2024 International Conference on Environment Engineering, Urban Planning and Design (EEUPD 2024)

E3S Web of Conferences Volume 617 (2025)

Nanjing, China
29 November - 1 December 2024

ISBN: 979-8-3313-1453-8

Printed from e-media with permission by:

Curran Associates, Inc.
57 Morehouse Lane
Red Hook, NY 12571

Some format issues inherent in the e-media version may also appear in this print version.

For additional information, please contact EDP Sciences – Web of Conferences
at the address below.

EDP Sciences – Web of Conferences
17, Avenue du Hoggar
Parc d'Activité de Courtabœuf
BP 112
F-91944 Les Ulis Cedex A
France

Phone: +33 (0) 1 69 18 75 75
Fax: +33 (0) 1 69 28 84 91

contact-edps@webofconferences.org

Additional copies of this publication are available from:

Curran Associates, Inc.
57 Morehouse Lane
Red Hook, NY 12571 USA
Phone: 845-758-0400
Fax: 845-758-2633
Email: curran@proceedings.com
Web: www.proceedings.com

TABLE OF CONTENTS

MULTI-DIMENSIONAL STUDY ON ECOLOGICAL ENVIRONMENT AND RESOURCE UTILISATION

Analysis of DOM Distribution Characteristics and Environmental Response Capacity in Different Areas of Urban Water 1
Yuanzhi Li

Urban Gardens and Plant Physiology: Green Infrastructure and Urban Sustainability 7
Yuchen Zeng

The Relationship Between Engagement with Natural Beauty and Pro-environmental Behaviors Using Structural Model Analysis 11
Hongqijie Wang, Pingping Liu

The Relationship Between Late Devonian Environmental Evolution and Organic Matter Enrichment 15
Yingying Ma

Emergy Analysis of Ecosystem in Qiandao Lake Asian Games Zone 21
Chao Wang, Yinrui Sun, Yuchen Zheng

Analysis and Research on the Relationship Between Bird Related Faults in Transmission Lines and Bird Activity Patterns 26
Weiyun Mao, Chen Shen, Rong Xiao, Kefu Zhang, KaiZhao Guo, Wenbin Zhao, Ya Wang, Dongsheng Zhang

Application of Magnetic Coagulation Sedimentation Tank and Sand-adding High-efficiency Sedimentation Tank in Sewage Treatment 33
Yufei Shangguan, Binbin Qiu, Zongkai Ma, Li He

Research Progress of Plant Available Carbon in the Earth's Carbon Cycle 40
Xiang-ping Sun, Hua-jiao Qiu, Wang Zhuo, Cai-sheng Qiu, Guo-qian Zhong, Xin-ling Zhao, Zhi-min Zhao, Song-hua Long, Wook Won Sung

The Role of Plant Charcol in the Carbon Geochemical Cycle 45
Xiang-ping Sun, Zhuo Wang, Guo-qian Zhong, Wook Won Zhong

Impact of Dissolved Oxygen Limiting Strategies on the Simultaneous Partial Nitrification, Anammox and Denitrification (SNAD) Process 50
Zhouzhenyang Xia, Jiajun Luo, Zhifan Liu, Menglin Ou, Haiyan Fu

Drought Disaster Risk Assessment and Zoning in Gansu Province 55
Ting Yang, Yufei Cheng, Junde Wang

Solidification of Water Plant Sludge and Domestic Waste Incineration Bottom Ash 62
Qi-Guang Cao, Peng Lu, Dong-Qing Yang, Guo-Li Xie, Jun Li, Ming-Yue Zheng, Nan Li-Xing, Hui Xiao-Zhang, Liang Chen, Song Li

Study on the Uneven Settlement Patterns at the Bottom of Seepage Control Structures in Deep Backfilled Reservoir Basins 69
Yayun Meng, Xianfei Wang, Xiangfei Ye, Yaolai Liu, Li Yuan, Shuairan Peng

Study on the Community Structure and Diversity of Zooplankton in Dingzi Bay by Using Mathematical and Intelligent Analysis Tools ... 76
Liang Zhang, Juan Huang, Yumeng Liu, Xueji Gu, Chuanting Zhao

Research on Testing Methods of Life Detection Equipment for Water-bearing Soil Disasters ... 82
Yanyun Nan, Jin Yan, Shizhe Xie

Spatiotemporal Patterns and Influencing Factors of Ecosystem Services in Shandong Province ... 87
Liangjie Guo, Junyue Wang, Yongbo Zhai, Yanlei Zuo, Wei Liu

Study on the Mineral Characteristics and Upgrading of Titanium Iron Ore Concentrate in the Panxi Region of China ... 93
Guoyin Xu, Purong Wang

Research on Process Mineralogy of Titanium-Bearing Blast Furnace Slag in Panzhihua-Xichang Area ... 100
Guoyin Xu, Purong Wang

Comparison of Coastal Resilience Policies: A Perspective on Effective Global Governance Strategies ... 107
Weiwei Dong, Guilan Lin

Research on Environmental Risk Assessment Techniques for Acid Gas Fields in Complex Terrains ... 115
Xuan Xu, Jia Zeng, Chunling Tang, Fangbing Gong

Study on the Migration Pattern of Organic Carbon in Slope Farmland in the Southeast of Heilongjiang Province ... 119
Hongfeng Wang, Jiuyi Wang, Jing Zhao, Guomeng Fu, Ze Shao, Xingchen Yu

Transboundary Marine Spatial Planning: Theoretical Implications, Practical Analysis, and Future Prospects ... 123
Xuefeng Li, Xin Teng, Xiangguo Zhang, Xinchun Sang

Analysis of Spatio-temporal change of Carbon Emissions and Carbon Sinks and Carbon Balance partition in the Poyang Lake Ecological Economic Region ... 129
Yunsong Yang

Research and Characteristic Analysis of Ocean Wave Simulation in Jiangsu Coast ... 137
Xiaohong Liang, Wenjun Xiao

BCF Better Reflects the Bioavailability of Selenium in the Soil of the Northeastern Guanzhong Plain Compared to MF ... 141
Shengfei Yang, Pei Zhang, Fa Guo, Duoxun Xu, Yuchen Yan, Tuo Liu

STUDY ON URBAN RENEWAL AND LOW CARBON TRANSPORT AND ENERGY SYSTEMS

Research on Regional Traffic Simulation Analysis and Traffic Capacity Improvement Technology Under the Background of Urban Renewal ... 145
Liang Wang, Anfan Shang, Yang Feng, Hao Fu, Ziqi Gao

Research on the Impact of New Energy Vehicles on Carbon Emissions of Urban Transportation ... 152
Hongru Lin

Addressing Global Climate Challenges: Low-carbon Transport Planning Practices and Optimization Strategies in Bangkok 157
Sucheng Yao, Kanjanee Budthimedhee

Spatial Pattern and Influencing Factor Analysis of Parcel Lockers in London 162
Xinwei Kang

Impact of Photovoltaic Power Station on Vegetation Carbon Sink in Guizhou Province 167
Yong Zhang, Yong Liu, Jing Xu, Ping Zhao

Research on Zero Carbon Sports Building Technology Strategy in Regions with Hot Summers and Cold Winters 173
Yunxia Ding, Xilan Qiu, Jinxing Cai

Analysis of Carbon Emission Characteristics Over the Entire Life Cycle of Indoor Substation Structures 180
Jingyun Wu, Zheng Huang, Hao Liu

Photovoltaic Potential of Elevated Metro Stations: A Case Study of Shanghai, China 186
Hui Mao, Lina Zhang, Gaijing Zhang

Current Situation, Trend and Peak-Reaching Path of Low-Carbon Development of Water Transportation in Zhejiang Province 191
Meng Lu, Yanhua Wei, Yilei Lu, Yiying Zhou

Renewable Energy Development in Funafuti: Photovoltaic System Design and Economic Performance 198
Yuan Zhuang

Substitution Efficiency Test of CH3127I Photoelectric Method for Traditional Radioactive CH3131I Method 204
Yao Tian, Sijun Fan, Weiye Yin, Shuai Zhou, Zhaojie Li, Shuai Yang, Chen Wang

A New Type of Cable Duct Channel and Design Method with Composite Support of Gravity Retaining Wall and Premixed Fluid Solidified Soil 211
Zhenhai Zhou, Linkang Zhou

Research on Implementation Pathways for Near-Zero Carbon Emissions in High-Density Urban Districts 216
Guoshu Yang, Anni Guo, Gaijing Zhang

Analysis and Research on the Influence of Civil Aircraft System on CO2 Emission 224
Kexin Sun, Yi Yang, Zhaojie Xuan

Price Mechanisms in Green Electricity Markets and Game Analysis of Participant Behavior 228
Jing Liu, Tiantian Chen, Yiqing Xiao, Jing Wang

Evaluation of the Application and Development Potential of Distributed PV in Industrial Parks– Take the 180MW Distributed Photovoltaic Project as an Example 235
Yetian Long, Shuming Lu, Shuaiming Chen, Chen Chen, Guanghua Chen, Xianjin Shan, Yuanyuan Hu

Public Transportation Commuting: Inequity and Polycentric Urban Structure 242
Wantong Fu

Analysis and Thinking on the Difficulties of Substation Location in Guizhou under the Background of New Energy 250
Linlin Li, Jiaxing Li, Shujuan Bai, Huafeng Chen

STUDY ON SUSTAINABLE URBAN AND RURAL DEVELOPMENT AND SPATIAL PLANNING

Study on Post-assessment System of Countryside with Distinctive Characteristics in Jiangsu Province 254
Xiu Li, Yanlin Liu, Lan Yu

An Empirical Study on Win-win Strategies for the Theme Park Sustainable Development and Tourist Satisfaction Based on IPA 259
Qifeng Yin, Xingmei Yang, Yanhong Liu

Simulation of Spatial and Temporal Dynamics in Habitat Quality amid Rapid Urbanization based on Random Forest Algorithm: A Lingui District, Guilin City Case Study 267
Xiaoyan Zhang, Hanxin Liu, Xin Yang, Gang Wu, Junjie Tan

Evaluating Urban Ecological Livability Based on the Coupling Coordination Degree Model: A Case Study of Guangzhou 274
Shan Huang, Baoqiang Wang

Research on the Guidelines for Green Ecological Planning and Construction of New Cities in Shanghai 284
Hongyan Pan, Guoshu Yang, Xilan Qiu

Constructing the Boundary-Domain Theory: A Spatial and Social Framework 288
Hua Zhu, Zhangpan Song, Xinghua Lai, Haiyang Li

Strategy and Practice of Artistic Transformation of Discarded Teaching Facilities in the University 292
Mingqi Shi, Jiayi Shi, Wei Cheng

Design Method and Construction Process of Slope Greening in A Basalt Mine 298
Xiaoming Chu, Yu Yang, Yunxiao Yu

A Study on the Morphology and Efficiency of Urban Blocks Around the Radial Plaza: Based on a Comparison of Dalian and Paris 303
Xinning Guo, Dongfengt Yang

Jiading New City Green Ecological Urban Planning and Practice — A Case Study of Yuanxiang Lake Central Activity Zone 312
Guoshu Yang, Hongyan Pan, Hua Sun

Urban Waterfront Space in Macao: Distribution Characteristics, Existing Problems and Design Strategies 317
Xuefeng Li, Qi Yue, Xinchun Sang, Xiangguo Zhang

Environmental, Social, and Governance (ESG) Committees and Corporate ESG Information Disclosure: From the Perspective of Organizational Information Processing 322
Hang Yang

Spatial Planning and Sustainable Development of Qiangang Village Based on Catalyst Theory 327
Xiuhong Lin, Yilin Wu

A Preliminary Investigation on the Relationship Between the Sustainable Construction Field and the Sustainable Development Goals: A Proposed Research Framework 332
Zhenquan Zhou, Deprizon Syamsunur, Xiang Fan, Jiawen Li

Hefei City Carding in the 21st Century 338
Menghan Zhu, Baoqiang Wang

Evaluation on the Network Structure of the Park Square System Based on the Social Network Analysis Method — An Case of Qindu District, Xianyang City, China 347
Yuyi Chen, Tianpeng Wang

Research on the Planning Strategy of Child Friendly Commuting Pathways Based on Society Ecosystems Theory: A Case Study of Xicheng Community in Zhuhai 352
Kang'en Chen, Helan Zhang, Yujuan Pan, Tao Xia

A Review of the Applications of CFRP Reinforcements in Civil Engineering 360
Jue Wang, Xincheng Su, Quanqing Gao

The Evolution of Chinese Architecture in the 21st Century 367
Tiantian Yao

A Study on the Age-Friendly Landscape Evaluation System for Rural Communities in Western Zhejiang: A Case Study of Quzhou 371
Jiahe Xu, Mimi Tian

Comprehensive Performance Evaluation for Sustainable Development of Small Towns: A Case Study of Tianhuangping Town, Anji County 378
Chao Li, Feng Yang, Maomao Yan, Ying Qu

Experimental Study on the Strength Characteristics of Loess Reinforced with Calcium-based Mixture, Chitosan and Sisal Fiber 383
Lei Lei, Hao Wan, Xiaolong Wei, Jun Liang, Han Zhang, Xiaoli Liu, Zhuo Chen, Kanliang Tian

Study on Mechanical Properties of Recycled Aggregate Concrete with Cement-Based Crystalline Materials 391
Jinhao Ou, Jiang Feng, Jiaqi Liu

How Digital Empowerment Promotes Rural Autonomous Governance? — A Case Study of Houwei Village, Xingdian Street, Pukou District, Nanjing 396
Xiu Li, Zhixuan Yang, Yinglu Liu

Author Index

Preface

In an era marked by rapid urbanization and environmental challenges, the intersection of environment engineering and urban planning and design has emerged as a crucial field of inquiry. It is against this backdrop that the 2024 International Conference on Environment Engineering, Urban Planning and Design (EEUPD 2024) was convened, bringing together scholars, researchers, and practitioners from around the globe to share insights, innovations, and best practices.

The EEUPD 2024 focused on the following central ideas: strengthening environmental sustainability, addressing global environmental challenges, improving the quality of life of residents, promoting social sustainable development, and facilitating scientific and technological innovation and research. It paid special attention to the integration of environmental sustainability and urban design, and explored how to promote win-win strategies for urban sustainable development and environmental protection through technological innovation.

The conference featured a diverse and engaging program, designed to foster knowledge exchange and collaboration. The highlight of the event was the keynote speeches delivered by several distinguished experts in the field. For instance, Professor El-Sayed Salama from Lanzhou University, a globally recognized scientist, presented on the topic of "Current Status of Microplastics (MPs) Pollution in the Different Environment," offering valuable insights into this pressing environmental issue. Professor Joon Sik Kim from Xi'an Jiaotong-Liverpool University, a transdisciplinary academic in international communities focusing on smart sustainable cities, watershed revitalization, and geographic information systems, talked about "The Complexity and Dynamics in Planning Practice of Smart Sustainable Cities". These keynote addresses not only set the tone for the conference but also provided a comprehensive overview of the latest advancements and challenges in the field.

In addition to the keynote speeches, the conference included oral presentations and poster sessions. The oral presentations allowed researchers to share their work in detail, engaging in lively discussions with peers. These sessions covered a wide range of topics, from environmental engineering technologies to innovative urban planning strategies. The poster sessions, on the other hand, provided an opportunity for more visual and interactive presentations. Participants were able to display their research findings through well-designed posters, facilitating one-on-one interactions and in-depth discussions. This format enabled researchers to receive valuable feedback and establish connections with other professionals in the field.

The proceedings of EEUPD 2024 represents a collection of cutting-edge research and innovative ideas presented during the conference. Each paper has undergone a rigorous peer-review process, ensuring the high quality and relevance of the contributions. The publication of the proceedings serves multiple important purposes. Firstly, it provides a permanent record of the conference, allowing researchers who were unable to attend to access the latest findings and developments. Secondly, the proceedings volume offers a valuable resource for academics, practitioners, and policymakers interested in the intersection of environment engineering and urban planning and design. By disseminating these contributions widely, we hope to inspire further research and practical applications that can contribute to sustainable urban development.

The success of this conference will have a long-term positive impact on the field of environmental engineering and urban planning, strengthen the sustainability of cities around the world, improve the quality of life of people, and provide practical strategies and technical support for addressing global environmental issues.

The Committee of EEUPD 2024

Committee Member

Conference General Chair
Prof. Hongwei Wu, University of Hertfordshire, UK

Technical Program Committee Chair
Assoc. Prof. Tao Zhang, China Agricultural University, China
Assoc. Prof. Dr. Md. Hasanuzzaman, University of Malaya, Malaysia

Publication Chair
Assoc. Prof. Yanbo Wang, Aalborg University, Denmark
Assoc. Prof. Dr. Mohd Rusli Yacob, Universiti Putra Malaysia, Malaysia
Assoc. Prof. Houssam Khelalfa, University of Jijel, Algeria
Assoc. Prof. Dr. Mohammadreza Vafaei, Universiti Teknologi Malaysia, Malaysia
Assoc. Prof. Dr. Fatimah Binti De'nan, Universiti Sains Malaysia, Malaysia

Committee Member
Prof. Dr. Ahmed F. Zobaa, Brunel University London, UK
Prof. Hüseyin Bilgin, EPOKA University, Albania
Prof. Rui Neves, Polytechnic Institute of Setubal, Portugal
Prof. Jorge de Brito, University of Lisbon, Portugal
Prof. Baoyu Gao, Shandong University, China
Prof. Rui Liang, China University of Mining and Technology, China
Prof. Kun Liang, Ningbo Institute of Materials Technology and Engineering Chinese Academy of Sciences, China
Assoc. Prof. Muhammad Hassan, Southwest Jiaotong University, China
Assoc. Prof. Grzegorz Lisak, Nanyang Technological University, Singapore
Assoc. Prof. Zawawi Bin Daud, Universiti Tun Hussein Onn Malaysia, Malaysia
Assoc. Prof. Vaishali Sahu, The NorthCap University, India
Assoc. Prof. ATMA Wafa, University of mustapha stambouli of Mascara, Algeria
Assoc. Prof. Maiyong Zhu, Jiangsu University, China
Assoc. Prof. Rongkui Su, Central South University of Forestry and Technology, China
Visiting Assoc. Prof. Rawshan Ara Begum, Macquarie University, Australia
Asst. Prof. Roya Peirovi-Minaee, Gonabad University of Medical Science, Iran
Asst. Prof. Tao Yang, City University of Hong Kong, China
Asst. Prof. Qiliang Wang, The Hong Kong Polytechnic University, China
Dr. Junzhuo Liu, University of Chinese Academy of Sciences, China
Dr. Wei Wu, City University of Hong Kong, China
Dr. Yiyuan Zhang, The Hong Kong Polytechnic University, China

Statement of Peer review

In submitting conference proceedings to *Web of Conferences*, the editors of the proceedings certify to the Publisher that

1. They adhere to its **Policy on Publishing Integrity** in order to safeguard good scientific practice in publishing.
2. All articles have been subjected to peer review administered by the proceedings editors.
3. Reviews have been conducted by expert referees, who have been requested to provide unbiased and constructive comments aimed, whenever possible, at improving the work.
4. Proceedings editors have taken all reasonable steps to ensure the quality of the materials they publish and their decision to accept or reject a paper for publication has been based only on the merits of the work and the relevance to the journal.

Title, date and place of the conference

2024 International Conference on Environment Engineering, Urban Planning and Design (EEUPD 2024)
November 29 - December 1, 2024
Nanjing, China

Proceedings editor(s):

Assoc. Prof. Yanbo Wang, Aalborg University, Denmark
Assoc. Prof. Houssam Khelalfa, University of Jijel, Algeria
Assoc. Prof. Dr. Mohammadreza Vafaei, Universiti Teknologi Malaysia, Malaysia
Assoc. Prof. Dr. Fatimah Binti De'nan, Universiti Sains Malaysia, Malaysia

Date and editor's signature

Yanbo Wang

Houssam Khelalfa

Mohammadreza Vafaei

Fatimah Binti De'nan

Analysis of DOM Distribution Characteristics and Environmental Response Capacity in Different Areas of Urban Water

Yuanzhi Li*

School of Chemical and Environmental Engineering, Wuhan Polytechnic University, Wuhan, 430023, China

Abstract: To accurately define the conditions of urban water bodies, research was conducted on the distribution characteristics and environmental responsiveness of DOM in different areas of urban water bodies. Water samples were collected from the severely polluted Zaohe River, Chanhe River, and Bahe River in Xi'an, including surface water and wastewater discharged from sewage outlets on these rivers. After processing the water samples, established measurement techniques were employed to analyze the DOM in the water samples. Taking Beijing as an example, remote sensing imagery was used to extract surface temperature and water body information within the study area, combined with land use and land cover data, to analyze the environmental pattern characteristics of urban water bodies. By selecting the summer period, during which the water bodies exhibit significant cooling effects, typical water areas were divided into surface and linear parts, and high-resolution imagery and fixed-point measured data were used to further explore the characteristics of urban water bodies and their environmental responsiveness.

1. Introduction

The distribution of Dissolved Organic Matter (DOM) varies across different media and carriers in nature. According to research, generally, the content of DOM in inland rivers and lakes is higher than that in coastal waters, while the DOM content in coastal waters is higher than that in offshore waters. In the vertical direction, although the distribution patterns of DOM in different oceans are not entirely consistent, on the whole, the DOM content in surface waters is relatively high, while that in intermediate waters varies greatly without a fixed pattern. The DOM content in deep waters is relatively low, but its chemical properties are more stable, indicating that its composition differs from the DOM in surface waters or rivers and lakes. DOM can absorb and scatter sunlight, thereby altering the transparency and color of water bodies [1]. These changes not only affect the aesthetic appeal of water bodies but may also impact the photosynthesis and living environments of aquatic organisms. Furthermore, DOM can combine with nutrients such as nitrogen and phosphorus to form complex organic complexes, influencing the bioavailability of these nutrients. This has significant implications for the eutrophication of urban water bodies. DOM serves as a flocculating agent for soil colloids, promoting the aggregation and gelation of colloidal particles and subsequently influencing soil structure and texture. These changes have important effects on soil aeration, water retention, and nutrient retention capacities. The abundant organic matter contained therein is one of the essential nutrient sources for crop growth. By providing nutrients and improving the soil environment, DOM can promote the growth and development of crops. As a crucial link connecting organic and inorganic carbon in life forms, DOM participates in various biogeochemical cycles. This is of great significance for maintaining ecosystem stability and promoting the carbon cycle.

Many scholars also pay attention to the ecological function of water landscape. For example, with the help of calculation methods such as ecological footprint, the ecological value of urban fresh water supply is evaluated or the ecosystem landscape service value of urban rivers is evaluated. With the help of RS and GIS tools, we can explore the change of watershed ecosystem service value from the time span, or establish a new evaluation framework and index system for the existing watershed system according to the detection and previous evaluation methods. On a smaller scale, some literatures have analyzed and studied the influence of water system on the humidity and heat of the surrounding environment [2]. Zhang Xinchun and Zhang Peihong carried out numerical simulation using Fluent and Airpak software to simulate and analyze the humidity and heat around the residential area, and studied whether there was a water system around the residential area and the effect of the location of the water system on the heat and humidity environment. However, the research focuses on environmental humidity, temperature and software simulation technology, and does not take into account the characteristics of the water landscape. Based on the above background, the distribution characteristics and environmental response capacity of DOM in different areas of urban water bodies were analyzed.

* Corresponding author: 13908692990@163.com

2. Study on the distribution characteristics of Dom in different areas of urban water bodies

2.1 Overview of the study area

The water samples used in this study come from Zaohe River, Xianhe River and Bahe River in Xi'an. The point coordinates are located by Owei satellite map, and the point distribution map is made by arcgis.

2.1.1 General situation of Zaohe River and selection of sampling sites.

Twenty-five sampling sites were selected in the Zaohe River basin, of which four surface water samples were collected from the upper Chang'an section (Z1, Z2, Z3, Z4), and eight surface water samples were collected from the western third ring section of the middle reaches (Z5, Z6, Z7, Z8, Z9, Z10, Z11, Z12). A total of 16 surface water samples were collected from Caotan to the Weihe Estuary (Z13, Z14, Z15, Z16). Six samples were collected from 2 outlets on the east bank and west bank of the ninth sewage treatment plant, 2 outlets of temporary sewage treatment project in Yanta section, Yuhua sewage treatment plant outlet in Yanta district of Xi'an, and the first stage outlet of Xi'an second sewage treatment plant (ZP1, ZP2, ZP3, ZP4, ZP7, ZP8). At the same time, three samples were collected from the direct discharge outlet of Changan District, the Zaohe Management Station of Xi'an Mianghui Canal Management Center, and the outlet No. 18 of Weiyang Palace site of Han Chang'an City (ZP5, ZP6, ZP9).

2.1.2 General situation of Huanghe River and selection of sampling points.

20 sampling sites were selected in the Xianhe River basin, of which 7 surface water samples were collected from the upper Yanta section (C3, C4, C5, C6, C7, C8, C9); 4 surface water samples were collected from the middle Baqiao section (C10, C11, C12, C13); and 3 surface water samples were collected from the downstream Weiyang section, and a total of 14 surface water samples were collected. Six samples were collected from the sewage outlet of Huahai Hotel, Eurasia International sewage outlet and Xi'an third sewage treatment Plant outlet (CP1, CP2, CP3, CP4, CP5, CP6).

2.1.3 General situation of Bahe River and selection of sampling sites.

Twenty sampling sites were selected in the Bahe River basin, including 7 surface water samples (B18, B17, B16, B15, B14, B13, B12) in the upper reaches of Lantian County, 7 surface water samples (B11, B10, B9, B8, B7, B6, B5) in the middle Baqiao section, and 4 surface water samples (B4, B3, B2, B1) in the downstream Weiyang section. a total of 18 surface water samples were collected. Two samples were collected from Xi'an Jinjiang International outlet and Baliu West Road Yushuikou (BP1, BP2).

2.2 Collection and treatment of samples

2.2.1 Sample collection and pretreatment.

All sampling points are located by Ovi map satellite. The samples were mixed for 3 times at each sampling point, and the mixed water samples were packed into a 1L polyethylene bottle washed with 10% hydrochloric acid and washed with raw water. The water temperature (t), pH value, turbidity (NTU), dissolved oxygen (DO) and dissolved total solids (TDS) were measured by U51 multi-parameter water quality tester. Immediately after sampling, the collected water samples were sent back to the laboratory and stored in a refrigerator at 4 °C [3]. The water samples were filtered by nylon filter membrane (Supor, Pall). The filtered water samples were stored in the refrigerator at 4 °C for the determination of total organic carbon (TotalOrganicCarbon, TOC) and three-dimensional fluorescence (3D-EEM).

2.2.2 Instruments and equipment.

Cold dryer, centrifuge, electronic balance, shaker, vacuum pump, filter bottle, filter cup, nylon filter membrane (Supor, Pall, 0.45 μ m), TOC detector (Shimadzu, TOC-L) and three-dimensional fluorescence spectrometer (Shimadzu, Fmur7000) were used in this experiment.

2.2.3 Basic information

The basic information, water quality parameters, and sampling time of the sampling points are shown in Table 1.

Table 1. Basic Information on Urban Water Bodies

Serial Number	Sampling point name	Sampling date	Water temperature (°C)	pH value	Dissolved oxygen (mg/L)	Chemical Oxygen Demand (COD, mg/L)
1	River A - Sampling Point 1	2024-5-1	20	7.2	6.5	50
2	Lake B - Sampling Point 2	2024-5-2	22	7.5	7.0	60
3	Pond C - Sampling Point 3	2024-5-3	25	7.0	5.5	80
4	Marine D-sampling point 4	2024-5-4	22	7.5	6.7	60
5	Farmland E-sampling point 5	2024-5-5	23	7.3	5.6	50

Chemical Oxygen Demand (COD): It reflects the content of oxidizable organic matter in water and is measured using chemical methods.

2.3 Determination of characteristic parameters of DOM Water quality Distribution

Water quality parameters (usually including water temperature, pH, turbidity (Turbidity), electrical conductivity, dissolved oxygen (DO), dissolved total solids (TDS), BOD, COD and other physical and chemical indexes), water quality parameters can be used to analyze the degree of water pollution. In this paper, U51 multi-parameter water quality analyzer was used to determine the indexes of water temperature (t), pH value, turbidity, dissolved oxygen and total dissolved solids.

2.3.1 Turbidity

Turbidity (in NTU) is an optical effect, an indication of the degree to which light is blocked when it passes through the water layer, and the ability of the water layer to scatter and absorb light. It is not only related to the content of suspended solids, but also to the composition, particle size, shape and surface reflection properties of impurities in water. The suspended matter and colloid in water, such as soil, silt, fine organic matter, inorganic matter, plankton and so on, can make the water become turbid and show a certain turbidity. Usually, the higher the turbidity, the more turbid the solution.

2.3.2 Total soluble solid

Total dissolved solids (Totaldissolvedsolids, TDS) refers to the total amount of dissolved components in water, including the total amount of various ions, molecules and compounds dissolved in water, but excluding suspended solids and dissolved gases. The higher the TDS value, the more dissolved matter (including both inorganic and organic matter) in the water. Too high TDS indicates that the water quality is not good.

2.3.3 Dissolved oxygen

Dissolved oxygen (DO) refers to the molecular oxygen dissolved in water, expressed in milligrams per liter of water. There are two sources of dissolved oxygen in water: one is. The water surface absorbs oxygen from the air. Second, algae absorb carbon dioxide and release oxygen through photosynthesis. The content of dissolved oxygen in water is directly related to water temperature, air pressure, solute (such as salt, etc.) and the partial pressure of oxygen in the atmosphere. The amount of dissolved oxygen in water is related to the size of its contact surface with atmosphere. The larger the contact surface is, the more dissolved oxygen is, which decreases with the increase of water temperature and increases with the increase of partial pressure of oxygen. In the normal state, the dissolved oxygen in the surface water should be close to saturation.

The characteristic parameters of DOM water quality distribution measured in the field are shown in the Table 2 below. The three outlets of ZP3, ZP4 and ZP5 were closed in September, and the water quality parameters of these three outlets in September were not measured. B16, B17 and B18 are unable to enter the sampling site in September because of road construction, so there is no September water quality parameter data for these three points (Table 3). The CP2 outlet was closed in September and the water quality parameters of the outlet in September were not measured (Table 4).

Table 2. Water Quality Parameters of Zaohe Point

Point number	May					September				
	T/°C	pH	NTU	DO (mg/L)	TDS (g/L)	T/°C	pH	NTU	DO (mg/L)	TDS (g/L)
ZP1	24.86	7.46	267	5.43	0.536	22.92	7.55	132	6.92	0.104
ZP2	24.69	7.54	77.6	5.4	0.521	22.74	7.35	4.5	6.85	0.508
Z1	24.74	7.88	110	6.04	0.492	22.79	7.57	8.0	8.05	0.481
ZP3	26.18	7.74	90.2	4.92	0.538	/	/	/	/	/
Z2	24.7	7.67	16.1	5.64	0.486	23.22	7.63	6.0	6.80	0.478
ZP4	26.46	7.7	12.8	4.6	0.531	/	/	/	/	/
Z3	26.28	7.87	106	6.38	0.481	23.89	7.83	54.2	6.49	0.439
ZP5	/	/	/	/	/	/	/	/	/	/
Z4	25.57	7.81	27.1	6.27	0.492	24.22	7.65	93.4	6.49	0.471
Z5	30.37	8.14	30.1	8.07	0.507	22.96	8.25	86..3	6.72	0.435
ZP6	27.32	7.59	12.1	7.9	0.427	24.35	7.36	0.6	6.49	0.424
Z6	29.74	7.67	12.1	7.9	0.427	24.35	7.36	0.6	6.49	0.428
ZP7	27.36	7.57	4	5.53	0.461	24.41	7.45	1.5	6.13	0.441
ZP8	27.94	9	13.3	5.67	0.427	23.66	7.70	0.0	6.50	0.477
Z7	29.58	7.93	36.8	8.25	0.438	23.61	7.63	16.2	6.79	0.447
Z8	31.88	7.49	39.4	5.71	0.274	24.00	7.69	29.0	8.39	0.436
Z9	27.41	7.96	12	9.09	0.472	24.35	7.78	27.6	7.12	0.438
ZP9	27.65	7.96	18.5	5.9	0.491	24.30	7.65	1.0	6.03	0.495
Z10	27.16	8.34	19.5	9.98	0.497	24.36	7.91	34.3	9.32	0.443
Z11	28.5	8.5	25.7	8.89	0.479	22.18	7.81	45.3	9.80	0.494
Z12	28.73	8.59	20.9	9.97	0.495	22.66	7.83	47.6	5.22	0.514
Z13	28.34	8.8	17.5	12.05	0.523	22.49	7.78	42.0	4.99	0.494
Z14	28.81	8.8	44.3	12	0.583	22.98	7.85	31.6	9.24	0.559
Z15	29.35	8.88	41.3	9.9	0.486	22.71	7.83	50.0	5.18	0.506
Z16	29.32	8.72	104	9.42	0.519	21.84	8.57	133	7.04	0.422

Table 3. Water Quality Parameters of Bahe River Points

Point number	May					September				
	T/°C	pH	NTU	DO (mg/L)	TDS (g/L)	T/°C	pH	NTU	DO (mg/L)	TDS (g/L)
B18	34.42	8.98	64	11.01	0.271	/	/	/	/	/
B17	31.13	8.75	49.9	12.21	0.306	/	/	/	/	/
B16	29.56	8.31	145	10.56	0.317	/	/	/	/	/
B15	31.21	8.65	22.2	11.28	0.37	20.94	8.96	25.5	10.07	0.236
B14	31.25	8.59	55.8	9.8	0.372	21.23	9.03	21.5	8.67	0.264
B13	30.26	8.5	27.2	9.27	0.357	22	9.34	14.4	7.77	0.215
B12	30.35	8.53	24.9	8.45	0.365	21.56	9.08	18.2	9.98	0.227
B11	30.32	8.81	12.7	7.82	0.357	22.98	7.95	28.1	4.01	0.288
B10	27.01	8.05	45.4	6.16	0.324	21.49	8.64	80.3	9.48	0.234
B9	25.88	8.5	17	10.55	0.287	21.53	8.33	102	7.83	0.262
B8	24.86	8.69	30.3	8.08	0.21	21.6	8.59	30.6	6.97	0.244
B7	30.37	9.12	12.2	9.35	0.184	21.5	8.44	104	6.89	0.271
BP2	22.91	8.05	8.7	11.2	0.622	20.05	8.25	0	5.68	0.633
B6	29.71	8.96	168	8.65	0.2	21.96	8.3	323	7.45	0.419
BP1	26.14	12.01	383	9.88	0.167	23.41	7.92	0	9.38	1.09
B5	28.84	7.05	590	7.86	0.087	23.03	8.68	109	8.76	0.292
B4	26.46	7.25	117	7.21	0.532	23.25	8.32	90.8	9.98	0.364
B3	23.45	9.06	50.5	7.8	0.242	23.26	8.27	479	7.32	0.352
B2	27.18	8.24	185	6.13	0.533	23.41	8.29	139	8.87	0.331
B1	29.72	8.67	65.7	7.12	0.535	22.88	8.63	106	7.63	0.427

Table 4. Water Quality Parameters of Chanhe River Points

Point number	May					September				
	T/°C	pH	NTU	DO (mg/L)	TDS (g/L)	T/°C	pH	NTU	DO (mg/L)	TDS (g/L)
C 3	22.87	7.35	51.5	9.85	0.205	26.61	8.36	30.6	7.27	0.307
CP3	25.03	8.28	150	5.06	0.623	20.96	8.39	96	5.9	0.391
C4	26.29	8.85	14.3	12.64	0.262	21.44	8.69	26.2	8.77	0.255
CP4	22.96	8.29	125	11.08	0.267	22.49	8.17	120	5.28	0.266
C5	26.25	8.44	43.7	10.51	0.213	21.23	8.59	51.4	9.99	0.27
C6	27.58	8.94	15.1	10.41	0.237	21.73	8.9	38.9	9.53	0.246
C7	23.34	7.25	17.8	8.22	0.264	21.95	8.9	38.9	9.53	0.246
C8	23.5	8.54	48.2	9.78	0.203	22.19	8.56	30.6	7.58	0.248
C9	23.7	8.29	61.3	9.75	0.163	22.52	8.73	26.9	7.45	0.247
C10	27.43	8.65	26.5	6.06	0.256	24.76	8.84	25.6	8	0.244
C11	25.71	8.62	26.3	7.96	0.244	25.03	8.78	31.7	6.77	0.244
CP5	27.39	8.01	74.5	4.22	0.368	23.04	8.21	56.6	10.06	0.403
C12	27.64	8.67	56.1	8.57	0.187	24.09	8.67	43.8	7.34	0.249
CP6	28.03	7.81	66.8	8.46	0.388	24.82	7.77	3.9	5.89	0.484
C13	23.73	8.07	33.4	7.76	0.515	24.86	7.92	17.4	5.59	0.583
C14	23.41	7.92	108	9.2	0.284	24.8	7.92	19	6.17	0.48
C2	25.96	7.88	32.9	8.87	0.532	24.06	8.3	22.1	6.15	0.406
CP2	28.15	7.98	426	7.34	0.217	/	/	/	/	/
CP1	30.21	8.59	127	9.92	0.3	24.03	8.33	57.1	6.62	0.379
C1	29.96	6.9	57.8	8.84	0.553	23.84	8.36	120	6.12	0.394

3. Study on Environmental response ability in different areas of Urban Water body

3.1 Landscape characteristics around urban water bodies

The purpose of this study is to explore the response law of the water body and the surrounding landscape to the thermal environment. The most important factor affecting the abnormal condition of the urban thermal environment comes from the difference of the underlying surface, considering the diversity of waterfront landscape in the study area. Green belts, artificial embankments, roads and buildings vary greatly in proportion and form, in order to facilitate measurement. In this paper, the surface cover type is used to characterize the characteristics of the landscape around the water body.

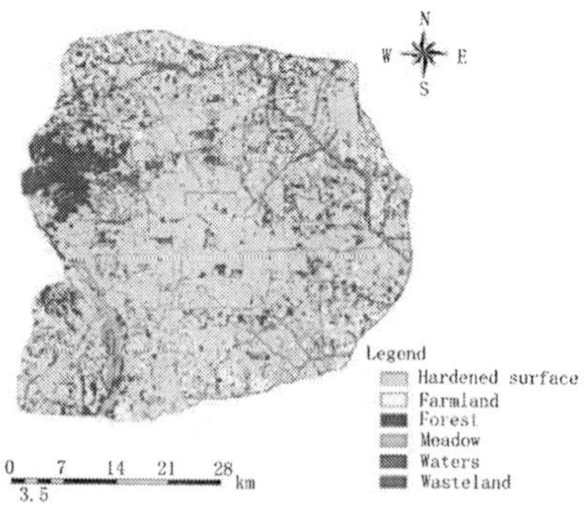

Figure 1. Land cover types within the Sixth Ring Road

The landscape characteristic data in Beijing sixth Ring Road come from the global change scientific data publication system "Beijing sixth Ring Road 5-year interval 30m resolution land cover data set". The data in 2023 are divided into six categories: hardened surface, farmland, forest, grassland, wasteland and water body, as shown in Figure 1.

It can be seen from the picture that most of the cover types in the sixth Ring Road are hardened surface, and the proportion is higher in the areas closer to the urban area. Forests, grasslands and farmland are mainly distributed in the edge areas, extending to the city center, and the distribution of wasteland is less [4-5]. Compared with the reality, the distribution of the water body in the land cover map is consistent with the water body extracted in this study, but the water body is missing in some places, which may be related to the basic image and interpretation, and the accuracy of other land types is better.

3.2 Calculation of water temperature

The statistical data of land surface temperature in the six rings in five years are shown in Table 5. The five-year data are divided into winter and summer, in which the average temperature is 41.50 °C in summer and 4.37 °C in winter.

Table 5. Water Temperature within Beijing's Sixth Ring Road (°C)

Time	MIN	MAX	RANGE	MEAN
September 2023	14.20	48.77	34.57	33.25
August 2023	21.21	50.10	28.89	34.15
November 2022	-9.57	19.27	28.83	8.88
July 2021	7.53	76.52	68.99	57.09
January 2021	-25.07	26.27	51.33	-0.14
	Summer average temperature			41.50
	Winter average temperature			4.37

3.3 Environmental response capability Index

There are different differences among the urban water landscape, which can be divided into two aspects: the water area itself and the surrounding landscape factors. The factors of water area itself mainly come from the nature of internal water, including water quantity, water quality and other factors, in which the difference of water quantity is difficult to compare at the same time in river-like waters and lake-shaped waters, so it is classified and discussed in the following chapters, in the global scope, the elements of water quality are selected to be analyzed, and the surrounding landscape differences are analyzed by buffer zone, which is measured by the proportion of surface cover [6].

The specific heat capacity of the water body is larger, and it shows more stable properties than other materials when the temperature changes, which is the basic principle of the response of the water body to the thermal environment. According to this characteristic, it is easy to infer that areas with large water volume should show lower temperatures in hot weather, but in reality, there are differences in water quality, especially in cities, which are affected by many factors. water quality is quite different from each other, especially organic pollutants [7-8].

In this study, a total of 42 water patches were extracted from the sixth ring, and the temperature status of each water body was obtained after superimposing the patch with the surface temperature. in order to facilitate measurement and comparison, the average temperature of each water body was arranged from small to large according to the water index of each patch. as shown in Table 6.

Table 6. Water Body Index and Summer Water Average Temperature

Parameter	Summer average temperature	Water body index	Parameter	Summer average temperature	Water body index
Kunyu River	39.38	-0.10	Donggan Canal	36.07	0.06
Xiaozhonghe	36.32	-0.05	Lotus flower pond	34.93	0.06
Orson dragon shaped water system	36.20	-0.05	Yuanbo Lake	36.45	0.06
Moat	40.47	-0.05	West Lake North	29.87	0.06
Qinghe	38.21	-0.05	Nanhaizi	36.80	0.07
Liangshui River	36.65	-0.05	Lianhua Lake	34.92	0.07
West Hunan	32.84	-0.02	South Lake	35.33	0.07
Shahe	34.74	-0.01	Lotus Lake	35.88	0.07
Red Scarf Park	36.96	-0.01	North Lake	35.21	0.08
The North Canal	34.26	-0.01	Jinyu	31.64	0.08
West of the Forbidden City moat	38.15	0.00	Beihai Zhonghai	33.97	0.09
Qianhai	37.10	0.01	South China Sea	33.92	0.09
Waterfront Park	33.75	0.01	West Lake Central	31.22	0.09
To the east of the Forbidden City moat	38.19	0.02	Purple Bamboo Courtyard	33.53	0.09
Longtan Lake East	34.62	0.02	Daning Reservoir	34.32	0.10
Xiaoyue Lake	35.37	0.03	Yuyuantan West	32.04	0.11
Taoran Pavilion	35.09	0.03	Zhongwu	38.27	0.12
Houhai	35.80	0.04	Fuhai	32.24	0.13
Xihai	36.90	0.04	Yuyuantan East	32.85	0.14
Longtan Lake West	35.42	0.04	Kunming Lake	30.78	0.16
Drainage canal	35.40	0.05	Taihu Lake	34.47	0.16

In the table, the waters with different patches of the same name are identified by location, and the unknown names are named by place names or park names. It can be seen from the table that the average value of the water index is the highest in Taihu Lake in the east of the study area, and the lowest is the Kunyu River on the west side of the center. From the shape of the water body, the patches in river form are generally in the low value area, and the lowest ten patches contain seven fluvial patches [9-10]. The area with high water body index is often the park water body close to the urban area, such as the Summer Palace and Old Summer Palace, most of the water bodies are in the forefront, including Kunming Lake, Zhongwu, Fuhai, West Lake, West Hubei and so on.

4. Conclusions

After an in-depth study of the distribution characteristics and environmental response capacity of dissolved organic matter (DOM) in different areas of urban water bodies, we can see the subtle changes and far-reaching effects in this complex ecosystem. Through the integration of multi-disciplinary perspective and advanced analysis technology, we can not only better understand the behavior pattern of DOM in urban water, but also provide a strong scientific basis for the protection and treatment of water environment. With the acceleration of urbanization, the pressure and challenges faced by urban water bodies are becoming more and more severe. As an important part of water environment, the source, migration, transformation and degradation of DOM directly affect the quality of water body, the stability of ecosystem and human health. Therefore, in-depth study of the distribution characteristics and environmental response capacity of DOM is of great significance to ensure the health and safety of urban water environment.

In this process, we are fully aware of the importance of interdisciplinary integration. The integration of knowledge and technology in many fields such as ecology, environmental science, chemistry and geography reveals the complex behavior pattern of DOM in urban water. At the same time, we also see the key role of advanced analysis technologies in the research, they not only improve the accuracy and reliability of data, but also provide us with more in-depth insight and broader vision. In the future, we will continue to deepen the research on the distribution characteristics and environmental response capacity of DOM in urban water bodies. On the one hand, we will further expand the scope of research and pay attention to more types of water bodies and a wider range of environmental factors; on the other hand, we will strengthen cooperation and exchanges with other fields to jointly promote the development of water environment protection and treatment.

References

1. V. A D, P. T D, A. O O, et al. Distribution and Variations of Elemental Sulfur in the Upper Part of the Black Sea Anoxic Water Column[J]. Oceanology, 2024, 64(2):250-259.
2. Xu T, Jia A, Huang Y, et al. Distribution and Main Controlling Factors of Gas and Water in Tight Sandstone Gas Fields: A Case Study of Sudong 41-33 Block in Sulige Gas Field, Ordos Basin, China.[J]. ACS omega, 2024, 9(19):20807-20818.
3. Okoli E, Nwaogazie L I, Ugwoha E. Assessment of the Spatial Distribution of Water Quality for Surface and Groundwater at Bonny Island, Nigeria[J]. Journal of Engineering Research and Reports, 2024, 26(6):50-59.
4. Isdianto A, Amanda S S, Yamindago A, et al. Water Quality Impact Christmas Tree Worms (Spirobranchus spp.) Distribution and Community Structure on Hard Corals At Sempu Island Nature Reserve, Malang, Indonesia[J]. Journal of Ecological Engineering, 2024, 25(5):175-186.
5. V. N K, G. V P, K. A R, et al. Distribution of Np, Pu, and Am in Water, Suspended Matter, and Bottom Sediments of Peter the Great Bay[J]. Radiochemistry, 2024, 66(1):115-123.
6. Aiswarya L, Siddharam, Sandeepika M. Soil Moisture Distribution Pattern under Drip Irrigation in Sandy Loam Soil Using Gravimetric Method[J]. Asian Journal of Soil Science and Plant Nutrition, 2024, 10(2):198-204.
7. Mazbah M U, uliu X, ingfeng H. anadium in a coastal Lagoon: distribution, sources, speciation, and ecological risks in sediments and water at Xiamen in China[J]. Chemistry and Ecology, 2024, 40(4): 369-387.
8. A. D T, G. S K, V. E I, et al. The Vertical Distribution of Microplastics in the Water Column of Lake Ladoga[J]. Water Resources, 2024, 51(2):146-153.
9. Castro D G D, Silva D L L A, Lopes N D M, et al. Effect of urbanization and water quality on microplastic distribution in Conceição Lagoon watershed, Brazil.[J]. Environmental science and pollution research international, 2024, 31(20):28870-28889.
10. Moudi M, Song X, He Y, et al. Enhancing Equitable Water Distribution in Agriculture: A Novel Optimal Framework for Irrigation Equity Index Improvement Under Diverse Adaptation Strategies[J]. Water Resources Management, 2024, 38(8):2669-268.

Urban Gardens and Plant Physiology: Green Infrastructure and Urban Sustainability

Yuchen Zeng*

School of Biological and Environmental Engineering, Jingdezhen College, Jingdezhen, Jiangxi, 333400, China

Abstract. The balance between supply and demand of green infrastructure (GI) is of great significance for maintaining the stability of urban or regional ecosystems, improving the modernization level of territorial spatial governance system and governance capacity, and promoting high-quality urban development [1]. With the acceleration of urbanization, the importance of urban garden and plant physiology has become increasingly prominent. This study aims to explore the role of green infrastructure in urban sustainability and to analyze how plant physiology supports the development of urban gardens. Based on a comprehensive analysis of relevant literature, this paper argues that green infrastructure is a key factor to achieve sustainable urban development, and the application of plant physiology can improve the ecological benefits and aesthetic value of urban gardens.

1 Introduction

With the rapid development of modern economy and the diversity of social life, the characteristics, styles and functions of urban landscapes are gradually changing and improving, but no matter how they change and progress, the functional needs must meet the people-oriented, that is, to meet the functional needs [2]. Urbanization is taking place at a rapid pace around the world. However, urbanization also brings many environmental and social problems: air pollution, water pollution, solid waste pollution, ecological damage, traffic congestion, etc. These environmental problems not only affect the quality of life of city residents, but also pose a threat to the entire ecosystem, and in order to solve these problems, cities need to adopt sustainable development strategies, of which green infrastructure is an important aspect. Urban gardens and plant physiology can provide a scientific basis for the construction and management of green infrastructure, so they are of great significance for achieving urban sustainability.

2 Green infrastructure for urban gardens

2.1 Definition and composition of green infrastructure

In the 1984 report on ecological city planning of the Man and the Biosphere (MAB) programme, UNESCO first proposed EI, which is similar to green infrastructure, as one of the five principles of ecological cities [3]. Green Infrastructure (GI) plays an essential role in ensuring the well-being of people and the green and sustainable development of cities. It is a concept that combines natural ecosystems and artificial infrastructure to improve the quality of human life and promote sustainable development by protecting, restoring, and managing natural ecosystems, providing ecosystem services. As a new type of green building technology, green infrastructure has great potential for development [4]. The components of green infrastructure are detailed in Table 1.

Table 1. Components of green infrastructure.

constituent	Examples of green infrastructure
Green space	Natural ecosystems such as parks, forests, wetlands, and grasslands, as well as artificial green spaces such as urban green spaces and community gardens.
Blue space	Water ecosystems such as rivers, lakes, and oceans, as well as artificial waters such as urban water systems and rain gardens.
Grown space	Unused or polluted land such as abandoned land, as well as artificial land such as urban waste land and landfills.
Green transportation	Eco-friendly modes of transport such as bike paths, pedestrian paths, and light rail, as well as sustainable modes of transport such as public transport and shared mobility.
renewable energy	Renewable energy sources such as solar, wind, and water, as well as energy infrastructure such as distributed energy resources and smart grids.

Note: The source is from Baidu Library https://wenku.baidu.com/ and other resources

* Corresponding author: 1394881973@qq.com

2.2 The role of green infrastructure in urban sustainability

Green infrastructure plays an important role in urban sustainability, and the fundamental idea lies in the organic integration of natural ecosystems and urban construction in order to realize the efficient use of resources, the protection of the ecological environment and the enhancement of human well-being, which plays a key role in improving the urban ecology, such as purifying the air, regulating the climate, and enhancing the image of the city, so as to promote the sustainable development of the city in all aspects.

2.2.1 Eco-efficiency

Studies have shown that large-scale green infrastructure can play a role in many aspects, such as lowering local temperatures, reducing air pollution, and alleviating the contradiction between urban development and natural ecology, providing people with positive and optimistic social benefits [5]. The most representative example is the two skyscraper "tree towers" erected in the downtown area of Milan, Italy, which is known as the "world fashion and design capital", which is also the world's first pair of green apartments, the skyscraper Bosco Verticale designed by Michel Brunello. As shown in Figure 1, the building in Milan is a green infrastructure that is widely based on vertical forests and is Leadership in Energy and Environmental Design (LEED) Gold certified. It plays a pivotal role in improving both local climate conditions and air quality.

Figure 1. Bosco Verticale. (Note: This image is derived from the Spread World Architecture https://www.archdaily.cn/)

2.2.2 Economic benefits

Choosing the right green infrastructure for the local area will greatly improve its ecological efficiency throughout its life cycle [4]. Many designers purchase and transport special equipment from distant areas in order to achieve cost savings in order to achieve the purpose of aesthetics and performance, but on the other hand, the transportation link will generate additional carbon emissions, which greatly reduces the economic benefits of green infrastructure. Each type of green infrastructure has its own characteristics, aesthetic potential, and maintenance needs, and should be selected based on a combination of the characteristics of the building and site climate [6], such as using vegetation and natural landscapes to control rain runoff to reduce the construction and maintenance costs of drainage systems, so as to achieve a win-win situation of economic and ecological benefits, and reduce costs by reducing the need for traditional infrastructure. In addition, green infrastructure promotes economic development by maintaining the local ecological environment by attracting investment and businesses.

2.2.3 Social benefits

The widespread application of green infrastructure in architecture is mainly due to its environmental restoration effect, but few studies have explored the aesthetic preferences of urban residents for green infrastructure and how the underlying cultural and historical factors affect people's perceptions of other urban natural systems [7], but the aesthetic preferences of urban residents for green infrastructure also affect the various social benefits it brings. The vast majority of green infrastructure provides places for recreation, physical exercise and social activities, improving the living environment and physical and mental health of residents, enhancing connections and interactions between residents, and strengthening community cohesion. It also creates employment opportunities for local residents, especially for some economically underdeveloped areas or disadvantaged groups, providing more development opportunities and promoting social equity.

3 Planning and design principles for green infrastructure

3.1 Principles of design

As a technical approach to construct urban ecosystems, improve human settlements, and improve the quality of urban open space, green infrastructure is of great significance for the optimization of the layout and service function of parks and green spaces in the context of China's new urbanization [8]. It is a nature-based solution for cities, integrating natural ecosystems and landscapes into urban planning and design to improve the sustainability and resilience of cities. In addition, the design of green infrastructure should follow the following principles, as detailed in Table 2.

Table 2. Design principles for green infrastructure.

Principles of design	The content of the principles
Coherence	The planning of green infrastructure should carry out the overall planning of the construction of urban and rural green infrastructure, strengthen the connectivity between different regions, and form a green infrastructure network.

Systematic issue	The planning of green infrastructure should take into account the stability of elements between different scales and different fields, and connect the functions of different green infrastructure spatially to form a complete system.
Connectivity	The planning of green infrastructure should connect different ecological resources, and at the same time connect government departments, organizations and individuals to realize the rational use and layout of natural resources.
Promptness	The planning of green infrastructure is a dynamic process, which should be revised according to the actual environment and characteristics of the city, and the optimal use of resources can be achieved through the formulation of a sound planning strategy.`

3.2 Planning for green infrastructure

As an emerging aesthetic symbol, green infrastructure with natural plants faces some challenges when integrating into the modern built environment [4]. The planning needs to consider a variety of factors to improve the sustainability of the city through reasonable planning and implementation, and the detailed steps are shown in Table 3.

Table 3. Planning steps for green infrastructure.

Steps for planning	Content
Define and structure the goals	Formulate matching construction goals for green infrastructure at different levels and scales, focus on different green infrastructure functions, and plan the natural resource elements that need to be incorporated.
Data collection and collation	Through scientific methods, the landscape data within the region are collected, and the evaluation criteria are formulated for data screening and sorting, so as to classify and grade the existing natural resource elements.
Connect facility features	Identify the core protected areas, and connect the natural resource elements at different levels around the core protected areas, and complete the layout and planning.
Set the network hierarchy	The elements of green infrastructure are hierarchically graded, the priority level of ecological protection goals and corresponding planning measures are determined, and specific planning strategies are formed.
Looking for feedback	In the process of green infrastructure planning, we widely collect feedback from relevant departments, organizations and individuals, and continuously improve the green facility planning plan.

4 Plant physiology and urban gardening

4.1 Basic Concepts in Plant Physiology

Plant physiology is a science that studies the laws of plant life activities and their interrelationship with the environment, and reveals the nature of plant life phenomena [7]. Starting from the 20th century, with the great development of physics, chemistry and cell biology and other disciplines, plant physiology has penetrated into the level of cells, organelles and biomolecules [8-9].

4.2 Application of plant physiology in urban gardens

Plant physiology is the science of studying the laws of plant life activities, and its application in urban gardens can help us better understand and manage plants to improve the quality and effectiveness of urban gardens. For this reason, four of them that most help to carry out garden plant management methods are listed, as detailed in Table 4.

Table 4. Application of plant physiology in urban gardens.

Applications	Element
Plant Selection	Based on the physiological characteristics of plants, plant species are selected to suit local climatic and soil conditions. For example, drought-tolerant plants are selected in arid areas and cold-resistant plants are selected in cold areas to improve plant survival and growth.
Plant configuration	According to the physiological characteristics of plants, plants are reasonably configured to improve their ecological benefits and landscape effects. For example, in urban gardens, light-loving plants can be configured in sunny places and shade-tolerant plants in shady places to improve the growth rate of plants and landscape effect.
Plant care	According to the physiological characteristics of plants, scientific maintenance and management of plants is carried out to improve the growth rate and resistance of plants. For example, rational watering according to the water demand and transpiration of plants; rational fertilization according to the nutrient demand of plants; and rational pruning according to the growth cycle of plants.
Plant protection	According to the physiological characteristics of plants, pest and disease control is carried out on plants in order to improve their health status. For example, suitable pesticides and control methods are selected according to the plant's resistance to pests and diseases, and the timing and concentration of pesticides are rationally adjusted according to the plant's growth cycle.

5 The relationship between green infrastructure and urban sustainability

5.1 Impact of green infrastructure on urban ecosystems

Green infrastructure has the stability of eco-efficiency, and the eco-efficiency it produces presents the characteristics of stability and sustainability. [10]. Green infrastructure can reduce the occurrence of urban flood disaster through the water storage and drainage function of vegetation, and improve the service function of urban ecosystem. It plays a vital role in air purification, water purification, climate regulation and many other aspects. It also promotes sustainable urban development while improving the quality of urban ecosystems.

5.2 Contribution of green infrastructure to urban socio-economic development

Green infrastructure is a flexible and versatile green building technology, and the use of different technologies for different environments can improve the benefits of green infrastructure [5]. And it contributes to urban socio-economic development in the following ways:

(1) Improve urban ecological quality: improve urban air quality, water quality and soil quality, reduce noise and pollution, improve the quality of urban ecological environment, and enhance the quality of life and health of residents.

(2) Promote economic development: create job opportunities and promote the development of related industries, such as construction, gardening and environmental protection. At the same time, the construction and maintenance of green infrastructure also requires a large amount of capital and technical investment, which can stimulate domestic demand and promote economic growth.

(3) Enhance city image: Improve the cityscape, enhance the image and attractiveness of the city, attract more talents and enterprises to move in, and promote the development and prosperity of the city.

(4) Enhance urban resilience: improve the city's ability to cope with natural disasters and climate change, reduce disaster risks and losses in the city, and safeguard the sustainable development of the city.

(5) Promote social harmony: provide more public space and recreational facilities, promote communication and interaction among residents, and enhance community cohesion and social harmony.

In conclusion, the contribution of green infrastructure to urban socio-economic development is multifaceted. It can improve the ecological quality of cities, promote economic development, enhance urban image, strengthen urban resilience and promote social harmony, and it is one of the most important means of realizing sustainable urban development.

6 Conclusion

Urban gardens and plant physiology play an important role in green infrastructure and urban sustainability. Through rational urban garden design and scientific application of plant physiology, the initiatives of stock quality improvement and renovation and incremental structure optimization are proposed in response to the heterogeneity and orderliness of the spatial configuration of various types of supply and demand in the municipal jurisdiction [1]. It can improve the ecological environment quality of the city and promote the sustainable development of the city. Therefore, the research and application of urban gardens and plant physiology should be strengthened to make greater contributions to the sustainable development of cities.

References

1. Zhao Haixia, Gu Binjie, Wang Junqi et al. Characteristics of the evolution of supply-demand fitness relationship of green infrastructure and its law--Taking Nanjing as an example [J/OL]. Journal of Ecology, 2024, (02): 1-13[2024-01-02]
2. Zhu Jasheng. Application of low carbon concept in urban garden landscape design [J]. Stone, 2023, (10): 76-78.
3. LIU Hailong, LI Dihua, HAN Xili. A review on the concept of eco-infrastructure and its research progress [J]. Urban Planning, 2005, (09): 70-75.
4. Wang Qiyao. Development of green infrastructure in modern architecture [J]. China Building Decoration and Decoration, 2023, (20): 71-73.
5. Gao M. Green Infrastructure and Ecological Environmental Protection in Cities [J]. Environment and Development, 2018, 30 (07): 181+183.
6. ZHANG Yifei,LI Jiahui,WANG Yiwei. A study of green infrastructure performance in Chinese cities - a nature-based solution [J]. Ecological Economy, 2022, 38 (11): 100-107.
7. DU Ting,WU Xiangguo,FANG Zhaoxin et al. Analyzing the development direction and way of green wall materials in China [J]. Concrete and Cement Products,2006(2):48-51.
8. WANG Peng,YAO Peng,JIANG Zixia. Spatial planning and design practice of urban comprehensive parks under the perspective of green infrastructure [J]. Forestry Science and Technology Newsletter, 2022, (09): 14-19.
9. WANG Yongfei, MA Sanmei, LI Hongye. Plant physiology [M]. Beijing: Science Press, 2019.
10. CHEN Xiaoya, HE Zuhua, FAN Pei et al. Review and prospect of plant physiology [J]. Journal of Agronomy, 2018, 8 (01): 7-11.

The Relationship Between Engagement with Natural Beauty and Pro-environmental Behaviors Using Structural Model Analysis

Hongqijie Wang[1,2] and Pingping Liu[1,2,*]

[1] CAS Key Laboratory of Mental Health, Institute of Psychology, Chinese Academy of Sciences, 16 Lincui Road, Beijing 100101, China
[2] Department of Psychology, University of Chinese Academy of Sciences, 19A Yuquan Road, Beijing 100049, China

Abstract: Addressing environmental issues requires enhancing pro-environmental behaviors (PEBs). This study examines the relationship between engagement with natural beauty (ENB) and PEBs, focusing on the mediating roles of awe and connectedness to nature (CN). A total of 387 participants from 17 provinces in China completed self-report measures via an online survey platform, assessing ENB, awe, CN, and PEBs. All scales used were well-established, with internal consistencies ranging from 0.79 to 0.89. Structural equation modeling showed that ENB significantly predicts PEBs, with awe and CN as parallel mediators. Additionally, ENB predicts PEBs through a serial mediation pathway involving both awe and CN. These findings enhance the understanding of the ENB-PEB relationship, offering new theoretical perspectives and practical implications. Specifically, promoting exposure to and appreciation of natural beauty can increase ENB, encouraging more pro-environmental behaviors.

1. Introduction

Environmental issues continue to threaten human survival and development, yet current actions to mitigate these issues remain insufficient [1]. The appreciation of natural beauty may offer a solution. According to the Stress Reduction Theory and the Attention Restoration Theory, nature's fascination is a core factor in its restorative effects [2]. Therefore, we propose that the perception of natural beauty could promote pro-environmental behaviors (PEBs).

Engagement with natural beauty (ENB) refers to an individual's tendency to perceive and emotionally engage with the beauty of nature [3]. Previous studies have confirmed that ENB positively predicts PEBs [3, 4, 5], but the underlying mechanisms remain unclear. This study aims to clarify the link between ENB and PEBs, focusing on the possible mediating roles of awe and connectedness to nature.

ENB is known to foster positive emotions [6], and awe, as a positive emotion, is likely associated with ENB. Awe is often elicited by vast, overwhelming stimuli, with natural landscapes being a common trigger [7]. Prior research indicated that individuals with higher ENB reported stronger dispositional awe [8]. Additionally, connectedness to nature (CN)—the subjective feeling of being connected to nature—is strongly related to ENB [4, 9]. Diessner et al. found that ENB predicts PEBs through the mediation of CN [5].

Awe also promotes PEBs. Individuals with high dispositional awe report more ecological behaviors [10]. As awe is a self-transcendent experience characterized by decreased self-salience and increased connectedness with others and the environment, it may positively correlate with CN, as confirmed by previous research [1, 11, 12]. Given the stable association between CN and PEBs [13, 14], awe may also promote PEBs through CN.

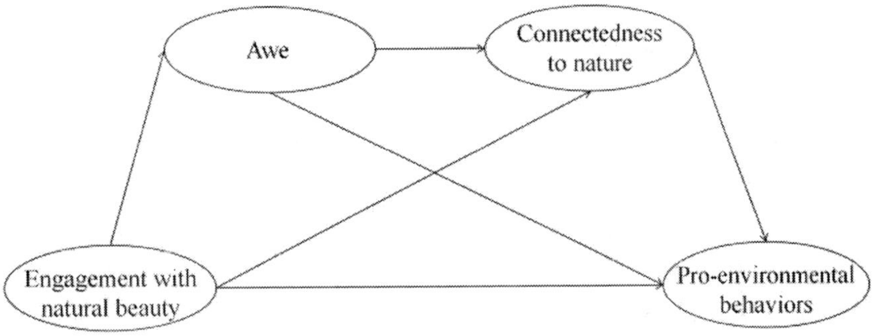

Fig. 1. Hypothesized conceptual model.

* Corresponding author: liupp@psych.ac.cn

To address gaps in previous research on the mechanisms linking ENB and PEBs, the current study explores how individuals' perception of natural beauty (ENB) promotes PEBs through awe and CN. We developed an integrated model incorporating these variables, proposing effective approaches to benefit environmental well-being.

Based on the above analysis, we propose the following hypotheses (see Figure 1):

Hypothesis 1: ENB positively predicts PEBs.

Hypothesis 2: Awe and CN serve as parallel mediators in the relationship between ENB and PEBs.

Hypothesis 3: Awe and CN serve as a chain mediation pathway in the relationship between ENB and PEBs.

2. Method

2.1 Participants and procedures

We recruited 421 Chinese participants via the online survey platform "Wenjuanxing," covering 17 provincial-level regions. After excluding 34 invalid responses due to failed attention checks, 387 valid questionnaires were retained, including 150 males. Participants' average age was 35.94 years (SD = 12.65), ranging from 16 to 71 years. All participants provided informed consent before completing measures of ENB, awe, CN, and PEBs. Demographic data were collected, and participants received 2 RMB for participation. The study was approved by the Institutional Review Board at the Institute of Psychology, Chinese Academy of Sciences.

2.2 Measures

All measures used a Likert scale from 1 (*completely disagree*) to 7 (*completely agree*), with higher scores indicating greater levels of each construct.

Engagement with Natural Beauty. ENB was assessed using Diessner et al.'s (2008) 4-item scale [15] (α = 0.83).

Awe. Awe was measured with 5 items adapted from Shiota et al.'s (2006) Dispositional Positive Emotion Scales [16] (α = 0.83).

Connectedness to Nature. CN was measured using the 10-item Connectedness to Nature Scale by Mayer and Frantz (2004) [13] (α = 0.89).

Pro-environmental Behaviors. PEBs were assessed with 6 items from established scales [17], tailored to Chinese daily life (α = 0.79).

2.3 Data Analysis

Descriptive statistics were computed using SPSS 26.0. Measurement models were validated with Mplus 8.0, followed by structural equation modeling. Bootstrap procedures tested mediating effects.

3. Results

3.1 Descriptive statistics and correlation analysis

Pearson correlation analyses were conducted among ENB, awe, CN, and PEBs (see Table 1). Significant positive correlations were found among all variables. Specifically, ENB was positively correlated with PEBs (r = 0.35, p < 0.01), supporting Hypothesis 1.

3.2 Measurement model evaluation

All item factor loadings for the latent variables (ENB, awe, CN, and PEBs) were above 0.40 and significant (p < 0.001), indicating adequate representation of the constructs [18]. Confirmatory factor analysis for the overall measurement model showed good fit: χ^2/df = 2.726, CFI = 0.925, RMSEA = 0.067 (90% CI: 0.060, 0.074), SRMR = 0.065. These indices suggest a well-fitted measurement model [18], allowing for further structural model analysis.

Table 1. Descriptive statistics and correlation analysis.

Constructs	*M*	*SD*	1	2	3	4
ENB	5.90	0.80	1			
Awe	5.39	0.87	0.50**	1		
CN	5.89	0.72	0.60**	0.62**	1	
PEBs	5.14	1.04	0.35**	0.46**	0.47**	1

Note. ENB = engagement with natural beauty, CN = connectedness to nature, PEBs = pro-environmental behaviors. ** p < 0.01.

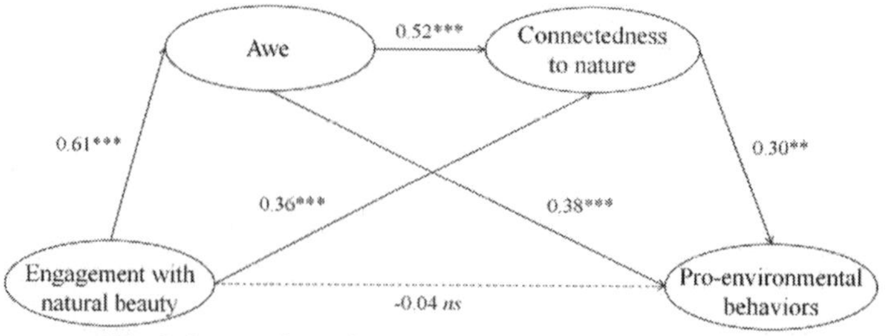

Fig. 2. Structural equation model of ENB, awe, CN, and PEBs.

Note. All coefficients are standardized. Solid lines indicate significant paths, dashed lines indicate non-significant paths. ** p < 0.01, *** p < 0.001.

3.3 Structural model analysis

Controlling for demographic variables, we tested a structural equation model based on the research hypotheses. The model showed good fit: $\chi^2/df = 2.349$, CFI = 0.919, RMSEA = 0.059 (90% CI: 0.052, 0.066), SRMR = 0.062. Therefore, the model is acceptable (see Figure 2). All paths were significantly positive except for the path from ENB to PEBs, which was not significant ($\beta = -0.04$, $p = 0.64$).

3.4 Mediation analysis

To test the mediation effects of various indirect paths, we used the Bootstrap method with 5000 resamples to calculate the 95% confidence intervals. If the confidence interval does not include zero, the mediation effect is significant. The results (see Table 2) show that all indirect paths have significant mediation effects, and the total effect of ENB on PEBs is significant.

Table 2. Mediation analysis.

Pathways	Estimate	95% CI	
		Lower 2.5%	Upper 2.5%
ENB → awe → PEBs	0.233***	0.122	0.381
ENB → CN → PEBs	0.105*	0.028	0.211
ENB → awe → CN → PEBs	0.093*	0.029	0.171
Total effect: ENB → PEBs	0.392***	0.250	0.507

Note. ENB = engagement with natural beauty, CN = connectedness to nature, PEBs = pro-environmental behaviors, CI = confidence interval. * $p < 0.05$, *** $p < 0.001$.

4. Discussion

Path analysis and mediation tests showed that ENB significantly affects PEBs ($b = 0.392$, $p < 0.001$), supporting Hypothesis 1. Awe and CN act as parallel mediators between ENB and PEBs, supporting Hypothesis 2. Additionally, there is a serial mediation effect through awe and CN, supporting Hypothesis 3.

This study found a positive relationship between ENB and PEBs, which is consistent with prior studies [3, 4, 5]. Since PEBs are forms of prosocial behaviors [19], their positive prediction by ENB may stem from its prosocial nature: ENB correlates with prosocial behaviors [6]. However, the direct effect of ENB on PEB was not significant, indicating full mediation by awe and CN.

This study extends Diessner et al. (2018) [5] by validating the multiple mediation effects of awe and CN. ENB predicts PEBs through CN, suggesting emotional and physiological responses to natural beauty foster a connection with nature, increasing PEBs.

Awe acts as both a parallel mediator and a serial mediator through the "awe-CN" pathway. Natural beauty often induces awe, and those who frequently perceive it are more likely to experience awe, supporting the Stress Reduction Theory: natural beauty alleviates stress and restores positive emotions [20]. The positive prediction of PEBs by awe through CN aligns with Yang et al. (2018) [11]. Awe fosters self-transcendence, allowing individuals to integrate into a larger world beyond themselves (such as society and the natural environment), thereby increasing prosocial behaviors [7].

This study has several limitations and suggests future research directions. The cross-sectional design validates the model but does not establish causality, necessitating longitudinal studies. Additionally, while focusing on trait ENB, it is unclear if immediate perceptions of natural beauty follow the same patterns. Future research should explore how state ENB affects pro-environmental behaviors and their mechanisms.

This study constructs and validates a multiple mediation model between ENB and PEBs, addressing gaps in understanding their mechanisms. Practically, it confirms the close association between ENB and PEBs, suggesting enhancing public appreciation of natural beauty could be effective. Specifically, daily engagement with nature and immersion in its beauty have the potential to enhance individuals' levels of ENB, ultimately contributing to environmental well-being.

5. Conclusion

This study develops and confirms a multiple mediation model linking ENB to PEBs. The results indicate that ENB significantly predicts PEBs, with awe and CN as parallel mediators. Additionally, the "awe-CN" pathway serves as a serial mediator.

References

1. Wang, C., Geng, L., & Casallas, J. D. R. (2022). Mindfulness to climate change inaction: The role of awe, "Dragons of inaction" psychological barriers and nature connectedness. Journal of Environmental Psychology, 84, 101912. https://doi.org/10.1016/j.jenvp.2022.101912
2. Devera, A. (2023). Restorative attributes of nature-based interventions that promote psychological well-being in adults: A systematic literature review (Doctoral dissertation, California State University, Northridge).
3. Diessner, R., & Niemiec, R. M. (2023). Can beauty save the world? Appreciation of beauty predicts pro-environmental behavior and moral elevation better than 23 other character strengths. Ecopsychology, 15(2), 93-109. https://doi.org/10.1089/eco.2022.0047
4. Barrows, P. D., Richardson, M., Hamlin, I., & Van Gordon, W. (2022). Nature connectedness, nonattachment, and engagement with nature's beauty predict pro-nature conservation behavior. Ecopsychology, 14(2), 83-91. https://doi.org/10.1089/ eco.2021.0036

5. Diessner, R., Genthôs, R., Praest, K., & Pohling, R. (2018). Identifying with nature mediates the influence of valuing nature's beauty on proenvironmental behaviors. Ecopsychology, 10(2), 97-105. https://doi.org/10.1089/eco.2017.0040
6. Zhang, J. W., Piff, P. K., Iyer, R., Koleva, S., & Keltner, D. (2014b). An occasion for unselfing: Beautiful nature leads to prosociality. Journal of Environmental Psychology, 37, 61-72. https://doi.org/10.1016/j.jenvp.2013.11.008
7. Jiang, T., & Sedikides, C. (2022). Awe motivates authentic-self pursuit via self-transcendence: Implications for prosociality. Journal of Personality and Social Psychology, 123(3), 576-596. https://doi.org/10.1037/pspi0000381
8. Güsewell, A., & Ruch, W. (2014). Are musicians particularly sensitive to beauty and goodness? Psychology of Aesthetics, Creativity, and the Arts, 8(1), 96-103. https://doi.org/10.1037/a0035217
9. Gu, X., Zheng, H., & Tse, C. S. (2023). Contact with nature for emotion regulation: the roles of nature connectedness and beauty engagement in urban young adults. Scientific Reports, 13(1), 21377. https://doi.org/10.1038/s41598-023-48756-4
10. Zhao, H., Zhang, H., Xu, Y., Lu, J., & He, W. (2018). Relation between awe and environmentalism: The role of social dominance orientation. Frontiers in Psychology, 9, 2367. https://doi.org/10.3389/fpsyg.2018.02367
11. Yang, Y., Hu, J., Jing, F., & Nguyen, B. (2018). From awe to ecological behavior: The mediating role of connectedness to nature. Sustainability, 10(7), 2477. https://doi.org/10.3390/su10072477
12. Yaden, D. B., Haidt, J., Hood Jr, R. W., Vago, D. R., & Newberg, A. B. (2017). The varieties of self-transcendent experience. Review of General Psychology, 21(2), 143-160. https://doi.org/10.1037/ gpr0000102
13. Mayer, F. S., & Frantz, C. M. (2004). The connectedness to nature scale: A measure of individuals' feeling in community with nature. Journal of Environmental Psychology, 24(4), 503-515. https://doi.org/10.1016/j.jenvp.2004.10.001
14. Chen, X., Tang, J., & Liu, P. (2024). How place attachment affects pro-environmental behaviors: The role of empathy with nature and nature relatedness. Current Psychology, 43, 20571-20583. https://doi.org/10.1007/s12144-024-05820-0
15. Diessner, R., Solom, R. D., Frost, N. K., Parsons, L., & Davidson, J. (2008). Engagement with beauty: Appreciating natural, artistic, and moral beauty. The Journal of Psychology, 142(3), 303-332. https://doi.org/10.3200/JRLP.142.3.303-332
16. Shiota, M. N., Keltner, D., & John, O. P. (2006). Positive emotion dispositions differentially associated with Big Five personality and attachment style. The Journal of Positive Psychology, 1(2), 61-71. https://doi.org/10.1080/17439760500510833
17. Kaiser, F. G., & Wilson, M. (2000). Assessing People's General Ecological Behavior: A Cross-Cultural Measure. Journal of Applied Social Psychology, 30(5), 952-978. https://doi.org/10.1111/ j.1559-1816.2000.tb02505.x
18. Guo, J., Zhang, J., & Pang, W. (2021). Parental warmth, rejection, and creativity: The mediating roles of openness and dark personality traits. Personality and Individual Differences, 168, 110369. https://doi.org/10.1016/j.paid.2020.110369
19. Zelenski, J., & Desrochers, J. (2021). Can positive and self-transcendent emotions promote proenvironmental behavior? Current Opinion in Psychology, 42, 31-35. https://doi.org/10.1016/ j.copsyc. 2021.02.009
20. Luo, L., & Jiang, B. (2022). From oppressiveness to stress: A development of Stress Reduction Theory in the context of contemporary high-density city. Journal of Environmental Psychology, 84, 101883. https://doi.org/10.1016/j.jenvp. 2022. 101883

The Relationship Between Late Devonian Environmental Evolution and Organic Matter Enrichment

Yingying Ma*

Beijing Haidian Kaiwen Academy, Beijing 100000, China

Abstract. This study investigates the interplay between environmental evolution and organic matter (OM) accumulation during the Late Devonian, with a focus on the Frasnian-Famennian (F-F) transition in the Lengshuijiang section, South China. By analyzing geochemical proxies such as Total Organic Carbon (TOC), Phosphorus/Aluminum (P/Al) ratios, Chemical Index of Alteration (CIA), and Co/Ni ratios, we aim to elucidate the factors that controlled OM enrichment during this period. Our findings suggest that the processes driving OM accumulation, including primary productivity and preservation mechanisms, were closely linked to significant environmental changes. The data reveal that periods of high TOC coincided with anoxic or euxinic conditions, driven by global sea-level changes and oceanic stagnation, contributing to increased environmental stress and the F-F mass extinction. Additionally, fluctuations in CIA values indicate alternating warm and humid to cooler, more arid climatic phases, which likely exacerbated environmental stress and disrupted marine ecosystems. This research highlights the interconnectedness of climate, ocean chemistry, and biological evolution during periods of significant environmental upheaval, offering valuable insights into the processes that drove one of Earth's most profound biotic crises.

1 Introduction

The accumulation of organic matter (OM) in sedimentary archives is crucial for reconstructing ancient environmental conditions and assessing hydrocarbon resources.

Currently, the two most commonly accepted models for enhanced OM accumulation are the primary productivity model and preservations model. The primary productivity model suggests that high biological productivity in surface waters leads to the deposition of large quantities of organic material on the seafloor. This productivity is typically driven by the prosperity of primary producers, which are high sensitive to nutrient influxes. The nutrient supple, mainly associate with P, Ba, Zn and other biolimited elements, can result from volcanic activity, continental weathering, and other processes that enhance the bloosom of phytoplankton and other microorganisms[1]. The subsequent death and settling of these organisms contribute to significant OM input on the seabed[2]. This model implies that nutrient-rich environments, such as those found in upwelling zones, are likely to exhibit substantial OM accumulation [3].

In contrast, the preservation model emphasizes the role of depositional conditions in inhibiting the decomposition of OM after it has been deposited. Anoxic (oxygen-depleted) and euxinic (sulfide-rich) conditions are particularly favorable for OM preservation, as they prevent the aerobic degradation of organic material. During the Late Devonian, widespread anoxia, potentially driven by global sea-level changes and oceanic stagnation, created conditions that favored the preservation of OM [4]. Evidence for such anoxic conditions is found in the extensive black shale deposits from this period [2][5]. The Late Devonian period, known for its significant climatic and environmental transformations, presents an unique opportunity to examine the mechanisms driving OM enrichment. This study focuses on the interplay between primary productivity and preservation mechanisms within the context of Late Devonian environmental evolution.

The Late Devonian environment was marked by notable fluctuations, including a significant cooling event known as the Devonian glaciation and oceanic anoxia expansion [6][7]. These climatic changes affected sea levels, ocean circulation, and weathering processes, which in turn influenced the redox conditions of marine and lacustrine environments. Redox reactions, mostly driven by changes in oxygen availability, played a crucial role in determining biohabitability of organisms and the preservation potential of OM in sediments[8]. The transition from oxic to anoxic conditions facilitated the preservation of OM by reducing rates of decomposition and oxygen respriation[9].

This research paper aims to explore the relationship between Late Devonian environmental evolution and organic matter accumulation, with a focus on the primary productivity and preservation models based geological and geochemical investigations on the Upper Devonian shale-carbonate succession in South China. By examining geochemical proxies in sedimentary records across the Frasnian-Famenian transition, this study seeks to

* Corresponding author: 20190703005@hd.kaiwenacademy.cn

elucidate the factors that controlled OM enrichment during this period and their paleoenvironmental implications. The findings will contribute to a deeper understanding of the processes that drive OM accumulation and preservation in ancient sedimentary environments and environmental perturbations in South China that potentially led to biotic turover, offering insights into the conditions that prevailed during one of the most dynamic periods in Earth's history.

2 Method

2.1 Sample Collection and Preparation

Samples were collected from the Lengshuijiang section, focusing on the Late Devonian interval. This section is dominated by shale-carbonate succession, and located in the Hunan Province, South China and has been systematically studied for biostratigraphy, sedimentology, and chemostratigraphy across the F-F extinction event[10][11]. The collected shale samples were carefully cleaned, dried, and pulverized for geochemical analysis.

2.2 Total Organic Carbon (TOC) Analysis

To determine the Total Organic Carbon (TOC) content, approximately 200 mg of each shale sample was subjected to a two-step acid treatment to remove any carbonate material. The samples were treated with 10% (volume) hydrochloric acid (HCl) at 60 °C for 12 hours twice, ensuring complete carbonate removal. After acid treatment, the samples were thoroughly washed with distilled water to remove any remaining HCl. The washed samples were then dried overnight at 50 °C and weighed. The TOC content was subsequently analyzed using a Euro-3000 analyzer at the Sample Solution Inc. Wuhan.

2.3 X-ray Fluorescence Spectrometry (XRF)

XRF was employed to determine the concentrations of selected major elements and certain trace elements (including Ba, Ni, Cr, and V). For the determination of the Loss on Ignition (LOI), approximately 500 mg of pulverized sample was oxidized at 1000 °C for 1 hour. The oxidized sample was then cooled and thoroughly mixed with 5000 mg of lithium borate.

2.4 Chemical Index of Alteration (CIA)

The Chemical Index of Alteration (CIA) was calculated to evaluate the intensity of chemical weathering for the F-F interval of Lengshuijiang section. CIA values were calculated using the following formula:

$$CIA = \left[\frac{Al_2O_3}{(Al_2O_3+CaO^*+Na_2O+K_2O)}\right] \times 100$$

where CaO^* represents the CaO content in the silicate fraction. To correct for the presence of phosphate, the available P2O5 data was used to adjust the CaO values, calculated as CaO* = mole CaO – mole P2O5 × 10/3 [12]. If the corrected CaO* value was higher than the Na2O content, it was assumed to be equivalent to the Na_2O content [12]. This method of calculating the CIA provides insights into the intensity of chemical weathering and is based on established protocols[13][14][15][16].

3 Discussion

The stratigraphic variations in TOC, P/Al, CIA, and Co/Ni, as illustrated in Figure 1, provide valuable insights into the paleoenvironmental conditions during the Late Devonian period at the Lengshuijiang section. These geochemical proxies are crucial for understanding the factors influencing organic matter (OM) accumulation and preservation within this sedimentary sequence.

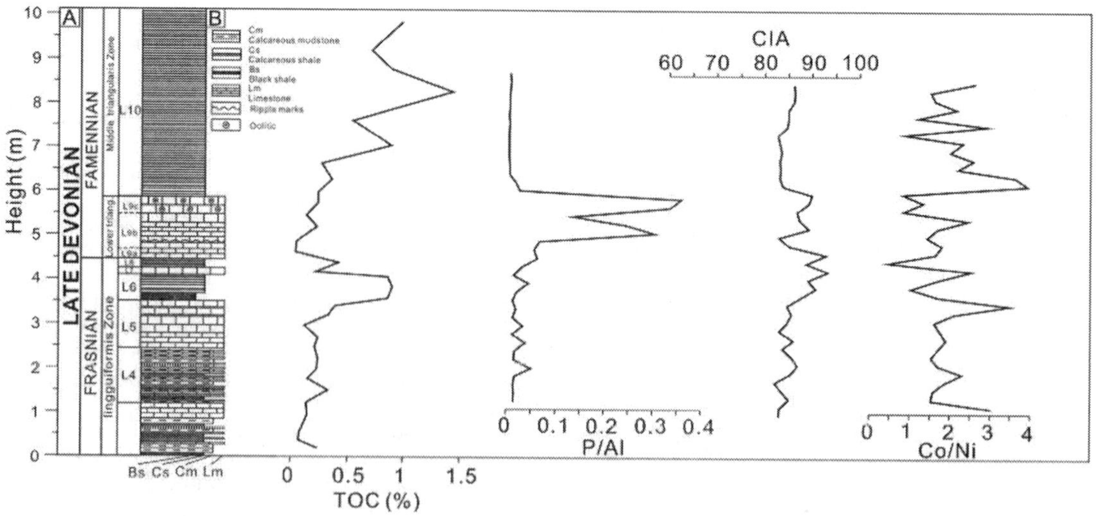

Figure 1: Lithology, TOC, P/Al, CIA, and Co/Ni variations in the stratigraphy column of the Lengshuijiang section.

3.1 Total Organic Carbon (TOC) Variations

The TOC profile displayed in Figure 1 shows significant fluctuations throughout the stratigraphy, with elevated concentrations observed at approximately 3–4 meters and 7–8 meters in height, where TOC values reach up to 1.5%. These elevated TOC intervals likely correspond to periods of enhanced primary productivity or improved preservation conditions. According to the primary productivity model, increased biological activity in surface waters leads to a higher deposition rate of organic material on the seafloor[2][3][17][18]. The preservation model further emphasizes the role of anoxic or euxinic conditions in minimizing organic matter decomposition [19]. The co-occurrence of high TOC values with elevated Co/Ni ratios within these intervals supports the hypothesis that anoxic conditions were prevalent during these periods and conducive to organic matter preservation. In contrast, lower TOC values, ranging from 0.1% to 0.3%, are observed at approximately 0–2 meters and 5–6 meters in height. These intervals may indicate either reduced primary productivity or more oxic conditions, where the degradation of organic matter outpaced its accumulation[20]. The correlation between decreasing Co/Ni ratios and lower TOC values suggests that more oxic conditions, which promote organic matter decomposition, were likely responsible for the observed reduction in TOC[19].

3.2 Phosphorus/Aluminum (P/Al) Ratio

The Phosphorou is one of the most essential bio-limited elements to sustain life largely through phosphates, compounds containing the phosphate ion, PO_4^{3-}.The P/Al ratio serves as a proxy for paleoproductivity. The observed variations in P/Al across the stratigraphy are indicative of fluctuating nutrient levels, which directly influence primary productivity. In Figure 1, high P/Al ratios are observed between approximately 5-6 meters in height above the F-F boundary, where the ratios reach up to 0.35. These drastically elevated P/Al ratios indicate periods of increased nutrient availability during the F-F transition, which likely drove higher biological productivity in surface waters. Such conditions could be attributed to enhanced continental weathering or upwelling, which supplies the necessary nutrients for phytoplankton growth. The intervals with elevated P/Al ratios, corresponding with higher TOC values, suggest that increased nutrient availability and productivity were key factors driving organic matter accumulation during these periods. In contrast, low P/Al ratios are observed at approximately 0–4 meters and 7-9 meters in height, where the ratios drop to around 0.05. These lower ratios may reflect periods of reduced nutrient influx, leading to diminished primary productivity. Additionally, under anoxic conditions, phosphorus tends to diffuse from sediments back into the water column, potentially explaining the observed decrease in P/Al ratios in the organic-rich intervals.

3.3 Chemical Index of Alteration (CIA)

The CIA values provide insights into the intensity of chemical weathering and, by extension, the prevailing climatic conditions during sediment deposition. In Figure 1, high CIA values are observed between approximately 3–4 meters and 5-6 meters in height across the F-F transition, where the CIA values range from 85 to 90. These values suggest a warm and humid tropical climate, which would have intensified chemical weathering processes [13][14]. Such a climate would have promoted greater nutrient fluxes into the marine system, thereby enhancing primary productivity and contributing to organic matter accumulation. Low CIA values, ranging from 70 to 75, are observed at approximately 0–2 meters and 5–6 meters in height. These values correspond to colder and more arid conditions, where weathering processes would have been less intense, reducing nutrient supply and limiting primary productivity that climatic fluctuations played a significant role in controlling the delivery of nutrients to the marine environment, thereby influencing organic matter accumulation and preservation.

3.4 Co/Ni Ratio

The Co and Ni are commonly used redox-sensitive elements. The Co/Ni ratio, as shown in Figure 1, is utilized as a proxy for redox conditions, with higher values. In the stratigraphy, high Co/Ni ratios (ranging from 3.0 to 3.8) are observed between approximately 3–4 meters and 7–8 meters in height. These intervals suggest the presence of anoxic or euxinic conditions, which are favorable for the preservation of organic matter [19]. The correlation between these high Co/Ni ratios and elevated TOC values further supports the interpretation that reducing conditions were prevalent during these periods, leading to enhanced preservation of organic matter. Conversely, low Co/Ni ratios (ranging from 0.5 to 1.5) are observed at approximately 0–2 meters and 5–6 meters in height. These intervals likely reflect more oxic conditions, where the preservation of organic matter was less efficient due to increased aerobic degradation. The decrease in Co/Ni ratios correlates with lower TOC values, indicating that the redox conditions during these intervals were less conducive to the preservation of organic matter.

3.5 Synthesis and Paleoenvironmental Implications

The interplay between primary productivity (P/Al), preservation potential (Co/Ni), and weathering intensity (CIA) reveals a complex environmental evolution during the Late Devonian at the Lengshuijiang section. The high TOC intervals coincide with periods of high primary productivity and/or strong OM preservation under anoxic conditions. Conversely, lower TOC values reflect either reduced productivity or more oxic conditions that favored OM decomposition. The climatic fluctuations, as inferred from the CIA values, suggest alternating warm/humid and cold/arid phases, which would have significantly impacted the nutrient supply and, consequently, the primary productivity. When compared with the Bakken Shale in North America, similar euxinic conditions are observed during key transgressive phases, where trace metals such as molybdenum (Mo) and vanadium (V) indicate the episodic expansion of euxinia. Both sections reveal that euxinia played a critical role in OM preservation, with high Mo and V concentrations signaling intense anoxia in the Bakken Shale, paralleling the Co/Ni-driven redox insights from the Lengshuijiang section. The geochemical data collectively suggest that both primary productivity and preservation mechanisms were critical in controlling OM accumulation during this period, with climate playing a pivotal role in modulating these processes. The presence of photic zone euxinia (PZE) in the Bakken Shale further underscores the importance of surface water oxygen depletion, which may have occurred during similar intervals of enhanced primary productivity in both basins. This study contributes to a better understanding of the factors driving OM enrichment in ancient sedimentary environments, particularly during periods of significant climatic and environmental change such as the Late Devonian. The findings emphasize the importance of integrating multiple geochemical proxies to reconstruct past environmental conditions and their influence on sedimentary OM accumulation.

3.6 Implications for Environmental Change and the Frasnian-Famennian (F-F) Mass Extinction

The occurrence of shale sandwiched by carbonate indicates rapid sea level rise during the F-F transition. This could be related to the progressive onlapping of euxinic waters found here and in North America, which has major implications for the massive organic matter accumulation and, thereby, the Late Devonian mass extinction. The environmental changes became unstable through our analysis of the Lengshuijiang section. These phenomena are not only important for understanding organic matter (OM) accumulation but also hold important implications for the biotic crises of the Late Devonian, particularly the Frasnian-Famennian (F-F) mass extinction. This extinction event, one of the "Big Five" mass extinctions, is characterized by a severe loss of marine biodiversity, including significant declines in reef-building organisms and a variety of pelagic species. The data presented in this study suggest that the environmental stressors leading to the F-F extinction were closely linked to the same processes that influenced OM accumulation. High TOC intervals, associated with anoxic or euxinic conditions as indicated by elevated Co/Ni ratios, coincide with periods of increased environmental stress. Similar euxinic conditions are observed in the Bakken Shale, where euxinia is indicated by Mo and V enrichments, especially during the Hangenberg Event, marking another significant extinction event. In both cases, euxinia is coupled with high OM accumulation, reflecting oxygen-depleted environments that would have been detrimental to marine life. These anoxic conditions, likely driven by global sea-level changes and oceanic stagnation, would have created inhospitable environments for many marine organisms, contributing to widespread extinctions. Moreover, the fluctuations in CIA values throughout the stratigraphy reflect significant climatic shifts during the Late Devonian, alternating between warm, humid periods and cooler, more arid phases. The Bakken Shale shows similar evidence of climatic shifts, as inferred from geochemical markers, which suggest alternating periods of high productivity and euxinia, contributing to environmental instability. These climatic oscillations could have exacerbated the environmental stress, leading to disrupted ecosystems and contributing to the mass extinction. The warm, humid conditions associated with high CIA values would have intensified chemical weathering, leading to nutrient influxes that may have initially boosted primary productivity. However, as these nutrients were consumed, the subsequent anoxia and euxinia likely led to rapid declines in marine oxygen levels, creating conditions unfavorable for many forms of life. In the Bakken Shale, these conditions are mirrored by trace metal enrichments and TOC peaks, further suggesting that nutrient-driven productivity initially surged but ultimately contributed to anoxic collapse. The F-F mass extinction appears to be a culmination of these environmental factors. The stress from fluctuating redox conditions, combined with climatic instability, likely created a scenario where the resilience of marine ecosystems was compromised, leading to the widespread loss of biodiversity. The organic matter accumulation patterns we observe are thus both a record of these environmental changes and a contributor to the conditions that led to this significant extinction event. These findings highlight the interconnectedness of climate, ocean chemistry, and biological evolution, particularly during periods of significant environmental upheaval. Both the Bakken Shale and the Lengshuijiang section provide critical case studies on how localized euxinia can have global biotic implications, especially during mass extinction events like the F-F crisis. These transformative events in Earth history are encapsulated in the shale spanning from South China and North America, and this study synthesizes this massive organic matter accumulation mechanism genetically related to sea level, climate, ocean chemistry and mass extinction. The insights gained from this study provide a more nuanced understanding of the processes that drove one of the most profound biotic crises in Earth's history, offering lessons for understanding the potential impacts of modern environmental changes.

4 Conclusion

This study provides a comprehensive analysis of the relationship between Late Devonian environmental evolution and organic matter (OM) accumulation in the Lengshuijiang section. By examining the variations in geochemical proxies, including TOC, Phosphorus/Aluminum (P/Al) ratios, Chemical Index of Alteration (CIA), and Co/Ni ratios, the research elucidates the complex interplay between primary productivity, redox conditions, and climatic influences during this period. The findings indicate that periods of enhanced OM accumulation are closely linked to both increased primary productivity, as suggested by elevated P/Al ratios, and improved preservation conditions under anoxic environments, as reflected in high Co/Ni ratios. The correlation between high TOC values and these geochemical indicators underscores the dual importance of productivity and preservation mechanisms in driving OM enrichment. Furthermore, the CIA values provide crucial insights into the climatic backdrop of the Late Devonian, revealing that shifts between warm, humid conditions and cooler, arid phases played a significant role in modulating chemical weathering intensity and, consequently, nutrient fluxes into the marine system. These climatic fluctuations likely had a direct impact on primary productivity, influencing the overall patterns of OM accumulation observed in the stratigraphy. This study contributes to a deeper understanding of the environmental processes that governed OM accumulation during one of Earth's most dynamic periods. The integration of multiple geochemical proxies offers a robust framework for reconstructing paleoenvironmental conditions and provides valuable insights into the factors that control OM preservation in ancient sedimentary environments. Future research could build on these findings by exploring additional proxies and expanding the study to other regions to further refine our understanding of the Late Devonian environmental evolution.

References

1. Wang, R., Lang, X., Ding, W., Liu, Y., Huang, T., Tang, W., & Shen, B., The coupling of Phanerozoic continental weathering and marine phosphorus cycle. Scientific Reports, 10, 5794 (2020).
2. Demaison, G. J., & Moore, G. T., Anoxic environments and oil source bed genesis. In Organic Geochemistry I, Vol. 2, pp. 9-31(1981). Pergamon Press Ltd.
3. Cowie, G. L., & Hedges, J. I.,The role of anoxia in organic matter preservation in coastal sediments: Relative stabilities of the major biochemicals under oxic and anoxic depositional conditions. Advances in Organic Geochemistry 1991, Organic Geochemistry, 19(1-3), 229-234. https://doi.org/10.1016/0146-6380(92)90039-Z(1992).
4. Liu, M., Chen, D., Jiang, L., Stockey, R.G., Aseal, D., Zhang, B., Liu, K., Yang, X., Yan, D., Planavsky, N.J., Oceanic anoxia and extinction in the latest Ordovician. Earth and Planetary Science Letters 588, 117553(2022).
5. Tessin, A., Hendy, I., Sheldon, N., & Sageman, B.,Redox-controlled preservation of organic matter during "OAE 3" within the Western Interior Seaway. Paleoceanography, 30(6), 702-717 (2015). https://doi.org/10.1002/2014PA002729.
6. Sahoo, S.K., Gilleaudeau, G.J., Wilson, K., Hart, B., Barnes, B.D., Faison, T., Bowman, A.R., Larson, T.E., Kaufman, A.J., Basin-scale reconstruction of euxinia and Late Devonian mass extinctions. Nature 615, 640-645(2023).
7. Zhao, H., Shen, J., Algeo, T.J., Racki, G., Chen, J., Huang, C., Song, J., Qie, W., Gong, Y., Mercury isotope evidence for regional volcanism during the Frasnian-Famennian transition. Earth and Planetary Science Letters 581, 117412(2022).
8. Fang, C., Liu, M., Zhang, C., Tang, H., Li, J., Xing, G., Li, F., Xu, N., Wu, T., Liu, B., Middle Ordovician climatic and oceanic destabilization in a slope-setting of the Yangtze platform, South China, and its role as a regional brake on the Ordovician radiations. Palaeogeography, Palaeoclimatology, Palaeoecology 648, 112265(2024).
9. Percival, L. M. E., Bond, D. P. G., Rakociński, M., Marynowski, L., Hood, A. v. S., Adatte, T., Spangenberg, J. E., & Föllmi, K. B., Phosphorus-cycle disturbances during the Late Devonian anoxic events. Global and Planetary Change, 184, 10307(2020). https://doi.org/10.1016/j.gloplacha.2019.103070.
10. Ma, X. P., & Bai, S. L., Biological, depositional, microspherule, and geochemical records of the Frasnian/Famennian boundary beds, South China. Palaeogeogr Palaeoclimatol Palaeoecol, 181, 325–346(2002).
11. Ma, X. P., Gong, Y., Chen, D., & Racki, G., The Late Devonian Frasnian–Famennian event in South China—patterns and causes of extinctions, sea level changes, and isotope variations. Palaeogeogr Palaeoclimatol Palaeoecol, 448, 224–244(2016).
12. McLennan, S.M., Weathering and global denudation. The Journal of Geology 101, 295-303(1993).
13. Nesbitt, H. W., & Young, G. M., Early Proterozoic climates and plate motions inferred from major element chemistry of lutites. Nature, 299, 715-716 (1982).
14. Fedo, C. M., Nesbitt, H. W., & Young, G. M., Unraveling the effects of potassium metasomatism in sedimentary rocks and paleosols, with implications for paleoweathering conditions and provenance. Geology, 23, 921-924(1995).
15. Young, G. M., & Nesbitt, H. W., Paleoclimatology and provenance of the glaciogenic Gowganda Formation (Paleoproterozoic), Ontario, Canada: A chemostratigraphic approach. Geology, 111, 264-274(1999).
16. Rieu, R., Allen, P.A., Plötze, M., Pettke, T., Climatic cycles during a Neoproterozoic "snowball" glacial epoch. Geology 35, 299–302(2007).
17. Zou, C., Zhu, R., Chen, Z.-Q., Ogg, J.G., Wu, S., Dong, D., Qin, Z., Wang, Y., Wang, L., Lin, S., Cui, J., Su, L., Yang, Z., Organic-matter-rich shales of China. Earth-Science Reviews 189, 51-78(2019).
18. Fang, C., Zhang, C., Meng, G., Xu, J., Xu, N., Li, H., Liu, M., Liu, B., Constraints on the Accumulation of Organic Matter in the Upper Permian Dalong Formation from the Lower Yangtze Region, South

China. Acta Geologica Sinica - English Edition 98, 150-167(2023).

19. Tribovillard, N., Algeo, T. J., Lyons, T., & Riboulleau, A., Trace metals as paleoredox and paleoproductivity proxies: An update. Chemical Geology, 232, 12-32(2006).
20. Ge, X., Chen, D., Zhang, G., Huang, T., Liu, M., El-Shafeiy, M.,Marine redox evolution and organic accumulation in an intrashelf basin, NE Sichuan Basin during the Late Permian. Marine and Petroleum Geology 140, 105633(2022).

Emergy Analysis of Ecosystem in Qiandao Lake Asian Games Zone

Chao Wang[a], Yinrui Sun[b], Yuchen Zheng[c,*]

[a] Beijing CQC information Security Technology Co., Ltd, Fengtai, Beijing, 100070 China
[b] Hangzhou Qiandao Lake Global Tourism Co., Ltd., Chunan, Hangzhou, 310000 China
[c] China quality Certification Center, Fengtai, Beijing, 100070 China

Abstract: Under the background of double carbon targets, the sustainable development of city of major sports events become the focus of global attention. This study focuses on assessing the sustainability of the overall operations within the Qiandao Lake Asian Games Zone. The evaluation utilized an emergy analysis method, resulting in the creation of an emergy chart for the system, and the development of an evaluation index system. The findings indicate that: The environmental load rate of the Qiandao Lake competition area is 50.3, the sustainability index ESI is 0.02, and the system sustainable development emergy index EISD is 0.038. This shows that the environmental pressure in the system is high, the environmental load rate is high, and the environment is in a relatively unsustainable state of development. This is mainly due to the fact that the Qiandao Lake competition area, on the basis of tourism and sightseeing, also has the functions of hosting events and recuperating and vacationing. The operation and maintenance process of sports venues will bring a lot of energy and economic consumption. In the subsequent development, the system can increase the investment of internal renewable resources by expanding the area, reduce the investment of non-renewable resources through green lighting transformation, green transportation planning, photovoltaic system construction, etc. in the competition area, and improve the energy value use efficiency. Strengthen the system's sustainable development capabilities.

1. Introduction

The sports industry holds a pivotal role in China's ongoing economic development. Under the guidance of organizations like the United Nations and the International Olympic Committee, environmental protection initiatives linked to large-scale sports events have gained global attention[1]. Qiandao Lake, also known as Xin'anjiang Reservoir, is situated in Chun'an County, Hangzhou City, Zhejiang Province, with a small portion extending to the northwest of Jiande City, Hangzhou. It is an artificial lake created for the construction of the Xin'anjiang Hydropower Station, designed to impound the upper reaches of the Xin'anjiang River (positioned between 29°11'–30°02' north latitude and 118°34'–119°15' east longitude). The Hangzhou Asian Games Chun'an Competition Area is located within the Chun'an Qiandao Lake Tourism Resort zone, the basic area location map is shown in Fig. 1(a), encompassing a total area of approximately 8.4 square kilometers (8,400,000 m^2), with a water area of 4.2 square kilometers and a land area of 4.2 square kilometers. This area primarily hosts major events such as cycling, triathlon, and open water activities, including track cycling, mountain cycling, road cycling, BMX, triathlon, and open water swimming, among other minor events.

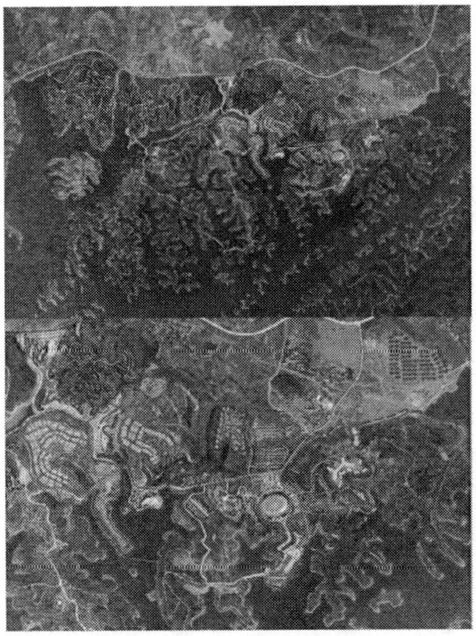

Fig. 1 Asian Games Base Location Map (a) Qiandao Lake Competition Area Plan (b).

*Corresponding author: zhengyuchencqc@163.com

As per measurements from Google Earth, the competition area spans about 2 million m^2, and the floor plan of the competition area is shown in Fig. 1(b). Utilizing emergy analysis theory and methods, the study conducts an emergy analysis on the ecosystem of the Qiandao Lake Competition area, aiming to provide theoretical support for the sustainable development of this region.

2. Material and methods

The emergy analysis for the sustainable development assessment of the Qiandao Lake Competition Area involves specific steps tailored to the research objectives. In the first step, the system boundary is defined as the Qiandao Lake Asian Games competition area, and relevant data on the natural environment, geography, and socioeconomics are systematically collected. Input and output parameters are then classified. The second step entails calculating emergy values for each component of the system using the collected data, with results recorded and analyzed. Subsequently, an emergy diagram is constructed based on the data and calculation outcomes, and an emergy index system is developed, categorizing aspects into society, ecology, and economy. Finally, the environmentally sustainable development level of the competition area is assessed using this constructed emergy index system. The ecosystem emergy diagram,shown in Fig. 2, illustrates inputs like solar and wind energy, and non-renewable resources such as electricity and water, alongside outputs including wastewater, solid waste, carbon fixation, oxygen release, stored water resources, and tourism income, providing a comprehensive foundation for evaluating the environmental sustainability of the Qiandao Lake Competition Area.

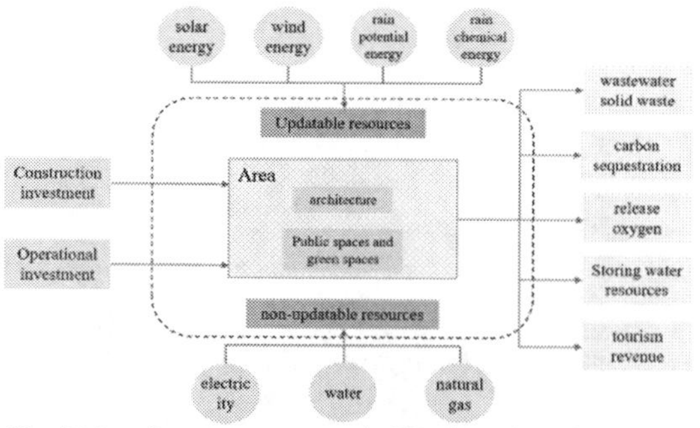

Fig. 2 Map of ecosystem energy in Qiandao Lake Asian Games area.

2.1 Calculation of energy

The local renewable natural resources absorbed by the system include solar energy, wind energy, rainwater potential energy and rainwater chemical energy. The original data is calculated as follows:

(1) Solar emergy

$$\text{Emergy} = \text{Area}\times\text{Solar radiation} \quad (1)$$

Where Area (m^2) and Solar radiation (J/m^2).

(2) Wind energy

$$\text{Emergy} = \text{Area}\times\text{Air density}\times\text{Drag coefficient}\times\text{Speed}^3 \quad (2)$$

Where Area (m^2), Air density (Kg/m^3), and Speed (m/s).

(3) Rain chemical energy

$$\text{Emergy} = \text{Area}\times\text{Rain density}\times\text{rainfall}\times G \quad (3)$$

Where Area (m^2), Rain density (Kg/m^3), rainfall (m) and G is Gibbs free energy (J/Kg).

(4) Rain potential energy

$$\text{Emergy} = \text{Area}\times\text{Rain density}\times\text{rainfall}\times g\times\text{Altitude} \quad (4)$$

Where Area (m^2), Rain density (Kg/m^3), rainfall (m), g is gravitational acceleration (m/s^2) and Altitude (m).

(5) Carbon sequestration[2]

$$\text{Emergy} = 1.63\times Rc\times\text{Area}\times X\times B \quad (5)$$

Where Area (m^2), Rc is carbon ratio in CO_2, X is Conversion coefficient of green area and B is Annual net plant productivity ($t/(hm^2\cdot a)$).

(6) Oxygen release[2]

$$\text{Emergy} = 1.19\times\text{Area}\times X\times B \quad (6)$$

Where Area (hm^2), X is Conversion coefficient of green area and B is Annual net plant productivity ($t/(hm^2\cdot a)$).

(7) Storing water resources[3]

$$\text{Emergy} = \text{Area}\times\rho\times G\times(Ro-Rg) \quad (7)$$

Where Area is the water area (m^2); ρ is the density of water (Kg/m^3); G is Gibbs free energy (J/kg); Ro is the average annual rainfall (m); Rg is the average annual evaporation (m).

2.2 Emergy index system

Table 1. Emergy index system of sustainable development in Qiandao Lake Competition area.

Index	Emergy index	Expression	Representative
Emergy flow	Renewable resource energy flow /sej	R	The resource can be renewed
	Unrenewable resource energy flow /sej	N	The resource cannot be renewed
	Input emergy	I	Import resources from the outside
	Output emergy	O	Outport resources from the outside
	Total emergy	U=R+N+I	Total emergy
Social evaluation	Emergy density ED / (sej/m^2)	U/S	Evaluation of the intensity of emergy value
Natural evaluation	Energy Exchange Law (EER)	EER = I/O	
	Energy Output Ratio (EYR)	EYR = U/I	System economic benefit
	Environmental Load Ratio (ELR)	ELR = (I*N + N)/R	Evaluation of the environmental pressure
Sustainable evaluation	Sustainable Indicator	ESI = EYR/ELR	Measurement of the sustainable development capacity of urban ecosystems
	Energy Sustainable Development Indicator	EISD = (EYR*EER)/ELR	Actual sustainable development capacity

In line with emergy analysis theory and the specific conditions of Qiandao Lake, a sustainable development index for the competition area is established. Various indicators, including emergy flow, social and natural system evaluation, and sustainable evaluation, are carefully selected and summarized, as presented in Table 1. This framework provides a comprehensive set of metrics to gauge and assess the sustainable development of the Qiandao Lake competition area.

3. Results and discussions

3.1 Results

Utilizing the original data on ecosystem energy in the competition area and the conversion rate of solar energy value, a system emergy analysis table was computed, as detailed in Table 2. Drawing upon the emergy analysis table (Table 2) and the emergy index system (Table 1) specific to the Qiandao Lake Competition area, the parameters for the sustainable development index of the competition area's ecosystem were calculated, as detailed in Table 3. This comprehensive analysis provides valuable insights into the sustainable development dynamics within the Qiandao Lake competition area.

3.2 System evaluation index analysis

The results of this study and the values of other ecosystem evaluation indicators are summarized in Table 4.

Table 2. Qiandao Lake competition area system emergy value analysis.

Project	Raw data	Solar conversion	Reference	Results
		Renewable natural resources		
Solar energy	9.28E+15	1.00E+00	[4]	9.28E+15
Chemical energy of rainwater	1.65E+13	7.00E+03		1.15E+17
Potential energy of rainwater	5.29E+12	1.28E+04		6.77E+16
Wind energy	2.62E+13	8.00E+02		2.10E+16
	Total			**2.13E+17**
		Non-updatable resource		
Electrical power	3.33E+13	1.60E+05	[5]	5.33E+18
Natural gas	3.03E+13	1.78E+05	[6]	5.39E+18
Water	2.02E+11	8.06E+04	[4,7]	1.63E+16
	Total			**1.07E+19**
		Input energy		
Construction	2.58E+09	6.20E+11	[8]	1.60E+21
Management	1.29E+08	6.20E+11		8.00E+19
Energy consumption	1.22E+07	6.20E+11		7.57E+18
	Total			**1.69E+21**
		Output energy		
Wastewater	2.02E+11	6.66E+05	[9]	1.35E+17
Solid waste	3.02E+08	1.80E+06		5.44E+14
Carbon sequestration	1.01E+09	7.49E+09	[8]	7.55E+18
Oxygen release	2.70E+09	7.49E+09		2.02E+19
Water resource	2.56E+12	4.00E+04	[3]	1.03E+17
Tourist income	1.40E+09	6.20E+11	[4,8]	8.68E+20
	Total			**8.96E+20**

Table 3. Emergy index analysis table of sustainable development in Qiandao Lake Competition area.

Index	Emergy index	Expression	Area
Emergy flow	Renewable resource energy flow /sej	R	2.13E+17
	Unrenewable resource energy flow /sej	N	1.07E+19
	Input emergy	I	1.69E+21
	Output emergy	O	8.96E+20
	Waste emergy	W	1.35E+17
	Total emergy	U=R+N+I	1.70E+21
Social evaluation	Emergy density ED / (sej/m^2)	U/S	8.49E+14
	Energy Exchange Law (EER)	EER=I/O	1.88E+00
Natural evaluation	Energy Output Ratio (EYR)	EYR=U/I	1.01E+00
	Environmental Load Ratio (ELR)	ELR=N/R	5.03E+01
Sustainable evaluation	Sustainable Indicator	ESI=EYR/ELR	0.0200
	Energy Sustainable Development Indicator	EISD=(EYR*EER)/ELR	0.0377

Table 4. Summary of ecosystem evaluation indicators.

	Qiandao Lake	Qilihai Wetland (2014)[10]	Emei Mountain [11]	China (2016)[12]
Emergy density ED / (sej/m^2)	8.49E+14	5.254E+12	4.29E+18	6.58E+12
Energy Output Ratio (EYR)	1.01	1.048	-	16.90
Environmental Load Ratio (ELR)	50.3	20.788	8.88	3.13
Sustainable Indicator	0.02	-	0.13	5.40
Energy Sustainable Development Indicator	0.038	-	0.12	-

https://doi.org/10.1051/e3sconf/202561701005

3.2.1 Social system evaluation index analysis

Emergy density, representing the emergy utilized per unit area, serves as an indicator reflecting the intensity and level of economic development for the assessed entity. A higher emergy density signifies a more developed economy and a higher standing in the hierarchical order. As shown in Table 4, in the Qiandao Lake Competition area, the energy density measures 8.49e+14 sej/m², significantly surpassing that of China's ecological economy system and Qilihai Wetland in 2016, while being lower than that of the Mount Emei Scenic Area. This observation suggests a positive impact of the Qiandao Lake ecosystem development on the residents' quality of life within the system. As the system continues to mature through subsequent development, there is the potential for further enhancement in the quality of economic development and living standards for the area's inhabitants.

3.2.2 Analysis of natural system evaluation index

The average value yield of the natural system evaluation index EYR serves as an indicator reflecting the system's economic development's dependence on external factors and the output benefits achieved. A higher EYR value suggests a more efficient energy utilization of the system. In Table 4, the emergy productivity of the Qiandao Lake Competition Area and Jinshanping Yanyun Town is comparable to that of the Qili Sea but notably lower than that of China's ecosystem in 2016. This indicates that, in comparison to the tourism development zone, the energy utilization efficiency of the Qiandao Lake Competition area is at a normal level, yet there exists substantial potential for development.

The environmental load factor (ELR), representing the ratio of energy from non-renewable resources to energy from renewable natural resources, reflects the direct coordination relationship between the social economy and the ecological environment of the system. A system environment load rate of less than 3 is considered a low load, a rate between 3 and 10 is considered a medium load, and a rate greater than 10 is considered a high load[4]. The system's environmental load ratio of 50.3 for the Qiandao Lake Competition area is considerably higher than other systems, signifying a substantial environmental pressure within the Qiandao Lake Competition area system. This is attributed to the distinctive characteristics of sports venue operation and event hosting in addition to the tourism and vacation properties of the Qiandao Lake Competition area. Furthermore, compared to areas such as Mount Emei and other scenic spots, the overall size of the competition area is currently smaller. Future attention should be directed towards the efficient use of resources or the development of surrounding areas to promote regional sustainable development.

3.2.3 Sustainable comprehensive evaluation index analysis

The analysis of the system's sustainable development capability is primarily based on two key indices, namely ESI and EISD. The Sustainability Indicator (ESI) is the ratio of the system output rate to the environmental load rate. When ESI is less than 1, it signifies a consumer economy where the system heavily depends on external resource and labor inputs, resulting in a high environmental load ratio. An ESI value between 1 and 10 indicates a more active system economy with greater development potential, while an ESI greater than 10 suggests a low level of resource development and utilization, indicative of poor economic development[11].

Both the Qiandao Lake Competition Area and Mount Emei Scenic Area have an ESI less than 1, indicating a current state of a consumption-based economy for these areas. The systems rely on external resource inputs, resulting in a high environmental load rate. This suggests that the systems are in a relatively unsustainable development state. Attention should be directed towards increasing the input of internal renewable resources, reducing the reliance on non-renewable resources, and enhancing overall resource utilization.

EISD, a composite index of the system's sustainable development ability, factors in both social and economic benefits and ecological environmental pressures. A higher EISD value implies greater social and economic benefits under a unit environmental pressure, indicating a better sustainable development ability for the system. As shown in Table 4, the EISD value for Mount Emei Scenic Area is 0.12, while the EISD value for the Qiandao Lake Competition Area is 0.038. These values provide a quantitative assessment of the sustainable development capability, with higher EISD values signifying better overall sustainability for the system.

4. Conclusions

The preliminary evaluation of the sustainable development of the Qiandao Lake Competition Area using the emergy analysis method yields valuable insights for the system's future development and operation. The analysis reveals that the renewable natural resource input in the Qiandao Lake Competition area is 2.13e+17 sej, non-renewable resource input is 1.07e+19 sej, input energy is 1.69e+21 sej, and output energy is 8.96e+20 sej. Calculations show that the ESI value for Qiandao Lake Competition Area is less than 1, categorizing it as a consumption-based economy heavily reliant on external resources with a high environmental load ratio. This places the system at risk of being in a state of high load after operation, indicating an overall unsustainable development state. In subsequent development, the focus should be on ensuring efficient resource use within the system or expanding the development area to reduce environmental load and promote sustainability.

While Qiandao Lake Competition Area shares the nature of sightseeing tourism with Mount Emei and Qilihai Scenic Area, it stands out with additional functions such as sports events, health, and vacation. The operation and maintenance of sports venues contribute to significant energy and economic consumption. Post-Asian Games, areas like the Asian Games Village may be converted into apartments, commercial houses, villas, etc., potentially increasing the consumption of non-renewable resources in subsequent regions. To address this, the system should enhance the input of internal renewable resources and consider feasible measures such as green lighting transformation, green transportation planning, photovoltaic power generation system construction, heat pump air-conditioning system construction, and green electricity purchase. These steps aim to improve energy use efficiency and strengthen sustainable development capabilities. However, it's important to note that there are shortcomings in the current research, including limited representation of various data indicators due to the system being in the preliminary operation stage. Future efforts should focus on continuous monitoring and timely updates to enhance the accuracy and completeness of research results.

Author Contributions

All authors contributed to the conception and design of the article, data collection and analysis were jointly performed by all authors, W.C wrote the first draft, and the manuscript was written by Z.Y.C. and all authors commented on previous versions of the manuscript. All authors have read and agreed to the published version of the manuscript.

Declaration of Interest Statement

The author declare that they have no known competiong financial interest or personal relationships that could have appeared to influence the work reported in this paper.

References

1. Chen, Y.Y.; Guo, Z.; Liu, B. Research progress and practical application of carbon footprint in sports activities (in Chinese). Sports Journal, 2022, 29(01): 17-24.
2. Wu, W.T.; Xia, G.Y.; Bao, Z.Y. The Assessment of the Carbon Fixation and Oxygen Release Value of the Urban Green Space in Hangzhou. Chinese Gardens, 2016, 32(03), 117-121.
3. Yang, T.; Zhang, D.Q.; Shen, C.Y.; et al. Valuation of River Basin Ecosystem Services Based on Emergy Analysis:A Case Study of Dongjiang River Basin. 2023, 44(1), 9.
4. Tang, M.; Hong, J.; Wang, X.; & He, R. Sustainability accounting of neighborhood metabolism and its applications for urban renewal based on emergy analysis and SBM-DEA. Journal of Environmental Management, 2020, 275, 111177.
5. Odum, H. Environmental Accounting–EMERGY and Environmental Decision Making. John Wiley, New York, 1996.
6. Brown, M.T., Protano, G., Ulgiati, S. Assessing geobiosphere work of generating global reserves of coal, crude oil, and natural gas. Ecol. Model, 2011, 222, 879–887.
7. Odum, H.T. Emergy of global processes. Handbook of Emergy Evaluation: a Compendium of Data for Emergy Computation Issued in a Series of Folios: Center for Environmental Policy, Environmental Engineering Sciences. University of Florida, Gainesville (FL). 2000.
8. Zhao, K.X.; Li, W. Application of Emergy Analysis Method in Evaluating the Sustainable Development of Community Environment. International Urban Planning, 2023, 38(05), 74-82.
9. Sun, X.R.; Shui, W.; Zheng, J.Y.; Tao, Y.; Li, H. Evaluation of metabolic sustainability in Xiamen based on emergy analysis. Acta Ecologica Sinica, 2021, 41(11): 4342-4353.
10. Zhang, L.; Gao, X.; Cai, W.M. Ecological efficiency evaluation of Qilihai wetland tourism land in Tianjin (in Chinese). Wetland Science, 2017, 15(04): 489-496.
11. Yang, H.; Shen, L.C. Emergy analysis of tourism ecosystem-Taking Emei Mountain Scenic Area as an example.Eco-economy, 2020, 36(04): 129-132.
12. Ma, WJ.; Liu, J. Evaluation of sustainable development of eco-economic systems in China based on emergy analysis. Chinese Journal of Applied Ecology, 2020, 31(6): 2029-2038.

Analysis and Research on the Relationship Between Bird Related Faults in Transmission Lines and Bird Activity Patterns

Weiyun Mao[1], Chen Shen[2], Rong Xiao[2], Kefu Zhang[2], KaiZhao Guo[2], Wenbin Zhao[3], Ya Wang[3,*], Dongsheng Zhang[4]

[1] State Grid Shanghai Electric Power Research Institute, Shanghai, China
[2] State Grid Shanghai Electric Power Company, Shanghai, China
[3] Shanghai Electric Power University, Shanghai, China
[4] Shanghai Ocean University, Shanghai, China

Abstract: With the implementation of national greening policies, bird related faults in transmission lines have become increasingly serious, and have become the third largest transmission line fault problem after lightning strikes and external damage. In previous studies on electric birds, strong stimuli such as electricity, sound, and light were mostly used for bird prevention, and there was a lack of research on the relationship between bird activity patterns and line fault data. This study is based on public collection of bird watching data and compares the situation of power facility failures in Shanghai. The results show that the high incidence areas of bird related failures coincide with bird activity areas, and bird related failures in transmission lines are closely related to bird habits and bird diversity, both of which are directly affected by the surrounding environment. By analyzing these survey data, the distribution and activity patterns of these birds were studied, providing a basis for bird line symbiosis and achieving a combination of ecological protection and power grid construction.

1. Introduction

Bird related faults in transmission lines have become the third largest transmission line fault problem after lightning strikes and external damage, and are also a key focus of power grid animal protection [1-2]. The negative interaction between animals and infrastructure leading to line tripping, hardware damage, and additional maintenance is a reality faced by power companies around the world. With the expansion of the power grid, this interaction has become increasingly common. It is estimated that in some parts of the world, 10-23.5% of power outages in the power system are caused by bird events, and in China, it is also as high as about 19%. According to the impact of birds on the power grid, it can be mainly divided into various types such as bird droppings, bird nests, pecking birds, etc. Unlike line faults caused by external forces, bird related faults are often closely related to the distribution and habits of birds. Previously, measures to prevent bird harm in the power system mainly relied on physical stimuli such as electricity, light, and sound, but the effectiveness of these methods was not satisfactory. Therefore, identifying these birds and mastering their activity patterns can help us carry out targeted bird pest prevention based on bird habits, change "human bird confrontation" to "human bird harmony", and promote ecological prevention and control governance that combines dredging and blocking. This not only ensures power safety but also guarantees the rights of bird survival and reproduction.

The diversity of bird species related to power grids has always been a hot topic for scholars both domestically and internationally. Common bird species that frequently appear in regions such as the United States and Africa and pose a threat to overhead transmission lines include woodpeckers, egrets, starlings, bald eagles, bald eagles, crows, storks, etc. Research on the distribution and regional activities of birds related to the power grid has been conducted in power grids across the country. With the deepening understanding of bird species and habits, measures to prevent bird damage based on bird behavior are gradually receiving attention. A survey of birds along the 400 kV Qaidam-Lhasa DC TL was carried out by using line transect method and point count method, and then the comprehensive improvement measures for such bird-related outages were put forward [3]. Scholars have studied the relationship between bird activity and transmission line flashover and provided treatment methods [4-5]. Polat Ö et al proposed regulatory framework about protecting bird species, impacts of birds on utility equipments and remedial actions that can be taken for problem specific cases. Distribution companies need apply a systematic approach for reducing electrocutions and collisions by developing and implementing Avian Protection Plans [6]. University of Calgary, Canada Nicole (Nikk) Heck, a renowned scholar in the Department of Environmental Science, used voltage levels, geographical environments, and bird habits to based on the arrangement of wires, a zoning level was established to predict the degree of danger of bird impact on transmission lines. The model has been

* Corresponding author: 1040986374@qq.com

applied to the local power system and environmental protection department with good results Israeli scholars discuss power transmission [7]. A study was conducted on bird related faults along the route and the surrounding geographical environment, and it was found that rivers, fish ponds, and forests are located around the route. The more obvious the environmental features that can attract birds to inhabit, the stronger the ability of the surrounding area to carry birds. The more birds that inhabit the area, the higher the likelihood of line faults [8-10]. Classification of the threat level of birds to transmission lines based on their characteristics. However, there has been no systematic report on the activity patterns of birds in the power system in China.

Shanghai is located at the mouth of the Yangtze River and is also the middle section of the East Asian migratory belt, with abundant bird resources. In order to better understand the impact of birds on the Shanghai power system, this article collected the latest photos of bird activities inside and around power equipment taken by bird watching enthusiasts, analyzed the species, habits, and activity patterns of these birds, drew a geographical distribution map of bird activities in Shanghai and a distribution map of bird related faults in the Shanghai power grid, compared the distribution characteristics of the two, combined with the distribution of green spaces in the Shanghai area, analyzed the relationship between bird related faults and the surrounding environment, and provided scientific basis for Shanghai Power to carry out targeted bird prevention work.

2. Information collection and statistics

2.1 Collection of Bird Activity Information

By publicly soliciting bird activity information around power equipment from bird watching enthusiasts, collecting photos of bird activities, as well as information on the time, location, bird species, and bird behavior of the photos taken. For different bird watching enthusiasts shooting data of the same bird species at the same location at different times, only the first submitted data is used to avoid duplicate information and ensure the accuracy and effectiveness of the data as much as possible. For bird identification, refer to the Chinese Bird Observation Manual, for bird classification information, refer to the Chinese Bird Classification and Distribution Directory, for bird residence type and ecological type information, refer to the bird directory information published by the Shanghai Wild Bird Association and the ecological Nanhui official account, and for bird protection level, refer to the Chinese Key Wildlife Conservation Directory and the IUCN red list.

2.2 Statistics of Bird Damage Fault Information

Table 1 shows the statistics of bird related faults in Shanghai from 2018 to 2022. Through the data, we also found that the frequency of bird infections is also very high. Increasing year by year, the reason may be that with the vigorous development of green environmental protection, the ecological environment has been greatly improved. The activities of birds have also become more frequent. Therefore, our protection measures for birds should mainly focus on Fengxian District, Jinshan District, Chongming Island, and Pudong New Area are high-risk areas for bird disasters.

In order to eliminate errors caused by annual data statistics, this article used the total number of bird malfunctions from 2018 to 2022 to draw a geographical distribution map. Using local polynomial interpolation, trend surface analysis takes 40, polynomial order takes 1, and smoothing method is used to plot with a smoothness of 0.5.

Table 1. STATISTICS OF BIRD RELATED MALFUNCTIONS IN SHANGHAI FROM 2018 TO 2022

Region	2018	2019	2020	2021	2022	total
Pudong	1	2	2	8	13	26
city center	0	0	0	0	0	0
Northern part of the city center	1	1	4	1	1	8
Southern part of the city	0	2	2	2	7	13
Songjiang	0	2	3	3	4	12
Jiading	1	1	0	0	4	6
Qingpu	3	4	4	1	6	18
Jinshan	7	6	7	10	19	49
Fengxian	15	16	7	7	13	58
Chongming	12	2	2	10	11	37
Changxing	0	0	2	1	4	7

3. Results and Discussion

Bird watching enthusiasts provided a total of 92 survey records, covering all representative areas of Shanghai. A total of 33 bird species were recorded, including 4 national second-class protected animals (red falcon, peregrine falcon, common falcon, and black winged kite).

Table 2. LIST OF BIRD SPECIES RECORDED NEAR POWER FACILITIES IN SHANGHAI

Serial Number	latin name	English name	Record the number of times	Residence type	Ecological type
1	*Pycnonotus sinensis*	Light-vented Bulbul	3	R	songbird
2	*Lanius schach*	Long-tailed Shrike	7	R	songbird
3	*Alcedo atthis*	Common Kingfisher	1	R	scansorial birds
4	*Ceryle rudis*	Pied Kingfisher	2	R	scansorial birds
5	*Upupa epops*	Eurasian Hoopoe	1	R	scansorial birds
6	*Turdus mandarinus*	Chinese Blackbird	2	R	songbird
7	*Cuculus canorus*	Common Cuckoo	1	S	scansorial birds
8	*Eurystomus orientalis*	Oriental Dollarbird	2	Mp	scansorial birds

9	*Motacilla cinerea*	Grey Wagtail	1	W	songbird
10	*Spilopelia chinensis*	Spotted Dove	3	R	Land poultry
11	*Streptopelia orientalis*	Oriental Turtle Dove	1	R	Land poultry
12	*Streptopelia tranquebarica*	Red Turtle Dove	3	S	Land poultry
13	*Dicrurus macrocercus*	Black Drongo	1	Mp	songbird
14	*Spodiopsar cineraceus*	White-cheeked Starling	3	R	songbird
15	*Gracupica nigricollis*	Black-collared Starling	3	R	songbird
16	*Acridotheres cristatellus*	Crested Myna	7	R	songbird
17	*Egretta garzetta*	Little Egret	1	S/W	wading bird
18	*Lonchura striata*	White-rumped Munia	1	R	songbird
19	*Passer montanus*	Eurasian Tree Sparrow	4	R	songbird
20	*Falco peregrinus*	Peregrine Falcon	2	Mp/W	raptor
21	*Falco tinnunculus*	Common Kestrel	7	R	raptor
22	*Copsychus saularis*	Oriental Magpie-robin	4	R	songbird
23	*Muscicapa griseisticta*	Grey-streaked Flycatcher	1	Mp	songbird
24	*Phoenicurus auroreus*	Daurian Redstart	1	W	songbird
25	*Pica serica*	Oriental Magpie	7	R	songbird
26	*Cyanopica cyanus*	Azure-winged Magpie	1	R	songbird
27	*Riparia riparia*	Sand Martin	1	Mp	songbird
28	Cecropis daurica	Red-rumped Swallow	1	Mp/S	songbird
29	Hirundo rustica	Barn Swallow	9	S	songbird
30	Chloris sinica	Oriental Greenfinch	1	R	songbird
31	Buteo japonicus	Eastern Buzzard	2	W	raptor
32	Elanus caeruleus	Black-shouldered Kite	6	Mp	raptor
33	Jynx torquilla	Wrvneck	1	Mp	scansorial birds

Note: Residence type: R-resident bird, W-winter migratory bird, S-summer migratory bird, Mp migratory bird.

From the above data, it can be seen that among the 33 bird species surveyed, there are 18 resident birds, 5 summer migratory birds, 5 winter migratory birds, and 8 migratory birds. The observed ecological groups of birds are songbirds, land birds, climbers, raptors, wading birds, and waterfowl. Forest birds such as songbirds (19 species), climbers (6 species), raptors (4 species), and land birds (3 species) that live on land are more likely to use the power system for habitat. Meanwhile, according to Fig. 1, songbirds were observed to be the largest percentage of bird ecological taxa. From records to frequency, among all 92 records, birds with a frequency of more than 5 occurrences include the house swallow (9 times), myung (7 times), brown backed shrike (7 times), magpie (7 times), red falcon (7 times), and black winged kite (6 times), confirming that Passeriformes forest birds and raptors use more power lines and poles than other facilities. Only one species of wading bird (egret) was recorded briefly staying on the power line among waterbirds, while there was no record of waterfowl at all. This is because power facilities are generally built on land, close to the habitat of forest birds, and water birds use power facilities relatively less frequently.

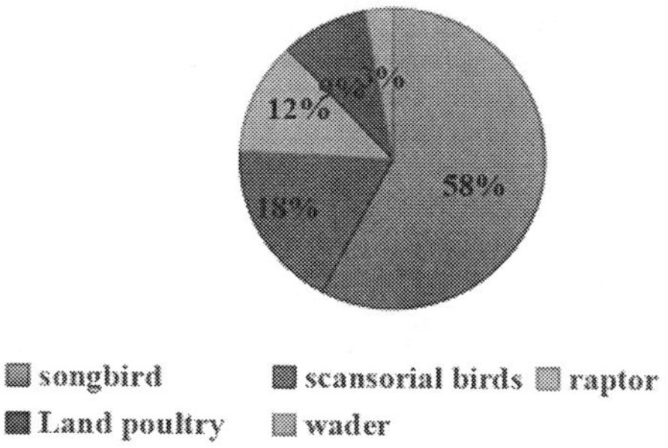

Figure 1. The number and percentage of bird species in different ecological types

Previous bird damage data in the power system shows that different bird behaviors have varying degrees of harm to electricity. The flashover of insulators caused by bird droppings pollution is mainly caused by birds in the family Egret, Stork, Order Falconiformes, Order Scops, order Passeriformes, family Crow, family Beraceae, and family Convolvulaceae. The main bird species include night herons, egrets, magpies, gray magpies, big billed crows, bald nosed crows, black necked starlings, etc. Bird nest malfunctions involve birds such as the stork shaped, falcon shaped, and sparrow shaped orders. The main bird species include night herons, herons, Oriental white storks, black storks, red falcons, magpies, gray magpies, big billed crows, bald nosed crows, and black necked starlings. The birds that cause short circuiting faults are generally large in size, with wings spread over 1.5 meters. The main bird species include the Oriental White Stork, Black Stork, Great White Egret, Heron, Great Bustard, Black necked Crane, Spotted Geese, Giant Hawk, and Common Hawk. Bird pecking fault refers to the damage or flashover of composite insulators caused by birds pecking at the umbrella skirt or sheath. The birds that cause bird pecking faults mainly include magpies and big billed crows, and suspected bird species include gray magpies, bald nosed crows, pearl necked spotted doves, black curly tails, gray starlings, *etc.*

The flashover of insulators caused by bird droppings pollution is mainly caused by birds in the family Egret, Stork, Order Falconiformes, Order Scops, order Passeriformes, family Crow, family Beraceae, and family

Convolvulaceae. The main bird species include night herons, egrets, magpies, gray magpies, big billed crows, bald nosed crows, black necked starlings, etc. Bird nest malfunctions involve birds such as the stork shaped, falcon shaped, and sparrow shaped orders. The main bird species include night herons, herons, Oriental white storks, black storks, red falcons, magpies, gray magpies, big billed crows, bald nosed crows, and black necked starlings. The birds that cause short circuiting faults are generally large in size, with wings spread over 1.5 meters. The main bird species include the Oriental White Stork, Black Stork, Great White Egret, Heron, Great Bustard, Black necked Crane, Spotted Geese, Giant Hawk, and Common Hawk. Bird pecking fault refers to the damage or flashover of composite insulators caused by birds pecking at the umbrella skirt or sheath. The birds that cause bird pecking faults mainly include magpies and big billed crows, and suspected bird species include gray magpies, bald nosed crows, pearl necked spotted doves, black curly tails, gray starlings, etc.

The data in table 2 shows that the majority of birds reside on power lines, with the majority of their stops on power lines and a small portion on the towers of power poles. Among them, larger bird species, including raptors, magpies, black necked starlings, etc., are more inclined to use the tower area of electric poles, which may be related to their competitive advantage. Smaller birds tend to use wires for rest. Short circuits caused by bird nest materials are the main cause of power failures. Our survey shows that most birds use electrical wires for rest, while other behaviors include nesting, excretion, and feeding. The phenomenon of building nests and raising chicks on power poles has been recorded 6 times, including 4 times for magpies, 1 time for black necked starlings, and 1 time for suspected red falcons to occupy magpie nests for raising chicks. A small number of birds defecate on the wire (recorded up to 5 times) and feed (recorded up to 3 times: all are black winged kites feeding on rodents). This survey did not find any cases of birds pecking at insulators or short circuiting bird bodies.

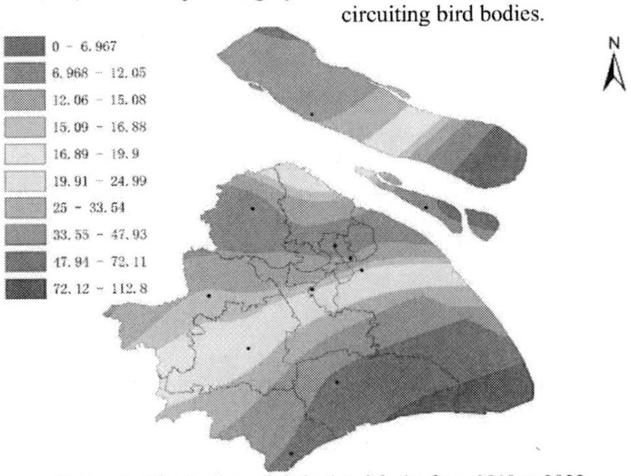

Figure 2. Distribution of bird related faults from 2018 to 2022

The spatial distribution of bird disasters in the Shanghai area can be visually observed from the above figure. All bird malfunctions are caused by bird nests, with Fengxian District, Jinshan District, Chongming Island, and Pudong New Area being the main high-risk areas, while bird disasters in the central urban area are significantly less frequent. These high-risk areas are mostly suburban counties or areas with a relatively large suburban area. These areas have high forest coverage, abundant food, and rivers crisscrossing, providing a good living environment for birds. Therefore, our protection measures for birds should mainly focus on Fengxian District, Jinshan District, Chongming Island, and Pudong New Area are high-risk areas for bird disasters.

We can find that Fengxian District, Jinshan District, Chongming Island, and Pudong New Area are the main high-risk areas, while the city center. Bird disasters occur much less in urban areas. Most of these high-risk areas are suburban counties or relatively large suburban areas. These areas have high forest coverage, abundant food, and criss crossing rivers, providing them with a good living environment, After more than a decade of habitat and reproduction, the number of birds has significantly increased. Birds have fixed habitat habits and environmental factors. Please note that nests within the same tower will be constructed repeatedly. In order to speed up the process of building a nest, birds use one bird to build a nest. It has evolved into a situation where multiple birds collectively build nests. Through the data, we also found that the frequency of bird infections is also very high. Increasing year by year, the reason may be that with the vigorous development of green environmental protection, the ecological environment has been greatly improved, The activities of birds have also become more frequent. The tripping of bird nests mostly occurs from night to early morning, and the occurrence of faults during the day is relatively rare. The reason for this is that birds inhabit bird nests at night, and the humidity is high at night. Rainy and foggy weather can easily cause insulation short circuits. Therefore, our protection measures for birds should mainly focus on Fengxian District, Jinshan District, Chongming Island, and Pudong New Area are high-risk areas for bird disasters.

Table 3. Annual Statistics of Bird Special Trip Data (Unit: Times)

year	2018	2019	2020	2021	2022	total
Number of trips	40	36	33	43	82	234

Table 4. Monthly Statistics of Bird Special Trip Data (Unit: Times)

Annual/ Month	1	2	3	4	5	6	7	8	9	10	11	12
2018	/	9	12	5	1	1	4	4	1	/	3	/
2019	1	9	12	4	1	/	1	4	1	1	1	1
2020	2	3	11	4	2	/	4	1	1	2	1	2
2021	8	9	12	2	3	5	2	1	/	/	/	1
2022	10	9	16	10	6	4	11	6	4	2	1	3
Total	22	39	63	25	13	10	22	16	7	5	7	7

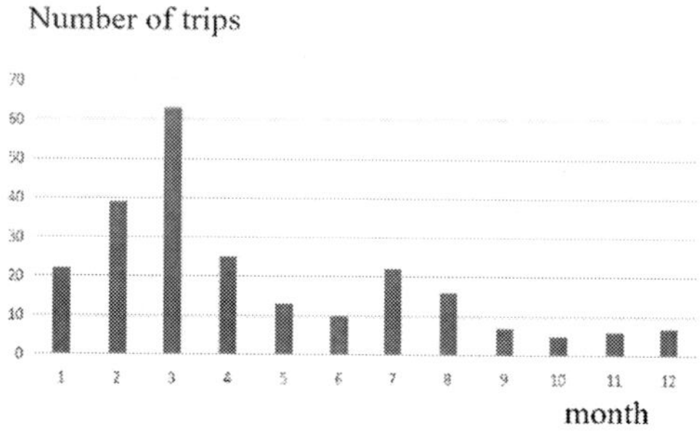

Fig 3. Monthly Distribution of Bird Damage Trips

As in Table 3, From January 2018 to December 2022, there were a total of 234 malfunctions caused by birds, accounting for 2.55% of the total.

From an annual perspective, the data for the first four years remained relatively stable, but in 2022, the number of bird infestations suddenly increased, as in Table 4 almost doubling that of the previous four years.

From the figure above a monthly perspective, there are two peak periods, one from February to April and the other from July to August, which may be related to an increase in bird nesting and breeding activities.

Fig. 4 is the bird distribution of Table 2. In the figure, different bird species are labeled according to their appearance locations, allowing for a visual observation of their activity locations.

The survey records cover 10 districts in Shanghai. A total of 33 bird species were recorded, including 4 species of national second-class protected animals (red falcon, peregrine falcon, common falcon, and black winged kite).Therefore, in Fig. 3, there are more bird species in coastal areas such as Pudong New Area (20 species) and Chongming District (14 species), while there are fewer bird species in central urban areas with higher urbanization levels (Xuhui District with more recorded data has 10 bird species). In the urban area, only 10 bird species have been recorded, all of which are common resident birds except for the swallow. The area outside the outer ring road includes all 33 recorded bird species.

Comparing the distribution map of bird related faults in Shanghai from 2018 to 2022 (Fig. 2) and the distribution map of bird activity patterns (Fig. 3), it can be found that there is a certain degree of overlap in the key areas of the two maps, without considering the existence of a small amount of data errors. This is because frequent bird activity is bound to easily cause bird related faults in the transmission line.

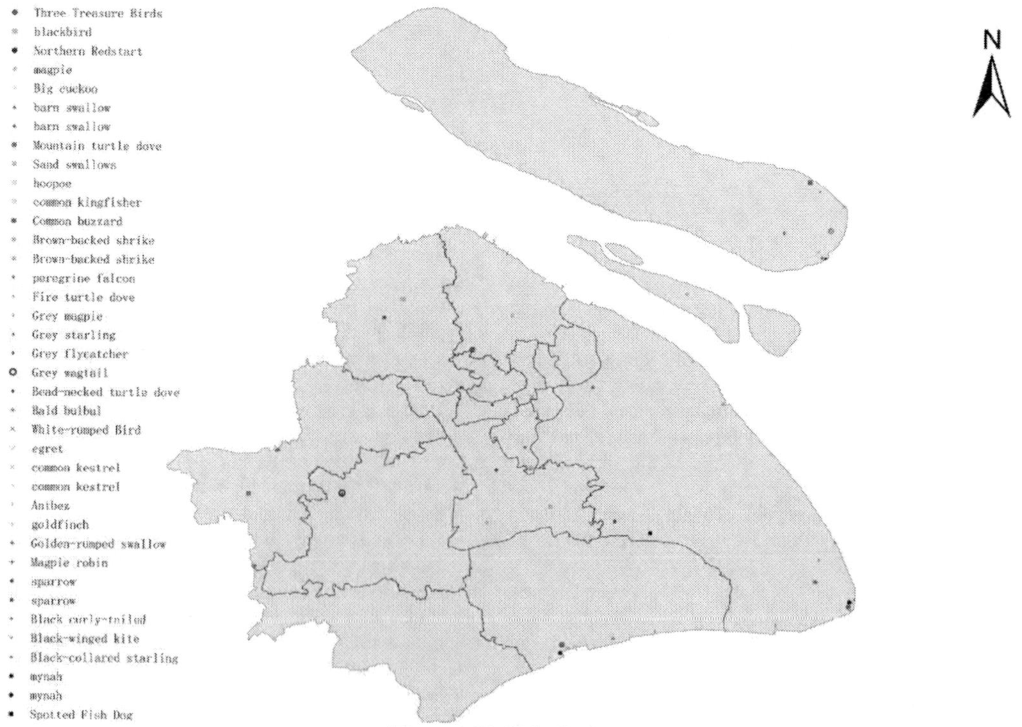

Figure 4. Bird Distribution

In the history of Shanghai power grid transmission lines with voltage levels of 110kV and above, there have been no bird damage faults. Occasionally, bird damage faults occur in 35kV and 10kV distribution lines. With the continuous improvement of Shanghai's green space level and the grounding of urban overhead lines, the overall level and distribution of bird damage have changed. The number of power grid and power station accidents caused by bird damage has increased.

Unlike Shanghai Power Grid, Beijing Power Grid is mainly distributed in plain areas with vast and flat terrain, lush vegetation, and suitable for bird habitat. Bird activities are particularly frequent in water source areas such as rivers, lakes, reservoirs, and agricultural areas. In the urban and suburban areas of Beijing, small and medium-sized birds such as sparrows, magpies, and crows are the main species. However, in the remote suburban areas such as Miyun District, Shunyi District, and Tongzhou District, large and medium-sized birds such as sparrows and black storks are the main species. The activities of medium and large bird species pose a significant threat to the safe operation of power lines, with 92% of bird damage incidents occurring in suburban areas in the past five years.

The Wuhan Nanchang double circuit 1000 kV transmission line project crosses the lakeside area of Poyang Lake, where a large number of water birds overwinter. Due to human economic activities and extreme weather in recent years, there has been a shortage of food for wintering birds in Poyang Lake. Since 2017, there have been a large number of cranes, geese, ducks, and other animals transferring to the surrounding farmland to feed, resulting in damage accidents caused by bird collisions with ultra-high voltage transmission lines.

The bird related faults often manifest as natural laws, bird habits, and human production and life, as well as being closely related to the insulation configuration, tower structure, size, and other factors of the transmission line. Therefore, starting from bird activities, preventive measures can be improved in the following aspects: (1) Reasonable wiring and planning of transmission line routes; (2) Select tower locations and optimize tower structures based on the nesting environment and nest size of repetitive nesting birds; (3) Bird droppings pollution needs to consider the surrounding vegetation conditions, avoid bird activity areas, and arrange transmission lines.

Research has found that bird activity patterns, bird hazard characteristics, and their correlation with power lines have strong regional characteristics that require long-term development and dynamic tracking. We need to transform 'human bird confrontation' into 'human bird harmony', promote ecological prevention and control governance combining dredging and blocking, and ensure power safety and bird survival and reproduction rights.

4. Conclusions

Bird activities exhibit seasonality, regionalism, species diversity, and are influenced by human activities characteristics. These characteristics determine that the impact of bird activities on power grid safety is a complex issue that requires us to conduct comprehensive analysis from multiple perspectives. The results show that:

(1) The bird related faults in transmission lines are closely related to the habits and diversity of birds;

(2) Bird related faults in transmission lines and bird activities are directly affected by the surrounding environment;

(3) The bird prevention work of transmission lines should fully consider the impact of the surrounding environment and the activity characteristics of birds.

The construction principle is to prioritize prevention, protect diversity, and minimize risks. All power infrastructure should follow relevant technical, ecological, social, and economic standards based on appropriate routes and locations. Considering the identified risks, sensitive and/or priority areas, as well as the number of accidents involving small animals affecting power facilities, collision avoidance and prevention measures should be taken at the best available locations and routes, or feasible mitigation measures should be taken. If this cannot be achieved, measures to offset the impact should be taken.

In future research, we should continue to analyze the habitat conditions and bird diversity around the power system, and investigate how birds affect power system safety based on actual power hazards, providing scientific basis for the practice of harmonious power systems between humans and birds. It is worth noting that infrastructure related to clean energy may have adverse effects on nature in sensitive or protected areas, and it is necessary to plan ahead to avoid repeating the mistakes of the fossil fuel era and reduce these errors in the future.

Acknowledgments

State Grid Corporation of China Science and Technology Project Funding (Project Number: B 30 940 230 00J).

References

1. Wang W, Ma Z, Fan Y, et al. Analysis and Protection Studies of Bird Droppings Falling on the Electric Field Distribution near the 330 kV Transmission Line V-type Composite Insulators[J]. IEEE Access, 2024.
2. Desta B Z, Wogari M M, Gubanski S M. Analyses of unexplained faults in transmission lines in the power grid of Ethiopia[J]. Electric Power Systems Research, 2024, 231: 110293.
3. Ding Y, Zhou J, Li C, et al. Bird-related fault analysis and prevention measures of±400 kV Qinghai-Tibet DC transmission line[J]. Energy Reports, 2021, 7: 426-433.
4. Wang W, Ma Z, Fan Y, et al. Analysis and Protection Studies of Bird Droppings Falling on the Electric Field Distribution near the 330 kV Transmission Line V-type Composite Insulators[J]. IEEE Access, 2024.
5. Bi M, Li K, Zhang S, et al. Electric field simulation analysis of ICOG corona ring and study on flashover characteristics of bird droppings[J]. Electric Power Systems Research, 2024, 233: 110459.

6. Polat Ö, Yumak K, Atilla N E, and Bağriyanik M. An overview of bird related issues in electrical power systems[C]//IOP Conference Series: Materials Science and Engineering. IOP Publishing, 2016, 161 (1): 012091.
7. Nicole (Nikki) Heck.A landscape-scale model to predict the risk of bird collisions with electricpower transmission lines in Alberta[D]. Dissertation of Univerity of Calgary, 2007:73-78.
8. Huang Z, Zhou Q, Liao Z, et al. Probability Prediction of Bird Damage Based on Multiple Linear Regression[C]//2020 7th International Conference on Information, Cybernetics, and Computational Social Systems (ICCSS). IEEE, 2020: 645-649.
9. Zhu P, Zhao Q, Liao Z, et al. Analysis of bird caused transmission line fault based on Bayesian linear regression[J]. Nanjing Xinxi Gongcheng Daxue Xuebao, 2022, 14 (2): 227-232.
10. Qiu Z, Zhu X, Liao C, et al. Detection of bird species related to transmission line faults based on lightweight convolutional neural network[J]. IET Generation, Transmission & Distribution, 2022, 16 (5): 869-881.

Application of Magnetic Coagulation Sedimentation Tank and Sand-adding High-efficiency Sedimentation Tank in Sewage Treatment

Yufei Shangguan[1,*], Binbin Qiu[1], Zongkai Ma[1], Li He[1]

[1] POWERCHINA Chengdu Engeneering Corporation Limited, Chengdu, 611130, China

Abstract: Magnetic coagulation sedimentation tank and sand-adding high-efficiency sedimentation tank are more and more widely used in the upgrading of sewage plants, industrial wastewater treatment and river water treatment, which are on the basis of ordinary high-density sedimentation tanks, the addition of magnetic powder and micro-sand as the core of flocculation to strengthen the flocculation process. This paper discusses the development history, research direction and application results of magnetic coagulation sedimentation tank and sand-adding high-efficiency sedimentation tank, and summarises the advantages and disadvantages of the two in engineering applications. In the case of the same footprint and treatment effect, compared with the magnetic coagulation and sedimentation tank, the sand-adding high-efficiency sedimentation tank is more adaptable, less difficult to operate and maintain, and more economical in operating costs.

1. Introduction

With the increasing attention to water environment issues in China, the discharge standards for wastewater treatment plants have become stricter. The discharge standards for wastewater treatment plants in most regions have been raised to Class 1A of "Discharge standard of pollutants for municipal wastewater treatment plant" or Class IV of "Environmental quality standards for surface water".

Consequently, more and more wastewater treatment plants have undergone or are undergoing upgrading and reconstruction to meet these standards. The advanced treatment unit with high-density sedimentation tank and denitrifying filter as the main body is currently the mainstream process for upgrading and reconstruction. The high-density sedimentation tank, typically represented by the Densadeg process developed by Degremont (the predecessor of Suez Environment), is an improved flocculation clarification tank with external sludge recycling. It combines the principles of inclined tube clarification and sludge thickening. However, the following issues have been encountered during the upgrading and reconstruction process using high-density sedimentation tanks: 1) Large variation ranges in influent water quantity and quality; 2) Unstable effluent quality; and 3) Tight plant site conditions.

In recent years, magnetic coagulation sedimentation tanks and sand-adding high-efficiency sedimentation tanks, which add magnetic powder and micro-sand as flocculation cores to enhance the flocculation process, have achieved good results. Both technologies improve the formation and separation characteristics of flocs during the flocculation stage, forming large and dense flocs, thereby obtaining better settlement effects. They are characterized by strong anti-shock load capacity, high surface hydraulic load, small footprint, high removal efficiency, and stable effluent quality.

Through research and case analysis, this paper studies the treatment principles of magnetic coagulation sedimentation tanks and sand-adding high-efficiency sedimentation tanks, as well as the pros and cons of their application in engineering, and analyzes the selection of wastewater treatment processes under different conditions.

2. Development of Magnetic Coagulation Sedimentation Tanks

2.1 Magnetic Coagulation and Magnetic Separation Technology

Magnetic coagulation sedimentation technology involves synchronously adding magnetic powder to coagulation sedimentation processes, allowing it to flocculate and integrate with pollutants to enhance the coagulation and flocculation effects, resulting in larger and more compact flocs for high-speed settlement. The flow chart of magnetic coagulation and sedimentation tank is shown in Fig. 1. Magnetic powder, typically Fe_3O_4, and coagulant dosage are adjusted according to the influent water quality and load. Magnetic separation technology refers to a separation technique that utilizes differences in the magnetic potential of elements or components and employs an external magnetic field, thereby enhancing the separation process.

* Corresponding author: shangguan940807@163.com

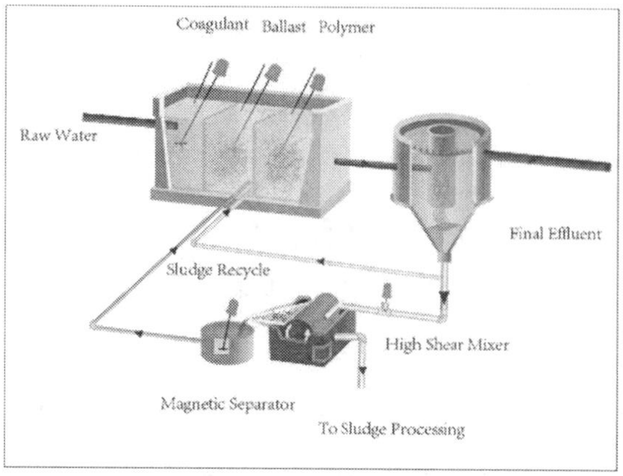

Fig. 1. Process flow chart of magnetic coagulation and sedimentation

2.2 Research of magnetic coagulation and precipitation

Although magnetic coagulation technology has been widely used in engineering, it has little effect on the removal of dissolved COD, ammonia oxides, benzene, phenols and other difficult to degrade substances, and the removal of TP is overly dependent on the role of aluminium and iron salts in the coagulant. At present, the research aspects of magnetic coagulation and precipitation mainly focus on the following four aspects.

Development of New Magnetic Species and Magnetic Species Modification. Magnetic species materials in addition to common Fe_3O_4 and ferromagnetic materials, new magnetic species materials are also being used in research. For example, the use of electric arc furnace powder as a magnetic species to remove suspended particles in the suspension, which can greatly shorten the settling time[1]. The use of hydroxyferric iron powder as a magnetic species in the pollution of rivers and lakes, the removal of total phosphorus is greater than 80% [2].

Magnetic Flocculant Optimisation. Magnetic flocculant, refers to the preparation of magnetic species and flocculants combined in one magnetic flocculant, not only to strengthen the treatment effect but also simplify the process. At present, most of magnetic flocculants are made by chemical co-precipitation of nano-Fe_3O_4 with organic, inorganic and biological coagulants. For example, PAC wrapped in magnetic powder surface, and then combined with PAM, greatly improving the clarification rate [3]. The use of hydrolysis to prepare $CoFe_2O_4$ particles and decomposited in chitosan, to form magnetic $CoFe_2O_4$ composite particles, through which the removal of indigo dye in printing and dyeing wastewater can be up to 95.3% [4].

Magnetic Adsorbent Optimisation. In the process of magnetic species modification and optimisation, the adsorption of functional materials and magnetic materials are combined to prepare magnetic adsorbents, both to meet the requirements of the magnetic composite specific gravity, strong magnetism and to improve the ability of the magnetic species to adsorb pollutants.

The preparation of magnetic nanocomposites includes chemical precipitation, sol-gel, solvothermal, and other methods [5]. The types of adsorption functional materials include inorganic materials, organic small molecule materials and organic polymers. Among them, inorganic adsorption functional materials include SiO_2, Al_2O_3, graphene and activated carbon [6-8]. The modification of inorganic materials is characterised by easy operation, controllability, high stability and good adsorption performance. Organic polymer modification can be divided into natural polymers and synthetic polymer modification, natural polymers from a wide range of sources, low cost, loaded without secondary pollution [9-12]. Synthetic polymer surface contains a wealth of active genes, the removal of organic matter and heavy metal ions in wastewater is effective [13-14].

Magnetic Coagulation Coupling Technology. Magnetic coagulation technology, biological treatment, catalytic oxidation, membrane treatment and other combined technology are used to deal with different water quality conditions of pollutants. For example, the use of magnetic coagulation and EHBR purification pond to deal with black smelly water, magnetic coagulation to quickly remove SS, non-dissolved COD and TP, the lower end of the use of unused ponds to establish a strengthened coupled biofilm reactor, to remove dissolved COD and NH_4^+-N [15]. The use of powdered activated carbon and magnetic coagulation to treat industrial wastewater, the effluent water quality COD is less than 65 mg/L, SS is less than 4 mg/L. The activated carbon is added in a section of the coagulation area, which has a significant effect on the removal of chromaticity and does not require large-scale transformation of the process [16]. The use of magnetic coagulation coupled with EDTC(heavy metal trap) to treat nickel-containing wastewater, the residual nickel concentration in the effluent is lower than 0.1mg/L, and the addition of magnetic species can reduce the use of EDTC by 20%, shorten the sedimentation time, and improve the effluent water quality [17].

3. Magnetic coagulation sedimentation tank application

Magnetic coagulation sedimentation tank, as a highly efficient and rapid water treatment technology, can effectively shorten the settlement and flocculation time. It has the advantages of good effluent water quality, small footprint, rapid settlement, etc. As shown in Table 1, magnetic coagulation sedimentation tank has been widely used in the field of water treatment in recent years, especially in industrial wastewater treatment and domestic sewage treatment plant upgrading.

A sewage treatment plant with the scale of 12×10^4 m^3/d in Guangdong Province, in which upgrading process used magnetic coagulation and sedimentation process, completed the implementation of the conditions in a very small footprint, with the surface load of 19.93 $m^3/(m^2\cdot h)$ and the peak surface load of 25.91 $m^3/(m^2\cdot h)$. After upgrading, it ensured efficient and stable removal of SS and TP [18].

Table 1. Application examples of magnetic coagulation and sedimentation tanks

Project Scale	Type of sewage	Processing	Processing effect	Citations
$12\times10^4m^3/d$	Municipal sewage	MBBR + magnetic coagulation and sedimentation	Under the condition of very small occupation, the quality of effluent water is better than the Class IV of "Environmental quality standards for surface water".	[18]
$15\times10^4m^3/d$	Municipal sewage	MBBR + magnetic coagulation and sedimentation	Scale expaned from $10\times10^4m^3/d$ to $15\times10^4m^3/d$, effluent quality met the Class IV of "Environmental quality standards for surface water".	[19]
$20\times10^4m^3/d$	Printing and dyeing wastewater	Carbon loaded magnetic coagulation and sedimentation	Running costs was low, effluent COD, TN, TP and other 14 indicators met the direct discharge standards.	[20]
$2.0\times10^4m^3/d$	River polluted water	Magnetic coagulation + biological aerated filter + ecological wetland	Suspended solids, COD, ammonia nitrogen and total phosphorus removal achieved the desired effect to meet the Class III of "Environmental quality standards for surface water".	[21]
$1.0\times10^4m^3/d$	Municipal wastewater from cold areas	Modified AO process + magnetic coagulation sedimentation tank	Stable effluent quality and stable nitrogen removal efficiency under low temperature.	[22]
$6000m^3/d$	Printing and dyeing wastewater	Magnetic coagulation and precipitation (pre-treatment)	With 50% space saving, the pre-treatment effluent effect was stable and the water quality is significantly better than before.	[23]
$8.0\times10^4m^3/d$	Municipal sewage	Magnetic coagulation and efficient sedimentation + denitrification deep-bed filtration tank	All the main indicators in the effluent (except TN) met the Class III of "Environmental quality standards for surface water", and the effluent quality was stable.	[24]

A wastewater treatment plant did expansion and upgrading at the same time. The scale expanded from 10×10^4 m^3/d to 15×10^4 m^3/d, and the effluent quality met the Class IV of "Environmental quality standards for surface water". Expansion and upgrading was used with MBBR and magnetic coagulation sedimentation tank, and the project was completed with good effect of nitrogen and phosphorus removal, occuping less land, and strong shock-resistant ability [19].

Carbon loaded magnetic coagulation sedimentation tank was used as a comprehensive printing and dyeing wastewater advanced treatment process in the sewage treatment plant in Zhejiang Province. After completed, the operating cost was low, and the effluent COD, TN, TP, and other 14 indicators met the direct discharge standards [20].

The 'magnetic coagulation sedimentation tank + biological aerated filter + ecological wetland' process was used to treat polluted river in Zhejiang Province. In which the pollutants in the water body need to be purified and the quality of the water environment need to be improved. After the implementation of the project, suspended solids, COD, ammonia nitrogen and total phosphorus removal rate achieved the desired effect, and met the Class III of "Environmental quality standards for surface water" [21].

Magnetic coagulation process was used to improve the effect of pre-treatment in a printing and dyeing enterprise wastewater treatment station in Shunde City of Guangdong Province. Compared with original inclined tube sedimentation tank, project saved 50% of the area, COD removal rate was stable at 40% to 50%, and the effluent water quality was significantly better than before [23].

The 'magnetic coagulation sedimentation tank + denitrification deep-bed filter' process was uesd in the depth treatment of a sewage treatment plant in Baoding City of Hebei Province. The data of effluent quality after deep treatment showed that the concentrations of COD, SS, TP, NH_3-N and TN in the effluent of the wastewater treatment plant were under good control, and the main indexes of the effluent (except for TN) met the Class III of "Environmental quality standards for surface water", which was better than the project standard [24].

Since 2015, magnetic coagulation technology has been widely used. In the upgrading of sewage treatment plant, magnetic coagulation sedimentation tank can ensure the stability of the effluent under the conditions of land tension. In industrial wastewater treatment, magnetic coagulation technology is often used as enhanced pretreatment or depth treatment. In polluted river water treatment, magnetic coagulation sedimentation tank can meet the requirements of large changes in the amount of water and fast settling speed. But after several years of operation, the magnetic coagulation sedimentation tank has also exposed some problems as follows.

(1) With the increase of operation year, the loss of magnetic powder increases, and the operation cost gradually rises.

(2) Magnetic powder separation equipment is easy to fail, which affects stable operation. Magnetic coagulation using high shear and magnetic separator for the recovery of magnetic powder in the remaining sludge, high shear and magnetic separator are power equipment, high shear and magnetic separator on the market quality varies, the failure rate is high. Once the magnetic separation equipment failure on the whole plant to meet the standard operation has a huge impact.

(3) Magnetic coagulation pipeline clogging and wear and tear problems. Magnetic coagulation of the magnetic powder medium, due to its large specific gravity, making the magnetic sludge pipeline clogging problems occur frequently. The production method of magnetic powder determines that the spherical degree of magnetic powder is poor, and the wear and tear of pipe fittings and equipment is more prominent.

(4) The residual sludge in the magnetic coagulation and sedimentation tank contains magnetic powder particles, which leads to the limitation of sludge disposal.

4. Sand-adding high-efficiency sedimentation tank technology development

In view of these problems in magnetic coagulation sedimentation tanks, in recent years, the application of sand-adding high-efficiency sedimentation tanks, in which the added magnetic powder is adjusted to micro-sand, has become more and more common. The high-efficiency sedimentation tank with sand is a new type of process formed on the basis of ordinary high-density sedimentation tank by changing the sludge circulation into micro-sand circulation, with Actiflor process developed by Veolia as a typical representative Actiflor high-efficiency sedimentation process. As shown in Fig. 2, the process consists of coagulation tank, dosing tank, ripening tank, inclined tube(plate) sedimentation tank, and micro-sand circulation and sludge discharge system.

The process through the addition of micro-sand (particle size of 100~150μm), so that the pollutants in the polymer flocculants and microsand polymerisation into large particles of easy-to-sediment flocs, thus accelerating the sedimentation rate of pollutants in the sedimentation tank, and combined with the principle of inclined plate sedimentation, greatly reducing the area of the sedimentation tank and sedimentation time, and can get good effluent results[25].

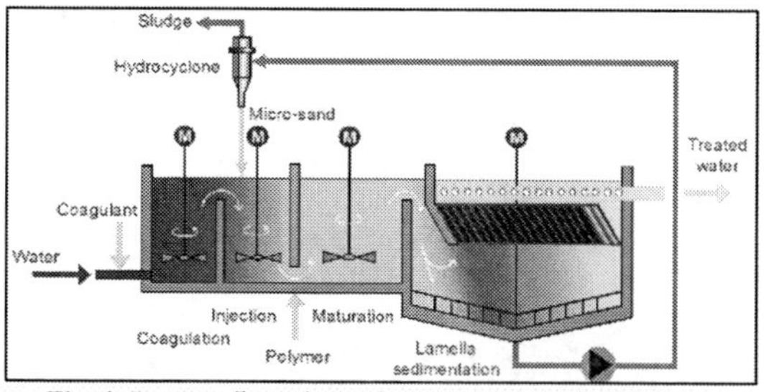

Fig. 2. Process flow chart of sand-adding high-efficiency sedimentation tank

Coagulation pool. Coagulant is added to the raw water and mixed with the sewage rapidly under the action of rapid stirrer to form small flocs and then enter the flocculation pool.

Input pool. Micro-sand and small flocs formed by coagulation are quickly mixed under the action of rapid stirrer, and micro-sand is used as the core to form flocs with higher density and heavier weight, which is conducive to the rapid sedimentation in the sedimentation tank.

Maturing pool (flocculation pool). The flocculant prompts the small flocs to form larger flocs through adsorption, electrical neutralisation and bridging between each other, and the role of the slow stirrer not only enables the agent and flocs to be fully mixed but also does not damage the flocs already formed. The slow speed stirrer can make the chemicals and flocs mix fully without destroying the large flocs already formed.

Inclined plate sedimentation tank. Sand-containing flocs achieve high-speed sedimentation in the inclined pipe(plate) sedimentation tank, the water in the upper part of the sedimentation tank is collected by the catchment tank, and the sludge containing micro-sand is settled at the bottom of the tank, and is collected by the scraper to the central area at the bottom of the sedimentation tank.

Micro-sand circulation and sludge discharge. The sludge at the bottom of the sedimentation tank is pumped out by the micro-sand circulation pump to the hydrocyclone for mud-sand separation. Due to the difference in density between micro-sand and sludge, the sludge is separated from the micro-sand by the centrifugal force of the hydrocyclone. The separated micro-sand is directly reused in the dosing tank.

5. Application of sand-adding high-efficiency sedimentation tanks

As shown in Table 2, in recent years, the sand-adding high-efficiency sedimentation tank has been gradually applied in engineering due to its features such as high treatment load, saving civil construction area, short construction period, strong shock load resistance, stable operation effect, and convenient operation and maintenance (compared with magnetic coagulation sedimentation tank), etc. In actual production application, the design parameters are different according to different treatment objectives.

The total scale of Jiaxing Joint Wastewater Treatment Plant is 60×10^4 m^3/d, and the proportion of industrial wastewater in the influent is about 50 percent or more. The 'sand-adding high-efficiency sedimentation tank + denitrification deep-bed filter + ozone oxidation' progress was used to improve the effluent water quality to Class 1A of "Discharge standard of pollutants for municipal wastewater treatment plant". After upgrading, the new unit operating cost of the sewage plant is low, and the effluent quality can stably meet the requirements.[26].

The scale of a sewage treatment plant was 10×10^4 m^3/d in Guangdong Province, in which the main process is 'modified A^2/O + secondary sedimentation tank + ultraviolet disinfection'. After the modified A^2/O process was transformed into a combination of 'MBBR + sand-adding high-efficiency sedimentation tank', and the effluent quality met the Class IV of "Environmental quality standards for surface water". The project covered an additional area of 3300 m^2, the land-saving effect was remarkable [27].

Zhengzhou New District Wastewater Treatment Plant had a total scale of 100×10^4 m^3/d. The advanced treatment system of the wastewater plant adopted sand-adding high-efficiency sedimentation tanks and V-filter tanks. A total of 2 sand-adding high-density sedimentation tanks were set up, and the surface load was 31.0~43.7 $m^3/(m^2\cdot h)$. The effluent water quality was better than Class 1A of "Discharge standard of pollutants for municipal wastewater treatment plant" [29].

Table 2. Application of sand-adding high-efficiency sedimentation tank

Project Scale	Type of sewage	Processing	Processing effect	Citations
$60\times10^4m^3/d$	Municipal wastewater+industrial wastewater	Sand-adding high-efficiency sedimentation tank+Denitrification deep-bed filtration tank	The operating cost of the sewage plant is low, and the effluent quality can stably meet the requirements.	[26]
$10\times10^4m^3/d$	Municipal sewage	MBBR + Sand-adding high-efficiency sedimentation	Effluent quality met the Class IV of "Environmental quality standards for surface water", with remarkable land-saving effect.	[27]
$40\times10^4m^3/d$	Municipal sewage	Sand-adding high-efficiency sedimentation tank	The effluent quality met Class 1A of "Discharge standard of pollutants for municipal wastewater treatment plant".	[28]
$100\times10^4m^3/d$	Municipal sewage	Sand-adding high-efficiency sedimentation tank + V-filter tank	The effluent water quality was better than Class 1A of "Discharge standard of pollutants for municipal wastewater treatment plant"	[29]
$1.5\times10^4m^3/d$	Municipal wastewater+rainwater	Advanced AAO tank + Sand-adding high-efficiency sedimentation tank	Effluent quality after sewage treatment was better than Class 1A of "Discharge standard of pollutants for municipal wastewater treatment plant"; rainwater treatment was stable and met the standard.	[30]
$2.4\times10^4m^3/d$	Lake water	Biological aerated filter + Sand-adding high-efficiency sedimentation tank + Filter+ Ultrafiltration + Ultraviolet disinfection	The total phosphorus of the final effluent can be controlled to be less than 0.02mg/L.	[31]

Xinwei River sewage treatment plant in Wuhan City of Hubei Province, the use of semi-underground construction form, coupled with $1.5\times10^4m^3$ rainwater storage tanks and primary enhanced treatment facilities. Sewage treatment adopted the process of 'advanced AAO tank + sand-adding high-efficiency sedimentation tank'. The system had strong shock load resistance and stable operation, and the effluent water quality reaches and is better than the national standard of Class A discharge standard was better than Class 1A of "Discharge standard of pollutants for municipal wastewater treatment plant". The rainwater treatment adopted the process of 'aerated sand sedimentation tank + sand-adding high-efficiency sedimentation tank', which had the characteristics of flexible operation and fast start-up, and the treated effluent can reach the quality stably [30].

The Star Wishing Lake of Shanghai Disneyland supporting construction, had a volume of about 15×10^5 m^3. The water treatment process is 'biological aerated filter → sand-adding high-efficiency sedimentation tank → filter → ultrafiltration → ultraviolet disinfection'. The scale of the sand-adding high-efficiency sedimentation tank was 2.4×10^4 m^3/d, adding micro-sand with particle size of 80~100μm, and the upward flow rate of the inclined plate sedimentation tank was 46 m/h. After mixing, flocculation, micro-sand addition cycle and sedimentation treatment, the total phosphorus content of the lake water can be effectively reduced, and with the ultrafiltration advanced treatment process, the total phosphorus of the final effluent can be controlled to be less than 0.02mg/L [31].

Compared with the magnetic coagulation sedimentation tank, the sand-added high-efficiency sedimentation tank is more adaptable, less difficult to operate and maintain, and more economical in operation cost under the same conditions of land occupation and flocculation effect. It is gradually widely used in the fields of municipal wastewater treatment, industrial wastewater treatment, and lake and river management.

6. Conclusion

(1) Magnetic coagulation sedimentation tank and sand-adding high efficiency sedimentation tank, have achieved the removal of pollutants on the basis of saving land area and improve the ability to resist shock load, stable operation, high treatment load, used in the upgrading of wastewater treatment plants more and more widely.

(2) Magnetic coagulation sedimentation tank process in the actual selection process, because the remaining sludge contains magnetic particles, not accepted by most of the sludge blending and disposal units (thermal power plants, waste incineration stations, etc.), need to be considered in a comprehensive manner of sludge disposal. And the magnetic coagulation sedimentation tank of the magnetic powder recovery system in the equipment wear and tear faster, need to strengthen the usual maintenance.

(3) Magnetic coagulation sedimentation tank and sand high-efficiency sedimentation tank, increasing the recovery and addition of magnetic powder and micro-sand process, in the actual operation of the project will increase the workload and difficulty of operation and maintenance, should be combined with the actual situation of the project for the selection of processes.

References

1. Tang H, Wang L, Sun W, et al. Electric arc furnace dust as magnetic carrier particles for removal of micro-fine particles from suspensions[J]. Separation and Purification Technology, 2017, 176 (Complete): 220-230. DOI: 10.1016/j.seppur.2016.12.024.
2. B I D V A, A M M, C F G, et al. Chemical interferences when using high gradient magnetic separation for phosphate removal: Consequences for lake restoration[J]. Journal of Hazardous Materials, 2011, 192 (3): 995-1001.DOI: 10.1016/j.jhazmat. 2011.05.090.

3. Yuan, Zhao, Wenyan, et al. Harvesting Chlorella vulgaris by magnetic flocculation using Fe3O4 coating with polyaluminium chloride and polyacrylamide[J]. Bioresource Technology, 2015. DOI: 10.1016/j.biortech.2015.09.087.
4. Juliana, M. N, et al. Synthesis of a novel CoFe2O4/chitosan magnetic composite for fast adsorption of indigotine blue dye[J]. Carbohydrate Polymers, 2019.DOI: 10.1016/j.carbpol.2019.04.054.
5. GUO ZH, LIU ZH Tao, SHEN B X, et al. Preparation of MBFX-8/magnetic Fe3O4@CMC composite flocculant and removal of Cu2+[J]. Journal of Environmental Engineering, 2018, 12 (2): 6. DOI: CNKI: SUN: HJJZ.0.2018-02-010. (in Chinese)
6. Gao F. An Overview of Surface-Functionalized Magnetic Nanoparticles: Preparation and Application for Wastewater Treatment[J]. ChemistrySelect, 2019, 4 (22): 6805-6811. DOI: 10.1002/slct.201900701.
7. Gao Hairong, Jin Shuailong, Chen Jianjun, et al. Preparation of magnetic composites of black algae and ultrasonic adsorption of Pb^ (2+) in water[J]. Water Treatment Technology, 2024, 50 (6): 75-79. (in Chinese)
8. Yao Q L, He M, Kong Y R, et al. Y_2O_3-functionalized graphene-immobilized Ni–Pt nanoparticles for enhanced hydrous hydrazine and hydrazine borane dehydrogenation[J]. Rare Metals, 2023 (10): 3410-3419.
9. J.H. Wang, S.L. Lei. Removal of Cu (II) from water by sulfonic acid-based modified magnetic adsorbent[J]. Environmental Chemistry, 2019, 38 (8): 8. DOI: 10.7524/j.issn.0254-6108.2018102603. (in Chinese)
10. Kim, Bohwa, Jeon, et al. Downstream integration of microalgae harvesting and cell disruption by means of cationic surfactant-decorated Fe3O4 nanoparticles[J]. Green chemistry, 2016, 18 (14): 3981-3989.
11. HU Fang, ZHENG Shunji, SHI Changwei, et al. Progress in the preparation of Fe3O4/agroforestry waste and its adsorption of dyes[J]. Printing and Dyeing, 2022, 48 (06): 79-84. (in Chinese)
12. Hua J, Wang M, Lin Shuting, et al. Degradation of typical phenylarsonic acid pollutants by Fe3O4/FeS2 activated H2O2[J]. 2021, 41 (06): 2646-2656.DOI: 10.3969/j.issn.1000-6923.2021.06.018. (in Chinese)
13. LU Navy, LI Yuanshuai, LI Xiaoli. Preparation of magnetic carbon nanotube composites and their adsorption properties[J]. Environmental Science and Technology, 2019, 42 (4): 7. DOI: CNKI: SUN: FJKS.0.2019-04-017. (in Chinese)
14. QI, XIUXIU, LI, NAJUN, XU, QINGFENG, et al. Water-soluble Fe3O4 superparamagnetic nanocomposites for the removal of low concentration mercury (II) ions from water[J]. RSC Advances, 2014, 4 (88): 47643-47648. DOI: 10.1039/c4ra05935b.
15. Yu Pianhong. Magnetic coagulation/enhanced coupled biofilm process for river water quality enhancement[J]. China Water Supply and Drainage, 2019, 35 (18): 5. DOI: CNKI: SUN: GSPS.0.2019-18-025. (in Chinese)
16. JI Qingqing, CHEN Li, ZHOU Wenbin, et al. Application of activated carbon+magnetic coagulation and precipitation in wastewater plant to improve the standard pilot[C]//Environmental Engineering 2019 National Academic Annual Conference. 2019: 15-19. (in Chinese)
17. Qiu Yiqin. Magnetic flocculation coupled with heavy metal trap EDTC for deep removal of acidic complexed nickel and recycling of magnetic species[D]. Guangdong University of Technology, 2017.DOI: CNKI: CDMD: 2.1017.842190. (in Chinese)
18. SHEN Hao, YANG Zhongqi, WANG Fan, et al. MBBR+magnetic coagulation and sedimentation process for the renovation of land-constrained wastewater plants[J]. China Water Supply and Drainage, 2023, 39 (8): 112-119. (in Chinese)
19. YANG Ping, XU Kangkang, ZHOU Jiazhong, et al. MBBR+magnetic coagulation for the upgrading and expansion of A^ (2) /O micro-exposure oxidation ditch wastewater plant[J]. China Water Supply and Drainage, 2023, 39 (10): 67-74. (in Chinese)
20. ZHANG Jianliang, XIE Wei, CAI Jianfeng, et al. Carbon loaded magnetic coagulation and high efficiency sedimentation process for Keqiao Jiangbin wastewater plant[J]. China Water Supply and Drainage, 2023, 39 (18): 120-125. (in Chinese)
21. ZHU Yu, CHEN Jun, ZHAO Qin.BAF+magnetic coagulation technology applied to urban river water quality improvement project design[J]. Green Technology, 2023, 25 (24): 141-146. (in Chinese)
22. LI Yufei, SU Chang, CHEN Kexin, et al. Analysis and application of technologies in upgrading wastewater plants in alpine areas[J]. Water Treatment Technology, 2024, 50 (2): 140-143.DOI: 10.16796/j.cnki.1000-3770.2024.02.026. (in Chinese)
23. Zeng Xiangzhuan, Huang Yixun, Wang Junxian, et al. An example of pretreatment project of magnetic coagulation and precipitation for dyeing and printing wastewater[J]. Water Supply and Drainage, 2024 (6): 99-104. (in Chinese)
24. SHI Kaiying, GUO Zhongwei, ZHANG Wensheng. Application of magnetic coagulation and sedimentation + denitrification deep-bed filtration in the renovation of wastewater treatment plant[J]. Environmental Science and Technology, 2024 (3): 47-51. (in Chinese)
25. Sun Wei, et al. Pollution characteristics of CSOs and selection of core treatment process[J]. Water Supply and Drainage, 2021, 47 (S01): 138-144. (in Chinese)
26. Chen Xucheng. Design and experience of upgrading Jiaxing joint wastewater treatment plant[J]. China Water Supply and Drainage, 2020, 36 (4): 6. DOI: CNKI: SUN: GSPS.0.2020-04-012. (in Chinese)
27. YANG Leisan, LI Junfei, ZHOU Weizhi, et al. A case study of upgrading wastewater treatment plant based

on systematic concept[J]. China Water Supply and Drainage, 2021 (037-018). (in Chinese)

28. WANG Qiguang. Design of large-scale aeration biofilter wastewater treatment plant upgrading[J]. Urban Road and Bridge and Flood Defence, 2024 (1): 149-152. (in Chinese)
29. Tan Yunfei. Application of sand-filled high-density sedimentation tank in Zhengzhou new district wastewater plant[J]. Urban Road and Bridge and Flood Defence, 2018 (8): 4. DOI: 10.16799/j.cnki.csdqyfh.2018.08.045. (in Chinese)
30. CHEN Zhizhen, LIN Chunxiao, SHEN Dan, et al. A case study of semi-subterranean wastewater treatment plant design with coupled primary rainwater storage and treatment facilities[J]. Water Purification Technology, 2023, 42 (8): 165-171. (in Chinese)
31. ZHOU Chuanting, WANG Mengyu, ZHU Xiayan, et al. Engineering scheme for water environment maintenance and greening irrigation system in Shanghai Disney[J]. China Water Supply and Drainage, 2022, 38 (18): 4. DOI: 10.19853/j.zgjsps.1000-4602.2022.18.010. (in Chinese)

Research Progress of Plant Available Carbon in the Earth's Carbon Cycle

Xiang-ping Sun[a,b,*], Hua-jiao Qiu[a], Wang Zhuo[b], Cai-sheng Qiu[a], Guo-qian Zhong[a], Xin-ling Zhao[a], Zhi-min Zhao[a], Song-hua Long[a], Wook Won Sung[b]

[a] Institute of Bast Fiber Crops, Chinese Academy of Agricultural Sciences, Changsha 410205, Hunan, China
[b] Department of Marine Environmental Engineering, Gyeongsang National University, 2 Tongyeonghaean-ro, Tongyeong, Gyeongnam 53064, Republic of Korea

Abstract. To our knowledge, study has yielded that biological carbon sequestration is the use of photosynthesis of plants to improve ecological. The purpose of the present study was to investigate Research progress of plant available carbon in the earth's carbon cycle. One way to overcome these problems is to that plants can fix carbon dioxide as they grow through photosynthesis, and soil can sequester the carbon as decomposed biomass. This was demonstrated in a number of studies that showed that there are four mechanisms that affect the carbon sink of terrestrial ecosystems, namely climate change, CO2 fertilization, nitrogen deposition fertilization and the regeneration of forests after early destruction. Therefore, It is interesting to note that, how to fully utilize the carbon fixed by plants to extend the service time of plant carbon sinks or reduce the consumption of fossil energy is a worthwhile research issue in the process of achieving the "dual carbon" goal.

1 Introduction

This is particularly important when investigating that plants absorb carbon dioxide from the air, combine the carbon dioxide with sunlight, and convert it into something that plants can use the energy used, that is, carbon-based sugars [1,6]. We speculate that this might be due to that plants can fix carbon dioxide as they grow through photosynthesis, and soil can sequester the carbon as decomposed biomass [2,3]. This does seem to depend on that carbon is the basic material that constitutes the "skeleton" of the plant body and is also an important material for plants to carry out photosynthesis [3,4]. Alternatively, it could simply mean that carbon in the soil can provide "food" for microorganisms [4]. Therefor, sufficient carbon is beneficial to create a good root growth environment. It is unclear whether this is a suitable for that there are four mechanisms that affect the carbon sink of terrestrial ecosystems, namely climate change, CO_2 fertilization, nitrogen deposition fertilization and the regeneration of forests after early destruction [3-5]. Under certain assumptions, this can be construed as that CO_2 is the main gas containing carbon and the main form of carbon participating in the material cycle [6]. At this stage of understanding, we believe that plants fix CO_2 during photosynthesis. However, differences in pathways lead to different photosynthetic plants choosing to absorb ^{13}C [5,7]. This may explain why the ratio is different, so that the stable carbon isotopes in C3 and C4 plants [8]. There are differences in the ratio ($\delta^{13}C$), However, $\delta^{13}C$ can be used to characterize plants [9]. Nonetheless, we believe that it is well justified to that indicators of differences in physiological and ecological characteristics of strains [8-10]. As discussed, this is due to the fact that plants use carbon dioxide during the process of photosynthesis, where plants convert energy from the sun into chemical carbohydrate molecules [11]. The empirical findings in this study provide a new understanding of that it is worth conducting extensive research on how to more reasonably and fully utilize plant resources, especially the carbon sink resources of plant waste [19,20]. Notwithstanding these limitations, the study suggests that the main sources of greenhouse gas carbon dioxide, such as deforestation and the use of fossil fuels [20]. In spite of its limitations, the study certainly adds to our understanding of exploring more subtle influencing factors, such as the interactions between plants and animals [21]. A new study has come to a paradoxical conclusion that plants accumulate more carbon in areas where carnivores and herbivores are present, compared to isolated environments that are less susceptible to grazing and trampling [17,22]. This aspect of the research suggested that inorganic carbon fixation includes the dissolution of CO_2 in the atmosphere by seawater, the absorption of CO_2 by saline alkali soil in arid areas, and the formation of carbonaceous rocks [18]. In summary, this paper argued that the transfer efficiency of organic carbon as an energy carrier from plants to litter and then to soil in terrestrial ecosystems is still unclear [19]. The analysis leads to the following conclusions biological carbon sequestration is the use of photosynthesis of plants to improve ecological [20]. The main conclusion that can be drawn is that the system's

* Corresponding author: sunxiangping@caas.cn

carbon absorption and storage capabilities make it the oldest carbon sequestration system on earth [18-20]. In the study, this method is the most economical and has the least side effects for fixing CO_2 in the atmosphere. However, the main source of carbon in plants is CO_2 in the air [27]. The present findings confirm that there is also some water-soluble organic carbon in the soil, and the plant roots directly absorb the small molecular organic carbon elements dissolved in water from the organic matter in the soil, and input them into the plant through electrochemical reactions to form the plant's internal tissue and energy source [28]. Ideally, these findings should be replicated in a study where the rate at which plants and photosynthetic microorganisms absorb carbon from the atmosphere through photosynthesis and through biological respiration [29]. The present findings confirm that it can replace traditional organic fertilizer, and its effective substance content is 5 to 20 times that of traditional organic fertilizer [18]. This conclusion follows from the fact that dynamic changes in the global carbon cycle, climate change and human are closely related to the impacts of similar activities [15-19]. Results provide a basis for that carbon compounds in plants are transferred to carbon compounds in animals through the food chain [30]. These findings provide a potential mechanism for that the respiration of plants and animals converts part of the carbon in the body [29]. We have shown that transformed into carbon dioxide and emitted into the atmosphere [17,30]. The main objective of our investigation has been to obtain some knowledge of that in-depth revelation of the microbial driving mechanism of biogeochemical cycles still needs to be study. The emphasis of this study lies in that to stabilize the combination of carbon isotope tracing technology and multi-omics technology.

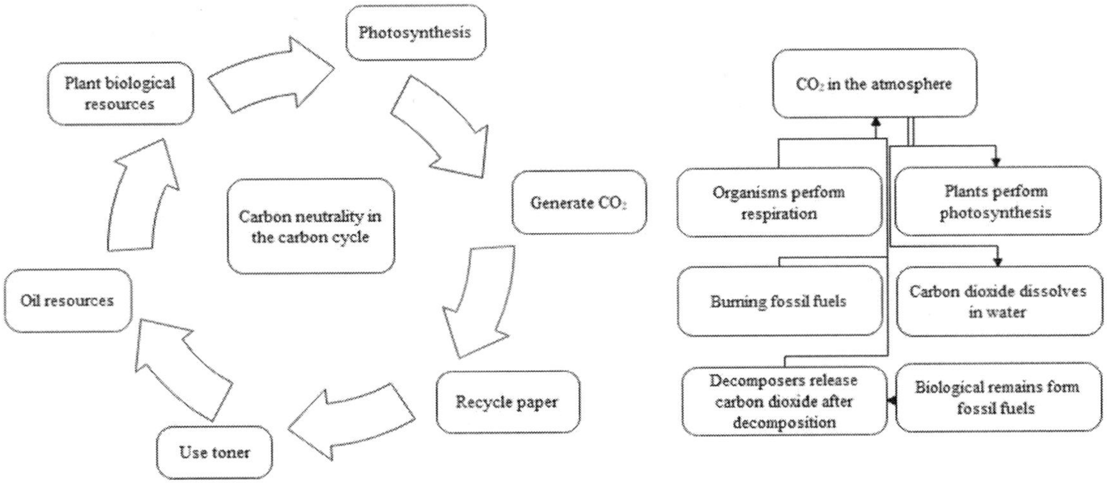

Fig. 1. plant carbon cycle mechanism [5-10].

2 Research progress on plant carbon cycle mechanism

Generally speaking, the percentage of solar energy utilized by plants is 1% to 3% (Fig. 1). This provides a good starting point for discussion and further research. This suggests that under the same climate conditions, different plants have different productivity and carbon sequestration potential (Fig. 1) [7]. This is an important finding in the understanding of the changes in microbial communities can alter microbial carbon utilization efficiency, and the microbial diversity under dry conditions has a significant impact on microbial carbon utilization efficiency (Fig. 1) [18]. The positive impact can be offset. In line with previous studies, forests are major absorbers of carbon and major storers of carbon in biobanks [10]. The positive impact can be offset [18]. This result ties well with previous studies where in under the same climate conditions, different plants have different production power and carbon sequestration potential (Fig. 1) [21,22]. Because of this potential limitation, we treat through selective breeding and planting technology, it is possible to improve high plant productivity and increased carbon sequestration efficiency (Fig. 1).

In our study, bacterial community structure and diversity are the strongest drivers of CUE (Fig. 1). This does seem to depend on that CUE has a positive impact [26,27]. A popular explanation is that the process of biochar degrading into CO_2 on the food chain is relatively short, branches, animal fur, and bones with high levels of lignin and cellulose undergo degradation for several years to several decades (Fig. 1) [30]. We showed that the life cycle of crop carbon and animal carbon in agricultural ecosystems is relatively short [7,9]. The present findings confirm that returning organic waste such as straw and livestock manure to the field is the main way for agricultural carbon neutrality (Fig. 1). This allows the conclusion that in the biological cycle of carbon, carbon dioxide in the atmosphere is absorbed by plants, converted into organic matter through photosynthesis, and then converted from organic matter into carbon dioxide through biological respiration and bacterial decomposition and enters the atmosphere [30].

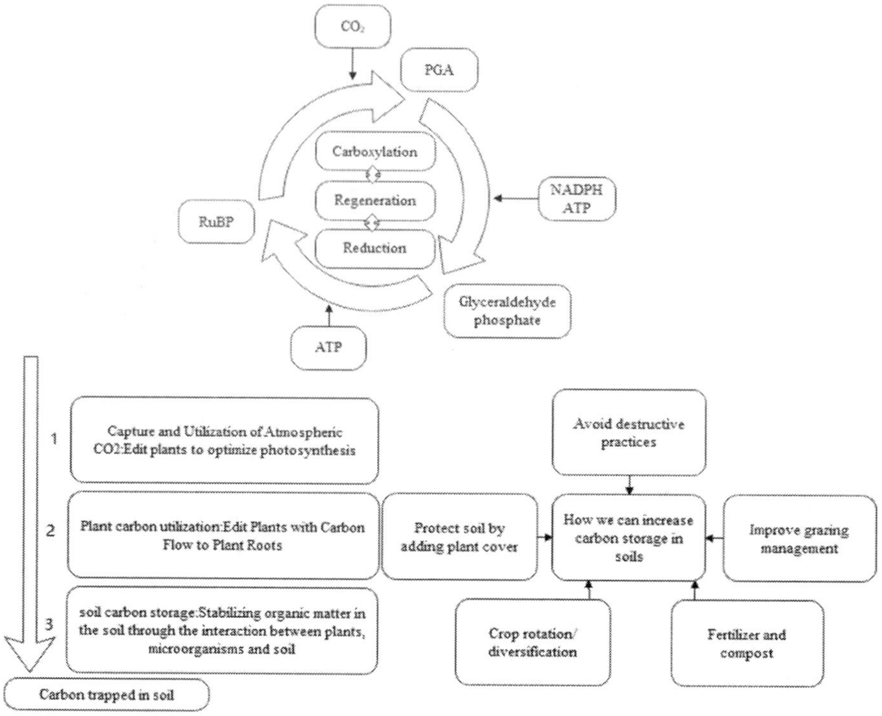

Fig. 2. role of plant carbon in the carbon geochemical cycle [21-25].

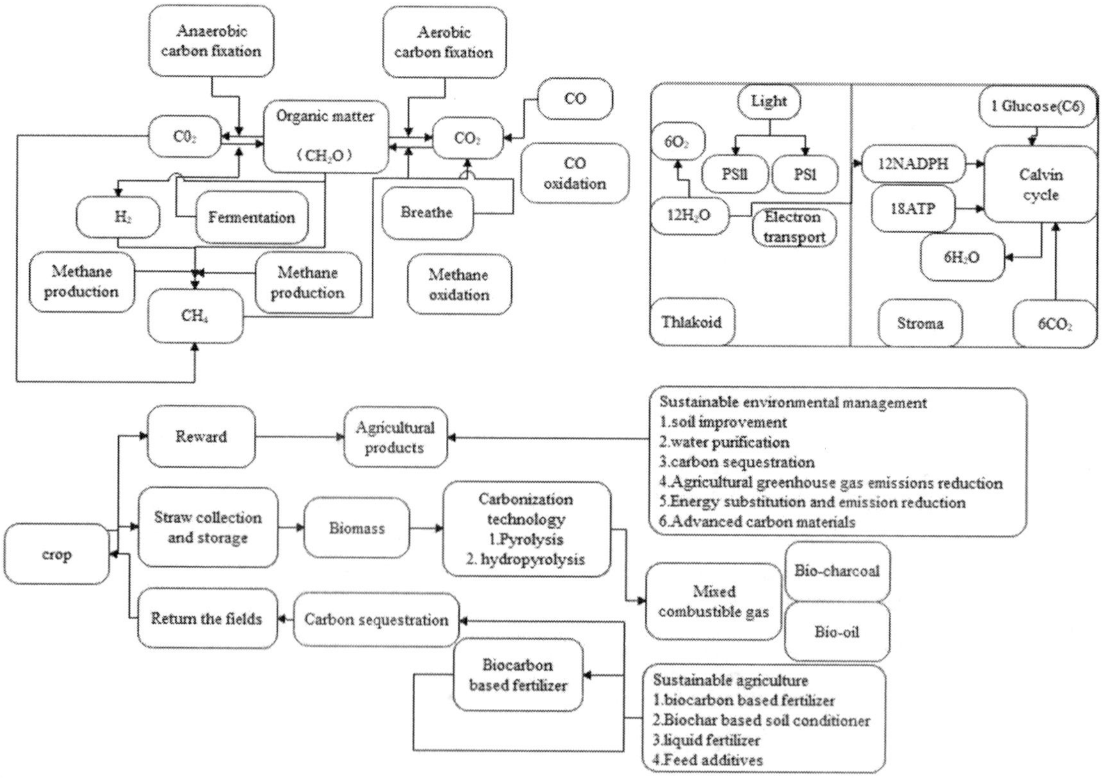

Fig. 3. Plant available carbon in carbon geochemical cycle [30-33].

3 The role of plant carbon in the carbon geochemical cycle

More broadly, research is also needed to determine plant carbon supply pathway: root system secretes organic matter (Fig. 2) [1,23]. The question raised by this study is root autolysis and decomposition (Fig. 2). A limitation of this study is that dry matter decomposition on the ground [4,25]. Taken together, these results suggest that the carbon supply dynamics of plants are the result of a combination of factors such as plant type, various environmental factors, and artificial management measures (Fig. 2) [19]. In general, therefore, it seems that various factors that affect plant photosynthesis and plant residue decomposition can all affect the carbon supply dynamics of plants (Fig. 2) [17,28]. These findings suggest that in general photosynthesis is the main energy source and all food needed by humans origin (Fig. 2) [29].

The energy required for movement comes from converting solar energy into biomass energy photosynthesis (Fig. 2) [28]. Taken together, these results suggest that the potential of plant carbon sinks in nature is huge, far greater than emissions from human activities [27,28]. Therefor, destruction of forest vegetation can cause a rapid increase in carbon emissions, while the carbon fixation effect produced by planting trees and grass is slow and long-term (Fig. 2) [16]. As discussed, this is due to the fact that Soil carbon pools occupy a central position in the carbon cycle (Fig. 2).

4 Research prospects of plant available carbon in carbon geochemical cycle

At this stage of understanding, we believe that the carbon cycle on the earth is mainly manifested in that green plants in natural ecosystems absorb carbon dioxide from the air, convert it into carbohydrates through photosynthesis, and release oxygen (Fig. 3) [12,14,15]. At the same time, because of this potential limitation, we treat carbon dioxide is released into the atmosphere through biogeochemical cycle processes and human activities (Fig. 3) [13,19,23]. After animals and plants die, most of their remains are emitted into the atmosphere in the form of carbon dioxide through the decomposition of microorganisms. However, It is unclear whether this is a suitable for under conditions of environmental stress or human interference, the importance of underground carbon exchange (Fig. 3). Alternatively, it could simply mean that plants absorb CO_2 from the atmosphere through photosynthesis, store the carbon in the plant body, and fix it as organic compounds [22]. That among them, part of the organic matter is absorbed through the plant's own respiration (Fig. 3) [29,33].

It is the starting point of the carbon biogeochemical cycle (Fig. 3). CO_2 fixation pathways during photosynthesis of different photosynthetic plants (Fig. 3). The difference results in a different proportion of selective absorption of ^{13}C. There are differences in $^{\delta13}C$ of different plants. It is important to note, that the present evidence relies on that carbon storage in natural ecosystems and carbon release is basically balanced on a longer time scale (Fig. 3). Unless the intensity of the terrestrial ecosystem carbon pool increases, any one Sooner or later, a carbon sink will be balanced by a carbon source (Fig. 3). This does seem to depend on the greatest uncertainties in the current global carbon cycle arise primarily from terrestrial ecosystems (Fig. 3).

5 Conclusion

Identical conclusions were obtained in studies where green plants fix the absorbed solar energy in carbohydrates through photosynthesis, and these compounds are then transmitted along the food chain and oxidized to release energy within various levels of organisms. Prior investigations have implemented diverse approaches to that biological carbon sequestration has huge potential, and different management measures have different impacts on ecosystems has a significant impact on the carbon sequestration capacity of the system, but its carbon sequestration amount is still lacking hold accurate understanding of continuity and stability.

Acknowledgments

The Agricultural Science and Technology Innovation Program (No. ASTIP-IBFC-04).

References

1. Wu H, Cui H, Fu C, et al. Unveiling the crucial role of soil microorganisms in carbon cycling: A review[J]. Science of The Total Environment, 2023: 168627.
2. Horwath W R. Soil carbon formation and persistence[M]//Soil Microbiology, Ecology and Biochemistry. Elsevier, 2024: 329-367.
3. Birner B, Rödenbeck C, Dohner J L, et al. Surprising stability of recent global carbon cycling enables improved fossil fuel emission verification[J]. Nature Climate Change, 2023, 13(9): 961-966.
4. Jones C, Mcconnell C, Coleman K, et al. Global climate change and soil carbon stocks; predictions from two contrasting models for the turnover of organic carbon in soil[J]. Global Change Biology, 2010, 11(1):154-166.
5. Stevens D P A. The impact of Land Surface Model vegetation parameterization on the terrestrial water and carbon cycles[J]. 2023.
6. Gakaev R. Functional classification of forests: Study of carbon sequestration[C]//BIO Web of Conferences. EDP Sciences, 2023, 76: 06004.
7. Schuur E A G, Bockheim J, Canadell J G, et al. Vulnerability of Permafrost Carbon to Climate Change: Implications for the Global Carbon Cycle[J]. Bioscience, 2008, 58(8):701-714.
8. Hashimoto S, Ito A, Nishina K. Divergent data-driven estimates of global soil respiration[J]. Communications Earth & Environment, 2023, 4(1): 460.

9. Spackman C, Allison C D. Carbon Sequestration on Rangelands: A Primer[J]. Agriculture and its Role in the Global Carbon Cycle \He, Liyin California Institute of Technology ProQuest Dissertations Publishing, 2023. 30604349. DOI:10.7907/34d6-nj32
10. Brown T, Hampp J. Ultra-long-duration energy storage anywhere: Methanol with carbon cycling[J]. Joule, 2023, 7(11): 2414-2420.
11. Mavrovic A, Sonnentag O, Lemmetyinen J, et al. Reviews and syntheses: Recent advances in microwave remote sensing in support of terrestrial carbon cycle science in Arctic–boreal regions[J]. Biogeosciences, 2023, 20(14): 2941-2970.
12. Liu J, Gao W, Liu T, et al. A bibliometric analysis of the impact of ecological restoration on carbon sequestration in ecosystems[J]. Forests, 2023, 14(7): 1442.
13. Guddaraddi A, Goutam P K, Sonaniya P, et al. Role of Biological Carbon Mitigation in Carbon Sequestration: A Review[J]. International Journal of Environment and Climate Change, 2023, 13(11): 749-757.
14. Prajapati S K, Choudhary S, Kumar V, et al. Carbon Sequestration: A Key Strategy for Climate Change Mitigation towards a Sustainable Future[J]. 2023.
15. Zaret M. Microbes, herbivores, and grassland carbon cycling responses to biodiversity loss and nutrient pollution[D]. University of Minnesota, 2023.
16. Mao Z, Liu H, Zhang Z. Enhancing the ocean carbon sink capacity of Shandong province, China, under the "dual carbon" goal[J]. Frontiers in Marine Science, 2023.
17. Fendrich A. Carbon cycle as affected by soil erosion in Europe[D]. Université Paris-Saclay, 2023.
18. Hotchkiss E R, DelSontro T. Organic Carbon Cycling and Ecosystem Metabolism[M]//Wetzel's Limnology. Academic Press, 2024: 939-997.
19. Zhao D, Cai J, Xu Y, et al. Carbon sinks in urban public green spaces under carbon neutrality: A bibliometric analysis and systematic literature review[J]. Urban Forestry & Urban Greening, 2023: 128037.
20. Wani O A, Kumar S S, Hussain N, et al. Multi-scale processes influencing global carbon storage and land-carbon-climate nexus: A critical review[J]. Pedosphere, 2023, 33(2): 250-267.
21. Exbrayat J F, Bloom A A, Carvalhais N, et al. Understanding the land carbon cycle with space data: current status and prospects[J]. Surveys in Geophysics, 2019, 40: 735-755.
22. Piao S, Wang X, Wang K, et al. Interannual variation of terrestrial carbon cycle: Issues and perspectives[J]. Global Change Biology, 2020, 26(1): 300-318.
23. Lal R. Carbon cycling in global drylands[J]. Current climate change reports, 2019, 5: 221-232.
24. Jiang Y, Sun Y, Liu Y, et al. Exploring the correlation between waterbodies, green space morphology, and carbon dioxide concentration distributions in an urban waterfront green space: A simulation study based on the carbon cycle[J]. Sustainable Cities and Society, 2023, 98: 104831.
25. Prajapati S K, Kumar V, Dayal P, et al. THE ROLE OF CARBON IN LIFE'S BLUEPRINT AND CARBON CYCLE UNDERSTANDING EARTH'S ESSENTIAL CYCLING SYSTEM: BENEFITS AND HARMS TO OUR PLANET[J]. International Journal, 2023, 1(1): 21-32.
26. Wu H, Cui H, Fu C, et al. Unveiling the crucial role of soil microorganisms in carbon cycling: A review[J]. Science of The Total Environment, 2023: 168627.
27. Horwath W R. Soil carbon formation and persistence[M]//Soil Microbiology, Ecology and Biochemistry. Elsevier, 2024: 329-367.
28. Zhang J, Zhang C, Ma W, et al. Improving the Model Performance of the Ecosystem Carbon Cycle by Integrating Soil Erosion–Related Processes[J]. Atmosphere, 2023, 14(12): 1724.
29. Ma C, Xu Z. Research Status, Problems and Direction of Soil Organic Carbon in Zoige Peat Wetland[J]. Research in Ecology, 2023, 5(3): 1-10.
30. Wang S, Chen D, Liu Q, et al. Dominant influence of plants on soil microbial carbon cycling functions during natural restoration of degraded karst vegetation[J]. Journal of Environmental Management, 2023, 345: 118889.
31. Mu C, Abbott B W, Norris A J, et al. The status and stability of permafrost carbon on the Tibetan Plateau[J]. Earth-Science Reviews, 2020, 211: 103433.
32. Zhou W, Li J, Yue T, et al. Research progress of the grassland carbon cycle and grassland degradation in China[J]. Remote Sensing Monitoring and Evaluation of Degraded Grassland in China: Accounting of Grassland Carbon Source and Carbon Sink, 2020: 1-16.
33. Soong J L, Fuchslueger L, Marañon-Jimenez S, et al. Microbial carbon limitation: The need for integrating microorganisms into our understanding of ecosystem carbon cycling[J]. Global change biology, 2020, 26(4): 1953-1961.

 https://doi.org/10.1051/e3sconf/202561701009

The Role of Plant Charcol in the Carbon Geochemical Cycle

Xiang-ping Sun[a,b,*], Zhuo Wang[b], Guo-qian Zhong[a], Wook Won Zhong[b]

[a] Institute of Bast Fiber Crops, Chinese Academy of Agricultural Sciences, Changsha 410205, Hunan, China
[b] Department of Marine Environmental Engineering, Gyeongsang National University, 2 Tongyeonghaean-ro, Tongyeong, Gyeongnam 53064, Republic of Korea

Abstract: The role of plant charcol is to promote healthier and more vigorous growth in plants. Therefore, the carbon cycle is mainly manifested in green plants which absorbing carbon dioxide from the air, which converting it into glucose through photosynthesis, at the same time, releasing oxygen in the biosphere. However, artificial carbon fixation is the direct or indirect artificial removal of atmospheric carbon dioxide through spatiotemporal scale conversion in the study. This article reviews the flow patterns and mechanisms of plant-derived carbon in the carbon cycle. Which attempting to explain carbon moves from one reservoir to another through a variety of mechanisms in the carbon cycle. Which is expected to provide reference significance for the research on the biological cycle process of carbon and the carbon neutralization of carbon peak in the study.

1 Introduction

To mitigate the crisis caused by global warming, there is an urgent need to reduce greenhouse gas emissions and lower the concentration of carbon dioxide in the atmosphere [1-3]. However, the use of plant photosynthesis to fix atmospheric carbon dioxide in plants and soils is an important link in the carbon cycle that has been shaped by nature over billions of years[4]. Therefore, green plants play a very important role in the carbon cycle[5]. Which are one of the major life forms on earth and are the most important and fundamental producers in ecosystems, which playing a key role in the ecosystems of the biosphere, and in the cycling of matter and in the flow of energy. Plants provide us with food, fiber, shelter, medicines and fuel in the study[6]. They are organisms that use sunlight to produce food, such as plants, which absorb carbon dioxide from the atmosphere and use it to build carbohydrates, blood lipids, proteins, and other important biomolecules in the study[7]. Therefore, carbon absorption by producers organisms that uses sunlight to make food, such as plants, absorb carbon dioxide from the atmosphere and use it to build carbohydrates, blood lipids, proteins and other important biomolecules[8]. Therefore, photosynthesis converts inorganic matter into organic matter through light reaction and carbon assimilation, and which is the most important chemical reaction on earth. Perennials have more developed root systems than annuals, favouring long-term carbon storage. Decomposers release carbon. Plants and animals that die without being consumed by other organisms will be decomposed by other organisms called "decomposers"[8-10].

In the study, plants absorb carbon dioxide and convert it into energy for growth. When a plant dies, carbon dioxide is released from the plant's decomposition. The role of carbon in plants is to promote healthier and more vigorous growth[10-11]. Therefore, adding organic manure or broken down plant parts (carbon-rich or brown manure in compost) to the soil around growing plants essentially causes them to be nourished by the fertilizer, allowing the plants to grow vigorously in the study[12]. However, agricultural carbon peaking and carbon neutrality are one of the important contents of "double carbon" development, and which have a solid supporting role in achieving the "double carbon" goal in China. Especially, there is still a lack of research on it, and there is no systematic report on it. This article mainly elaborates on the importance of plants in the carbon cycle, studies the role of plant carbon in the carbon geochemical cycle, and provides technical assistance and reference opinions for carbon neutrality and carbon peaking.

2 Research progress of plant charcol

In order to survive, some plants have evolved another way of sequestering carbon. They utilize PEPcase (phosphoenolpyruvate carboxylase), which has a high affinity for carbon dioxide, for carbon fixation[1]. C4 plants are plants that assimilate carbon through the C4 pathway. They have both C3 and C4 pathways. (Table 1).

In dry and hot days of C3 plants, the stomata will close, which will prevent carbon dioxide from entering the leaves and oxygen from escaping the leaves in they study (Table 1). Oxygen will gradually accumulate, and Rubp carboxylase will fix the oxygen to produce a two-carbon compound. The plant will decompose the two-carbon compound into carbon dioxide and water, which will produce photorespiration and reduce crop yields (Table 1).

* Corresponding author: sunxiangping@caas.cn

Table 1. Plants carry out photosynthesis pathways C3 and C4[6-10]

Classification	Photosynthesis type	Classification of plants using photosynthesis	Plant type
C3	C3 way	Plants that carry out photosynthesis through the C3 pathway are called C3 plants	Wheat, rice, soybean, bitter bamboo, tobacco, cotton, ginkgo, cedar, fir, silver fir
C4	C4 way	Plants that carry out photosynthesis through the C4 pathway are called C4 plants	Rice, Sorghum, Sugarcane, Setaria, Imperatae, Millet (Milliet), Paspalum Paspalum, Broadleaf Paspalum, Peapaspalum, Single Branch Barnyardgrass, Multigenus Barnyardgrass, Terminal Bluestem, Nandi Setaria, Narok Setaria, Gorgeous foxtail, multi-tasseled tiger's tail, four-tasseled foxtail, two-tasseled tiger'stail, burrata, sweetroot, red amaranth, spiny amaranth, purslane, dog's tongue

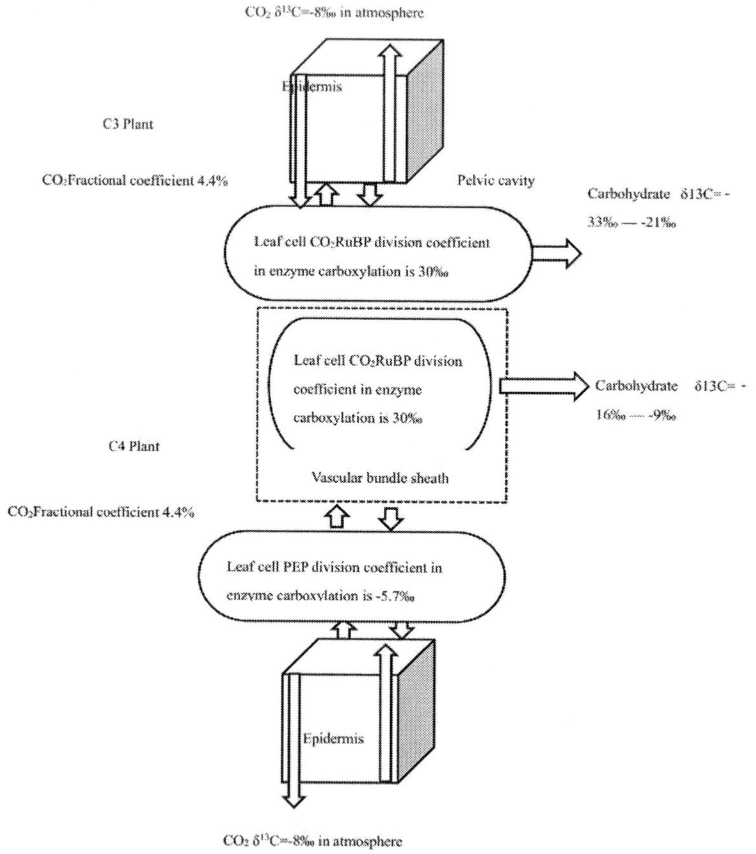

Fig 1. The difference of using CO2 between the C3 and C4 plan't[13-16]

The carbon isotope fractionation effect during photosynthesis in C3 and C4 (Fig 1). Green plant photosynthesis fixes carbon (Fig 1). Therefore, respiration is the process of releasing carbon. Therefore, C4 plants have one more carbon dioxide fixation than C3 plants (Fig 1). However, they have a high carbon dioxide binding capacity and can increase crop yields. In the C4 pathway, PEPcase carboxylase can distinguish carbon dioxide and oxygen very well. It only participates in the fixation of carbon dioxide to generate malic acid or aspartic acid (Preston, 2006). However, since the stable carbon isotopes of C3 plants are -22‰ to 30‰, the average is -27‰ in the study (Fig 1). The stable carbon isotopes of C4 plants are -9‰-19‰, with an average of -14‰ (Fig 1). Although C4 plants are powerful, the C4 pathway consumes a large amount of energy during carbon fixation (Fig 1). In contrast, the division of labor in C3 plant cells is less clear than that of C4 plants, and the efficiency of CO_2 utilization is lower (Fig 1). To a certain extent, C3 plants can be considered to be "prokaryotes" among plants, while C4 plants are more like "eukaryotes"(Fig 1). Therefore, the carbon cycle in the biosphere is mainly manifested in green plants absorbing carbon dioxide from the air, converting it into glucose through photosynthesis, and releasing oxygen (O_2).

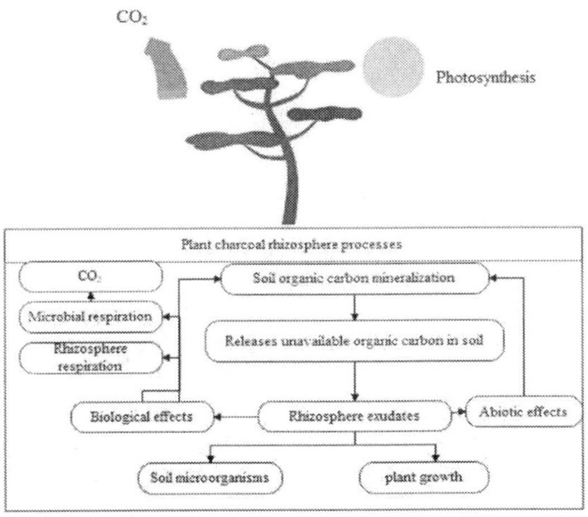

Fig 2. The life activities of plants using carbon[17-20]

The carbon cycle includes the following main steps: Plants absorb carbon photosynthesis from the atmosphere during growth. These plants are then devoured by animals, causing organisms to accumulate carbon in their bodies in the study (Fig 2). Therefore, these organisms eventually die, and their decomposition releases carbon back into the atmosphere. As time goes by, some carbon that has not returned to the atmosphere will become fossil fuels in the study (Fig 2). These fossil fuels are then used for human activities, releasing additional carbon into the atmosphere[21]. However, during photosynthesis, plants absorb carbon dioxide and sunlight to produce fuel - glucose and other sugars - used to build plant structures. This process forms the basis of the rapid (biological) carbon cycle (Fig 2). More than 90% of plant dry matter comes from light synthetic carbon fixation reaction and organic matter assimilated by photosynthesis are the material basis for crop yield in the study[22]. Therefore, improving crop light energy utilization efficiency is the key to improving crop yield. important way to measure. Therefore, photosynthesis is where plants absorb carbon dioxide to produce carbohydrates (Fig 2). However, most of the carbon in the body is in the form of carbon dioxide due to respiration (Fig 2).

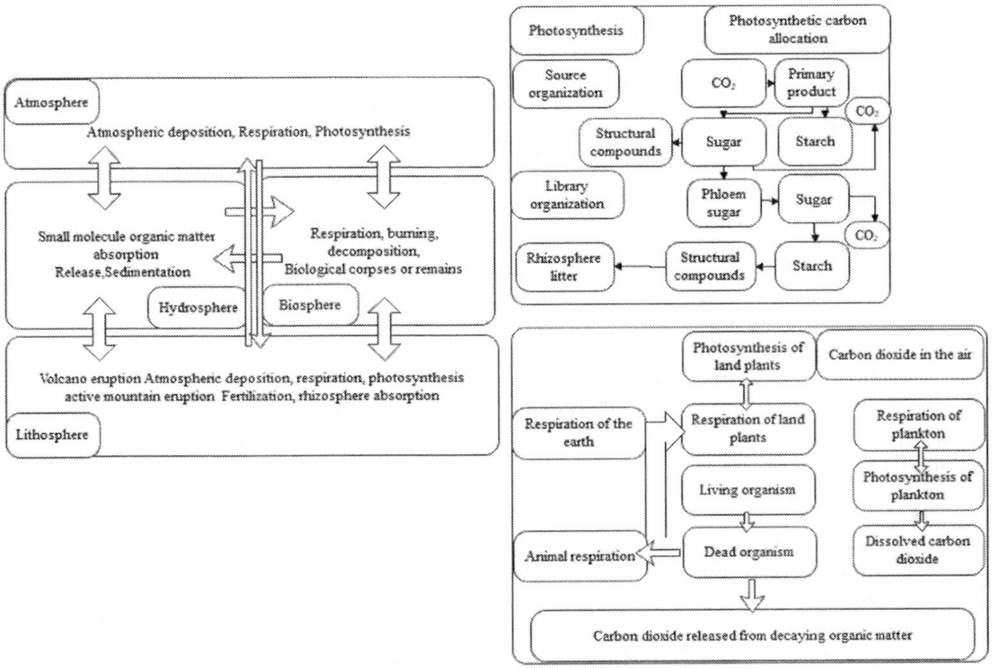

Fig 3. Based on plant carbon of carbon geochemical cycle[16-22]

In the study, carbon moves from one reservoir to another through a variety of mechanisms (Fig 3). In the food chain, plants transfer carbon from the atmosphere into the biosphere through photosynthesis, respiration, excretion, and decomposition release carbon back into the atmosphere or soil, continuing the cycle (Fig 3)[16-17]. However, C4 plants are able to utilize low concentrations of carbon dioxide, while C3 plants cannot. Plant residues are the main initial source of soil organic matter (Fig 3). However, under the mediation of soil microorganisms, it is transformed into soil through a complex decomposition process (Fig 3) [15-16]. Soil organic matter exists stably. The small molecular compounds produced by the decomposition of plant residues again form macromolecules of humus. At the same time, a process that has so far not been demonstrated in natural soil systems (Fig 3) [14]. Lignin is a type of phenylpropane polymer, and its relative content Controls the degradation rate of litter. However, some studies have shown that lignin is not as stable in soil as expected (Fig 3)[18-20].

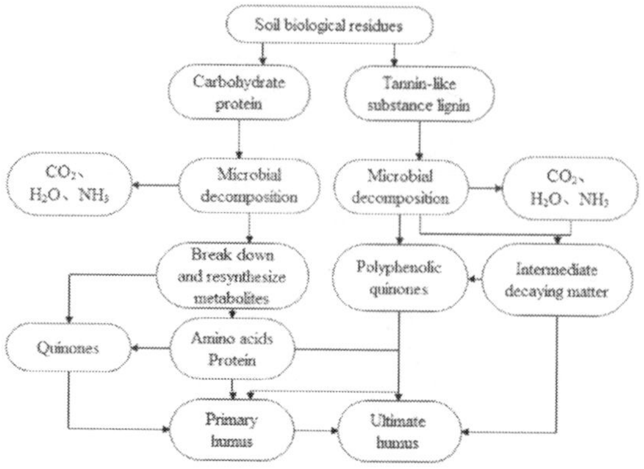

Fig 4. Carbon geochemical cycle involving plant carbon [12-16,19,20]

In the study, restoration of original oxidized form (CO_2) by mineralization of organic form (Fig 4)[23]. There are three different mechanisms for organic mineralization and carbon dioxide emission (Fig 4). They are: breathing unintentional (forest fire) and intentional (fuel) burning in the study. However, organic matter is decomposed by microorganisms[24]. The respiration of plants and animals and the unintentional and intentional burning of plants and their parts lead to the decomposition of organic carbon molecules and the subsequent release of carbon dioxide into the atmosphere (Fig 4) [25-26]. In the study, amending soil with organic carbon not only benefits plant growth, but also drains well, prevents water pollution, benefits microorganisms and insects, and eliminates the need for chemical fertilizers (Fig 4)[23-25]. While our dependence on fossil fuels is what got us into chaos in the first place, utilizing organic gardening techniques is one way to combat the collapse of global warming[26]. Of course, it is also the reason why we promote the use of organic fertilizer compost. Carbon dioxide in the air or organic carbon in the soil, the role of carbon and plant growth is extremely valuable, in fact, without this process, life as we know it would not exist (Fig 4).

3 Conclusion

Carbon completes a complete cycle, moving from the air to plants and/or animals and then back to the atmosphere. There are many ways to improve the carbon cycle, and some progress has been made in theoretical research, which provides some new clues for improving the photosynthetic efficiency of plants. Some of the successful examples also show us that by improving the carbon cycle, they mainly focus on One is to introduce the C4 cycle pathway into C3 plants, the other is to change the relationship between carbon cycle and modification of key enzymes; third, inhibition of plant photorespiration. Plant-derived carbon will play an irreplaceable role in the carbon geochemical cycle. It provides a good method for us to study the carbon cycle and implement carbon peak and carbon neutral pathways.

Acknowledgments

The Agricultural Science and Technology Innovation Program (No. ASTIP-IBFC-04).

References

1. Adeleye AT, Akande AA, Odoh CK, Philip M, Fidelis TT, Amos PI, Banjoko OO. Efficient synthesis of bio-based activated carbon (AC) for catalytic systems: A green and sustainable approach. J Ind Eng Chem, 2021, 96: 59-75.
2. Salani G M, Rizzo E, Brombin V, et al. Carbon Soil Mapping in a Sustainable-Managed Farm in Northeast Italy: Geochemical and Geophysical Applications[J]. 2024.
3. Altwala A, Mokaya R. Predictable and targeted activation of biomass to carbons with high surface area density and enhanced methane storage capacity. Energ Environ Sci, 2020, 13: 2967-2978.
4. Upchurch G R, Marino B D. Plant Cuticle: A Geochemical and Biophysical Sensor of the Carbon Cycle[J]. AGU Fall Meeting Abstracts, 2007.
5. Martinez-Braceras J D T J. Physical and geochemical record of an early Eocene carbon-cycle perturbation on a turbiditic continental margin[J]. Sedimentology: Journal of the International Association of Sedimentologists, 2021, 68(2).
6. Balan S A, Welsh H, Amundson R. Comparative C, N, and S cycling along a Californian grassland chronosequence[J]. Geoderma, 2023, 439: 116682.
7. Jerden J, Mejbel M, Antonio Nilson Zamunér Filho, et al. The impact of geochemical and life-cycle variables on carbon dioxide removal by enhanced rock weathering: Development and application of the Stella ERW model[J]. Applied Geochemistry, 2024, 167.
8. Cao W. Advances in Geochemical Monitoring Technologies for CO2 Geological Storage[J]. Sustainability, 2024, 16.
9. Cen C, Xu Z. Forest ecosystem responses to environmental changes: the key regulatory role of biogeochemical cycling[J]. Journal of Soils and Sediments, 2010, 10: 210-214.
10. Mcgivern B B, Cronin D R, Ellenbogen J B, et al. Microbial polyphenol metabolism is part of the thawing permafrost carbon cycle[J]. Nature Microbiology, 2024, 9(6): 1454-1466.
11. Li Y, Tang S, Xi C Z. Microbial distribution characteristics related to carbon cycle and their potential impact on methanogenesis of coal reservoirs in underground in situ environments[J]. Environment and history, 2024, 30(1): 40-51.
12. Jaffe A L, Bardot C, Anne-Hélène.Le Jeune, et al. Variable impact of geochemical gradients on the functional potential of bacteria, archaea, and phages from the permanently stratified Lac Pavin[J]. Microbiome, 2023, 11(1): 14.

13. Olson J S, Garrels R M, Berner R A, et al. The natural carbon cycle[J]. Atmospheric carbon dioxide and the global carbon cycle, 1985, 239: 175-214. Rodionov A, Amelung W, Peinemann N, et al. Black carbon in grassland ecosystems of the world[J]. Global Biogeochemical Cycles, 2010, 24(3).
14. Jenerette G D, Park I W, Andrews H M, et al. Biogeochemical cycling of carbon and Nitrogen in chaparral dominated ecosystems[J]. Valuing Chaparral: Ecological, Socio-Economic, and Management Perspectives, 2018: 141-179.
15. Lal R. Agricultural activities and the global carbon cycle[J]. Nutrient cycling in agroecosystems, 2004, 70: 103-116.
16. Shi X, Wu B, Qiao S, et al. Distribution, burial fluxes and carbon sink effect of sedimentary organic carbon in the eastern China seas[J]. SCIENCE CHINA Earth Sciences, 2024, 67(10).
17. Forbes M S, Jerden J, Mejbel M, Antonio Nilson Zamunér Filho, et al. The impact of geochemical and life-cycle variables on carbon dioxide removal by enhanced rock weathering: Development and application of the Stella ERW model[J]. Applied Geochemistry, 2024, 167.
18. Pogodaeva T, Khodzher T. Organic Carbon in the Bottom Sediments of Lake Baikal: Geochemical Processes of Burial and Balance Values[J]. Water (20734441), 2023, 15(16).
19. Song Z, Müller K, Wang H. Biogeochemical silicon cycle and carbon sequestration in agricultural ecosystems[J]. Earth-Science Reviews, 2014, 139: 268-278.
20. Yu L, Li S, Wu K. A natural analogue for carbon capture and storage: Petrographic and geochemical changes in sandstone after CO2 emplacement in the Yinggehai Basin, South China Sea[J]. AAPG Bulletin, 2023, 107(4): 643-683.
21. Oelkers E H, Gislason S R. CARBON CAPTURE AND STORAGE: FROM GLOBAL CYCLES TO GLOBAL SOLUTIONS[J]. Geochemical Perspectives, 2023, 12(2): 179-346.
22. Kirichenko Y, Rickli J D, Bontognali T R R, et al. Insights into stable strontium isotope fractionation in marine gypsum and its geochemical implications[J]. Geochimica et Cosmochimica Acta, 2024, 378: 270-285.
23. Inoue Y, Okiyama R, Hagiwara Y, et al. Raman spectroscopic evaluation of precision of oxygen isotope ratio of carbon dioxide[J]. GEOCHEMICAL JOURNAL, 2023.
24. Cao W. Advances in Geochemical Monitoring Technologies for CO2 Geological Storage[J]. Sustainability, 2024, 16.
25. Harsha M L, Whisenhant E A, Hebert D, et al. Photodissolution of Rare Earth Elements and Dissolved Organic Carbon from Subbituminous Coal: Effects of Environmental Variables and Implications for Biogeochemical Cycling[J]. Environmental Science & Technology Letters, 2023, 10(9): 747-754.
26. Pandao M R, Lalitha G R, Kishore A J, et al. A Comprehensive Review on Biochar[J]. International Journal of Environment and Climate Change, 2023, 13(11): 3453-3461.

Impact of Dissolved Oxygen Limiting Strategies on the Simultaneous Partial Nitrification, Anammox and Denitrification (SNAD) Process

Zhouzhenyang Xia[1], Jiajun Luo[2], Zhifan Liu[1], Menglin Ou[1] and Haiyan Fu[1,*]

[1]School of Environmental Scienece and Engineering, Xiamen University of Technology, Xiamen 361024, China
[2]Wanjia Qingshui (Xiamen) Environmental Technology Co., LTD, Xiamen 361022, China

Abstract. The effect of dissolved oxygen (DO) limitation strategies on the nitrogen removal performance of the SNAD system under high DO damage was explored. Nitrogen removal efficiency (NRE) decreased rapidly in a high DO environment (2.0-3.0 mg/L). The NRE temporarily recovered to 87.32% after the DO limiting strategy was adopted. However, the NRE decreased to 59.35% with the rapid increase in nitrate concentration and the average anaerobic ammonia oxidation rate decreased to 1.37 mg-TN/(gVSS-h). In order to suppress the gradually increasing nitrate concentration, DO was further limited to 0.3-1.0 mg/L. This strategy effectively stabilised the nitrate concentration and achieved the recovery of anammox bacteria (AnAOB) activity, but could not stably restore the system denitrification performance. At the same time, the pH in the system dropped below 7, which hindered the restoration of the nitrogen removal efficacy. The DO limitation strategy can temporarily restore AnAOB activity. Achieving stable nitrogen removal may require synergistic adjustments of alkalinity and hydraulic retention time, among other strategies.

1 Introduction

As the carbon peak and carbon neutrality policies become increasingly stringent, the quality of wastewater discharge is becoming more strictly regulated across all sectors. However, the traditional nitrification-denitrification processes have several disadvantages, including the requirement for significant inputs of organic carbon sources, increased energy consumption and high operating costs [1]. Additionally, the treatment process generates a considerable quantity of residual sludge, which presents a challenge to the attainment of carbon emission reduction targets in wastewater treatment facilities [2]. Due to the absence of an organic carbon source, minimal aeration requirements, and the generation of negligible quantities of greenhouse gases such as NxO during the reaction process. In recent times, the Anammox process has emerged as a prominent area of research.

Anammox oxidises NH_4^+-N to N_2 using NO_2^--N as the electron acceptor. Thus, anammox process has the potential to achieve a 60% reduction in aeration, eliminating the necessity for an organic carbon source, and reducing sludge production by 90% [3]. However, the anammox process will be subject to several limitations in the practical application of the project. These include poor attachment and fixation ability, suitability for growth at higher temperatures, unstable substrate nitrite generation, and dissolved oxygen (DO).

The simultaneous partial nitrification, anammox and denitrification (SNAD) process represents a significant component of the anammox process. The SNAD process is primarily employed for high ammonia and low C/N ratio wastewater [4]. The SNAD process commences with the oxidation of a portion of NH_4^+-N into NO_2^--N by AOB. Subsequently, anammox bacteria (AnAOB) utilises NO2--N and a fraction of NH_4^+-N as substrates for anaerobic ammonia oxidation, which converts ammonia nitrogen into nitrogen gas for removal and generates a minor quantity of nitrate. Finally, denitrifying bacteria (DNB) is responsible for the removal of nitrate, utilising organic carbon as an electron donor. Concurrently, organic carbon is converted into CO2 in this process [5]. The denitrification and decarbonisation of the SNAD system is achieved.

AnAOB play a significant role in the nitrogen removal process within the SNAD system. Studies have shown that AnAOB is sensitive to the concentration of DO, and higher DO environments reduce the affinity of the AnAOB communities [6]. AnAOB activity is inhibited when dissolved oxygen exceeds 0.5 mg/L. AnAOB activity can be restored rapidly when DO limitation is lifted. The recovery of AnAOB may not occur at high DO when DO is above 2 mg/L. Due to partial nitrification at the front end, Anammox bacteria are inevitably influenced by DO [7]. The impact of the constrained DO approach on the performance of the SNAD process has yet to be fully elucidated. In this study, we limited the dissolved oxygen concentration by flowmeters. Exploring the effect of the DO limiting strategy on the performance of the SNAD process.

* Corresponding author: fuhy@xmut.edu.cn

2 Materials and methods

2.1 Reactor set-up and operation

The reactor unit is constructed from polymethyl-methacrylate (PMMA) with an effective volume of 20.0 L. The reactor utilises biological ropes wrapped around a turntable as a carrier. The detailed specifications regarding reactor components are shown in [8]. The reactor employed an air pump to provide aeration. The study used three flowmeters to regulate the DO environment in the sections. In the specific experimental operation, the hydraulic retention time (HRT) of the reactor was 24 h, and the reflux ratio was 100%. The temperature of the reactor was maintained at 30 ± 2 °C. In phase I (1-20d), the air intake was adjusted to 2 L/min by flowmeters. In phase II (21-44d), the air intake was adjusted to 500 ml/min by flowmeters. In phase III (45-68d), the air intake was reduced to 300 ml/min.

2.2 Wastewater and inoculation

The sludge for this experiment was obtained from a mainstream anammox reactor that had been in stable operation for 300 days. The initial flocculated sludge exhibited a mixture of suspended solids (MLSS) concentrations of 14,500 mg/L, and a mixture of volatile suspended solids (MLVSS) concentrations of 8,000 mg/L. The study was conducted using a simulated wastewater with a C/N ratio of 1, with the sole source of NH_4^+-N being NH_4Cl, and the sole source of organic matter being glucose. 1.26 g/L $NaHCO_3$ was introduced as an inorganic carbon source and source of alkalinity, and 0.1360 g/L KH_2PO_4, 0.0200 g/L $MgSO_4 \cdot 7H_2O$, 0.1360 g/L $CaCl_2$, 0.3500 ml/L trace elements were added. The trace element contents are shown in [8].

2.3 Sampling and analytical methods

The concentration of NH_4^+-N was determined in accordance with the Nano reagent spectrophotometric method. The concentration of NO_3^--N was determined by UV spectrophotometry. The concentration of chemical oxygen demand (COD) was determined by rapid extinction spectrophotometry. The concentrations of MLSS and MLVSS were calculated by gravimetric method. The pH of the samples was determined with a PHBJ-260, and the DO content was determined with a JPB-607A DO meter.

2.4 Calculation methods

The formulae used in this study are as follow:

$$V_{AOB}\ (mg\text{-}NH_4^+\text{-}N/(\text{gVSS}\cdot\text{h})) = ([NH_4^+\text{-}N]_{\text{inf}} - [NH_4^+\text{-}N]_{\text{eff}})/\text{MLVSS}\times\Delta t \quad (1)$$

$$V_{Anammox}\ (mg\text{-}NH_4^+\text{-}N/(\text{gVSS}\cdot\text{h})) = 2.04\times([NH_4^+\text{-}N]_{\text{inf}} - [NH_4^+\text{-}N]_{\text{eff}} + [NO_2^-\text{-}N]_{\text{eff}})/\text{MLVSS}\times\Delta t \quad (2)$$

V_{AOB} denotes ammonia oxidation rate, $V_{Anammox}$ denotes Anammox rate, and MLVSS denotes combined liquor volatile suspended solids concentration. Since no sludge was discharged during the reactor operation, the MLVSS concentration was taken as 8000 mg/L. Δt represents the change in time, which was set to 24 h due to the regular daily sampling.

3 Results and discussion

3.1 Nitrogen and Dissolved Oxygen Analysis

The nitrogen removal performance of the SNAD system with time is shown in **Fig. 1**. In phase I, a high dissolved oxygen environment (2.0-3.0 mg/L) was established by increasing the aeration in **Fig. 2**. During the initial 10-day period, the ammonia nitrogen and total nitrogen treatment performances remaining above 90%. On the 11th day, there was an abrupt elevation in the concentration of NO_3^--N in the system, rising from 11.96 mg/L to 27.05 mg/L, decreasing to 80.71% in the total nitrogen treatment performance. Following a 10-days period of continuous operation, the total nitrogen treatment performance of the system reached a value of approximately 80%. The NO_3^--N concentration was sustained at 24.25 mg/L throughout this period. The mean pH value declined from 7.78 to 7.39, yet the internal environment of the system remained in a weakly alkaline state in **Fig. 3**. The elevated dissolved oxygen levels resulted in the uncontrolled proliferation of AOB and NOB, leading to a significant nitrification reaction and the production of a considerable amount of NO_3^--N. The nitrosation reaction within the system was unable to provide a stable electron acceptor, NO_2^--N, for anaerobic ammonia oxidation.

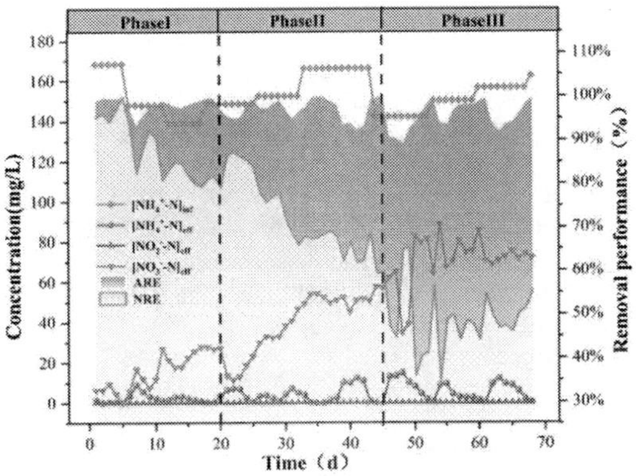

Fig. 1. Nitrogen concentration and removal efficiency in the SNAD.

In phase II, the effect of weakening NOB and strengthening AnAOB was achieved by limiting the DO to 0.5-1.0 mg/L in **Fig. 2**. In the phase II (21-25d), the NO_3^--N effluent concentration was markedly diminished, with the mean value declining from 24.25mg/L to 16.14mg/L. This suggests that the system NOB was effectively weakened. The NRE increased to a value exceeding 80%, reaching a maximum of 87.32%. In phase

II (21-25d), there was an increase in the effluent NO_3^--N from 23.38 mg/L to 57.65 mg/L, and a corresponding decrease in the NRE from 77.91% to 59.35% in **Fig. 1**. The results demonstrated a precipitous decline in nitrogen removal performance and a concomitant surge in the effluent nitrate concentration, which marked the onset of the collapse phase. The significant alteration in the DO environment resulted in the weakening of the AnAOB. This phenomenon led to a rapid decline in system NRE. The ARE was maintained at a high level throughout both phase I and phase II of the experiment, with an average ARE of 97.76%. In phase II, there was a significant fluctuation in the pH value of the system, as shown in **Fig. 3**. The high DO may have prompted further screening and intensification of AOB and NOB, enabling them to rapidly adapt to the significantly altered dissolved oxygen environment while maintaining a powerful nitrification. This results in a considerable depletion of alkalinity within the system, which further inhibits the nitrogen removal performance.

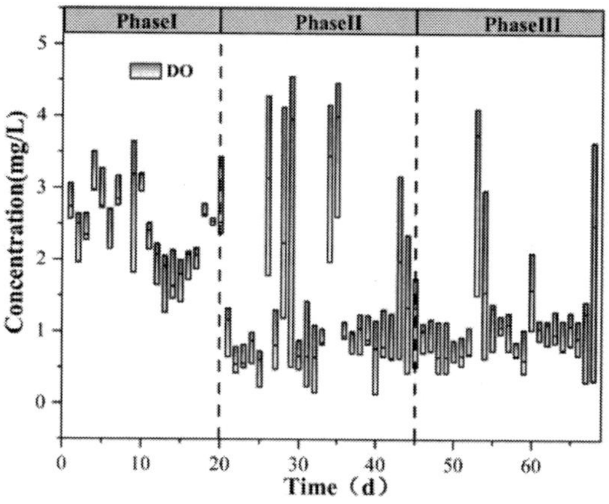

Fig. 2. DO changes in the SNAD.

In phase III (45-68 d), reducing DO to 0.3-1.0 mg/L to suppress the continuous increase of NO_3^--N concentration in the system. The effluent NH_4^+-N was 6.23 mg/L, and the ARE remained above 90%. In phase III (44-49 d), the effluent NO_3^--N was markedly diminished from 57.65 mg/L to 39.11 mg/L. The NRE subsequently declined to 35.86%. In phase III (50-68d), the average NO_3^--N was at 75.21 mg/L. The NRE increased from 35.86% to 55.32%. The concentration of NH_4^+-N exhibited a fluctuating increase. The average pH value decreased to 6.80 in **Fig. 3**. The results demonstrate that the nitrogen removal performance of the system can be temporarily restored by the DO limiting strategy. The nitrogen removal ability is not fully achieved. This phenomenon may be attributed to the weakly acidic environment. Additionally, elevated salinity also impedes the activity of AOB and AnAOB, allowing heterotrophic bacteria to progressively predominate the autotrophic denitrification process. The DO limiting strategy may increase the alkalinity within the system.

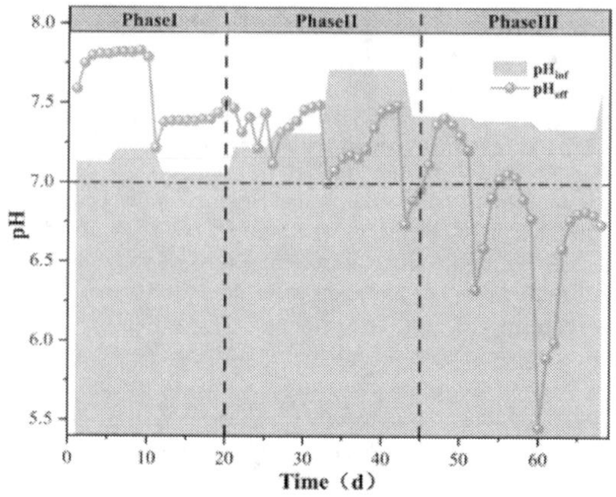

Fig. 3. pH changes in the SNAD.

3.2 COD Removal and Activity Analysis

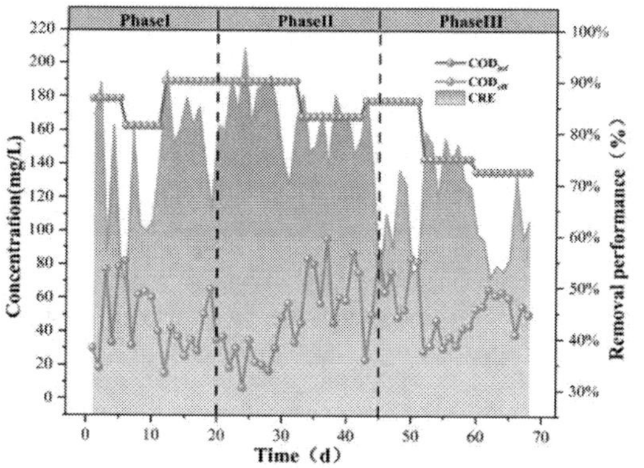

Fig. 4. COD Removal Performance in the SNAD.

The removal of organic matter is also a key indicator for evaluating the performance of the Anammox system in nitrogen and carbon removal. **Fig. 4** shown the COD removal performance. The influent COD was maintained between 140 and 180 mg/L. In a high DO environment (2.0-3.0 mg/L), the average effluent COD was at 45.48 mg/L and the average CRE was at 74.09%. Anammox represents the primary mechanism for nitrogen removal during the SNAD process. DNB is responsible for the removal of a minor proportion of NO_3^--N generated by anammox, in addition to the organic carbon sources [9]. This suggests that the DNB was not significantly impacted by the high DO environment. In phase II, the DO was at 0.5-1.0 mg/L. During the initial eight-day period, the effluent COD exhibited a pronounced decline, reaching a minimum value of 6.69 mg/L. The results indicate that the DO limitation strategy can transiently improve the performance of SNAD process. The phenomenon may be attributed to the promotion of DNB in low DO. DNB are heterotrophic parthenogenetic bacteria, exhibiting a proclivity for nitrate reduction under anaerobic conditions. In phase II, the effluent COD returned to original level, with average effluent COD of 58.11 mg/L and average CRE of 67.01%. In phase III, the

DO reached 0.3-1.0 mg/L. During the initial seven-day period, the average COD increased 70.41 mg/L with a maximum of 86.96 mg/L. On the 8th day, the effluent COD decreased to 28.43 mg/L. Subsequently, the system rapidly recover in carbon removal performance, accompanied by stabilisation of the effluent COD. The findings indicated that the DO factor exerted a discernible impact on the removal of organic matter by the SNAD process. The DNB can adjust expeditiously to the changing DO environment.

Changes in the activity of Anammox, along with changes in the nitrite accumulation rate (NAR), were calculated using equations (1) and (2), shown in **Fig. 5**. The impact of the DO limiting strategy was subjected to further analysis. During the initial five-day period, the average anammox rate exceeded 1.77 mg-TN/(gVSS-h). In phase I (6-20 d), the average anammox rate decreased to 1.51 mg-TN/(gVSS-h), which indicated the collapse of the SNAD system under conditions of elevated DO. On 34th day, the aanammox rate increased from 1.56 mg-TN/(gVSS-h) to 1.73 mg-TN/(gVSS-h). This suggests that the activity of AnAOB was reinstated under conditions of low dissolved oxygen (0.5-1 mg/L). The inhibition of AnAOB by a high DO environment was reversible [10]. However, the recovery of AnAOB activity was slower. In phase III, the DO strategy adjustment resulted in a rapid decline in the ammonia oxidation rate and anammox rate, reaching 0.67 mg-NH_4^+-N/(gVSS-h) and 1.37 mg-TN/(gVSS-h), respectively. This suggests that the minor change in DO can also influence the activity of AnAOB. The anammox rate exhibited a rapid decline, followed by a gradual increase over time. However, the nitrogen removal performance of the system did not recover. This phenomenon may be attributed to a change in pH, which may have provided insufficient nitrite substrate for AnAOB. This was supported by changes in NAR within the system. The results suggest that limiting the DO strategy can achieve the recovery of AnAOB activity, but further studies may be required to ascertain the recovery of the SNAD process.

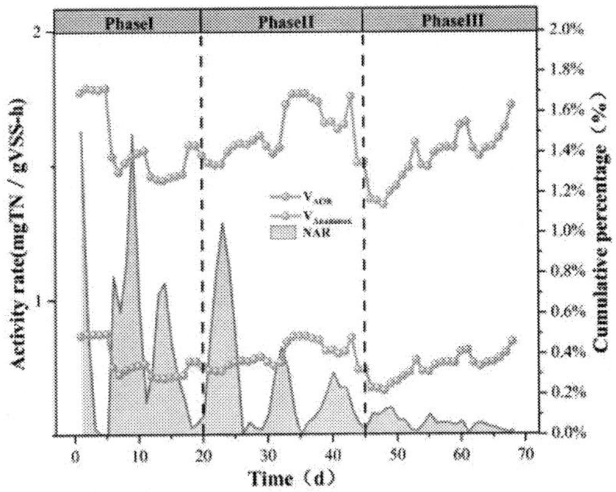

Fig. 5. Changes in anammox activity, and NAR.

4 Conclusions

This study used flowmeters to regulate the DO environment in the sections. The DO limiting strategy could briefly recover the performance of SNAD system and AnAOB activity. Due to the nitrate concentration increased rapidly, the system was unable to achieve stable nitrogen removal. However, the inhibition of AnAOB by a high DO environment was reversible When the DO environment was stabilised, anammox activity and ammonia oxidation rate increased. However, the effects of salinity and pH factors were stronger and limited the recovery of nitrogen removal performance.

Acknowledgments

This research was financially supported by Fujian Provincial Industry-university Research Collaborative Innovation (2021Y4005), and Key Technology Collaborative Innovation Platform Project for Improving Urban and Rural Ecological Environment Quality in Fu-Xia-Quan National Independent Innovation Demonstration Zone (Grant No. 2022-P-024).

References

1. Bin Ma, Yanbing Liang, Yujian Zhang, Yan Wei.: Achieving advanced nitrogen removal from low-carbon municipal wastewater using partial-nitrification/anammox and endogenous partial-denitrification/anammox. Bioresource Technology 383, 129227 (2023).
2. Huilong OuYang, Yi Wang, Shirong Dong, Yuening Wang, Ting Yang, Lijuan Gao, Baohua Chai, Pengyu Zhuang, Xiaomei Kou, Wentao Li.: Performance and biofilm characteristics of an anammox biofilter for treating low nitrogen wastewater at room temperature. Biochemical Engineering Journal 204, 109225 (2024).
3. Yeshi Cao, Mark C M van Loosdrecht, Glen T Daigger.: Mainstream partial nitritation-anammox in municipal wastewater treatment: status, bottlenecks, and further studies. Applied Microbiology and Biotechnology 101, 1365-1383 (2017).
4. Dong Wang, Guowen Wang, Fenglin Yang, Changfa Liu, Liang Kong, Ying Liu.: Treatment of municipal sewage with low carbon-to-nitrogen ratio via simultaneous partial nitrification, anaerobic ammonia oxidation, and denitrification (SNAD) in a non-woven rotating biological contactor. Chemosphere 208, 854-861 (2018).
5. Bo Li, Bruce J. Godfrey, Raymond RedCorn, Pieter Candry, Britt Abrahamson, Zhiwu Wang, Ramesh Goel, Mari-K.H. Winkler.: Mainstream nitrogen removal from low temperature and low ammonium strength municipal wastewater using hydrogel-encapsulated comammox and anammox. Water Research 242, 120303 (2023).
6. M. Han, H. De Clippeleir, A. Al-Omari, B. Wett, S. E. Vlaeminck, C. Bott, S. Murthy.: Impact of carbon

to nitrogen ratio and aeration regime on mainstream deammonification. Water Science and Technology 74 (2), 375–384 (2016).

7. Chuanqi Zhao, Gang Wang, Xiaochen Xu, Yuesuo Yang & Fenglin Yang.: Long-term operation of oxygen-limiting membrane bioreactor (MBR) for the development of simultaneous partial nitrification, anammox and denitrification (SNAD) process. Environmental Technology 39(17), (2018).
8. Jiajun Luo, Yicheng Wu, Haiyan Fu, Muxing Fu, Mian Liu, Huibin Guo, Lei Jin, Shuguang Wang.: Shift in microorganism and functional gene abundance during completely autotrophic nitrogen removal over nitrite (CANON) process. Journal of Environmental Management 359, 121009 (2024).
9. Yeqi Du, Deshuang Yu, Xiaoxia Wang, Jianyuan Zhen, Chunxue Bi, Xiuzhen Gong, Ji Zhao.: Achieving simultaneous nitritation, anammox and denitrification (SNAD) in an integrated fixed-biofilm activated sludge (IFAS) reactor: Quickly culturing self-generated anammox bacteria. Science of the Total Environment 768, 144446 (2021).
10. Jinli Zheng, Chenrong Gu, Jiawei Zeng, Yangzhe Liu, Wenru Liu, Xiao Ma, Ruifu Zhang, Xiao-Ming Ji.: Linking ecological niche and metabolic hierarchy in low-oxygen-driven anammox systems. Journal of Water Process Engineering 67, 106230 (2024).

Drought Disaster Risk Assessment and Zoning in Gansu Province

Ting Yang, Yufei Cheng*, Junde Wang

Gansu Academy of Water Conservancy, Gansu 730000, China

Abstract. The special geographical location and complex climatic conditions of Gansu Province determine that the problem of drought disasters in Gansu Province will exist for a long time. Assessment and zoning of drought disaster risk is the basis for decision-making of drought risk management, the basic link of disaster prevention and mitigation, and the urgent need for sustainable economic and social development, which has important theoretical significance and application value. In this study, 87 county-level administrative regions in Gansu Province were taken as the evaluation unit, and the typical year method, percentile method, ArcGIS spatial analysis method were used to assess the risk of agricultural drought disaster, the risk of drinking water due to drought and the risk of urban drought disaster, and the drought disaster risk degree was calculated, and then the comprehensive risk of drought in Gansu Province were obtained, and the spatial distribution of drought risk in Gansu Province was analyzed. The research results can provide an important basis for formulating drought and disaster reduction programs, regional disaster prevention and mitigation strategies and plans, and regional economic and social sustainable development plans.

1 Introduction

Drought is one of the most harmful and unpredictable natural disasters in the world, which has long restricted regional economic development and brought huge economic losses to the region, and it may have a huge impact on society, economy, agriculture and ecological environment[1-2]. In the context of extreme climate, due to the decrease of regional precipitation and the intensification of global warming, drought events continue to occur frequently, and it is urgent to grasp the temporal and spatial distribution and risk degree of drought to mitigate the adverse effects of drought and ensure regional food security[3-5]. At present, there is no universal definition of drought disaster risk, and in research it usually refers to the possibility of regional catastrophic drought events in the future and causing losses, mainly including losses in human production, livelihood, ecological environment and natural resources[6-9].

Gansu Province is located in the northwest inland region of China, where it meets the three major deserts of Badain Jarin, Tengger and Ulan Buhe[10]. Agriculture, livelihoods and ecology are more susceptible to drought events due to precipitation and evaporation, and the province is the most drought-prone region in China, with economic losses caused by drought being much more severe each year than other regions[11-12]. If the drought disaster intensifies, it will seriously affect the distribution and yield of crops, lead to the degradation of vegetation, and then accelerate regional desertification. Therefore, on the basis of the analysis of regional water resources, this study took 87 counties (districts) as the evaluation unit, and used the typical year method and percentile method to evaluate the risk of drought disaster in agriculture, human drinking water and urban areas in Gansu Province, so as to explore the spatial distribution of drought disaster risk in Gansu Province, aiming to provide a theoretical basis for the optimal allocation and scientific evaluation of regional water resources, and to be of great significance for the formulation of regional drought early warning and disaster prevention and mitigation programs.

2 Data and methodology

2.1 Study Area

Gansu Province is located in the hinterland of northwest China, geographical location: 92°13' ~ 108°46'E, 32°31'N ~ 42°57'N, the total area is 425,800 km2. The topography is complex and diverse, mainly mountains, plateaus, deserts, Gobi, etc., and the terrain is inclined from southwest to northeast. The climate types are complex and diverse, the rainfall is small, the evaporation is large, the distribution of water resources is uneven, the annual rainfall is between 36.6~734.9mm, mostly concentrated in summer, accounting for more than 50% of the total precipitation of the year, and more than 80% of the area belongs to arid and semi-arid areas.[13-15].

2.2 Research data

In this paper, the long series of water resources data from 1956 to 2016 are used, which are derived from the third water resources survey and evaluation of Gansu Province. From 1990 to 2022, the data on water supply, water demand, crop sowing area, drought-affected area, total

* Corresponding author: feifei105@163.com

rural population, and population with drinking water difficulties due to drought are mainly derived from drought control planning, Gansu Statistical Yearbook, Gansu Water Conservancy Statistical Yearbook, and survey results of drought disasters from 2008 to 2020.

2.3 Research Methods

2.3.1 Drought hazard risk assessment

(1) The hydrological frequency analysis method was used to calculate the series of water resource frequencies, and the P-III. type fitting curve was used to obtain the water resources under different drought frequencies in 87 county-level administrative regions in Gansu Province in 5 years, 10 years, 20 years, 50 years and 100 years. The rate of agricultural drought damage and the rate of people having difficulty in drinking due to drought were selected as the risk assessment indicators of agricultural drought and people's difficulty in drinking due to drought. The indicator is calculated as follows:

$$I_d = \frac{A_d}{A} \tag{1}$$

where: I_d is the rate of agricultural damage due to drought, %; A_d is the area affected by drought, hm^2; A is the sown area of crops, hm^2.

$$P_d = \frac{N_d}{N_p} \tag{2}$$

where: P_d is the rate of drinking difficulty due to drought, %; Nd refers to the population with difficulty in drinking water due to drought; N_P is the total rural population.

With the construction of water conservancy projects, the water supply capacity of each region has been greatly improved, and the impact of drought disaster has been reduced. The calculation formula is as follows:

$$k_p = \frac{x + w_i}{x + w_{2022}} \tag{3}$$

where: x is the amount of water resources corresponding to a typical year; w_i is the amount of water stored and transferred from large and medium-sized reservoirs in the first year; w_{2022} is the annual water supply of large and medium-sized reservoirs in 2022.

(2) Using the percentile method, the drought disaster risk level was divided into five levels: high risk, medium high risk, medium risk, medium low risk and low risk (Table 1). The percentiles are calculated as follows:

$$P_x = L + \frac{\frac{x}{100} \times N - F_h}{f} \times i \tag{4}$$

where: P_x is the xth percentile; N is the total number of times; L is the lower limit of the g roup to which P_x belongs; f is the number of times P_x is in the group; F_h is the number of accumulations less than L; i is the group spacing.

Table 1. Criteria for grading drought disaster risk.

Risk level	Percentiles
Low	P≤50%
Medium-low	50%<P≤65%
Middle	65%<P≤80%
Medium-high	80%<P≤95%
High	P>95%

For cities and towns, considering whether there are two water sources, emergency backup water sources and other factors, according to the two sources and one backup, two sources, one source and one backup, one source stability, one source unstable and other different water sources, the urban drought disaster risk level is judged (Table 2 for the grade standard).

Table 2. Criteria for grading urban drought disaster risk.

Risk level	Watersourcestatus
Low	Two sources and one backup
Medium-low	Two sources
Middle	One source, one backup
Medium-high	One source is stable
High	One source isunstable

2.3.2 Drought hazard risk zoning

(1) Drought disaster risk refers to the likelihood that the occurrence and development of drought will have an impact and harm on social, economic and natural environmental systems[16]. The calculation of drought disaster risk mainly considers the risk of agricultural drought disaster and the risk of people's difficulty in drinking due to drought, and the specific calculation formula is as follows:

$$R = \sum_{i=0}^{n} (p_i - p_{i+1}) \left(\frac{L_i + L_{i+1}}{2} \right) \tag{5}$$

where: R is the risk degree of agricultural drought disaster or the risk degree of difficulty in drinking water due to drought; P_i is the frequency of drought (when it occurs once in 100 years, P_i is 0.01); L_i is the effect of P_i (agricultural disaster rate or drought water difficulty rate) for this calculation unit.

(2) According to the minimum and maximum values of the risk degree in the region, the risk zone is divided into high-risk area, medium-high risk area, medium-risk area, medium-low risk area, and low-risk area (Table 3).

$$\Delta R = (R_{\max} - R_{\min}) / 5 \tag{6}$$

where: ΔR is the equal spacing of the risk degree in the region; R_{max} is the maximum value of risk degree in the region; R_{min} is the minimum value of the risk degree in the region.

Table 3. Criteria for zoning drought disaster risk classification.

Degree of risk	Risk zoning
$R_{min} \leq R < R_1$	Low-riskareas
$R_1 \leq R < R_2$	Lowtomedium risk area
$R_2 \leq R < R_3$	Medium-risk area
$R_3 \leq R < R_4$	Mediumand high-risk areas
$R_4 \leq R \leq R_{max}$	High-risk areas

3 Results & analysis

3.1 Drought hazard risk assessment

3.1.1 Risk assessment of drought disasters in agriculture

According to the drought disaster risk classification standard, the agricultural drought disaster risk map under different drought frequencies in the province was drawn (Fig. 1).

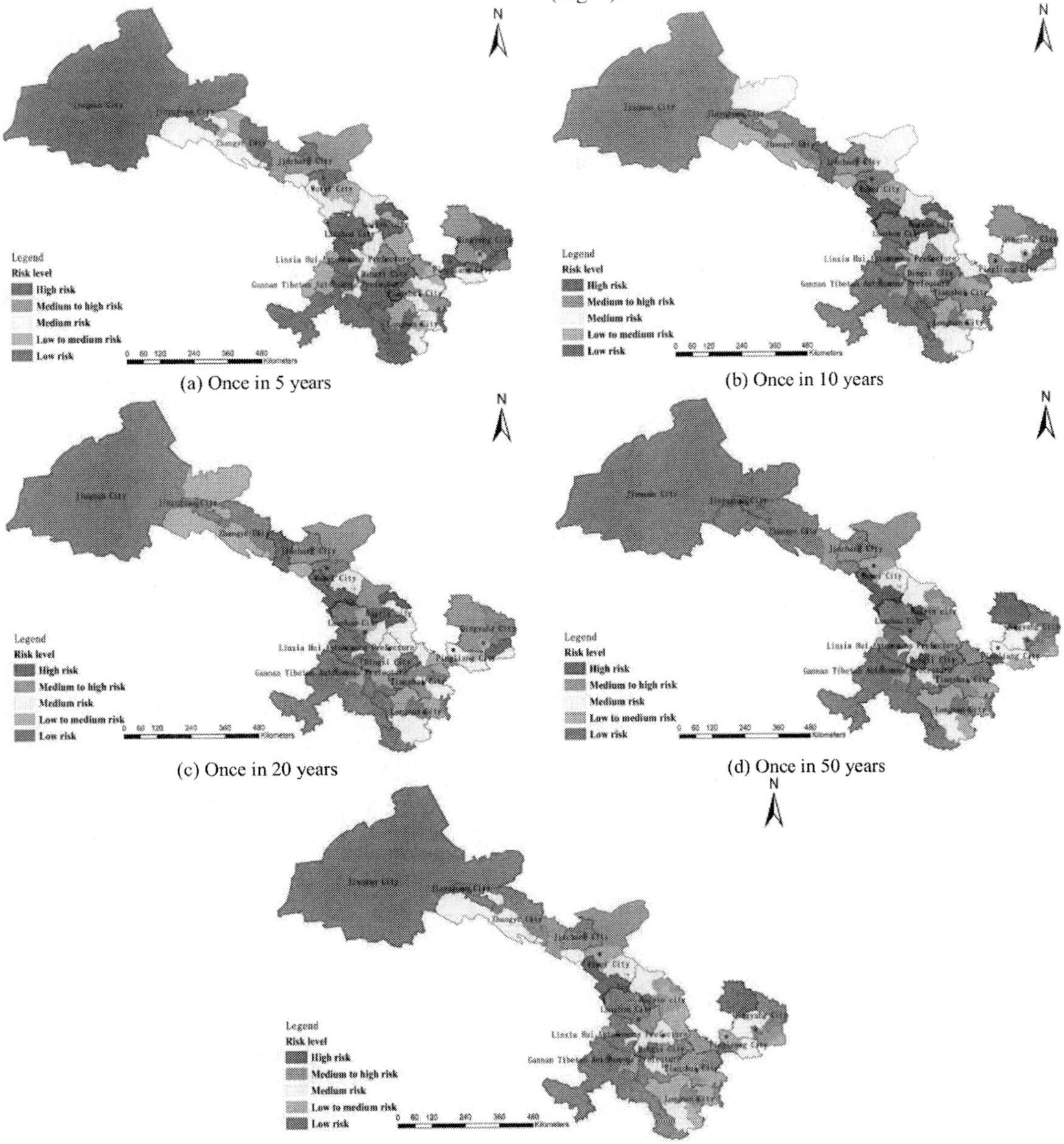

(a) Once in 5 years

(b) Once in 10 years

(c) Once in 20 years

(d) Once in 50 years

(e) Once in 100 years

Fig. 1. Drought disaster risk map of Gansu Province under different droughtfrequencies.

https://doi.org/10.1051/e3sconf/202561701011

Under the frequency of drought once in 5 years and once in 100 years, the risk level of agricultural drought disaster in the province showed an increasing trend, as shown in Fig. 2. When the frequency of drought occurs once in 5 years, the low risk level is the main one, accounting for 51.72% of the total number of counties (counties and districts) in the whole province, and a few counties and districts are medium to low to medium and high grades, and the high risk level only accounts for 5.75%. In the 10years and 20years drought frequency, the proportion of low risk is still the main one, accounting for 50.57% of the province, the proportion of medium and low risk decreases, the proportion of medium risk and medium and high risk increases, and the proportion of high risk remains unchanged. When it occurs once in 50 years and 100 years, the proportion of high-risk counties and districts is 5.75%, mainly distributed in Wuwei Tianzhu Tibetan Autonomous County, Baiyin District of Baiyin City, Dingxi Weiyuan County, Pingliang Huating City, Qingyanghuan County, and the proportion of medium and high-risk counties is 14.94% and 16.09% respectively, mainly distributed in Zhangye Shandan County, Jinchang Yongchang County, Wuwei Minqin County, Baiyin Jingyuan County, Lanzhou Yuzhong County, Dingxi Longxi County, Tianshui Wushan County, Gangu County, Qinzhou District, Qin'an County, Pingliang Jingning County, Zhuanglang County, Qingyang Ning County.

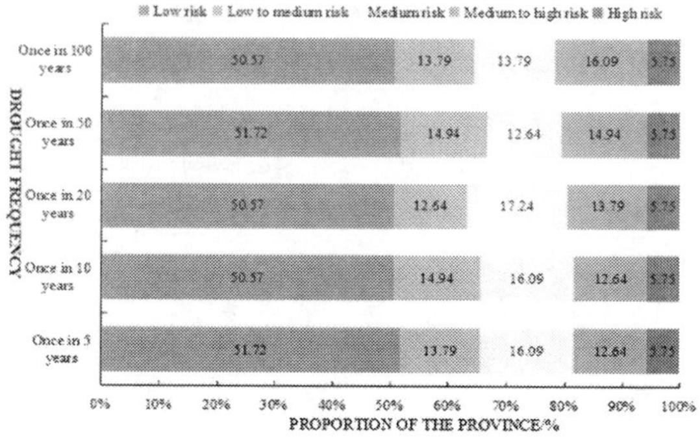

Fig. 2. Proportion of agricultural drought disaster risk level in Gansu Province under different drought frequencies.

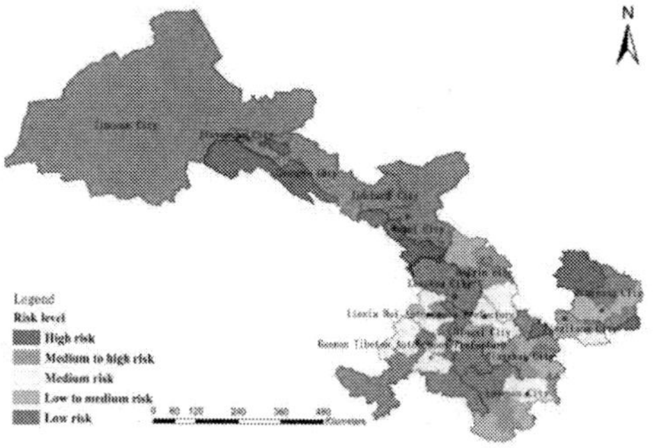

(a) Once in 5 years

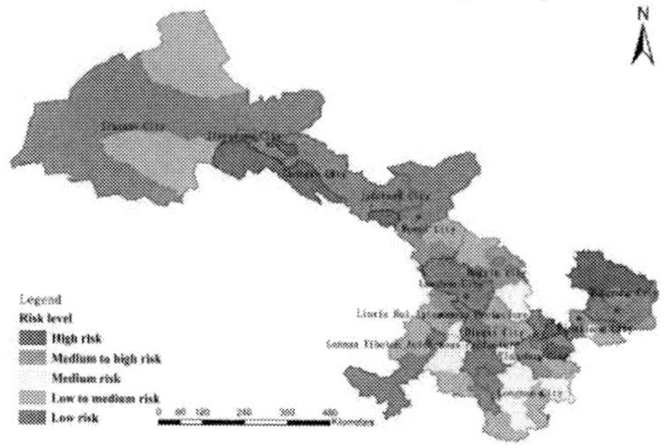

(b) Once in 10 years

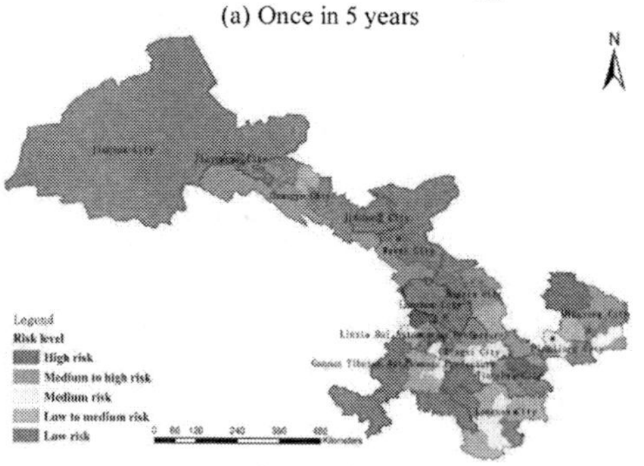

(c) Once in 20 years

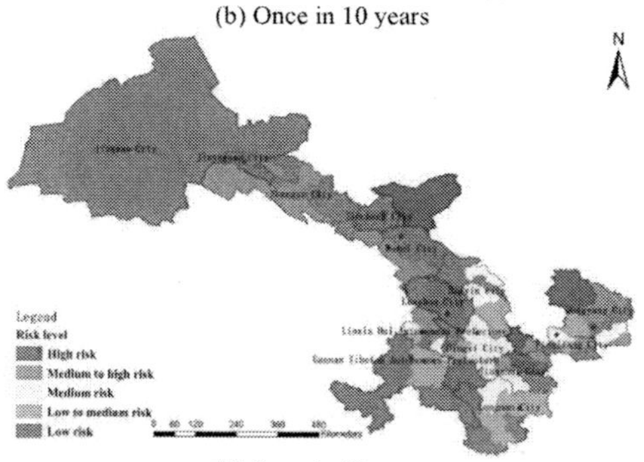

(d) Once in 50 years

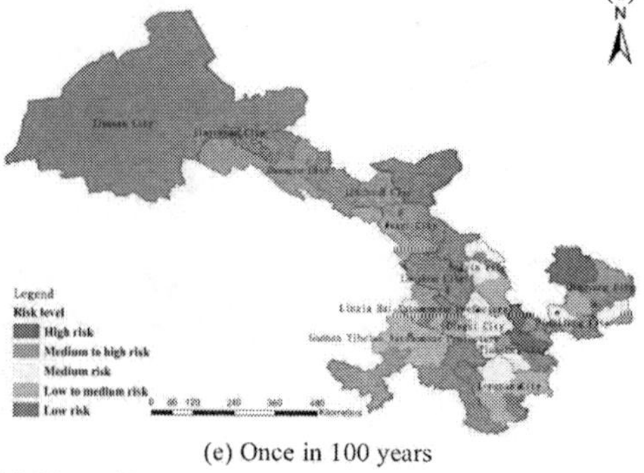

(e) Once in 100 years

Fig. 3. Risk map of people with drinking difficulties due to drought in Gansu Province under different drought frequencies.

3.1.2 Risk assessment of people's drinking difficulties due to drought

According to the risk level standard of drinking difficulties due to drought, the risk level map is drawn, as shown in Fig. 3.

From the frequency of droughts once in 5 years to once in 100 years, the risk level of drinking difficulties due to drought still shows an increasing trend (see Fig. 4). The frequency of 5-year and 10-year drought was mainly low-risk level, and the proportion of high-risk level was 5.75%, mainly distributed in Zhangye Sunan Yugur Autonomous County, Wuwei Tianzhu County, Tianshui Qingshui County, Pingliang Zhuanglang County, Qingyanghuan County, Zhengning County and Huachi County. When the frequency of drought occurs once in 20 years, once in 50 years and once in 100 years, the proportion of low risk is the largest, the proportion of medium and low risk increases, and the proportion of medium risk, medium high risk and high risk level remains unchanged, which are 12.64%, 14.94% and 5.75%, respectively, and the high risk is mainly distributed in Wuwei Minqin County, Linxia Yongjing County, Dingxi Lintao County, Tianshui Qin'an County, Tianshui County, Pingliang Jingning County and Qingyanghuan County.

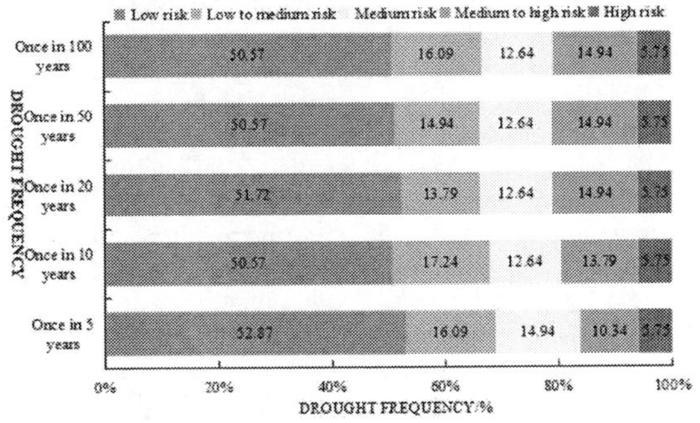

Fig. 4. Proportion of people with drinking water difficulties due to drought in Gansu Province under different drought frequencies.

3.2 Drought hazard risk zoning

The work of drought disaster risk zoning can grasp the degree and comprehensive risk of drought disaster in different regions such as agriculture, drinking water and urban areas, and provide a scientific basis for understanding the characteristics and levels of drought disaster risk in different regions, and formulating disaster prevention and mitigation policies and regional risk response measures[17]. Based on the drought disaster risk level of agriculture and human drinking water, the drought disaster risk zoning level was determined according to the risk zoning level standard (Table 3), and the comprehensive drought disaster risk zoning was determined according to the most unfavorable principle in combination with the urban risk level, as shown in Fig. 5.

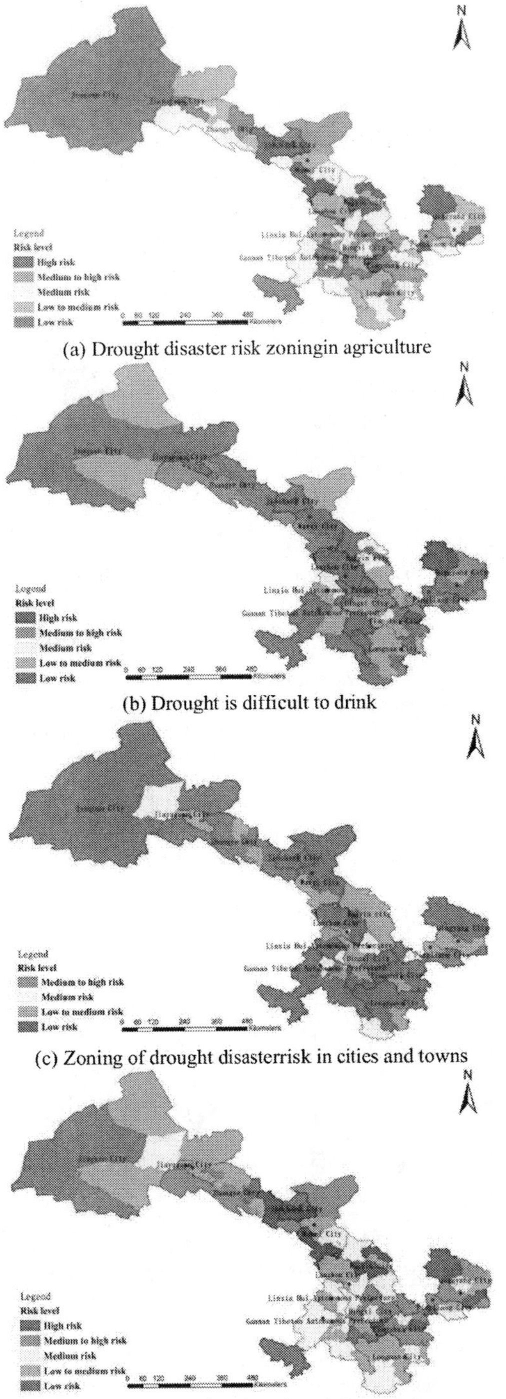

(a) Drought disaster risk zoningin agriculture

(b) Drought is difficult to drink

(c) Zoning of drought disasterrisk in cities and towns

(d) Zoning of comprehensive risk of drought disasters

Fig. 5. Risk zoning of different types of drought disasters.

It can be seen from the zoning map of agricultural drought disaster risk that there are 18 high-risk and medium-high risk areas, accounting for 20.69%, mainly distributed in Zhangye Shandan County, Jinchang

https://doi.org/10.1051/e3sconf/202561701011

Yongchang County, Wuwei Tianzhu and Minqin County in Hexi Region, Baiyin District and Jingyuan County in Longzhong Region, Dingxi Anding and Weiyuan Counties, Tianshui Wushan and Gangu Counties in Longxi Region, Huating and Jingning County in Pingliang in Longdong Region, Huan County, Ning County and Zhenyuan County in Qingyang. 22 were medium-risk areas, accounting for 25.29%; There were 47 low-risk and medium-low risk, accounting for 54.02%, mainly distributed in the areas west of Hexi, Gannan and Longnan.

It can be seen from the risk zoning map of drinking difficulties due to drought that there are 9 high-risk and medium-high risk areas, accounting for 10.34%, mainly distributed in Sunan County of Zhangye City in Hexi Region, Tongwei County of Dingxi City in Longxi Region, Qingshui County of Tianshui City in Longxi Region, Huanxian County, Huachi and Zhengning County of Qingyang City in Longdong Region, and Zhuanglang County of Pingliang City. There are 5 medium-risk areas, mainly distributed in Jingyuan County, Yongjing County, Guanghe County, Dongxiang Autonomous County and Gangu County; There were 73 low-risk and low-risk areas, accounting for the largest 83.91%.

It can be seen from the urban drought disaster risk zoning map that more than half of the regions are in medium-low risk and low-risk areas, with a total of 77 areas, accounting for 88.51%. There are 7 medium-risk areas, mainly distributed in Yumen County; There are 3 medium and high-risk areas, mainly distributed in Xigu and Anning Districts of Lanzhou City and Xifeng District of Qingyang City, and there are no high-risk areas.

It can be seen from the comprehensive risk zoning map of drought disasters that there are 28 high, medium and high-risk areas, accounting for 32.18%, mainly distributed in Zhangye, Jinchang, Wuwei, Baiyin, Qingyang, Pingliang, Tianshui, Dingxi and other cities. There were 24 medium-risk areas, accounting for 27.58%, and 35 medium-low-risk and low-risk areas, accounting for 40.23%.

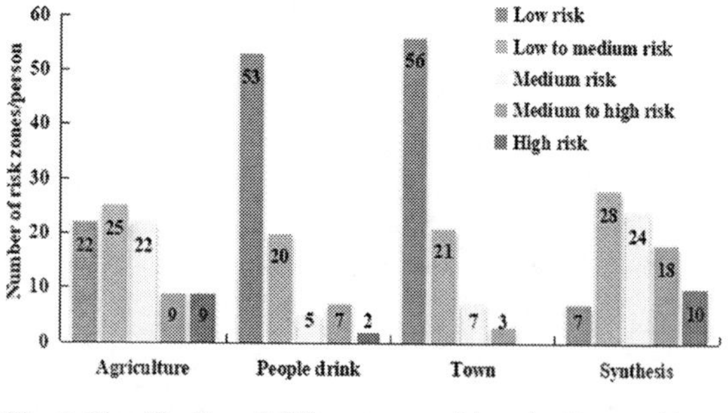

Fig. 6. Classification of different types of drought disaster risks.

From the perspective of the number of different types of drought disaster risk levels (Fig. 6), the comprehensive risk of drought disaster > the risk of agricultural drought disaster> the risk of urban drought disaster due to drought > the risk of drinking water in urban areas. During the "13th Five-Year Plan" period, governments at all levels and relevant departments in Gansu Province insisted on taking urban and rural drinking water safety as the bottom line task, and establis hed a rural water supply engineering system based on centralized water supply projects and supplemented by decentralized water supply projects, which historically solved the problem of drinking water safety in Gansu.

4 Conclusions and discussions

In this paper, the following conclusions are drawn through the risk assessment and risk zoning of drought disasters in Gansu Province.

(1) The frequency of drought disasters ranges from 5 to 100 years, and the impact of drought disasters gradually increases, and the risk level gradually increases. When it occurs once in 100 years, the high-risk areas of agricultural drought disasters are mainly distributed in Tianzhu County of Wuwei City, Baiyin District of Baiyin City, Weiyuan County of Dingxi City, Huating County of Pingliang City, and Huanxian County of Qingyang City. The high-risk areas of drinking water difficulties due to drought are mainly distributed in Minqin County of Wuwei, Qin'an County and Qingshui County of Tianshui City, Jingning County of Pingliang City, and Huanxian County of Qingyang City.

(2) In general, the comprehensive risk zoning of drought disaster > the risk of agricultural drought disaster> the risk of drinking water due to drought > urban drought disaster risk zoning. The agricultural drought disaster risk area was divided into high and medium and high risk areas, accounting for 20.69%, and the high-risk areas were mainly distributed in the three major regions of Hexi, Longxi and Longdong. There are 2 high-risk areas in the risk area of drinking water difficulties due to drought, which are mainly distributed in Longdong area; The counties and districts with high and medium and high risk of drought disaster accounted for 32.18%, mainly in the Hexi Corridor, Longzhong and Longdong arid areas.

Under the conditions of rapid temperature rise and extreme drought, Gansu Province needs to further improve its ability to resist the risk of agricultural drought disasters. Due to the impact of desertification and land salinization, the risk of agricultural drought disasters in Hexi continues to increase, so it is necessary to strengthen the prevention of agricultural drought disasters, strengthen water-saving irrigation, and implement cross-basin water diversion projects, so as to ensure water conservancy for agricultural production and ensure food security. With the increase of the risk level of drinking water difficulties due to drought, it is necessary to pay attention to the areas prone to severe drought, especially the high-risk and medium-high risk counties (districts) will face the problem of drinking water difficulties, and take effective measures to ensure the safety of drinking water in rural areas in a timely manner when necessary, so as to ensure the "last mile" of water supply.

Acknowledgments

Project fund information: This article is from the 2023 Water Conservancy Science and Technology Promotion Project, Research on Drought Disaster Risk Assessment Based on Multiple Objectives (Ganshui Jian Guan Fa

[2023] No. 67 23GSLK014), 2024 Soft Science Special Project (24JRZA018), ProvincialKeyR & DProject (23YFFA0018).

References

1. Karim R M, Rahman A M. Drought risk management for increased cereal production in Asian Least Developed Countries[J]. Weather and Climate Extremes, 2015, 7(C):24-35.
2. Hedayat H, Kaboli H S. Drought risk assessment: The importance of vulnerability factors interdependencies in regional drought risk management[J]. International Journal of Disaster Risk Reduction, 2024, 100: 104152.
3. Zeng Z, Wu W, Li Z, et al. Quantitative assessment of agricultural drought risk in Southeast Gansu Province, Northwest China[J]. Sustainability, 2019, 11(19): 5533.
4. Hao H, Bo Z, Shang-qian M A, et al. Temporal and spatial variations of meteorological drought and drought risk analysis in Hedong area of Gansu province[J]. Chinese Journal of Agrometeorology, 2020, 41(07): 459.
5. Han L, Zhang Q. The Characteristics of Agricultural Meteorological Disasters and Responses to Climate Change in Gansu Province[C]//2018 7th International Conference on Agro-geoinformatics (Agro-geoinformatics). IEEE, 2018: 1-4.
6. Sharafi L, Zarafshani K, Keshavarz M, et al. Drought risk assessment: Towards drought early warning system and sustainable environment in western Iran[J]. Ecological Indicators, 2020, 114
7. Sharafi S, Ramroudi M, Nasiri M, et al. Role of early warning systems for sustainable agriculture in Iran[J]. Arabian Journal of Geosciences, 2016, 9: 1-17.
8. Akbari M, Alizadeh Noughani M. Early warning systems for desertification hazard: a review of integrated system models and risk management[J]. Modeling Earth Systems and Environment, 2024, 10(4): 4611-4626.
9. Belal A, El-Ramady R H, Mohamed S E, et al. Drought risk assessment using remote sensing and GIS techniques[J]. Arabian Journal of Geosciences, 2014, 7(1):35-53.
10. Wang Y, Deng F, Cai Y, et al. Spatial and Temporal Change in Meteorological Drought in Gansu Province from 1969 to 2018 Based on REOF[J]. Sustainability, 2023, 15(11):
11. Zeng Z, Wu W, Li Z, et al. Quantitative assessment of agricultural drought risk in Southeast Gansu Province, Northwest China[J]. Sustainability, 2019, 11(19): 5533.
12. HAN J, JIAO M, CAO Y. Disaster evolution and loss assessment of meteorological disasters in east Gansu[J]. HUBEI AGRICULTURAL SCIE NCES, 2022, 61(13): 12.
13. ZHANG Juan, YAO Xiaojun, LI Jing, et al. Research on agricultural drought in Gansu Province based on multi-source remote sensing data[J]. Arid Land Geography, 2023, 46 (01): 11-22.
14. LI Liang, Pich Linvolak, CAI Huanjie. Spatiotemporal characteristics of drought in Gansu Province based on standardized precipitation evapotranspiration index [J]. Agricultural Research in the Arid Areas, 2019, 37 (03): 256-266.
15. HAN Lanying, ZHANG Qiang, YANG Yang, et al. Comprehensive loss characteristics of major meteorological disasters in Gansu Province under climate change[J]. Journal of Arid Land Resources and Environment, 2019, 33 (07): 107-114. DOI:10.13448/j.cnki.jalre.2019.205.
16. YAO Yubi, ZHANG Qiang, LI Yaohui, et al. Drought disaster risk assessment technology, its scientific problems and prospects[J]. Resources Science, 2013, 35 (09): 1884-1897.
17. Yuan Zhe, Xu Xiangyu, Xu Jijun, et al. Regional drought disaster risk zoning method and application based on disaster damage fitting analysis: A case study of Hunan Province[J]. China Flood Control and Drought Relief, 2023, 33(03):24-29.) DOI:10.16867/j.issn.1673-9264.2023082.

Solidification of Water Plant Sludge and Domestic Waste Incineration Bottom Ash

Qi-Guang Cao, Peng Lu *, Dong-Qing Yang, Guo-Li Xie, Jun Li, Ming-Yue Zheng, Li-Nan Xing, Xiao-Hui Zhang, Liang Chen, Song Li, Ya-Peng Song, Hong-Yu Yue, Yuan Yuan, Hong-Mei Chen, Lang Ming, Jia-Le Li, Zhen Wang, Yu-Xin Zhang

Beijing Polytechnic, Beijing 100176, China

Abstract. This study investigates the solidification of water plant sludge mixed with domestic waste incineration bottom ash through cement addition, The solidification experiment was carried out by adding the same amount of cement in different dry ratio of sludge and incineration bottom ash. The density of solidified body, unconfined compressive strength of solidified body were tested. pH value, total nitrogen and total phosphorus in the extract of original material and solidified body were also tested. It is found that with the increase of the amount of incineration bottom ash, the density and the unconfined compressive strength of the solidified body increased, the coefficient of variation gradually decreased, and the stability of the solidified body is improved. At the same time, the variation coefficients of the solidified bodies were high and the consistency between repeats is low with six different ratios of incineration bottom ash and sludge. With the amount of sludge added increases, the solidification cannot effectively inhibit the microbial activities in the sludge, which may have a harmful impact on the strength of the solidified body.

1. Introduction

Water plant sludge is the final solid waste in municipal sewage treatment, which mainly comes from suspended substances intercepted during sewage treatment; Biomass increased in the process of biological treatment system and substances obtained from sewage due to adsorption of biological or chemical flocs [1-3]. The weight of sludge proportion is about 5% of the total wastewater treatment capacity. According to the sewage treatment process section, it can be divided into grid residue, grit chamber sediment, primary sedimentation tank sludge and surplus activated sludge [4]. In 2019, China's sludge output has exceeded 60 million tons (based on 80% moisture content), and it is estimated that the annual sludge output in the year of 2025 will exceed 90 million tons [5, 6]. Anaerobic digestion, thermal hydrolysis, composting and incineration are China's predominant sludge treatment technologies, with an estimated output of 6,944, 8,342, 11,250 and 27,122 t/dw in 2019, respectively [7]. Sludge from primary sedimentation tank and excess activated sludge in sewage treatment plant are bottlenecks of sludge treatment [8]. The main characteristics of sludge are high water content (up to more than 90%), high organic content, easy to decay and smell, fine particles, small bulk density, and colloidal liquid [9].

The average content of organic matter in dry sludge is about 40%, the calorific value is about 14000kJ/kg, and the average content of nitrogen and phosphorus in nutrients is 15 mg/g and 1.7 mg/g respectively [10]. Due to the addition of a large number of chemical agents such as flocculants and coagulant aids in the sewage treatment process, the salt content of the sludge and surplus sludge in the primary sedimentation tank is very high. The sludge is rich in heavy metals and other harmful substances in the sewage. It is not suitable for agricultural application even if the sludge contains nutrients such as nitrogen and phosphorus required for plant growth. The reduction, harmless and resource utilization of municipal sludge has always been a difficult problem in the field of solid waste treatment.

Cement is widely used for sludge solidification and stabilization treatment, and has good solidification effect on sludge [11-13]. Magnesium oxysulfate cement showing excellent immobilization efficiency for Zn, Pb and As [14]. Hydraulics such as calcium hydroxide, calcite and hydrated calcium silicate make the solidified products have good mechanical properties [15-17]. However, when using cement to treat sludge containing a large amount of organic pollutants, the organic pollutants hinder the hydration reaction of cement and seriously hindered the solidification effect [18, 19].

To sum up, it has been reported that there are different ratios of solidified agent and auxiliary materials when solidified process. At the same time, the moisture content of raw materials also has great differences, which makes it difficult to popularize the sludge cement solidification process in practice. This study intends to explore the solidification effect and engineering feasibility of sludge from the perspective of different dry base ratios.

* Corresponding author: lupeng531@163.com

2. Materials and methods

The sludge used in the experiment is taken from the East District Sewage Treatment Plant of Beijing Yi-Zhuang Environment Co., Ltd. It is the sludge cake obtained from the plate and frame pressure filtration of the excess sludge after the treatment of the conditioner. The water content is about 60%. The water content calculation and detection indicators are required to take the wet base weight as the denominator [20].

The sample of waste incineration bottom ash is taken from a domestic waste incineration plant in Beijing (with a daily treatment capacity of 1600 tons). The incineration residue is discharged after the water quenching and cooling by the waste leachate. The newly generated bottom ash appears dark brown and has a smell. Place the collected bottom ash samples in a ventilated place in the laboratory for natural air drying (avoid direct sunlight) for one week. During this process, the bottom a changes from dark brown to gray white.

According to the provisions of JGJ55-2000 (J64-2000), the impurities such as metal, glass fragments, ceramics, iron wires, bottle caps, etc. shall be removed. The stainless steel screen with 8.00mm and 4.00mm aperture is used to divide the dried domestic waste incineration bottom ash into two different particle size ranges. Materials with a particle size greater than 8.00 mm contain more impurities, so the materials on the screen shall be discarded.

The cement used is quick-setting quick-hardening sulphoaluminate cement (license No.: XK08-001) produced by Shandong Jinlucheng Engineering Materials Co., Ltd. Adopt 100mm×100mm×100mm triple standard mold, and the dry density of ordinary concrete is 2000~2800kg/m ^{3}, Take the intermediate density value of 2.4×10^3 kg/m^3, Therefore, the theoretical mass of the three solidified blocks in the triple mold is 5.5kg. When weighing the sample, take into account the uncertainties in the operation process, weigh each component according to the total amount of 7kg after mixing with water, and make three parallel solidified body test blocks for each treatment. Use the concrete curing box to keep standard constant temperature and humidity for curing.

After the unconfined compressive strength of the solidified body is determined, the solidified body is crushed and passed through a 60-mesh sieve. The sludge, incineration bottom ash and broken solidified powder were extracted by the Reversing Method of Solid Waste Leaching Toxicity Leaching Method (GB5086.1-1997[21]), and the pH value, total nitrogen and total phosphorus content in the raw material extraction solution were determined. The data obtained from the test were statistically analyzed by SPSS software.

3. Operation process and convergence

3.1 Determination of sludge moisture content

In order to calculate the dry basis ratio of materials used for solidification treatment, it is necessary to measure the absolute water content of sludge first. The moisture content of sludge is determined by drying in the Method for Determination of Soil Moisture (NY/T 52-1987[22]), and the results are as follows (**Table 1**):

Table 1. Determination of absolute moisture content of sludge

Sample No.	absolute moisture content (dry base, /%)	moisture content (/%)
1	162.6	61.9
2	180.4	64.3
3	186.1	65.1
Average	176.4	63.8
Error	12.3	1.6

3.2 Material ratio and solidify method

See the table (dry basis ratio) for the quality and proportion of each material. Calculate the proportion of each material according to the dry basis ratio of sewage sludge and incineration bottom ash respectively, and weigh them after converting them into actual mass according to the moisture content of each material (Table 2).

Table 2. Weight of sludge and incineration bottom ash in solidify process

Treatment	A	B	C	D	E	F
Sludge and incineration bottom ash proportion (dry basis ratio)	2:3	3:7	1:4	1:9	1:4	1:9
Sludge weight /kg (wet basis ratio)/kg	2.95	2.22	1.48	0.74	1.6	0.8
Coarse bottom ash /kg	0.86	1	1.15	1.29	1.15	1.29
Fine bottom ash /kg	0.86	1	1.15	1.29	1.15	1.29
cement /kg	1.7	1.7	1.7	1.7	1.7	1.7
water /kg	1.01	1.1	1.2	1.1	—	0.1
Proportion of sludge on dry basis /kg	0.24	0.18	0.12	0.06	0.04	0.02

Note: Treatment A to treatment D is mixed solidification of air dried sludge and incineration bottom ash.
Treatment E and treatment F is mixed solidification of wet sludge and incineration bottom ash in SBR simulator. The moisture content of wet sludge is calculated as 90%

Treatment A, B, C and D used the dried sewage sludge (water content 60%) in solidify procession. Treatment E and F added the wet sludge (water content 90%) that produced in the SBR simulation device in solidify procession. Use shovel, scraper and other tools to evenly mix various materials on the hardened ground. After fully mixing, put the standard test block mould of 100mm×100 mm×100mm coated with release agent, put it on the oscillation table for oscillation treatment for 3-5 minutes, discharge the gas in the solidified body, and promote the further mixing of the solidified body materials. Stop the oscillation, and use the cement scraper to smooth the upper surface of the test block. Put the standard mold with solidified substance into the curing box, set the temperature of the curing box to 20±0.5°C, and the relative humidity to 95%. After 24 hours of curing, use the high-pressure air pump to

 https://doi.org/10.1051/e3sconf/202561701012

connect the nozzle matching the interface at the bottom of the mold, start the air pump to blow the solidified body out of the mold, and operate carefully to avoid collision during the demoulding process of the test block. Then put the removed test block back into the cement test block curing box for curing for 30 days. Take out and weigh the weight of each block, calculate the density of the solidified body, and then test the unconfined compressive strength of it.

4. Analysis and discussion

The density and compressive strength test results of the solidified body are shown in Fig. 1. It can be clearly seen that among the six different treatments, treatment E and treatment F with wet sludge have the highest strength of the solidified body, but the results are very unstable and the reproducibility of it is very poor. This is related to the lowest proportion of dry sludge in the actual solidification process.

It can be seen from the table that with the increase of the amount of incineration bottom ash, the density of the solidified body increases, at the same time, the coefficient of variation decreases gradually, and the stability of the solidified body is improved.

It can be seen from Table 3 that the unconfined compressive strength of the solidified body is gradually increasing with the decrease of sludge addition and the cement addition increases. Li Lei et al. [17] considered that the addition of cement is the main factor to improve the dry-wet durability of the solidified body. The experimental results verify the above view. The strength of the solidified body in this experiment is higher than that of LIU Ji-zhuang [15] *et al.*

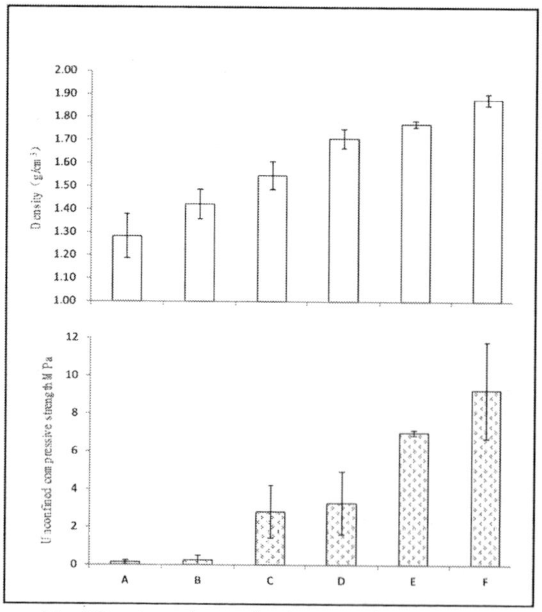

Fig. 1 Density and unconfined compressive strength of different treatments

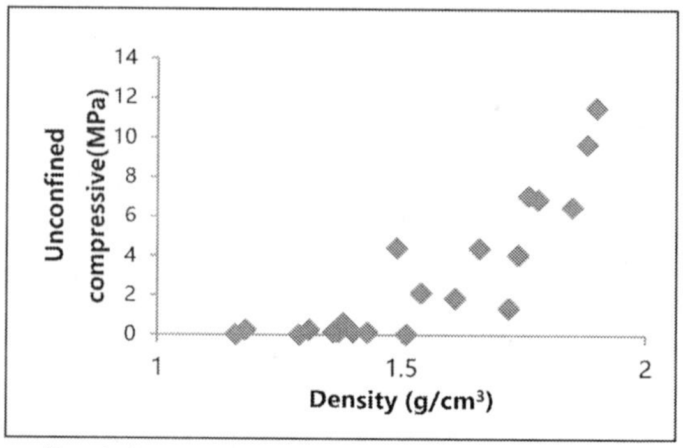

Fig. 2 Correlation between density and unconfined compressive strength

The correlation analysis of the obtained density and unconfined compressive strength shows that the density of the solidified body is significantly correlated with the unconfined compressive strength (Fig. 2) and that correlation coefficient is 0.855 by two-tailed test. The increase in the density of solidified solids is mainly related to the addition of incineration bottom ash.

The paper combed the existing cement solidification methods in China and recalculated the sludge moisture content according to the engineering regulations [20]. It can be seen from Table 3 that, no matter what method is used, the strength of sludge and cement mixed solidification is not satisfactory, and the strength can only meet the requirements of landfill treatment, far from the requirements of building materials application. At the same time, the quantity of solidified sludge is far below the expected value that the dry basis mass ratio of sludge can rarely exceed 50% of the mass of solidified body.

Silicate solidification materials and alkaline solidification materials play a major role in the formation of solidified sludge strength. Cement solidification is still the most important way to obtain sludge solidification strength. Chang Fangqiang *et al.* [2] showed that under the same conditions, the cement solidification strength is about 2 to 5 times to that of lime, and the strength is only less than geopolymer solidification method [5]. When cement is used to treat sludge containing a large amount of organic pollutants, the organic pollutants will hinder the hydration reaction of cement and seriously decreased the solidification effect [23].

The study of Helene *et al.* [23] on solidification of dredged sludge with cement found that a single solidification material made the strength of solidified sludge increase slowly, and could not achieve the expected effect. Zhu Hong *et al.* [24] used calcium oxide to solidify sewage sludge, but did not explain the final results. Dai Rui [9] studied the solidification effect of quicklime and fly ash on sewage sludge. With the increase of the amount of the two curing agents, the unconfined compressive strength of the mixture increases, but the maximum compressive strength is less than 45kPa. Chen Meng *et al.* [25] used cement and fly ash as the main solidify agents to study the effect of sodium hydroxide on the sludge solidification effect. The results show that the reasonable addition of sodium hydroxide can reduce the organic content of solidified sludge and enhance the strength of solidified sludge. At the same time, the strength of solidified body increases with the increase of fly ash content, but the unconfined compressive strength does not exceed 400kPa. Liang Bing *et al.* [26] studied the engineering characteristics of municipal sludge solidification under the joint action of waste incineration bottom ash and solidification agent by using the self-prepared solidification agent with the main components of calcium silicate and calcium oxide. The maximum strength of the mixed solidified body of waste incineration bottom ash and sewage plant sludge reached 30 kPa. Li Shiwen *et al.* [27] determined the optimal dosage of curing agent by orthogonal experiment as follows: cement 0.12 kg/kg, mineral waste residue 0.10 kg/g, fly ash 0.02 kg/kg, after 6 days curing the compressive strength of curing block was 360kPa.

At present, only a few researchers in China made sludge solidified body reach the final compressive strength that higher than 1MPa. YUE Jun-chao [28] determined the optimum dosage of curing agent for every 100 g of dewatered sludge (water content 80%) mixed with 20 g of cement, 15 g of fly ash 0.4 g of calcium chloride and 0.2 g of naphthalene water-reducing agent. The unconfined compressive strength of solidified sludge after curing for 7 days, 14 days and 28 days can reach 0.83 MPa, 1.22 MPa and 1.52 MPa respectively. Che cheng-dan [8] reported that when the cement content is 0.07 kg/kg, the fly ash content is 0.02 kg/kg, and the coal slag content is 0.06 kg/kg, that is, m (sludge): m (cement): m (fly ash): m (coal slag)=1:0.07:0.02:0.06, the unconfined compressive strength of the sludge solidified block cured for 7 days is about 1.8 MPa, which is the highest curing strength obtained by the cement solidification method in China at present.

Table 3. Comparison of sludge solidification effect in China. (Calculated according to 100g of solidified wet sludge, and the water content of sludge is calculated according to the technical manual of solid waste treatment engineering [20])

Dry basis mass ratio of solidified sludge	Water content	Sludage (dry basis)	Name of main solidified agent	Amount of main solidified agent(g)	Auxiliary agent	Amount of auxiliary agent(g)	Maximum strength (MPa)	researcher
--	81.1%	Not provided	sulphoaluminate cement	3.9g	Fly ash	6g	<0.09 MPa	SUN Qin-ling [30]
			portland cement	3g				
0.10	86.65%	13.35g	Dicalcium silicate and calcium oxide	20g	Incineration bottom ash	100g	<0.12 MPa	LIANG Bing[26, 31]
0.13	82.5%	17.5g	Dicalcium silicate and calcium oxide	20g	Incineration bottom ash	100g	<0.28 MPa	HU Xue-Tao[32]
0.13	94.94%	3.76g	Ultrafine powder after crushing of construction waste	15g	Fly ash	4.7g	Not analyzed	DONG Liang-fei [33]
					CaO	5g		
					Polyacrylamide potassium salt(KPAM)	0.7g		
					Sulfonated oil	0.3g		
0.18	92.1%	7.9g	425 cement	15	clay	20g	<0.04 MPa	CHANG Fang-qiang[2]
					coal cinder	10g		
0.19	94.15%	9.51g	Ternary mixed cement	40g	--	--	<5.0MPa	ZHANG Jing-bo[34]
0.33	79.4%[1]	25g	cement	10g	Amine persulfate	0.3g	<0.45 MPa	YANG Ai-wu[35]
0.33	Calculated by 80%	20g	quick lime	20g	Fly ash	20g	0.0431 MPa	DAI Rei[9]
0.35	80%	20 g	cement	20g	Fly ash	15g	1.52MPa	YUE Jun-chao[28]
					calcium chloride	0.4g		
					Naphthalene series water reducer	0.2g		
0.36	79.01%	20.99g	cement	11g	Local soil	27 g	0.215 MPa	WANG Yu-feng[10]
0.39	80%	15.2g	cement	12g	coal ash	10g	0.318 MPa	CAO Sheng[36]
					lime	2g		
0.47	78.5%	21.5g	cement	12g	Fly ash	2g	0.36 MPa	LI Shi-wen[27]
0.51	55%[2]	45g	cement	8g	Fly ash	30g	<0.40MPa	CHEN Meng[25]
					NaOH	0.4g		
0.58	81.1%	16.65g	portland cement	3	Fly ash	4.7	0.127 MPa	ZHOU Jian[37]
					lime	1.2		
					Component A	3		
0.60	70%[3]	30 g	cement	20g	--	--	<0.3 MPa	LIU Ji-zhuang[15]
					Acrylamide	0.6g		
					Sea-based soft clay	39.825g		
0.66	71.24%	28.76g	cement	7g	Fly ash	2g	1.8MPa	CHE Cheng-dan[16]
					coal cinder	6g		

The highest proportion of sludge (dry base) in the experiment is treatment A, which is 0.24, and the unconfined compressive strength reaches 0.14 MPa. The solidification strength is lower compared with the results of domestic scholars with similar dry base ratio (Table 3). It is necessary to adjust the proportion of reagents in future experiments, especially increasing the stability of organics in sludge.

Compared with Portland cement, the content of silicon dioxide generated from the incineration of sewage sludge is 1.5 times that of Portland cement. The content of aluminum oxide (Al_2O_3) and Ferric oxide (Fe_2O_3) is on the high side. The content of aluminum oxide is about 3 times of that in Portland cement, and the content of Ferric oxide is more than 3 times of that in Portland cement. The content of calcium oxide (CaO) is low, about one quarter of that in Portland cement [29]. In future, the method of properly increasing calcium oxide can be considered to promote the strength of solidified body.

It can be clearly seen from Table 4 that the content of total nitrogen and total phosphorus in the leaching solution of incineration residue is low, and the pH value is alkaline. However, the pH value in the sludge extraction solution is weak acidic, and the total nitrogen and total phosphorus content in the extraction solution is very high, which is more than 25 times of the content in the incineration bottom ash. The high content of total nitrogen and total phosphorus indicates that the sludge contains a large amount of organic matter. After the solidified process, the content of total nitrogen and total phosphorus in the leaching solution decreases significantly, which is related to the stabilization effect of cement and incineration bottom ash on sludge, as well as the dilution of organic substances in sludge by other components of the solidified materials.

Table 4 Nitrogen and phosphorus in extraction solution of raw material and solidified body

item	nitrogen content (mg/L)	Phosphorus content(mg/L)	pH (22°C)
Incineration bottom ash	0.65	12	9.22
Sludge	16.42	303	6.48
A	1.29	20	10.57
B	0.81	22	10.79
C	0.28	7	11.13
D	0.15	3	11.58

In the curing process of the solidified body in the curing box, hypha grows on the surface of the solidified body of high sludge ratio treatment of A and B (Fig. 3).

However, the solidified sludge in the laboratory simulation device has no such phenomenon. It was preliminarily determined as actinomycetes from the pH value and nutritional conditions of the solidified body. This is similar to the cement solidification experiment results of Feng Bin [38]. In his experiment, there are obvious white and yellow mold spots on the surface of the solidified body. The growth of actinomycetes may have a certain impact on the strength of the solidified body, which needs further analysis and research in subsequent experiments.

The characteristics of bacterial colonies from culture experiments of sludge sampling and the solidified body surface sampling were similar. It can be preliminarily determined that the mycelium growing on the surface of the solidified body comes from the sludge of the sewage treatment plant.

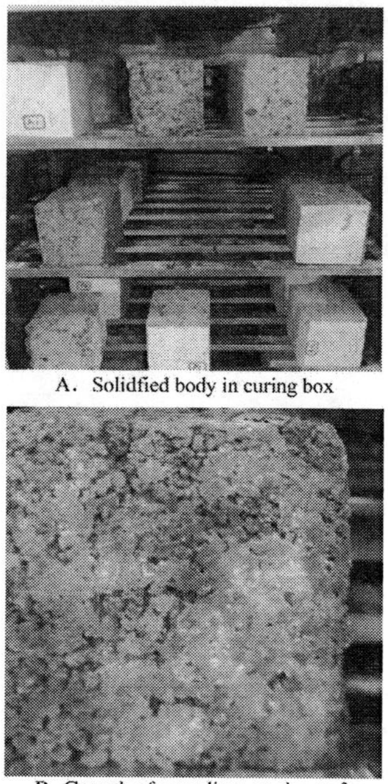

A. Solidfied body in curing box

B. Growth of mycelium on the surface of Solidfied body solidified body

Fig. 3 Mycelium formation in solidified body

5. Conclusion

According to the current reports, few sludge solidification forms account for more than 50% of dry basis; And the strength of the solidified body formed by sludge and cement mixture is very low, which cannot meet the requirements of building materials. Therefore, cement solidification of sewage treatment plant sludge can only be used as a temporary means in case of emergency. It has no engineering application and promotion value from the current technical level.

When the sludge dry base ratio increased, the coefficient of variation of the density and compressive strength of the solidified body increased, and the consistency of the properties of the solidified body became worse. The unconfined compressive strength of the solidified body is significantly related to the density, which is mainly due to the increase of the proportion of the incineration bottom ash. It is necessary to explore the stabilization of organic matter in sludge to improve the strength of the solidified body after the solidification treatment using sludge and municipal solid waste incineration bottom ash mixture.

It is preliminarily determined that mycelium produced during the solidification of sewage treatment plant sludge and incineration bottom ash was from sewage plant sludge, and the influence of these microorganisms on the strength or other aspects of solidification needs to be determined by further experiments. Such as using antibacterial agents or other additives to improve solidification strength.

Acknowledgments

Major scientific and technological projects of Beijing Polytechnic 2020Z003-KXZ; Teaching Innovation Team Project of Beijing Polytechnic CJGX2024-011-04.

References

1. REN Hong-ying, 2003, Study on utilization of municipal sewage sludge in Beijing, Beijing University Of Technology.
2. CHANG Fang-qiang, TU Fan, and LUO Cai-Song, 2010. Sewagesludge sludge solidification and its influencing factors. Journal of Fujian University of Technology, **8**(03): p. 258-261.
3. Mohamed Ammar, Ezz Yousef, Muhammed A. *et al.*, 2023. A Comprehensive Review of the Developments in Electrocoagulation for the Removal of Contaminants from Wastewater. Separations, **10**: p. 337.
4. CHEN Xiao-juan and L. Xiao-fang, 2012. Discussion on the treatment, disposal and resource utilization of urban sludge. Environmental protection and circular economy, **32**(01): p. 41-45.
5. DAI Xiao-hu, 2020. Present situation and development trend of sludge treatment and disposal in China. Science, **72**(06): p. 30-34+4.
6. ZHENG Weicheng, SHAO Yuchao, QIN Shulin *et al.*, 2024. Future Directions of Sustainable Resource Utilization of Residual Sewage Sludge: *A Review.* Sustainability, **16**: p. 6710.
7. Toyin Dunsin Saliu and Sébastien Sauvé, 2024.A review of per- and polyfluoroalkyl substances in biosolids: geographical distribution and regulations. Frontiers in Environmental Chemistry, **5**.
8. Shenbin, H., 2019. Analysis on Utilization Effect of Surplus SludgeDisposal in Municipal Sewage Treatment Plants. Environment science and management, **44**(10): p. 87-91.
9. DAI Rui, ZHANG Heng, XU Wei, *et al.*, 2015. Experimental study on the city sewage sludge solidification. Brick-Tile, (09): p. 11-13.
10. WANG Yu-feng, 2009, Solidification treatment experimental study on the sewage sludge of urban sewage treatment *plant,* Chang'an University.
11. Valls, S. and E. Vàzquez, 2000. Stabilisation and solidification of sewage sludges with Portland cement. Cement and Concrete Research, **30**(10): p. 1671-1678.
12. Hamdy A. Abdel-Gawwad, Samah A. Sanad, and Mona S. Mohammed, 2020. A clean approach through sustainable utilization of cement kiln dust, hazardous lead-bearing, and sewage sludges in the production of lightweight bricks. Journal of Cleaner Production, **273**: p. 123129.
13. Hsu Chin-Wei and Chen Chun-Tao, 2020.Strength development of cement pastes with alkali-activated dehydrated sewage sludge. Construction and Building Materials, **255**: p. 119243.
14. WANG Lei, ZHANG Yuying, CHEN Liang., *et al.*, 2022. Designing novel magnesium oxysulfate cement for stabilization/solidification of municipal solid waste incineration fly ash. Journal of Hazardous Materials, **423**: p. 127025.
15. LIU Ji-zhuang, Experimental study on solidification property of municipal sludge. New Building Materials, 2014. **41**(12): p. 53-55.
16. CHE Cheng-dan, ZHU Nan-wen, LI Yan-lin, *et al.*, 2008 New technology for solidifying sewage sludge. Journal of Safety and Environment (03): p. 56-59.
17. LI Lei, ZHU Wei, and L. Cheng, Study of wet and dry properties of solidified sludge. Rock and Soil Mechanics, 2009. **30**(10): p. 3001-3004+3012.
18. Stegemann, J.A. and Buenfeld N.R., 2002.Prediction of unconfined compressive strength of cement paste with pure metal compound additions. Cement and Concrete Research, **32**(6): p. 903-913.
19. Zhang Zhong-zhuang, Wang Xiao-yu, Chu Song-zhu., *et al.*, 2024. Effect of water loss during curing on hydration reaction and hydrates conversion in calcium aluminate cement-bonded castables. Journal of Iron and Steel Research International, **31**: p. 2133-2141.
20. NIE Yong-feng., 2013 Handbook on Solid Waste Management and Tecnlogy.: Chemical Industry Press Co. Ltd.
21. National Environmental Protection Agency, 1997, Test method standard for leaching toxicity of solid wastes-Roll over leaching procedure.
22. Administration of standardization, 1987. Method for the determination of soil water content, standardization administration.
23. HELENE T., SERGE L., and JACQUES L., 2001. Mechanical improvement and vertical yield stress prediction of clayey soils from eastern Canada treated with lime or cement. Revue Canadienne De Géotechnique, **38**(3): p. 103－111.
24. ZHU Hong and MENG Yu, 2010. Sludge solidification in municipal sewage sludge. China Resources Comprehensive Utilization, **28**(02): p. 38-39.
25. CHEN Meng, YANG Guo-lu, and YU Liang-ying, 2013. Exprimental study on solidification and stabilization of polluted silt using fly ash and sodium hydroxide. Journal of Huazhong University of Science and Technology(Natural Science Edition) **41**(10): p. 123-127.

26. LIANG Bing, *et al.*, 2017. Engineering characteristics analysis of sewage sludge solidified by different gradations municipal solid waste incineration bottom ash. Chinese Journal of Environmental Engineering, **11**(02): p. 1117-1122.
27. LI Shi-wen, NING Xun-an, DENG Zhong-liang, *et al.*, 2010., Study on the technology of papermaking sludge solidification/stabilization. Chinese Journal of Environmental Engineering, **4**(08): p. 1911-1915.
28. YEU Jun-chao, 2022., Research on the optimal configuration of different sludge curing agent and their application in the solidification of municipal sludge. Journal of Green Science and Technology, **24**(06): p. 14-18.
29. HAN Li-li, 2011. Discussion on the application of ecological cement made from municipal sludge. China water & wastewater, **27**(02): p. 107-108.
30. SUN Qing-lin, LI Yi, ZHOU Jiong, *et al.*, 2016., Research on fast solidification agent of sewage sludge from municipal sewage treatment plant. China water & wastewater, **32**(21): p. 23-27.
31. LIANG Bing, HU Xue-tao, and CHEN Yi-jun, 2016. Engineering characteristics of municipal sludge solidified by municipal solid waste incineration bottom ash as a additive. Chinese Journal of Environmental Engineering **10**(11): p. 6705-6710.
32. HU Xue tao, LIANG Bing, CHEN Yijun 2016, Influence of municipal solid waste incineration bottom ash with different grain sizes on engineering characteristics of sewage sludge solidification. Chinese Journal of Environmental Engineering, 2015. **9**(11): p. 5567-5572.
33. DONG Liang fei, ZHANG, Xiao ya, GONG Jia hao *et al.*, 2020., Research and development of new sludge solidifying agent based on waste recycling. journal of Changzhou University (natural science edition), **32**(05): p. 50-57.
34. ZHANG Jingbo, CHEN Yanchang, SONG Peng, *et al.*, 2018. Study on the curing of sludge with composite cement. Shandong Chemical Industry, 2018. **47**(16): p. 197-198+201.
35. YANG Ai-wu and HU Yao, 2018. Study of engineering properties and micromechanism of new municipal sludge solidified soil. Rock and Soil Mechanics, **39**(S1): p. 69-78.
36. CAO Sheng, DONG Xiao-nan, and XING Yanfeng, 2014, Research on solidification method for sludge from a sewage treatment plant. Environmental science and management, 2014. **39**(08): p. 90-94.
37. ZHOU Jian, Zeng Xiaojun, Zhou Jiong *et al.,* 2014. Combined effects of sludge rapid solidification using response surface methodology. Chinese Journal of Environmental Engineering, **8**(02): p. 408-414.
38. FENG Bing, 2015, Experimental study on co-solidification of sewage sludge and MSWI bottom ash and stability analysis of solidified sludge landfill, Zhejiang University.

Study on the Uneven Settlement Patterns at the Bottom of Seepage Control Structures in Deep Backfilled Reservoir Basins

Yayun Meng[1], Xianfei Wang[1], Xiangfei Ye[1], Yaolai Liu[1], Li Yuan[2,*], Shuairan Peng[2]

[1] PowerChina Zhongnan Engineering Corporation Limited, Changsha, Hunan, China
[2] Hohai University, Nanjing, Jiangsu, China

Abstract. When a pumped storage power station adopts a full reservoir basin seepage control scheme, significant uneven settlement may occur at the bottom if there is a deep backfilled area, which could lead to cracks in the seepage control structure and threaten the system's safety. This study analyzes the basic patterns of uneven settlement in the bottom of the seepage control structure in deep backfilled basins. Using a finite element deformation gradient method, risk areas were identified, and the influence of material parameters in the backfilled area on uneven settlement was considered. The results showed that uneven settlement is particularly evident near the boundary of the excavation-backfill line and dam body boundary, with deformation gradient values exceeding the critical 1.0% mainly in corner areas of the excavation-backfill boundary, although the overall affected area is relatively small. From the perspective of material parameters, proper compaction in the backfilled areas, especially near the excavation-backfill boundary and dam body boundary, can effectively reduce uneven settlement at the reservoir bott.

1. Introduction

Pumped storage power stations are generally located in urban load centers or adjacent to scenic spots, and combining the functions of power stations with the surrounding environment is gradually becoming a new design trend. For example, in order to reduce the damage to the ecological environment caused by engineering excavation slag, part of the slag will be backfilled to the bottom of the reservoir, which can reduce the impact of engineering slag on the neighboring ecological environment while reducing the amount of land requisition and immigration works [1]. For projects with complex geological conditions and deep backfill, especially when the geological structure is developed, the reservoir is prone to leakage of large flows [2], but the water volume of the pumped storage power station reservoir directly affects the economic benefits of the power station, so it is crucial to choose the appropriate form of seepage control.

If the reservoir area rock body karst or fissure development, low water table, impermeable layer depth is large, need to use the whole reservoir basin seepage control program[3]. The whole reservoir basin seepage control program than the seepage control curtain program reliability is good, but can also take into account the reservoir bank protection, the dam surface seepage control panel and the reservoir perimeter seepage control panel smooth connection, combined with the reservoir bottom surface plate or composite geomembrane and other seepage control materials, the formation of joint seepage control body [4-6]. However, if there is a deep backfilled area at the reservoir bottom, significant uneven settlement may occur after water impoundment[7]. Such pronounced uneven deformation can lead to cracking of the bottom panels and local tearing of the composite geomembrane[4,8], resulting in localized damage to the seepage control system and consequently affecting the normal operation of the entire seepage barrier for the reservoir basin. For cracks caused by uneven settlement of earth and rock dams, the deformation inclination method [9] and the equivalent beam method [10] are the most commonly used estimation methods, but both calculations require a large amount of monitoring information [11]. Additionally, coupling the deformation gradient method with finite element deformation analysis can serve as a practical approach for crack estimation during the engineering design phase[12].

2. Numerical simulation

2.1 Calculation model

To study the impact of deep backfilled areas on the safety of seepage control structures in pumped storage power stations, this paper analyzes the uneven settlement patterns at the bottom of deep backfilled reservoirs based on finite element calculations. The finite element deformation gradient method is employed to identify potential crack locations in the reservoir bottom, providing guidance for the engineering design phase.

Gongyi Houji River Pumped Storage Power Station is located in Gongyi City under the jurisdiction of Zhengzhou City, the engineering grade is first class, and the engineering scale is large (1) type. Power station on the

* Corresponding author: yuanli97@hhu.edu.cn

reservoir dam for concrete panel rockfill dam, dam crest elevation 860.00m, the maximum height of 142.0m (at the dam axis), the dam crest length of 421m. dam fill material zoning from upstream to downstream in turn for the bedding area, special bedding area, transition area, the main rockfill area, downstream rockfill area, the downstream side of the dam to set up the pressure slope body, the top elevation of the slope body is 785.00m. dam and bank seepage control C30 reinforced concrete panels, panels with 0.4m thickness, panel slit width of 12m, panels around the bottom of the reservoir and the bottom of the reservoir drainage corridor using the perimeter of the seam connection, and the bottom of the reservoir geomembrane composed of closed seepage control body. The bank of the reservoir ring road elevation 860.00m below to the bottom of the reservoir elevation 810.00m according to 1:1.4 slope ratio rule of excavation, the bottom of the reservoir in the elevation of 810.00m parts of the excavation into a platform, higher than 810.00m parts of the excavation, lower than 810.00m parts of the use of stone slag backfill to the elevation of 810.00m.

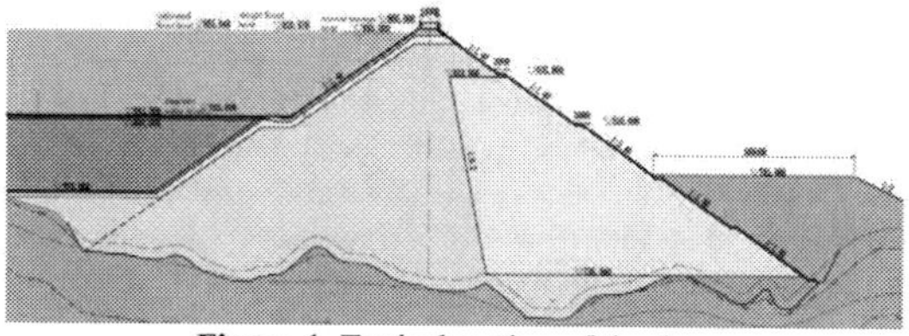

Figure 1. Typical section of the dam

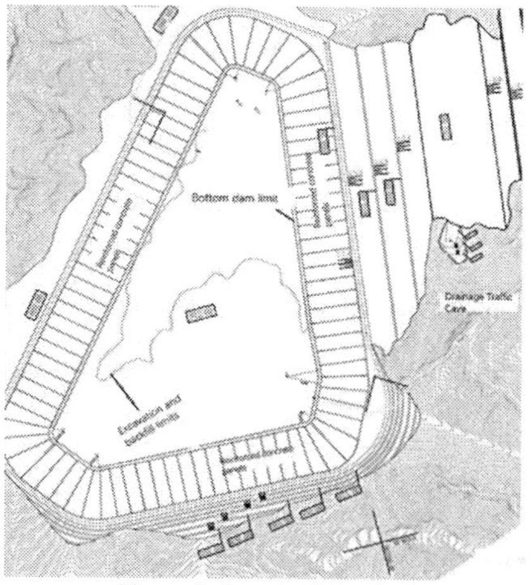

Figure 2. Layout of the dam

The cross-section and plan layout of the designed dam are used to establish the overall three-dimensional finite element model of the dam, and all the units use eight-node hexahedral isoparametric units.(Typical section and plan view of the dam are shown in Fig. 1 and Fig. 2) The overall coordinate system of the model adopts the right-handed coordinate system, taking the direction of water flow as the positive direction of X; vertically upward as the positive direction of Z-axis, and along the dam axis from the right bank to the left bank as the positive direction of Y-axis, and the model has a total of 64,973 nodes and 57,525 units. There are a total of 142 panels set up in the dam and reservoir bank panels with joint units.

The model does not consider bedrock, and the dam and the bottom of the backfill area at the bottom of the reservoir are set as three-way fixed constraints, and the loads contain the self-weights of the dam, backfill, and compression slopes, as well as the water loads in the reservoir basin. The final 3D finite element graph is shown in Fig. 3.

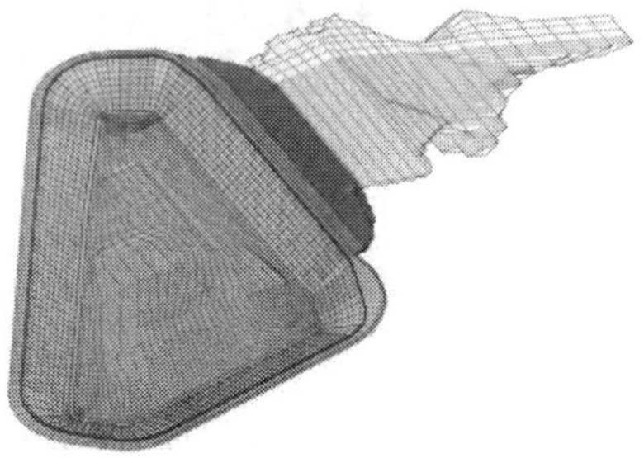

Figure 3. Overall 3D finite element model of the dam

2.2 Material parameters

The dam material calculation model used for the calculation is Duncan-Chang E-B model, with the parameters shown in Table 1. C30 concrete is used for the panel and other structures, with a density of 2500 kg/m3, and a linear elasticity model is used, with a modulus of elasticity of 30 Gpa and a Poisson's ratio of 0.167.

Table 1. Parameters of Duncan-Chang E-B model for dam material

materials	dry density (g/cm3)	E-B model parameter						
		φ_0 (°)	$\Delta\varphi$ (°)	R_f	K	n	K_b	m
padding zone	2.35	53.0	6.9	0.79	1340	0.42	685	0.14
transition zone	2.31	52.1	6.9	0.76	1260	0.39	640	0.12
main rock pile area	2.23	50.9	6.7	0.78	1160	0.32	550	0.03
Downstream rock pile	2.22	50.1	6.6	0.82	1130	0.39	540	0.08
backward sloping body	2.01	35.2	4.2	0.73	430	0.42	180	0.21
backfill slag	2.15	44.5	4.5	0.75	560	0.45	280	0.26

The contact surface between the panel and the dam and bank bedding uses Goodman's no-thickness contact surface unit, and the relationship between stresses and relative displacements on the upper and lower contact surfaces can be expressed as follows:

$$\{\sigma\} = [k_0]\{w\} \tag{1}$$

where:

$\{\sigma\} = [\tau_{yx} \quad \sigma_{yy} \quad \tau_{yz}]^T$ is the contact surface stress in three directions;

$\{w\} = [\Delta u \quad \Delta v \quad \Delta w]^T$ is the relative displacement of the two surfaces above and below any point on the contact surface;

$$[k_0] = \begin{bmatrix} k_{yx} & 0 & 0 \\ 0 & k_{yy} & 0 \\ 0 & 0 & k_{yz} \end{bmatrix} \tag{2}$$

For the three-dimensional problem, the tangential stiffness is generalized to the two-dimensional contact surface. The two tangential direction stiffnesses are:

$$K_{yx} = K_1 \bullet \gamma_w \left(\frac{\sigma_y}{p_a}\right)^{n'} \left(1 - \frac{R'_f \tau_{yx}}{\sigma_y tg\delta}\right)^2$$
$$K_{yz} = K_1 \bullet \gamma_w \left(\frac{\sigma_y}{p_a}\right)^{n'} \left(1 - \frac{R'_f \tau_{yz}}{\sigma_y tg\delta}\right)^2 \quad (3)$$

K_1 is defined as the factorless quantity, obtained from the straight shear test; γ_w is the bulk weight of water; where δ is the friction angle between the two contact materials; and n', R'_f are index to damage ratio, obtained from the straight shear test. The parameters used in this paper are shown in Table 2.

Table 2. Panel mat contact surface parameters

designation	K_1	n'	R'_f	δ
Panel mat contact surface	4800	0.56	0.74	36.6

The water stopping unit proposed by Ho hai University[13] is used for panel joint water stopping, and the parameters are shown in Table 3, where the unit is m; the unit of F is kN / m.

Table 3. $F-\delta$ relationships and parameters for water stops

Classifications		Copper water stop	Water Stop Plastic Sheet
Tensile stress	$F-\delta$	$\delta = F/(a+bF)$	$F = k\delta$
	parameters	$a=175, b=47.6$	$k=4000,\ \delta \le 0.0115$ $k=600,\ \delta > 0.0115$
Compressive stress	$F-\delta$	$\delta = F/(a+bF)$	$F = k\delta$
	parameters	$a=650, b=41$	$k=530,\ \delta \le 0.0115$ $k=196,\ \delta > 0.0115$
Normal shear stress	$F-\delta$	$\delta = F/(a+bF)$	$F = k\delta$
	parameters	$a=225, b=40$	$k=0$
Slope shear stress	$F-\delta$	$F = k\delta$	$F = k\delta$
	parameters	$k=608,\ \delta \le 0.0125$ $k=560,\ \delta > 0.0125$	$k=1400$

2.3 Deformation Inclination Finite Element Method

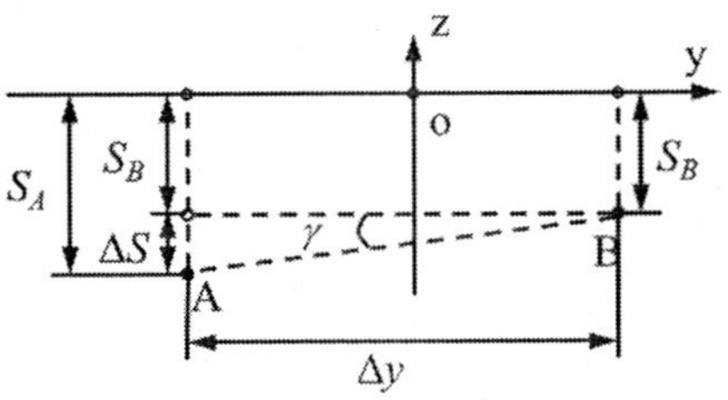

Fig. 4. Basic principle of deformation inclination method

The dam body will have uneven settlement due to the different mechanical properties of different types of dam materials, resulting in misalignment between the dam materials. When the misalignment exceeds a certain limit, the dam body will be damaged along the direction of misalignment, forming cracks. Therefore, some scholars have proposed the deformation inclination method for predicting dam cracks based on the observation data of dam settlement. According to engineering experience[9,14,15], it is generally believed that the critical deformation inclination is 1%, that is, if the deformation inclination exceeds 1%, tension cracks will appear. The basic principle of the deformation inclination method is shown in Figure 4.

At the same elevation of the body of the dam, there are two observation points A and B. The horizontal distance between the two points is Define the deformation inclination of points A and B on a given day:

$$\gamma \approx \tan\gamma = \frac{\Delta S}{\Delta y} \times 100\% \quad (4)$$

where: γ is the deformation inclination between A and B, ΔS is the cumulative settlement difference between A and B at the moment.

Peng Chong [12] combined the deformation inclination method with the finite element method to calculate the deformation inclination of each point by numerical methods. When the deformation of the dam body at a certain moment obtained by finite element calculation, the deformation inclination of each unit can be calculated according to the following formula:

$$\gamma_x = \frac{\partial S}{\partial x} \times 100\% = \sum_{i=1}^{n} \frac{\partial N_i}{\partial x} S_i \times 100\%$$
$$\gamma_y = \frac{\partial S}{\partial y} \times 100\% = \sum_{i=1}^{n} \frac{\partial N_i}{\partial y} S_i \times 100\% \qquad (5)$$

where: N_i is the shape function of the cell and S_i is the calculated settlement value at the cell node.

3. Uneven settlement pattern of deep backfilled reservoir bottom

Based on the model and parameters in Chapter II, the deformation patterns of the deep backfill area at the bottom of the reservoir basin containment body and the dam body were calculated during the as-built and impoundment periods, in which the backfill and the dam material were used in the construction loading process of simultaneous filling. The settlement distribution of the deep backfill area at the bottom of the basin containment body and the dam body under the normal water storage level condition is shown in Figure 5.

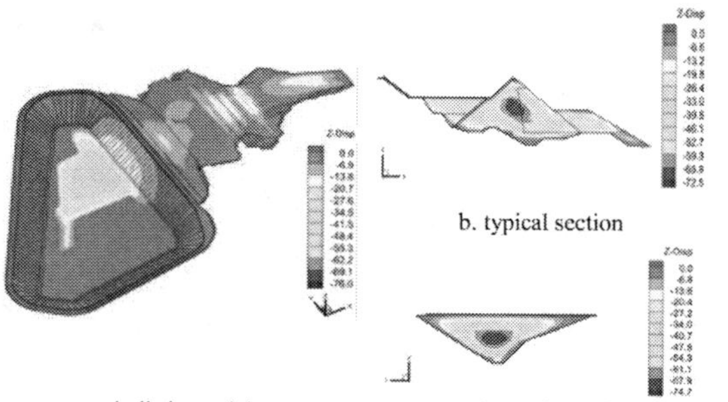

a. holistic model b. typical section c. dam axis section

Fig. 5. Settlement of the dam under normal storage level condition (cm)

It can be seen that due to the relatively small deformation modulus of the backfill material used in the deep backfill area at the bottom of the reservoir basin containment body, and the bedrock in the excavation area is set as a fixed constraint in the model, there is a clear boundary for the settlement at the bottom of the reservoir, i.e., there is a large inhomogeneous settlement at the boundary of the excavation and backfill.

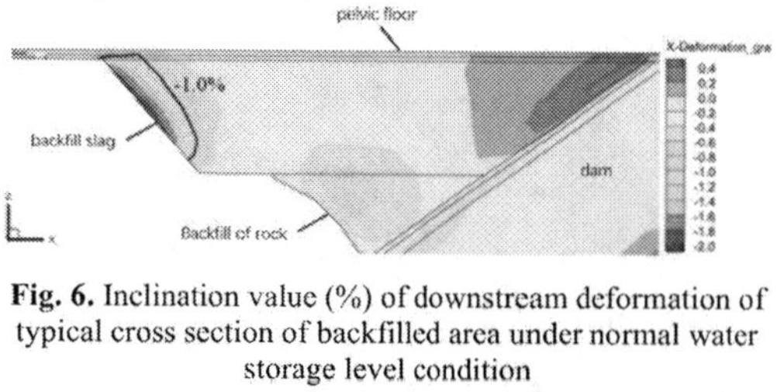

Fig. 6. Inclination value (%) of downstream deformation of typical cross section of backfilled area under normal water storage level condition

Figure 6 is the typical section on the backfill area of the river deformation inclination value calculation results, from the calculation results can be seen, backfill area backfill slag due to small parameters, deformation, there is an obvious uneven settlement in the left and bedrock contact position, its river deformation inclination of up to 2.0%; in addition to the backfill slag and the dam directly there is also an uneven settlement, and its river deformation inclination of about 0.4%. The deformation inclination value of backfill slag is more than 1% in a certain width. However, due to the deformation inclination value is greater than 1.0% of the region is basically located in the backfill slag inside, backfill area to withstand the compressive stress is larger, so generally will not form tension cracks threaten the backfill area and the overall stability of the dam body.

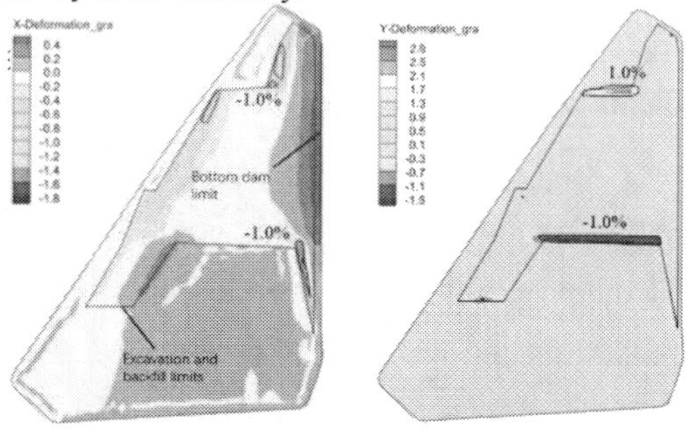

a. Inclination of deformation in the down-river direction b. Inclination of deformation in the cross river direction

Fig. 7. Inclination value (%) of deformation at the bottom of the reservoir under normal storage level condition

For panel dams with reservoir basin containment, the safety of the reservoir basin containment system directly affects the operational status and economic benefits of the dam, in which the composite geomembrane used at the bottom of the reservoir will be affected by the uneven settlement of the reservoir bottom. Figure 7 shows the deformation inclination value of the entire reservoir bottom in the normal storage level condition in the down-river direction and the transverse river direction, in which the deformation inclination value is larger in the region of the excavation backfill boundary and the reservoir bottom near the boundary of the dam body. This indicates that the bottom of the reservoir will have a large uneven settlement due to the large difference in modulus at the material demarcation location. The area where the deformation inclination value exceeds the critical value of 1.0% is mainly located near the excavation and backfill boundary, especially the corner area of the boundary, and the overall area is relatively small. Because the geomembrane at the bottom of the reservoir has a certain tensile strength and strong flexible deformation ability, and is not completely applicable to the critical deformation inclination value of 1.0% of the geotechnical dam material, from the perspective of the safety of the impermeable body, can be overlying geomembrane backfill material shear test to get the geomembrane damage to the critical deformation inclination value.

Uneven settlement in the down-river direction exists near the boundary of the dam at the bottom of the reservoir, and its deformation inclination value is maximum 0.4%, while the uneven settlement in the cross-river direction is not significant. Since the deformation inclination value at this location is less than the critical value of 1.0%, there is no damage problem at the location of the geomembrane-joint with the bank panel and the bedding layer underneath.

4. Influence of material parameters in the backfilled area on the uneven settlement of the bottom of a deep backfilled reservoir

On the basis of the calculation parameters in Chapter 2, the influence of the modulus of the material in the backfilled area on the uneven settlement of the bottom of the deep reservoir is studied by multiplying the coefficients of the Duncan-Chang E-B model parameters of the backfilled slag. Due to too many parameters of Duncan-Chang E-B model, according to engineering experience, the effect of K, Kb on the modulus is more significant, i.e., the K, Kb parameters of backfilled slag are discounted or increased for calculation, and the coefficients are taken from 0.25 to 2.0.

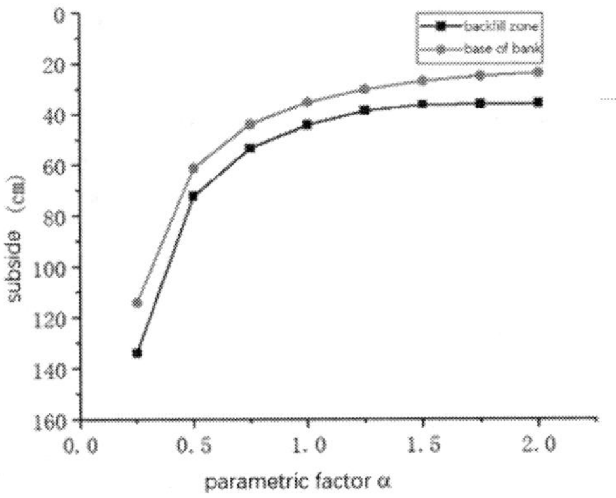

Fig 8. Maximum settlement (cm) of backfilled area and reservoir bottom with different parameters

Figure 8 shows the maximum settlement values of different parameters of the backfilled area and the bottom of the reservoir under the normal water storage level condition, in which the parameter coefficient α takes 1.0 as the actual backfilled slag parameter of the project. The settlement distribution law under different parameters is consistent, and the distribution of settlement at the bottom of the reservoir is similar to that in Fig. 5(a), with the maximum settlement appearing on the surface of the bottom of the reservoir with the largest thickness of the backfilled area; the maximum settlement of the backfilled area appears in the middle of the backfilled area with the largest thickness. When the parameter coefficient α is smaller, i.e., the modulus of the backfilled area is smaller, the settlement of the backfilled area and the bottom of the reservoir is larger, and the settlement of the backfilled area and the bottom of the reservoir decreases rapidly with the gradual increase of α , but the trend of the change gradually becomes slower.

Figure 9 shows the deformation inclination values of different parameters of the excavation and backfill boundary and the dam bottom boundary under the normal water storage level condition, in which the deformation inclination value is the largest deformation inclination value in the river direction and transverse direction, when the parameter coefficients are smaller, and the bottom settlement is bigger, the deformation inclination value of the excavation and backfill boundary and the dam bottom boundary is bigger; especially when the parameter coefficients are $\alpha \leq 0.5$ (at this time the parameters have been far less than the conventional backfill parameters), the material modulus of the backfill area plays a dominant role in the uniformity of bottom settlement.), the material modulus of the backfill area plays a dominant role in the uneven settlement of the reservoir bottom. With the gradual increase of coefficient α , the modulus of the backfill area gradually reduces the influence of the uneven settlement at the bottom of the reservoir. From the engineering practice, it can be seen that to avoid the risk of damage to the impermeable body of the reservoir basin above the deep backfill area, the backfill area can be appropriately milled; especially, milling and compacting in the vicinity of the excavation and backfill boundaries and the boundaries of the dam body at the bottom of the reservoir to improve the modulus of the material in the backfill area, which can reduce the phenomenon of the inhomogeneous settlement of the impermeable body at the bottom of the reservoir very effectively

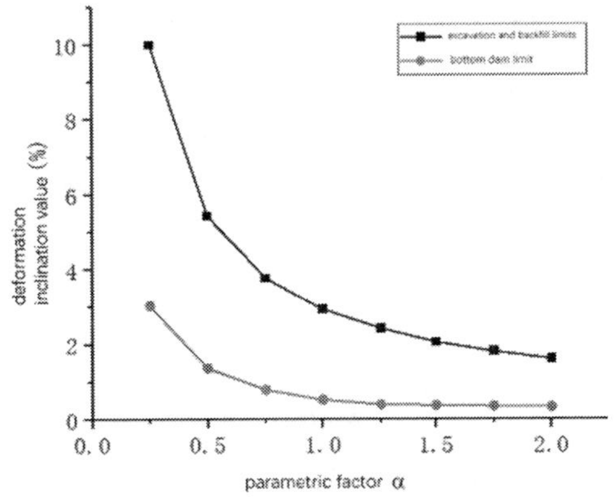

Fig 9. Typical positional deformation inclination values (%) for different parameters

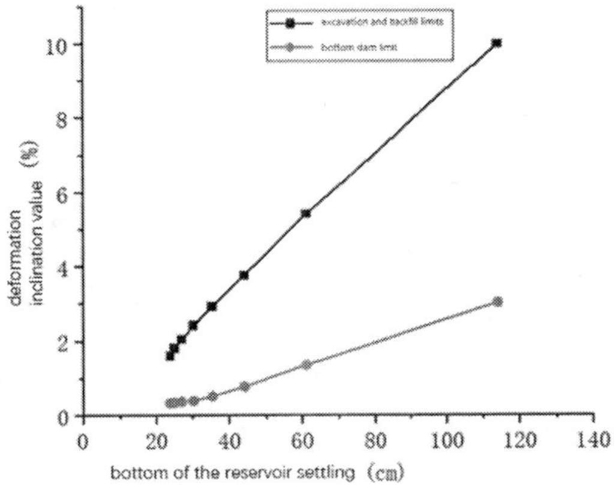

Fig 10. Relationship between the values of deformation inclination at typical locations and the maximum settlement of the reservoir bottom

Combining the results of Fig. 8 and Fig. 9, the relationship between the deformation inclination value and the maximum settlement at the bottom of the reservoir is organized as Fig. 10, which shows that the maximum settlement at the bottom of the reservoir is basically linearly related to the maximum deformation inclination value at different locations. If the critical deformation inclination value of 1.0% of bedding material is taken into account, the maximum allowable settlement at the bottom of the reservoir is 48 cm; accordingly, on the basis of the critical deformation inclination value of geomembrane obtained from the shear test, the maximum allowable settlement at the bottom of the reservoir can also be obtained, and the maximum allowable settlement index at the bottom of the reservoir can be established finally.

Based on the allowable settlement index of the reservoir bottom, on the one hand, finite element calculation can be utilized to invert the material parameters of the backfilled area, which can be used as an index of the mechanical performance of the materials in the backfilled area during the construction; on the other hand, it can be compared with the actual observation results of the reservoir bottom, which can always reflect the inhomogeneous settlement situation of the reservoir bottom, so as to guarantee the operation safety of the reservoir basin seepage control body in the long term service.

5. Conclusions

In this paper, for the problem of uneven settlement at the bottom of the deep backfilled reservoir basin impermeable body, the uneven settlement law of the backfilled area and the bottom of the reservoir is analyzed from the practical engineering point of view; and the location of the possible cracks at the bottom of the reservoir is studied based on the deformation inclination finite element method. Then, this paper also considers the influence of the material parameters of backfilled area on the uneven settlement of the bottom of the deep backfilled reservoir to provide guidance for engineering applications. The calculation results show that:

(a) The phenomenon of uneven settlement is more obvious near the excavation and backfill boundary and the dam body boundary at the bottom of the reservoir, and the area where the deformation inclination value exceeds the critical value of 1.0% is mainly located in the corner area of the excavation and backfill boundary, but the overall area is relatively small.

(b) The smaller the parameter of mechanical properties of materials in the backfilled area is, the larger the uneven settlement of the bottom of the reservoir is, but with the gradual increase of the parameter, the influence of the modulus of the backfilled area on the uneven settlement of the bottom of the reservoir gradually decreases.

(c) When the whole reservoir basin seepage control form can be used to backfill area for appropriate crushing, especially in the excavation backfill boundary and the reservoir bottom dam body boundary near the crushing compact, can effectively reduce the size of the reservoir bottom uneven settlement.

(d) The maximum settlement of the reservoir bottom and the maximum deformation inclination value of different locations is basically linear relationship, the actual engineering application can be determined through the test of the critical deformation inclination value of the reservoir bottom impermeable body, and finally establish the allowable settlement index of the reservoir bottom.

References

1. LI S, HU Y, ZHANG L. (2022) Coupling coordination relationship of pumped storage power station and eco-environment system [J]. Journal of Energy Storage, 52(105029).
2. WANG Z G, HE W. (2011) Classifying geo-hazards in relation to hydropower and water resources engineering [J]. Jouenal Of Geological Hazards And Environment Preservation, 22(04):35-40.
3. XIN J S, WANG F Q, HAO J G. (2023) Selection and design of whole reservoir basin seepage control types of pumped-storage power stations [J]. China Renewable Energy Engineering Institute, 49(11):67-72+78.
4. ZHANG X Q. (2011) Bituminous concrete impervious facing structure for the upper reservoir of zhang he wan pumped-storage power station [J]. Water Power, 37(04): 39-42.
5. SHI H X, CHANG S S. (2010) Seepage control system selection of upper reservoir of liyang pumped-storage Power Station [J]. Water Power, 36(07): 50-2+78.
6. SHI H X, HU Y L, CHANG S S. (2016) Application of geomembrane in li yang pumped storage power plant[J]. Chinese Journal of Geotechnical Engineering, 38(S1):15-20.
7. MA F, LI J, LI D H, et al. (2020) Study on seepage control technology for reservoir basin of pumped storage power station [J]. water resources and Hydropower Engineering, 51(S1) : 122-129.
8. KONG F H, HUA J J, WANG L H, et al. (2018) Stress and deformation analysis of the composite geomembrane used as the seepage prevention for the whole reservoir basin on deep overburden [J]. China Rural Water and Hydropower, , (05):152-155.
9. Nanjing Hydraulic Research Institute, et al. (1979) Cracks in soil dams and Observational analysis [M]. Water Resources and Electric Power Press, Nanjing.
10. LI J C. (1983) Simplified methods for evaluating cracks in earth dam [J]. Hydro-Science and Engineering,, (03):1-11.
11. GAO Z L, ZHANG H, XIONG M. (2020) Research on the cracking characteristics and causes of dam crest in high core wall rock—fill dam on the deep overburden layer [J]. China Rural Water and Hydropower, (05):103-108+113.
12. PENG C, ZHANG Z L, ZHANG B Y, et al. (2013) Deformation gradient finite element method for analyzing cracking in high earth-rack dam and its application [J]. Rock and Soil Mechanics, 34(05):1453-1458.

13. GU G C, SHU Y M, SHEN C S. (2004) Experience and innovation in earth and rock dam engineering [M]. China Electric Power Press, Nanjing.
14. ZHANG B Y, ZHANG M C, SUN X. (2008) Centrifugal modeling of transverse cracking in earth core dams [J]. Rock and Soil Mechanics, (05):1254-1258.
15. Sahu M, Jana S, Agarwal S, et al (2011). Point form velocity prediction in meandering open channel using artificial neural network [C].2nd International conference on environmental science and technology, (06): 209-212.

https://doi.org/10.1051/e3sconf/202561701014

Study on the Community Structure and Diversity of Zooplankton in Dingzi Bay by Using Mathematical and Intelligent Analysis Tools

Liang Zhang*, Juan Huang, Yumeng Liu, Xueji Gu and Chuanting Zhao

North China Sea Marine Forecasting and Hazard Mitigation Center of Ministry of Natural Resources, Qingdao 266061, China

Abstract. In order to know the community structure and diversity of zooplankton in Dingzi Bay, based on the investigation data in April 2024(spring), the species composition, dominant species, community structure and the correlation between species diversity and environmental factors of zooplankton were analyzed by using computer systems and digital intelligent analysis tools. The results showed that the water quality at the bottom of Dingzi Bay was poor, and a tolal of 10 species and 5 taxa of zooplankton larvae were found. The average individual density of zooplankton was 387.0 ind/m^3, and the average biomass (wet weight) was 188.6 mg/m^3. The average Shannon-Wiener index (*H'*), Margalef's species richness diversity (*d*) and Pielou's evenness index (*J'*) were 1.35, 0.92, 0.46, respectively. The dominant species were *Calanus sinicus*, *Tortanus derjugini*, Brachyura Zoea larva and fish egg. Cluster analysis showed that the zooplankton in the investigation area can be divided into 2 groups. The correlation analysis showed that water temperature was an important environmental factor affecting zooplankton indexes.

1 Introduction

Dingzi Bay is located in the southern coast of Shandong Peninsula, and its mouth is adjacent to the Yellow Sea. The bay is long and narrow like a T-shaped, in which Wulong River, Baisha River and Lianyin River flow into the bay [1]. In recent years, due to the influence of various human factors such as sea-related engineering development, reclamation and breeding, sea reclamation, the water environment in this area has deteriorated and biodiversity has declined [2]. Zooplankton is one of the important components of biodiversity in marine ecosystem [3]. The proportion and quantity of zooplankton species can reflect the changes of marine environment [4], which has great research value in biological oceanography, water eutrophication research and marine ecological environment protection. In recent years, the researches on Dingzi Bay mainly include water quality [5], sediments [6], ocean current [7] and microbenthic [1], but no research on zooplankton in Dingzi Bay have been reported. With the help of computer systems, mathematical and intelligent analysis tools, the characteristics of zooplankton community structure and biodiversity, the correlation between zooplankton species diversity and environmental factors were analyzed. The study can provide basic data and scientific basis for ecological environment protection and restoration in the sea area of Dingzi Bay.

2 Materials and methods

2.1 Sample Collection and Processing

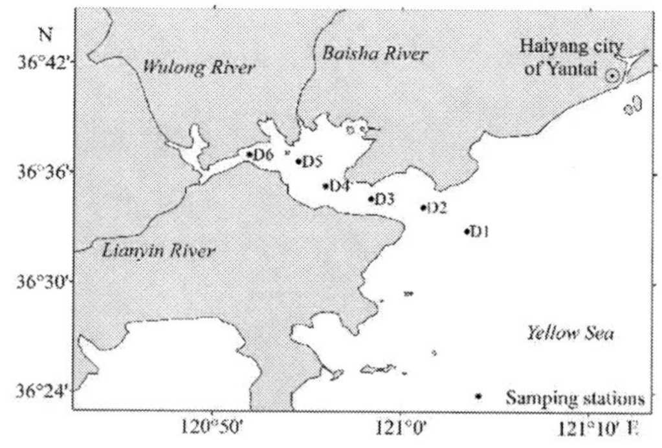

Fig. 1. The location of sampling stations.

Investigators conducted zooplankton surveys at 6 stations in Dingzi Bay in April 2024 (spring), samping stations as shown in Fig. 1. The sampling net was type I plankton net in shallow water. The zooplankton species identification and individual count were carried out according to the specification for marine monitoring—Part7 [8]. At the same time, relevant environmental factors were monitored according to the specification for marine monitoring—Part4 [9] and Part 7 [8]. The identification of zooplankton species was quality controlled through double person retesting, and environmental factors were

* Corresponding author: 36302162@qq.com

quality controlled through measures such as password parallel samples and internal control samples to ensure data consistency and reliability.

2.2 Data processing and analysis

2.2.1 Evaluation of seawater quality and eutrophication index

According to Marine Function Zoning of Shandong Province (2011—2020), the station D2~D6 is located in Dingzi Bay Tourism, leisure and entertainment area, and station D1 is located in Haiyang-Jimo agricultural and fishery area, So the second category of seawater water quality standards in the Seawater Water Quality Standard (GB3097—1997) is used as a reference, and the single factor quality index method [10] is used to evaluate seawater quality. The eutrophication evaluation of seawater quality in the sea area is carried out according to the Specifications for oceanographic survey—Part9 [11].

2.2.2 Individual density, biomass and dominance

The individual density, biomass (wet weight) and dominace (*Y*) of zooplankton were calculated with the help of computer systems, according to previous reach [12]. When $Y \geq 0.02$, it is determined that the zooplankton is the dominant species [13].

2.2.3 Species diversity and biodiversity index threshold (Dv)

Shannon-Weaver diversity index (H') [14], Margalef species richness index (d) [15] and Pielou evenness index (J') [16] were calculated by using the PRIMER 6.0 software package. The formulas were as follows:

$$H' = -\sum_{i=1}^{S} P_i \log_2 P_i \quad (1)$$

$$D = (S-1)/\log_2 N \quad (2)$$

$$J' = H'/\log_2 S \quad (3)$$

where S is the species number at each station, P_i is the percentage of the abundance of the i-th species to the total abundance, N is the total abundance at each station.

The biodiversity index threshold Dv ($Dv = H' \times J'$) determined by Chen Qingchao et al. (1994) was used to describe the biodiversity [17].

2.2.4 Division of community structure

Cluster analysis module in PRIMER 6.0 software package was used to perform cluster and MDS sequencing analysis on zooplankton communities. To reduce the interference of opportunistic species on community structure, first remove species with relative abundance less than 1% in the population, but retain species with relative abundance greater than 3% at any station. After the original abundance data is transformed and standardized to the fourth root, a similarity matrix is constructed based on the Bray Curtis similarity coefficient, and then hierarchical clustering analysis is used to connect the samples into groups step by step. The availability of analysis results was judged by the range of stress coefficient [18].

2.2.5 Bioenvironmental correlation analysis

The obtained zooplankton data and environmental factor data were analyzed with Pearson Correlate analysis in IBM SPSS Statistics 19.0 software package. The significance test is conducted using two-sided test.

3 Results and discussions

3.1 Hydrological Environment

The monitoring results of environmental factors in Dingzi Bay in April 2024 are shown in Table 1. The highest water temperature was at station D6, the lowest at station D2, showing a general trend of decreasing from the bottom of the bay to the outside of the bay. The highest salinity was at station D2, the lowest at station D6, and the salinity generally increased from the bottom of the bay to the outer bay. The highest COD and active phosphate appeared at station D6, the lowest at station D1. The highest inorganic nitrogen and active silicate appeared at station D5, and the lowest at station D2. The distribution trend of COD, active phosphate, inorganic nitrogen and active silicate decreased gradually from the bottom of the bay to the outside of the bay.

Table 1. The environmental factors in investigation sea area.

Environmental factors	Monitoring results		Quality Index		Over-standard rate / %
	Range	Average value	Range	Average value	
Depth / m	3.3~9.5	5.6	—	—	—
Water temperature / °C	15.91~19.00	3.67	—	—	—
Salinity	28.634~31.051	30.271	—	—	—
pH	8.21~8.27	8.25	0.17~0.34	0.27	0
DO / (mg/L)	5.88~7.86	7.12	0.39~0.82	0.53	0
COD / (mg/L)	0.66~1.54	1.03	0.22~0.51	0.48	0
Inorganic nitrogen / (μg/L)	48~377	145	0.16~1.26	0.48	17
Active phosphate / (μg/L)	1.37~24.1	8.57	0.05~0.80	0.29	0
Active silicate / (μg/L)	201~987	490	—	—	—
Chlorophyll a / (μg/L)	1.67~3.64	2.37	—	—	—

Note: "—" indicates that no relevant data is involved.

The second category of seawater water quality standard (GB3097—1997) was used to evaluate the environmental data, among which 17% of the inorganic nitrogen stations exceeded the standard, and the rest factors did not exceed the standard. According to the "2023 Shandong Ecological Environment Status Bullentin", the sea areas in Shandong Province that were inferior to the fourth category of seawater quality standard are mainly distributed in Laizhou Bay, Bohai Bay, Dingzi Bay and Jiaozhou Bay, and the main element exceeding the standard in Dingzi Bay was inorganic nitrogen [19].

3.2 Species composition and dominant species

The eutrophication evaluation of the sea water in the survey were showed that the nutrient index of the survey stations ranged from 0.1 to 2.6, with an average value of 0.7. The sea water at stations D5 and D6, witch at the bay bottom showed eutrophication. The distribution trend of environmental factors was mainly due to the influence of the injection of rivers into the bay, which carried a large amount of nutrients from land-based sewage, resulting in the reduction of water quality at the bottom of the bay. It can be seen that the water quality at the bottom of Dingzi Bay was poor. A total of 10 species (Table 2) were found in the sea area investigated, which were belong to Coelenterata, Arthropoda, Chaetognaths, respectively. Coelenterata (3 species) accounted for 30% of the total species, Arthropoda (6 species) accounted for 60%, and Chaetognaths (1 species) accounted for 10%. In addition, 5 taxa of planktonic larvae were found. There were 4 dominant species (Table 3), namely, *Calanus sinicus*, *Tortanus derjugini*, Brachyura Zoea larva and fish egg. *Calanus sinicus* was the first dominant species with 100% station frequency, followed by Brachyura Zoea larva and *Tortanus derjugini*. As a warm temperate species, *Calanus sinicus* is widely distributed in the coastal areas of Bohai, Yellow Sea and East Sea, and is the dominant species in these sea areas [20]. Planktonic larvae account for 50% of the dominant species. The main reason is that the water temperature starts to rise in spring, and the abundance of bait phytoplankton is high, which is suitable for planktonic larva mass reproduction. Planktonic larvae are an important part of marine zooplankton, and generally become the dominant species during the peak reproduction period [21]. *Tortanus derjugini* is mainly distributed in coastal low-salt waters, especially in the estuary area [20]. In this survey, the density of *Tortanus derjugini* was higher at stations D5 and D6, which were near the estuary.

Table 2. List of zooplankton species.

No.	Species name	Taxa
1	*Calanus sinicus*	Arthropoda
2	*Tortanus derjugini*	Arthropoda
3	*Acartia pacifica*	Arthropoda
4	*Centropages abdominalis*	Arthropoda
5	*Paracalanus parvus*	Arthropoda
6	Gammaridea	Arthropoda
7	*Obelia* spp.	Coelenterata
8	*Clytia hemisphaerica*	Coelenterata
9	*Rathkea octopunctata*	Coelenterata
10	*Aidanosagitta crassa*	Chaetognaths
11	Macrura larva	Plankonic larva
12	Brachyura Zoea larva	Plankonic larva
13	Polychaeta larva	Plankonic larva
14	Fish egg	Plankonic larva
15	Fish larva	Plankonic larva

Table 3. Dominant species of zooplankton.

No.	Species name	*Y*	Frequency of occurrence	Individual density/ (ind/m^3)
1	*Calanus sinicus*	0.69	100%	284.1
2	*Tortanus derjugini*	0.07	83%	36.8
3	Brachyura Zoea larva	0.10	100%	27.0
4	Fish egg	0.05	83%	20.6

3.3 Individual density and biomass distribution

The individual density of zooplankton ranged from 67.3 to 1 223.3 ind/m^3, with an average of 387.0 ind/m^3. The highest individual density appeared at station D6, and the lowest at station D4. The biomass varied from 33.3 to 58.0 mg/m^3, with an average of 188.6 mg/m^3. The highest biomass appeared at station D6, while the lowest at station D2. The station with the highest biomass and individual density both appeared at station D6, which was mainly due to the appearance of large number of high biomass of *Calanus sinicus*. The individual density was 796.7ind/m^3 at station D6, while was 80.6 ind/m^3 and 50.0 ind/m^3 at stations D2 and D4.

The biomass and individual density of zooplankton in the surveyed area showed similar trend, with the overall trend of higher at the bottom of the bay and outside the bay mouth, and lower in the middle of the bay.

3.4 Community species diversity

The diversity index and evaluation results of each station are shown in Table 4. According to the classification of biocommunity richness by Chen Qingchao et al. (1994) [17], the average threshold of zooplankton diversity index was 0.6, and the diversity description was general. Except for D1 and D3 stations, the diversity threshold was 0.1 and 0.3, and the diversity description was poor. The remaining stations classes were described as general.

3.5 Group Classification

The 6 zooplankton survey stations in Dingzi Bay can be divided into 2 groups with 60% similarity (Fig. 2), Group I contains D5 and D6 stations located at the bottom of the bay, and group II contains D1~D4 stations. The pressure coefficient in the MDS map is 0.02, which ranging from 0.01 to 0.05, and the group results are credible [18]. The similarity between the stations was high, and the division of groups was related to the distribution of the 4 dominant species found in this survey, which all had the highest individual density in group I, while the individual density was lower in group II. Studies have shown that the division of ecological groups was caused by the uneven distribution of dominant species [22].

Table 4. The diversity index and evaluation results of zooplankton.

Stations	*H'*	*J'*	*d*	*Dv*	Diversity description
D1	0.36	0.14	0.68	0.1	poor
D2	1.56	0.60	0.93	0.9	general
D3	0.87	0.34	0.76	0.3	poor
D4	1.55	0.49	1.20	0.8	general
D5	2.11	0.70	0.91	1.5	general
D6	1.65	0.50	1.06	0.8	general
Average	1.35	0.46	0.92	0.6	general

Table 5. Correlations between zooplankton and the content of environmental factors.

Pearson relevance	Water temperature	Salinity	pH	DO	COD	Active phosphate	Inorganic nitrogen	Active silicate	Chlorophyll a	Biomass
Salinity	-0.96*	1.00	0.68	-0.49	-0.96**	-0.93*	-0.86*	-0.95**	-0.85*	-0.57
Species number	0.83*	-0.75	-0.41	0.56	0.74	0.64	0.51	0.68	0.53	0.45
Biomass	0.43	-0.57	0.15	0.07	0.46	0.27	0.19	0.35	0.09	1.00
Individual density	0.48	-0.63	0.10	0.07	0.54	0.33	0.25	0.41	0.16	0.99**
H'	0.66	-0.61	-0.71	0.28	0.73	0.72	0.67	0.70	0.79	-0.02

Concentrate:**. Significantly correlated at the 0.01 level (bilateral). *. Significantly correlated at the 0.05 level (bilateral).

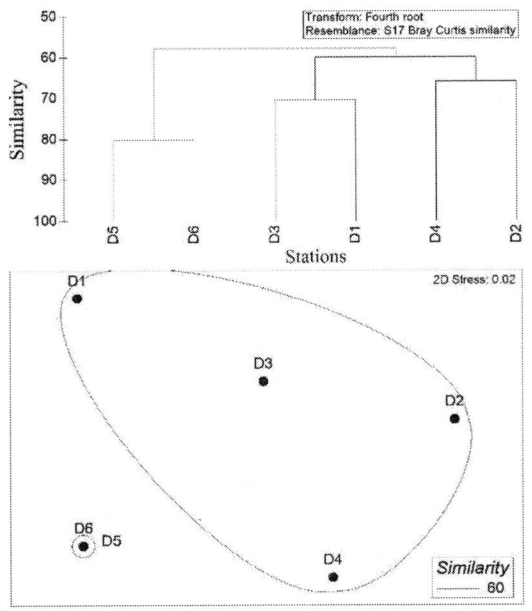

Fig. 2. The results of cluster and MDS analysis.

3.6 Bioenvironmental correlation analysis

As can be seen from Table 5, correlation analysis showed that the number of zooplankton species was significantly positively correlated with water temperature ($p<0.01$), and the biomass of zooplankton was significantly positively correlated with individual density ($p<0.05$). There was a significant negative correlation between salinity and DO, COD, active phosphate, inorganic nitrogen and active silicate. Salinity was negatively correlated with zooplankton species number, biomass, individual density and *H'*, while water temperature was positively correlated with other zooplankton indexes, but not significantly correlated.

Bioenvironmental correlation analysis showed that water temperature was positively correlated with the number of zooplankton species, biomass, individual density and diversity index, and there was a significant positive correlation with species number ($p<0.01$), indicating that water temperature was the key environmental factor affecting zooplankton indicators [23]. There was no significant positive correlation between the nutrients and zooplankton, which shows that the impact of nutrients on zooplankton is relatively weak. The analysis may be caused by the fact that nutrients first affect the growth of phytoplankton, and then the phytoplankton have an impact on zooplankton [24]. In this investigation, the water quality of inorganic nitrogen exceeded the standard, and the water of D5 and D6 stations at the bottom of the bay showed eutrophication, so the water quality at the bottom of Dingzi Bay was poor. The species diversity index is often used to evaluate the pollution degree of marine environmental quality [25], the survey zooplankton species diversity index evaluation shows that D1 and D3 stations were evaluated as heavy pollution, D2, D4, D6 stations were highly moderate pollution, and D5 was light to moderate pollution. The species diversity index assessment of marine environmental pollution degree is inconsistent with the evaluation results of water quality eutrophication, so the evaluation of species diversity index should be comprehensively considered with the evaluation results of water quality and other environmental factors.

4 Conclusions

In April 2024 (spring), a total of 10 species of zooplankton were found in Dingzi Bay, which belonged to Coelenterata, Arthropoda, Chaetognaths, and Arthropoda accounted for the highest proportion, and 5 taxa of plankonic larvae were found. The dominant species were *Calanus sinicus*, *Tortanus derjugini*, Brachyura Zoea larva and fish egg. The planar distribution of zooplankton biomass and individual density showed a trend of higher at bay bottom and outside bay mouth, and lower in middle bay. The zooplankton diversity in spring of Dingzi Bay was rated as general, and the zooplankton could be divided into 2 groups. The water quality at the bottom of Dingzi Bay is poor, and the zooplankton diversity index needs to be combined with environmental monitoring data to reflect the quality of the ecological environment.

Due to the unique characteristics of each bay, the results of this study are not recommended for direct application to similar bays. However, the methods used in this study can be applied to related research on zooplankton in other bays.

The changes in the community structure of zooplankton are the result of multiple factors. With seasonal changes, hydrological conditions such as climate, water temperature, salinity, and environmental factors in the sea area also undergo seasonal changes. The community structure of zooplankton also undergoes succession, mainly reflected in the number of species, the proportion of each group, and the differences in dominant species. At present, there is no research on zooplankton in Dingzi Bay. Due to limited survey data, this study only used spring survey data to study the community structure of zooplankton in Dingzi Bay. The next step is to conduct surveys in other seasons to analyze the impact of seasonal changes on the community structure and diversity of zooplankton in Dingzi Bay.

References

1. Ji, Y. L., Pu. S. C., Tao, H. H., et al. (2022) Study on the structure of microbenthic community in coastal wetlands with different vegetation types in Dingzi Bay Journal of Shandong Agricultural University (Natural Science Edition), 53 (03): 412-420. DOI:10.3969/j.issn.1000-2324.2022.03.012.
2. Yu, Z. H., Jin, X. S., Li, X. S. (2010) Analysis of ecological niche for major fish species in the central and southern Yellow Sea. Progress in Fishery Sciences, 31 (6): 1-8.
3. Fu, P., Yang, P., H, J. L., et al. (2023) Effects of the occurrence of Green tide (Ulva prolifera) on marco and mesozooplankton community structure in the sea area near Dingzi Bay.Marine Environmental Science, 42 (04): 604-611+621. DOI: 10.13634/j.cnki.mes. 2023.04.019.
4. Hays, G. C., Richardson, A. J., Robinson, C. (2005) Climate change and marine plankton. Trends in Ecology & Evolution, 20 (6): 337-344.
5. Wang, W., Yuan, S. S. (2019) Spatial and temporal distribution and variation characteristics of suspended matter in the sea water of Dingzi bay in the past 20 years. Ground water, 42 (6):116-118. DOI: 10.19807/j.cnki.DXS.2020-06-039.
6. Wang, W., Tang, S. K., Hu, Y. P., et al. (2022) Spatio-temporal variation and influencing factors of seafloor sediment grain size off the mouth of Dingzi Bay of Southern Shandong Peninsula. Marine Geology & Quaternary Geology, 42 (02):70-80. DOI:10.16562/j.cnki.0256-1492.2021050601.
7. CAO, X. F., Chen, Z. L., Yu, J. W., et al. (2023) Researches on the characteristics of tide and current in the coastal area of Dingzi Bay. Transactions of Oceanology and Limnology, 45 (01): 74-81. DOI:10.13984/j.cnki.cn37-1141.2023.01.011.
8. General Administration of Quality Supervision, Inspection and Quarantine of the PRC. (2008) The specification for marine monitoring—Part7: Ecological survey for off shore pollution and biological monitoring: GB17378.7—2007. China Standard Press, Beijing.
9. General Administration of Quality Supervision, Inspection and Quarantine of the PRC. (2008) The specification for marine monitoring—Part4: Sea water analysis: GB17378.4—2007. China Standard Press, Beijing.
10. Environmental Engineering Evaluation Center, State Environmental Protection Administration. (2005) Compilation of Technical Guidelines and Standards for Environmental Impact Assessment. China Environmental Science Press, Beijing.
11. General Administration of Quality Supervision, Inspection and Quarantine of the PRC, Standardization Administration of the PRC. (2007) Specifications for oceanographic survey—Part9: Guidelines for marine ecological survey: GB/T 12763.9—2007. Standards Press of China, Beijing.
12. Zhang, L., Wang, J. J., Wang, L., et al. (2022) Zooplankton community in the sea near Qingdao. Marine Sciences, 46 (08): 88-100. DOI: 10.11759/ hykx20210920001.
13. Wong, M. W., Townsend, D. W. (1999) Phytoplankton and hydrography of the Kennebec estuary, Maine, USA. Marine Ecology Progress Series, 178: 133-144.
14. Shannon, C. E., Weaver, W. (1949) The mathematical theory of communication. University of Illinois Press, Urbana.

15. Margalef, R. (1958) Information theory in ecology. General Systematics, 3: 36-71.
16. Pielou, E.C. (1969) An Introduction to Mathematical Ecology. Wiley-Interscience, New York.
17. Chen, Q. C., Huang, L. M., Yin, J. Q., et al. (1994) Study on zooplankton diversity in Nansha Islands // Nansha Comprehensive Scientific Expedition Team, Chinese Academy of Sciences. Study on Marine Biodiversity of Nansha Islands and its adjacent Waters I. China Ocean Press, Beijing. pp. 42-50.
18. Clarke, K. R., Gorley, R. N. (2001) PRIMER v5: user manual/tutorial. PRIMER-E Ltd, Plymouth.
19. Shandong Department of Ecology and Environment. (2024) Bulletin of Ecological Environment Status of Shandong Province in 2023 http://xxgk.sdein.gov.cn /xxgkml/hjzkgb/202406/P020240606452239592735 .pdf.
20. Yang, J. Q., Chen, Y. G., Zhang, D. R., et al. (2023) Adaption of Calanus sinicus to the environment factors in bays and estuaries in spring season. Journal of Applied Oceanography, 42 (03):460-468. DOI: 10.3969/J.ISSN.2095-4972.20220427002.
21. Liu, G. X., FANG, J., Chen, H. J. (2013) The community structure of zooplankton in the enteromorpha prolifera outbreak area of the south Yellow Sea during spring in 2009. Journal of Ocean University of China: Natural Science Edition, 43 (7): 72-77. DOI: 10.16441/j.cnki.hdxb.2013.07.013.
22. Luo, M., Miao, S. Y., Yu, H. B., et al. (2013) Community structure of zooplankton in the offshore water of Wanning at the end of spring. Marine Science, 37 (11): 79-84.
23. Wang, Y. H., Li, S. Q., Ye, S., et al. (2024) Correlation analysis of zooplankton community structure and environmental factors in the Oujiang River Estuary. Haiyang Xuebao, 2024, 46 (3) ： 98–110. DOI: 10.12284/hyxb2024042.
24. Jiang, H. C., Chen, H. G., Song, X. K., et al. (2015) Zooplankton community structure in Jincheng area of Laizhou Bay and its relationship with environmental factors. Acta Ecologica Sinica, 35 (22): 1-12.
25. Kong, F. X. (2000) Environmental Biology. Higher Education Press, Beijing.

Research on Testing Methods of Life Detection Equipment for Water-bearing Soil Disasters

Yanyun Nan[1,a], Jin Yan[1,*] and Shizhe Xie[2]

[1] National Earthquake Response Support Service, Beijing 100049, China
[2] China University of Geosciences Beijing, Beijing 100080, China

Abstract. Aiming at the current water-bearing soil disaster life detection equipment testing research is relatively small, mainly relying on key components and equipment as a whole of the factory test, the lack of standardised system testing problems, this paper designs the water-bearing soil disaster life detection equipment comprehensive performance testing methods and testing process. We combine the equipment with the actual application scenarios, analyse the equipment testing requirements, determine the equipment testing content and testing methods, and establish a standardised testing process. The establishment of standardised test methods is of great importance in assessing the overall capability of the equipment and improving its effectiveness and reliability in complex disaster environments.

1 Introduction

A water-bearing soil disaster is a geological disaster with a water content within a certain range, usually referring to mudslides, landslides and other geological disasters triggered by flash floods or heavy rainfall. Life detection equipment refers to devices used for detecting physiological signals of living beings, typically divided into contact-based life detection equipment and non-contact detection equipment. Water-bearing soil disasters are characterised by a large disaster area, a large number of trapped people, a deep burial depth, a short rescue window, a complex environment, etc. Improper detection will result in the loss of the best rescue time, which will cause more casualties in turn.

With the development of China's emergency rescue cause, as well as the implementation of the Ministry of Science and Technology emergency rescue equipment science and technology publicity, prompting the rapid development of water-bearing soil disaster life detection equipment progress, some new equipment application and birth. However, due to the complexity of the water-containing soil disaster environment, the actual environment simulation is difficult, resulting in the application of some detection equipment performance and design features in the disaster site adaptability there is still a certain gap. At present, the relevant equipment testing mainly focuses on equipment safety protection level testing[1-4]. Reliability testing is primarily conducted based on current national standards, while functional testing is mainly carried out according to project evaluation requirements, with the goal of meeting these requirements. However, there is a lack of a systematic testing process, testing content, and testing regulations for the equipment. Therefore, systematically studying the testing requirements, testing content, and testing methods for life detection equipment in water-bearing soil disasters is of great significance for enhancing the evaluation of R&D projects for such equipment and improving practical operational capability testing.

2 Testing requirements

In the event of a water-bearing soil disaster, timely and effective life detection is the key to successful rescue. Firstly, life detection equipment can quickly determine the location of trapped people and reduce the search time; secondly, through advanced sensors and algorithms, it can improve the accuracy of detection and avoid misjudgment; thirdly, it can reduce the number of times rescuers enter high-risk areas and improve the efficiency of rescue, while avoiding injury to rescuers at the disaster site. According to the detection principle, water-bearing soil disaster life detection equipment mainly includes acoustic wave detection equipment[5-9], radar detection equipment[10], infrared detection equipment[11-12] and video detection equipment[13] four types. Different types of equipment detection principle is different, but their test requirements are the same, can be used in common test methods and test procedures for testing.

2.1 Requirements for tester

The testers should be familiar with the various test equipment and measuring instruments before the test, and should know the various test techniques and test contents, as well as emergency disposal measures. At the same time, testers should receive systematic pre-test training and be able to perform the test in accordance with the test operating procedures.

* Corresponding author: 2220901019@cun.edu.cn, a 01107203nyy@163.com

2.2 Requirements for test environment

The design and testing of water-bearing soil disasters life detection equipment must take into account complex and changing geological conditions, operating conditions, possible extreme weather conditions and other factors. The laboratory test environment is one of the most important steps in ensuring that the equipment will work effectively in actual rescue operations, and the test site is an important part of determining whether the equipment can meet the requirements of actual rescue operations.

2.2.1 Laboratory test

Laboratory tests are mainly routine tests and environmental adaptability tests of equipment. Routine testing requires appropriate data measurement and data acquisition equipment. The environmental adaptability test is carried out by simulating the environment, adjusting the temperature and humidity levels in the laboratory, and simulating the temperature and humidity changes during different seasons, which is crucial for evaluating the performance of the equipment. In addition, the equipment operates in a complex environment and it is also necessary to simulate extreme weather conditions (e.g. heavy rain, high temperature, low temperature, etc.) to test the durability and stability of the equipment, as well as to test the efficiency and accuracy of the equipment by simulating a variety of interference sources (e.g. electromagnetic interference, terrain obstacles, etc.).

2.2.2 Site test

The main purpose of the test site is to test the performance of the equipment, i.e. to test the functions of the life detection equipment, such as recognition and detection capability. The test site can be a real scene or a simulated similar scene. The real scene refers to the scene where the disaster occurs or the scene where the water-bearing soil disasters has occurred and been preserved, and this kind of scene is the best test site for the life detection equipment of water-bearing soil disasters. Similar scenes are disaster scenes or disaster platforms similar to the actual situation built according to the study of water-bearing soil disasters cases, and such scenes should match the real situation in terms of soil moisture content setting, obstacle setting, calibration depth and position setting, and the performance and reliability of life detection equipment in complex environments can be comprehensively evaluated by setting up a reasonable test environment. This not only helps to improve the level of equipment research and development, but also provides strong technical support for actual rescue work.

In addition, the test site should be equipped with surveillance cameras and safety warning signs to ensure the safety of the test personnel, and at the same time ensure that the test site is isolated from the outside world to prevent the entry of unrelated personnel.

3 Test Content

In order to ensure the effectiveness of life detection equipment in actual rescue operations, it must be rigorously tested. Testing typically includes:

(1) Appearance and structure test. check the visual quality and structural integrity of the equipment.

(2) Function test. Evaluating the equipment's detection direction, detection precision and detection accuracy.

(3) Environmental adaptability test. Test the equipment's performance in extreme environments such as high and low temperature, humidity and heat, vibration and shock.

(4) Reliability test. Verify the long-term performance and reliability of the equipment.

4 Test methods

Appearance and structural tests can be carried out in the laboratory or in the field; functional, environmental and reliability tests are usually carried out in the field. The high and low temperature test, constant humidity and heat test, vibration test, shock test, protection level test, and temperature test in the environmental adaptability test and reliability test are mainly based on the applicable standards; the appearance and structure inspection is conducted with the goal of equipment integrity; the test conditions of the functional test are based on the basic test conditions of the national key research and development project on the evaluation requirements for the equipment of water-bearing soil disaster. According to the different working conditions and test items, the test procedure can be adjusted accordingly, and the specific test conditions are shown in Table 1.

4.1 Appearance and structure test

Appearance and structure inspection usually adopts the method of visual inspection, mainly from the following aspects to check the surface coating and product marking.

(1) The appearance of the surface of the test equipment should be neat and clean, without obvious external mechanical damage, coating cracks, deformation, corrosion and other phenomena.

(2) The text, symbols and signs on the surface of the test equipment should be clear, firm, without damage, wrinkles, falling off and other phenomena.

Structural inspection is usually combined with a visual manual adjustment method, if necessary, tools can be used, mainly to check whether the structure is reasonable, there is no mechanical damage and rotating stagnation and so on.

Portable equipment preparation should also check the size and weight of the test sample.

4.2 Function test

Functional tests using practical methods, which may be carried out simultaneously with visual and structural inspections, or using specially designed test items. Detection capability, detection precision and detection accuracy shall be tested in accordance with the following requirements:

(1) Detection capability is usually tested by the ability of the detection equipment to detect orientation (single orientation, multiple orientation) and depth;

(2) The detection accuracy is usually tested by the detection accuracy of soil with different water content and different burial depth;

(3) Detection accuracy is tested by the detection accuracy of soils with different water contents and different burial depths.

Table 1. Test conditions of life detection equipment for water-bearing soil disasters.

Test name	Test parameter	Test condition	Test mode
Visual and structural inspection	Surface	Laboratory or field test	Off working state
	Structure	Laboratory or field test	
Function testing	Probe direction	Multiaspect	Operative mode
	Detection accuracy	When the moisture content is not more than 20% and the buried pressure depth is not more than 30m, the positioning accuracy is better than 2m	
	Detection accuracy	≥60%	
Environmental adaptability testing	High temperature test	According to the GB/T2423.1-2008	Off working state
	Low temperature test	According to the GB/T2423.1-2008	
	Constant hot and humid test	According to the GB/T2423.3-2016	
	Vibration test	According to the GB / Z26139	
	Impact test	According to the GB / T 2423.5-2019	
	Protection grade test	According to GB / T 4208-2017	
Reliability test	Temperature	According to the GB / T 5080.7-1986	

4.3 Environmental adaptability test

Environmental testing usually includes climatic and mechanical tests. Climatic environment mainly includes low temperature, high temperature, constant humidity and heat, rain and dust; mechanical environment test mainly includes vibration and shock.

4.3.1 Climatic environment test

Climatic environment test should be divided into working environment test and storage environment test. Test equipment environmental adaptability test items in accordance with the technical requirements to be determined, if the technical requirements are not clearly specified, for field use of the test equipment or its field installation part, low temperature, high temperature, constant humidity and heat and packaging and transportation test is a mandatory test items, if necessary, can increase the alternating humidity and heat test. Regardless of whether the technical specifications are specified or not, the cast part of the high-altitude detection instrument shall be tested for low temperature and low air pressure; the instrument used in water-bearing soil and in the environment shall be tested for watertightness and dustproofness. The order of the tests shall be determined according to the actual situation. If it is necessary to get the conclusion of whether it is qualified or not as soon as possible, the severe test items should be placed at the front; if it is necessary to get as much information as possible about the adaptability of the tested equipment to various environments, the severe test items should be placed at the back. The climatic adaptability test should be carried out first under working conditions and then under storage conditions. The environmental adaptability test item should give priority to the use of relevant national/industry standards for testing, and determine the test method in conjunction with the actual situation of the environmental conditions under which the equipment under test will be used.

(1) IP level test

IP level test is mainly based on GB/T 4208-2017 "Shell protection level (IP code)". ip6x (solid foreign matter protection test) test in the dust box, the air pressure in the shell during the normal working cycle is lower than the surrounding atmospheric pressure for the test, observe the sample shell dust deposition. ipx5 test, using the standard requirements of the anti-spraying test device in all possible direction to test the sample shell water spray, observe the water in the shell.

(2) Temperature and humidity test

Temperature and humidity test is mainly based on "GB/T2423.1-2008 Environmental Tests for Electrical and Electronic Products Part 2: Test Method Test A": low temperature, "GB/T2423.2-2008 Environmental Tests for Electrical and Electronic Products Part 2: Test Method Test B: high temperature", "GB/T2423.3-2016 Environmental Tests Part 2 Test Method Test Cab: Constant Humidity and Heat'. Damp heat' is performed. The low temperature test is conducted under the condition of -40°C, and the specimens are connected to a laptop computer and turned on for 2 hours, and the specimens are observed to upload data. High temperature test at 55°C, the sample external laptop turned on for 2h, observe the sample upload data. Damp heat test at 40°C, 93% RH conditions, the sample external laptop power work for 2h, observe the sample upload data.

4.3.2 Mechanical environment test

(1) Vibration test

The vibration test is mainly based on "GB/T 2423.56-2018 Environmental Test Part 2: Test Methods Test Fh: Broadband Random Vibration and Guidelines". Under the conditions of different vibration frequencies, the sample is turned on, the external laptop is connected, and the X, Y and Z axes work for 1 hour each, and the data received in the external laptop is observed.

(2) Impact test

The impact test is mainly based on "GB/T 2423.5-2019 Environmental Testing Part 2: Test Methods Test Ea and Guidelines: Impact". The specimen is energised, the external laptop is connected and continuous pulses are applied in the positive direction along three perpendicular axes and the data received in the external laptop is observed.

4.4 Reliability test

The reliability test is mainly conducted according to 'GB/T 5080.7-1986 Equipment Reliability Test Verification Test Programme for Failure Rate and Mean Time to Failure under the Assumption of Constant Failure Rate'. Combine actual operating conditions, apply temperature and humidity stress, set cycle steps and count the number of responsible failures of the equipment during the cumulative test time. Analyse the performance of the equipment in extreme environments, such as stability at high temperature, low temperature and humidity.

5 Test process

Equipment testing is a systematic effort that can be divided into four phases: preparation phase, execution phase, data collection and analysis, and test report generation, as shown in Fig. 1.

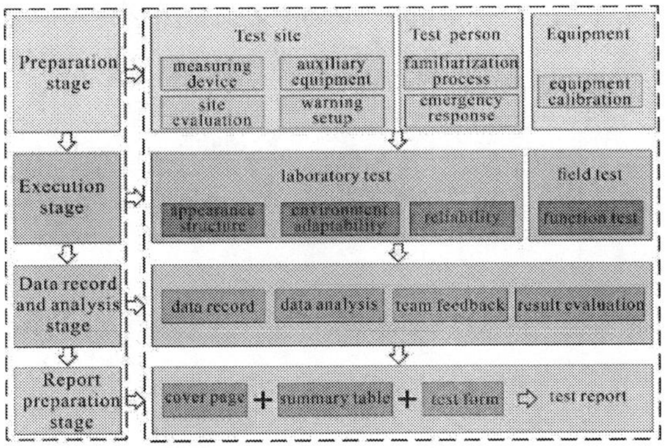

Fig. 1. Flowchart of life detection equipment test for disasters in water-bearing soils.

(1) Preparation stage

The preparation stage mainly involves the preparation of personnel, site and test equipment. The test personnel and environment should meet the test requirements. Before the equipment test, the test site should be configured for the measurement of standard equipment (such as thermometers, hygrometers, vibration tables, impact testers, etc.), auxiliary equipment (simulated mannequins, soil moisture content testers, etc.) and the equipment under test for calibration to ensure the accuracy of measurement data.

(2) Execution stage

Set up the appropriate environmental conditions according to the test plan, and measure the test content according to the test method. Functional test in the test point placed in the simulation doll, set up different depth and shape of burial pressure, depth and shape and the real occurrence of scenes similar to the situation of people buried in the pressure. After the setup is completed, start the life detection equipment to start the test.

(3) Data Recording and Analysis

Record the data of each test in detail, including but not limited to environmental conditions, signal strength, response time, positioning error, etc. Rescue personnel can be invited to participate in the testing process to collect their feedback on the use of the equipment, thus facilitating subsequent improvement while assessing the applicability of the equipment. If all test items in the test content pass the test, the equipment is assessed as applicable, otherwise it is assessed as not applicable.

(4) Report preparation

The test report consists of three parts: the cover page, the general table and the test table, which are numbered consecutively from the general table to X of X, located at the top right-hand corner of the page.

The cover page is suitable for printing on hard cover paper as a general text, including the front and back. The general table contains the file number, test unit name, address, contact department, person in charge, telephone number, postal code, test items, test time, test unit (official seal), issuer and the basic information of the test unit. Among them, the test unit address to write the test unit headquarters address, test items have more than one address should be filled in the test table; test item list of project names, should be relative to the name of the items in its post-test table; the issuer should be signed by a black pen or carbon table; test unit (seal) column should be stamped with the official seal of the statutory test unit, should not be stamped with the test special seal, classification test table of the technical assessment columns stamped with test special seal. The test form is divided into four categories according to the test content: appearance and institutional inspection test form, function test form, environmental adaptability test form and reliability test form, without file number and test unit information, and the date of the test special seal (item) is the test time of the item.

6 Conclusions

This paper proposes a systematic test method and standardised test procedure to address the problems encountered in the testing of life detection equipment for water-bearing soil disasters. By testing the appearance and structure, function, environmental adaptability and reliability of the equipment, it can not only comprehensively evaluate the performance of the equipment, but also help to discover potential problems, thus providing a scientific basis for the improvement and optimisation of the equipment.

(1) In terms of testing requirements, this paper emphasises the importance of professional training and safety awareness of testers to ensure a smooth testing process. At the same time, it describes the requirements for the construction of the laboratory test environment, including the control of temperature and humidity, the simulation of extreme weather conditions, and the establishment of multiple sources of interference, all of which help to more realistically reflect the performance of the equipment in the actual disaster environment.

(2) In terms of test content, this paper covers multiple dimensions such as visual and structural inspection, functional test, environmental adaptability test and reliability test, aiming at a comprehensive evaluation of the overall performance of the equipment. In particular, the detection direction, detection precision and detection accuracy in the functional test are explained in detail, and these indicators are crucial to improving the effectiveness and reliability of life detection equipment.

(3) Environmental adaptability test and reliability test are the key to evaluate the stable operation of the equipment under harsh environmental conditions. This paper proposes a systematic and standardised test method to ensure the authority and credibility of the test results.

(4) Finally, this paper proposes a complete testing process including four steps: preparation stage, execution stage, data recording and analysis, and report preparation, which provides a standard operating guideline for the testing of life detection equipment. Through this series of testing procedures, the reliability and safety of the equipment in practical use can be effectively ensured, while at the same time providing valuable experience and reference for future equipment development.

In conclusion, the testing methods and procedures proposed in this paper are of great practical significance for improving the practical applicability and rescue effectiveness of life detection equipment for water-bearing soil disasters. This testing procedure applies to the suitability testing of all equipment, and functional testing can be slightly adjusted according to the specific requirements of different equipment.

Acknowledgments

National Key Research and Development Project (Research on comprehensive Efficiency Testing Platform of Life Search and Rescue Equipments for Water-bearing Soil Disasters (2021YFC3090405) provides funds support for this paper.

References

1. Xiang Y., Jiang F. (2016) A Review of Testability Verification Technology of Equipment, 34(2): 65-70.
2. Wang J.P., Zhang Y.C., Zhu X.R., et al. (2019) High-cold Environment Rescue Equipment Testing and Research. China Emergency Rescue, (4): 34-40.
3. Wang B.S. (2020) Research on Test Method of Anti-EMP for Radar Countermeasure Equipment. Ship Electronic Engineering, (309): 160-163.
4. Wei, Z. X. (2021) Challenge and Countermeasure of Intelligent Equipment Experiment and Test. Measurement & Coutrol Technology, 40(2): 1-12.
5. Hou P.G., Li N., Song T. (2014) Research status and development of life detection technology. Transducer and Microsystem Technologies, 33(7): 1-8.
6. Kim H.G., Kim J.Y. (2017) Acoustic event detection in multichannel audio using gated recurrent neural networks with High-resolution spectral features.Etri J, 39(6): 832-840.
7. Emre C., Giambattista P., Toni H., et al. (2017) Convolutional recurrent neural networks for polyphonic sound event detection. IEEE-Acmt Audio Spe, 25(6): 1291-1303.
8. Wan T., Zhou Y., Ma Y., et al. (2019) Noise robust sound event detection using deep learning and audio enhancement. 2019 IEEE International Symposium on Signal Processing and Information Technology (ISSPIT). IEEE, 1-5.
9. Romanov S.A., Kharkovchuk N.A., Sinelnikov M.R., et al. (2020) Development of an non-speech audio event detection system. 2020 IEEE Conference of Russian Young Researchers in Electrical and Electronic Engineering (EIСon Rus). IEEE, 1421-1423.
10. Lu G.H., Wang J.Q., Yang G.S., et al. (2004) Current state of the technology of bioradar. Biomedical Engineering Foreign Medical Sciences, 27(6): 368-383.
11. Sun G., Nakayama Y., Dagdanpurev S., et al. (2017) Remote sensing of multiple vital signs using a CMOS camera-equipped infrared thermography system and its clinical application in rapidly screening patients with suspected infectious diseases. Int J Infect Dis, 55: 113-117.
12. Dong X., Martin J., Wang S., et al. (2018) A review of hyperspectral imaging for nanoscale materials research. Appl Spectrosc Rev, 54(4): 285-305.
13. Persaud D. (1982) Analysis of discrimination mechanisms in the mammalian olfactory system using a model nose. Nature, 352-355.

Spatiotemporal Patterns and Influencing Factors of Ecosystem Services in Shandong Province

Liangjie Guo[1], Junyue Wang[2], Yongbo Zhai[2], Yanlei Zuo[3], Wei Liu[1,*]

[1]College of Geography and Environment, Shandong Normal University, Jinan, Shandong, China
[2] Shandong Institute of Territorial and Spatial Planning, Jinan, Shandong, China
[3] Jinan Institute of Surveying and Mapping, Jinan, Shandong, China

Abstract. Ecosystem services (ESs) are essential for human survival and development and serve as key indicators in evaluating ecological civilization progress. This study investigated the spatiotemporal patterns of ESs in Shandong Province from 2000 to 2020, using the Geographical Detector model to assess the impact of diverse influencing factors. The results showd that: (1) From 2000 to 2020, grain production (GP), soil conservation (SC), and carbon sequestration (CS) showed a fluctuating increase of 39.92%, 106.88%, and 32.21%, respectively. Water yield (WY) remained relatively stable, while habitat quality (HQ) declined rapidly after 2010. (2) Land use types and elevation demonstrated high explanatory power regarding the spatial heterogeneity of each ES, with population density also having a significant impact on all services except SC. Precipitation emerged as the strongest explanatory factor for water conservation. Furthermore, all pairwise interactions between factors exhibited a bi-factor enhancement in 2020, indicating a substantial role of multi-factor interactions in ES dynamics. This study provides essential insights to support decision-making, addressing the challenges ESs face and promoting the sustainable, high-quality development of Shandong's socio-ecological-economic system.

1 Introduction

Ecosystem Services (ESs) encompass all benefits that ecosystems provide to humans, serving as a critical indicator of regional environmental quality with profound implications for the sustainable development of human society[1, 2]. Different types of ESs are influenced by a range of natural and socio-economic factors, and studying the spatiotemporal changes, drivers, and interrelationships of ESs is a key focus in contemporary ecological research. Such studies are essential for improving ecosystem sustainability management, providing a foundation to further refine ESs spatiotemporal dynamics theory, and advancing ecological civilization and human well-being[3].

Research on ESs has made significant progress, with scholars constructing evaluation index systems for ESs from multiple perspectives and employing a range of methods to explore their spatiotemporal variations in depth[4]. Despite considerable progress, understanding the temporal dynamics and interactive effects among influencing factors remains incomplete. This study addresses these limitations by extending the analysis period to 2000–2020, enabling a more detailed examination of ESs changes in Shandong Province. Furthermore, it applies the Geodetector model to quantitatively assess how multiple factors interact to shape the spatial distribution of ESs.

Shandong Province is a key component of economic development in both the Bohai Economic Rim and the Yellow River economic belt. As a significant industrial hub and a strategic anchor for economic growth in northern China, Shandong boasts a well-developed economy, dense population, and diverse topography. Conducting a spatiotemporal analysis of ESs and their influencing factors in this region is crucial for balancing economic development with environmental sustainability and accelerating green, low-carbon, high-quality development[5]. Based on the assessment and analysis of key ESs, this study uses Geodetector analysis to clarify the primary influencing factors. The focus is on examining the spatial distribution and variation patterns of Shandong's ESs under different driving factors, ultimately providing foundational data for decision-making aimed at promoting high-quality development and improving human well-being in the province.

2 Study area and Materials

2.1 Overview of the Study Area

Shandong Province, situated between latitudes 34°23′–38°17′N and longitudes 114°47′–122°42′E, serves as an essential economic hub within both the Bohai Economic Rim and the Yellow River Economic Belt. Bordered by the Bohai Sea to the north and the Yellow Sea to the east,

* Corresponding author: liuw_simlab@163.com

Shandong's topography is predominantly mountainous and hilly, interspersed with plains and basins. The province's central area is marked by high-elevation mountains, while low, flat plains define the southwest and northwest regions, and gently rolling hills characterize the eastern landscape (Fig. 1).

Covering a total area of approximately 157,965 km², Shandong is a major agricultural province where over 70% of the land is used for agriculture. Of this agricultural land, cultivated land comprises the largest portion, representing around 65% as of 2023.Shandong's climate is warm temperate monsoonal, with hot, rainy summers and cold, dry winters. The region experiences an average annual temperature of 11–14°C and an average annual precipitation of 680.5 mm, supporting diverse ecosystems. In 2023, Shandong Province's GDP was about 9,206.9 billion yuan, with a permanent population of 101.23 million, ranking fifth in the country in terms of population density.

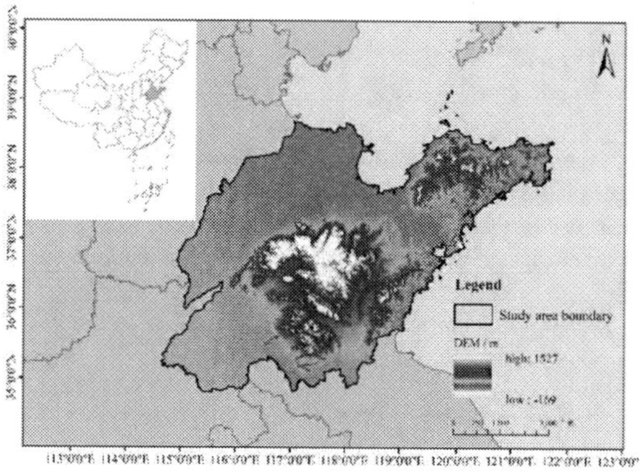

Fig. 1. Location and DEM Spatial Distribution in Shandong Province.

2.2 Data source and preprocessing

This study utilizes a range of datasets, including land use, the Normalized Difference Vegetation Index (NDVI), meteorological data, soil characteristics, digital elevation model (DEM), and socioeconomic indicators. Land use data from 2000 to 2020, alongside population and GDP statistics, were sourced from the Resource and Environmental Science Data Center. NDVI values were derived from specific bands of Landsat 5/7/8 satellite imagery, while meteorological data were collected from the National Meteorological Science Data Center. Soil attributes were obtained from the National Tibetan Plateau Data Center, and DEM data were drawn from the Geospatial Data Cloud's ASTER GDEM dataset. All datasets were resampled to a 1 km × 1 km spatial resolution using methods such as spatial interpolation and nearest-neighbor allocation to ensure consistency across data layers.

3 Research Methods

3.1 Assessment of ecosystem services

Based on Shandong Province's positioning as a major agricultural and economic province, as well as its natural environmental characteristics, this study selects grain production (GP), soil conservation (SC), water yield (WY), carbon sequestration (CS), and habitat quality (HQ) as typical ESs[6]. These are evaluated using remote sensing inversion methods (vegetation index), the RULSE model, the water balance equation, the CASA model, and the HQ module of the InVEST model (Table 1)[7].

Table 1. Ecosystem Services Assessment Methods.

Types	Methods	Source
Grain production	$GP_i=\frac{NDVI_i}{\sum_{i=1}^{n} NDVI_i}\times GP_{SUM}$	[7]
Soil conservation	$A_c=R\times K\times LS\times(1-C\times P)$	[7]
Water yield	$WY=P-ET$	[7]
Carbon sequestration	$W_{CO_2}=NPP\times 2.2\times 1.63$	[8]
Habitat quality	$Q_{xj}=H_j\left[1-\left(\frac{D_{xj}^z}{D_{xj}^z-k^z}\right)\right]$	[9]

3.2 Ecosystem Services Driving Factors

To analyze the driving factors of the five ESs in Shandong Province, this study used sampling points at 6 km intervals to extract data on various influencing factors. Using geographic detector analysis, this study revealed the impact of each factor on the spatial differentiation of ESs. This model did not require linear assumptions or constraints and quantified interactions and spatial correlations between independent variables through factor and interaction detectors.

First, based on research and regional characteristics of Shandong Province, this study constructed an indicator system for key ES driving factors (Table 2). The geographic detector model in the "GD" package in R was applied to discretize and categorize factors[10], allowing for precise quantification of their influence. The factor detector then calculated the q-value to measure the independent contribution of each factor to ESs, as shown in Equation (1):

$$q = 1 - \frac{\sum_{h=1}^{L} N_h \sigma_h^2}{N\sigma^2} \quad (1)$$

In this equation, h represents the number of categories for variable Y or influencing factor X within the study area; N and Nh represent the number of observational units in the entire study area and in category h, respectively; σ^2 and σ_h^2 denote the variance of the dependent variable in the entire study area and within category h, respectively. The q-value ranges from 0 to 1, where a higher q-value indicates a stronger explanatory power of the influencing factor X_i on the dependent variable Y, and a lower q-value indicates a weaker explanatory power.

Table 2. Indicator System for ESs Driving Factors.

Numbers	Types	Driving Factors	Codes
1	Natural environmental factors	DEM/ m	X_1
2		Slope (SLO)	X_2
3		Rainfall / mm	X_3
4		Temperature (TEM)/°C	X_4
5		NDVI	X_5
6		Land use type (LU)	X_6
7	Socio-economic factors	Population density (POP)	X_7
8		GDP	X_8

Use the interaction detector to assess the mutual influence among different driving factors, determining whether they act independently or exhibit synergistic effects. The types of interactions are shown in Table 3.

Table 3. Types of interactions and basis of judgment.

Interaction type	Basis of judgment
Non-linear reduction	$q\left(X_1 \bigcap X_2\right) < Min[q(X_1), q(X_2)]$
Single-factor non-linear reduction	$Min[q(X_1), q(X_2)] < q\left(X_1 \bigcap X_2\right) < Max[q(X_1), q(X_2)]$
Bi-factor enhancement	$q\left(X_1 \bigcap X_2\right) > Max[q(X_1), q(X_2)]$
Independent	$q\left(X_1 \bigcap X_2\right) = q(X_1) + q(X_2)$
Non-linear enhancement	$q\left(X_1 \bigcap X_2\right) > q(X_1) + q(X_2)$

4 Results and Analysis

4.1 Spatial and temporal dynamics of ecosystem services

4.1.1 Grain production

The GP in Shandong Province showed a continuous increasing trend, rising from 3.89×10[7] tons in 2000 to 5.44×10[7] tons in 2020. The most significant increase occurred between 2005 and 2010, with production rising from 4.23×10[7] tons to 5.39×10[7] tons, a growth rate of 27.54% (Fig. 2). Spatially, the northwest and southwest regions were areas of high GP, while the northern and eastern coastal areas were characterized by lower production levels (Fig. 3).

4.1.2 Soil conservation

From 2000 to 2020, the SC in Shandong Province exhibited significant interannual variations, but overall showed a fluctuating increasing trend, rising from 1.55×10^8 tons to 3.27×10^8 tons (Fig. 2). In terms of spatial distribution, the central and northeastern mountainous regions had higher SC capacity, while the surrounding peripheral areas exhibited relatively lower capacity. This was mainly related to human activities and topography. The mountainous areas had larger slope and slope length factors, and the land use was primarily composed of arable land, forests, and grasslands with good vegetation coverage, leading to higher SC service levels. In contrast, the plain areas, with their relatively flat terrain and land use dominated by construction land, had lower potential soil erosion and therefore lower SC service capacity (Fig. 3).

4.1.3 Water yield

The total WY in Shandong Province ranged from 1.14×10^{11} m³ to 1.56×10^{11} m³, with minimal interannual variation. The spatial distribution characteristics of WY did not exhibit significant regularity due to the combination of precipitation and evapotranspiration intensity under different spatial distribution patterns. In 2000, WY in Shandong Province showed a spatial pattern of gradual decrease from the southeast toward the north. By 2005, the total WY increased, with values decreasing from the southeastern coastal areas to the northwest. In 2010, WY slightly declined, displaying higher values in the central and northeastern regions that diminished outward. In 2015, a notable decrease occurred, with a distribution pattern that decreased progressively from the central north-south axis toward both sides. By 2020, the spatial trend returned to a gradual decline from the southeast coast to the northwest (Fig. 3).

4.1.4 Carbon sequestration

From 2000 to 2020, the CS in Shandong Province showed a fluctuating increasing trend, rising from 2.53×10^8 tons to 3.34×10^8 tons (Fig. 2). Regions with higher CS capacity were mainly located in the low-altitude plain areas (Fig. 3), where temperatures were suitable and vegetation grew well. In contrast, the central and northeastern mountainous regions, with high altitudes, had lower CS capacity. By 2020, the areas with higher CS capacity were primarily concentrated in the western part of Shandong Province.

4.1.5 Habitat quality

During the study period, HQ showed a declining trend. From 2000 to 2010, the HQ index in Shandong Province remained relatively stable between 0.53 and 0.54. However, after 2010, due to the expansion of construction land and road development, HQ significantly decreased, dropping from 0.53 in 2010 to around 0.08 in 2015 and 2020 (Fig. 2). Spatially, HQ was highest in the central and southern Shandong hills and the Jiaodong low mountain regions, while lower HQ values were observed in the northern areas (Fig. 3).

 https://doi.org/10.1051/e3sconf/202561701016

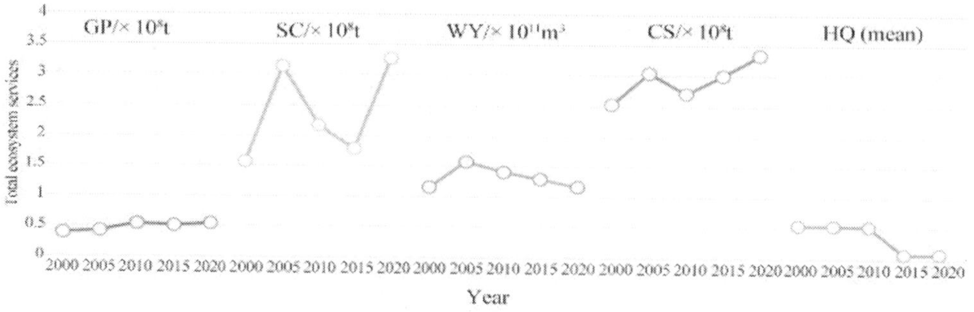

Fig. 2. Changes in Ecosystem Services in Shandong Province from 2000 to 2020.

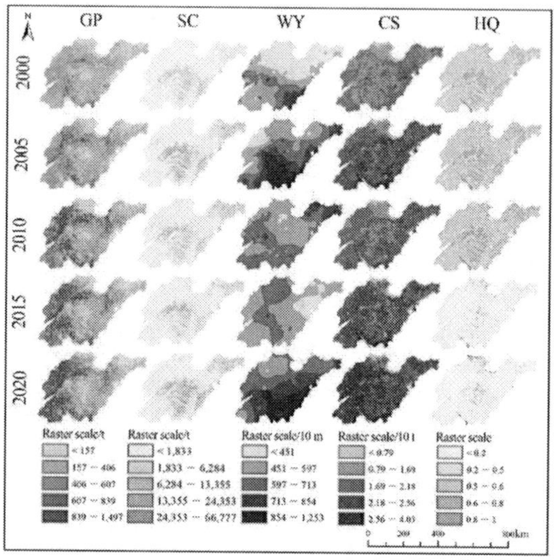

Fig. 3. Spatial Distribution of Ecosystem Services in Shandong Province from 2000 to 2020.

4.2 Ecosystem Service Driving Factor Detection

Natural environmental and socio-economic factors were selected as primary drivers to analyze the spatiotemporal variability of ESs. Through the application of the GeoDetector model, the influence of each factor on the spatial heterogeneity of GP, SC, WY, CS, and HQ in Shandong Province was effectively quantified (Fig. 4).

From the spatial distribution perspective, the results of the factor detector indicated that land use type, elevation, and population density had the most significant explanatory power for GP. Regarding SC, elevation, land use type, and slope had the greatest explanatory influence, with elevation being the dominant factor, showing a q-value of 0.47. This underscored the pivotal role of topography in shaping both soil erosion and conservation processes. For WY, precipitation emerged as the primary driver, with a q-value near 0.97, explaining almost the entirety of this ecosystem service's spatial variation. Elevation also contributed significantly to explaining WY. In terms of CS and oxygen release, the NDVI and land use type showed the highest explanatory power, with NDVI being the dominant factor and a maximum q-value of 0.45.For HQ, land use type and elevation had the highest explanatory power, though the contributions of all influencing factors to this service remained relatively modest.

In terms of temporal variation, from 2000 to 2020, the explanatory power of each driving factor for GP and HQ showed varying degrees of increase. Specifically, the q-value of land use type for GP rose from 0.15 to 0.56, and for HQ, it increased from 0.08 to 0.12. The explanatory power of the driving factors for other ESs remained relatively stable over time.

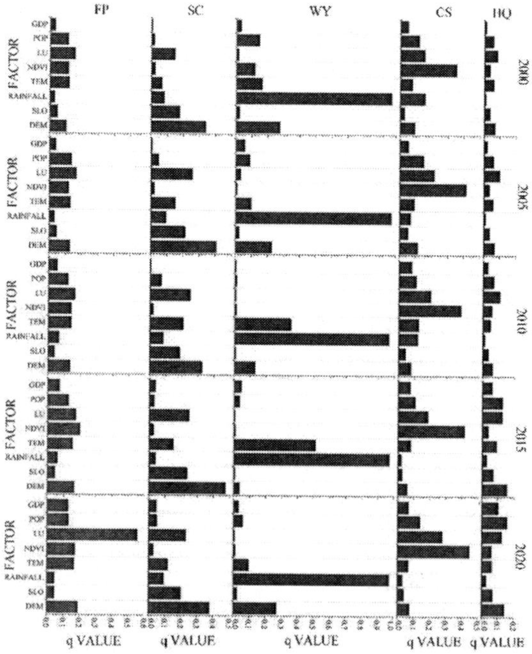

Fig. 4. Results of ESs Driving Factors in Shandong Province from 2000 to 2020.

4.3 Interactive detection analysis

This study, through the interaction detection of factors for ESs in 2020, revealed that the interaction between any two factors provided a greater explanatory power for ESs than individual factors (Fig. 5), demonstrating a bi-factor enhancement effect.

For GP, the main interacting factors influencing spatial variation were land use type combined with temperature (q value=0.73), and land use type combined with elevation (q value=0.70). For SC, the key interaction pairs were elevation and precipitation (q value=0.44), as well as elevation and land use type (q value= 0.42). For WY, precipitation interacted significantly with other influencing factors. For CS, the interaction between the NDVI and land use type emerged as the most influential factor (q value= 0.55). For HQ, elevation and population density were the primary interacting factors affecting spatial variability (q value= 0.29).

Overall, these inter-factor interactions significantly increased the explanatory power of models predicting the spatial distribution of ESs, underscoring the complex, multifaceted nature of ecosystem dynamics.

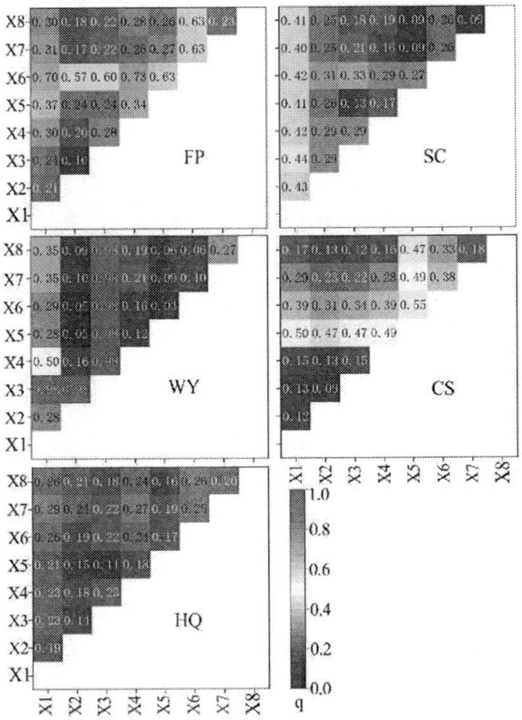

Fig. 5. Comparison of Interactive detection analysis Results of ESs Impact Factors in Shandong Province in 2020.

5 Conclusions

This study quantitatively assessed the five ESs (GP, SC, WY, CS, and HQ) in Shandong Province from 2000 to 2020 and used geographic detectors to quantify the contributions of natural and socio-economic factors to their spatial heterogeneity. The results indicated that:

(1) From 2000 to 2020, GP, SC, and CS increased by 39.9%, 110.7%, and 32.2%, respectively, while WY remained stable and HQ sharply declined after 2010. Spatially, GP had high values in western Shandong, SC in central and northeastern hills, and WY, CS, and HQ showed less variation.

(2) Land use type and elevation were strong explanatory factors for ES spatial heterogeneity. Population density also strongly influenced all services except SC, while precipitation was the primary factor for WY. In 2020, combined factors had greater explanatory power for ESs than individual factors.

Future research should focus on exploring multi-factor interactions driving dynamic changes in ESs and the mechanisms behind them using high-precision data. This will enhance regional planning and management, provide more precise evidence for the collaborative optimization and sustainable development of ESs, and better address their challenges.

Acknowledgments

Shandong Provincial Natural Science Foundation of China(ZR2024QD114); 2024 Experimental Teaching Reform Research Project of Shandong Normal University(2024MS22); Jinan City-School Integration Project(JNSX2023036).

References

1. Costanza, R. d'Arge, R. de Groot, R. The Value of the World's Ecosystem Services and Natural Capital[J]. Ecol Econ, 1998, 25(1): 3-15.
2. Chen Y.Q. Liu W. Zhao F. et al. (2024) Multi-Scale Analysis of Ecosystem Service Trade-Offs/Synergies in the Yangtze River Delta. Land, 13(9): 1462. DOI: 10.3390/LAND13091462
3. Li, H.R. Ma, S. Yin, Y.K. et al. (2024) Analysis of Spatiotemporal Changes and Influencing Factors of Ecosystem Service Multifunctionality: A Case Study of the Qinghai-Tibet Plateau Ecological Barrier Zone. Journal of Ecology and Rural Environment, 40: 1038-1046. DOI: 10.19741/j.issn.1673-4831.2023.0426
4. Liao, Z.Q. Fan, Y.T. Wang, J.X. et al. (2024) Spatiotemporal Evolution of Ecosystem Service Functions and Driving Mechanisms in the Yangtze River Delta. Environmental Science, 1-19. DOI: 10.13227/j.hjkx.202405081
5. Ren, Q.R. Liu, D.D. Liu, Y.F. et al. Spatio-temporal dynamics and socio-ecological determinants of ecosystem service interplays in Shandong Province's coastal region (2000–2020): Implications for environmental protection and sustainable ecosystem management[J]. Environ Res, 2024, 243: 117824. DOI: 10.1016/J.ENVRES. 2023.117824
6. Liu, W. Zhan, J.Y. Zhao, F. et al. Spatio-temporal variations of ecosystem services and their drivers in the Pearl River Delta, China[J]. J Clean Prod, 2022, 337: 130466. DOI: 10.1016/J.JCLEPRO. 2022. 130466
7. Liu, W. Zhang, F. Wei, Y.H. et al. (2023) The relationship between supply and demand of ecosystem services in the Pearl River Delta urban

agglomeration. Acta Ecologica Sinica, 43: 4461-4472. DOI: 10.5846/stxb202112203611

8. Wen, Y.Y. Sun, Q. Yan, Y.C. et al. (2020) Impacts of the terrestrial ecosystem changes on the carbon fixation and oxygen release services in the Guangdong-Hong Kong-Macao Greater Bay Area. Acta Ecologica Sinica, 40: 8482-8493. DOI: 10． 5846 / stxb202003310765
9. Sallustio, L. De Toni, A. Strollo, A. et al. (2017) Assessing habitat quality in relation to the spatial distribution of protected areas in Italy. J Environ Manage, 201: 129-137. https: //doi.org/10.1016/ j.jenvman.2017.06.031
10. Wang, J.Q. Xing, Y.Q. Chang, X.Q. et al. (2024) Spatial distribution of ecosystem services and analysis of driving factors in Northeast China. Environmental Science, 45: 5385-5394. DOI: 10.13227/j.hjkx.202311022

Study on the Mineral Characteristics and Upgrading of Titanium Iron Ore Concentrate in the Panxi Region of China

Guoyin Xu[a*], Purong Wang[b]

Panzhihua University, Panzhihua 617000, Sichuan, China

Abstract. By conducting sieve analysis and MLA process mineralogy automated analysis system research on a titanium iron ore concentrate sample from the Panxi region, this study found that the main metallic mineral in the Anning titanium concentrate is ilmenite, which is the primary target mineral for recovery in this study. In addition, there are minor metallic minerals such as titanomagnetite and hematite, as well as gangue minerals including chlorite, hornblende, sphene, and diopside. Ilmenite accounts for 89.42% of the total mineral mass, and its liberation degree has reached 95.26%, indicating that ilmenite has essentially achieved liberation. The sample contains 47.61% TiO_2, 2.63% SiO_2, 1.05% CaO, and 2.84% MgO. The particle size of the sample is mainly below -200 mesh, accounting for 73.75%, while the fine particle grade with a higher grade is -325 mesh, accounting for 28.95%. SiO_2 and TiO_2 are mainly concentrated in the -200 mesh particle size grade, while gangue minerals rich in Ca, Si, and Mg are mainly chlorite, titanite, sphene, and other silicate minerals. Gangue minerals associated with ilmenite include chlorite, sphene, and titanite, while those enclosed by ilmenite are mainly composed of chlorite, hornblende, sphene, and titanite. These gangue minerals, due to their low intergrowth particle size, are difficult to remove by mineral processing methods. Gangue minerals containing Ca, Mg, and Si have low intergrowth particle sizes and complex associations with ilmenite, making them difficult to completely remove.In the flotation tests, this study employed an open circuit flotation test process of "one roughing, two scavenging, and middling separate treatment". In the roughing stage, the dosage of sulfuric acid was 3000g/t, and the dosage of FS+CS was 1500+1500g/t, while the collector YS was used at 500g/t. Through this process, the study obtained two types of titanium concentrates: one with a yield of 36.72%, TiO_2 content of 49.62%, SiO_2 content of 0.95%, and TiO_2 recovery rate of 38.24%, which is a low-silica titanium concentrate; the other with a yield of 63.28%, TiO_2 content of 46.51%, SiO_2 content of 3.52%, and TiO_2 recovery rate of 61.76%, which is a regular titanium concentrate. The high-quality titanium concentrate obtained meets the requirements for silicon, calcium, and magnesium in the chlorinated titanium dioxide process, while the flotation tailings can be used as regular titanium concentrate.

1 Introduction

The Panxi region, as an important titanium resource-rich area in China, mainly produces titanium concentrate from the beneficiation process of vanadium-titanium magnetite. Titanium concentrate with more than 45% TiO_2 is primarily obtained through flotation, which is not only high in titanium content but also often contains elements such as vanadium and iron. Ilmenite flotation is a vital process in the extraction of titanium minerals from ores, particularly in the context of increasing demand for titanium dioxide (TiO_2) in various industries. Research in this area has focused on improving flotation efficiency, understanding the surface chemistry of ilmenite, and developing new reagents and methods.

Understanding the surface chemistry of ilmenite is crucial for optimizing flotation processes. Studies have investigated the effects of pH, ionic strength, and the presence of various ions on the flotation behavior of ilmenite [1]. The development of effective collectors and frothers is a significant area of research. Various organic and inorganic reagents have been tested to enhance the flotation of ilmenite [2]. Innovative flotation techniques, such as column flotation and microbubble flotation, have been explored to improve the separation efficiency of ilmenite from gangue minerals. Research has focused on optimizing various flotation parameters, such as air flow rate, pulp density, and collector dosage, to maximize ilmenite recovery and grade [3].

The research progress in ilmenite flotation has been substantial, with advancements in understanding surface chemistry, reagent development, process optimization, and environmental sustainability. Continued research in these areas is essential for improving the efficiency and sustainability of ilmenite extraction processes [4-7].

Titanium dioxide (TiO_2), also known as titanium white pigment, is one of the key raw materials for the production of titanium white. It is obtained from titanium concentrate through either the sulfate process or the chloride process, and is widely used as a white pigment in the coatings, plastics, papermaking, and ink industries.

* Corresponding author: [a*]1252638725@qq.com; [b]240037447@qq.com

The raw material for the sulfate process titanium white pigment is ilmenite or acid-soluble titanium slag, which reacts with sulfuric acid to undergo acidolysis, forming titanyl sulfate, then hydrolyzing to form titanic acid, and finally obtaining titanium dioxide after calcination and grinding. The advantage of the sulfate process is that it is easy to obtain raw materials, has a simple process, and has a lower investment cost. However, this method generates a large amount of sulfur-containing exhaust gases, acidic wastewater, and ferrous sulfate waste, causing significant environmental pollution, and the product quality is not as stable and high-quality as that of the chloride process.

The chloride process for titanium white pigment mainly uses rutile as the raw material, which reacts with chlorine gas to form titanium tetrachloride, and further oxidizes to obtain titanium dioxide. This process includes raw material preparation, chlorination, purification of titanium tetrachloride, oxidation, and post-treatment. The advantage of the chloride process is its high product purity, uniform particle size distribution, and superior optical properties, making it suitable for the production of high-quality titanium white pigment. However, this process has strict requirements for the content of silicon, calcium, and magnesium in the raw materials, such as the TiO_2 content needing to reach more than 90%, and the total content of CaO and MgO needs to be below 1.5%. During the smelting of high-titanium slag, silicon will react with titanium to form silicon-titanium alloy at high temperatures, consuming some titanium resources, so the silicon content also needs to be strictly controlled.

The goal of this study is to process titanium concentrate from the Panxi region through flotation technology to obtain high-quality titanium concentrate with low silicon, calcium, and magnesium, suitable for chloride process production. Currently, there is relatively little research on how to further improve the quality of ordinary titanium concentrate through flotation [8-9].

2 Experimental Raw Material Analysis and Reagents

2.1 Analysis of the Properties of the Original Ore

The original ore used in the experiment was selected from the flotation titanium concentrate of the Anning Titanium Industry concentrator in the Panxi region. To ensure the representativeness of the test samples, the samples taken were strictly processed and then sampled using the grid reduction method.

2.1.1 Multi-element Chemical Analysis of the Original Ore

The chemical composition and content of the original ore were analyzed using X-ray diffraction, and the analysis results are shown in Table 1.

From Table 1, it can be seen that the grade of TiO_2 in the original ore is 47.61%, the grade of Fe_2O_3 is 45.65%, the content of P_2O_5 is 0.03%, and the content of SO_3 is 0.29%. The contents of SiO_2 and CaO are 2.63% and 1.05%, respectively, which are much higher than the requirements for the content of calcium and magnesium in the raw materials for the preparation of titanium white pigment by the chloride process.

Table 1. Results of multi-element chemical analysis of raw ore / %.

Al_2O_3	CaO	Cl	Fe_2O_3	TiO_2	MgO	Na_2O	P_2O_5	SO_3	SiO_2	ZrO_2
0.80	1.05	0.03	45.65	47.61	2.84	0.09	0.03	0.29	2.63	0.01

2.1.2 Occurrence State of Main Elements

Table 2. The occurrence state of main elements in the titanium concentrate /%.

Mineral	Al (%)	Ca (%)	Fe (%)	K (%)	Mg (%)	Na (%)	P (%)	S (%)	Si (%)	Ti (%)	V (%)
Ilmenite	11.78	0.00	93.59	18.88	13.65	0.00	0.00	0.00	6.32	97.28	0.00
Chlorite	20.96	0.17	0.97	0.00	33.64	0.00	0.00	0.00	16.41	0.11	7.97
Forsterite	0.06	0.05	0.04	0.07	3.31	0.13	0.00	0.00	2.05	0.00	0.79
Diopside	2.15	16.36	0.21	0.00	15.12	0.00	0.00	0.00	15.41	0.00	0.00
Andesine	9.59	3.80	0.00	0.00	0.00	68.71	0.00	0.00	9.59	0.00	0.00
Kaersutite	7.87	14.17	0.62	14.39	8.91	0.00	0.00	0.00	15.95	0.00	0.00
Magnetite	0.00	0.00	1.21	0.00	0.00	0.00	0.00	0.00	0.00	0.00	0.00
Kaolinite	0.20	0.01	0.00	0.00	0.05	0.23	0.00	0.00	0.09	0.00	0.00
Fayalite	0.30	0.00	0.03	0.00	0.00	0.00	0.00	0.00	0.46	0.00	0.63
MgO-Al_2O_3	16.43	0.00	0.28	0.00	12.15	0.00	0.00	0.00	2.36	0.01	2.37
Cao-TiO_2	0.00	3.62	0.05	0.00	0.14	0.00	0.00	0.00	0.00	0.17	0.00
Titanite	1.62	30.89	0.09	0.00	1.35	0.00	0.00	0.00	10.53	0.95	32.46
Rutile	0.44	6.78	0.17	0.00	0.00	0.00	0.00	0.00	2.43	0.94	14.11
Pyrite+MgO-Al_2O_3	0.60	0.00	0.04	0.00	0.89	0.00	0.00	4.66	0.08	0.00	0.00
Garnet	0.00	0.37	0.01	0.00	0.00	0.00	0.00	0.00	0.10	0.00	0.18
Hematite (Ti)	2.12	0.00	1.23	0.00	1.75	0.00	0.00	0.00	0.00	0.30	21.79
Titanaugite	4.02	6.02	0.33	28.84	4.73	0.00	0.00	0.00	5.16	0.19	15.57
Orthoclase	0.25	0.05	0.00	4.86	0.00	0.00	0.00	0.00	0.35	0.00	0.20
Muscovite	1.36	0.29	0.01	6.95	0.06	1.00	0.00	0.00	0.93	0.00	0.75
Anorthite	7.60	7.45	0.09	0.00	0.00	10.71	0.00	0.00	5.02	0.02	3.18
Quartz	0.00	0.00	0.00	0.00	0.00	0.00	0.00	0.00	0.60	0.00	0.00
Augite	5.59	2.56	0.06	0.00	2.96	8.61	0.00	0.00	2.44	0.00	0.00
Pyrrhotite	0.00	0.00	0.93	0.00	0.00	0.00	0.00	95.34	0.00	0.03	0.00
Biotite	1.38	0.00	0.04	14.20	1.31	0.00	0.00	0.00	1.63	0.00	0.00
Apatite	0.00	7.42	0.00	0.00	0.00	0.00	100.0	0.00	0.00	0.00	0.00
Al_2O_3	4.30	0.00	0.00	0.00	0.00	0.00	0.00	0.00	0.00	0.00	0.00
Albite	1.38	0.00	0.00	11.81	0.00	10.61	0.00	0.00	1.91	0.00	0.00
Zicron	0.00	0.00	0.00	0.00	0.00	0.00	0.00	0.00	0.21	0.00	0.00
Unknown	0.00	0.00	0.00	0.00	0.00	0.00	0.00	0.00	0.00	0.00	0.00
Total	100.0	100.0	100.0	100.0	100.00	100.0	100.0	100.0	100.0	100.0	100.0

According to the data provided in Table 2, it can be understood that the distribution characteristics of the main elements in the titanium concentrate are as follows: the silicon (Si) element is mainly concentrated in several minerals such as chlorite, diopside, hornblende, and sphene; the calcium (Ca) element is primarily found in diopside, hornblende, sphene, and plagioclase; while the magnesium (Mg) element is mainly enriched in ilmenite, chlorite, diopside, and magnesium-aluminum spinel. Therefore, in this flotation study, special attention should be paid to the separation effects between ilmenite and the three minerals of chlorite, diopside, and hornblende, and suitable mining reagents and corresponding process flows should be selected accordingly.

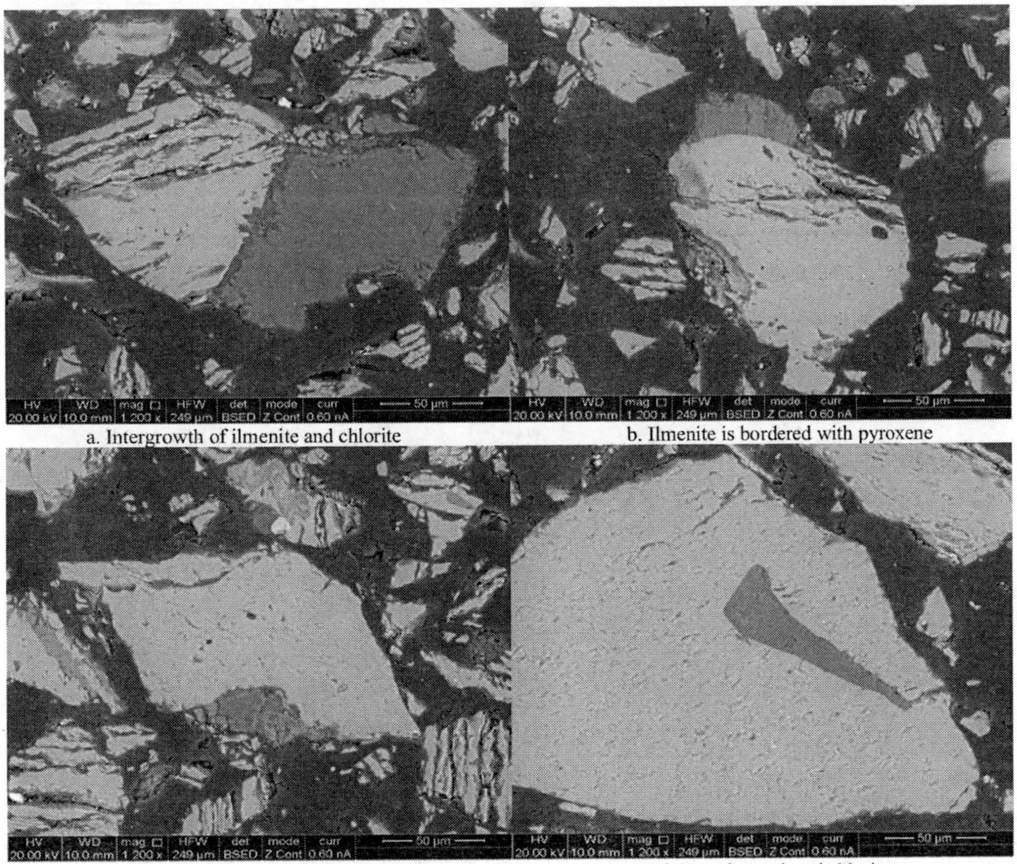

a. Intergrowth of ilmenite and chlorite

b. Ilmenite is bordered with pyroxene

c. Interaction between ilmenite and sphenite

d. Ilmenite enclosed chlorite

Fig. 1. Symbiosis relationship between titanomagnetite and other minerals (the bright area is titanomagnetite).

2.1.3 Characteristics of intercalation between the main minerals in titanium concentrate

As shown in section a-b of Fig. 1, albite, chlorite, titanopside and diopside are the main gangue minerals rich in calcium (Ca), magnesium (Mg) and silicon (Si) elements. In the target mineral ilmenite, the gangue minerals associated with it are mainly chlorite, sphenite and titanite, and these gangue minerals can have the opportunity to be mostly removed in mineral processing. At the same time, the gangue minerals chlorite, hornblende, sphene and titanite are surrounded by ilmenite. Due to the small distribution size of these gangue minerals, they are difficult to be effectively separated and removed by beneficiation technology. Therefore, this may lead to a high content of silicon, calcium and magnesium in titanium concentrate even after flotation improvement.

2.2 Test instruments and main drugs

2.2.1 Test main instruments and equipment

The equipment used in the tests are shown in Table 3.

2.2.2 Test principal reagent

In addition to various collectors, sulfuric acid, diesel oil are industrial pure, other test agents are analytical pure. The main reagents used in the test are shown in Table 4. The test inhibitors were acidified water glass, sulfuric acid, water glass, sodium fluosilicate and prepared into 5% concentration solution.

Table 3. Main test equipment.

Equipment	Model number	Application
Drying cabinet	101-3B electric blast drying oven	Test sample drying
Rod mill	XMB-Φ240×300 rod mill	Ore grinding
Electronic balance	Model 20001 electronic balance	Mineral weighing
Single cell flotation machine	XFD-63-0.5L	Actual mineral test
Single cell flotation machine	XFDIV-0.5L	Actual mineral test
Electronic analytical balance	AUY220 electronic balance	Reagent weighing
Vacuum filter	DL-5C	Filtered pulp
Sample machine	HF-3	Sample preparation
Thermostat sleeve	ZDHW	Pharmaceutical preparation

Table 4. Main reagents tested.

Pharmaceutical name	source	Official rank	use
Vitriol	Xinyu Chemical Co., LTD	Analytically pure	Regulator
Anhydrous Na_2CO_3	Chengdu Kelong chemical reagent factory	Analytically pure	Regulator
YS, YS-1 collector (sodium oleate, naphthenic acid, etc.)	Lab-made	Technical pure	Collector
Benzoyl/salicylhydroxamic acid	Tieling mineral processing pharmaceutical factory	Technical pure	Collector
Eutitanium benzyl arsenate	Hubei Xinghe Chemical Co., LTD	Technical pure	Collector
MOH	Jingjiang River in Hubei Province	Technical pure	Collector
Sodium hexametaphosphate	Chengdu Kelong chemical reagent factory	Analytically pure	Inhibitor
Sodium silicate	Tieling mineral processing pharmaceutical factory	Analytically pure	Inhibitor
CMC (Carboxymethyl cellulose)	Renqiu Jingda Chemical Co., LTD	Analytically pure	Inhibitor
CS (Modified Sodium pyrophosphate)	Lab-made	Analytically pure	Inhibitor
FS (oxalic acid, lignin mixture)	Lab-made	Analytically pure	Inhibitor
Sodium fluorosilicate	Tianjin Hengxing Chemical reagent manufacturing Co., LTD	Analytically pure	Inhibitor

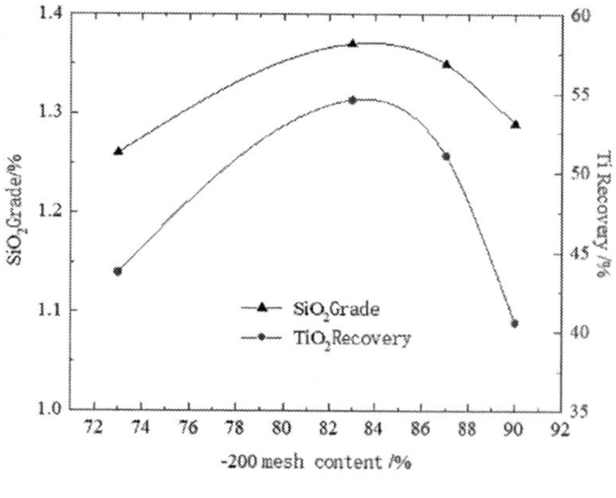

Fig. 2. Grinding fineness test results.

2.3 Test

In the quality improvement flotation test, XFD 0.5L single cell flotation machine was used, the flotation concentration was 40%, and the flotation process of one coarse and one fine was adopted to explore the best parameters of the flotation process (stirring time, amount of inhibitor types and amount of collector), the main purpose was to improve the grade of TiO_2 to more than 50%. Reduce the content of SiO_2, CaO and MgO (CaO+MgO≤1.5%), and provide qualified concentrate for the next preparation of high titanium slag for chlorination.

3 Quality improvement flotation test

3.1 Grinding fineness test

According to the results of process mineralogy, the gangue minerals associated with ilmenite mainly include chlorite, sphenite and titanite. This means that SiO_2, CaO and MgO have similar behavior in the flotation process. In the experiment, in order to balance cost and efficiency, TiO_2 enrichment and SiO_2 removal were mainly concerned. In the experiment, 500 grams of raw ore was taken, stirred and aerated for 2 minutes, then a single flotation was carried out, and the bubble scraping time was 3 minutes.

 https://doi.org/10.1051/e3sconf/202561701017

The specific results are shown in Fig. 2. It can be observed from Fig. 2 that with the extension of grinding time, the recovery rate of titanium element gradually increases. With the increase of grinding fineness, the titanium concentrate mineral yield, TiO_2 recovery rate and SiO_2 grade all showed a similar change trend: it rose at first, reached a peak when the fineness reached about 83%, and then began to decline. At the same time, the grade of TiO_2 generally showed an upward trend, but the change was not large. Therefore, the grinding fineness should not be too fine, and it can be maintained at about 83% of the -200 purpose grade.

3.2 Collector type test

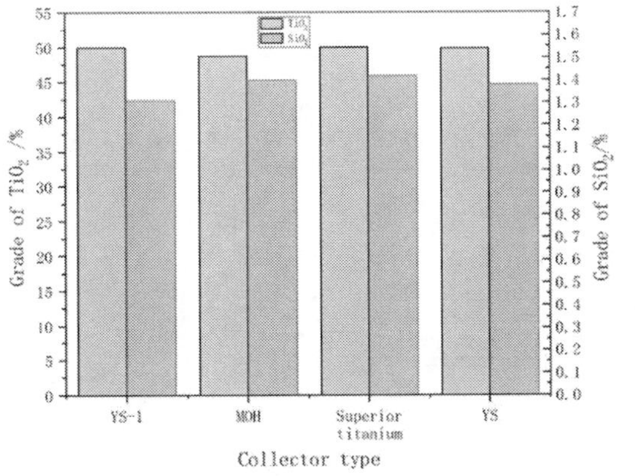

Fig. 3. Flotation results with different collectors.

The aim of this study is to reduce the content of silicon, calcium and magnesium in ilmenite. Therefore, the selection of suitable collector is very important for the flotation process of ilmenite. The selectivity and ability of collector directly affect the efficiency and selectivity of flotation, and then determine the recovery rate of ilmenite and the quality of concentrate. The flotation performance of ilmenite can be significantly improved by carefully selecting the collector and optimizing its application conditions. The common collectors of ilmenite are fatty acids, phosphoric acids, hydroxamic acids, arsonates and their combinations. These collectors exhibit different selectivity and harvesting effects by interacting with different functional groups on the surface of ilmenite. In practice, in order to enhance the effect of the collector, the commonly used collector is often mixed and mixed. On the basis of previous studies, the effects of self-made YS and YS-1 collector (composed of sodium oleate and naphthenic acid) on the flotation of ilmenite were compared with those of benzohydroxamic acid, salicylhydroxamic acid, benzyl arsenic acid, eutitanium, MOH and other common collectors. By comparing the flotation results of several different collectors, the most suitable collector for the mine was selected. The dosage of collector was fixed at 600g/t, the test process was shown in Fig. 4, and the test results were shown in Fig. 3.

The results in Fig. 3 show that under the same chemical regime, the titanium concentrate yield from Eutitanium is the highest, followed by YS and the lowest by MOH, but the titanium concentrate from Eutitanium has the highest grade of SiO_2, followed by MOH, all of which are greater than 1%. Overall, YS was selected as the collector.

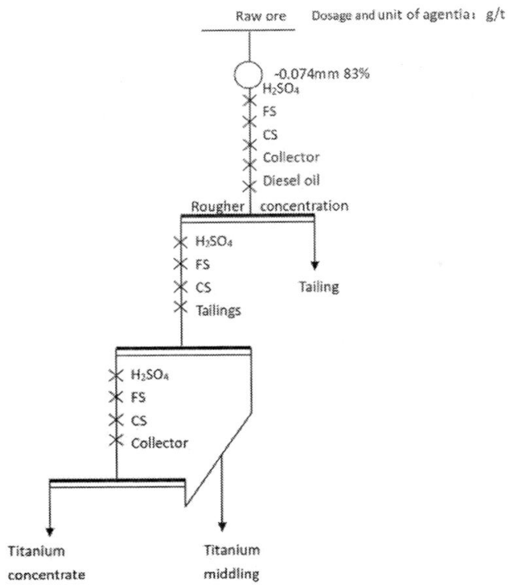

Fig. 4. Flotation process.

3.3 Sulfuric acid dosage test

Different kinds of collectors have specific requirements on pH. Under the premise of using YS as collector, sulfuric acid dosage test was carried out. The test results are shown in Fig. 5. The results show that with the increase of the amount of sulfuric acid, the flotation titanium concentrate mineral yield and TiO_2 recovery rate gradually decline, fluctuate at 1800g/t, and then gradually decline, TiO_2 grade basically shows an upward trend, SiO_2 grade basically shows a downward trend, comprehensive consideration, sulfuric acid dosage of 1800g/t is determined to be more appropriate.

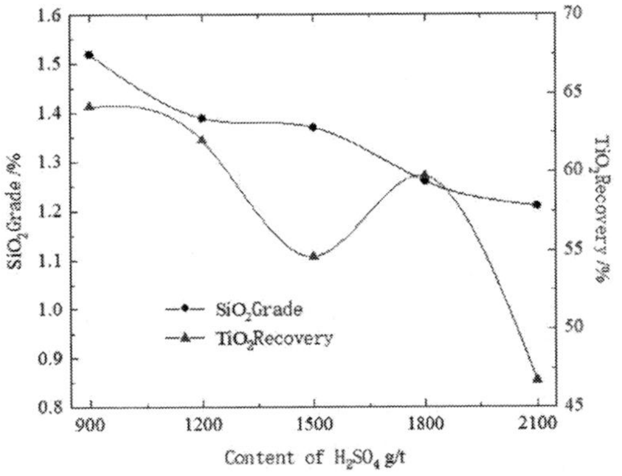

Fig. 5. Test results of sulfuric acid dosage.

3.4 Inhibitor type test

The main gangue minerals in the ore are SiO_2, CaO and MgO, so acidified water glass, sulfuric acid, water glass, sodium fluosilicate and lignin are selected as inhibitors. In the previous quality improvement test, FS+CS was found to have a good inhibition effect on silicate minerals. In order to further verify its inhibition effect on different silicate minerals, FS+CS was compared with sodium fluorosilicate, sodium silicate and oxalic acid. The test process is shown in Fig. 3. The dosage of control inhibitors is 1600g/t, other conditions remain unchanged, and the test results are shown in Fig. 6.

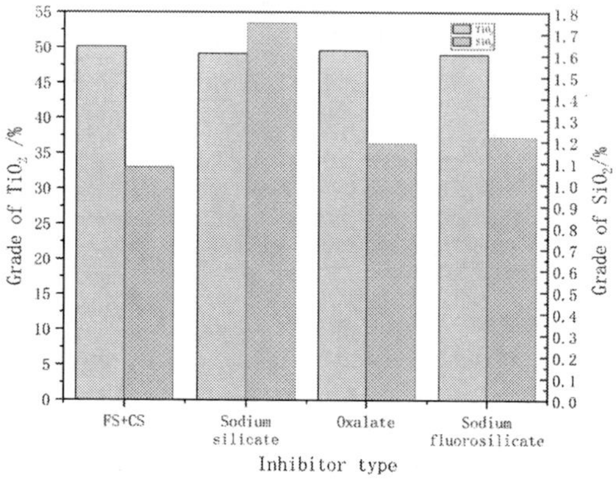

Fig. 6. Experiment of inhibitor types.

As can be seen from Fig. 6, TiO_2 grade of titanium concentrate obtained with FS+CS is the highest, reaching 50.11%, and the selective inhibition effect on gangue minerals is significantly better than the other three inhibitors. Oxalic acid also has a good inhibition effect on ilmenite, but SiO_2 is higher than FS+CS, and the inhibition effect on water glass is the worst. Therefore, FS+CS is selected as the inhibitor.

3.5 Intermediate ore treatment method test

Take 500 g of raw ore, stir for 10 min, adjust the pH value of pulp with sulfuric acid to 3, add inhibitor sulfuric acid and collector YS (5% concentration solution), stir for 3 min respectively, the amount of sulfuric acid is 2 000 g/t, the amount of MOH is 800 g/t, the process adopts "one crude - two refined, medium ore treated separately" open-circuit test flow. The test process is shown in Fig. 7, the bubble scraping time is 3 min, and the test results are shown in Table 5.

It can be seen from Table 5 that through the open circuit test process of "one crude - two refined, and separate treatment of middle ore" as shown in Fig. 7, the yield of 36.72%, including $TiO_2$49.62%, can be obtained when the dosage of coarse sulfuric acid is 3000g/t, FS+CS1500+1500g/t, and collector YS500g/t. Low silicon titanium concentrate with SiO_2 0.95% and TiO_2 recovery of 38.24% and yield of 63.28%, ordinary titanium concentrate with $TiO_2$46.51%, SiO_2 3.52% and TiO_2 recovery of 61.76%.

Table 5. Results of open-circuit test for single treatment of medium ore /%.

Product	Yield	Grade		Recovery	
		TiO_2	SiO_2	TiO_2	SiO_2
Titanite1	36.72	49.62	0.95	38.24	13.54
Titanite 2	63.28	46.51	3.52	61.76	86.46
Run-of-mine	100.00	47.65	2.58	100.00	100.00

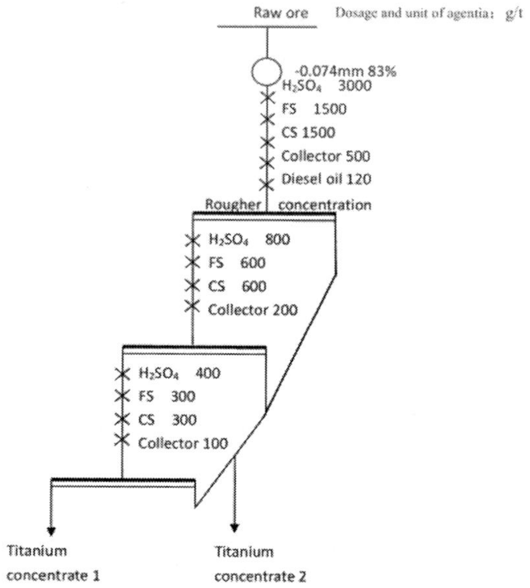

Fig. 7. Flotation flowsheet.

4 Test product analysis

For Anning ilmenite, low-silicon titanium concentrate 1 containing TiO_2 49.62%, SiO_2 0.95%, TiO_2 recovery rate 38.24% and TiO_2 46.51%, SiO_2 3.52% can be obtained through the process of "one crude, two refined, and medium ore treatment alone". TiO_2 recovery of 61.76% of ordinary titanium concentrate 2. In order to further analyze the quality of titanium concentrate, the particle size sieve analysis and chemical multi-element composition analysis of titanium concentrate 1 and 2 were carried out. The specific analysis results are shown in Tables 6, 7, 8 and 9.

Table 6. Sieve analysis results of titanium concentrate 1 /%.

Size fraction	Yield	Grade		Distribution rate	
		TiO_2	SiO_2	TiO_2	SiO_2
+100	2.90	50.08	0.84	2.93	2.53
-100~+200	23.01	50.63	1.03	23.49	24.58
-200~+325	46.34	49.85	0.75	46.58	36.05
-325	27.75	48.27	1.28	27.01	36.84
Total	100.00	49.60	0.96	100.00	100.00

Table 7. Sieve analysis results of titanium concentrate 2 /%.

Size fraction	Yield	Grade		Distribution rate	
		TiO_2	SiO_2	TiO_2	SiO_2
+100	1.06	47.02	2.49	1.07	0.75
-100~+200	14.67	48.47	2.03	15.27	8.42
-200~+325	38.15	47.12	3.56	38.62	38.41
-325	46.12	45.46	4.02	45.04	52.43
Total	100.00	46.55	3.54	100.00	100.00

Table 8. Results of chemical multielement analysis of titanium concentrate 1 /%.

Element	TiO_2	FeO	Fe_2O_3	SiO_2	CaO	MgO	Al_2O_3
Content	49.62	32.67	11.16	0.95	<0.2	1.21	0.44
Element	S	P	V_2O_5	Co	MnO	Na_2O	K_2O
Content	0.012	0.0061	0.013	0.003	0.051	0.019	0.017

Table 9. Results of chemical multielement analysis of titanium concentrate 2 /%.

Element	TiO_2	FeO	Fe_2O_3	SiO_2	CaO	MgO	Al_2O_3
Content	46.51	18.23	14.67	3.52	1.53	3.69	2.57
Element	S	P	V_2O_5	Co	MnO	Na_2O	K_2O
Content	0.13	0.099	0.072	0.058	1.02	0.28	0.13

Tables 6, 7, 8 and 9 concentrate analysis results show that when the finity of grinding is -200 mesh accounting for 83%, sulfuric acid is used as pH adjustment agent, FS+CS is used as silicate mineral inhibitor, and new chemical YS is used as ilmenite collector. Through the process of "one crude and two refined, medium ore is treated separately", the yield of 36.72% can be obtained, containing $TiO_2$49.62%. Low silicon titanium concentrate with SiO_2 0.95% and TiO_2 recovery of 38.24% and yield of 63.28%, ordinary titanium concentrate with TiO_2 46.51%, SiO_2 3.52% and TiO_2 recovery of 61.76%.

5 Conclusion

1) Chemical multi-element analysis was carried out on the raw ore, and the ore sample contained TiO_2 47.61%, SiO_2 2.63%, CaO1.05%and MgO_2.84%; The particle size of the ore sample is mainly concentrated in -200 mesh, accounting for 73.75%, and the micro-fine grade of -325 mesh is higher, accounting for 28.95%. The gangue minerals rich in Ca, Si and Mg impurity elements are mainly silicate minerals such as chlorite, titanaugite and sphene, which are concentrated in -200 order. The gangue minerals which interexisted with the target mineral ilmenite were mainly chlorite, sphenite and titanite, the gangue minerals coated by ilmenite were mainly chlorite, hornblende, sphenite and titanite, and these gangue minerals could not be removed by mineral processing because of their low distribution size.

2) The addition of inhibitors FS+CS, sodium fluosilicate, water glass and oxalic acid can effectively inhibit the flotation of main gangue minerals SiO_2, CaO and MgO, and improve the grade of titanium concentrate. The inhibition effect of FS+CS is better, and the comprehensive cost and test results confirm that FS+CS is a quality improvement flotation inhibitor.

3) YS-1, MOH, Eutitanium and YS are all selected as titanium collectors, but considering the selectivity of silicon, calcium and magnesium, YS has a better effect.

4) Through the process of "one crude and two refined, and separate treatment of middle ore", low silicon titanium concentrate 1 containing $TiO_2$49.62%, SiO_2 0.95%, TiO_2 recovery rate of 38.24% and ordinary titanium concentrate 2 containing $TiO_2$46.51%, SiO_2 3.52%, TiO_2 recovery rate of 61.76% can be obtained. Moreover, MgO+CaO can be fully filled at the same time ≤1.5%, which provides a basis for further enrichment.

Acknowledgements

Study and application of key technology for titanium extraction from acknowledgements of blast-furnace slag containing titanium in Panzhihua, Sichuan Province, No. 2022YFS0458.

References

1. Zhang, Y., & Liu, Y. (2018). Effect of pH on the flotation of ilmenite. Minerals Engineering, 122, 1-8.
2. Zhang, L., & Wang, X. (2022). Application of microbubble flotation in ilmenite separation." Separation and Purification Technology, 287, 120-130.
3. Liu, Y., & Zhang, Y. (2021). "Optimization of flotation conditions for ilmenite using response surface methodology." Minerals, 11(5), 487
4. Zhang, H., & Wang, Y. (2023). "Application of machine learning in optimizing ilmenite flotation processes." Minerals Engineering, 203, 107-115.
5. Kumar, A., & Singh, R. (2023). "Sustainable practices in ilmenite flotation: A review." Resources, Conservation and Recycling, 185, 106-115.
6. Cai, J., Deng, J., Wang, L. et al. Reagent types and action mechanisms in ilmenite flotation: A review. Int J Miner Metall Mater 29, 1656–1669 (2022)
7. Zhai, J., Chen, P., Sun, W., Chen, W., & Wan, S. (2020). A review of mineral processing of ilmenite by flotation. Minerals Engineering, 157, 106558.
8. Su Xin, Hu Houqin, Wang Hongbin. Experimental study on flotation of titanium concentrate from a concentrator in Panxi Area. Iron and Steel, 2023, 44(06):88-92. (in Chinese)
9. HU Houqin, Luo Rongfei, WU Xuehong. Experimental study on quality improvement of ilmenite flotation concentrate. Mining and Metallurgy, 2022, 31(04):19-24. (in Chinese)

Research on Process Mineralogy of Titanium-Bearing Blast Furnace Slag in Panzhihua-Xichang Area

Guoyin Xu[a*] and Purong Wang[b]

Panzhihua University, Panzhihua, Sichuan, China

Abstract. Titanium-bearing blast furnace slag is an important resource among the by-products of the steel industry, and the recycling of titanium elements holds significant economic and environmental importance. This study conducts a detailed process mineralogy analysis of Panzhihua-Xichang titanium-bearing blast furnace slag samples, exploring the mineral composition, occurrence state of titanium elements, the dissemination characteristics and particle size distribution of the main titanium-bearing minerals, as well as micro-area composition analysis of titanium elements. The research findings indicate that ilmenite and perovskite are the primary titanium-bearing minerals in the samples, with titanium elements mainly residing in these two minerals. Through analysis of the particle size distribution and degree of liberation, potential pathways for enhancing the recovery rate of titanium elements are proposed, providing a theoretical basis for the resource utilization of titanium-bearing blast furnace slag.

1 Introduction

With the acceleration of industrialization, the demand for energy and raw material mineral resources continues to rise. Mineral resources have become one of the key elements concerning social progress, economic prosperity, and national strength. Titanium, an important strategic metal, is widely used in aviation, chemical, medical, and many other fields. China is a country rich in titanium resources, accounting for 28% of the global total and leading the world in reserves. Notably, the Panzhihua and Chengde areas possess vast amounts of vanadium-titanium magnetite, with Panzhihua holding 9 billion tons and Chengde holding 8.4 billion tons [1]. However, due to the low titanium grade in Chinese vanadium-titanium magnetite, compounded with elements such as iron, vanadium, manganese, cobalt, nickel, gallium, copper, etc., the mineral processing is quite challenging, making it subpar titanium resources [2]. At present, the primary utilization of domestic vanadium-titanium magnetite is through blast furnace smelting, producing iron water and slag. During the subsequent blast furnace smelting, the majority of titanium enters the slag, which is why this type of slag, rich in titanium, is referred to as high-titanium blast furnace slag. It is an important man-made titanium resource, with a TiO_2 content typically ranging between 20% and 26% [3]. To date, approximately 60 million tons have accumulated, and the annual increase is about 3.6 million tons. Currently, high-titanium blast furnace slag is only used as a general industrial by-product for auxiliary materials, with a rather extensive utilization method that is difficult to recycle, resulting in low titanium utilization rates and minimal processing as a secondary mineral resource [4]. Therefore, studying the process mineralogy characteristics of Panzhihua high-titanium blast furnace slag and developing its titanium extraction technology as a secondary mineral resource is of great practical significance for further implementing industrial measures and conducting technical research to effectively utilize Panzhihua's secondary mineral resources [5]. The process mineralogy characteristics of Panzhihua high-titanium blast furnace slag determine its titanium extraction pathways. Current titanium extraction technology plans mainly include two methods: direct titanium extraction [6-8] and post-metallurgical modification for titanium extraction [9-12]. However, both methods face issues related to environmental protection, production costs, and equipment to varying degrees. Up to now, no titanium extraction technology has been able to achieve large-scale industrial production. Therefore, future research on Panzhihua high-titanium blast furnace slag should focus on diversifying the technological route and product variety.

2 Ore Scanning Electron Microscopy Morphology and Composition Analysis

2.1 Ore Sample Preparation

The ore samples were first crushed using a PE-150*250 jaw crusher, then further crushed to below -2 mm using an MPG-Φ200*75 roll crusher in a closed circuit, followed by mixing, splitting, and bagging to create samples for process mineralogy analysis.

* Corresponding author: [a*]1252638725@qq.com, [b]240037447@qq.com

2.2 Scanning Electron Microscopy Morphology of Main Minerals

Under the electron microscope, perovskite appears as grayish-white granular, cruciform, and dendritic crystals (Fig. 1 a, b), with most perovskite particles occurring in a wrapped form by ilmenite, and monomeric liberated perovskite particles are relatively rare (Fig. 1 d), with particle sizes generally below 10μm. In the samples, two types of ilmenite are mainly observed, namely ilmenite and rich ilmenite. Ilmenite is light gray amorphous (Fig. 1 a, c), anhedral crystals, and is the primary mineral wrapping perovskite, with titanium element content generally below 10% in its elemental composition. Rich ilmenite, also known as deep green rich ilmenite, often exhibits a dark gray strip or forked shape (Fig. 1 b), with anhedral crystals, coexisting with perovskite, mostly wrapped by ilmenite, and its elemental composition generally contains more than 10% titanium. Since both types of ilmenite belong to the pyroxene class and have similar properties, the data involving mineral content and elemental occurrence are not carefully distinguished when involved in the study, and they are all uniformly treated as ilmenite.

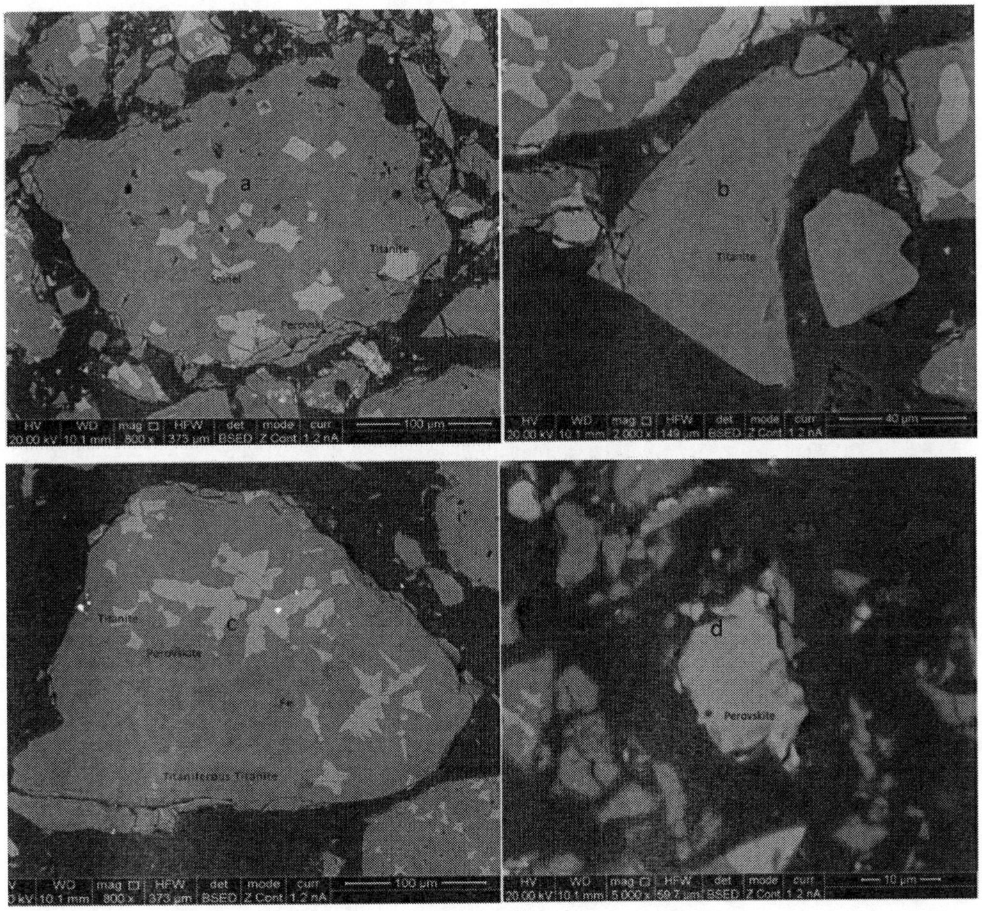

Fig. 1. (a、b、c、d) SEM images of ilmenite, perovskite.
(a, c: Granular, cruciform, and dendritic perovskite with amorphous ilmenite and strip, forked rich ilmenite, b: Monomeric ilmenite, d: Monomeric perovskite)

2.3 Mineral Composition and Content

From Table 1, it can be seen that the minerals in the samples are primarily ilmenite, followed by perovskite, with the combined content of the two exceeding 98%. Other minerals have lower contents. This implies that the subsequent processes of mineral processing or metallurgical separation should focus on the separation of ilmenite and perovskite.

2.4 Occurrence of Titanium Elements

From Table 2, it can be seen that the titanium elements in the samples are mainly found in ilmenite and perovskite, with a very low proportion found in other minerals. Therefore, the subsequent titanium extraction processes should primarily focus on ilmenite and perovskite.

Table 1. Mineral composition and content of titanium-bearing blast furnace slag / %.

Titanium-bearing Pyroxene	Perovskite	Metallic Iron	Ferric Oxide	Calcium aluminate	Spinel	Calcium Ferrite	Calcium Oxide
85.83	12.76	0.21	0.40	0.08	0.20	0.19	0.07

Table 2. Distribution of titanium elements in minerals in titanium-bearing blast furnace slag / %.

Titanium-bearing Pyroxene	Perovskite	Titanium nitride	Vanadium plate titanium ore	Iron oxide	Calcium ferrite	Spinel	Metallic iron	Ilmenite
57.79	41.67	0.14	0.11	0.09	0.08	0.05	0.04	0.02

Table 3. Elemental composition of perovskite micro-area / %.

Serial number	O	Ca	Ti	V
1	29.95	38.57	31.48	0
2	30.70	37.62	31.46	0.22
3	30.51	38.46	30.62	0.41
4	32.62	36.91	30.02	0.45
5	30.77	37.97	30.62	0.64
6	32.22	37.01	30.24	0.53
7	31.67	37.83	30.50	0
8	29.27	38.54	32.19	0
9	30.26	37.99	31.53	0.22
10	31.35	37.61	30.64	0.4
Mean value	30.93	37.85	30.93	0.29

Table 4. Elemental composition of ilmenite micro-area / %.

Serial number	O	Na	Mg	Al	Si	K	Ca	Ti	V	Mn
1	31.58	0.86	7.24	10.06	23.85	0.76	19.73	5.26	0.66	0
2	31.93	0.83	6.71	10.34	24.09	1.08	18.23	4.95	0.39	1.45
3	30.97	1.01	7.22	10.70	24.10	0.60	18.29	5.44	0.46	1.21
4	30.26	1.43	7.36	10.80	24.02	0.91	18.42	5.41	0.14	1.25
5	30.26	1.41	7.59	10.28	24.01	0.98	18.51	4.93	0.52	1.51
6	29.98	0	7.30	8.58	23.81	1.11	23.60	2.65	0.92	2.05
7	29.09	0.61	7.85	9.36	23.56	1.21	23.77	2.54	0.42	1.59
8	29.83	0.41	7.84	9.07	24.04	0.98	23.84	1.94	0.34	1.71
9	30.83	0.30	7.13	8.64	22.83	1.21	24.06	2.65	0.6	1.75
10	29.35	0.33	7.06	8.55	23.92	1.38	25.04	2.45	0.35	1.57
Mean value	30.41	0.72	7.33	9.64	23.82	1.02	21.35	3.82	0.48	1.41

Table 5. Elemental composition of rich ilmenite micro-area / %.

Serial number	O	Mg	Al	Si	Ca	Ti	V	Mn
1	31.17	13.44	15.68	15.66	12.61	9.75	0.5	1.19
2	30.49	14.00	15.22	15.63	12.73	9.85	0.63	1.45
3	32.11	14.92	15.50	15.99	12.31	8.34	0.83	0
4	31.50	14.12	15.52	15.28	12.55	10.60	0.43	0
5	31.56	12.78	16.72	13.45	13.09	11.58	0.82	0
6	32.54	13.89	16.02	13.77	12.58	10.63	0.57	0
7	32.38	13.76	14.97	14.69	12.14	11.06	1.00	0
8	33.31	13.22	15.44	13.73	12.46	11.48	0.36	0
9	31.92	13.79	15.70	14.75	12.45	10.98	0.41	0
10	31.61	13.88	16.35	14.47	12.92	10.09	0.68	0
Mean value	31.86	13.78	15.71	14.74	12.58	10.44	0.62	0.26

2.5 Titanium Element Grade of Main Titanium-Bearing Minerals

Using an energy-dispersive spectrometer, a rough statistical analysis of the elemental composition in the micro-areas of the main titanium-bearing minerals, ilmenite, rich ilmenite, and perovskite, was conducted. As can be seen from Table 3 to Table 5, the average titanium content in perovskite is 30.93%, in ilmenite it is 3.82%, and in rich ilmenite it is 10.44%. This indicates that the average maximum theoretical grades of titanium recovery that can be achieved for perovskite, ilmenite, and rich ilmenite in the samples are approximately 31%, 4%, and 10%, respectively.

2.6 Elemental Surface Concentration Distribution of Main Titanium-Bearing Minerals

The elemental surface scanning images of titanium-bearing minerals such as ilmenite and perovskite are shown in Fig. 2. As can be seen from Fig. 2, the concentration changes of various elements on the surface exhibit several different trends. O, Na, K, Mn appear as relatively uniform dispersed patterns on the surface, with Na, K, Mn mainly distributed in ilmenite and rich ilmenite. In contrast, the V element clearly tends to enrich in perovskite. Other elements, such as Ca, Ti, Mg, Al, Si, etc., show a concentration change on the surface that is positively correlated with the content of the corresponding elements in the minerals. The highest contents of Ca and Ti elements are found in perovskite, followed by ilmenite, and then rich ilmenite. This is reflected in Fig. 3, where the highest concentration of Ca is found in perovskite, followed by ilmenite, and then rich ilmenite. The highest concentration of Ti is found in rich ilmenite, followed by ilmenite, and then perovskite. The content of Mg and Al elements in rich ilmenite is significantly higher than in ilmenite, hence the concentration of Mg and Al elements in the rich ilmenite area is higher than in ilmenite in Fig. 2. However, the content of Si element in rich ilmenite is significantly lower than in ilmenite, hence the concentration of Si in the ilmenite area is higher than in rich ilmenite in Fig. 2.

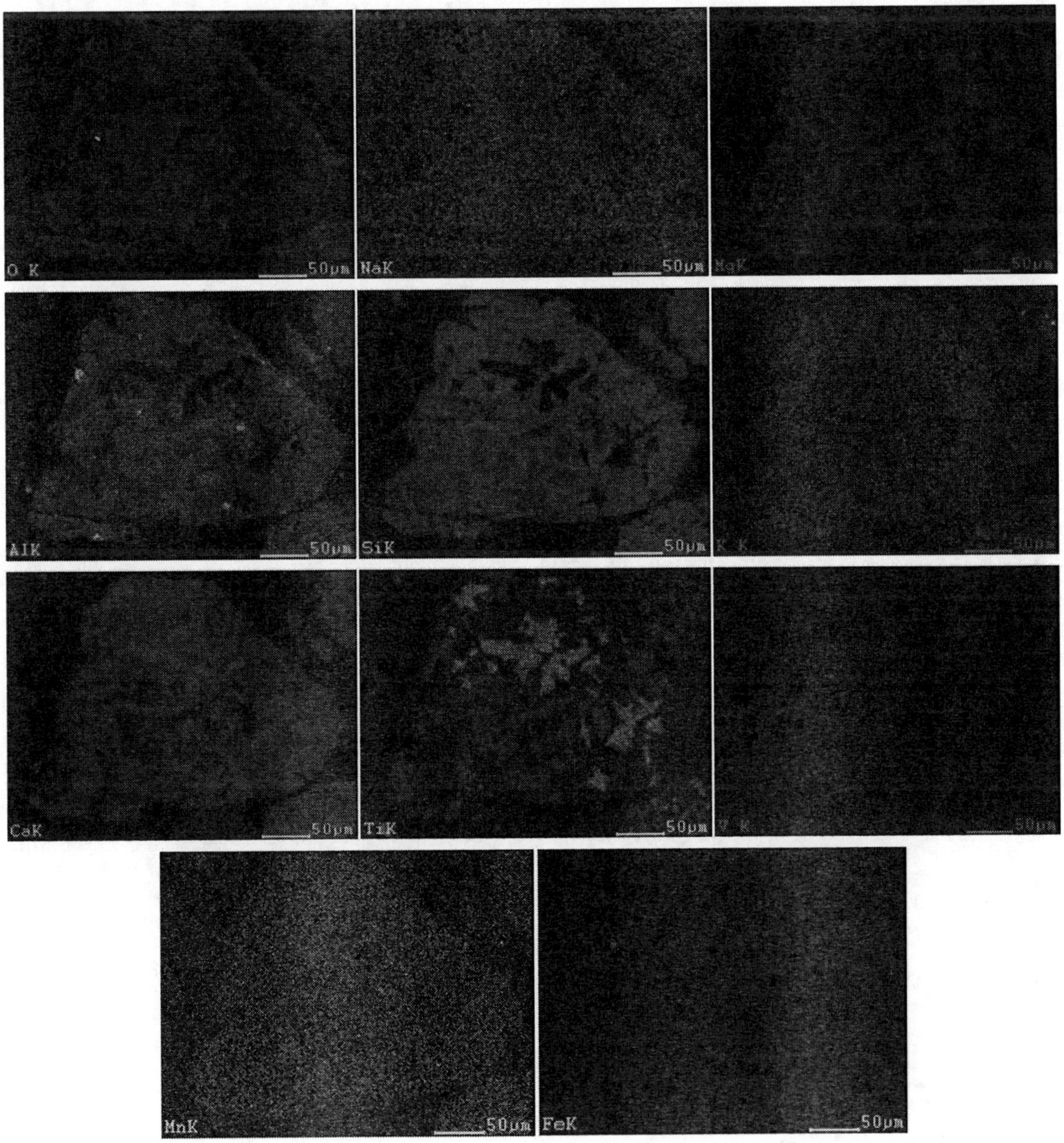

Fig. 2. Scanning electron microscopy images of micro-area elements of main titanium-bearing minerals.

Table 6. Liberation degree of perovskite.

Mineral	Monomer	Coenobium			
		> 3/4	3/4-1/2	1/2-1/4	< 1/4
Perovskite	10.08	2.02	3.88	9.29	74.73
Titanaugite	16.87	79.57	3.04	0.46	0.05

Table 7. The proportion of intergrowths and embedded minerals of titanium-bearing pyroxene and perovskite in the sample / %.

Embedded Mineral	Titanaugite	Perovskite	Metallic iron	Iron oxide	Vitreous	Quartz	Calcium aluminate
Titanaugite	0	95.82	0.56	0.37	0.11	0.03	0.29
Perovskite	99.12	0	0.12	0.06	0.01	0	0.09
Interlocking mineral	Spinel	Calcium ferrite	Olivine	Vanadium plate titanium ore	Titanium nitride	Calcium oxide	Chlorite
Titanaugite	1.21	0.64	0.18	0.37	0.2	0.2	0.02
Perovskite	0.32	0.1	0.05	0.04	0.03	0.06	0

Table 8. Overall particle size distribution of the sample / %.

Particle(μm)	-850+600	-600+355	-355+212	-212+106	-106+3.4
Perovskite	9.56	17.21	17.63	19.85	35.75

Table 9. Perovskite particle size in the sample / %.

Particle(μm)	-90+63	-63+45	-45+22	-22+9.6	-9.6+1
Perovskite	0.91	4.21	35.08	44.64	15.16

Table 10. Titanium pyroxene particle size in the sample / %.

Particle(μm)	-850+212	-212+150	-150+90	-90+63	-63+45	-45+22	-22+9.6	-9.6+1
Titanaugite	42.59	11.03	13.17	5.77	5.32	7.87	13.42	0.83

3 Liberation Degree and Intergrowth Relationship of Main Titanium-Bearing Minerals

3.1 Analysis of Liberation Degree of Main Minerals

It is not difficult to see from Table 6 that under the current grinding particle size, both perovskite and titanium-bearing pyroxene have low degrees of liberation, but there are differences in the specific situations of the two. The intergrowths of perovskite are mainly poor to extremely poor intergrowths of 1/2-1/4 and less than 1/4, and the content of medium to rich intergrowths of 3/4-1/2 and greater than 3/4 is very low, indicating that perovskite is severely wrapped and has a finer particle size, which requires grinding to a relatively very fine particle size in order to possibly improve its degree of liberation. The intergrowths of titanium-bearing pyroxene are mainly rich intergrowths greater than 3/4, which are easier to liberate, and only require appropriately increasing the grinding fineness to effectively improve their degree of liberation.

As can be seen from Table 7, the intergrowth relationship between perovskite and titanium-bearing pyroxene is the closest, with 95.82% of titanium-bearing pyroxene appearing in the form of intergrowth with perovskite, and 99.12% of perovskite appearing in the form of intergrowth with titanium-bearing pyroxene. Combined with the actual situation in the electron microscope morphology where perovskite is mostly produced wrapped by titanium-bearing pyroxene, it is not difficult to conclude that only by continuing to increase the grinding fineness can the degree of liberation of perovskite be effectively improved, allowing perovskite and titanium-bearing pyroxene to be separated.

3.2 Particle Size Distribution of Main Titanium-Bearing Minerals

From Table 8, Table 9 and Table 10, it can be seen that the overall particle size of the sample is relatively coarse, with the largest particle size between 710-850μm, indicating a grinding fineness below 850μm. The particle size of perovskite in the sample is finer, with the main particle size below 45μm, which means that if the degree of liberation of perovskite is to be significantly improved, the grinding fineness of the sample must be at least below -45μm or even finer. The particle size of titanium-bearing pyroxene is coarser, with most of its particle size distributed above 90μm, and some distributed between 9.6-90μm, that is, grinding fineness greater than 90μm can achieve a higher degree of liberation for titanium-bearing pyroxene.

Table 11. Chemical Element Composition of Minerals in Micro-Areas / %.

SPOT	Mineral	O	Na	Mg	Al	Si	K	Ca	Ti	V	Fe
1	Perovskite	29.04	0	0	0	0	0	39.10	31.54	0.32	0
2	Titanaugite	31.12	1.28	7.41	9.39	24.80	0.84	21.67	3.14	0.35	0
3	Titanium-rich pyroxene	31.61	0	13.88	16.35	14.47	0	12.92	10.09	0.68	0
4	Spinel	29.89	0	19.88	41.73	0	0	0	3.46	1.37	3.67

3.3 Micro-Area Composition of Main Titanium-Bearing Minerals

Fig. 3. Energy spectrum of titanium-bearing blast furnace slag.

From Fig. 3 and Table 11, it can be seen that the microchemical elements of the main titanium-bearing mineral perovskite in the titanium-bearing blast furnace slag are Ca and Ti, and the surface of the pyroxene is mainly Al, Si and Mg. Therefore, the flotation of perovskite can be targeted by using a selective flotation collector.

4 Conclusions

A. When dealing with ores containing titanium-bearing pyroxene and perovskite, our main goal is to achieve effective separation of the two. Titanium-bearing pyroxene is the main component of the ore, while perovskite is present in lesser amounts. The titanium content in titanium-bearing pyroxene usually does not exceed 10%, hence our mineral processing goal is to separate titanium-bearing pyroxene from perovskite. Since titanium-bearing pyroxene often wraps around perovskite, this results in a significant amount of titanium-bearing pyroxene being mixed into perovskite during the mineral processing, which reduces the grade of titanium and affects the efficiency of titanium extraction.

B. The theoretical maximum grades of titanium recovery that can be achieved for perovskite and titanium-bearing pyroxene are approximately 31% and 4%, respectively, while the titanium recovery for rich titanium-bearing pyroxene is around 10%. On the mineral surface, Ca and Ti elements are most abundant in perovskite, followed by Ca elements in titanium-bearing pyroxene, and Ti elements in rich titanium-bearing pyroxene. Na, K, Mn elements are mainly distributed in titanium-bearing pyroxene and rich titanium-bearing pyroxene, while V elements tend to enrich in perovskite. These distribution patterns of elements can be used to guide us in selecting targeted flotation collectors to achieve effective separation of titanium minerals.

C. To improve the degree of liberation of perovskite, we need to grind perovskite to a very fine particle size. However, doing so not only increases unnecessary grinding costs but also significantly affects the recovery rate of titanium due to fine grinding. To address this issue, we can consider adopting a multi-stage grinding and mineral processing technique, which can gradually separate titanium-bearing pyroxene and perovskite at different stages, thereby improving the recovery rate of titanium. Additionally, we can employ methods to enrich titanium elements through mineral processing, and then further purify titanium minerals through wet or pyrometallurgical processes to achieve a higher grade and recovery rate of titanium.

Acknowledgements

Study and application of key technology for titanium extraction from acknowledgements of blast-furnace slag containing titanium in Panzhihua, Sichuan Province, No. 2022YFS0458.

References

1. Cai, Y., Song, N., Yang, Y., Sun, L., Hu, P., & Wang, J. (2022). Recent progress of efficient utilization of titanium-bearing blast furnace slag. International Journal of Minerals, Metallurgy and Materials, 29, 22-31.
2. Lei, Y., Sun, L., Ma, W., Ma, X., Wu, J., Li, S., & Morita, K. (2018). An approach to employ titanium-bearing blast-furnace slag to prepare Ti and Al–Si alloys. Journal of Alloys and Compounds, 769, 983-990.
3. Li, J., Zhang, Z., Zhang, M., Guo, M., & Wang, X. (2011). The influence of SiO2 on the extraction of Ti element from Ti-bearing blast furnace slag. steel research international, 82(6), 607-614.
4. Fan, G., Wang, M., Dang, J., Zhang, R., Lv, Z., He, W., & Lv, X. (2021). A novel recycling approach for efficient extraction of titanium from high-titanium-bearing blast furnace slag. Waste Management, 120, 626-634.
5. Shi, J., Qiu, Y., Yu, B., Xie, X., Dong, J., Hou, C., ... & Liu, C. (2022). Titanium extraction from titania-bearing blast furnace slag: a review. Jom, 1-14.

6. Chen, K., Li, Y., Meng, X., Meng, L., & Guo, Z. (2019). New integrated method to recover the TiO2 component and prepare glass-ceramics from molten titanium-bearing blast furnace slag. Ceramics International, 45(18), 24236-24243.
7. Liu, Y., Chen, X., Mao, S., Xiao, Y., & Li, J. (2024). Extraction of Valuable Metals from Titanium-bearing Blast Furnace Slag by Acid Leaching. Journal of Wuhan University of Technology-Mater. Sci. Ed., 39(2), 376-385.
8. Pu, Z., Luo, Y., Wang, W., Zhang, G., Wang, M., Tu, J., ... & Jiao, S. (2022). A universal study of liquid metal cathodes for direct extraction of titanium within a closed loop. Journal of Cleaner Production, 368, 133135.
9. Zhen, Y. L., Zhang, G. H., & Chou, K. C. (2016). Carbothermic reduction of titanium-bearing blast furnace slag. High Temperature Materials and Processes, 35(3), 309-319.
10. He, M., Teng, L., Gao, Y., Rohani, S., Ren, S., Li, J., ... & Liu, W. (2022). Simultaneous CO2 mineral sequestration and rutile beneficiation by using titanium-bearing blast furnace slag: Process description and optimization. Energy, 248, 123643.
11. Wang, L., Liu, W., Hu, J., Liu, Q., Yue, H., Liang, B., ... & Li, C. (2018). Indirect mineral carbonation of titanium-bearing blast furnace slag coupled with recovery of TiO2and Al2O3. Chinese Journal of Chemical Engineering, 26(3), 583-592.
12. Zhang, R., Hou, Y., Fan, G., Huang, D., Ding, X., & Dang, J. (2022). Gas-based reduction and carbonization of titanium minerals in titanium-bearing blast furnace slag: a combined thermodynamic, experimental and DFT study. International Journal of Hydrogen Energy, 47(12), 7586-7599.

Comparison of Coastal Resilience Policies: A Perspective on Effective Global Governance Strategies

Weiwei Dong* and Guilan Lin

Third Institute of Oceanography, Ministry of Natural Resources, Xiamen 361005, China

Abstract. Coastal areas are key ecosystems that concentrate global biodiversity and are particularly vulnerable to the impacts of climate change. Countries around the world have developed diverse policies to address coastal resilience challenges, responding to threats from sea level rise, extreme weather events, and coastal development. This paper provides a comparative analysis of coastal resilience governance policies across different countries and regions, summarizing successful governance practices. The study reveals that developed countries like the Netherlands and the United States have effectively enhanced disaster resilience and ecological protection by combining hard engineering measures with nature-based solutions. In contrast, countries such as Japan and Indonesia have focused more on community participation and nature-based solutions, demonstrating the advantages of low-cost and sustainable governance. Successful policy tools exhibit several common characteristics, including the integration of engineering measures with nature-based solutions, broad community involvement, multi-level collaborative governance, and adaptive management. Moreover, the study highlights the challenges and prerequisites of cross-regional policy transfer. This paper provides feasible recommendations for policymakers to enhance the adaptability and resilience of coastal areas in response to the long-term risks posed by climate change.

1 Introduction

Coastal areas are among the most biodiverse yet fragile ecosystems on the planet. Not only do they harbor rich biological resources, but they also provide numerous ecosystem services on a global scale. However, these areas face multiple challenges due to climate change and human activities, including sea level rise, frequent extreme weather events such as storm surges and typhoons, and ecological degradation caused by human activities and coastal development [1-2]. Sea level rise threatens the safety of low-lying coastal cities and rural areas, especially under the accelerating context of climate change. Human activities in coastal areas have also exacerbated ecological degradation, such as the loss of mangroves and the destruction of coastal wetlands, weakening the natural protective barriers in these regions and further increasing their exposure to risks [3]. Globally, over 40% of the population lives within 100 kilometers of the coast, making coastal resilience a matter not only of ecological protection but also of social and economic sustainability [4-5]. This is especially true for developing countries and small island states, where coastal vulnerability is pronounced due to inadequate infrastructure to deal with extreme climatic events and heavy dependence on marine resources. Enhancing coastal resilience is therefore crucial for addressing these challenges and ensuring long-term sustainability.

Coastal resilience refers to the ability of coastal areas to absorb shocks and maintain ecological, social, and economic functions in the face of multiple environmental pressures, such as climate change, sea level rise, and human activities [6]. The core of this concept is to enhance the adaptive capacity of coastal areas, enabling them to maintain sustainable development in the face of sudden environmental disturbances. The concept of resilience originated in the field of ecology, where it was used to describe the ability of ecosystems to recover to a balanced state after disturbances [7]. As the impacts of climate change intensified, the concept of resilience was gradually extended to socio-ecological systems, emphasizing the interconnections between ecological, social, and economic components, thus promoting the interdisciplinary integration of resilience research [8]. Within this context, Cutter et al. proposed the concept of "socio-ecological system resilience," further expanding the connotation of resilience to include the ability of communities to respond to environmental risks, highlighting the dynamic adjustment and continuous transformation capabilities of human society in coping with disasters [9]. This transformation goes beyond merely restoring to the original equilibrium state; it emphasizes achieving better management and development during the response to risks, thereby enhancing the overall adaptive capacity of the system.

The evolution of coastal governance policies can be traced back to the mid-20th century. Early coastal management predominantly relied on hard engineering measures, such as seawalls and breakwaters, to cope with storm surges and coastal erosion [10]. However, these

* Corresponding author: dongweiwei@tio.org.cn

singular engineering solutions often neglected the complexities of coastal ecosystems, leading to further environmental degradation [11]. With increasing recognition of the interrelationship between the environment and economy, this engineering-focused approach was gradually questioned, prompting a shift toward more integrated management policies. From the 1980s onwards, ecosystem-based management concepts began to permeate coastal governance. At the international level, organizations such as the United Nations Environment Programme (UNEP) and the World Bank promoted the concept of Integrated Coastal Zone Management (ICZM), emphasizing the need to consider economic, social, and environmental factors in coastal management to achieve sustainable development [12]. ICZM, as an emerging policy framework, seeks to achieve the sustainable use and protection of coastal resources through stakeholder participation and cross-sectoral collaboration [13]. This framework has effectively enhanced the comprehensiveness of management measures and gained widespread international recognition and promotion. Entering the 21st century, global climate change has posed new challenges to coastal environments, driving the trend toward adaptive management in coastal governance. Adaptive management dynamically responds to changing environmental conditions, emphasizing decision-making under uncertainty and adjusting actions based on actual outcomes to reduce uncertainty and improve management effectiveness [14]. Recently, Nature-based Solutions (NbS) have become an important trend in coastal management. NbS aim to mitigate coastal risks by leveraging natural ecosystems and their functions, such as the restoration of mangroves, salt marshes, and coral reefs, which act as natural barriers [15]. This approach not only reduces infrastructure costs but also achieves multiple ecosystem service benefits. Recent research has shown that NbS not only have significant effects in disaster reduction but also enhance local community resilience and livelihoods, as well as increase the carbon sequestration capacity of ecosystems [16-17]. Furthermore, the socio-ecological systems perspective has also attracted increasing attention. This perspective emphasizes the dynamic interaction between human society and the natural environment, promoting sustainable coastal governance by identifying vulnerabilities and enhancing adaptive capacities [18]. Such research highlights the importance of social, equity, and participation aspects in coastal management, promoting community-based management models that enable local communities to play a more active role in governance, thereby enhancing policy acceptability and effectiveness.

Effective policy interventions play a crucial role in enhancing coastal resilience, not only in disaster risk reduction but also in safeguarding ecosystem services and improving residents' quality of life [19]. Policies can strengthen coastal infrastructure, protect natural barriers (such as mangroves, dunes, and coral reefs), and promote community adaptive capacity to enhance the resilience of coastal areas, thereby achieving multiple ecological, economic, and social goals [20]. In summary, policy research on coastal resilience is significant not only for ecological protection but also for the well-being and economic sustainability of coastal communities. This study aims to provide actionable recommendations for policymakers by conducting a comparative analysis of governance experiences across different countries and regions to address the increasingly severe coastal challenges under climate change.

The core questions of this study are to compare how different countries and regions around the world address coastal challenges through policy and to analyze the effectiveness and limitations of these policies in building resilience. By systematically comparing representative cases from various countries, this paper seeks to identify which policy tools are most effective in enhancing coastal resilience and whether these experiences can be adapted and transferred across regions. Specifically, this study will explore the following key questions: 1) What are the core components of coastal resilience policies in different countries and regions? 2) Which policy tools are most effective in enhancing coastal resilience? 3) Can successful policies be transferred to other regions with different ecological, economic, and social contexts? The structure of this study is as follows: Section 2 introduces the comparative case study methodology adopted in this paper; Section 3 provides a detailed analysis and summary of effective governance strategies in different countries; Section 4 discusses the commonalities and differences of effective governance policies, challenges faced, and the possibility of policy transfer; and Section 5 concludes with recommendations for future policy development.

2 Research methodology

2.1 Comparative Case Study

The primary research method employed in this paper is the comparative case study [21]. Coastal resilience policies involve complex social, ecological, and economic contexts across different countries and regions. Using a comparative case study approach allows for a deeper understanding of these policies within their unique contexts. By conducting detailed analyses of representative countries and regions, this study aims to better reveal the effectiveness and limitations of different policy tools and governance strategies in real-world applications, thereby providing reference points for global policy transfer and innovation.

2.2 Data Sources and Analysis Methods

Data collection primarily involved a systematic literature review, including academic papers, government reports, and documents from international organizations related to coastal resilience policies. These documents provided background information and theoretical support, helping to understand the evolution and implementation of policies in various countries.

For data analysis, the study conducted a systematic analysis of the policy content of typical case countries, identifying policy measures, outcomes, and limitations. The focus was on analyzing the differences and commonalities of different policy frameworks and revealing key factors that may influence policy transfer.

2.3 Case Selection Criteria

To ensure comprehensiveness and representativeness, this study adopted the following criteria for selecting case countries or regions: (1) Geographic diversity: Countries and regions with different geographic locations and climatic conditions were selected to ensure that the study covered a wide range of environmental conditions, such as the low-lying coastal plains of the Netherlands, the mountainous coastlines of Japan, the Gulf Coast of the United States, and the tropical coasts of Indonesia. (2) Diversity of policy types: Case selection considered different types of policies, including engineering-dominated policies, nature-based solutions, and community-participation frameworks, to allow for comparative analysis of the effectiveness of different policy types in enhancing coastal resilience. (3) Implementation effectiveness: Countries and regions that have achieved significant improvements in resilience were prioritized, providing successful governance strategies that could guide policy development in other regions. (4) Economic development level: Both developed and developing countries were included to compare the influence and constraints of different economic development levels on coastal governance strategies.

3 Comparative Analysis of Global Resilience Policies

3.1 Successful Experiences in Effective Governance Strategies

Numerous countries around the world have accumulated valuable experiences in coastal governance. Table 1 lists some representative examples. In Europe, the Netherlands' "Living with Water" strategy combines large-scale hydraulic engineering projects (such as the Delta Works) with natural measures (such as dune restoration), achieving a balance between flood control and ecological conservation, thereby significantly enhancing coastal resilience [22]. The United Kingdom, through its "Managed Realignment" projects, has returned portions of coastal land to the sea to enhance the self-regulating capacity of natural habitats, reduce flood risks to surrounding communities, and protect biodiversity [23].

In East Asia, Japan's "Satoyama" concept advocates for harmonious coexistence between humans and nature, enhancing the adaptive capacity of coastal communities to disasters through community-led management measures while fostering community cohesion and social capital [24-25]. In China's Zhejiang Province, the "Blue Coast Project" integrates coastal wetland restoration with the construction of artificial reefs, effectively improving coastal defense, increasing fishery production, and boosting eco-tourism income, while also raising environmental awareness among local residents [26]. In Southeast Asia, Indonesia has addressed coastal erosion and storm surges by rebuilding mangroves, which not only effectively reduce wave energy but also provide livelihoods through fisheries and eco-tourism for local communities [27]. In Bangladesh, natural infrastructure measures have restored mangroves and embankments, effectively reducing the impacts of floods and cyclones while enhancing the economic resilience of coastal communities [28]. The Philippines' "Community Mangrove Reforestation Program" has mitigated coastal erosion through mangrove restoration, and community participation has increased environmental awareness and economic income for residents [29].

In Oceania, Australia's Queensland has implemented the "Coastal Ecological Restoration Project," restoring coastal wetlands and coral reefs, thereby improving resilience to tropical storms, enhancing fisheries resources, and supporting the sustainable development of tourism [30]. In New Zealand, Maori-led initiatives have enhanced coastal defense by restoring natural wetlands and estuarine areas, promoting eco-tourism and cultural activities in the process [31].

In the Americas, the Gulf Coast of the United States has adopted ecosystem-based management, effectively reducing hurricane damage through wetland and mangrove restoration, while providing multiple ecosystem services such as water purification and biodiversity protection [32-33]. In the Caribbean, the Dominican Republic has enhanced coastal ecosystems' protective functions through the "Coral Nursery and Artificial Reef Project," restoring fishery resources and improving resilience to storms and wave action [34].

These cases demonstrate that enhancing coastal resilience requires a comprehensive approach that combines engineering measures, nature-based solutions, and community capacity-building while adapting to the ecological and social conditions of different regions to achieve sustainable governance outcomes.

Table 1. Strategies and effectiveness of different countries in coastal zone resilience enhancement.

Country/Region	Strategy	Main Management Measures	Outcomes
Netherlands	Coexistence with Water Strategy	Combination of large-scale hydraulic engineering and natural measures (e.g., Delta Works, dune restoration)	Achieves balance between flood control and ecological protection
Japan	Satoyama Initiative	Community involvement with traditional knowledge and community-led management measures	Enhances coastal resilience and strengthens community cohesion and social capital

USA	Wetland Restoration	Restoration of wetlands and mangroves	Significantly reduces storm damage; provides water purification and biodiversity conservation services
Indonesia	Mangrove Reforestation	Replanting of mangroves	Reduces wave energy, providing coastal protection
Bangladesh	Nature-Based Infrastructure	Restoration of mangroves and flood prevention measures	Reduces disaster risks and enhances community livelihood sources and climate resilience
Australia	Coastal Ecosystem Restoration	Restoration of coastal wetlands and coral reefs	Reduces coastal erosion, improves fisheries resources, and promotes tourism
United Kingdom	Managed Realignment	Converting farmland back to coastal wetland and allowing tidal flooding	Reduces flood risks, providing habitat for birds and other wildlife
Belize	Coral Reef Restoration and Artificial Reefs	Deployment of artificial reefs	Enhances protection against storms and sea-level rise, promoting fisheries and resource recovery
Philippines	Mangrove Rehabilitation	Community involvement in mangrove restoration	Reduces disaster risks, enhancing fisheries resources and ecotourism opportunities
New Zealand	Coastal Wetland Restoration	Restoration of wetlands and river catchments	Strengthens flood and coastal protection, supporting ecotourism and cultural heritage
China (Zhejiang Province)	Blue Bay Initiative	Restoration of coastal wetlands and construction of artificial fish reefs	Enhances coastal defense, increasing fisheries and ecological tourism revenue

3.2 Analysis of Governance Strategies in Typical Case Countries

3.2.1 Coastal Governance Strategies in the Netherlands

The Netherlands is a leading example of coastal governance, renowned for its "Living with Water" philosophy and integrated water management policies. The Dutch resilience strategy combines hard engineering with nature-based solutions to address sea level rise and flood risks [22]. Examples include the Delta Works and dune restoration projects. This approach not only reduces flood risks but also enhances ecological and socio-economic value. The "Building with Nature" concept in the Netherlands exemplifies this idea, emphasizing the interactions between physical, ecological, and governance systems to ensure flood safety [35]. The Netherlands also emphasizes public participation, employing a Multilayered Safety (MLS) approach to introduce more adaptive and diversified governance strategies to cope with flood risk uncertainties. This approach improves policy acceptance and implementation efficiency while promoting stakeholder engagement in policymaking [36].

3.2.2 Resilience Policies in Japan and Indonesia

In the Asia-Pacific region, Japan and Indonesia exhibit different policy paths in addressing coastal issues. Japan, as a developed country, places significant emphasis on community participation, advocating for the use of traditional ecological knowledge to enhance the adaptive capacity of coastal communities. For instance, the "Satoyama" concept emphasizes harmonious coexistence between humans and nature, using community-led management approaches to improve resilience against coastal hazards [37]. On the other hand, Indonesia, as a developing country, relies more on nature-based solutions, particularly in the restoration and conservation of mangroves. In recent years, Indonesia has established community protection zones and promoted mangrove planting, significantly improving mangrove ecosystem recovery and the living conditions of coastal residents [38]. Mangroves not only provide crucial ecological barriers against storm surges but also offer livelihoods through fisheries and eco-tourism. Indonesia's experience demonstrates the feasibility and sustainability of low-cost, community-based nature restoration measures under resource constraints.

3.2.3 Coastal Governance Practices in the United States and Brazil

In the Americas, the coastal governance strategies of the United States and Brazil each reflect distinct management philosophies and practices. The United States has adopted ecosystem-based management measures, emphasizing natural solutions such as wetland restoration and mangrove planting to mitigate the effects of hurricanes and storm surges, while also enhancing carbon sequestration and biodiversity [39]. These measures not only provide ecosystem services but also significantly reduce the damage caused by natural disasters to coastal communities. Additionally, U.S. governance strategies emphasize collaboration between federal, state, and local governments to ensure coordinated policy implementation [40]. In contrast, Brazil's coastal governance primarily relies on the combination of federal laws and local actions, particularly in the protection of coastal ecosystems and control of coastal development [41]. In recent years, Brazil has introduced concepts such as ecological compensation and nature reserves in coastal ecosystem management, promoting sustainable development of mangrove and coral reef protection through local community cooperation [42]. Brazil's approach to coastal management emphasizes biodiversity conservation through legal frameworks and community cooperation, thereby enhancing coastal resilience and effectively responding to the impacts of climate change.

3.3 Insights from Effective Governance Strategies

By analyzing coastal resilience policies in different countries and regions, several effective governance strategies can be derived:

3.3.1 Integration of Hard Engineering and Nature-Based Solutions

The Delta Works in the Netherlands and wetland restoration projects in the United States demonstrate that integrating "gray" (e.g., dikes, seawalls) and "green" (e.g., wetlands, mangroves) infrastructure is an effective strategy for addressing flood threats and protecting ecosystems. This approach not only reduces the impacts of flooding and storm surges but also provides additional ecosystem services such as water purification and fisheries resources [4,43]. However, the successful implementation of such strategies requires substantial financial support and technical capacity, which may be challenging for some developing countries. Thus, while this combined approach is successful, its applicability depends on the availability of economic and technical resources.

3.3.2 Importance of Community Participation

Japan's "Satoyama" concept successfully integrates community involvement into policy implementation, enhancing coastal adaptation by leveraging social capital and community knowledge and skills [44]. Similarly, Indonesia's mangrove restoration projects show that public participation and nature-based solutions can have significant impacts in resource-constrained environments [45]. These cases indicate that community participation plays a critical role in the sustainability of policies, particularly in regions with limited resources [46]. The success of community participation depends on how effectively governments can mobilize and utilize community resources.

3.3.3 Necessity of Multi-Level Collaborative Management

The United States' response to storm surges and hurricanes along the Gulf Coast demonstrates the importance of collaboration between federal, state, and local governments to ensure effective policy implementation and efficient resource allocation [40]. A multi-level governance structure not only enhances equity in resource distribution but also improves rapid response capacity during crises. The success of such multi-agency collaboration is facilitated by the mature governmental system and sound legal framework in the United States.

3.3.4 Scientific Support and Adaptive Management

The Netherlands' adaptive water management strategy involves continuous scientific monitoring and policy adjustments to flexibly respond to changing environmental risks [47]. This science-based adaptive management model emphasizes the importance of "learning by doing," particularly in the context of uncertainties brought by climate change. The success of this strategy lies in its comprehensive monitoring systems and scientific evaluation mechanisms, which enable policy optimization through dynamic adjustments [48].

4 Discussion

4.1 Differences and Commonalities in Policy Frameworks

A comparison of coastal resilience policies across various countries reveals significant differences in policy objectives, implementation methods, and stakeholder involvement, though some commonalities are evident. For instance, countries such as the Netherlands and the United States focus on combining hard engineering with nature-based solutions to address climate change threats. Their policy goals tend to be more explicit and quantifiable [1]. In contrast, countries like Japan and Indonesia emphasize community resilience building and the transmission of traditional knowledge, reflecting policy goals that are more socially oriented. In terms of implementation, the Netherlands and the United States employ large-scale infrastructure projects, such as levees and breakwaters, supplemented with ecosystem-based solutions like wetland restoration. On the other hand, Japan and Indonesia rely more on community involvement and ecological conservation, using practices such as mangrove restoration and vegetation protection to enhance resilience. Regarding stakeholders, developed countries like the Netherlands and the United States involve a diverse range of actors, including government agencies, academia, and the private sector, with clearly defined roles and high levels of participation [44]. In contrast, developing countries like Indonesia rely more heavily on community and NGO partnerships, emphasizing community-led participation mechanisms that are particularly viable in low-income, resource-constrained regions [45].

4.2 Effectiveness and Challenges

While the implementation of coastal resilience policies has yielded notable successes globally, countries also face common challenges. The Netherlands' integration of hard engineering and nature-based solutions has achieved dual goals of flood risk reduction and ecological protection. However, the high infrastructure costs underscore the issue of sustainable funding sources, which may hinder the replicability of such high-cost models in developing countries [49]. The success of Japan and Indonesia in community engagement demonstrates the effectiveness of low-cost, community-based solutions for resilience building. However, ensuring long-term policy continuity and sustained community involvement remains challenging [50]. With climate change increasing the frequency of disasters, issues like public fatigue and uncertain financial support may threaten the sustainability of these projects [51]. In the United States, ecosystem-based approaches like wetland restoration not only mitigate hurricane impacts but also provide multiple ecosystem services. However, policy implementation still encounters persistent coordination challenges between federal and local governments, especially in terms of funding allocation, policy priorities, and management responsibilities [32].

4.3 Possibilities and Constraints in Policy Transfer

Despite valuable lessons from successful policies, the transferability of these experiences across regions faces several constraints and challenges.

4.3.1 Socioeconomic Disparities

Economic capacity is a key factor determining policy success. High-investment engineering solutions like the Netherlands' Delta Project require substantial financial resources and advanced technical support. While effective in developed countries, such measures are often unaffordable for many developing nations. Conversely, Indonesia's experience demonstrates that community-based ecological restoration offers a low-cost and effective alternative, especially in resource-limited environments. Thus, policy transfer requires adjustments based on local economic conditions, with nature-based solutions proving more viable in financially constrained regions.

4.3.2 Cultural and Social Contexts

Policy success is not only determined by economic factors but is also influenced by cultural and social contexts. In some countries, community engagement and the use of traditional knowledge have proven successful, as seen in Japan's "Satoyama" concept. However, in other regions, cultural and social structures may be less conducive to such community-led management approaches. For example, in more centralized states, limited public input in policy-making processes can diminish community participation. Therefore, policy transfer must account for cultural and social variations, ensuring alignment with local behavioral norms and values.

4.3.3 Institutional and Governance Differences

The collaborative management model between federal and local governments in the United States has proven effective for policy implementation. However, this approach relies on a mature institutional framework and clearly delineated government responsibilities. For countries lacking well-established governance systems, cross-departmental and intergovernmental policy coordination poses significant challenges. Thus, institutional development and governance capacity must be strengthened in such countries to ensure the effective implementation of transferred policies.

5 Conclusion

This study's comparative analysis of coastal resilience policies across several countries highlights the successful practices and challenges in enhancing coastal resilience. Findings indicate that combining hard engineering and nature-based solutions, promoting community involvement, multi-level collaborative governance, and adaptive management are core components of effective resilience policies. However, challenges remain, including high costs, interdepartmental coordination issues in policy execution, and the sustainability of community engagement. In developing countries, economic constraints and governance limitations hinder the feasibility of high-cost, technology-intensive strategies. Future research could further focus on the adaptability of cross-regional policy transfer, exploring effective ways to apply and adjust successful governance practices across varying economic, social, and ecological contexts. Additionally, studies may examine how to better integrate science, technology, and adaptive management into policy design and decision-support systems to address the complexities and uncertainties inherent in coastal resilience governance.

References

1. Toimil, A., Losada, I. J., Nicholls, R. J., Dalrymple, R. A., & Stive, M. J. (2020). Addressing the challenges of climate change risks and adaptation in coastal areas: A review. Coastal Engineering, 156, 103611.
2. Masselink, G., & Lazarus, E. D. (2019). Defining

coastal resilience. Water, 11(12), 2587.

3. Abelson, A., Reed, D. C., Edgar, G. J., Smith, C. S., Kendrick, G. A., Orth, R. J., ... & Nelson, P. (2020). Challenges for restoration of coastal marine ecosystems in the anthropocene. Frontiers in Marine Science, 7, 544105.
4. Neumann, B., Vafeidis, A. T., Zimmermann, J., & Nicholls, R. J. (2015). Future coastal population growth and exposure to sea-level rise and coastal flooding - a global assessment. PLOS ONE, 10(3), e0118571.
5. Hinkel, J., Jaeger, C., Nicholls, R. J., Lowe, J., Renn, O., & Peijun, S. (2015). Sea-level rise scenarios and coastal risk management. Nature Climate Change, 5(3), 188-190.
6. Rölfer, L., Celliers, L., & Abson, D. (2022). Resilience and coastal governance: Knowledge and navigation between stability and transformation. Ecology and Society, 27(2).
7. Holling, C. S. (1973). Resilience and stability of ecological systems. Annual Review of Ecology and Systematics, 4(1), 1-23.
8. Folke, C. (2006). Resilience: The emergence of a perspective for social–ecological systems analyses. Global Environmental Change, 16(3), 253-267.
9. Cutter, S. L., Barnes, L., Berry, M., Burton, C., Evans, E., Tate, E., & Webb, J. (2008). A place-based model for understanding community resilience to natural disasters. Global Environmental Change, 18(4), 598-606.
10. Bird, E. C. F. (1985). Coastal changes: A global review. Wiley.
11. Elko, N., & Briggs, T. R. (2020). National coastal management challenges and needs. Shore & Beach, 88(4), 34.
12. Cicin-Sain, B., & Knecht, R. (2013). Integrated coastal and ocean management: concepts and practices. Island press.
13. Sorensen, J. C. (1997). National and international efforts at integrated coastal management: Definitions, achievements, and lessons. Coastal Management, 25(1), 3-41.
14. Williams, B. K., & Brown, E. D. (2016). Technical challenges in the application of adaptive management. Biological Conservation, 195, 255-263.
15. Vasseur, L. (2021). How ecosystem-based adaptation to climate change can help coastal communities through a participatory approach. Sustainability, 13(4), 2344.
16. Raymond, C. M., Frantzeskaki, N., Kabisch, N., Berry, P., Breil, M., Nita, M. R., ... & Calfapietra, C. (2017). A framework for assessing and implementing the co-benefits of nature-based solutions in urban areas. Environmental Science & Policy, 77, 15-24.
17. Seddon, N., Chausson, A., Berry, P., Girardin, C. A. J., Smith, A., & Turner, B. (2020). Understanding the value and limits of nature-based solutions to climate change and other global challenges. Philosophical Transactions of the Royal Society B: Biological Sciences, 375(1794), 20190120.
18. Folke, C., Biggs, R., Norström, A. V., Reyers, B., & Rockström, J. (2016). Social-ecological resilience and biosphere-based sustainability science. Ecology and Society, 21(3).
19. Adger, W. N., Hughes, T. P., Folke, C., Carpenter, S. R., & Rockström, J. (2005). Social-ecological resilience to coastal disasters. Science, 309(5737), 1036-1039.
20. Cabana, D., Rölfer, L., Evadzi, P., & Celliers, L. (2023). Enabling climate change adaptation in coastal systems: A systematic literature review. Earth's Future, 11(8), e2023EF003713.
21. Cunningham, S. (2021). Introducing the Comparative Case Study. In Causal Inference: The Mixtape (pp. 511-525). New Haven: Yale University Press.
22. van Koningsveld, M., & Mulder, J. P. M. (2004). Sustainable coastal policy developments in The Netherlands: A systematic approach revealed. Journal of Coastal Research, 20(2), 375-385.
23. Esteves, L. S., & Thomas, K. (2014). Managed realignment in practice in the UK: results from two independent surveys. Journal of Coastal Research, (70), 407-413.
24. Takeuchi, K. (2010). Rebuilding the relationship between people and nature: The Satoyama Initiative. Ecological Research, 25(5), 891-897.
25. Berkes, F. (2017). Environmental governance for the anthropocene? Social-ecological resilience and collaborative learning. Sustainability, 9(7), 1232.
26. Zhang, Y., Li, H., Hou, X., Guo, P., & Guo, J. (2024). Coastline protection and restoration: A comprehensive review of China's developmental trajectory. Ocean & Coastal Management, 251, 107094.
27. Primavera, J. H. (2005). Mangroves, fishponds, and the quest for sustainability. Science, 310(5745), 57-59.
28. Smith, A. C., Tasnim, T., Irfanullah, H. M., Turner, B., Chausson, A., & Seddon, N. (2021). Nature-based solutions in Bangladesh: evidence of effectiveness for addressing climate change and other sustainable development goals. Frontiers in Environmental Science, 9, 737659.
29. Primavera, J. H., & Esteban, J. M. A. (2008). A review of mangrove rehabilitation in the Philippines: successes, failures and future prospects. Wetlands Ecology and Management, 16, 345-358.
30. Hein, M. Y., Vardi, T., Shaver, E. C., Pioch, S., Boström-Einarsson, L., Ahmed, M., ... & McLeod, I. M. (2021). Perspectives on the use of coral reef restoration as a strategy to support and improve reef ecosystem services. Frontiers in Marine Science, 8, 618303.
31. Hepburn, C. D. (2019). Local, community-led interventions to address global-scale problems and environmental extremes in coastal ecosystems. In Marine Extremes (pp. 141-158). Routledge.

32. Costanza, R., Pérez-Maqueo, O., Luisa Martinez, M., Sutton, P., Anderson, S. J., & Mulder, K. (2008). The value of coastal wetlands for hurricane protection. Ambio, 37(4), 241-248.
33. Barbier, E. B., Hacker, S. D., Kennedy, C., Koch, E. W., Stier, A. C., & Silliman, B. R. (2011). The value of estuarine and coastal ecosystem services. Ecological Monographs, 81(2), 169-193.
34. Shaver, E. C., McLeod, E., Hein, M. Y., Palumbi, S. R., Quigley, K., Vardi, T., ... & Wachenfeld, D. (2022). A roadmap to integrating resilience into the practice of coral reef restoration. Global Change Biology, 28(16), 4751-4764.
35. Borsje, B. W., de Vries, S., Janssen, S. K., Luijendijk, A. P., & Vuik, V. (2017). Building with nature as coastal protection strategy in the Netherlands. In Living Shorelines (pp. 137-156). CRC Press.
36. Molenveld, A., & van Buuren, A. (2019). Flood risk and resilience in the Netherlands: In search of an adaptive governance approach. Water, 11(12), 2563.
37. Duraiappah, A. K., Nakamura, K., Takeuchi, K., Watanabe, M., & Nishi, M. (Eds.). (2012). Satoyama–Satoumi ecosystems and human well-being: Socio-ecological production landscapes of Japan. United Nations University Press.
38. Martin, E., Ulya, N. A., Yunardy, S., Agustina, K., Meidalima, D., & Chuzaimah, C. (2024). Navigating Mangrove Protection: A Jurisdictional Approach to Climate Action in South Sumatra, Indonesia. Climate Law, 14(1), 67-94.
39. Arkema, K., Bennett, R., Dausman, A., & Materman, L. (2019). United States: Blending Finance Mechanisms for Coastal Resilience and Climate Adaptation. Green Growth That Works: Natural Capital Policy and Finance Mechanisms from Around the World, 213-236.
40. Jordan, S., & Benson, W. (2013). Governance and the Gulf of Mexico coast: how are current policies contributing to sustainability?. Sustainability, 5(11), 4688-4705.
41. Horta, P., Pinho, P. F., Gouvêa, L., Grimaldi, G., Destri, G., Mueller, C. M., ... & Cotrim da Cunha, L. (2020). Climate Change and Brazil's coastal zone: socio-environmental vulnerabilities and action strategies. Sustainability in Debate/Sustentabilidade em Debate, (3).
42. Asmus, M. L., & Scherer, M. E. G. (2016). Ecosystem-based knowledge and management as a tool for integrated coastal and ocean management: A Brazilian initiative. Journal of Coastal Research, 75(sp1), 1282-1286.
43. Renaud, F. G., Sudmeier-Rieux, K., & Estrella, M. (Eds.). (2013). The role of ecosystems in disaster risk reduction. United Nations University Press.
44. Olsson, P., Folke, C., & Berkes, F. (2004). Adaptive comanagement for building resilience in social–ecological systems. Environmental Management, 34(1), 75-90.
45. Pelling, M., & Blackburn, S. (2013). Megacities and the Coast. Rutledge, Inc. NY.
46. Nyaranga, M. S., Hao, C., & Hongo, D. O. (2022). The role of public participation in governance towards achieving sustainable development. Part 2. RUDN Journal of Public Administration, 9(1), 29-41.
47. Van Buuren, A., Keessen, A. M., Van Leeuwen, C., Eshuis, J., & Ellen, G. J. (2015). Implementation arrangements for climate adaptation in the Netherlands: characteristics and underlying mechanisms of adaptive governance. Ecology and Society, 20(4).
48. van Valkengoed, A. M., & Steg, L. (2019). Meta-analyses of factors motivating climate change adaptation behaviour. Nature Climate Change, 9(2), 158-163.
49. Brière, C., Janssen, S. K., Oost, A. P., Taal, M., & Tonnon, P. K. (2018). Usability of the climate-resilient nature-based sand motor pilot, The Netherlands. Journal of coastal conservation, 22, 491-502.
50. Patton, M. Q. (2014). Qualitative research & evaluation methods: Integrating theory and practice. Sage publications.
51. IPCC. (2014). Climate change 2014: Impacts, adaptation, and vulnerability. Cambridge University Press.

Research on Environmental Risk Assessment Techniques for Acid Gas Fields in Complex Terrains

Xuan Xu*, Jia Zeng, Chunling Tang, Fangbing Gong

Petro China Suothwest Oil & Gas field Company Safety, Environment & Technology Supervision Research Institute, No.12 Tianfu Rd North, Chengdu, 610041, Sichuan, China

Abstract. Analyzed the current status and difficulties of the environmental risk analysis system, proposed that environmental risk simulation should be conducted from two dimensions: concentration distribution and receptor injury probability analysis, to support the environmental risk management system analysis. Explored the precise assessment of atmospheric environmental risks in the development of acidic gas fields under complex terrain conditions, combined with the toxicity characteristics of hydrogen sulfide, and provided regional and graded emergency management strategies and suggestions.

1. Introduction

Natural gas, as a clean and efficient fossil fuel, is primarily composed of low molecular weight alkanes. Acidic natural gas mainly refers to natural gas containing high concentrations of hydrogen sulfide (H_2S) and carbon dioxide (CO_2). Acidic gas fields in China are mainly distributed in complex mountainous areas. In these complex mountainous areas, there are significant differences and diversities in terrain, ecosystems, land use, hydrological characteristics, and soil types. At the same time, due to the dense population and scattered rural households in the southwestern region of China, with limited transportation conditions in the mountains, these factors all pose challenges to the environmental risk management of acidic gas fields.

2. Environmental Risk Analysis of Acidic Gas Fields

2.1 Accident Scenario Analysis

In the environmental risk analysis of acidic gas fields, the main accident scenarios include leaks, fires and explosions, and the subsequent secondary pollution events they trigger. Leaks can lead to the release of H_2S gas, causing severe toxic effects, especially in densely populated areas. After a blowout occurs, the raw gas from the formation is ejected through the wellhead, and hydrogen sulfide is released into the atmosphere along with methane, posing a threat to human health and environmental safety. This study has identified the blowout as the most credible accident with the greatest impact on atmospheric environmental risk.

2.2 Simulation

Environmental risk simulation research is a scientific method for assessing and predicting the potential impact of pollution events on ecosystems and human health. Yang Dongdong et al. used CFD to simulate the diffusion process of hydrogen sulfide on offshore drilling platforms.[1] Gao Kaigeconducted a simulation of hydrogen sulfide leakage diffusion at a high-sulfur well site in Northeastern Sichuan based on the FLACS software[2] . Xin Baoquan[3] and Xu Xiaoli[4] used DNV's PHAST software to study the diffusion characteristics, concentration distribution, and downwind distance of H_2S under different leakage conditions.

Considering the properties of hydrogen sulfide gas after a leak and the real-time conditions at the leak site, and based on the high-resolution terrain and underlying surface conditions around the leak point, an appropriate diffusion model is selected to simulate and analyze the concentration and affected area of hydrogen sulfide. Researchers have discussed the application of the CHARM model in atmospheric diffusion over complex terrain and its application in emergency environmental risk management, with positive feedback [5,6]. This study uses the CHARM model for simulation. The wind field of the CHARM model projects observational data from the observation points upwards to the grid parts without observational data, using a power exponent extrapolation in the projection calculation. After the wind speed is projected upwards, the wind speed and direction at each position in the wind field are decomposed into horizontal (u) and vertical (v) vectors. The data for each horizontal plane in the entire wind field is formed by weighted interpolation from the vertical files, followed by corrections for terrain and other parameters. The main simulation process of the CHARM diffusion module is: setting up the grid—establishing the wind field—simulating diffusion. Additionally, this module is capable

*Corresponding author's e-mail: 421095176@qq.com

of considering the surface deposition of chemicals, chemical reactions, and the treatment of special pollutants (hydrogen fluoride).

In the preliminary environmental risk assessment of construction projects in China, the values for air temperature, stability, relative humidity, and wind speed under the "most adverse meteorological conditions" are all specified, as shown in Table 1. Compared to large-scale models such as CFD (Computational Fluid Dynamics) and CALPUFF (California Line Source Dispersion Model), the CHARM (Complex Hazardous Air Release Model) model does not require the driving of external reanalysis data. It can determine the "most unfavorable meteorological conditions" at the point of leakage. Then, through iterative calculations adjusted according to topographic data, the model minimizes three-dimensional divergence to obtain the dynamic influence of the terrain on the wind field, deducing the wind field that most closely resembles the most adverse diffusion conditions

Table 1. The most unfavorable meteorological conditions

Air temperature(°C)	Relative humidity (%)	Wind speed(m/s)	Stability
25	50	1.5	F

2.3 Simulation Results

Based on 20 years of local climate data, the prevailing wind direction in the mountainous area is north wind. Considering the relative positions of the leak point and the surrounding sensitive points, most of the sensitive points are located to the east, southeast, and southwest of the leak point. Therefore, this study plans to simulate four wind directions at the leak point: north (N), northeast (NE), northwest (NW), and west (W), with all other meteorological conditions being the same, to obtain the concentration distribution map of hydrogen sulfide.

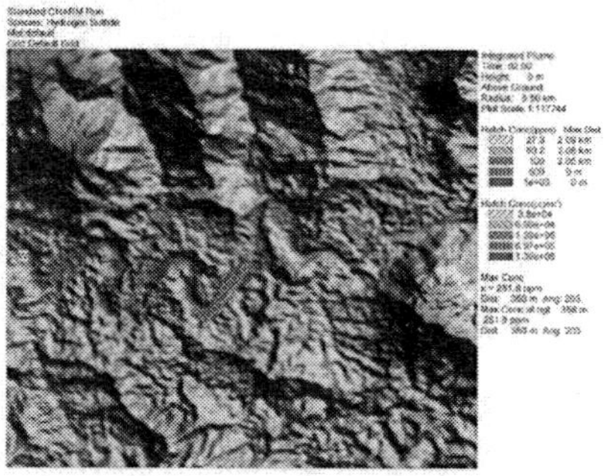

Figure 1. Diffusion Concentration Distribution Map under North Wind (N) Scenario.

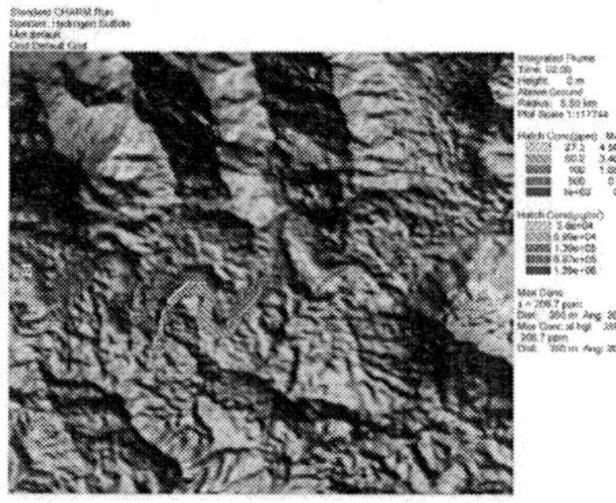

Figure 2. Diffusion Concentration Distribution Map under Northeast Wind (NE) Scenario.

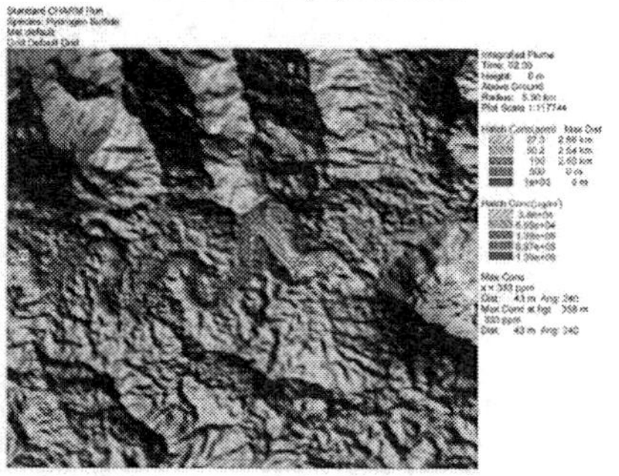

Figure 3. Diffusion Concentration Distribution Map under West Wind (W) Scenario.

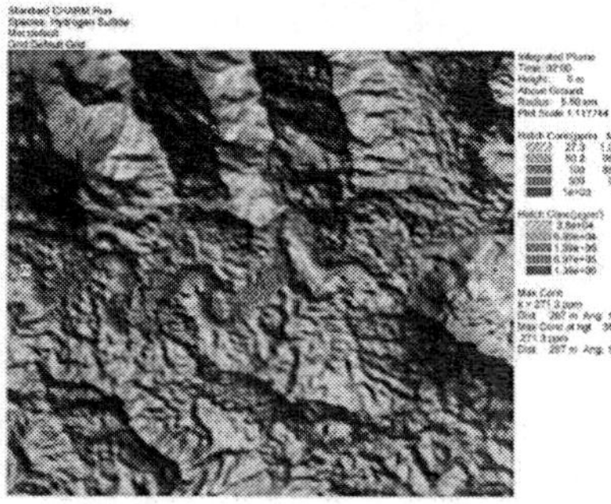

Figure 4. Diffusion Concentration Distribution Map under Northwest Wind (NW) Scenario.

Based on the predicted concentration distribution map, it can be seen that terrain and other underlying surface parameters have a significant impact on the wind field, with pollutants diffusing along the riverbed towards the downwind direction, as shown in Figure 1-Figure 4. Under conditions where the river channel is in the same direction as the wind (northeast wind, NE), the diffusion range of hydrogen sulfide is the greatest. Under conditions where the river channel intersects with the wind direction (northwest wind, NW), hydrogen sulfide is blocked by the high mountains to the east and south, and under the influence of the northwest wind, it accumulates on the southeast side of the leak point where the terrain is lower. Under west wind conditions, hydrogen sulfide is forced by the terrain of the high mountains to the east and continues to diffuse eastward after circumventing the mountains to the north along the river channel.

2.4 Exposure Assessment of Receptors

Exposure assessment of receptors is the most critical part of consequence evaluation. Goyak et al.[7] determined the starting point for setting exposure limits for hydrogen sulfide and emphasized the progressive biological effects of acute exposure. Qin Ge et al. [8] studied the histopathological characteristics of hydrogen sulfide poisoning. Wang Yao et al. [9] discussed the relevant risk factors for acute hydrogen sulfide poisoning combined with central nervous system injury. Zhi Yan et al. conducted a full life cycle environmental risk management of toxic and hazardous substances [10]. Jiang Hualiang et al. combined the diffusion of toxic and hazardous substances to study the early warning system for chemical industrial parks [11].

In the exposure assessment of receptors, this paper adopts the grid points most affectedunder the influence of various dominant wind directions as our receptors for analysis (the peak of P-value as shown in Table 2). In practical work, this criterion will vary with the actual population distribution. This study extracted the changes in hydrogen sulfide concentration at the receptor under accident conditions. Figure 5 to Figure 8 display the changes in hydrogen sulfide concentration of the receptors that are most affected under various wind direction conditions. It can be observed that under the influence of the highest peak value (as shown in Figure 5), the highest instantaneous peak value of hydrogen sulfide concentration is less than 350 ppm, and the duration is less than 2 minutes. To further investigate the exposure injuries of the receptor when the hydrogen sulfide concentration is high, this study then conducted an exposure assessment by combining the methods and injury probability parameters recommended by the Netherlands Organisation for Applied Science Research (TNO).The selection of receptors followed the principle of considering the most unfavorable conditions, using the points of maximum hydrogen sulfide concentration for each wind direction as representative receptor points.

Table 2 Exposure Assessment under Predicted Maximum Concentration Points in Different Wind direction Scenarios

wind direction	Threshold Limit Value (TLV) for hydrogen sulfide exposure/ppm	Duration/min	A	B	n	Y	P
W	100	15	-11.5	1	1.9	0.73	9.7×10^{-6}
NW	100	21	-11.5	1	1.9	1.06	4.16×10^{-5}
N	100	17	-11.5	1	1.9	0.85	1.69×10^{-5}
NE	100	18	-11.5	1	1.9	0.91	2.16×10^{-5}

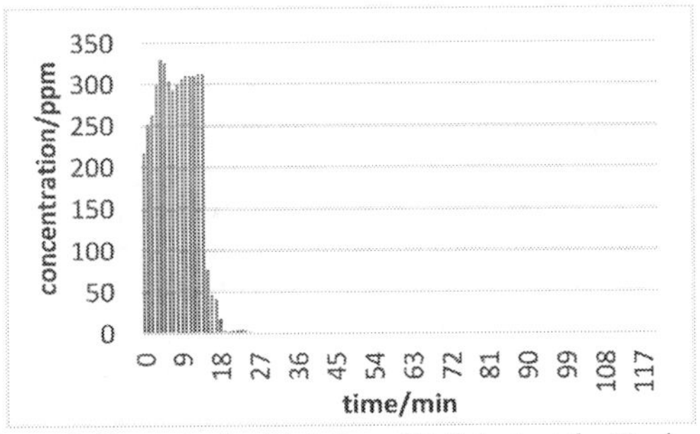

Figure 5. Concentration Variation of Pollutants at the Maximum Impact Receptor under West Wind (W) Scenario.

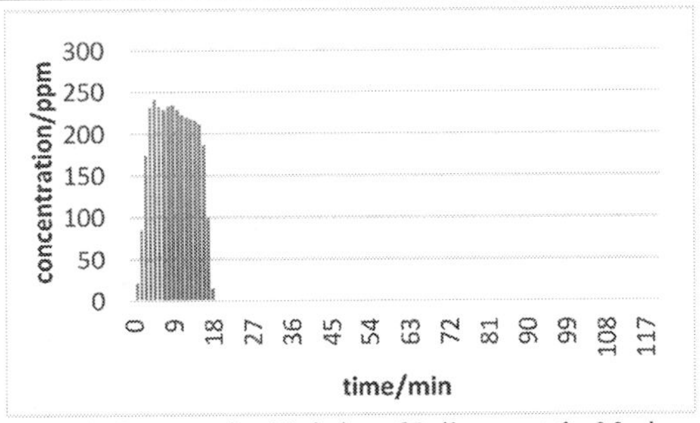

Figure 6. Concentration Variation of Pollutants at the Maximum Impact Receptor under North Wind (N) Scenario.

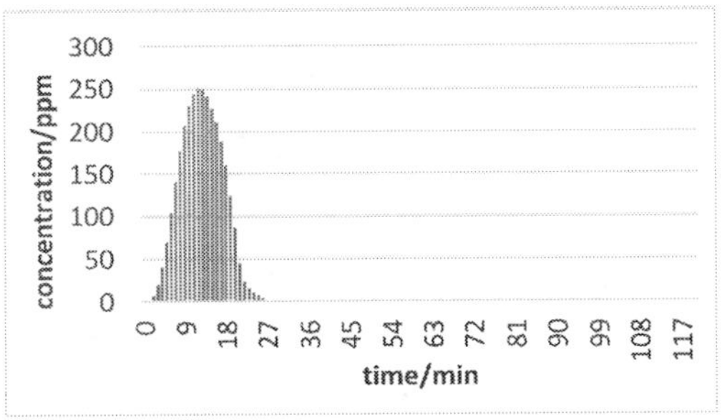

Figure 7. Concentration Variation of Pollutants at the Maximum Impact Receptor under Northwest Wind (NW) Scenario.

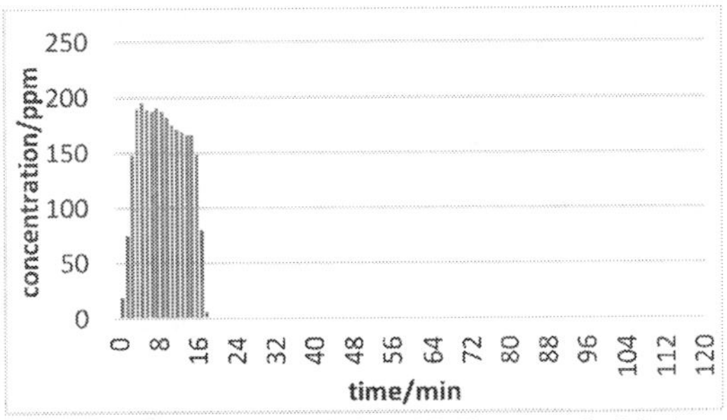

Figure 8. Concentration Variation of Pollutants at the Maximum Impact Receptor under Northeast Wind (NE) Scenario.

In the event of a complete blowout, for the four wind direction scenarios, the range of fatality probabilities is between 9.7×10^{-6} and 4.16×10^{-5}, which is considered an acceptable level, as detailed in Table 2. It is noteworthy that under the northwest wind condition, the receptor representative point downwind has the highest hydrogen sulfide fatality probability at 4.16×10^{-5}. The reason for this, according to the study, is that under northwest wind conditions, hydrogen sulfide is blocked by the high mountains to the east and south, and accumulates in the low-lying areas in front of the mountains. This results in a smaller diffusion range and higher concentrations of hydrogen sulfide, leading to a higher fatality probability for receptors. After exposure assessment analysis, the fatality probabilities caused by the blowout accident case in this study are all acceptable.

3. Conclusions

Based on the needs of emergency management in the development of acidic gas fields, an innovative comprehensive environmental risk analysis system that includes concentration distribution and receptor exposure assessment is proposed. Taking a high-sulfur gas well blowout accident as a case study, the CHARM model is used for simulation and prediction, extracting information from the three dimensions mentioned above. Combined with the characteristics of acute hydrogen sulfide injury, regional and hierarchical emergency measures and strategies are provided. This provides a scientific basis for orderly organizing mass evacuations and other emergency measures under accident scenarios, refining and perfecting emergency management.

Currently, there are still many deficiencies in the risk assessment of the atmospheric environment and emergency management in natural gas exploration and development. For instance, the precise determination of accidental gas leaks, the nesting connection between real-time wind direction and simulated wind fields, and the feedback coupling of monitoring data to emergency measures all need to be supplemented and improved. There is an urgent need for more and deeper research results to support these efforts, with the hope that future research will continue to enhance the depth and breadth of study.

References

1. Yang Dongdong, Chen Guoming, Zhu Yuan, et al. Dynamic assessment of consequences for poisoning accidents caused by H2 S release on offshore platforms[J]. CIESC Journal, 2020, 71(08): 3839-3848.
2. Gao Kaige, Li Yong, Zhu Xianjun, et al. Safety evaluation of well control of high-sulfur gas wells and measures to improve blowout emergency capability[J]. Chemical Industry and Engineering Progress, 2020, 39(S2):83-88.
3. Xin Baoquan, Lu Wei, Zhang Xiangjian, et al. Quantitative analysis method and index response law of hydrogen sulfide leakage[J]. The Chinese Journal of Process Engineering, 2018, 18(S1):82-88.
4. Xu Xiaoli, Mao Weizhi, Zhao Zihao, et al.Application of Hydrogen Sulfide Dispersion Simulation Technique on Offshore Production Facility[J]. Ship Engineering, 2021, 43(03):142-145.
5. Duan Yong, Hu Cuijuan, Li Guangru. The Algorithm of CHARM and Its Application in Environmental Risk Assessment[J]. Environmental Engineering, 2014, 32(06).
6. Mark W E. Complex Hazardous Air Release Model (CHARM), Technical Reference Manual[Z]. 2009.
7. Katy O Goyak, R Jeffrey Lewis. Application of Adverse Outcome Pathway Networks to Integrate Mechanistic Data Informing The Choice of A Point of Departure for Hydrogen Sulfide Exposure Limits, Critical Reviews in Toxicology, 2021.
8. Qin Ge, Wu Xiaojun, Zhao Peng, et al. Research Progress on Metabolic Mechanisms and Testing Methods of Hydrogen Sulfide Poisoning[J]. Environmental Chemistry.2024, 1-13.
9. Wang Yao, Zhang Jinsong, Fan Bowen, et al. Research on the related risk factors and predictive value of acute niversity (Natural Science Edition) 2023, 43(12):1650-1655+1680.
10. Zhi Yan, Wu Huihui, Yu Fang, et al. Design of a Comprehensive Life-cycle Environmental Risk Management System for Hazardous Chemicals. Asian Journal of Ecotoxicology. 2024, 19(5), 64-74
11. Jiang Hualiang, Yin Qin, Ren Xiangyu, et al.Identification and Monitoring of Dispersion Pathways in the Early Warning System for Toxic and Hazardous Gases in Chemical Park. Environmental Pollution & Control. 2024, 46(06): 784-788.

Study on the Migration Pattern of Organic Carbon in Slope Farmland in the Southeast of Heilongjiang Province

Hongfeng Wang[1,2,a], Jiuyi Wang[1,2,b], Jing Zhao[1,2,c], Guomeng Fu[1,2,d], Ze Shao[1,2,e], Xingchen Yu[3,*]

[1] Harbin Center for Integrated Natural Resources Survey, China Geological Survey, Harbin, Heilongjiang, China
[2] Observation and Research Station of Earth Critical Zone in Black Soil, Ministry of Natural Resources, Harbin, Heilongjiang, China
[3] Yantai Center of Coastal Zone Geological Survery, China Geological Survey, Yantai, Shandong, China

Abstract. This study addresses the issue of soil organic carbon (SOC) migration in slope farmland in the southeastern part of Heilongjiang Province by constructing a machine learning-based predictive model for organic carbon migration. Based on multi-year monitoring data, a three-tier model architecture consisting of a data layer, a computational layer, and an application layer was established, and an improved random forest algorithm was used to dynamically simulate the organic carbon migration process. The model integrates terrain, meteorological, and soil characteristic parameters to achieve accurate predictions of organic carbon fluxes at different spatial and temporal scales. The results indicate that the model exhibits good predictive ability under different slope gradients and rainfall conditions, providing a scientific basis for regional soil and water conservation and farmland quality improvement.

1. Introduction

Soil organic carbon loss in slope farmland is a key factor affecting the stability of agricultural ecosystems. The hilly region in the southeastern part of Heilongjiang Province, an important grain-producing area, experiences severe organic carbon loss in slope farmland, which directly threatens regional agricultural sustainability. Existing studies mainly rely on experimental observations and empirical models for predictions, but prediction accuracy is insufficient under complex terrain conditions, and the simulation of response mechanisms to factors such as rainfall and terrain remains weak[1]. This study, based on machine learning methods and multi-year monitoring data, aims to construct a high-precision organic carbon migration prediction model to reveal the migration patterns and provide scientific guidance for regional soil and water conservation.

2. Data collection and processing methods

2.1 Multi-source Data Collection System Design

In this study, typical slope farmland areas were selected in the southeastern part of Heilongjiang Province to establish a multi-source data collection system, including soil sampling, automatic meteorological stations, and surface runoff monitoring plots. Soil samples were collected at three depths: 0-20 cm, 20-40 cm, and 40-60 cm, with 45 sampling points in total, as shown in Figure 1. The automatic meteorological stations recorded data on rainfall, temperature, and soil moisture every 30 minutes[2]. The surface runoff plots, each with an area of 2m × 5m, were set up at three different slope gradients: 5°, 15°, and 25°, with three repetitions at each slope, to monitor runoff volume and sediment content. Additionally, Landsat 8 satellite imagery was used to obtain NDVI data for the study area from 2020 to 2023, with a temporal resolution of 16 days.

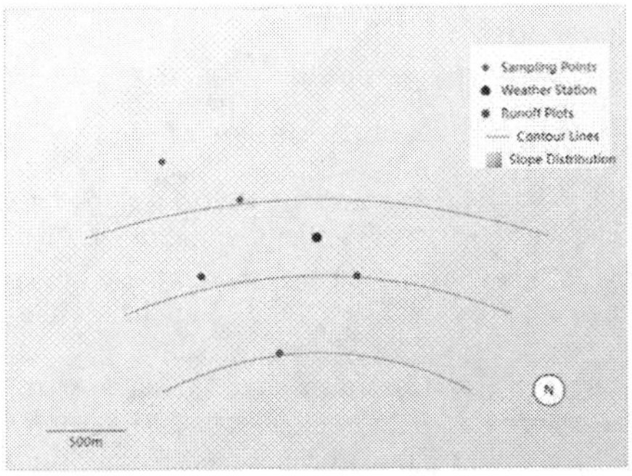

Figure 1 Sampling Point Distribution Map of the Study Area

2.2 Data Preprocessing Methods

The raw data collected were systematically preprocessed to ensure data quality. Soil organic carbon content was determined using the potassium dichromate method, with three parallel experiments conducted, and the relative standard deviation was controlled within 5%. Meteorological data were processed using a moving average method to handle outliers, with validation against

* Corresponding author: *yuxingchen@mail.cgs.gov.cn, a wanghongfeng@mail.cgs.gov.cn, b wjiuyi@mail.cgs.gov.cn, c zhaojing01@mail.cgs.gov.cn, d fuguomeng@mail.cgs.gov.cn, e shaoze@mail.cgs.gov.cn

adjacent meteorological station data for consistency[3]. After removing outliers from the runoff sediment data, linear interpolation was used to fill in missing values. Remote sensing images were atmospherically corrected, geometrically corrected, and cloud-detected to generate NDVI datasets. The typical data statistics after preprocessing are shown in Table 1.

Table 1 Statistical Characteristics of Preprocessed Typical Data

Monitoring Indicator	Sample Size	Mean Value	Standard Deviation	Coefficient of Variation
Soil Organic Carbon (g/kg)	135	18.45	3.27	0.177
Soil Bulk Density (g/cm^3)	135	1.32	0.15	0.114
Annual Rainfall (mm)	1095	623.5	85.6	0.137
Runoff Volume (L/m^2)	86	2.83	0.95	0.336

2.3 Database Design and Implementation

An integrated organic carbon migration database system for slope farmland was built using PostgreSQL, enabling unified management and fast retrieval of multi-source heterogeneous data. The database consists of three main modules: spatial data, attribute data, and metadata, and adopts an object-oriented data organization approach. Spatial data stores location information for sampling points, meteorological stations, and runoff plots; attribute data includes monitoring data and analysis results; and metadata records key information about data collection and processing[5]. After one year of operation, the system has accumulated a data storage volume of 2.3TB, with an average query response time of less than 1.5 seconds, meeting the research requirements.

3. Organic carbon migration model construction

3.1 Model Architecture Design

Based on the field monitoring data from slope farmland in the southeastern part of Heilongjiang Province, a three-tier model architecture was established, consisting of a data layer, a computational layer, and an application layer. The data layer integrates basic data such as soil, meteorological, and terrain data, which are stored and managed in a PostgreSQL database[4]. The computational layer includes three functional modules: migration process simulation, flux calculation, and predictive analysis, with core algorithms implemented in Python. The application layer provides data visualization and decision support interfaces, enabling real-time display of monitoring data and output of model prediction results.

3.2 Organic Carbon Migration Mechanism Modeling

For the organic carbon migration characteristics in the study area, a migration flux calculation model based on mass conservation was established. The soil organic carbon migration flux F can be expressed as: $F = K \times C \times Q \times S$ Where:K is the migration coefficient (d^{-1}),C is the soil organic carbon content (g/kg), Q is the runoff volume (mm),S is the slope factor.Model parameters were obtained through field observations, where the K value ranged from 0.015 to 0.028 d^{-1}, the runoff volume Q was monitored automatically at 30-minute intervals, and the slope factor S was derived from a 1:1000 terrain map. The model's calculation results showed that the average annual migration flux of surface soil organic carbon in the study area is 156.3 kg/hm^2. As shown in Figure 2, there is a positive correlation between the migration coefficient K and the flux F. As the K value increases, the flux F also increases, indicating a strong linear dependency. This suggests that higher K values correspond to greater annual migration fluxes of soil organic carbon.

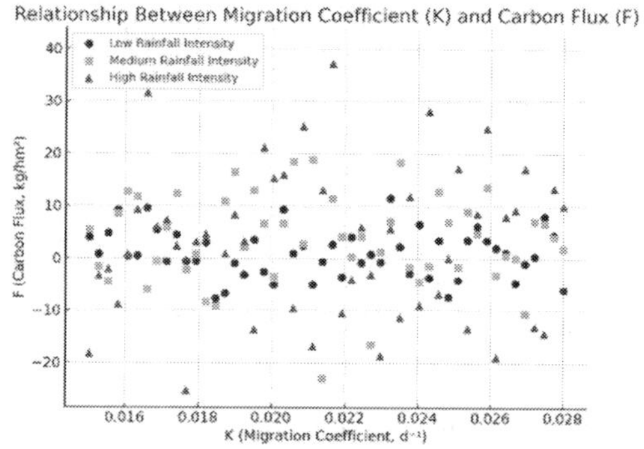

Figure 2 Migration Coefficient (K) vs. Flux (F)

3.3 Core Algorithm Selection

This study systematically improved the standard Random Forest algorithm. By introducing an adaptive weight mechanism based on terrain gradients, dynamic weight allocation for features across different slope intervals was achieved, enhancing the model's sensitivity to terrain variations. In terms of tree structure, the optimal split point selection method based on the CART algorithm significantly improved the model's responsiveness to terrain changes. Additionally, by integrating the Gradient Boosting method, a cascade-structured multi-layer decision tree was constructed, which enhanced the model's predictive capability for extreme rainfall events. The improved algorithm outperformed the standard Random Forest in multiple aspects: the RMSE on the validation set was reduced by 26.3%, model training time decreased by 35%, and predictive stability across different slope gradients improved by 42%. These advancements enable the model to demonstrate greater adaptability and predictive accuracy under complex terrain conditions.

3.4 Model Optimization and Integration

By incorporating Moran's I spatial autocorrelation analysis and terrain correction factors, the model's

https://doi.org/10.1051/e3sconf/202561701021

adaptability to hilly areas was optimized, with the spatial autocorrelation coefficient improved to 0.72. Using 10-fold cross-validation, the standard deviation of the validation set decreased to 0.086, and the 95% confidence interval estimated by Bootstrap resampling was ±8.3%. The model integration adopted a weight-adaptive method, dynamically adjusting the weights under extreme weather conditions, which improved the prediction accuracy by 18.6%. After optimization, the model's prediction error in the 5°-25° slope range was controlled within 15%, with the average error in the 5°-8° slope range being 8.7%, as shown in Figure 3.

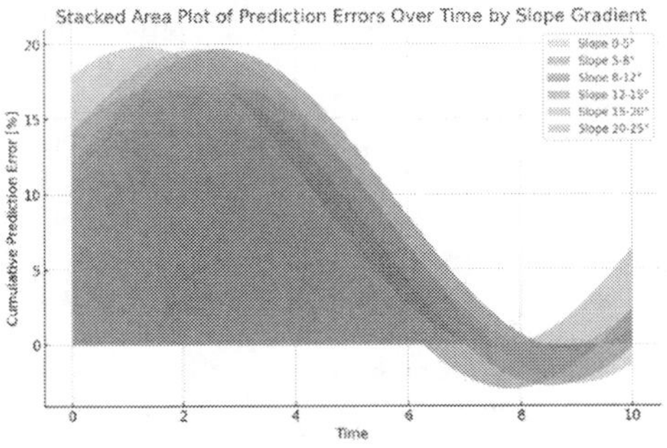

Figure 3 Cumulative Prediction Error Change with Time for Different Slope Gradients

4. Model validation and result analysis

4.1 Experimental Design and Implementation

Three representative slope farming test areas were selected in the southeastern part of Heilongjiang Province, located in low mountain hill areas with altitudes ranging from 200 to 450 meters. The experimental period was from May to October 2023. In each test area, three slope gradients (5°-8°, 8°-15°, 15°-25°) were set up as observation plots, each with an area of 100 m². A stratified sampling method was used to collect soil samples at 20 cm intervals from 0-60 cm in depth, for a total of 387 samples. Soil organic carbon content was measured using the potassium dichromate volumetric method, runoff was monitored with self-recording flow meters, and rainfall was recorded by automatic weather stations.

4.2 Model Performance Evaluation

The model's prediction performance was evaluated using statistical indicators, including the coefficient of determination (R^2), root mean square error (RMSE), and mean absolute error (MAE). The validation results from the three test areas showed that the R^2 values for predicting organic carbon content were 0.89, 0.87, and 0.85, with RMSE values ranging from 1.12 to 1.45 g/kg and MAE values ranging from 0.86 to 1.13 g/kg. The prediction accuracy varied under different slope conditions, with the highest accuracy in the 5°-8° slope range, where the R^2 reached 0.91. As shown in Figure 4, the relationship between predicted and measured values for different slope conditions is illustrated. Particularly in the 5°-8° slope range, the scatter plot is more concentrated, indicating the highest prediction accuracy in this range, consistent with the R^2 value of 0.91. The model validation in this study has certain limitations. The current validation area is primarily concentrated in the 200-450 m elevation range, with validation samples mainly from black soil regions. The model's applicability in high-altitude areas and under extreme rainfall conditions remains to be tested. Future research will expand the validation scope to higher altitude regions and establish validation plots in areas with different soil types to enhance the model's generalizability.

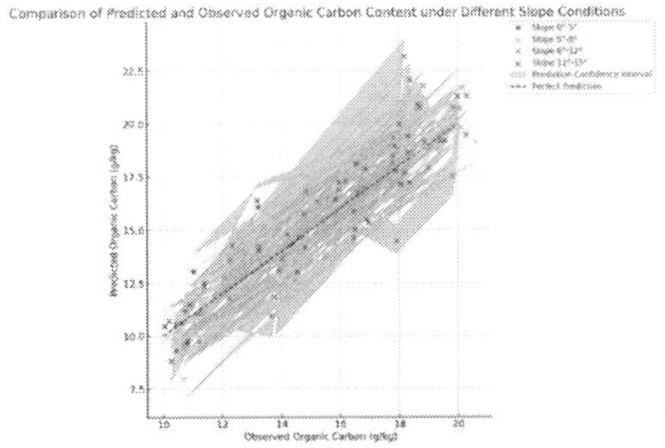

Figure 4 Comparison between Predicted and Measured Organic Carbon Content under Different Slope Conditions

4.3 Comparison with Existing Models

The model developed in this study was compared with the RUSLE and WEPP models to assess prediction accuracy and applicability. Under the same experimental conditions, the prediction error of this model was 23.5% lower than the RUSLE model and 18.7% lower than the WEPP model, as shown in Figure 5. Notably, this model demonstrated clear advantages when simulating the dynamic characteristics of organic carbon migration during rainfall events[6]. Simulation results from the 2023 flood season showed that this model's time resolution could reach 30 minutes, while existing models generally operate on a daily scale, making it more suitable for detailed management and decision-making.

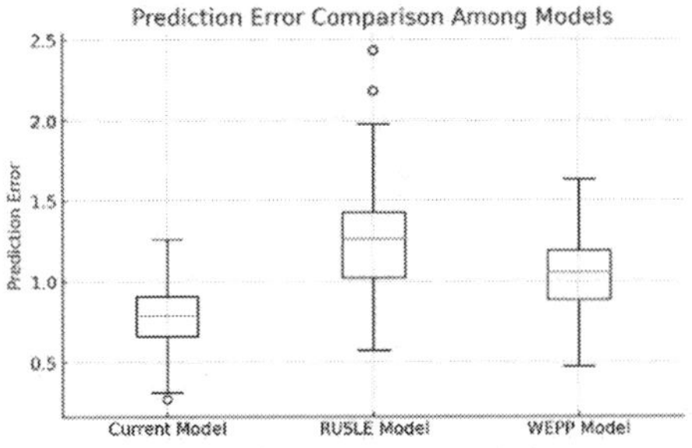

Figure 5 Comparison of Model Prediction Errors

4.4 Practical Applications

Based on the research findings, this study proposes systematic application recommendations. At the policy level, it is suggested to establish a zoned management system, implementing differentiated protection measures for various slope ranges. Policies on soil and water conservation subsidies should be improved, with a focus on supporting the transformation of sloping farmland with slopes greater than 15°. A comprehensive soil organic carbon monitoring network should be developed to enable dynamic supervision. In agricultural practices, contour planting combined with straw return technology should be implemented in areas with slopes of 5-15°. For slopes of 15-25°, the emphasis should be on promoting no-tillage technology and constructing ecological terrace systems. During heavy rainfall events exceeding 50 mm/h, runoff diversion measures should be promptly adopted to prevent soil erosion. To promote technological adoption, it is recommended to develop mobile monitoring and early warning systems, establish standardized operating procedures, and enhance the effectiveness of implementation through technical training and demonstration-based promotion.

5. Conclusion

The findings of this study have significant implications at both regional and global scales. In terms of carbon sequestration, the annual average carbon sequestration potential of the study area reaches 0.8-1.2 t/hm². By promoting no-tillage practices, an additional 20-30% increase in carbon sinks can be achieved. In mitigating climate change, optimizing tillage methods can effectively reduce greenhouse gas emissions by 15-20%, while improving soil drought resistance and enhancing the adaptive capacity of agricultural systems. The results provide critical scientific support for achieving the "carbon peak and carbon neutrality" goals and have a positive impact on promoting green agricultural transformation and ecological civilization construction. The methodological innovations and practical application experiences of this study offer valuable insights for soil and water conservation and sustainable agricultural development in similar regions worldwide.

Acknowledgments

National Field Verification of Land Change Survey (Harbin Center) "(project number: DD20230517)

China Geological Survey "Natural Resources Monitoring (Harbin Center)" (project number: DD20243006)

References

1. Guo Z, Li J. Effects of cultivating different crops on soil erosion of slope farmland[J]. Bangladesh Journal of Botany, 2021: 925-932.
2. Temna Y, Kokhan S. Nexus Soil Organic Carbon and Water Erosion[C]//International Conference of Young Professionals *GeoTerrace-2024*. European Association of Geoscientists & Engineers, 2024, 2024(1): 1-5.
3. Kołodyńska-Gawrysiak R, Gawrysiak L, Poesen J, et al. The impact of natural closed depressions on soil organic carbon storage in eroded loess landscapes of East Poland[J]. Land Degradation & Development, 2023, 34(17): 5273-5292.
4. Datta A, Nayak D, Smith J U, et al. Climate smart agricultural practices improve soil quality through organic carbon enrichment and lower greenhouse gas emissions in farms of bread bowl of India[J]. Soil Research, 2022, 60(6): 455-469.
5. Yang J, Masoudi A, Li H, et al. Soil total nitrogen mediated the impact of climatic factors on urban soil organic matter under different land uses[J]. Journal of Soil Science and Plant Nutrition, 2024: 1-18.
6. Franzluebbers A J. Soil organic carbon sequestration calculated from depth distribution[J]. Soil Science Society of America Journal, 2021, 85(1): 158-171.

Transboundary Marine Spatial Planning: Theoretical Implications, Practical Analysis, and Future Prospects

Xuefeng Li, Xin Teng, Xiangguo Zhang and Xinchun Sang*

National Ocean Technology Center, Tianjin 300112, China

Abstract. Marine spatial planning, as a highly effective policy tool for mitigating conflicts and fostering collaboration, has garnered significant attention in coastal countries and regions worldwide in recent years. With the global rise and advancement of marine spatial planning, transboundary marine spatial planning has emerged as a novel concern for nations aiming to safeguard the marine environment, foster ocean sustainable development, and foster maritime collaboration. Nevertheless, significant challenges persist in the implementation of transboundary MSP due to legislative gaps, difficulties in acquiring and effectively managing cross-border data, variations in governance systems, and other related issues. Therefore, the objective of this article is to enhance comprehension and awareness regarding transboundary marine spatial planning through a systematic review of its developmental process and an extensive elucidation of its implications in transboundary contexts. Furthermore, a comprehensive analysis of MSP in the Baltic Sea region is conducted as a case study. Finally, drawing on both theoretical and empirical research, this study presents future prospects for the advancement of transboundary MSP in terms of enhancing legal frameworks and regulations, facilitating scientific data sharing, and establishing robust governance systems.

1 Introduction

In recent years, the marine biodiversity and ecological functions have been significantly impacted by global climate change, ocean pollution, overfishing, and habitat degradation. Human activities continue to exacerbate the deterioration of coastal marine ecosystems worldwide. Simultaneously, there is a persistent increase in demand for marine resources and spatial utilization by coastal countries and regions, resulting in escalating conflicts between different maritime activities as well as conflicts between maritime activities and the marine environment. The effective alleviation of these conflicts, attainment of the dual objectives of ocean economic growth and marine environmental protection, and establishment of an ecosystem-based marine governance system have emerged as pressing concerns for coastal countries and regions worldwide.

Marine/Maritime Spatial Planning (MSP) is a public process of analyzing and allocating the spatial and temporal distribution of human activities in marine areas to achieve ecological economic, and social objectives that are usually specified through a political process. Global MSP has established a comprehensive planning system and framework, demonstrating an increasing trend of regional collaboration. Furthermore, coastal countries and regions are actively implementing and enhancing their MSPs. Initially pioneered by European nations (Belgium, Netherlands, Germany, Norway), China (Marine Functional Zoning), Australia (Regional Marine Planning), and the United States, MSP-related initiatives have now been undertaken by nearly 80 countries across six continents (Europe, Africa, Asia, Australia/Oceania, North America, and South America) encompassing four oceans (Atlantic Ocean, Pacific Ocean, Indian Ocean, and Arctic Ocean)[1,2]. The dynamic nature of the marine environment, the transboundary distribution and impact of ocean resources and activities, as well as the concept and approach of land-sea coordination in marine management, collectively establish that the essence of MSP lies in transcending boundaries. Thus, the process of transboundary MSP is recognized as a mechanism that facilitates enhanced integration and coordination among existing management frameworks, with the aim of promoting ecosystem-based approaches, safeguarding valuable ecosystem services, reinforcing effective fisheries management, addressing marine pollution concerns, and advancing cross-border marine protected area planning. Nevertheless, significant challenges persist in the implementation of transboundary MSP due to legislative gaps, difficulties in acquiring and effectively managing cross-border data, variations in governance systems, and other related issues.

Therefore, this article provides a systematical overview on the progression of MSP and the implications of transboundary MSP. Furthermore, a comprehensive analysis of the Baltic Sea MSP is conducted as a case study. Finally, drawing on both theoretical and empirical research, this study presents future prospects for the advancement of transboundary MSP in terms of enhancing legal frameworks and regulations, facilitating

* Corresponding author: 375705336@qq.com

scientific data sharing, and establishing robust governance systems.

2 Theoretical implications of Transboundary MSP

2.1 Revisiting the progression of MSP

By integrating concepts derived from MSP, academic research, national policies, and international practices, the evolutionary trajectory of MSP can be delineated into three pivotal stages[3].

The embryonic stage of the concept emerged during 1996-2005 when numerous developed maritime nations began recognizing the significance of marine resource utilization, marine environmental protection, and integrated ocean management. They acknowledged the role of MSP, established a framework for marine protected areas, elucidated the multifunctional nature of marine space, and devised comprehensive plans for the ocean along with zoning strategies to address conflicts in maritime usage and safeguard the marine environment. Despite being an effective measure for conserving marine resources and managing coastal zones comprehensively, MSP has not yet attained a unified conceptualization or standardized process.

The second stage represents the initial development period (2006-2013): In 2006, UNESCO organized the inaugural International Workshop on Marine Spatial Planning, which focused on implementing ecosystem-based coastal zone management through marine spatial planning. This workshop provided foundational concepts and strategic frameworks for the advancement of marine spatial planning. In 2009, IOC-UNESCO published their guidebook "Marine spatial planning: A step-by-step approach toward ecosystem-based management", which presented a comprehensive summary and synthesis of technical approaches to MSP[1]. This seminal publication not only provided a widely recognized definition for MSP but also delineated specific operational procedures, thereby enhancing its theoretical value and offering invaluable guidance. Consequently, it established a robust foundation for the global promotion of MSP. During this period, the international community has primarily focused on topics such as ecosystem-based ocean management, stakeholder engagement, facilitation of marine protected area establishment through MSP, assessment of cumulative environmental impacts from ocean activities, and examination of climate change implications for expanding protected areas.

The third stage represents a period of rapid growth (2014-present): Throughout this phase, MSP has acquired enhanced connotations and functions, encompassing the mitigation of climate change, promotion of blue growth, emphasis on institutional design, advancement in land-sea coordination, support for sustainable ocean governance, and an increased focus on cross-border cooperation. In 2014, Directive 2014/89/EU of the European Parliament and of the Council was introduced to establish a comprehensive framework for MSP. The primary objective of this directive is to facilitate sustainable growth in the marine economy, promote environmentally sound development in marine regions, and ensure responsible utilization of marine resources. The directive mandates the 22 coastal member states of the European Union to engage in collaborative efforts within their maritime areas pertaining to MSP and strengthen cross-border cooperation with non-member countries. Additionally, it necessitates the development of national-level MSP plans by March 31, 2021. IOC-UNESCO and the Directorate-General for Maritime Affairs and Fisheries of the European Commission (DG-MARE) jointly released the "Joint Roadmap to Accelerate Maritime/Marine Spatial Planning (MSP) Processes Worldwide" in 2017, highlighting the imperative for policymakers to incorporate transboundary collaboration into MSP formulation and implementation. In 2019, the MSPGlobal initiative was launched by IOC-UNESCO and the European Commission with the objective of enhancing global transboundary MSP and fostering an enabling external environment for decision-makers, stakeholders, scientists, and the public to actively engage in MSP while promoting multi-level integrated governance.

2.2 Implications of transboundary MSP

The concept of cross-border MSP encompasses three distinct dimensions: cross-administrative boundary MSP, cross-geographical boundary MSP, and cross-governance boundary MSP[4].

2.2.1 Cross-administrative boundary MSP

Cross-administrative boundary MSP refers to the process and framework for resource allocation in shared maritime areas that span multiple jurisdictions. It involves collaborative management across various administrative boundaries such as supra-national organizations, national borders, state or city borders, and inter-departmental boundaries. This type of transboundary resource allocation can also extend vertically across national, state or city, and inter-departmental boundaries. While each country has its own defined administrative jurisdiction range, the interconnectedness of marine ecosystems necessitates transnational cooperation in large-scale oceanic spatial planning. In most cases, the driving force behind cross-administrative boundary MSP is the need to address environmental degradation or ecosystem decline at a broader scale, requiring multiple countries or administrative authorities to collaborate in order to protect ecosystems, mitigate destructive activities, and combat climate change.

2.2.2 Cross-geographical boundary MSP

Cross-geographical MSP extends beyond the ocean space and encompasses land, rivers, coastlines, and atmosphere boundaries to achieve an integrated approach. It is a comprehensive and ecological process that crosses geographical boundaries between land and sea, necessitating integration with various types of planning.

Regional management tools such as land use planning on the land side, watershed management planning, MSP on the sea side, and coastal zone integrated management planning may concentrate or overlap within specific coastal areas.

2.2.3 Cross-governance boundary MSP

Cross-governance boundary MSP involves cross-border cooperation in institutional norms, governance systems, data information levels, and stakeholder involvement for allocating maritime space resources. It encompasses multiple forms and processes of maritime space governance across different categories, scales, and time sequences overlapping each other. The application of cross-border thinking in MSP goes beyond administrative, geographical, and ecological boundaries; due to social, cultural, policy differences in neighboring jurisdictions, it is challenging to achieve complete consensus on preferences. Therefore, it requires crossing different domains, governance systems, and stakeholders' boundaries to facilitate cross-border cooperation among stakeholders regarding issues such as allocation of cross-border marine public resources and production and distribution of cross-border maritime public goods.

3 Case Study: Transboundary MSP in the Baltic Sea

3.1 Background

MSP is extensively practiced worldwide, and the pursuit of transboundary MSP has emerged as a burgeoning domain of collaboration among maritime nations. Europe currently leads in the implementation of extensive transboundary MSP projects, offering valuable insights for cooperation and governance in global ocean regions. Since 2000, numerous transboundary MSP initiatives have been successfully executed across various European seas, with a particular focus on the Baltic Sea.

The Baltic Sea, characterized by its semi-enclosed nature and shallow waters, presents significant environmental challenges. The inherent geographical features of this sea necessitate urgent cooperation among coastal countries for effective marine development and conservation. Transboundary MSP facilitates collaborative efforts in environmental protection, including the reduction of pollutant emissions and preservation of marine ecosystems.

3.2 Transboundary MSP Projects

The Baltic Sea region (BSR) has successfully implemented several transboundary MSP projects, including ARTWEI, Baltic LINes, Baltic SCOPE, BALTSEAPLAN, BaltSpace, BONUS BASMATI, Capacity4MSP, C-Scope, Coastal and Maritime Spatial Planning, ECODUMP, ESaTDOR, HERRING, KNOWSEAS, MSP-LSI, Pan Baltic Scope, PartiSEApate, PERICLES, Plan Bothnia, Plan4Blue, SeaGIS, Seanergy 2020, SPICOSA, TOPCONS, etc. (Table 1)[5].

Baltic LINes (Coherent Linear Infrastructures in Baltic Maritime Spatial Plans): aims to enhance the transnational consistency of shipping routes and energy corridors in the MSP of the Baltic Sea region. It seeks to establish an optimal framework for future blue growth activities, including maritime transportation, offshore energy development, coastal tourism, etc., over the next 10-15 years, thereby bolstering investor confidence. Key project activities encompass developing MSP requirements pertaining to BSR shipping and energy sectors; coordinating BSR MSP data infrastructure for efficient exchange and dissemination of shipping route and energy corridor data; drafting guidelines for seamless MSP data exchange and dissemination; ensuring transnational consistency in linear infrastructure planning; providing recommendations for formal BSR agreements on cross-border consultation regarding linear infrastructure formation during the MSP process.

Baltic SCOPE (Towards coherence and cross-border solutions in Baltic Maritime Spatial Plans): for the first time, the authorities of MSP and regional marine organizations in the BSR have collaborated to seek planning solutions for cross-border issues and enhance the process of MSP. Their primary objective is to propose joint solutions that address cross-border MSP, promoting greater coordination and consistency among national plans. To achieve this goal, two MSP case studies were conducted with a focus on shipping, energy production, fisheries, and environmental functions in these areas and their potential trade-offs. The results comprise reports, research papers (on maritime traffic, energy production, fisheries, and environment), collaborative maps as well as online tools.

Capacity4MSP (Strengthening the capacity of MSP stakeholders and decision makers): is a project platform designed to enhance the capacity of stakeholders, policymakers, and decision-makers in MSP through robust dialogue activities and the application of acquired knowledge. This initiative aims to elevate the visibility and impact of projects, foster potential synergies, as well as deepen and expand expertise by integrating, amplifying, and transferring project outcomes towards innovative practical solutions.

Pan Baltic Scope: aims to achieve coordinated and consistent MSP among countries in the BSR, as well as establish a sustainable cross-border mechanism for cooperation on MSP. They have established a planning forum as a central platform for collaboration on non-specific planning issues between planning authorities and regional organizations. Cross-border activities are conducted at various geographical levels to meet national MSP processes' requirements and support the successful implementation of the EU MSP Directive. They are developing tools and methods including ecosystem-based approaches, cumulative impacts assessment, green infrastructure integration, economic and social analysis, as well as data sharing mechanisms. The project also addresses land-sea interactions.

Through the joint efforts of multiple MSP projects, they have developed shared tools and methodologies, leveraging insights from prior projects, to foster targeted cross-border cooperation in bolstering national planning solutions. Their ecosystem-based approach and strategic environmental assessment handbook offer practical guidance and modular implementation to cater to the requirements of planners. The incorporation of green infrastructure into MSP has been thoroughly examined, including the mapping of ecological values and provision of ecosystem services in the BSR. Additionally, they have developed a comprehensive pan-Baltic map highlighting significant fishing grounds areas. Their models demonstrate that climate change will exert substantial impacts on species distribution, thereby posing threats to both ecosystem functionality and the blue economy within the next 80 years. Consequently, they are identifying key regions as potential climate refuges or "last line of defense". The open-source desktop application and online version of the BSII-CAT cumulative impact assessment tool are now available for use, along with the free online tool PlanWise4Blue, to assess the economic and cumulative impacts of MSP. The BASEMAPS map and data service provides a comprehensive overview of planning progress in each state, enabling users to visualize ocean utilization within the plans and explore designated areas for MSP based on type and sector.

3.3 Practical experience

The European Union provides a legal and policy framework to promote the development and collaboration of national MSP, with intergovernmental cooperation mechanisms serving as organizers and coordinators. The network governance model, along with other relevant stakeholders such as organizations and institutions, offers supplementary support for transboundary MSP projects in terms of both procedural aspects and substantive content. The aforementioned elements are intricately interconnected and functionally linked, in conjunction with the financial support from the European Union and the establishment of a collaborative data platform for the Baltic Sea region. These factors have collectively contributed to the robust development and overall advancement of transboundary MSP in the Baltic Sea region, offering valuable insights for cross-border cooperation in MSP across other regions[6].

Table 1. Compilation of main transboundary marine spatial planning initiatives in the Baltic Sea region[5].

Initiative	Period	Countries	Leader Partner
ARTWEI: Action for the Reinforcement of the Transitional Waters' Environmental Integrity	2010-2013	Denmark, Germany, Lithuania, Poland, Russia, Sweden	Coastal Research and Planning Institute (Lithuania)
Baltic LINes: Coherent Linear Infrastructures in Baltic Maritime Spatial Plans	2016-2019	Denmark, Estonia, Finland, Germany, Latvia, Lithuania, Netherlands, Poland, Russia, Sweden	Federal Maritime and Hydrographic Agency (Germany)
Baltic SCOPE: Towards coherence and cross-border solutions in Baltic Maritime Spatial Plans	2015-2017	Denmark, Estonia, Finland, Germany, Latvia, Poland, Sweden	Swedish Agency for Marine and Water Management (Sweden)
BALTSEAPLAN: Planning the future of the Baltic Sea	2009-2012	Denmark, Estonia, Germany, Latvia, Lithuanian, Poland, Russia, Sweden	Federal Maritime and Hydrographic Agency (Germany)
BaltSpace: Towards Sustainable Governance of Baltic Marine Space	2015-2018	Denmark, Germany, Latvia, Lithuania, Poland, Sweden	Sodertorn University (Sweden)
BONUS BASMATI: Baltic Sea Maritime Spatial Planning for Sustainable Ecosystem Services	2017-2020	Denmark, Finland, Latvia, Germany, Sweden	Aalborg University (Denmark)
Capacity4MSP: Strengthening the capacity of MSP stakeholders and decision makers	2019-2021	Belgium, Denmark, Estonia, Finland, Germany, Latvia, Lithuania, Poland, Sweden, Russia	VASAB Secretariat
C-Scope: Combining Sea and Coastal Planning in Europe	2007-2013	Belgium, United Kingdom	Belgian Coordination Centre on ICZM (Belgium)
Coastal and Maritime Spatial Planning: Coastal and Maritime Spatial Planning in Pärnu Bay Area in Estonia and Coastal Municipalities in Latvia	2012-2015	Estonia, Latvia	Estonian Ministry of the Interior (Estonia)
ECODUMP: Application of ecosystem principles for the location and management of offshore dumping sites in SE Baltic Region	2011-2014	Lithuania, Poland, Russia	Klaipeda University Coastal Research and Planning Institute (Lithuania)
KNOWSEAS: Knowledge-based Sustainable Management for Europe's Seas	2009-2013	Bulgaria, Denmark, Finland, France, Germany, Ireland, Italy, Netherlands, Norway, Poland, Portugal, Spain, Sweden, Turkey, United Kingdom	The Scottish Association for Marine Science LBG (United Kingdom)

MSP-LSI: Maritime spatil planning and land-sea interactions	2018-2019	Croatia, Denmark, Germany, Netherlands, Poland, Slovenia, Sweden	The University of Liverpool (United Kingdom)
Pan Baltic Scope	2018-2019	Denmark, Estonia, Germany, Latvia, Poland, Sweden	Swedish Agency for Marine and Water Management (Sweden)
PERICLES: Preserving and Sustainably Governing Cultural Heritage and Landscapes in European Coastal and Maritime Regions	2018-2021	Denmark, Estonia, France, Greece, Malta, Netherlands, Portugal, United Kingdom	Aalborg University (Denmark)
SPICOSA: Science and policy integration for coastal System Assessment	2007-2011	EU Member States, Israel, Norway, Turkey, Ukraine	French Institute for Exploitation of the Sea (France)

4 Future prospects for advancing transboundary MSP

While the majority of MSP activities are typically confined to jurisdictional waters, there is a growing recognition of the necessity for cross-boundary thinking in order to facilitate ecosystem-based integrated management. This becomes particularly crucial in shared maritime areas where addressing ecological concerns such as pollution may necessitate transcending administrative boundaries. Moreover, effectively managing coastal human activities often entails navigating spatial boundaries that intersect and overlap geographically, including those between oceans, land, rivers, and coastlines. Furthermore, when acquiring, managing, and sharing data for the purposes of MSP governance and decision-making processes, it also involves traversing technological boundaries as well as those among diverse stakeholders.

4.1 Policy and legal advocacy

The legal foundation for transboundary MSP is currently limited, resulting in the predominantly voluntary nature of such planning among most countries and often manifesting as informal project cooperation. The degree of integration between cross-border policies and legislative arrangements in adjacent jurisdictions is a key factor in determining the feasibility of transboundary MSP. The higher the level of integration and coherence between policy and legal structures, the greater the likelihood of achieving transboundary MSP. In terms of policy convergence, it is imperative to integrate and interconnect policies across neighboring countries, diverse departments, and multiple domains. The transboundary and cross-sovereign nature of marine resources necessitates the distinctiveness of the legal system governing these resources. Relying solely on individual laws falls short; a comprehensive and systematic legal framework must be established. At the international level, fostering the establishment of novel and stable national practices that foster innovation, while facilitating the evolution of customary international law, will provide a robust legal and policy framework for future advancements in transboundary MSP.

4.2 Data and information exchange

The collection and organization of spatial data pose the most prevalent challenge in transboundary MSP. Inconsistencies in scales often lead to incompatible situations during the data collection process, thereby exacerbating the difficulty of data management. Moreover, as a cross-border endeavor spanning different jurisdictions and spatial domains, the lack of comparability across various scales renders futile efforts invested in data collection. When engaging in cross-border collaboration for data management, it is imperative to gather, systematize, and synchronize information based on the specific requirements of data, following a set of technical guidelines pertaining to administrative boundaries, biological characteristics, as well as past, present and future human activities. The cross-border data standardization platform and spatial analysis system facilitate users in observing, comprehending, querying, and interpreting data in various formats while visualizing the information. This ensures that cross-border datasets adhere to a unified measurement framework and standardized data protocols.

4.3 Establishment of governance systems

Both countries and regions have unique administrative and management systems, which vary significantly among nations that may engage in cross-border cooperation for marine spatial planning. Additionally, the priorities of different countries may be incompatible, posing challenges for connecting, integrating, and restructuring cross-border marine spatial planning. Therefore, when establishing a broader governance structure, it is imperative to consider the existing economic, political, cultural, and legal frameworks and leverage insights gained from other domains in order to effectively implement cross-border marine spatial planning. It is crucial to fully harness the influence of stakeholders and foster extensive collaboration through relevant governmental and non-governmental entities. A coordinating body can be established to facilitate stakeholder engagement and commitments. This cross-border coordination mechanism can operate at both supranational and subnational levels, or even extend beyond the project itself to establish a more enduring coordinating institution.

5 Conclusions

In recent years, the field of MSP has witnessed a flourishing global development trend in both research and practice. The significance of transboundary MSP has garnered substantial attention and recognition at the international level, particularly within Europe. The proper management of transboundary issues has emerged as a pivotal component of ecosystem-based MSP. Furthermore, the international community is increasingly recognizing the imperative for MSP to encompass diverse cross-border factor, for instance cross-administrative boundary MSP, cross-geographical boundary MSP, and cross-governance boundary MSP. Effective transboundary MSP necessitates collaborative efforts among multiple countries and regions, encompassing the expedited enactment of relevant legislation and policy implementation; enhanced optimization of data and information management to facilitate data sharing; establishment of robust cross-border coordination mechanisms; and improvement of governance systems across borders[7]. These measures collectively aim to safeguard marine ecosystems, foster sustainable development of oceans, address global climate change, and resolve transboundary issues that cannot be resolved by individual nation.

References

1. T. Rafael, H. Cabral, J. Mourato, et al. (2024) Marine spatial planning: a systematic literature review on its concepts, approaches, and tools (2004-2020). Marit. Stud., 23: 6.
2. X. Li, C. Ma, Q. Yue, et al. (2024) Research on environmental assessment and management strategies of the Fildes Peninsula in Antarctica. World Reg. Stud., 33(3):15-28.
3. S. Li. (2022) The origins, evolution and prospects of Marine spatial planning: Empirical analysis of global-scale data. Pac. j., 30(11): 92-106.
4. X. Ma, C. Zhao. (2019) Integration, evolution and realization: Methodology for transboundary marine spatial planning. J. Ocean Univ. China (Soc. Sci.), 5:69-80.
5. IOC-UNESCO. (2021) MSPglobal - compendium of existing and emerging cross-border and transboundary MSP practices. Paris, UNESCO.
6. J. Moodie, F. Sielker. (2022) Transboundary marine spatial planning in European sea basins: Experimenting with collaborative planning and governance. Plan. Pract. Res, 37(3): 317-332.
7. S. Li, S. Jay. (2023) Addressing transboundary challenges: Exploring the interactive relations between collaborative governance and transboundary marine spatial planning in Europe. Mar. Policy, 158: 105880.

Analysis of Spatio-temporal change of Carbon Emissions and Carbon Sinks and Carbon Balance partition in the Poyang Lake Ecological Economic Region

Yunsong Yang*

School of Economics and Management, Nanchang Institute of Science & Technology, Nanchang 330108, China

Abstract. Research on spatio-temporal evolution characteristics of carbon sources/sinks and carbon balance zoning is of significance in promoting the coordinated development of regional low-carbon development, improving the scientificity and operability of carbon reduction and sink enhancement policies, and realizing the goal of "double carbon". In this study, Spatio-temporal evolution characteristics of carbon sources/sinks and carbon balance zoning in the Poyang lake ecological economic region (PYLEER) have been analyzed by correcting the long time series DMSP / OLS, NPP / VIIRS night lighting data, land use data, and GIS. The results were as follows: ① From 2000 to 2021, the total carbon emission of PYLEER continued to rise but the growth rate was declining, showing an overall convergence trend, but not reaching the peak carbon emission, and the spatial pattern of carbon emission was stable. The change of carbon sink has gone through two stages, with a slow growth trend from 2000 to 2005 and a decreasing trend from 2005 to 2021. ② From 2000 to 2021, the carbon balance changed greatly, with the carbon balance spanning 1612.90-20269.18t. The carbon compensation rate was less than 1, showing carbon source, and the carbon compensation rate was decreasing, indicating certain carbon reduction pressure. The economic contribution coefficient of carbon emission generally showed a trend of decreasing first and then increasing, and the spatial pattern changed greatly. The spatial pattern of carbon ecological carrying coefficient relatively was stable. ③ From 2000 to 2010, the carbon balance zoning of PYLEER produced some changes. And Jiujiang city and Jingdezhen city changed from carbon sink development area to economic development area, and Xinyu city had become comprehensive optimization area. To 2021, the carbon balance zoning has changed again compared with 2010, with Yingtan becoming a comprehensive optimization zone, Jiujiang, Yichun, and Jingdezhen becoming carbon sink development zones, and Ji'an and Fuzhou becoming low-carbon zones.The above research results are of guiding significance to the formulation of scientific carbon emission reduction and carbon neutralization policies in PYLEER.

1 Introduction

Climate warming has become an immense challenge to human survival and societal progress. Acknowledging that carbon emissions arising from human activities and industrial energy consumption are the main driving force behind global warming [1], considerable focus has been placed on this issue by the international community, especially regarding "carbon reduction" and "enhancement of carbon sinks". Currently, more than 130 countries and regions have pledged their commitment towards attaining the goal of "carbon neutrality" [2-3]. As the world's largest developing country in the world, China bears an important responsibility to build a community with a shared future for mankind and promote sustainable development of human civilization. In September 2020, Chinese leader Xi Jinping solemnly announced the "dual carbon" goal at the United Nations General Assembly that China will strive to peak its carbon dioxide emissions before 2030 and achieve "carbon neutrality" before 2060 [4].

The PYLEER is established upon the foundation of the Poyang Lake Urban Circle, with a primary objective to concurrently pursue ecological preservation and economic advancement. In December 2009, the Chinese government officially endorsed the "Planning for the PYLEER " which designates this area as an exemplar for harmonious ecological civilization, socio-economic progress within China and worldwide, as well as pioneering low-carbon economic development in China. Therefore, it holds immense practical significance to conduct an in-depth investigation into the temporal and spatial change of carbon emissions/sinks within the PYLEER and undertake scientific zoning for carbon balance. This endeavor can effectively advance the low-carbon development of the region, enhance the scientific rigor and operational feasibility of policies pertaining to carbon reduction and sequestration, ultimately facilitating the achievement of "dual carbon" goals.

* Corresponding author: yys7401@163.com

2 Materials and Methods

2.1 Study Area

The PYLEER located in the middle and lower reaches of the Yangtze River [5], encompasses Nanchang, Jiujiang, Jingdezhen, Yingtan, Xinyu, Fuzhou, Yichun, Shangrao, and Ji'an cities in Jiangxi Province. It spans a total area of 51,117 km^2 which accounts for 30.62% of the Jiangxi province's total area. Moreover, it represents 49.01% of the province's population and contributes to 57.04% of its GDP [6]. The topography primarily consists of low mountains, hills and plains with a tendency towards lower elevation in the central region gradually rising towards higher elevations on its periphery. The core region is characterized by the Poyang Lake Plain which boasts an intricate network of waterways facilitating robust economic activity within southern China while providing favorable conditions for harmonizing ecological preservation with economic development [7]. The PYLEER is a crucial area for ecological function protection in China and holds global significance as an ecological area . In recent years, the region has experienced rapid development, leading to increased resource consumption and subsequent detrimental effects on the local ecological environment.

2.2 Data Sources

2.2.1 Night light data

The night light data includes DMSP/OLS and NPP/VIIRS. Among them, DMSP/OLS night light image from 2000 to 2013, with spatial resolution of 1km. NPP/VIIRS night light image from 2012 to 2021, with spatial resolution of 0.5km. The data are all from the National Geophysical Data Center (NGDC) under the National Oceanic and Atmospheric Administration of the United States. Taking the municipal administrative region as the correction unit, DMSP/OLS data as the reference, and NPP/VIIRS data as the correction object, the NPP/VIIRS data from 2014 to 2021 were corrected. Reconstruct long time series, comparable DMSP/OLS and NPP/VIIRS datasets.

2.2.2 Land use data

The land use data is derived from the national land cover data released by the Resource and Environmental Science Data Center of the Chinese Academy of Sciences, with spatial resolution of 30m. According to ArcGIS10.7, the land use types are re-divided into five categories: cultivated land, woodland, grassland, water area and unused land according to the Land Use Status Classification, and re-sampled as 1km land grid.

2.2.3 Other data

The spatial resolution of Land Scan population data is 1km, from the Oak Ridge National Laboratory (ORNL) of the U.S. Department of Energy; GDP and energy consumption statistics of each city are from Jiangxi Provincial Statistical Yearbook and China Energy Statistical Yearbook from 2001 to 2021. The provincial and municipal vector administrative data were sourced from the 1:4,000,000 database of the National Basic Geographic Information Center.

2.3 Calculation Method

2.3.1 Estimation of carbon emissions

a) Calculation of carbon emissions of energy consumption

The carbon emissions of the PYLEER were calculated based on energy consumption statistics from the 9 cities within the zone, utilizing carbon emission coefficients determined by the Intergovernmental Panel on Climate Change (IPCC) in 2006 for various energy sources. Following Su Yongxian's approach for calculating carbon emissions, CO_2 emissions from 9 primary energy sources were computed using their respective consumption statistics [8]. The formula employed is as follows:

$$CO_2 = \frac{44}{12} \times \sum_{i=1}^{9} K_i E_i$$

where i is the type of energy, E_i is the i-th energy source consumption, K_i represents the CO_2 emission coefficient of the i-th energy source (The carbon emissions per unit of standard coal consumption).

b) Carbon emissions simulation of energy consumption

The original data was grouped using the natural break method, and linear regression analysis without an intercept was applied to establish a correlation between carbon emissions derived from nighttime light data and statistical energy consumption data [9]. This approach aimed to generate a spatial distribution map of carbon emissions.

2.3.2 Estimation of Carbon sinks

Carbon sinks primarily involves the absorption of carbon by vegetation, soil, and water, which is closely linked to land use patterns. The calculation of carbon absorption from different land use types such as cropland, forestland, grassland, built-up land, and water bodies is performed using the coefficient method. This entails obtaining the carbon absorption volume for each land use type by multiplying its corresponding carbon absorption coefficient with the area it occupies. The total carbon absorption volume across all land use types is then determined [10]. The formula can be expressed as follows:

$$C_s = \sum_{m=1}^{5} e_m = \sum_{m=1}^{5} T_m \delta_m$$

where C_s represents the total carbon absorption, e_m is the carbon absorption attributed to land use type m, T_m represents the area of land use type m(hm^2); and δ_m indicates the carbon absorption rate ($t \cdot hm^{-2}$), According to existing studies, the coefficients for carbon absorption in cropland, forestland, grassland, water bodies, and unused land (The amount of carbon emissions absorbed by each unit of M-type land per year)are 0.007, 3.810, 0.910, 0.252, and 0.005 ($t \cdot hm^{-2} \cdot a^{-1}$) respectively [11].

2.3.3 Carbon Compensation Rate

The carbon compensation rate (CCR) is the ratio of carbon absorption to carbon emissions in a city, serving as an indicator of the city's pressure for reducing carbon emissions [12]. The formula is expressed as follows:

$$CCR = CS_i/C_i$$

where CS_i is the carbon uptake and emission of a specific urban area; when $CCR > 1$, the urban area functions as carbon sink; conversely, when $CCR < 1$, it is carbon source.

2.3.4 Economic contribution coefficient of carbon emissions

The economic contribution coefficient of carbon emissions (ECC) is carbon productivity from the perspective of economic benefits [13], and the formula is as follows:

$$ECC = (G_i/G)/(C_i/C)$$

where G_i and C_i represent the GDP and carbon emissions of a city, respectively, while G and C are the GDP and carbon emissions of PYLEER. When $ECC > 1$, it signifies that the city boasts relatively high energy efficiency and carbon productivity; conversely, when $ECC < 1$, it suggests that the city has comparatively lower energy efficiency and carbon productivity.

2.3.5 Carbon ecological support coefficient

Carbon ecological support coefficient (ESC) is the quotient between the proportion of carbon absorption of a city in the carbon absorption of PYLEER and the proportion of carbon emissions of the city in the carbon emissions of PYLEER, which represents the carbon sink capacity of the city.

$$ESC = (CS_i/CS)/(C_i/C)$$

where CS_i and C_i respectively represent the carbon absorption and carbon emission of a city, while CS and C respectively are the carbon absorption and carbon emission of PYLEER. When $ESC > 1$, it indicates that the city possesses a relatively high capacity for carbon sink; when $ESC < 1$, it suggests that the city has a relatively low capacity for carbon sink.

3 Results and Discussion

3.1 Temporal and spatial evolution characteristics of carbon emissions

As shown in Fig. 1, carbon emissions from energy consumption in PYLEER increased from 2000 to 2021. From 48.698012 million t in 2000 to 217.675143 million t in 2021, an increase of 4.47 times. From 2000 to 2014, the rapid development of industrialization and urbanization in the study area led to a significant increase in energy demand and a significant increase in carbon emissions; by 2015, the implementation of relevant carbon emission reduction policies led to a decline in carbon emissions; by 2000, the increase in carbon emissions affected by the epidemic was small.

The carbon emissions in each year showed that carbon emissions in all cities have increased to a certain extent. The high carbon emission zones in each year are mainly concentrated in Nanchang, Jiujiang, Shangrao and Yichun. The above cities are large in scale, high in GDP value, developed in industry, convenient in transportation, densely populated, with strong human activities and high energy consumption, and Jiujiang and Shangrao are rich in mineral resources, which are important energy bases in China. The low carbon emission areas are Jingdezhen, Yingtan and Xinyu. The development of these cities is mostly based on tourism, mineral resource exploitation and industrial scale are relatively small, and urban carbon emissions are relatively low. The change trend of carbon emissions in each city is basically consistent with the change trend of the whole region. In terms of the increase in carbon emissions, the increase in high carbon emission areas is significant, while the increase in low carbon emission areas is relatively small.

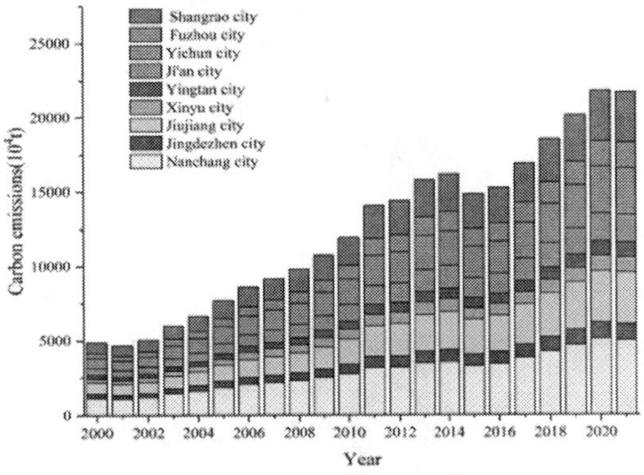

Fig. 1. Trend of carbon emissions in PYLEER from 2000 to 2021.

3.2 Temporal and spatial evolution characteristics of carbon sinks

3.2.1 Analysis of characteristics of carbon sequestration changes

The carbon sink in the PYLEER underwent two distinct stages. From 2000 to 2005, there was a gradual growth trend in the carbon sink, with an increase from 2.88799 million tons in 2000 to 2.88921 million tons in 2005. However, from 2005 to 2021, it exhibited a declining trend, decreasing from 2.88921 million tons in 2005 to 2.7704 million tons in 2021. This decline can be primarily attributed to the rapid urbanization level improvement, extensive expansion of urban areas, and encroachment on land space designated for carbon sink purposes. In terms of individual cities' carbon sinks within this region: Nanchang city experienced a marginal upward trend but at a relatively slow rate; Jingdezhen, Xinyu, and Yichun witnessed a decrease without significant burden reduction; Jiujiang, Yingtan, Fuzhou, Shangrao and Ji'an displayed fluctuating patterns albeit with minimal amplitude (Fig. 2).

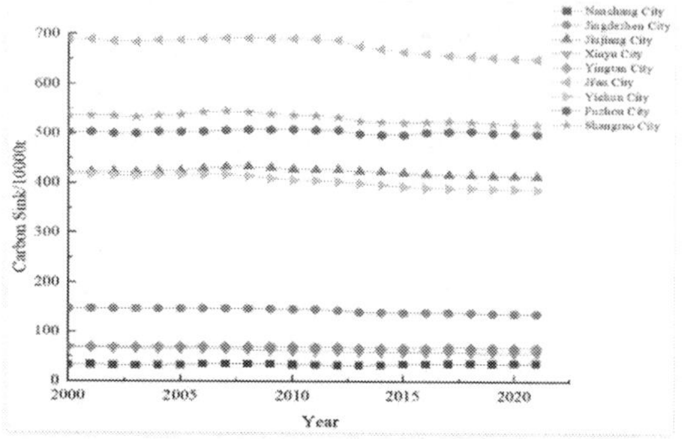

Fig. 2. Carbon sink in PYLEER from 2000 to 2021.

3.2.2 Spatial distribution characteristics of carbon sinks

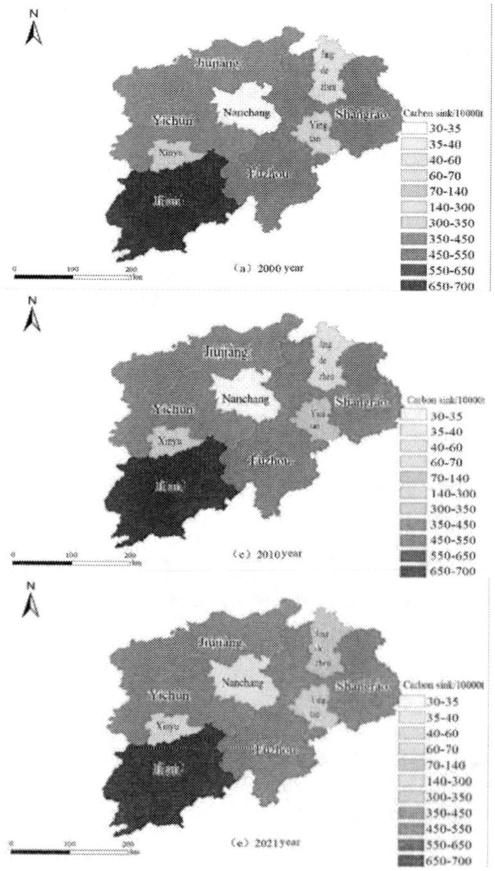

Fig. 3. Characteristics of carbon sink in PYLEER from 2000 to 2021.

The distribution of low-carbon areas in each period can be observed from Fig. 3, which shows that these areas are primarily concentrated in the eastern cities of Ji'an, Fuzhou, and Shangrao. Throughout the study period, there is a relatively stable spatial distribution pattern for carbon sinks. Among all periods examined, Ji'an City exhibits the largest carbon sink capacity, followed by Fuzhou City and Shangrao City. Jingdezhen City and Xinyu City have comparatively smaller carbon sink capacities, while Nanchang City has the smallest carbon sink capacity. It is worth noting that different cities possess distinct energy structures leading to significant variations in their carbon absorption capabilities. Overall, the spatial distribution patterns of carbon sinks remain relatively consistent in 2000, 2010 and 2021.

3.3 Carbon balance analysis

3.3.1 Analysis of characteristics of carbon balance time series change

The temporal changes in carbon balance within the PYLEER (Fig. 4) reveal a wide range of fluctuations, spanning from 1.61290 to 20.26918 million tons. Nanchang exhibits the largest variation in carbon balance among all cities, ranging from 0.88 to 4.5 million tons, indicating its relatively weak capacity for carbon balance due to rapid urban expansion, industrial development and population growth over the past two decades as a provincial capital city. Jiujiang, Yichun and Shangrao show relatively dispersed distribution of carbon balance with most values concentrated between 0.25-3.4 million tons. Fuzhou and Ji'an had negative carbon balances initially but gradually deteriorated over time. Jingdezhen, Yingtan and Xinyu exhibit strong capacity for carbon balance with concentrated distributions mainly within the range of 0.15-10 million tons.

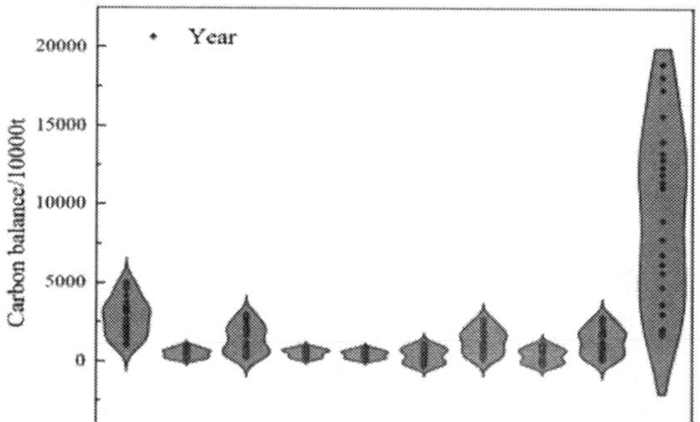

Fig. 4. Temporal change of carbon balance in PYLEER from 2000 to 2021.

3.3.2 Temporal and spatial evolution characteristics of CCR

The analysis of Fig. 5 reveals that throughout the entire study period, the CCR in the PYLEER remained below 1, indicating a carbon source status. Moreover, there was a declining trend in the carbon offset rate, suggesting significant pressure for reducing carbon emissions. Examining individual cities, with the exception of Ji'an City in 2000 and Fuzhou City in both 2000 and 2005, all other cities acted as carbon sinks. Nanchang City, Jingdezhen City, Jiujiang City, Xinyu City, Yingtan City, Yichun City, Shangrao City were identified as consistent sources of carbon emissions in the study period. Furthermore, it is evident that their capacity for absorbing carbon fell short of compensating for their respective emissions.

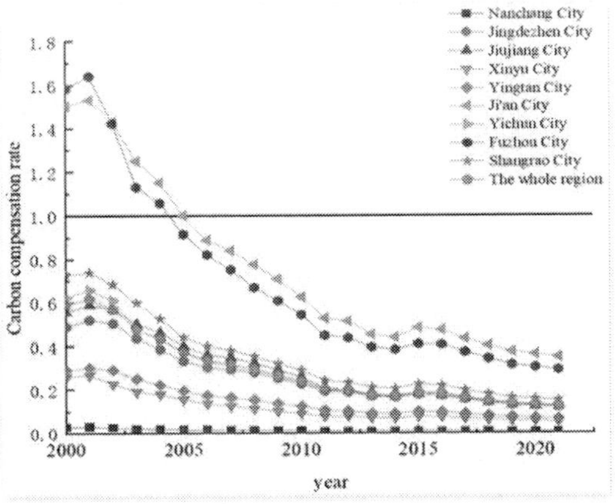

Fig. 5. Temporal and spatial evolution characteristics of CCR in PYLEER from 2000 to 2021.

3.3.3 Temporal and spatial evolution characteristics of ECC

The data presented in Fig. 6 (a、b) reveals that in the year 2000, ECC varied from 0.8 to 1.69, indicating an imbalance between the carbon emission contribution and economic contribution of each city. Notably, Nanchang and its surrounding areas, including Yingtan, and Shangrao, exhibited high values. This suggests that these cities possess higher levels of carbon productivity and energy use efficiency compared to other cities. Conversely, Ji'an and Fuzhou displayed relatively lower coefficients.

The ECC of 7 cities exhibit greater than 1 in 2000. By 2010, the number of cities with a coefficient exceeding 1 had decreased to 4, and the ECC for all 9 cities was lower compared to that in 2000. This indicates a declining trend in both carbon productivity and energy utilization efficiency within PYLEER from 2000 to 2010. Nanchang and Yingtan have transitioned out of the higher value zone, while Jiujiang, Xinyu, and Jingdezhen have joined the lower value zone. In contrast, by 2021, the number of cities with ECC greater than 1 increased to reach 7. Furthermore, compared to 2010, there has been an overall increase for all nine cities between 2010 and 2021. These findings suggest an upward trend in carbon productivity and energy use efficiency across most regions during this time frame. These phenomena are mainly due to the inclusion of carbon intensity targets in national development plans in 2011, the government has implemented more stringent environmental regulations, facilitated industrial transformation and upgrading, and promoted increased carbon productivity and energy efficiency. The spatial distribution of the high-value and low-value areas for ECC remained consistent with that of 2010. However, Nanchang, which had previously dropped out of the high-value area, re-entered it again, while Fuzhou, originally part of the low-value area, also moved into the high-value area queue. These changes indicate significant shifts in the spatial pattern of ECC from 2000 to 2021. Furthermore, there is an overall trend observed where ECC initially decreased before subsequently increasing.

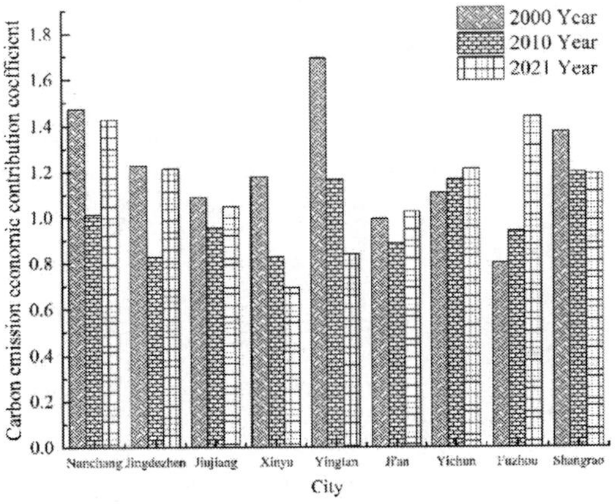

Fig. 6. Temporal and Spatial distribution of ECC in the PYLEER from 2000 to 2021.

3.3.4 Spatial and temporal evolution characteristics of ESC

Fig. 7. Spatial distribution of ESC in PYLEER from 2000 to 2021.

The data presented in Fig. 7 reveals that Ji'an exhibits the highest ESC (2.45), while Nanchang displays the lowest value (0.05), with Ji'an's coefficient being nearly 49 times greater than that of Nanchang. The four cities with ESC greater than 1 exhibit higher capacities for carbon sequestration. In 2010, there was minimal change observed in the spatial distribution of ESC, indicating a stable pattern during the period from 2000 to 2010. Ji'an boasts the highest coefficient (2.94), while Nanchang exhibits the lowest value (0.04). The highest value is approximately 73.5 times greater than the lowest value, highlighting substantial disparities within PYLEER regarding their respective ESC. Compared to data from 2000, all six cities have witnessed an increase of ESC during the period from 2000 to 2010 due to effective implementation of environmental protection measures and low-carbon policies within PYLEER. The spatial distribution of ESC remained consistent between 2021 and 2010, suggesting stability within PYLEER spatial pattern throughout this time frame.

3.4 Carbon balance zoning

According to ECC and ESC, and with reference to previous studies [14], the cities in PYLEER can be classified into four categories: low-carbon conservation areas (ECC > 1 and ESC > 1), economic development areas (ECC < 1 and ESC > 1), carbon sink development areas (ECC > 1 and ESC < 1), and comprehensive optimization areas (ECC < 1 and ESC < 1). Fig. 8 illustrates that Yichun and Shangrao cities were categorized as low-carbon conservation areas in 2000. These regions exhibit high ECC, energy utilization efficiency, carbon productivity, strong capacity for carbon sink formation while maintaining a relatively balanced approach towards both economic development and natural resource protection. Ji'an and Fuzhou are both part of the economic development zone. The ESC of both cities exceeds 2.4, which is significantly higher than that of other cities, indicating a strong carbon sink capacity. However, their economic development lags behind with low economic contribution coefficient to carbon emissions and inefficient energy utilization and carbon productivity. To enhance energy efficiency and carbon productivity while ensuring the functionality of ecological carbon sinks, it is advisable to learn from the development experience of low-carbon conservation areas, introduce advanced low-carbon technologies, and foster low-carbon industries. This will effectively harness the potential of carbon sinks. The main regions for carbon sink development include Jiujiang, Jingdezhen, Nanchang, Yingtan, and Xinyu. These areas exhibit high levels of carbon emission economic contribution coefficient as well as energy utilization efficiency and carbon productivity; however, they possess a lower carrying capacity for carbon ecology resulting in weaker overall carbon sink capabilities compared to Ji'an and Fuzhou's stronger absorption capacities despite their relatively higher levels of economic development and technological advancement where energy resources are efficiently utilized leading to high energy utilization efficiency along with elevated levels of carbon productivity.

By 2010, there was a notable shift in the carbon balance zoning compared to that of 2000. Jiujiang and Jingdezhen transitioned from being carbon sink development zones to economic development zones, indicating a decrease in the economic contribution coefficient of carbon emissions and an increase in the carbon ecological carrying coefficient over the past decade in these two cities. In contrast, Xinyu introduced a new comprehensive optimization zone by 2010, highlighting lower carbon emission economic contribution coefficient and carbon ecological carrying coefficient, reduced energy utilization efficiency and carbon productivity, as well as weaker carbon sink capacity. There is ample room for improvement in both economic construction and ecological environmental protection. It is crucial to equally prioritize enhancing both carbon sink capacity and carbon productivity while safeguarding land designated for such purposes; simultaneously controlling the rise of carbon emissions and adhering to low-carbon development.

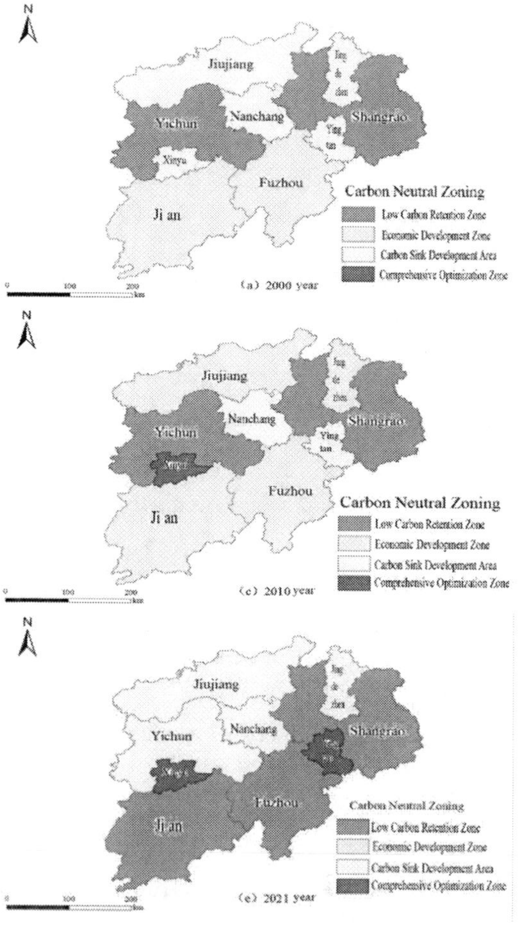

Fig. 8. Carbon balance zoning in the PYLEER from 2000 to 2021.

By 2021, there will be certain changes in the carbon balance zone compared to 2010. Yingtan City will evolve into a comprehensive optimization zone, while Jiujiang, Yichun, and Jingdezhen will transform into carbon sink development zones. Ji'an and Fuzhou will transition into low-carbon conservation zones. This highlights the need for Nanchang, the provincial capital, and its surrounding cities to prioritize urban carbon absorption levels and enhance their capacity for urban carbon sinks. Both Xinyu and Yingtan require urgent improvements and transformations in economic construction as well as ecological protection efforts. The economic development of Ji'an, Fuzhou, and Shangrao is relatively balanced with natural resource preservation measures that necessitate maintenance.

4 Conclusions

During the period from 2000 to 2021, the total carbon emissions in the PYLEER continued to increase but with a decreasing growth rate, indicating a convergence trend; however, peak carbon emissions have not been reached. Moreover, regions with relatively high carbon emissions were concentrated in cities characterized by developed transportation networks, high population density, large urban areas, high levels of urbanization, and abundant mineral resources reserves. The spatial pattern of carbon emissions remained stable throughout the entire study period.

The carbon sink has undergone two stages of change: a gradual growth trend from 2000 to 2005, followed by a declining trend from 2005 to 2021. This can be attributed to the recent national urbanization construction policy, which has resulted in a significant increase in construction land area and a decrease in carbon sink land such as forests and grasslands.

From 2000 to 2021, there will be significant changes in the carbon balance of the PYLEER. The overall carbon compensation rate during the study period remains below 1, indicating it is a carbon source. Additionally, the rate of carbon offset shows a downward trend, suggesting pressure for reducing carbon emissions. Between 2000 and 2010, both carbon productivity and energy use efficiency exhibited a decline; however, between 2010 and 2021, most cities showed an increasing trend in both indicators. Furthermore, throughout this period, the spatial distribution of the ecological carrying coefficient demonstrates higher values in southern regions compared to northern regions while maintaining relative stability.

From 2000 to 2010, there were certain changes in the carbon balance zone of the PYLEER. Jiujiang and Jingdezhen transitioned from being carbon sink development zones to economic development zones, while Xinyu City was newly designated as a comprehensive optimization zone. By 2021, further changes occurred in the carbon balance zone compared to 2010. Yingtan City became a comprehensive optimization zone, whereas Jiujiang, Yichun, and Jingdezhen transformed into carbon sink development zones. Additionally, Ji'an and Fuzhou evolved into low-carbon conservation zones.

Acknowledgments

This research was funded by Jiangxi Provincial Social Science Fund [Research on carbon emission difference and coordinated emission reduction countermeasures in Poyang Lake urban agglomeration under the dual-carbon target], grant number [22JL15D]; Jiangxi Provincial University Humanities and Social Sciences Project [Space and temporal evolution mechanism of carbon source / carbon sink and carbon balance prediction in poyang Lake ecological economic zone] , grant number [JJ22111] ; Science and technology project of Jiangxi Provincial Education Department [Space and temporal evolution mechanism of carbon source / carbon sink in poyang Lake ecological economic zone], grant number [GJJ2202918] scientific rigor and operational feasibility of policies pertaining to carbon reduction and sequestration, ultimately facilitating the achievement of "dual carbon" goals.

References

1. Firat, E.; Edmund, N.U.; Lucy, D.P. Determinants of carbon emissions: nexus among carbon emissions, coal, agriculture, trade and innovations. Environment, Development and Sustainability. 2024, 26: 17237-17251.
2. Ramakant, T.; Dileep, K.G.; Prashant, K.C., et al. Long-term spatio-temporal vegetation dynamics to climate change in koppen climatic regions of india. Earth Systems and Environment. 2024, (8):1327-1346.
3. Claudien, H.S.; Ayalkibet, M. S.; Mchura, T.M.; et al. Decoupling for a greener future: a spatio-temporal analysis of CO2 emissions and economic growth. Environmental Science and Pollution Research. 2024, 31:56886 -56900.
4. Qu, Y.D.; Li, P.F.; Chen, Y.Q.; et al. Discussion on the key areas of future agricultural discipline in the background of transforming the knowledge production mode. Universities and Disciplines. 2023, 4(1), 39-47.
5. Zhang, T.; Yu, D. Temporal and spatial dynamics of wetland landscape patterns in the Poyang lake eco-economic zone. Journal of Hydroecology. 2022, 43(4), 1-7.
6. Xiao, S.; Duo, L.H.; Zou, Z.L.; et al. Spatio-temporal matching of blue space with typical economic factors: a case study of Poyang lake ecological economic zone. Geography and Geo-Information Science. 2023, 39(1):31-36.
7. Zheng, X.P.; Chen, Z.A.; Wei, X.J. The impact of construction land expansion on the landscape fragmentation of cultivated land in the Poyang lake ecological economic zone. Journal of Agricultural Resources and Environment. 2024, 41(2):280-291.
8. Su, Y.X.; Chen, X.Z.; Ye, Y.Y.; et al. The characteristics and mechanisms of carbon emissions from energy consumption in China using DMSP/OLS night light imageries. Acta Geographica Sinica. 2013, 68(11), 1513-1526.
9. Du, H.B.; Wei, W.; Zhang, X.Y.; et al. Spatio-temporal evolution and influencing factors of energy-related carbon emissions in the Yellow river basin: based on the DMSP/ OLSand NPP/VIIRS nighttime

light data. Geographical Research. 2021, 40(7), 2051-2065.

10. Li, Y.Y.; Wei, W.; Zhou, J.J.; et al. Changes in Land Use Carbon Emissions and Coordinated Zoning in China. Environmental Science, 2023, 44(3):1267-1276.
11. Li, J.B. Carbon emission effect and control strategy of urbanization. Nanjing: Nanjing University Press, 2021.
12. Li, J.B.; Chen, H.M.; Zhang, C.L.; et al. Spatiotemporal evolution characteristics of carbon sources and carbon sinks and carbon balance zoning in the Yangtze river delta region. Environmental Science, 2023, https://doi.org/10.13227/j.hjkx.202308149.
13. Zhao, R.Q.; Zhang, S.; Huang, X.J.; et al. Spatial variation of carbon budget and carbon balance zoning of central plains economic region at county-level. Acta Geographica Sinica. 2014, 69(10), 1425-1437.
14. Lu, J.Y.; Huang, X.J.; Dai, Q.; et al. Spatio-temporal Scale Analysis on the Equality of Energy Consumption Carbon Emission Distribution in China. Journal of Natural Resources. 2012, 27(12), 2006-2017.

Research and Characteristic Analysis of Ocean Wave Simulation in Jiangsu Coast

Xiaohong Liang[1,*], Wenjun Xiao[2]

[1] Jiangsu Coastal Research Center, Nanjing 210036, China
[2] East China Sea Forecast and Disaster Reduction Center, Ministry of Natural Resources, Shanghai 200136, China

Abstract. In this paper, the wave disaster data of Jiangsu Province from 2003 to 2022 is used to analyze the disaster situation, and the third generation wave model SWAN is used to simulate the wave characteristics of Jiangsu sea area. The research and analysis results of wave disaster characteristics of Jiangsu Sea area show that: The whole coastal area of Jiangsu is mainly dominated by northeast and southeast wave, and the large wave occurs from August to September, and the distribution of large wave has obvious seasonal and regional differences.

1. Introduction

Waves not only affect the safety of Marine transportation, fishing, oil and gas exploitation and other Marine activities, but also induce coastal and Marine disasters and threaten the safety of engineering structures[1]. Jiangsu Province is located in the mid-latitude Marine and continental transition zone and climate transition zone, and is also affected by the weather systems of westerlies, subtropical and low-latitude easterlies. The coastal areas are extremely vulnerable to tropical cyclones, cold air and extratropical cyclones and other disastrous weather systems. However, the sea area of Jiangsu Province has special topography, fragile Marine environment, large beaches and shallow seas, and poor cover conditions, resulting in frequent wave disasters. At the same time, the Marine industry in Jiangsu is well developed. The Marine pillar industries are Marine fishery, coastal tourism and transportation, etc. In addition, emerging industries such as Marine biomedics, Marine renewable energy utilization and Marine power industry are developing rapidly[2]. Frequent wave disasters cause huge losses to the lives and property of the people and the development of Marine economy. According to the statistics of China Marine Disaster Bulletin (2003-2012) and Jiangsu Marine Disaster Bulletin (2013-2022), 183 people died (including missing) due to wave disasters in Jiangsu Province in the past 20 years, and the direct economic loss[3-4] was more than 137 million yuan.

Wave disasters generally occur at sea. The direct victims are firstly Marine traffic and production facilities, including offshore platforms, offshore power stations, ships and aquaculture equipment, etc., and secondly, the staff[5] on various offshore facilities. The casualties of Marine disasters in Jiangsu mainly come from the wave process[6] caused by typhoons, extratropical cyclones and cold air. The study on the spatio-temporal distribution characteristics of wave disasters is of great significance for disaster emergency defense and spatial layout of Marine economic activities. Many scholars and Marine workers have carried out research[7-13] on the characteristics of wave disasters in Jiangsu Province, but mainly based on limited statistical analysis of data. In order to further refine the analysis on the spatio-temporal distribution characteristics of wave disasters in Jiangsu Province, this paper has sorted out the measured wave data of Jiangsu sea for many years. The third generation wave model SWAN was adopted to simulate the wave characteristics in Jiangsu Sea area, and the wave height and wave direction frequency distribution in Jiangsu Sea area in different months were quantitatively analyzed, providing important references for Marine production operations, engineering construction, disaster prevention and other fields in Jiangsu Sea area.

2. Data and research methods

2.1 Data Sources

The main data used in this article:

The hourly tide level and effective wave height data of Jiangsu coastal stations, including the hourly tide level data of Lianyungang, Yangkou Port and Lusi tide level stations, as well as the effective wave height data of Yancheng Wai and Subei shoal buoys, are from the on-site observation of tide stations and wave buoys arranged by the Ministry of Natural Resources in Jiangsu coastal waters.

2.2 Numerical Model

In this study, the third-generation wave model SWAN is adopted, and the grid covers the whole sea area of the

* Corresponding author: 58403777@qq.com

https://doi.org/10.1051/e3sconf/202561701024

China Sea, the calculation range is 105°E-136°E, 9°N-41°N, and it is gradually encrypted from the outside to the inside. The maximum resolution of the grid in the near-shore section is 50m, and the time step is 10min. The water depth data of SWAN model comes from maritime surveying and mapping charts and measured data, and the wind field data comes from the re-analysis wind field of ERA-5 for 10 years from 1991 to 2000, and the numerical simulation wind field for 20 years from 2001 to 2020 calculated by mesoscale model WRF. The flow field data are derived from the 30-year hourly tidal flow results calculated by numerical simulation of three-dimensional Ocean circulation and ecological Model FVCOM (Finite-Volume Coastal Ocean Model).

3. Numerical simulation of ocean waves in Jiangsu Province

3.1 Model selection and calculation range

In this study, SWAN, the third-generation wave model, is adopted. Due to its reasonable design of various energy source terms in shallow waters, SWAN is especially suitable for nearshore. Moreover, due to its enrichment of deep-water energy source terms and the introduction of higher-order computing format[14], compared with the traditional wave model that adopts explicit format, the computing step size can be large even in shallow waters. The model is suitable for the Jiangsu sea area where the shoals are widely distributed and the water depth and topography are complicated. In this paper, in order to simulate the influence of different weather systems more accurately, the grid covers the whole sea area of the China Sea, the calculation range is 105°E-136°E, 9°N-41°N, and it is gradually encrypted from the outside to the inside. The maximum resolution of the grid in the near-shore section is 50m, and the time step is 10min. The water depth of the model is combined with the measured data of maritime surveying and mapping.

4. Analysis of simulation results

4.1 Wave high frequency distribution

Based on the sea wave calculation results of nearly 30 years obtained by simulation, the calculation results of hourly effective wave height in each 0.5°×0.5° grid are extracted, and the occurrence times of I, II, III, IV, V and VI wave heights in the grid are calculated respectively, according to the classification standards of near-shore wave strength (Table 1). The monthly average frequency of class I, II, III, IV, V and VI wave heights in the sea area of Jiangsu Province is plotted. This paper focuses on the distribution characteristics of the average frequency of wave heights in May and August .

Table 1. Classification criteria of near-shore wave intensity

Scale of Wave strength	Class I	Level II	Level III	Level IV	Level V	Level VI
Effective wave height /m	9.0 the Hs or less	6.0≤Hs < 9.0	4.0≤Hs < 6.0	2.5≤Hs < 4.0	1.3≤Hs < 2.4	Hs < 1.3

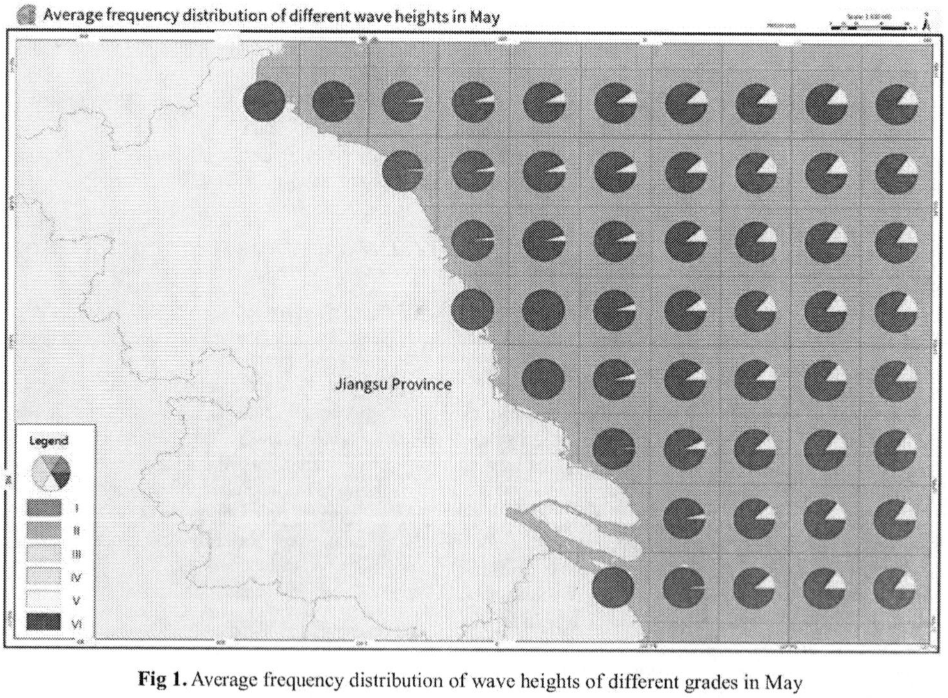

Fig 1. Average frequency distribution of wave heights of different grades in May

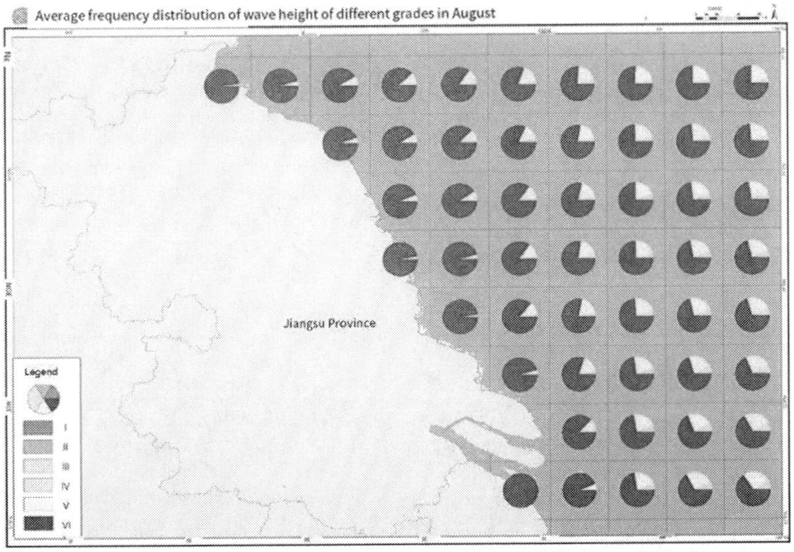

Fig 2 Average frequency distribution of wave heights of different grades in August

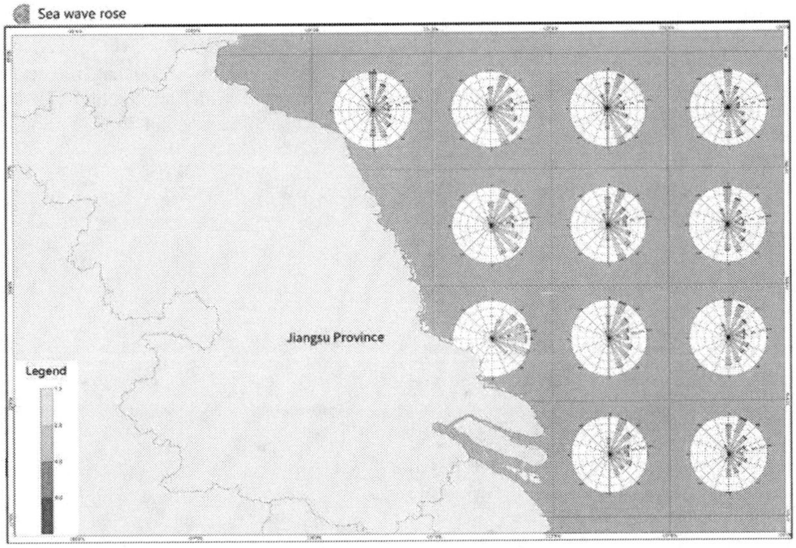

Fig 3 Wavy rose chart

As can be seen from the monthly average frequency distribution of wave heights of different grades, the average wave heights of the coastal waters of Jiangsu are mainly light to medium waves in each month (Fig 1), accounting for more than 80%, and the proportion of big waves increases gradually from west to east, and the proportion of big waves basically reaches about 10% when the east reaches 124°E. The monthly changes are obvious: The large waves in August and September accounted for a relatively large proportion (Fig 2), indicating that the large waves and even the huge waves were mainly caused by the influence of summer typhoons. From the distribution point of view, the large waves of summer typhoons were mainly distributed in the sea east of 122°E, and the proportion of large waves in the south was larger than that in the north. The important disaster bearing bodies in these areas include mariculture areas and coastal tourism areas, etc. It is necessary to take good wave prevention measures during typhoons to avoid major disaster losses; While April and May are the months in which light waves account for a large proportion of the year, and the coastal waters are basically light waves, but the proportion of waves in the waters east of 122°E in Lianyungang and Yancheng offshore and 122.5°E in the waters east of Nantong offshore is relatively large. Although the frequency of large waves is low, special attention should be paid to the development dynamics of these areas during cyclones, cold air and other small-scale processes, and early warning and monitoring should be carried out. To ensure the safety of production and operation personnel.

4.2 Wave rose diagram

Through statistical analysis of the simulation results in each 1°×1° grid, the number of times of different grade wave heights in each direction (16 directions) was obtained, and the occurrence percentage of each direction and grade wave heights, as well as the frequency distribution of different grade wave heights and the joint occurrence frequency distribution of wave heights in the wave direction were calculated, and the rose diagram on each grid point was drawn. It can be seen from the figure that the northeast and southeast waves are the main waves along the coast of Jiangsu Province. Large waves to large waves mainly appear in the east to southeast direction. (Fig 3)

5. Conclusion

(1) The northeast and southeast waves are the main waves in the whole Jiangsu sea area. Large waves to large waves mainly appear in the east to southeast direction. The average wave height of Jiangsu coastal waters in each month is mainly light to medium waves, accounting for more than 80%, and the proportion of big waves gradually increases from west to east, and the proportion of big waves basically reaches about 10% when it reaches 124°E in the east.

(2) The population loss caused by wave disasters in Jiangsu sea area is the biggest cold wave and cyclonic wave, and the key defense months are April-May; The proportion of waves in the waters east of 122°E in Lianyungang and Yancheng coastal areas and 122.5°E in the waters east of Nantong coastal areas was larger in Apron-May.

References

1. ELKO N, FEDDERSEN F, FOSTER D, et al. The future of nearshore processes research[J]. Shore and Beach, 2015, 83(1): 13-38.
2. Cheng Manman, Chen Wei. Research on development status of Marine industry in Jiangsu Province based on principal component analysis [J]. Economic Theory and Practice, 2020(11):12-19.]
3. State Oceanic Administration. China Marine Disaster Bulletin[EB/OL].2003-2012.https://www.mnr.gov.cn.
4. Jiangsu Provincial Department of Natural Resources. Communique [EB/OL]. Marine disasters in jiangsu province in 2013-2022. http://zrzy.jiangsu.gov.cn/.
5. Sun Yuntan. Research on Marine disaster emergency management in China [D]. Qingdao: Ocean University of China, 2010.
6. Han Xue, Huang Zuying. [J]. Marine Development and Management, 2015(9):75-77.]
7. WANG Y, MAO X, JIANG W. Long term hazard analysis of destructive storm surges using the ADCIRCSWAN model: A case study of Bohai Sea, China[J]. International Journal of Applied Earth Observation & Geoinformation, 2018, 73: 52-62.
8. LI J, PAN S, CHEN Y, et al. Numerical estimation of extreme waves and surges over the northwest Pacific Ocean[J]. Ocean Engineering, 2018, 53 (1) : 225-241.
9. Peng Ji, Tao Aifeng, Qi Kelen, et al. Analysis of wave disaster characteristics in China in recent ten years [C]// Chinese Society of Ocean Engineering. Proceedings of the 16th China Ocean (Coastal) Engineering Symposium (Volume 1). Beijing: China Ocean Press, 2013:814-817.
10. Chen Qicambridge, Han Bo, Yang Qing-Hua, et al. [10] Analysis of wave disaster characteristics of shipwrecks in the eastern waters of Zhejiang [J]. Marine Forecast, 2023, 40(1):28-38.
11. ZHANG Kunlan, LI Guimin, Yang Chao, et al. [11] Statistical analysis of wave characteristics along the coast of Hebei [J]. Science and Technology Innovation, 2021, 27:72-75.
12. Shi Shi 'an. Numerical simulation and statistical analysis of typhoon wave in Jiangsu Coast [D]. Yangzhou: Yangzhou University, 2023.
13. LIANG X H, Peng M, Han X. Analysis of the influence of the Jianghuai Cyclone out to sea on the East coast on 2010-04-21 [J]. Marine Development and Management, 2013, 30(3): 64-67.
14. ROGERS W, HWANG P, WANG D. Investigation of wave growth and decay in the SWAN model: three regional-scale applications[J]. Journal of Physical Oceanography, 2003, 33(2): 366–389.

BCF Better Reflects the Bioavailability of Selenium in the Soil of the Northeastern Guanzhong Plain Compared to MF

Shengfei Yang[1,*], Pei Zhang[1], Fa Guo[1], Duoxun Xu[1], Yuchen Yan[1], Tuo Liu[2]

[1] Xi'an Center of Geology and Mineral Resources Survey, China Geological Survey, Xi'an 710100, Shaanxi, China
[2] Xi'an Center of Geological Survey, China Geological Survey, Xi'an 710054, Shaanxi, China

Abstract. A large area of selenium-rich soil has been identified in the Guanzhong Plain; however, the bioavailability of selenium remains poorly understood, which poses significant challenges to the development and utilization of these resources. This study investigates selenium bioavailability by focusing on surface soils, wheat grains, and rhizosphere soils in the northeastern Guanzhong Plain. The results reveal that the average selenium content in surface soil exceeds the background levels for both the Guanzhong Plain and China. Humic acid-bound selenium and strongly organically bound selenium are the predominant forms in rhizosphere soil. The bioconcentration factor (BCF) better reflects the bioavailability of selenium in the soil of the northeastern Guanzhong Plain compared to the mobility factor (MF). Moreover, selenium bioavailability in the study area is notably higher than in surrounding regions. These findings offer valuable insights into the sustainable development and efficient utilization of selenium-rich land resources in this region.

1. Introduction

Selenium is widely recognized as a crucial trace element indispensable for sustaining the health and well-being of both humans and animals [1]. China has been identified by the World Health Organization as one of the countries facing selenium deficiency [2]. Adequate selenium intake is essential for the population, and consuming naturally selenium-rich crops provides a simple and effective means of achieving this [3]. Selenium-rich soil provides the foundation for growing and producing crops naturally enriched with selenium. In recent years, with the gradual progress of land quality geochemical surveys, a large area of selenium-rich soil has been discovered in the Guanzhong Plain. The study area is a representative region for the distribution of selenium-rich soil.

Selenium bioavailability refers to the capacity of selenium in soil to be absorbed by plants and incorporated into the food chain. This process is influenced by various factors, including soil properties, selenium speciation, plant species, and environmental conditions. In recent years, extensive research has been conducted to explore the factors affecting selenium bioavailability and its evaluation methods. Studies have shown that soil pH, organic matter content, and selenium speciation are critical determinants of selenium bioavailability. For example, soil pH significantly affects selenium solubility and mobility. In acidic soils, selenium is more readily available for plant uptake, while in alkaline soils, selenium tends to become more fixed, reducing its bioavailability [4]. Furthermore, the oxidation state of selenium is another crucial factor influencing its bioavailability. Selenite (Se(IV)) is more bioavailable and readily absorbed by plants than selenate (Se(VI)) [5]. Additionally, there are differences in selenium uptake among plant species, with crops such as wheat and rice demonstrating higher selenium accumulation capacity [6]. To more accurately assess selenium bioavailability, researchers have utilized the Diffusive Gradients in Thin Films (DGT) technique, which simulates the root absorption process of plants and provides more precise data on bioavailability [7].

The selenium content in crops is influenced by the selenium levels in the soil, as well as the forms in which selenium is present in the soil [8]. Generally, the bioavailability of soil elements can determine the absorption amount of soil elements by crops [9]. The bioavailability of soil selenium is primarily influenced by soil properties such as pH, organic matter (SOM), cation exchange capacity, and oxide content, as well as geographical factors like parent material, elevation, average annual precipitation, and temperature [10]. BCF and MF are two parameters used to characterize selenium (Se) in soil. The BCF value is mainly calculated based on the total elemental content in the soil, whereas the MF value focuses on the bioavailable fraction of elements in the soil [11-13].

The Guanzhong Plain is a vital grain-producing region in Northwest China, where wheat is the primary staple crop. Despite its importance, research on the sources and bioavailability of selenium (Se) in this region is limited, hindering the full potential of selenium-rich soils and selenium-enriched agricultural products. This study explores the distribution of selenium in the selenium-rich

* Corresponding author: 125286220@qq.com

 https://doi.org/10.1051/e3sconf/202561701025

soils of the northeastern Guanzhong Plain and examines the selenium content in local wheat. It identifies key factors influencing selenium enrichment in both soils and wheat and investigates potential sources of selenium. Additionally, the study proposes thresholds for the bioavailability of selenium in soil. The findings provide valuable theoretical insights for the sustainable use of selenium-rich soils and the cultivation of naturally selenium-enriched crops, offering significant practical value for agricultural development.

2. Conclusions

2.1 Distribution characteristics of selenium in surface soil and wheat grain

2.1.1 Distribution characteristics of selenium in surface soil

The distribution of selenium (Se) in the surface soils of China shows significant regional variability, a pronounced Se-deficient zone from Northeast to Southwest China has revealed, and the selenium-rich areas are mainly concentrated in the southeastern and southwestern regions, including parts of Hunan, Guangxi, Guangdong, Fujian, Hunan and Zhejiang, where the selenium content in soils often exceeds the average levels found globally[14]. The main factors affecting soil selenium enrichment include the type of soil, climatic conditions, pH of soil, and the organic content.

The selenium content in surface soil ranges from 0.132 to 1.052 mg/kg, with an average of 0.286 mg/kg(Table 1), which exceeds the background values for both the Guanzhong Plain and the national average for China [15]. The eastern alluvial plain region is characterized by high selenium content. According to the selenium-rich soil classification standards (Se ≥ 0.3 mg/kg) outlined in the China Geological Survey's 'Natural Selenium-Rich Land Demarcation and Identification DD2019-10 (Trial)', the soil in the area has a selenium enrichment rate of 40%.

Table 1. Characteristics of selenium content in surface soil (mg/kg)

	Study area	Guanzhong Plain	China
Maximum value	1.052	2.628	
Minimum value	0.132	0.034	
Average value	0.286	0.174	0.22

2.1.2 Distribution characteristics of selenium in wheat grain

The selenium content in wheat grains ranges from 0.024 to 0.859 mg/kg, with an average of 0.176 mg/kg. This is not only higher than the average selenium content in wheat grains from other counties in the Guanzhong Plain, but also exceeds the selenium levels across China(Fig 1).

The significant difference between the minimum and maximum selenium concentrations in wheat grains from Baoji and Xianyang reflects considerable variability in selenium content across these regions. In contrast, the narrower range of selenium concentrations in wheat grains from Weinan and the study area indicates a more uniform distribution of selenium. As a result, the study area and Weinan are more favorable for the development of selenium-enriched land compared to other regions.

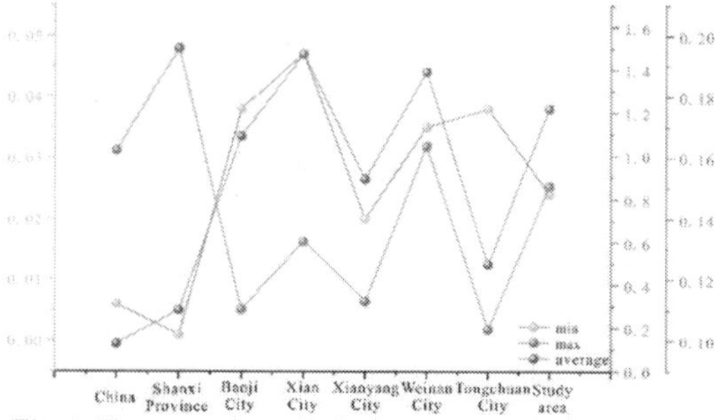

Fig. 1. The curve chart of selenium content of wheat grain in different regions

According to the "Selenium Content Standards for Selenium-Rich Foods and Related Products (DB61/T556-2018)", the selenium content rate in wheat grains in the study area is 53.33%, and the selenium-enriched rate of 40.91%. According to the Guidelines for the Delimitation and Identification of Naturally Selenium-Enriched Land (trial implementation), the study area is categorized into selenium-enriched zones (Se content ≥ 0.3 mg/kg) and non-selenium-enriched zones (Se content < 0.3 mg/kg). In selenium-enriched zones, the Se content in wheat grains ranges from 0.056 to 0.859 mg/kg, with an average of 0.254 mg/kg. In contrast, wheat grains in non-selenium-enriched zones have Se contents ranging from 0.047 to 0.062 mg/kg, with an average of 0.048 mg/kg, which is significantly lower than that in selenium-enriched zones.

2.1.3 Speciation of selenium in soil

2.2 Factors influencing selenium content in soil and wheat grains

2.2.1 The factors influencing selenium in soil

Studies have indicated that total carbon (TC), nitrogen (N), sulfur (S), phosphorus (P), soil organic carbon (SOC), pH, and certain heavy metals are correlated with selenium (Se) content in soil [6]. In the study area, the correlation coefficients between selenium and these elements and indicators are relatively high, while, the correlation with pH is weak, with the exception of a small portion of strongly alkaline soils, the soils in the study area are predominantly alkaline. This is likely due to the limited range of pH variation, which weakens the correlation with selenium content. Similarly, neighboring regions also exhibit a weak correlation with soil selenium.

2.2.2 The factors influencing selenium in wheat grain

The factors associated with content of Se in soil and wheat grain are generally similar. In the study area, the factors most strongly correlated with wheat grain Se content include soil Se, humic acid-bound Se, strongly organically bound Se, and P content.

The strongly bound organic form is the predominant species of selenium (Se) in wheat root soil, accounting for 19.390%–67.227%, with an average of 45.711%. The residue-bound form is the secondary species, representing 14.065%–60.763%, with an average of 31.090%(Fig 2). This distribution pattern is similar to the distribution of selenium forms in other selenium-rich soils, such as those in the Guanzhong Plain, Ziyang County in Shaanxi Province, and Enshi City in Hubei Province.

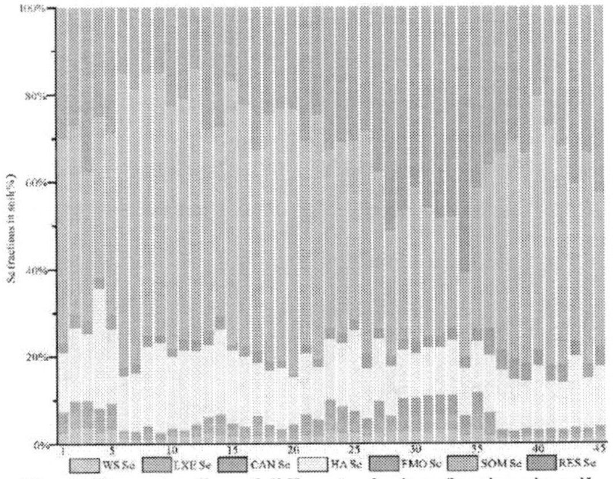

Fig. 2. The proportion of different selenium fractions in soil

(Water-soluble Se (WS-Se), Ion-exchangeable Se (EXE-Se), Carbonate-bound Se (CAN-Se), Humic-acid bound Se (HA-Se), Fe/Mn-oxide bound Se (FMO-Se), Strongly organic-bound Se (OM-Se), Residual Se (RES-Se))

Water-soluble selenium (WS-Se), ion-exchangeable selenium (EXE-Se), and carbonate-bound selenium (CAN-Se) exist in free or weakly adsorbed forms on soil particle surfaces, these forms are directly available for plant uptake and exhibit greater bioavailability compared to organically bound selenium. Humic acid-bound selenium (HA-Se) and strongly organically bound selenium (SOM-Se) are the predominant forms of selenium in soil of the study area, although they cannot be directly absorbed by plants, they act as potential reservoirs of bioavailable selenium, during plant growth, strongly organically bound selenium and Fe-Mn oxide-bound selenium can transform into bioavailable forms accessible to plants [6]. This also further explains the discrepancy in the study area, where the available selenium content in the soil is low, yet the selenium content in wheat grains remains high.

2.3 The bioavailability of selenium in soils

2.3.1 The characterization methods for bioavailability of soil

The bioavailability of soil elements can be expressed using the Bioconcentration Factor (BCF) and Mobility Factor (MF). The BCF value indicates the ability of plants to absorb and accumulate specific elements, a higher BCF for a particular crop reflects a stronger capacity for absorbing that element. The MF reflects the relative mobility of an element within the soil, with higher MF values indicating a greater capacity for the element to be absorbed by plants. This implies a higher bioavailability of the element to plants[11-13].

The bioconcentration factor (BCF) values in the study area range from 5.411% to 81.128% (Fig 3), with most values concentrated between 19.508% and 45.902% and the average values was 24.654%, higher than the surrounding counties. The BCF values for wheat in non-selenium-enriched zones range from 11.481% to 45.901%, with an average of 22.282%, concentrated primarily around 20.001%. In selenium-enriched zones, the BCF values for wheat range from 5.411% to 81.128%, with most values falling between 22.473% and 60.724% and an average of 40.207%, nearly twice that of non-selenium-enriched zones. These BCF values demonstrate that the Se content in wheat grains is predominantly influenced by the Se content in the rhizosphere soil. This observation is consistent with studies from other regions, which show that grain Se content in selenium-rich soils is significantly higher than in selenium-deficient or low-selenium soil.

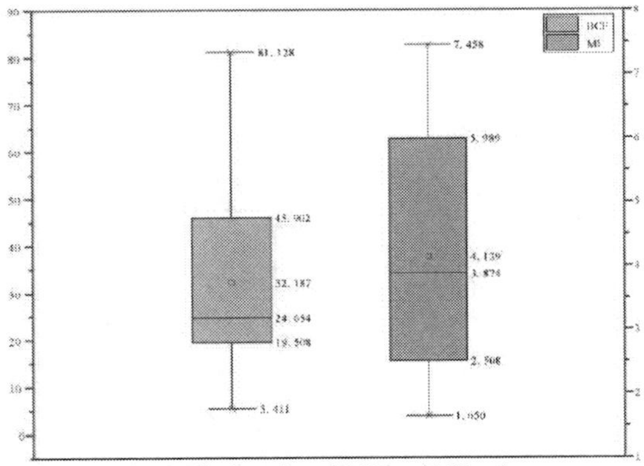

Fig. 3. The boxplot of BCF and MF values

The MF values in the study area range from 1.678% to 8.059%, with an average of 3.651%, which is lower than the national average for 16 soil types and the average for Shaanxi Province [6]. The MF values for wheat in non-selenium-enriched zones range from 1.830% to 7.690%, with an average of 4.565%, while, in selenium-enriched zones, the MF values for wheat range from 1.678% to 8.059%, with an average of 4.058%. Although the MF values are higher in the selenium-enriched area, the difference compared to the non-enriched area is not significant, and the average MF value is even lower. This suggests that the MF value is not a reliable indicator of selenium bioavailability in the soil of the study area

2.3.2 Characterization of selenium bioavailability in soil

Through the calculation and comparison of BCF and MF values, it is further demonstrated that the accumulation of selenium (Se) in wheat grains is mainly influenced by the selenium content in the soil. This finding is consistent with the results of studies conducted in Enshi, Hubei, Jizhou, Tianjin, and Yanqi, Xinjiang.

Research on the application of selenium fertilizer to the soil has demonstrated that an increase in soil selenium content leads to a significant enhancement of selenium levels in wheat grains [16-18], that further show the total selenium content in the soil is not only strongly correlated with selenium content in the wheat grains, but also a key determinant of selenium accumulation in the grains.

In summary, the soil in the northeastern Guanzhong Plain is alkaline, and combined with relatively stable pH and high organic matter content, the BCF values reflects the bioavailability of selenium in the soil is better than MF.

Acknowledgments

This paper is a partial result of the Shaanxi provincial natural science basic research program general project, "Study on the main controlling factors and enrichment patterns of selenium-enriched wheat in the Guanzhong Plain" (2023-jc-yb-255).

References

1. J. Zeng, D. Guan, Z. Dai, et al. Selenium contents, speciation and bioaccessibility of Se-enriched rice grains from Chinese markets, Soil & Environmental Health, 1, 100037 (2023)
2. Z.Q. Hu, X. Xiong, J.W. Bu, et al. Form, Bioavailability, and Influencing Factors of Soil Selenium in Subtropical Karst Regions of Southwest China, Appl. Sci. 14, 5192 (2024)
3. L.Q. Zhang, B.Y. LU. Quality and safety of selenium-rich agricultural food and considerations on its industrial development. Quality and Safety of Agro-products, **2**:3-9 (2020)
4. C.M. Yang, H. Yao, YJ Wu, et al. Status and risks of selenium deficiency in a traditional selenium-deficient area in northeast China. Sci. Total Environ, 762: 144103(2021)
5. Q.T. Dinh, M.K.Wang, T.A.T.Tran, et al. Bioavailability of selenium in soil-plant system and a regulatory approach. Critical Reviews in Environmental Science and Technology, 49(6): 443-517(2023)
6. N.N. Liu, M. Wang, F. Zhou, et al. Selenium bioavailability in soil-wheat system and its dominant influential factors: A field study in Shaanxi province, China. Sci. Total Environ, 170-141664 (2021)
7. H. Zhang, X.B. Feng, C.X.Jiang, et al. Understanding the paradox of selenium contamination in mercury mining areas: high soil content and low accumulation in rice [J]. Environmental Pollution, 188: 27-36(2023)
8. M.K. Wang, Z.W.Cui, M.Y.Xue, Et al. Assessing the uptake of selenium from naturally enriched soils by maize using diffusive gradients in thin-films technique (DGT) and traditional extractions, Environment, **689**:1-9 (2019)
9. S. Huang. Geochemical characteristics and genesis of Se rich soil in Guizhou area ,Yunnan Chemical Technology, **45**(7) : 147-148 (2018)
10. M.Y. Song, H.X. Li, Y.C. Wei, et al. Geochemistry of the selenium, Zhitang Town, Longyou County, Zhejiang Province, Guizhou Geology, **22**(3):176-180 (2005)
11. F. Zhou, Q. Peng, M. Wang, et al. Advances in the evaluation of selenium bioavailability in soil-plant system, Chin Sci Bull, **67**: 461-472 (2022)
12. L.X. Wang, T. Yu, F.Y. Li, et al. A Summary of Research Progress on Bioavailability Assessment Method of Selenium in Soil and Its Influencing Factor, Rock and Mineral Analysis, **42**(2):239-253 (2023)
13. S.H. Kim, S.J. Bae, Y.S. Hwang. Comparative bioaccumulation, translocation, and phytotoxicity of metal oxide nanoparticles and metal ions in soil-crop system, Sci. Total Environ, 856-158938 (2023)
14. M. Peng, H.X. Penguin. Progress and main achivements on land quality geochemical survey project. Computing Techniques for Geophysical and Geochemical Exploration, 44(6): 805-814(2022)
15. Y. Hou, Z.F. Yang, T.Yu, et al. Soil geochemical parameters in Chin, Beijing: Geological Publishing House (2020)
16. Radawiec, A., Szulc,W., Rutkowska, B. Selenium Biofortification of Wheat as a Strategy to Improve Human Nutrition, Agriculture, **11**, 144 (2021)
17. J.Y. Zhao, Y.L. Liu, X.L. Tian, et al. Simulation and prediction for the spatial heterogeneity of soil selenium bioavailability at different stratigraphic scales, Chemosphere, 344-140295 (2023)
18. K.M. Bao, Y. Wang, X.P. Du, et al. Comparison of Selenium Accumulation in Edible Parts of Wheat and Broad Bean, Agronomy, **13**, 1939. (2023) https://doi.org/10.3390/agronomy13071939

E3S Web of Conferences 617, 02001 (2025) https://doi.org/10.1051/e3sconf/202561702001
EEUPD 2024

Research on Regional Traffic Simulation Analysis and Traffic Capacity Improvement Technology Under the Background of Urban Renewal

Liang Wang[a], Anfan Shang[b], Yang Feng[*], Hao Fu[c] and Ziqi Gao[d]

Northwest Engineering Corporation Limited of PowerChina, Xi'an, Shaanxi 710065, China

Abstract. With the continuous advancement of urban renewal, existing road resources are difficult to meet new transportation needs. In order to alleviate various traffic problems arising from urban renewal, this article starts from macro and micro perspectives, based on macroscopic fundamental diagram theory and urban traffic microcirculation theory, and studies an innovative macro traffic control strategy and micro traffic organization method for urban renewal areas. The evaluation is conducted through VISSIM dynamic traffic simulation. This article takes the Xiaozhai commercial district in Xi'an as a case study to verify the feasibility and effectiveness of the proposed method. The results show that the macro region boundary control algorithm based on the macroscopic fundamental diagram can always maintain the cumulative number of vehicles in the region around the optimal number of vehicles. Meanwhile, the micro traffic organization method based on the concept of microcirculation can effectively utilize road resources for diversion, significantly improving the service level of intersections and roads in the control area. The average operating delay of vehicles in the road network is reduced by 13.29%, the queue length at intersections is reduced by 19.20%, and the efficiency of regional traffic operation is effectively improved. This study ultimately developed a practical, portable, and promotable urban renewal area traffic governance method system, which has significant industry, social, and economic benefits.

1 Introduction

In the past urban construction process, the development focus was overly inclined to the new urban areas, which promoted economic development and improved the level of urbanization through rapid urban expansion. The renewal and traffic management of the old urban areas lagged behind. The old urban areas had inherent deficiencies such as lack of transportation space resources, backward traffic organization models, and unsystematic road networks. With the rapid advancement of urban renewal in the old city, the functions of its plots changed significantly, and the population increased rapidly Aggregation pole The transportation supply and demand contradiction is increasingly prominent and further intensified[1], and the transportation governance problem in the old city renewal area is imminent.

At present, many experts and scholars have conducted research on urban traffic management methods. In terms of traffic organization, Karimi et al.[2] have made efforts to alleviate urban traffic congestion by minimizing the total number of vehicles in the area, At the same time, the total number of vehicles in the congested areaCheng timeThe minimum goal is to find the optimal one-way road configuration scheme for the target. In order to improve the utilization of one-way roads and divert traffic pressure from main roads, LONG et al.[3] proposed an optimization method for one-way traffic organization from the perspective of traffic load and road balance, with the goal of reducing road saturation and vehicle detours, and established a bi-level programming model. In terms of signal control, dynamic signal control is achieved by sensing real-time changes in traffic flow in the road network, optimizing signal timing schemes in real time, and continuously achieving balance in traffic flow distribution in the road network. Nathan [4] established a target function with the minimum total delay at each intersection in the regional road network, and optimized the phase difference, signal cycle, and effective green time at each intersection. Cao Hongwang et al.[5] considered the dynamic characteristics of regional traffic flow, comprehensively analyzed the calculation methods of external and internal entrance delays, and proposed a regional coordination control method based on chaos genetic algorithm. The macroscopic fundamental diagram (MFD) is a measurable and objectively existing macroscopic traffic flow model used to describe the relationship between variables such as traffic flow, average density, and average speed in road traffic networks. The model was originally proposed by Daganzo[6] proposed and experimentally proved the existence of MFD at the urban scale, which provides important practical significance for the analysis and

[*] Corresponding author: [*]253181610@qq.com, [a]wangliang@nwh.cn, [b]02084@nwh.cn, [c]fuhao@nwh.cn, [d]229734178@qq.com

management of regional transportation. The improvement and development of MFD have gained recognition from many scholars and experts, and it has begun to be applied to traffic network control. In recent years, many scholars have simulated regional transportation through VISSIM[7] and SUMO[8] simulation software, drawn regional macro-basic maps, and proposed a series of regional boundary control methods.

Currently, existing traffic management technologies are fragmented and not systematic, with limited application scenarios, and cannot be effectively applied to regional traffic management in the context of urban renewal. How to scientifically and effectively improve the overall operational level of urban renewal regional road networks remains a major research focus in the field of traffic management and control, and many data, scenarios, and technologies have not yet been broken through. Therefore, this article will systematically study the characteristics of changes in traffic travel demand in new business formats and new spaces after urban renewal in old urban areas. In response to problems such as traffic congestion, parking difficulties, serious interference between import and export, crowding of pedestrian rights, and delays in signal control, we will explore methodologies from both macro-regional traffic boundary control and micro-traffic organization and management perspectives, propose scientific and efficient traffic congestion management and traffic capacity improvement methodologies in the context of urban renewal in old urban areas, and apply VISSIM traffic simulation software to verify the effectiveness of the proposed methods. This will help alleviate urban renewal congestion and improve traffic capacityRear areaProvide technical support for transportation issues such as organization and management.

2 Methods for improving the traffic capacity of urban renewal areas

2.1 Method for Controlling Regional Boundaries Based on the Macroscopic Fundamental Diagram

Understanding the dynamic changes in road network traffic flow in real time has always been the foundation for research on road network structure design and road network performance optimization. Usually, road network traffic flow includes characteristics such as time distribution, spatial distribution, composition type, and formation reason of traffic flow. However, in actual traffic environments, it is difficult to collect sufficient traffic information to describe and analyze some or all of the characteristics of network traffic flow in real time. The macroscopic basic graph of the traffic network, as an objective and measurable attribute of the road network, can describe the relationship between macroscopic traffic parameters of road network traffic flow, such as average traffic flow and average density of the road network, indirectly describing the time and spatial distribution characteristics of road network traffic flow. This article will start from the overall structure of the road network, systematically analyze the relationship between traffic flow variables of multiple sections or multiple roads in the regional road network, and achieve the drawing of the macroscopic basic graph of the research area based on this.

2.1.1 Macroscopic Fundamental Diagram

This article uses the cumulative vehicle-departure vehicle relationship diagram (MFD) to present the overall traffic situation of the road network. The MFD can be seen as a function between the outflow *F(D(t))* of the road network and the cumulative number of vehicles N in the road network. According to Figure 1, the MFD presents a continuous unimodal curve, which roughly divides regional traffic flow into free flow, saturated flow, oversaturated flow, and congested state. The left side of the peak indicates that as the cumulative number of vehicles in the road network increases, the number of vehicles leaving the road network also increases; while the right side of the peak indicates that as the cumulative number of vehicles in the road network increases, the number of vehicles leaving the road network gradually decreases. Therefore, by fluctuating the traffic flow of the road network in the urban renewal area near the optimal cumulative number of vehicles N*, it can maximize the number of vehicles leaving the area, thus ensuring that the overall traffic operation efficiency of the road network in the urban renewal area is at its best and remains stable.

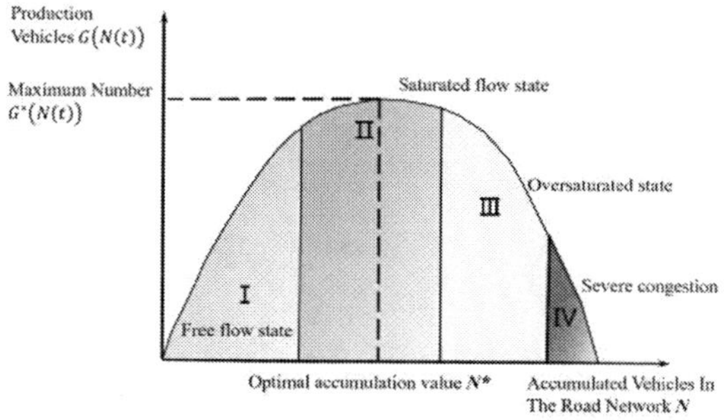

Fig. 1. Macroscopic fundamental diagram.

The time is discretized into equal intervals, *D(t)* represents the number of vehicles operating within the regional road network at the end of period *t*, $d_{in}(t+1)$ represents the number of vehicles entering the control area from outside the control area during the *t+1* period, and $d_{out}(t+1)$ represents the number of vehicles leaving the control area during the *t+1* period. The number of vehicles operating within the road network in the next period *D(t+1)* is given by formula (1):

$$D(t+1)=D(t)+d_{in}(t+1)-d_{out}(t+1) \quad (1)$$

The optimal cumulative number of vehicles D^* for the regional road network is obtained by fitting the macroscopic basic diagram and using the peak value of the number of vehicles driving out on the vertical axis, which is the maximum number of vehicles driving out. The difference between the number of vehicles obtained at a certain time and the given optimal cumulative threshold for vehicles is used as a dynamic input, and the increment

$\Delta D^*(t+1)$ of the optimal number of vehicles considering internal occurrence in the region at time t+1 can be calculated using formula (2).

$$\Delta D^*(t+1) = D(t+1) - D^* \tag{2}$$

2.1.2 Regional boundary control model

Let the number of vehicles entering the control area through the entrance section i outside the control area during the $t+1$ period be $d_{i,in}(t+1)$. Considering the significant differences in service levels at different intersections, the weight of each entrance is determined by the ratio of the number of vehicles entering the control area through a certain boundary entrance to the total number of vehicles entering the control area. Then, the incremental $\Delta D^*(t+1)$ obtained from formula (3) is allocated to each entrance section of the boundary, that is, the number of vehicles allocated to the entrance section i $\Delta D_{i,in}(t+1)$, as shown in formula (3).

$$\Delta D_{i,in}(t+1) = \Delta D^*(t+1) \cdot \frac{d_{i,in}(t+1)}{d_{in}(t+1)} \tag{3}$$

The flow rate $q_{i,in}(t+1)$ of vehicles entering the road network through the entrance section i during the $t+1$ cycle can be calculated by comparing the number of vehicles $d_{i,in}(t+1)$ and the corresponding travel time ΔT. The more reasonable reduction time $\Delta T_{i,in}(t+1)$ for the entrance green light in each cycle can be obtained, as shown in formula (4).

$$\Delta T_{i,in}(t+1) = \frac{\Delta D_{i,in}(t+1)}{q_{i,in}(t+1)} \tag{4}$$

On this basis, in order to avoid the boundary control method causing excessive queue length and vehicle overflow at the entrance, which may affect the traffic efficiency of other sections outside the control area, and to avoid the problem of insufficient pedestrian crossing time caused by the reduction of effective green light duration at the entrance direction, this paper proposes two constraint conditions: (1) queue length constraint at the periphery of the boundary; (2) The shortest green light duration constraint is shown in formulas (5) and (6).

$$T_d = \frac{n_i - (L_i \times R_i)/l}{q_{i,in}(t+1)} \tag{5}$$

Among them, L_i represents the maximum queue length allowed on the i entrance section, R_i represents the number of lanes on the i entrance section, l represents the average vehicle length, and n_i represents the number of existing queue vehicles on the i entrance section.

$$T_{min} = 7 + \frac{W}{V} - I_i \tag{6}$$

In this context, W denotes the width of the road segment, V stands for the speed of pedestrians crossing the street, and I_i represents the duration of the green light interval.

In summary, the green light time $T_{i,in}(t+1)$ of the entrance section i in the next cycle can be obtained by subtracting $\Delta Ti,in(t+1)$ from the green light time $T_{i,in}(t)$ of the previous cycle, and considering the constraints of peripheral queue length and the shortest green light time, the effective green light time that should be adjusted in each system control cycle can be obtained, as shown in formula (7).

$$T_{i,in}(t+1) = T_{i,in}(t) - \max\left(0, \Delta T_{i,in}(t+1)\right) + \max\left(0, T_d\right) \geq T_{min} \tag{7}$$

2.2 Research on the micro-traffic organization method in the renewal area based on the concept of "micro-circulation"

A well-developed urban micro-circulation system for transportation can not only share traffic flow, alleviate traffic pressure, and improve road capacity, but also serve residents' daily travel. "Urban micro-circulation for transportation" refers to the use of roads such as branch roads, narrow alleys, small streets, and sidewalks outside the main road network to achieve traffic diversion from the main road network by dredging the micro-circulation network. In urban transportation, traffic flow is generally concentrated on the main roads, which are often extremely congested during peak traffic hours. At this time, organizing traffic micro-circulation to bypass the main roads with traffic flow from branch roads, diverting traffic from the main roads, and alleviating congestion can effectively ensure smooth regional traffic operations. This section considers the current situation of the road network in the transformed plot and the impact of the transformation on the road networkTraffic attraction impactThe combination of "point-line-surface" is used to optimize and reconstruct the bottleneck points of traffic congestion and road network organization, ultimately improving the road network traffic capacity and traffic operation efficiency of the entire urban renewal area.

Microcirculation transportation is committed to building a more environmentally friendly, healthy, and convenient urban transportation system. Firstly, establish convenient non-motorized and pedestrian traffic systems, such as setting up bicycle lanes, pedestrian walkways, and pedestrian crossings, to provide a safe and convenient environment for non-motorized vehicles and pedestrians. Secondly, optimize the public transportation network, encourage citizens to use public transportation, and reduce their dependence on private cars. In addition, through the design of green belts, leisure squares and other humanized urban spaces, create a livable environment and promote the choice of walking and cycling. Finally, take measures such as traffic restrictions and increasing parking fees to control the traffic of motor vehicles and reduce traffic congestion and environmental pollution. The role of microcirculation transportation is mainly reflected in reducing traffic congestion, improving air quality, enhancing urban accessibility, promoting physical health, and reducing energy consumption. Therefore, the implementation of the microcirculation transportation concept helps optimize the urban transportation structure, promote the development of green travel modes, and improve the quality of life of residents.

3 Experimental Design and Data Collection

For the Xintiandi urban renewal project of Yanta culture, the land use nature of the area has undergone significant changes after the completion of the large underground parking lot. The area is adjacent to the Xiaozhai business district, with a long-term high volume of pedestrian and vehicle traffic. The opening of a large parking lot with 1,314 parking spaces will inevitably cause further tension in the surrounding traffic environment. Therefore, it is urgent to alleviate traffic congestion and improve traffic capacity in the context of urban renewal in the old city. This study selects the Xiaozhai urban renewal area in Xi'an as the research object, with the specific scope bounded by Xiaozhai West Road and Xiaozhai East Road as the northern boundary, Changming Road The southern boundary is defined by Cuihua Road, the eastern boundary by Zhuque Street, and the western boundary by the southern section of Zhuque Street, covering a total area of approximately 2.43 square kilometers. This area is primarily composed of high-grade arterial roads, with secondary roads interspersed between them for traffic distribution. There are a total of 16 intersections within the area, with signal control used except for the intersection between Changming Road and Chang'an South Road. Figure 2 illustrates the location of 13 entrances and exits within the area boundary.

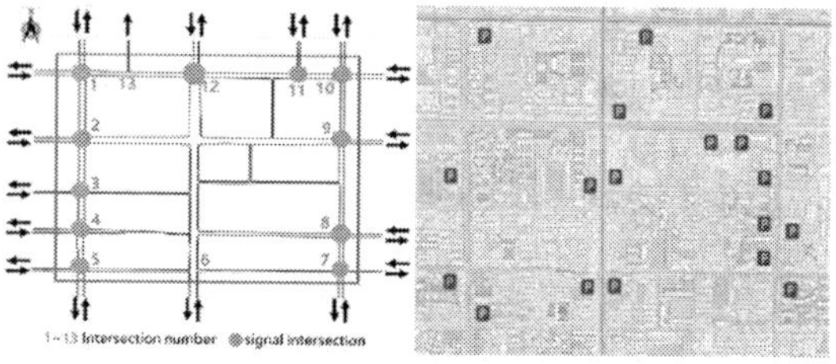

Fig. 2. Schematic diagram of key data collection points in the study area.

To obtain actual data support, 120 professional and technical personnel were organized to conduct a data survey on the control area for three consecutive hours during the peak period from 16:00 to 19:00 every day. The survey included: inflow and outflow traffic volume at the intersection of the control area boundary (recorded at intervals of one minute), traffic flow at intersections and road sections within the area, inflow and outflow traffic volume at important traffic occurrence points, signal timing schemes at each intersection, and the maximum acceptable queue length at each entrance section.

4 Result analysis

Traffic absorption of the renovated area after the parking lot is openedHair levelThe improvement measures proposed in this article include the signal control scheme for boundary intersections based on the macroscopic fundamental diagram algorithm, the coordinated control of green wave for arterial roads, the optimization of intersection signal timing, the design of lane functions, the improvement of intersection channelization, and the optimization of regional traffic organization, as shown in Table 1. Whether the above improvement measures are beneficial to alleviating traffic congestion and improving the efficiency of regional road network traffic operation needs further evaluation of the updated regional traffic operation level after the parking lot is opened.

Based on the above-mentioned research on regional traffic improvement measures, a 3D regional traffic simulation model was constructed based on the current VISSIM regional traffic simulation model after the parking lot in the urban renewal area was opened, and the above-mentioned regional traffic optimization measures were implemented. The simulation model is shown in Figure 3.

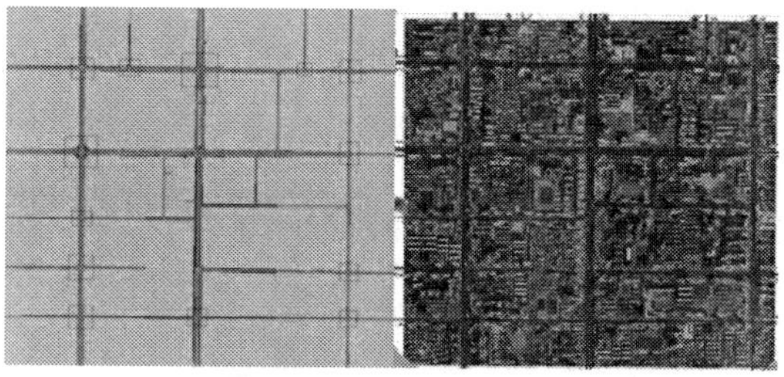

Fig. 3. Parking lot opened and the region Optimized VISSIM traffic simulation model.

Table 1. Traffic Improvement Measures in the Study Area.

Optimization object	Optimization measures
boundary signal intersection	boundary signal control based on macroscopic basic diagram
South Section of Zhuque Street	Coordinated control of green wave signal on main lines
Intersection of Chang'an Middle Road and Yanta West Road	Signal timing optimization, lane function improvement, channelization design
Intersection of Chang'an South Road and Yannan First Road	Signal timing optimization, lane function improvement, and channelization design
Yan Nan Yi Road - Cuihua Road Intersection	Optimization of signal timing and improvement of lane functions
Road network near the renovation area	Optimization of parking lot entrance and exit organization

To scientifically verify the effectiveness of improvement measures, evaluation and analysis are conducted at two levels: road network and intersection. The road network level evaluation includes four indicators: average delay, average number of stops, average operating speed, and parking delay, which are used to evaluate the improved research area through simulation. The intersection level evaluation includes three indicators: average queue length, maximum queue length, and average vehicle delay at intersections, which are used to evaluate the traffic operation level of the road network in the area where the parking lot is optimized. Tables 2 and 3 show the comparison of traffic operation levels in the area where the parking lot is optimized, and Figures 4 and 5 show the comparison of traffic operation levels in the area where the parking lot is optimized.

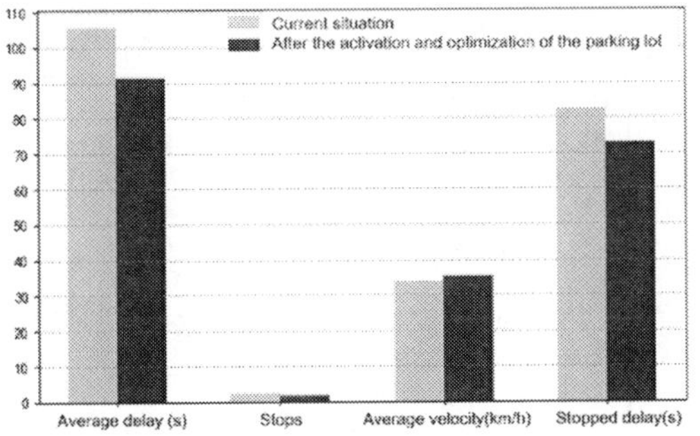

Fig. 4. Comparison of Traffic Operation Levels in Regional Road Networks.

Table 2 and Figure 4 clearly show that the increase in traffic pressure in the study area with the opening of parking lots did not have a negative impact on the overall traffic operation level of the regional road network. On the contrary, through the design of regional traffic optimization measures, the newly added traffic pressure was successfully alleviated and significant improvements were achieved. Although the opening of parking lots introduced additional traffic flow, the average delay time of the road network decreased from 105.36 seconds to 91.36 seconds through the above optimization strategies. In addition, the average number of parking times in the road network decreased to 1.79 after optimization, a decrease of 16.76% compared to the current situation, indicating that the optimization measures successfully reduced parking frequency, allowing motor vehicles to pass through the road network quickly and continuously, and played a positive role in overall traffic flow. After optimization, the average speed of the road network increased to 35.35 km/h, an increase of 4.61% compared to the current situation, reducing vehicle travel time and improving the traffic operation efficiency of the vehicle road network. Finally, in terms of parking delays, the opening and optimization of parking lots reduced parking delays from 82.41 seconds to 73.02 seconds, a decrease of 12.61%. This further indicates that through optimization strategies, the problem of parking delays in the road network has been successfully alleviated, ensuring the smooth operation of traffic.

Table 3 and Figure 5 provide detailed comparative data on the traffic operation level of regional intersections, covering important performance indicators such as average queue length, maximum queue length, and average intersection delay. Firstly, in terms of average queue length, the current situation is 15.21 meters, while the opening and optimization of the parking lot have reduced it to 12.29 meters, a significant decrease of 19.20%. Secondly, in terms of maximum queue length, after the opening and optimization of the parking lot, the maximum queue length has been reduced from the current situation of 127.24 meters to 110.67 meters, a decrease of 13.02%. Finally, in terms of intersection delay, the parking delay after the opening and optimization of the parking lot is 23.37 seconds, which is an improvement of 5.81% compared to the current situation of 24.81 seconds. Unlike the significant decrease in queue length, the reason is that the implementation of regional boundary control at peripheral boundary intersections has had a slight impact on the parking delay at the entrance of intersections, but this scheme has significantly improved the real-time traffic operation level within the region, while also improving the average delay at peripheral intersections. This indicates that our optimization strategy has achieved substantial improvements in reducing parking delays and is helpful for improving the traffic service level of vehicle intersections.

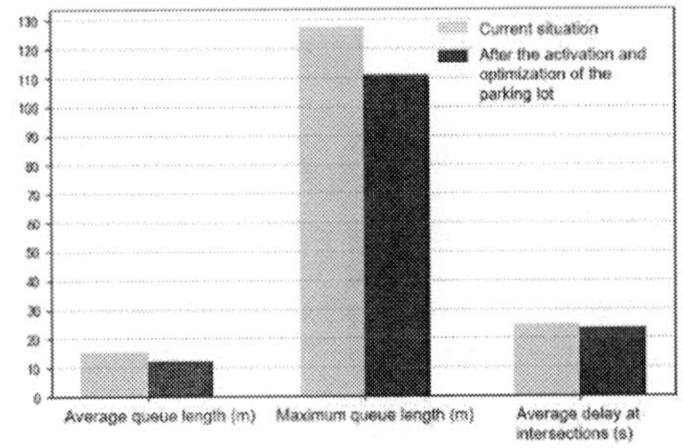

Fig. 5. Comparison of operation level at regional intersections.

Table 2. Comparison of Traffic Operation Levels of Regional Road Networks.

Comparison plan	Average delay (s)	Average number of parking times	Average speed (km/h)	Parking delay (s)
status	105.36	2.15	33.78	82.40
The parking lot is operational and has been optimized	91.36	1.79	35.35	73.02
Enhance the effect	13.29%	16.76%	4.61%	12.61%

Table 3. Comparison of Traffic Operation Levels at Regional Intersections.

Comparison plan	Average queue length (m)	Maximum Queue Length (m)	Average delay at intersection (s)
status	15.21	127.24	24.81
The parking lot is opened and optimized	12.29	110.67	23.37
Enhance the effect	19.20%	13.02%	5.81%

When adjusting signals, it is not only necessary to consider the operational efficiency within the control area, but also the queuing factors at boundary controlled intersections. The average queuing length at the entrance section of boundary controlled intersections is shown in Figure 6. As shown in the figure, although the boundary control algorithm has shortened the green light time for vehicles entering the area, the average queue length at the entrance section of the 11 boundary controlled intersections has not changed much. On the contrary, the average queue length at 7 intersections has decreased, indicating that the boundary control algorithm in this area does not affect the traffic operation efficiency outside the control area while optimizing the control area inside.

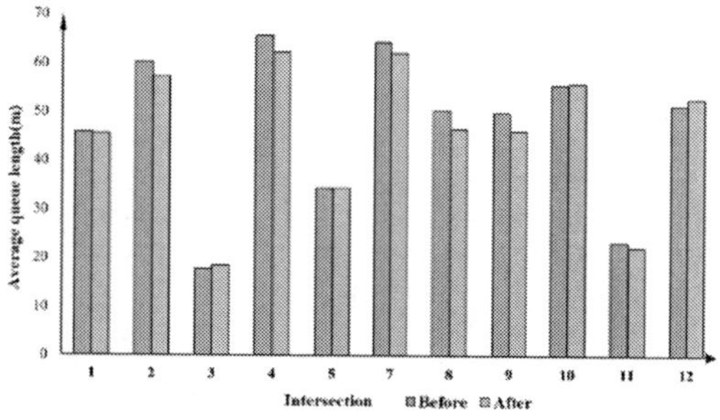

Fig. 6. Comparison of average queue length at the entrance section of boundary controlled intersections.

5 Conclusion

This article aims to address the problems of traffic congestion, parking difficulties, chronic road right occupation, and signal control delays in the renewal area. It explores methodologies from the perspectives of macro-regional traffic demand control and micro-traffic organization and management, and proposes a scientific and efficient method system for traffic congestion management and traffic capacity improvement in the context of urban renewal in old urban areas. By applying VISSIM traffic simulation software to build a regional simulation 3D model, the scientific and effectiveness of the proposed macro-micro traffic management technologies are successfully verified, and can provide strong experience and technical support for similar urban renewal projects in the future, forming a replicable technical paradigm that can be promoted.

(1) Proposed a region boundary control algorithm based on MFD

This algorithm ensures that the number of vehicles within the area is always maintained at an optimal level through boundary signal control, and is constrained by the length of the pedestrian crossing green light and the queue length at the entrance of the peripheral boundary intersection. It is suitable for control areas with large traffic generation, without affecting the traffic flow at the entrance of the area boundary intersectionDao generationThe impact of the proposed algorithm has made up for the deficiencies of the original boundary control method. Through evaluation indicators such as regional average operating speed and regional average vehicle delay, the applicability and effectiveness of the proposed algorithm in urban renewal areas have been successfully verified, effectively reducing traffic congestion within the area, improving average operating speed, and ultimately enhancing the traffic operation level and efficiency of the entire control area.

(2) Propose optimization strategies for key intersections and regional traffic organization based on the concept of "microcirculation"

Based on the concept of microcirculation, the optimization design and traffic organization of key intersections and road networks are carried out. By considering evaluation indicators such as intersection delay, average queue length, and average operating speed, the feasibility of the proposed scheme is successfully verified. In order to fully alleviate the radiation effect of parking difficulties, a three-level guidance scheme for traffic flow is adopted. By providing information on parking lot location and parking space, it can reduce the ineffective traffic flow on the road caused by blindly searching for parking spaces, which helps to alleviate the dynamic and static traffic contradictions at the city level. At the same time, it also helps to further promote the development of urban transportation informatization and intelligence.

(3) Built a regional traffic simulation model for urban renewal and established an optimization scheme evaluation mechanism

Through organizing professional and technical personnel to conduct on-site research and data collection, and using VISSIM simulation software to construct regional traffic simulation models before and after the optimization of the renewal area, scientific evaluations were conducted on improvement measures such as regional boundary control, signal timing scheme optimization, intersection channelization design, and regional traffic organization optimization. The simulation results showed that after implementing the above improvement measures in the urban renewal area, not only can the new traffic pressure brought by the opening of parking lots be overcome, but also the traffic operation level in the renovated area can be effectively improved.

However, the application of this boundary control method in practical scenarios is still difficult, requiring the control area to have complete intelligent monitoring equipment for real-time traffic volume statistics and calculation, and to write algorithms into signal and monitoring equipment for linkage control. Next, we will continue to study regional signal coordination and control strategies, explore intelligent monitoring devices and traffic information recognition algorithms, and help improve the traffic efficiency of urban renewal areas.

Acknowledgments

This article is supported by the Northwest Engineering Corporation of Power China (2023610002003846), Fundamental Research Funds for the Central Universities CHD (300102344604), the National Natural Science Foundation of China (52002030), and the Humanities and Social Sciences Foundation of the Ministry of Education (20XJCZH011).

References

1. Pan W, Du J. Towards sustainable urban transition: A critical review of strategies and policies of urban village renewal in Shenzhen, China[J]. Land Use Policy, 2021, 111: 105744.
2. Karimi H., Ghadirifaraz B., Shetab Boushehri S N., et al. Reducing traffic congestion and increasing sustainability in special urban areas through one-way traffic reconfiguration[J]. Transportation, 2021: 1-24.
3. Long E. R. Chemical contamination, acute toxicity in laboratory tests, and benthic impacts in sediments of Puget Sound: a summary of results of the joint 1997-1999 Ecology/NOAA survey[J]. 2003.
4. Nathan H. G., John D. C., Little H. G. Optimition of traffic signal settings by mixed integer linear programming. Part I: The network cording problem [J]. Transportation Science, 1975, 9(4): 321- 343
5. Cao Hongwang. Research on Regional Traffic Coordination Optimization Control Based on Vehicle Average Delay Model [D]. Jiangsu: Nanjing University of Posts and Telecommunications, 2020
6. Geroliminis N, Daganzo C F. Existence of urban-scale macroscopic fundamental diagrams: Some experimental findings[J]. Transportation Research Part B: Methodological, 2008, 42(9): 759-770.
7. Wang Yawei, Feng Yang, Wang Jiawen. Research on Boundary Control Method of Road Traffic Network Based on Macroscopic Basic Diagram [J]. Intelligent Computer and Applications, 2023, 13(2): 29-34+40
8. Ji Y, Daamen W, Hoogendoorn S, et al. Investigating the shape of the macroscopic fundamental diagram using simulation data[J]. Transportation Research Record, 2010, 2161(1): 40-48.

Research on the Impact of New Energy Vehicles on Carbon Emissions of Urban Transportation

Hongru Lin*

Qingdao Middle School (Qingdao academy), Qingdao, 266114, China

Abstract: With the acceleration of urbanization, the growth of urban travel demand has led to rising carbon emissions, and the development of low-carbon urban transport has become particularly urgent. This study focuses on the prediction of carbon emissions in the road traffic sector of City A in 2030, and explores the potential changes in carbon emissions under different scenarios. The research objects include private cars, taxis, buses and ordinary trucks, and two prediction scenarios are constructed: baseline scenario and policy scenario. The baseline scenario reflects the natural development trend of future carbon emissions under the existing technology and policy conditions, while the policy scenario considers the implementation of new energy vehicle promotion policies and evaluates the potential contribution of new energy vehicles to reducing road traffic carbon emissions. The multiple linear regression model and S-curve model are used to predict the total number of motor vehicles under different scenarios, and the "bottom-up" method of IPCC is used to calculate the carbon emissions of transportation. The results show that the promotion of new energy vehicles, especially pure electric vehicles, has a significant effect on reducing carbon emissions in the field of transportation.

1. Introduction

With the advancement of urbanization, residents' travel demand is increasing and travel modes are becoming more diversified, which leads to an increase in carbon emissions. Therefore, the development of urban low-carbon traffic is urgent. The proposal of "double carbon" goal has brought great challenges to the development of the transportation industry, and the problem of urban traffic pollution has become a hot topic for many scholars to study, so it is of great practical significance to strengthen the research on the influencing factors of carbon emissions of urban traffic problems for the realization of "double carbon" work. At present, the road traffic flow in the downtown area of a city is gradually approaching saturation, and congestion problems have occurred at individual intersections and sections during rush hours [1]. With the passage of time, urban road traffic pressure will further increase, and with the increase of the number of motor vehicles, carbon emissions will also increase. With the rapid development of the automobile industry, automobiles bring convenience to people's lives, but also bring increasingly serious environmental problems. Therefore, new energy vehicles, especially pure electric vehicles, have gradually attracted wide attention. Local governments in China have also adopted a series of measures to guide the low-carbon development of transportation, and the development of new energy vehicles is one of the important measures [2].

Taking A city as an example, the number of A new energy vehicles in A city in 2013 is only 1000. By the end of 2022, the number of a new energy vehicles had reached 945,000, accounting for 19.81% of the total number of vehicles [3]. New energy vehicles account for more than 50% of new car sales. Compared with traditional internal combustion engine vehicles, new energy has significant advantages in reducing carbon footprint. In this context, this study explains the contribution of popularizing new energy vehicles to reducing carbon emissions in the field of transportation by predicting the total carbon emissions of road traffic in City A under different conditions by 2030. In addition, by evaluating the expected emission reduction effects of different types of motor vehicles in 2030, this paper points out the most potential areas for new energy vehicles in future emission reduction.

2. Research Objects and Methods

2.1 Research Objects

Under the background of actively advocating the goal of achieving the peak of carbon emissions by 2030 in China, this study focuses on the prediction of carbon emissions in the field of road traffic in a city in 2030, aiming at exploring the potential changes of carbon emissions under different scenarios. The research objects cover private cars, taxis, buses and ordinary trucks, which constitute the main part of the total number of motor vehicles in a city. In this paper, two different prediction scenarios are constructed to assess the impact of new energy vehicle promotion on carbon emissions. Scenario 1 is set as the baseline scenario, that is, to predict the carbon emissions of road traffic in

*Corresponding author's e-mail: Hongru_Lin@outlook.com

City A without adopting the promotion policy of new energy vehicles [4]. This scenario reflects the natural development trend of future carbon emissions under the existing technology and policy conditions. In contrast, the second scenario is a policy scenario, which considers the implementation of new energy vehicle promotion policies, aiming to assess the potential contribution of new energy vehicles to reducing road traffic carbon emissions under the promotion of active policies. In this scenario, we assume that the market share of new energy vehicles will increase significantly, which will have a positive impact on carbon emissions [5]. In order to predict these two scenarios, this paper makes a detailed prediction analysis of the total number of four main ground vehicles in a city. Based on the existing vehicle registration data, vehicle usage and future development trends, we construct a prediction model and estimate the total number of motor vehicles under different scenarios. By comparing the prediction results of the two scenarios, this paper aims to reveal the potential impact of the new energy vehicle promotion policy on road traffic carbon emissions in City A.

2.2 Research methods

In this paper, the multiple linear regression model is used to predict the total number of all kinds of ground motor vehicles, and the optimization model (quadratic model, growth model, linear model, S curve model) is selected according to the fitting degree and significance of the model, so as to obtain the independent variables of the regression equation and the number of new energy vehicles [6].

Quadratic model: When the data exhibits a nonlinear relationship, the quadratic model can provide a better fit by introducing the square of the independent variable to capture the relationship.

Growth model: suitable for describing dynamic processes that grow over time and may include morphologies such as exponential or logarithmic growth.

Linear model: The most basic form of regression model, which assumes a linear relationship between the independent and dependent variables.

S-curve model: Also known as the logistic curve model, it is often used to describe a process in which growth has reached saturation, such as technology adoption or population growth [7].

In this paper, different models will be evaluated and compared according to their goodness of fit (such as R-squared), significance tests (such as F-test and t-test), and the distribution of residuals. The selected optimization model will be able to reflect the relationship between the independent variables and the number of new energy vehicles most accurately [8].

In the selection of transportation carbon emission accounting model, this paper refers to the "bottom-up" method issued by the Intergovernmental Panel on Climate Change (IPCC) for calculation, and calculates the carbon emission according to different types of vehicle ownership, mileage, fuel consumption per unit mileage and fuel carbon emission coefficient. The calculation formula is as follows:

$$Emission = \sum_{i,j} D_{i,j} \times S_{i,j} \times P_i \times G_i \times EF_{i,j} \qquad (1)$$

Emission is the total carbon emission of urban traffic in a certain period of time, and the unit is kg; i is the type of fuel consumed by urban traffic, mainly including diesel, gasoline, natural gas, etc. j is the vehicle type of urban traffic; $D_{i,j}$ is the annual mileage, in km, of a vehicle of category j using fuel i. $S_{i,j}$ indicates the energy consumption per unit mileage of vehicle of category j using fuel i, in L/km or kwh/km. P_i is the fuel density of the fuel i, in L/kg; G_i is the net calorific value of the fuel, in TJ/kg.$EF_{i,j}$ is the carbon emission factor of the fuel i, in kg/TJ or kg/Kwh. Where i= 1, 2, 3, 4 represent buses, private cars, taxis, freight cars, respectively. The carbon emission factors of different transportation modes are shown in Table 1.

Table 1. Carbon emission factors of different transportation modes (unit: tC/TJ)

Mode of transportation	Bus	A private car	Taxi	Freight car
Emission factor	0.066	0.239	0.276	0.141

3. Carbon Emission Projection under the Baseline Scenario

We divide the forecast of car ownership into a baseline scenario and a policy scenario. The baseline scenario is that in 2022, when a city no longer promotes new energy vehicles, we first examine the car ownership in 2030, and then examine its carbon emissions in 2030 [9].

This paper selects the carbon emissions and total carbon emissions of passenger and freight traffic in a city from 2013 to 2023, as well as the relevant data of various influencing factors, in which the total carbon emissions are calculated by "bottom-up method". Relevant data of influencing factors of transportation carbon emission in City A are shown in Table 2. The total carbon emission is represented by C; the total population is represented by X1; the gross national product is represented by X2; the number of motor vehicles is represented by X3; the passenger turnover is represented by X4; the freight turnover is represented by X5; the passenger volume of public transport is represented by X6; and the number of local tourists is represented by X7.

Table 2. Relevant Data of Influencing Factors of Transportation Carbon Emission in City A

Year	C/million tons	X1/thousand	X2/100 million yuan	X3/thousand vehicles	X4/million people	X5/million tons	X5/million people	X7/million people
2014	3.2	626.2	362.4	151	7.2	10.1	89.4	92574.3
2015	3.9	646.2	395.3	173	10.2	10.3	10.2	117902.6
2016	3.2	665.3	446.4	204	8.4	8.5	79.2	136663.9
2017	2.7	688.3	503.7	224	5.7	9.4	80.8	160660.6

Year	C/million tons	X1/thousand	X2/100 million yuan	X3/thousand vehicles	X4/million people	X5/million tons	X5/million people	X7/million people
2018	2.9	708.3	567.1	259	6.1	10.6	84.2	199020.0
2019	3.1	720.7	617.9	271	6.1	11.3	85.4	233720.9
2020	3.2	867.9	678.2	302	4.0	15.9	66.8	200802.9
2021	3.6	870.7	741.8	335	4.6	17.6	71.6	264390.6
2022	2.9	866	747.6	361	3.0	15.7	44.7	202412.1
2023	2.8	851.4	761.4	373	2.9	12.5	33.7	264718.9

Based on the data from 2014 to 2023, we can adopt the multiple linear regression model to predict the transportation carbon emissions and other related indicators in 2030. First, the existing data were statistically analyzed to determine the correlation between the various variables, and a multiple linear regression model including all significant independent variables was established. By adjusting the parameters of the model, we can ensure that the model has a high degree of fitting and prediction accuracy. Then, the value in 2030 is predicted according to the model. For example, if we focus on the number of cars per thousand people in 2030 (X3), we can use the model to predict the value of X3 for that year. In addition, for other indicators, such as C (carbon emissions), X1 (GDP per capita), X2 (gross regional product), X4 (population density), X5 (energy consumption) and X7 (disposable income of residents), we can also use similar methods to predict. In 2030, under the baseline conditions, the carbon emissions in the transportation sector of City A are 1.92 million tons.

4. Carbon Emission Projection under the Policy Scenario

In the scenario of "Promoting New Energy Vehicle Policy", the total number of motor vehicles in a city in 2030 will be composed of various types of new energy vehicles and various types of traditional fuel vehicles. In the context of promoting the policy of new energy vehicles, the number of new energy vehicles can be calculated according to the quota of new energy vehicles issued by the Transport Commission. At the same time, according to the official insurance number of China Automobile Technology Research Center, the proportion of new pure electric vehicles in new energy vehicles can be counted. We calculate the number of pure electric and plug-in hybrid buses, private cars, taxis and freight cars in proportion. The results are shown in Tables 2-6.

Table 3. Number of Buses (thousands)

Year	Fuel truck	Hybrid car	Pure electric vehicle	Total amount
2014	0.60	0	0	0.60
2015	0.69	0	0	0.69
2016	0.63	0.09	0.10	0.82
2017	0.69	0.10	0.11	0.90
2018	0.65	0.22	0.17	1.04
2019	0.68	0.23	0.17	1.08
2020	0.76	0.25	0.19	1.21
2021	0.79	0.31	0.24	1.34
2022	0.85	0.33	0.26	1.44
2023	0.88	0.34	0.27	1.49

Table 4. Private Car Ownership (thousands)

Year	Fuel	Hybrid	Pure electric	Total amount
2014	128.35	0	0	128.4
2015	144.62	1.61	0.82	147.1
2016	133.52	19.07	20.81	173.4
2017	146.61	20.94	22.85	190.4
2018	138.69	46.23	35.22	220.2
2019	145.12	48.37	36.86	230.4
2020	161.72	53.91	41.07	256.7
2021	168.00	65.49	51.26	284.8
2022	181.04	70.58	55.23	306.9
2023	187.06	72.92	57.07	317.1

Table 5. Number of Taxis (thousand)

Year	Fuel	Hybrid	Pure electric	Taxi
2014	0.91	0	0	0.9
2015	1.04	0	0	1.0
2016	1.08	0.00	0.15	1.2
2017	1.03	0.15	0.16	1.3
2018	0.98	0.33	0.25	1.6
2019	1.02	0.34	0.26	1.6
2020	1.14	0.38	0.29	1.8
2021	1.19	0.46	0.36	2.0
2022	1.28	0.50	0.39	2.2
2023	1.37	0.47	0.40	2.2

Table 6. Number of Freight Vehicles (thousand)

Year	Fuel	Hybrid	Pure electric	Freight car
2014	21.14	0	0	21.1
2015	24.22	0	0	24.2
2016	25.13	0.00	3.43	28.6
2017	24.15	3.45	3.76	31.4
2018	22.84	7.61	5.80	36.3
2019	23.90	7.97	6.07	37.9
2020	26.64	8.88	6.76	42.3
2021	27.67	10.79	8.44	46.9
2022	29.82	11.62	9.10	50.5
2023	31.85	10.97	9.40	52.2

By using the analysis software to rebuild the S-curve model and substituting the constants and coefficients into the formula, the total number of buses in 2010 is 1.81 thousand. Taking the pure electric bus as the dependent variable and the time series as the independent variable, the S-curve model is established by using the analysis software, and the constant and coefficient are substituted into the formula to predict that the number of pure electric buses will be 0.69 thousand in 2030. Taking the plug-in hybrid electric bus as the dependent variable and the time series as the independent variable, the S-curve model is established by using the analysis software. The constant of the equation is -0.049 and the coefficient is -1.367. By substituting the constant and coefficient into the formula, it can be predicted that the number of plug-in hybrid electric buses will be 0.79 thousand in 2030.

After eliminating the irrelevant independent variables, the significant influencing factors of the total number of private cars are the number of urban population and the built-up area. Based on the predicted urban population, built-up area and the constants and coefficients of the multiple linear regression equation, the total number of private cars in 2030 is predicted to be 481.41 thousand. Among them, with pure electric private cars as the dependent variable and time series as the independent variable, the linear model is established by using the analysis software, and the constant and coefficient are substituted into the formula to predict that the number of pure electric private cars in 2030 A is 142.51 thousand, with plug-in hybrid private cars as the dependent variable and time series as the independent variable. By substituting the constant and coefficient into the formula, it can be predicted that the number of plug-in hybrid private cars will be 184.66 thousand in 2030.

After eliminating the irrelevant independent variables, the significant factor affecting the total number of taxis is the consumption level of residents. By substituting the predicted consumption level of residents and the constants and coefficients of the multiple linear regression equation, the total number of taxis in a city in 2030 is 3.21 thousand. Taking the pure electric taxi as the dependent variable and the time series as the independent variable, the linear model is established by using the analysis software, and the constant and coefficient can be used to predict that the number of pure electric taxis will be 1.07 thousand in 2030 [10]. Taking the plug-in hybrid electric taxi as the dependent variable and the time series as the independent variable, a linear model is established by using the analysis software. The constant of the equation is -0.176 and the coefficient is 0.291. By substituting the constant and coefficient into the equation, it can be predicted that the number of hybrid taxis in City A will be 1.19 thousand in 2030.

Using the analysis software to re-establish the quadratic model, the constant and coefficient are substituted into the formula, and the total number of A trucks in 2030 is 40.51 thousand. Taking the pure electric truck as the dependent variable and the time series as the independent variable, the quadratic model is established by using the analysis software, and the constant and coefficient are substituted into the formula to predict that the number of pure electric trucks will be 14.91 thousand in 2030. Taking the plug-in hybrid electric truck as the dependent variable and the time series as the independent variable, the linear model is established by using the analysis software, and the constant and coefficient are substituted into the formula to predict that the number of plug-in hybrid electric trucks will be 15.61 thousand in 2030.

According to the formula

$$Emission = \sum_{i,j} D_{i,j} \times S_{i,j} \times P_i \times G_i \times EF_{i,j} \quad (2)$$

We can calculate that under the policy conditions, the total carbon emissions of urban transportation in City A in 2030 will be 1.06 million tons, which is 55.2% of the baseline conditions.

5. Relevant Recommendations

From the above results, the promotion of new energy vehicles has a significant inhibitory effect on urban traffic carbon emissions, and pure electric vehicles, as a representative of green travel, have a significant carbon reduction effect. We suggest that the government should take a leading position to promote pure electric vehicles from the field of public transport and gradually expand to the field of private cars in order to achieve wider social acceptance and application. The government should increase subsidies for pure electric vehicles and reduce the cost of purchasing and using pure electric vehicles through financial incentives. This will not only improve the public's recognition of pure electric vehicles, but also stimulate market vitality and promote the healthy development of the industry.

The most effective way to reduce carbon emissions is the transformation of traffic modes, which is extremely important to meet the needs of travellers and guide the adoption of energy-saving and environmentally friendly travel modes. Because of the advantages of high efficiency and environmental protection of public transport, giving priority to the development of public transport can not only alleviate traffic congestion, but also promote green travel. Improve the level of public transport service management, and then optimize the slow traffic network. There is a big difference in carbon emissions of 100 kilometres under different travel modes. Under the goal of "double carbon", it is urgent to promote the transfer of high-carbon vehicle travel to low-carbon green travel mode. By improving the service level and experience of green travel, we can attract the public to choose green travel and promote the realization of the goal of "double carbon" in urban transportation [11]

The government needs to strongly support the development of new technologies. The pure electric vehicle industry is in the stage of rapid development, and the continuous emergence of new technologies is of great significance for improving product performance and reducing costs. The government can encourage enterprises to increase research investment and promote technological innovation by setting up research funds and providing tax incentives. Infrastructure construction is the key to the popularization of pure electric vehicles [12]. The government should speed up the construction of infrastructure such as charging piles, improve the convenience of charging, and solve the problem of "long charging time and short range". At the same time, through rational planning of charging network, the coverage and accessibility of charging facilities are ensured to meet the needs of different regions and different users.

6. Conclusions

This study reveals the potential impact of new energy vehicle promotion policy on road traffic carbon emissions in City A by comparing the prediction results under the baseline scenario and the policy scenario. The results show that the promotion of new energy vehicles has a significant inhibitory effect on urban traffic carbon emissions, and pure electric vehicles, as the representative of green travel, have a particularly significant effect on carbon reduction. The government should take the leading position, give priority to the promotion of pure electric vehicles from the field of public transport, and gradually expand to the field of private cars, in order to achieve wider social acceptance and application. Increase subsidies for pure electric vehicles, reduce consumer purchase and use costs through financial incentives, improve public recognition, and stimulate market vitality. Support the development of new technologies, encourage enterprises to increase research investment, promote technological innovation, improve product performance and reduce costs. These measures can not only effectively reduce carbon emissions in the field of urban transportation, but also promote the development of urban transportation in a low-carbon, green and sustainable direction, and contribute to the realization of the "double carbon" goal.

References

1. Huang K, Zhou P, Liu Z, et al. The Calculation and Distribution of CAV Carbon Emissions on Urban Transportation Systems: A Comparative Analysis of Renewable and Non-Renewable Energy Sources[J]. Renewable Energy, 2024: 120884.
2. Wang K, Zheng L J, Zhang J Z, et al. The impact of promoting new energy vehicles on carbon intensity: Causal evidence from China[J]. Energy Economics, 2022, 114: 106255.
3. Li J, Jiang M, Li G. Does the new energy vehicles subsidy policy decrease the carbon emissions of the urban transport industry? Evidence from Chinese cities in Yangtze River Delta[J]. Energy, 2024, 298: 131322.
4. Zhang J, Jia R, Yang H, et al. Does electric vehicle promotion in the public sector contribute to urban transport carbon emissions reduction? [J]. Transport Policy, 2022, 125: 151-163.
5. Su C W, Yuan X, Tao R, et al. Can new energy vehicles help to achieve carbon neutrality targets? [J]. Journal of Environmental Management, 2021, 297: 113348.
6. Dixon J, Bukhsh W, Edmunds C, et al. Scheduling electric vehicle charging to minimise carbon emissions and wind curtailment[J]. Renewable Energy, 2020, 161: 1072-1091.
7. Barman P, Dutta L, Bordoloi S, et al. Renewable energy integration with electric vehicle technology: A review of the existing smart charging approaches[J]. Renewable and Sustainable Energy Reviews, 2023, 183: 113518.
8. Xie Y, Wu D, Zhu S. Can new energy vehicles subsidy curb the urban air pollution? Empirical evidence from pilot cities in China[J]. Science of The Total Environment, 2021, 754: 142232.
9. Pamucar D, Ecer F, Deveci M. Assessment of alternative fuel vehicles for sustainable road transportation of United States using integrated fuzzy FUCOM and neutrosophic fuzzy MARCOS methodology[J]. Science of the Total Environment, 2021, 788: 147763.
10. Lima M A, Mendes L F R, Mothé G A, et al. Renewable energy in reducing greenhouse gas emissions: Reaching the goals of the Paris agreement in Brazil[J]. Environmental Development, 2020, 33: 100504.
11. Zhang R, Hanaoka T. Deployment of electric vehicles in China to meet the carbon neutral target by 2060: Provincial disparities in energy systems, CO2 emissions, and cost effectiveness[J]. Resources, Conservation and Recycling, 2021, 170: 105622.
12. Zhao J, Xi X I, Na Q I, et al. The technological innovation of hybrid and plug-in electric vehicles for environment carbon pollution control[J]. Environmental Impact Assessment Review, 2021, 86: 106506.

Addressing Global Climate Challenges: Low-carbon Transport Planning Practices and Optimization Strategies in Bangkok

Sucheng Yao[a], Kanjanee Budthimedhee*

School of Architecture and Design, King Mongkut's University of Technology Thonburi, Bang Lamphu Lang, Khlong San, Bangkok, Thailand

Abstract. Carbon emissions are a critical issue in the global fight against climate change, with fuel consumption being a major contributor. The excessive use of private transportation in urban areas exacerbates fuel consumption and, consequently, carbon emissions. Bangkok, one of the most complex cities in terms of transportation, also faces significant climate challenges. This study focuses on transportation planning in Bangkok using a geodesign approach with the ArcGIS GeoPlanner tool. Based on Bangkok's government documents, higher-level urban transportation plans, and LISD (Low Impact Sustainable Development) guidelines, three key optimization strategies were identified: enhancing the coverage of integrated transportation, establishing electrified transportation infrastructure, and implementing regionally focused transportation zones. These strategies provide a reference for further research and for the practical optimization of transportation planning in other cities.

1 Introduction

The rapid global economic development has led to the continuous concentration of urban populations and the expansion of city scales. The global urban population grew from 751 million in 1950 to 4.2 billion in 2018, with the world's average urbanization rate reaching 55%. By 2050, the urban population is expected to increase by another 2.5 billion, with the average urbanization rate rising to 68%. This population growth has, to some extent, impacted urban climate change. The rise in anthropogenic carbon dioxide emissions has led to an increase in average temperatures, resulting in the greenhouse effect.

The greenhouse effect has caused a rise in extreme weather events, which have become a common issue in cities. In recent years, there has been a global increase in extreme weather. According to the US National Oceanic and Atmospheric Administration's 2022 Annual Climate Report, global concentrations of greenhouse gases, including carbon dioxide, have reached historic highs. Transportation, particularly from fuel emissions, is a major contributor to carbon emissions. Reducing the use of high-carbon-emission vehicles, such as private cars, can effectively lower urban carbon emissions. For instance, in Copenhagen, Denmark, the construction of a new bike lane led to a 20% increase in cyclists and a 10% reduction in car usage. However, sustainable low-carbon transportation modes, such as slow traffic (cycling and walking) and public transportation, are still less utilized than private transportation (private cars and motorcycles). Some studies suggest that this is due to the lack of safety, convenience, and comfort in sustainable urban transportation options.

The challenges faced in developing countries include high road congestion leading to inefficient energy consumption and carbon emissions [1]. Additionally, fast-moving motor vehicles, occupying roads alongside non-motor vehicles, compromise safety and convenience for tourists [2]. Furthermore, inadequate public transport connections hinder passengers from reaching their destinations easily [3].

To address these issues, this study uses a geodesign approach to analyze and plan urban transportation, focusing on the optimization of integrated transportation (cycling and public transportation), infrastructure improvements, and the establishment of transportation spaces. The goal is to explore transformation strategies during the urban transportation transition phase [4].

The innovations of this study compared to existing urban planning practices include:

1. The use of ArcGIS GeoPlanner, a rapid geodesign tool, enabling multi-disciplinary designers to collaborate and coordinate designs collectively.

2. Layering multiple data sets based on the existing local transportation conditions to increase traffic density suited to the area. This approach leverages the existing transportation density of the city to conduct spatial adaptability planning, significantly enhancing the economic efficiency of the planning process and supporting the gradual development of the city.

3. Integrating new technologies and renewable energy sources into transportation planning to enhance sustainability. This ensures that the planning is adaptable to the evolving urban transportation landscape.

* Corresponding author: *budthime@yahoo.com, [a]sucheng.yao1@kmutt.ac.th

2 Methodology

2.1 Study area

The study focuses on the Bangkok metropolitan area. Bangkok, the capital and administrative hub of Thailand, spans a total area of 605.7 square miles, with 270 square miles of built-up regions. The city is rich in natural surroundings, including seas, rivers, forests, and fertile land, contributing to its dynamic culture and strong economy, which make it Southeast Asia's premier travel destination. Key economic drivers include tourism, modern agriculture, and gem mining. In 2023, Bangkok's population is projected to reach 11,069,982, up from 1,365,000 in 1950, with an annual growth rate of 1.56%, or 170,284 people per year. This population surge highlights Bangkok's urban expansion within and beyond the city's hinterlands.

Bangkok experiences a tropical wet climate with a dry season from November to February, a peak period for tourism. The city's temperatures remain consistently high, with average maximum temperatures above 30°C (86°F) throughout the year. The rainy season, spanning from June to October, sees the heaviest rainfall between August and October. Despite the frequent rain, the city enjoys ample sunlight, with temperatures ranging from 32°C (89°F) to 35°C (96°F) year-round.

Bangkok's rapid development has led to challenges such as erratic weather patterns and ground subsidence, causing the city to sink by approximately two millimeters annually. Consequently, the Chao Phraya River is highly susceptible to flooding, and the World Bank predicts that 40% of the city could be underwater by 2030[5].

As one of the most congested large cities globally, Bangkok's status as a capital with well-developed urban infrastructure, along with its rich cultural heritage and historical landmarks, mirrors that of many major cities worldwide. Therefore, transportation planning and optimization strategies implemented in Bangkok could serve as a valuable reference for future urban transportation updates in other similar cities.

2.2 Database and planning tool

Based on documents from the Bangkok Metropolitan Administration (BMA)[5], such as the "Executive Summary Bangkok Master Plan on Climate Change 2021-2030," mid-to-long-term planning documents, and the LISD's "Climate Actions Directory," a low-carbon transportation dataset was developed. This dataset includes Bangkok's bicycle lanes, public transportation (bus routes, BRT lines, rail transit lines, and boat transport routes)[6].

ArcGIS GeoPlanner, a geodesign tool (Fig. 1), was used in this study. This urban planning tool is known for its simplicity and ease of use, allowing for rapid planning and practical application. Its user-friendly input and output interfaces provide an efficient design tool for various planners and related professionals. By leveraging these datasets and the capabilities of ArcGIS GeoPlanner, the research aimed to develop strategies to enhance low-carbon transportation options in Bangkok, contributing to the city's efforts to mitigate climate change impacts and improve urban mobility[7].

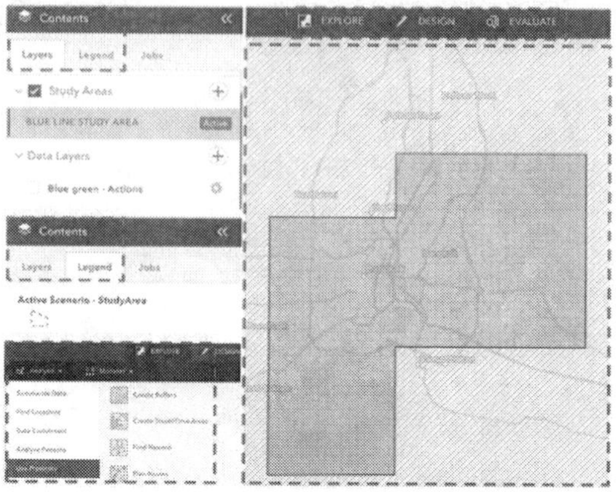

Fig. 1. ArcGIS GeoPlanner operation interface.

2.3 Planning process

Using ArcGIS GeoPlanner Version 3.8.1 (Build 180), datasets are inputted to create various layers for analysis and planning. Additionally, auxiliary datasets such as the Land Cover Map, Slope Map, and Carbon Emission Map are loaded to further support the analysis and planning process. During the planning process, it was discovered that some areas remained uncovered or unconnected by the transportation network even after integrating public transit. Certain regions faced insufficient connectivity, while others, despite having a high density of public transportation networks, still exhibited high carbon emissions. Additionally, there was a lack of adequate supporting infrastructure for electric vehicles. Addressing these issues requires an integrated geodesign approach.The final planning proposal will undergo expert evaluation and revision, leading to the establishment of a comprehensive transportation plan for Bangkok, followed by a summary of the optimization strategies(Fig. 2).

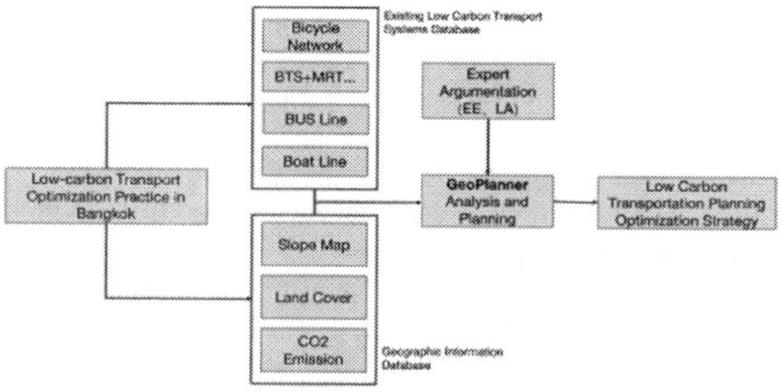

Fig. 2. Planning process.

3 Result

Bangkok's rich cultural heritage and historical landmarks, including temple architecture and canal landscapes, are dispersed throughout the city. These features provide a natural foundation for the tertiary industry, attracting numerous travelers and cultural enthusiasts. However, these cultural heritage sites, along with privatized land, limit the flexibility of modifying existing transportation infrastructure, such as road widening and upgrades.

Shenzhen, the fourth-largest city in China, serves as a noteworthy example by fully electrifying all its buses for transit services by the close of 2017 [8]. In Bangkok, the incorporation of BTS, MRT, and smart buses into urban public transportation is expected to create a more environmentally friendly public transport system, contributing to the reduction of carbon emissions in the city. However, despite these efforts, the transportation connections in Bangkok remain disjointed, leading to a surge in the number of cars and subsequent pollution, including elevated levels of PM 2.5.

Fig. 3. From top to bottom: bicycle line map, bus line map, train line map, boat line map.

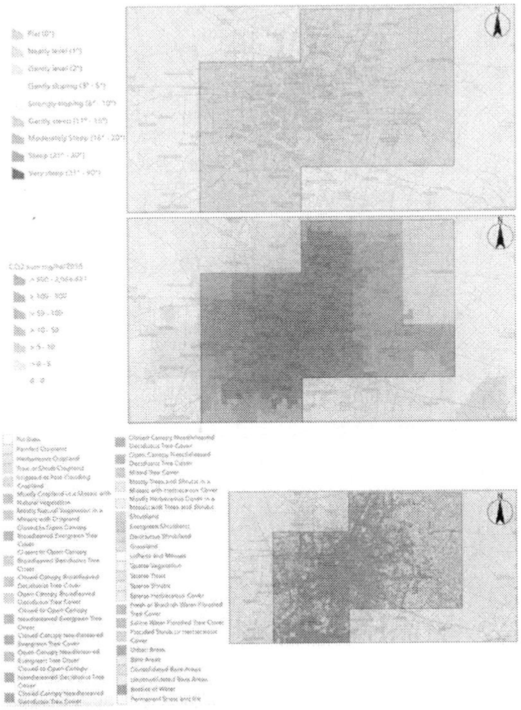

Fig. 4. From top to bottom: slope map, CO2 emission data, global land cover data.

Numerous studies have confirmed that well-developed slow-traffic infrastructure and public transportation systems contribute to reducing the reliance on private vehicles. Increased transportation coverage enhances the likelihood of residents using public transit, thereby lowering private travel rates. Based on the "Shenzhen Greenway Network Special Planning Summary (2011)," which mandates that greenways should be accessible within five minutes for nearby residents, this study evaluates the applicability of a similar 5-minute accessibility metric to the urban context of Bangkok, considering its climate, population, and area. Bicycle lanes and bus routes can be modified using existing infrastructure to increase network density. This includes expanding the coverage of a 5-minute walking radius (300-400 meters, taking the lower value) around bicycle lanes by increasing the density of the bicycle network. Similarly, the coverage area of a 5-minute (300m) walking radius around public transportation networks can be expanded by increasing the density of bus routes. Due to the higher infrastructure requirements of BTS, MRT, and water transport systems, this stage does not consider expanding these networks (Fig. 3, Fig. 4).

To enhance connectivity in specific areas, the planning process involved the following improvements: the existing bus lanes, totaling 845.97 km, were expanded by 156.67 km, representing a 15.6% increase. Similarly, the existing bicycle lanes (including those in planning) of 626.58 km were extended by 94.02 km, a 13% increase. These expansions were designed to connect existing dead-end roads, thereby improving the accessibility of public transportation in parts of Bangkok.

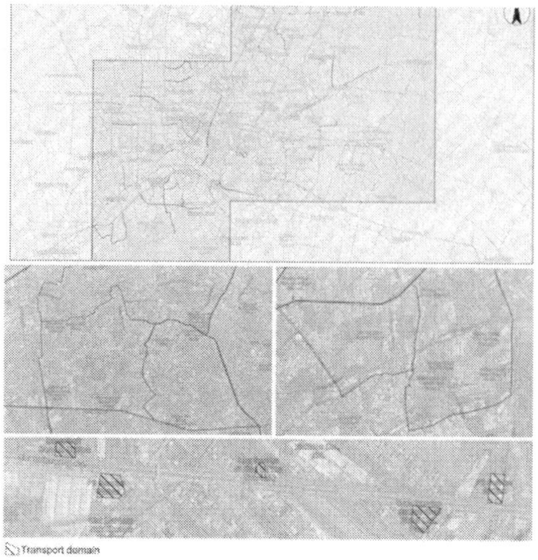

Fig. 5. From top to bottom: network plan, bicycle line planning, charging area planning.

Fuel consumption is a significant contributor to carbon emissions, so reducing fuel use can be an effective way to control emissions at the source. Although electric vehicles (EVs) are increasingly visible in Bangkok, fuel-powered cars and motorcycles remain the predominant modes of transport in the city. Based on existing documents, it is evident that transportation electrification is a crucial direction for optimizing transportation planning. To accommodate the growing shift towards electric public transit, freight vehicles, and private transportation, sufficient charging infrastructure is essential. This plan recommends equipping all gas stations, public and private parking lots with charging facilities. For underserved areas in the northeast, southeast, northwest, and southwest, charging stations should be installed in the parking lots of Lotus's and Big C supermarkets. Charging facilities should also be provided at some truck service areas, and bus depots should be equipped with bus charging stations. A high density of charging facilities can help alleviate range anxiety for private EV owners (Fig. 5).

This plan includes the renovation of 417 gas stations and parking lots across Bangkok's urban area, covering a total area of 122.34 hectares. These modifications will provide essential infrastructure to support the current and future development of electric vehicles and other electric transportation modes.

Transit-Oriented Development (TOD) is grounded in the planning theory aimed at fostering urban development by prioritizing public transportation. This approach involves the establishment of urban areas within a radius of 0.3km-0.8km, with residences situated up to 0.8km away from transportation hubs, all strategically guided by public transportation [9]. Combining the comprehensive low-carbon road network's 5-minute (300m) coverage area with an analysis using ArcGIS GeoPlanner, we identified a high-density low-carbon transportation coverage zone. Satellite imagery reveals that this area has a compact land use pattern, where almost all types of urban spaces (such as attractions, government agencies, residential areas, schools, and transportation infrastructure) are within walking distance. Based on the Climate Actions Directory and satellite data, this area could be designated as a Transit-Oriented Development (TOD) zone to reduce vehicle usage and related emissions (Fig. 6) [10].

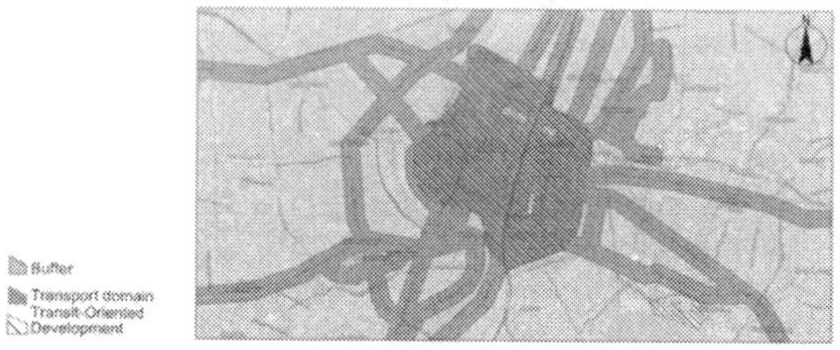

Fig. 6. Transit-Oriented Development (TOD) zone analysis and planning.

The area, spanning 2,905.71 hectares, is targeted for a reduction in private vehicle usage, with a primary focus on promoting public transportation as the main mode of travel.

4 Discussion

Based on the research outcomes and the evaluations provided by environmental engineering experts and landscape architecture specialists, compared to the pre-planning phase, the total length of bus routes has increased by 15.6%, bicycle lanes by 13%, and 417 new charging infrastructure points have been added. Additionally, a Transit-Oriented Development (TOD) area of 2,905.71 hectares has been designated. the following strategies have been summarized:

1. Integrating existing transportation networks to form a larger and more comprehensive system. On this foundation, increasing the coverage rates of various transportation modes will result in a more extensive and efficient network. Enhancing a single transportation mode's coverage has limited benefits, while a comprehensive approach is more economical and user-friendly.

2. Expanding the coverage of electrification infrastructure. Increasing the availability of charging facilities will promote the use of electric vehicles and encourage more people to transition to electrified private transportation.

3. Gradually developing different transportation spaces based on the existing network density. This approach will enable better control of transportation-related carbon emissions while ensuring user convenience.

ArcGIS GeoPlanner, the tool used in this study, is an efficient geodesign platform. It facilitates data sharing, making it easier for planners who are not GIS specialists to access geographic data, conduct analyses, and participate in the planning process. In this study, the collaborative use of ArcGIS GeoPlanner was not extensively highlighted at this stage. Future research aims to incorporate this tool more deeply into the design process, such as by involving local communities in planning efforts. This approach could enhance the resilience of specific areas while also raising low-carbon awareness among residents. Additionally, issues related to the promotion of electric vehicles and motorcycles, such as battery safety, require consideration from other professional fields. A multi-disciplinary, participatory planning approach will provide more robust references for implementing urban designs in specific cities.

There are some limitations to the implementation of this strategy. Firstly, Bangkok has previously attempted to implement bicycle lanes, but unfortunately, the effort was ineffective due to non-compliance with traffic laws by Thai pedestrians, motorcyclists, and drivers [11]. The feasibility of implementing bicycle lanes in practice needs further verification. Secondly, the strategy's effectiveness in other cities is uncertain, as we must account for the unique characteristics of each city. Bangkok, with its extensive and well-utilized waterways, incorporates boat transport as a supplement to its public transportation system. Other cities might explore region-specific forms of low-carbon public transportation. For instance, in Chinese cities, how electric bicycles can coexist with traditional bicycles is a question worth exploring.

5 Conclusion

This study addresses the optimization of urban transportation carbon emissions using geodesign methods, focusing on traffic planning optimization for the city of Bangkok. Three main optimization strategies have been identified: enhancing overall transportation coverage, expanding electrification infrastructure, and establishing key regional transportation zones. These strategies can be applied to transportation planning optimization practices in other cities.

Future urban transportation planning should consider the following recommendations: 1. Emphasize slow-moving and public transportation; 2. Improve city transportation connectivity and coverage to enhance accessibility; 3. Account for future changes in transportation and infrastructure.

Acknowledgments

2022 Suzhou Polytechnic Institute of Agriculture of Young Teachers Scientific Research Ability Enhancement Program Project. "Research on the Optimization of Rural Human Settlements in Southern Jiangsu under the Concept of Sponge City Construction." (QN[2022]05).

References

1. Srinet, P. Short, P. Doss, A, S, A. (2023). Towards sustainable transport in Bangkok. AIP Conf. Proc. 2946, 030009.
2. Chen, X, H. Yue, L, S, S (2017). Overtaking Interference on Shared Bicycle Path and Its Influence on Lane Design. Journal of Tong Ji University (Natural Science), 45 (12), pp.1810-1817.
3. Yang, Q, Y. (2021). Research on Urban Road Network Optimization Strategies for Travel Quality Improvements. PhD. Dalian University of Technology.
4. Steinitz, C. (2020). On Landscape Architecture Education and Professional Practice and Their Future Challenges. Land. 9 (7): 228.
5. BMA (2022). Executive Summary Bangkok Master Plan on Climate Change 2021-2030. Bangkok.
6. LISD. (2023). Climate Actions Directory. Edition1-Draft. Liechtenstein Institute for Strategic Development (LISD) Berlin. Berlin. pp. 127-145.
7. Kazak, J. K., Świąder, M., Arciniegas, G., Aslanoğlu, R., Wascher, D., Chrobak, G. (2023). The Application of GeoPlanner in Local Development Management. Acta Sci. Pol. Administratio Locorum. 22 (4): 525-535.
8. Lu, L., Xue, L., Zhou, W. (2018). How Did Shenzhen, China Build World's Largest Electric Bus Fleet?Insights, World Resources Institute, Washington, DC. Accessed at <https://www.wri.org/insights/how-did-shenzhen-china-build-worlds-largest-electric-bus-fleet>, accessed on May 6, 2022.
9. Calthorpe, P. (2009).The Next American Metropolis: EcologyCommunity, and the American Dream. China Architecture & Building Press.
10. Tohjiwa, A, D. Amartya, M, H, P, P. (2023). Implementation of TOD Concept on Gondangdia Station Area and the Influence on Menteng Heritage Area. IOP Conf. Ser.: Earth Environ. Sci. 1218 012004.
11. Panichagul, I. (2016), Not connected - encountering obstructions - disrupting the community's way of life" is the reason for the failure of bicycle lanes?ThaiPost.https://www.posttoday.com/politics/434552

Spatial Pattern and Influencing Factor Analysis of Parcel Lockers in London

Xinwei Kang*

Centre for Advanced Spatial Analysis, UCL, London

Abstract: When carriers deliver goods to customers, last-mile delivery is one of the supply chain's most costly and environmentally damaging components [1]. In recent years, a growing number of scholars have investigated parcel lockers as a solution to the problems of last-mile delivery [2]. The goal of this work is to satisfy urban consumers' demand for online shopping while lowering CO2 emissions. However, few studies have focused on the parcel locker itself. Therefore, for the first time, this paper introduces clustering methods into the study of last-mile delivery on parcel lockers; moreover, it builds a methodology to analyse the spatial distribution of parcel lockers and identifies parcel locker clustering based on the DBSCAN algorithm. This study employs Ordinary Least Squares Regression and Geographically Weighted Regression models with data on the socioeconomic, built environment and public transportation to quantify the relationship between the degree of parcel locker clustering in London and these factors at the global and local levels. The study's findings indicate that parcel lockers are spatially clustered in London, with a significant degree of clustering in central London. Fufrthermore, the degree of parcel locker clustering is affected by population density, restaurant density, supermarket density, underground station density and bus stop density. However, the influence of these factors on the degree of parcel locker clustering at different sites is varied. Based on this, we propose that the government and businesses collaborate to establish various renovation projects and strategies for various MSOAs to improve the spatial distribution of parcel lockers, helping London achieve its goal of becoming a zero-carbon city by 2050.

1. Introduction

1.1 Background

The emergence of e-commerce has created a high demand for specialised end-user delivery services. One of the major problems most businesses have is last-mile delivery, which is the most time-consuming and expensive phase of the delivery process. Hence, scholars worldwide have investigated alternative delivery options to lessen the costs and external impacts of last-mile delivery, such as increased traffic congestion and air pollution in populated areas [3]. Parcel lockers have grown as a potential last-mile delivery alternative, as well as a solution to end-to-end delivery issues generated by failed house deliveries [2]. The most frequently claimed benefits of parcel lockers are reduced risk of delivery failure and greater delivery consolidation, which reduces the number of drop-off sites and the distance vehicles travel.

London has committed to being zero-carbon by 2050, exceeding the UK's national goal of lowering greenhouse gas emissions by 80% from 1990 levels.

Transportation-related carbon emissions account for 25% of London's total emissions. Consequently, last-mile delivery must dramatically cut the carbon emissions it produces to fulfil London's 2050 goal. The parcel locker can cover a certain range of customers' parcel needs, and clustering of parcel lockers can help delivery vehicles find an optimal route within a cluster. Vehicles start from the original warehouse, visit multiple locations (locker locations) and return to the warehouse while minimizing the total number of trips, thus reducing CO2 emissions. Nevertheless, it is challenging to determine if the spatial distribution of parcel lockers in London satisfies the demand to decrease carbon dioxide emissions due to the scarcity of studies on the subject.

Therefore, this paper will use the geographical data on existing parcel lockers to efficiently explain and present the spatial pattern of parcel lockers in London, as well as promote building parcel lockers in particular neighbourhoods of London.

In some countries, such as the Nordics, Baltics, and Poland, there are large and dense networks of parcel locker stations and it is widely accepted as a preferred delivery method. In Germany, there is a large network of parcel locker stations in terms of numbers, but it lacks density compared to the aforementioned countries, and home delivery remains by far the most preferred delivery method. Provided that basic public data is available, there remains a need to investigate the generalizability and transferability of data-driven models at street-level resolution across different cities or regions. In countries or regions where parcel locker is more popular, it is easier

* Corresponding author: xinwei.kang.20@ucl.ac.uk

to verify the model and optimize the placement. However, compared to Germany, it may be necessary to improve the delivery model before gradually increasing parcel locker usage.

As a result, the concept can be applied not only in London but also UK-wide and aid international efforts to cut carbon emissions from last-mile delivery.

1.2 Research question and objectives

- What is the spatial distribution pattern of the parcel lockers?
- What factors impact the clustering of parcel lockers in London and how might this impact be quantified?
- Where do parcel locker services in London need to be further improved?

Moreover, how can these services be improved?

2. Literature review

2.1 The need for parcel lockers for sustainable last-mile delivery

Various infrastructural measures have been adopted recently to reduce pollution and traffic congestion in densely populated areas. Regarding urban logistics, the movement of products from the last transit point to the final drop point in the delivery chain is referred to as last-mile delivery [4], which is presently regarded as one of the most costly, inefficient and polluting parts of the whole supply chain. Therefore, last-mile delivery to parcel lockers is a breakthrough in alleviating the problem of CO2 emissions.

City logistics solutions, like the parcel locker, have been explicitly developed to lessen the size and impact of the increased delivery volume brought on by the internet. In high-density residential or commercial locations, parcel lockers take the form of delivery boxes where consumers may pick up their packages whenever it's convenient for them.

2.2 Spatial distribution and factor analysis

Previous studies have concentrated on grouping customers into separate clusters, each with its unique criteria. Then, to accommodate the local pickup and delivery demands, the parcel delivery businesses install parcel lockers in each cluster. While studies on parcel locker systems are still being conducted in Europe, those on parcel lockers are less common [5]. Furthermore, it is still not obvious how these parcel lockers should be distributed geographically to preserve or enhance urban commercial functions within the framework of sustainable urban development [6].

Considering both demographic and built environment factors, Lachapelle et al. [6] hypothesise that both influence parcel locker locations. However, limited research has examined the spatial distribution patterns of parcel locker sites, and the mjority of recent studies have focused on relatively limited regions.

This area of research can assist companies in determining if there is clustering in the spatial distribution of parcel lockers and what the primary driving forces are behind their construction. The locations and quantity of parcel lockers should also be prioritised to accommodate customer needs because companies may design the routes to allow for the greatest walking between parcel lockers if parcel lockers are densely clustered in urban areas, eliminating stopping at each place and reducing CO2 emissions [7].

It is critical to record and evaluate the site features of parcel lockers in order to accomplish the aims of urban and transportation planning. It will aid in ensuring their optimal and sufficient distribution throughout London.

3. Method

3.1 Spatial point pattern analysis - DBSACN

Several studies have employed the KDE function to compute the density of parcel lockers throughout the research area. However, from the parcel lockers summarised above, the location data of the parcel locker is point data. Therefore, to detect the spatial point pattern of its location, the DBSCAN clustering algorithm is suitable to be used to determine where in London the clusters of parcel lockers are occurring. This method can discover clusters of arbitrary shapes in space relatively quickly and identify noisy points, such as outlier data, without pre-setting the number of clusters to be formed.

3.2 Regression model

Different datasets were utilised to evaluate the factors impacting the degree of parcel locker clustering based on the variables mentioned above. The models were built using Ordinary Least Squares Regression (OLS). The Ordinary Least Squares model can adequately portray the influence of multiple factor indicators on London. The final models follow the structure of Equation 1.

On the other hand, the geographically weighted

$$y_i = \beta_0 + \sum_{k=1}^{p} \beta_i x_{ik} + \varepsilon \quad i = 1, 2, 3, \ldots, n \tag{1}$$

regression model can fully characterise the extent to which a variety of factors influence the spatial characteristics of London parcel locker and visualise the spatial differences in the factors, forming a framework of 'local characteristics - overall pattern - factor analysis'. Equation 2 shows the GWR model.

$$y_i = \beta_0(u_i, v_i) + \sum_{k=1}^{p} \beta_k(u_i, v_i) x_{ik} + \varepsilon_i \tag{2}$$

It is important to note that even if we utilise the GWR regression, the OLS regression is still significant in our study. The latter can assist select variables and reflect the influence of various socio-economic, built environment and public transportation factors on the global degree of parcel locker clustering. Additionally, by contrasting the GWR model's results with those of the OLS model, we can better highlight the GWR's distinctive contribution to the explanation of geographical disparities and the improvement of the overall effect of interpretation.

4. Results

4.1 Spatial point pattern analysis - DBSACN

After several iterations of the parcel locker spatial dataset, it is found that there is a large density of parcel lockers and a large cluster in central London. There are also some parcel locker clusters in the Wandsworth, Hammersmith and Fulham, and Kensington and Chelsea areas.

4.2 The influencing mechanism of clustering degree of parcel lockers

4.2.1 OLS Regression Results

Table 1. The results of OLS regression – after VIF

Dimensions	Indicators	Coefficient	Std. Error	t	P > |t|
Socio-economic	Population density	-0.0099	0.002	-5.259	0.000***
Build Environment	Workplace density	0.2084	0.013	15.640	0.000***
	Restaurant density	3.3458	0.641	5.219	0.000***
	Supermarket density	13.1400	4.853	2.708	0.007**
Public Transportation	Underground station densiy	20.5470	13.996	1.468	0.142
	Bus stop density	6.8387	1.411	4.843	0.000***
	Intercept	-0.7234	0.189	-3.817	0.000***
R2	0.557				
Adjusted R2	0.554				

Note: ***,**,* indicate p-value less than 0.001, 0.01, 0.05, respectively.

The OLS regression results for the six dependent variables are shown in Table 1. The R square value is 0.557, which means that the model explains 55.7% of the clustering of parcel locker distribution in London.

4.2.2 Results of the GWR model

R square for the GWR model is 0.663. It demonstrates that the GWR model's fit is more robust than the OLS regression model. The study used the optimal bandwidth obtained from the Gaussian kernel function to estimate the coefficients of the GWR model, using the adaptive bandwidth size. The coefficients of the independent variables differ significantly, as shown in Table 1, demonstrating that the indicator has different levels and trends of effect in each MSOA. Furthermore, except for workplace density, the impact of the indicators on the clustering degree of parcel lockers in London is both positive and negative, which also indicates the existence of spatial heterogeneity in the other five indicators.

A negative correlation between population density on the degree of parcel locker clustering was found in most MSOAs across London. It had the most significant impact on the areas in the centre of London, which means that the larger the density of the population of these areas, the smaller the degree of parcel lockers clustering (Figure 1 (a)). A portion of the gathering population is located in an urban centre with high real estate prices, and renting land in this location to build a parcel locker will raise the cost of its storage space, making it a possible challenge to build parcel lockers in densely populated areas of the city [8].

However, the coefficient estimated by the GWR suggested that the density of the population had a positive association with the degree of parcel locker clustering in Merton, Sutton and Croydon (Figure 1 (a)). The parcel locker density declines even faster than the population density decreases, which is the result of fewer parcel deliveries in less populated areas. Here, the installation of parcel lockers takes up more space than demand requires, resulting in waste.

Figure 1 (b) depicts the differing effect of restaurant density on the clustering degree of parcel lockers. It only increased the clustering degree of parcel lockers in northern London and had the most significant negative impact on Merton, Wandsworth and Lambeth. Furthermore, the London Eye is located in Lambeth, which attracts a steady flow of tourists. So, people are more likely to go to a restaurant after watching a game or visiting the attractions rather than ordering a takeaway delivery.

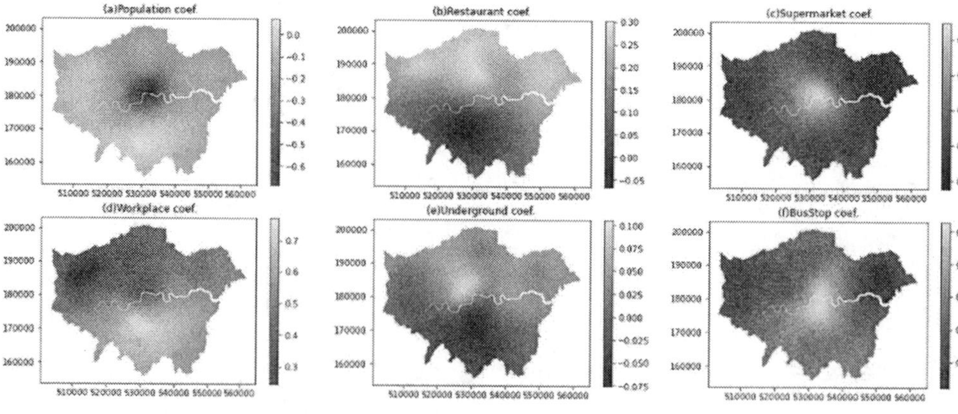

Fig. 1. Spatial distribution of regression coefficients in the GWR Mod

In addition, Figure 1 (c) shows that, contrary to the results for the population density, the density of supermarkets greatly promoted the degree of parcel locker clustering. However, its coefficients gradually decreased from central to suburb, indicating a trend of gradually decreasing from central London to the surroundings. It suggests that supermarket density had a higher impact on central London than other areas. Moreover, Figure 1 (d) shows workplace density is connected to the degree of parcel locker clustering in London. The coefficients in these MSOAs are all positive, ranging from 0.245 to 0.770, indicating that the higher the density of the workplace, the greater the degree of parcel locker clustering.

Figure 1 (e) exhibits differences from the OLS regression results - that the underground station density had no significant impact on the degree of parcel locker clustering in London. The results of GWR show the differing effect of underground station density on the parcel locker clustering degree. It contributes to the areas north of the river Thames in London while negatively impacting the areas south of the river Thames in London.

Some companies are installing parcel lockers in convenient locations, such as schools, subway stations and business buildings, as well as in areas close to residential neighbourhoods, to lower the high cost of home delivery. However, smaller businesses might be unable to pay the upfront land and installation fees in some places.

Figure 1 (f) shows a significant positive effect of the bus stop density on parcel locker clustering degree found in central London. However, there is a negative relationship between bus stop density and the degree of parcel locker clustering in some areas of London, such as Hillingdon, Harrow, Brent, Ealing, Havering and Barking and Dagenham. West London, although being an urban area of London, is typically a wealthy neighbourhood; therefore, people seldom want to travel by public transportation. Besides, bus stops are usually distant from where people live and residents do not want to carry their packages home from bus stops.

5. Discussion

The study discovered that there are problems with the uneven distribution of parcel lockers in London by assessing the spatial distribution of parcel lockers. The study calculates the number of points in each MSOA that belong to clusters as the dependent variable. The higher the degree, the closer the parcel lockers are to each other and the greater companies prioritise these areas as locations.

The factors that influence the degree of parcel lockers in London were then investigated using OLS regression and GWR. The adoption of the GWR model adjusts the OLS regression model's shortcoming of neglecting local features and reveals the various effects of each built environment indicator in different MSOAs. According to the GWR results, even indicators that are not significantly related to the parcel locker in the OLS regression may play a positive or negative role in some MSOAs. Hence, studying the degree of parcel locker clustering in London helps planners locate new parcel lockers.

The results and analyses explain the reasons why London parcel lockers form clusters. If the parcel lockers were 'better placed,' particularly close to public transportation, stores or supermarkets, over 15% of the customers who participated in the poll indicated that they would use them more frequently. Furthermore, some studies have investigated the effect of parcel lockers on the environment, claiming that parcel lockers have a beneficial effect on lowering greenhouse gas emissions [9].

The analysis of the spatial distribution of parcel lockers can provide some indication of urban development. From the results of our study, the distribution of parcel lockers embedded in residential space in the centre of london is an agglomeration. therefore places greater demands on the management of noise and pollution emissions. We found relatively low density of parcel lockers in the urban fringe. New town planning at the periphery ignores the accessibility of services. This may lead to a reduction in the efficiency of land resource allocation. These problems can be solved by optimizing the placement of parcel lockers. In addition, the planning of new towns should also take into account the accessibility of delivery service for vulnerable groups, such as the elderly and students. The question of how to balance equity and efficiency in the context of function-oriented new town planning is worth considering.

However, the assumption is not the number of lockers in each MSOA but the degree of parcel locker clustering in each MSOA, which will ignore some of the parcel lockers that are present but more dispersed. In terms of data, the dataset of parcel lockers is mainly picked from well-known companies in London, UK, while local pick-up/delivery points such as petrol stations or supermarkets were also considered, either as a primary delivery location or only after unsuccessful home deliveries. In some cases, delivery at these shared drop-off points was considered to be a lower quality of service.

6. Conclusion

In this study, the DBSCAN clustering algorithm was used to help determine the location of spatial clusters. The results show a significant parcel locker clustering in central London compared to other areas. Moreover, the regression results show that workplace density, restaurant density, supermarket density and bus stop density positively relate to the degree of parcel locker clustering. While the higher the population density, the lower the degree of parcel locker clustering. Besides, considering the spatial influence, the socioeconomic, built environment and public transportation factors have different impacts on the degree of parcel locker clustering in different MSOAs.

In summary, the models created for this work and the associated analysis carried out have the potential to provide a method for cities to produce a spatial overview of parcel locker clustering, which in turn can be used to identify 'parcel locker desert' areas. These models can also highlight where measures need to be taken to meet carbon targets and contribute to the creation of better cities and improved urban Last-mile delivery. This work can contribute to the creation of better cities and improved urban delivery, as well as working towards the global climate goal of limiting atmospheric temperatures to 1.5 degrees.

References

1. P. Klein, B. Popp, 'Last-Mile Delivery Methods in E-Commerce: Does Perceived Sustainability Matter for Consumer Acceptance and Usage?' *Sustainability*, **14**, 24, p. 16437. doi: 10.3390/su142416437. (2022)
2. S. Sina Mohri, N. Nassir, R.G. Thompson, H. Ghaderi, 'Last-Mile logistics with on-premises parcel Lockers: Who are the real Beneficiaries?' *Transp. Res. E: Log. & Trans. Rev.*, **183**, p. 103458. doi: 10.1016/j.tre.2024.103458. (2024)
3. V. Silva, A. Amaral, T. Fontes, 'Sustainable Urban Last-Mile Logistics: A Systematic Literature Review'. *Sustainability*, **15**, 3, p. 2285. doi: 10.3390/su15032285. (2023)
4. K. Ewedairo, P. Chhetri, F. Jie, 'Estimating transportation network impedance to last-mile delivery: A Case Study of Maribyrnong City in Melbourne'. *Int. J. Logist. Manag.*, **29,** 1, pp. 110–130. doi: 10.1108/IJLM-10-2016-0247. (2018)
5. A. Lagorio, R. Pinto, 'The parcel locker location issues: an overview of factors affecting their location', p. 9. (2020)
6. U. Lachapelle, M. Burke, A. Brotherton, A. Leung, 'Parcel locker systems in a car dominant city: Location, characterisation and potential impacts on city planning and consumer travel access'. *J. Transp. Geogr.*, **71**, pp. 1–14. (2018)
7. A. Ranjbari, C. Diehl, G. Dalla Chiara, A. Goodchild, 'Do parcel lockers reduce delivery times? Evidence from the field'. *Transp. Res. E: Log. & Trans. Rev.*, **172**, p. 103070. doi: 10.1016/j.tre.2023.103070. (2023)
8. E. Morganti, S. Seidel, C. Blanquart, L. Dablanc, B. Lenz, 'The Impact of E-commerce on Final Deliveries: Alternative Parcel Delivery Services in France and Germany'. *Transp. Res. Proc.*, **4**, pp. 178-190. (2014)
9. M. Peppel, S. Spinler, 'The impact of optimal parcel locker locations on costs and the environment'. *Int. J. Phys. Distrib. Logist. Manag.*, **52**, 4, pp. 324–350. doi: 10.1108/IJPDLM-07-2021-0287. (2022)

Impact of Photovoltaic Power Station on Vegetation Carbon Sink in Guizhou Province

Yong Zhang[a*], Yong Liu[b], Jing Xu[c], Ping Zhao[d]

Survey branch company PowerChina Guizhou Electric Power Engineering Co., Ltd, Guiyang, China

Abstract: Assessing the impact of photovoltaic power plants on carbon sequestration by regional vegetation is crucial for the scientific planning of clean energy initiatives. However, due to the extensive distribution and complex environments surrounding photovoltaic power stations, it is significant challenges that quantifying their impact in Guizhou Province using local field monitoring data, This study focuses on Guizhou Province and its photovoltaic power stations, examining changes in carbon sequestration from 2010 to 2022. By integrating remote sensing interpretation data related to photovoltaic installations, this research quantifies the emission reduction benefits associated with these power plants and determines the proportion of carbon sequestration attributable to vegetation affected by photovoltaic construction within overall emissions reductions. The findings reveal that: (1) The net primary productivity (NPP) across Guizhou Province exhibited an upward trend, with a notable increase observed in the western region while a localized decrease occurred in the eastern part; (2) The influence radius of photovoltaic power stations on vegetation carbon sequestration extends approximately 1500 m; however, over time this range has diminished to around 1000 m; (3) During 2010 to 2022, vegetation carbon sequestration impacted by photovoltaic development accounted for 8.39% per annum of total carbon emission reductions from solar energy generation, with a similar proportion noted for carbon dioxide emissions. By 2022, this percentage had decreased significantly to just 1.42%. This study provide a reference into accurately evaluating the emission reduction benefits conferred by photovoltaic power plants.

1. Introduction

In order to address the challenges posed by climate change and ecological degradation caused by rising carbon dioxide (CO^2) levels, China has established clear targets of achieving carbon peak by 2030 and carbon neutrality by 2060. According to the International Energy Agency (IEA), as of 2022, the global cumulative installed photovoltaic (PV) capacity reached 1185 GW, with China leading the world at 414.5 GW. As a key energy hub in China and a national pilot zone for ecological civilization, Guizhou Province plays a crucial role in building China Province's clean energy system. With ongoing enhancements to Guizhou Province's power infrastructure and development level, the province has achieved its target network framework of "three horizontal, two connected and one center", complete coverage of 220 kV power grid in counties, and new energy power generation accounting for 30.4% of total installed capacity - surpassing hydropower for the first time. This indicates that new energy development in Guizhou Province has reached a new milestone. According to statistical data, solar power generation capacity increased from 3% in 2018 to 18% in 2022 fourfold increase within just four years - reaching an installed capacity of 14.2 million kilowatts. Together with thermal power and hydropower, solar power forms the primary sources of energy generation in Guizhou Province. In order to effectively use renewable and pollution-free solar energy resources, large-scale construction of photovoltaic power stations. In the early stage of the construction of photovoltaic power stations, mechanical excavation will directly damage the surface [1], seriously affecting the local climate [2] and ecological environment. However, after the completion of construction, solar panels changed the ecological environment of the station by absorbing solar radiation. Studies have shown that photovoltaic panels can increase temperature and humidity, reduce wind speed and other effects, which can effectively reduce wind erosion and maintain soil. Secondly, large-scale photovoltaic power stations will form a positive feedback mechanism of albedo - precipitation - vegetation [3], and enhance the ecosystem service functions related to vegetation biodiversity [4, 5], vegetation soils [6], carbon storage, water conservation [7] and pollination of nearby farmland [8]. It has a positive effect on vegetation growth in the station.

However, under the current large-scale, diversified geographical conditions and complex underlying surface of photovoltaic power plant construction conditions, how to accurately and quantitatively assess the ecological environmental effects of photovoltaic power plant construction and the estimation of carbon reduction and sink gain is a key issue that needs to be solved. Therefore, the objective of this study is to (1) Analyze the spatio-

* Corresponding author: [a*]zhangyong20210819@163.com, [b]xujing-gzy@powerchina.cn, [c]liuyong-gzy@powerchina.cn, [d]zhaoping-gzy@powerchina.cn

temporal development of NPP of photovoltaic power station in Guizhou Province; (2) Quantifying the impact of PV power plant construction on vegetation carbon sink; (3) To discuss the estimation of carbon reduction increase for the construction of photovoltaic power stations taking into account the influence of vegetation carbon sink.

2. Materials and Methods

2.1 Study area and data

The study area is Guizhou Province in southwest China (Figure 1), with land elevation ranging from 584 m to 2887 m with a mean of approximately 1715 m and a slope ranging from 1° to 82° with a mean of approximately 18°. We downloaded medium resolution satellite image collection from Analytical Insight of Earth (AI Earth) platform to describe the carbon sequestration capacity spatiotemporal variation of vegetation (Available online: https://engine-aiearth.aliyun.com (accessed on 1 September 2024)): MODIS/006/MOD17A3HGF: Global 8-day Net Primary Productivity (NPP) at 500 m from 2010 to 2022. The temporal resolution of MODIS NPP data is annually. This data is based on MODIS /TERRA satellite remote sensing parameters, and the global vegetation net primary productivity dataset is simulated by BIOME-BGC ecological process model. Some studies have verified the validity of MODIS NPP. The determination coefficient R^2 between MODIS NPP and measured NPP is 0.75 [9]. This dataset has been widely used in global and regional vegetation productivity studies. The data product level is L4, and it has been processed by atmospheric correction, radiation correction, geometric correction.

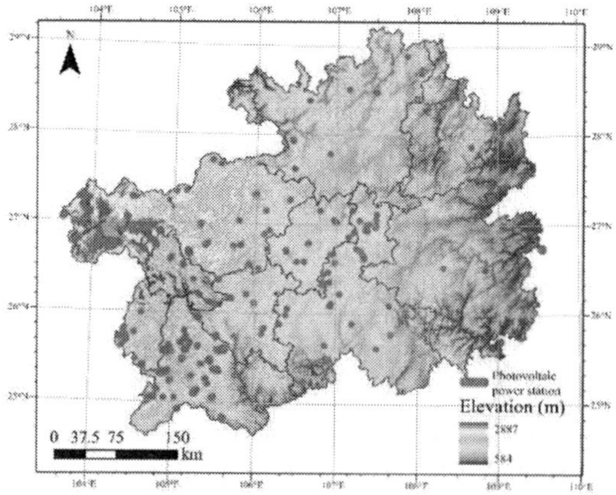

Figure 1. Study area location and corresponding DEM: Gray area is the photovoltaic power station, and red lines are municipal administrative divisions.

Meanwhile, we downloaded the PV power station vector [10] obtained from Sentinel-2 satellite image automatic interpretation based on deep learning in 2022 (Available online: https://doi.org/10.6084/m9.figshare.25347880.v1 (accessed on 5 September 2024)).

2.2 Method

2.2.1 Quantify the impacts of PV on vegetation carbon sequestration

The impacts of photovoltaic (PV) were quantified by comparing the NPP between the PV area and Vegetation area in Guizhou Province. We created 0.5 km wide non-photovoltaic (NPV) buffers zone, 0–0.5 km (buffers zone 1), 0.5–1 km (buffers zone 2), 1–1.5 km (buffers zone 3), and 1.5–2 km (buffers zone 4) away from the PV boundaries (Figure 2). We hypothesize that these buffer vegetation carbon sinks may be affected by PV construction

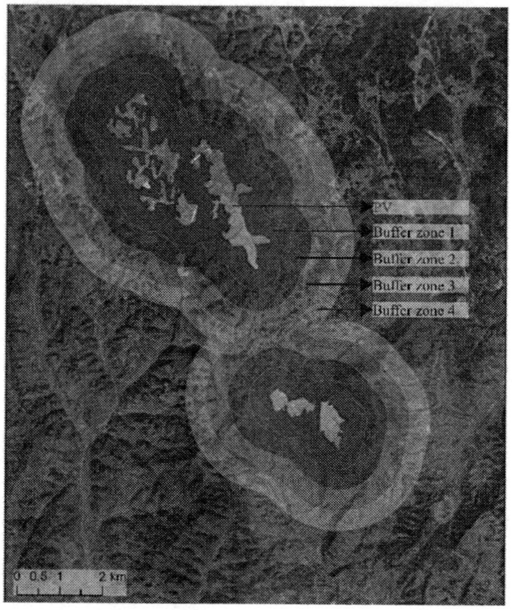

Figure 2. A photovoltaic (PV) area sample in Guizhou Province (104.76 E, 26.79 N). The light yellow area indicates the PV area. The belts denote non-photovoltaic (NPV) areas defined by 0.5 km wide buffers that vary in distance from the photovoltaic power stationa and include 0–2 km buffers zone 1, 2, 3, and 4.

The impacts of PV on vegetation carbon sequestration is quantified as the difference between PV and buffers zone and Guizhou Province in a time window centered on the year of PV construction (ΔNPP) (equation (1)):

$$\Delta NPP = NPP_{PV} - NPP_{G} \tag{1}$$

where NPV_{PV} is the average annual NPP of PV and adjacent buffer zones, and NPP_G is the average annual NPP of Vegetation area in Guizhou Province. Then, the influence of vegetation carbon sink was calculated (equation (2)):

$$\Delta C = NPP \times M \tag{2}$$

where ΔC is the vegetation carbon sink affected by the PV plant construction, M is the area affected by PV construction.

2.2.2 Sen's slope and Manner-Kendall (M-K) mutation test

We used the trend analysis method to obtain the temporal and spatial characteristics of NPP. The Sen trend degree is calculated by the following formula (equation (3)):

$$\beta = Median\left(\frac{x_j - x_i}{j-i}\right) \tag{3}$$

where i and j are the independent variables, representing the annual numbers; x_i and x_j are the NPP value of the i year and j year; β is trend degree, it's value is used to judge the rise or fall of the vegetation cover change in time series. When β is greater than 0, the vegetation cover is increasing from year i to year j; on the contrary, it is decreasing.

Mann-Kendall [11, 12] trend test process is to determine the size relationship (set as Z) between x_i and x_j in all dual values (x_i, x_j, j > i) of the sequence X = (x_i, x_j,⋯,x_n). The check statistic is calculated by the following formula:

$$Z = \begin{cases} \dfrac{S-1}{\sqrt{VAS(S)}} & (S>0) \\ 0 & (S=0) \\ \dfrac{S+1}{\sqrt{VAS(S)}} & (S<0) \end{cases} \tag{4}$$

$$S = \sum_{i=1}^{n-1}\sum_{j=i+1}^{n} \mathrm{sgn}(x_j - x_i),$$

$$\mathrm{sgn}(x_j - x_i) = \begin{cases} +1 & (x_j - x_i > 0) \\ 0 & (x_j - x_i = 0) \\ -1 & (x_j - x_i < 0) \end{cases} \tag{5}$$

$$VAR(S) = \left(n(n-1)(2n+5) - \sum_{i=1}^{m} t_i(t_i-1)(2t_i+5)\right)/18 \tag{6}$$

where Z is the standardized test statistic; S is the test statistic; n is the number of serial samples; m is the number of repeated data sets in the sequence; ti is the number of duplicate data in group i. At a given significance level, the critical value $Z_{1-\alpha/2}$ is found in the normal distribution table, when $|Z| <= Z_{1-\alpha/2}$, the trend is not significant; when $|Z| > Z_{1-\alpha/2}$, the trend is significant. In this paper, when the absolute value of Z exceeds 1.28, 1.64 and 2.32, it means that it passes the significance test of 90%, 95% and 99%, respectively. The Mann–Kendall test can be readily applied to time series data in Python through the utilization of libraries such as SciPy. These libraries provide functions that compute Kendall's Tau and yield p-values, thereby facilitating trend analysis. In this study, the Mann–Kendall test was conducted using the Python programming language [13].

2.2.3 Photovoltaic power generation potential and emission reduction assessment

To evaluate the photovoltaic carbon emission reduction potential of Guizhou province, solar energy should be converted into total energy output, and then the photovoltaic carbon emission reduction of Guizhou Province should be estimated according to the conversion formula from power generation to coal consumption and carbon dioxide emissions. The total energy output largely depends on the type of photovoltaic panels used, including monocrystalline silicon, polysilicon, amorphous silicon and thin film polysilicon. The power generation efficiency of different photovoltaic panels varies. According to the actual research situation, we choose the most widely used monocrystalline silicon in Guizhou Province, and its conversion rate is 22.5%. In addition, the photovoltaic power generation potential is also affected by the annual total solar radiation per unit area, the area of photovoltaic power plants and the area receiving solar radiation, while the power generation potential should also consider the ratio of discarded light (R), the power loss caused by dust (PL). According to the conversion standard of power generation and coal consumption and carbon dioxide emission, 301.5 g·C/kW·h will be calculated according to China's thermal power plant power supply, and carbon dioxide emission per unit of thermal power generation will be calculated according to 828 g/kW·h (Data source: China Electric Power Industry Annual Development Report 2022). The formula for solar power generation potential is as follows [14]:

$$S_{EGP}=S_A\times A_F\times A_{SR}\times P_E\times(1-R)\times(1-P_L) \tag{7}$$

where S_{EGP} is solar power generation potential (kW·h); S_A is Area of photovoltaic power station; A_F is the area factor coefficient of receiving solar radiation. The value of A_F is 0.35; A_{SR} is the annual total solar radiation per unit area. The value of A_{SR} is 1289.9 kW·h/m^2 (Available online: https://www.cma.gov.cn/); P_E is the power generation efficiency of solar panels. The value of P_E is 22.5%; R is the ratio of discarded light. The value of R is 7.5%; P_L is the power loss caused by dust. The value of P_L is 0.65%.

3. Results & Discussion

3.1 Temporal and spatial characteristics of NPP

Figure 3 shows temporal and spatial characteristics of NPP. First of all, the NPP of the six regions showed an increasing trend with the increase of years, among which the NPP growth rate of Guizhou province was the smallest, and the NPP growth rate of buffer zone 4 was the largest. Moreover, the annual NPP growth rate of the six regions is ranked as follows: Guizhou Province (4.6 g C*m^{-2}/a) < Photovoltaic area (6.9 g C*m^{-2}/a) < Buffer zone 1 (7.7 g C*m^{-2}/a) < Buffer zone 2 (8.4 g C*m^{-2}/a) < Buffer zone 3 (8.7 g C*m^{-2}/a) < Buffer zone 4 (9.0 g C*m^{-2}/a).

Then, in order to further explore the variation of NPP in space, the Mann-Kendall trend analysis of NPP is carried out from the pixel scale, and according to data Z value, the results are divided: Significant improved (Or severely degraded) (|Z| > 1.64), Slight degraded (Or improved) (1.28 <= |Z| <= 1.64), Stable (|Z| < 1.28). There were significant spatial differences in NPP in Guizhou province by Figure 4. It was shown that NPP increase area is located in the west of Guizhou Province, and the reduction in NPP is notably more pronounced in the east of Guizhou Province. And, during the period of 2010-2022. the areas with NPP significant improvement covered approximately 25.62% of Guizhou Province, the areas with NPP slight improvement covered approximately 10.00% of Guizhou Province, the areas with NPP slight reduction covered approximately 2.62% of Guizhou Province, the areas with NPP severely reduction covered approximately 2.62% of Guizhou Province, and the area characterized by stable and unchanged NPP constituted the largest proportion of the entire study area, accounting for approximately 59.14%. In addition, further statistics of the NPP changes of the photovoltaic power station region, during the period of 2010-2022, it is found that the areas with NPP severely reduction covered approximately 2.60% of photovoltaic power stations, the areas with NPP slight reduction covered approximately 1.42% of photovoltaic power stations, the areas with stable NPP region covered approximately 54.61% of photovoltaic power stations, the areas with NPP slight improvement covered approximately 12.06% of photovoltaic power stations, and the areas with NPP significant improvement covered approximately 29.31% of photovoltaic power stations.

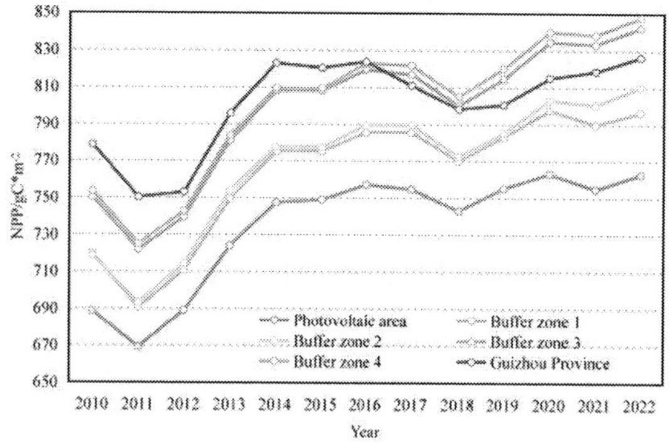

Figure 3. Temporal tracks of NPP during 2010—2022.

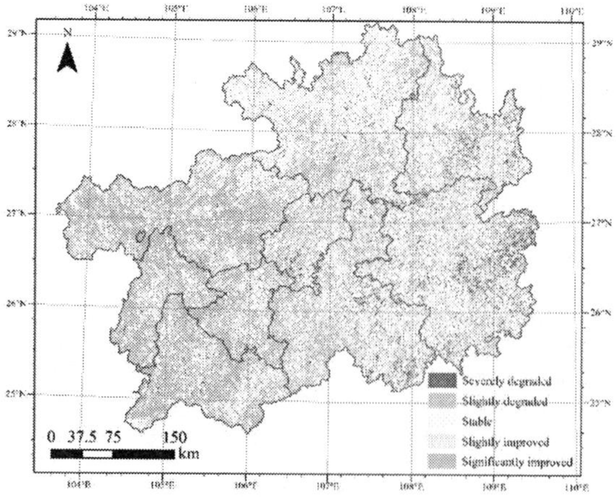

Figure 4. Distribution of NPP changes in Guizhou Province from 2010 to 2022.

3.2 PV impacts on local vegetation carbon sequestration

PV impacts on local vegetation carbon sequestration were quantified by the difference in by comparing the NPP between the PV area and Vegetation area in Guizhou Province. It can be seen from Figure 5 that the farther the buffer zone is from the photovoltaic power station, the closer the buffer zone NPP is to the NPP of Guizhou Province. Table 1 shows the annual average values of ΔNPP in the six regions are ranked from small to large as follows: Photovoltaic area (-65.85 g $C{\cdot}m^{-2}{\cdot}a^{-1}$) < Buffer zone 1 (-37.32 g $C{\cdot}m^{-2}{\cdot}a^{-1}$) < Buffer zone 2 (-32.95 g $C{\cdot}m^{-2}/a$) < Buffer zone 3 (-3.44g $C{\cdot}m^{-2}{\cdot}a^{-1}$) < Buffer zone 4 (0.35 g C $*m^{-2}{\cdot}a^{-1}$). And the 13-year average values of ΔC in the six regions are ranked from small to large as follows: Photovoltaic area (-6196.46 t $C{\cdot}a^{-1}$) < Buffer zone 1(-24639.06 t $C{\cdot}a^{-1}$) < Buffer zone 2 (-23764.95 t $C{\cdot}a^{-1}$) < Buffer zone 3 (-2807.23 t $C{\cdot}a^{-1}$) < Buffer zone 4 (317.56 t $C{\cdot}a^{-1}$).

Table 1. Net carbon sequestration of vegetation affected by PV

Area	*Photovoltaic area*	*Buffer zone 1*	*Buffer zone 2*	*Buffer zone 3*	*Buffer zone 4*
ΔNPP ($gC*m^{-2}/a$)	-65.85	-37.32	-32.95	-3.44	0.35
ΔC (t C/a)	-6196.46	-24639.06	-23764.95	-2807.23	317.56

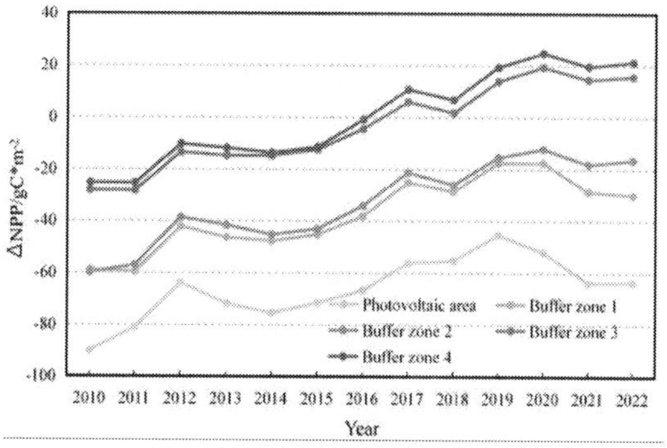

Figure 5. Temporal tracks of ΔNPP during 2010—2022.

3.3 Photovoltaic power generation potential and emission reduction assessment

Table 2 shows that during 2010 to 2022, the photovoltaic power generation in Guizhou increased by 8755.63 g W·H, averaging 2269.64 g $W{\cdot}H{\cdot}a^{-1}$, reducing standard carbon consumption by 684.30 $kt{\cdot}a^{-1}$ and carbon dioxide emission by 1879.26 $kt{\cdot}a^{-1}$. Combined with the results of the net carbon sequestration of vegetation affected by PV, we can get that the total photovoltaic emission reduction ratio of the net carbon sequestration of vegetation affected by PV each year is 8.39%, and the proportion of carbon dioxide is similar. However, in 2022, the proportion of total PV emission reduction in the net carbon sequestration of vegetation affected by PV was reduced to 1.42%.

Table 2. Statistical overview of photovoltaic power generation potential and emission reduction

Statistic	*2010*	*2011*	*2012*	*2013*	*2014*	*2015*	*2016*	*2017*	*2018*	*2019*	*2020*	*2021*	*2022*
Power generation potential (gW·h)	8.60	16.74	29.10	51.76	55.48	76.92	157.40	381.78	494.57	922.89	1678.17	2374.04	2648.42
Standard carbon (kt C)	23.62	45.96	79.91	142.15	152.37	211.24	432.25	1048.47	1358.24	2534.50	4608.69	6519.74	7273.28
Carbon dioxide content (kt CO^2)	8.60	16.74	29.10	51.76	55.48	76.92	157.40	381.78	494.57	922.89	1678.17	2374.04	2648.42

3.4 PV development and feasibility of power generation potential estimation in Guizhou Province

To evaluate the feasibility of estimating the power generation potential of Guizhou province based on the spatial distribution of PV interpreted by remote sensing, we compared the simulated power generation potential in Guizhou Province with electricity generation published in the news for the last four years. Figure 6 was shown that there was an extremely high correlation and that the relative error was less than 7% in 2022.

According to the "14th Five-Year Plan for the Development of new Energy and renewable Energy in Guizhou Province", the installed capacity of photovoltaic power generation in Guizhou Province will reach 3100 GW in 2025, and the installed capacity of photovoltaic power generation in Guizhou Province will reach 1420 GW by the end of 2022. In the process of rapid development of photovoltaic, we should pay more attention to its impact on the ecological environment. To better achieve the goal of peaking carbon emissions and carbon neutrality.

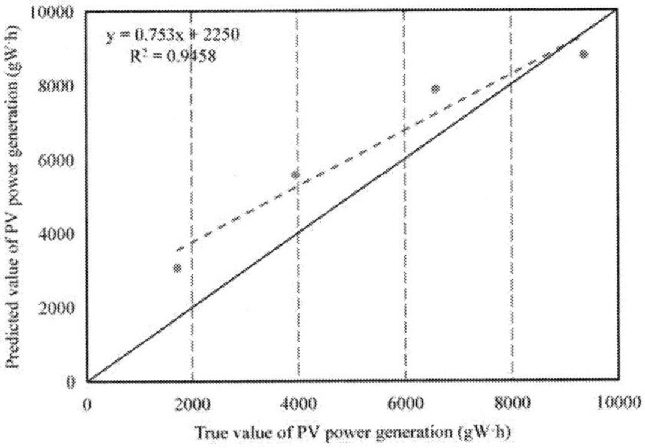

Figure 6. A scatter plot of simulated power generation potential in Guizhou Province with power generation published in the news over the past four years.

4. Conclusions

We analyzed the spatial-temporal variation trend of NPP of photovoltaic power stations in Guizhou Province, quantified the net carbon sequestration of vegetation and the power generation potential and carbon emission reduction of PV power stations reduced by the construction of PV power stations, and based on this, obtained the carbon emission reduction of PV power stations and the proportion of vegetation impact. The results showed that :(1) During 2010-2022, NPP significantly increased in the western Guizhou Province, while NPP locally deteriorated in the eastern Guizhou province. NPP of PV power stations showed an overall upward trend; (2) According to the changes of different buffer ΔNPP, we know that the impact range of PV on vegetation carbon sequestration is about 1500 m in the past decade, but the impact range is reduced to about 1000 m over time. And the farther away from the photovoltaic power station, the less vegetation is affected; (3) According to the estimated power generation potential of photovoltaic power stations, we understand that the power generation of PV power stations has increased by 8755.63 gW·H in the past ten years; (4) Each year, the net carbon fixation by vegetation influenced by PV accounts for 8.39% of the total emissions reduction achieved through PV power generation, with a similar proportion observed for carbon dioxide. However, by 2022, the share decreased to 1.42%. The quantification method can be applied to other countries or regions with available PV power station information. Further studies using satellite data at finer resolution than MODIS data could reveal the impact with more spatial detail. These observations can be combined with numerical simulations to advance the mechanistic understanding of PV power station impacts on the local native vegetation growth. The enhanced understanding of the impacts of PV power stations contributes to informing the environmental consequences of solar energy development and guiding clean energy planning for sustainable development.

Acknowledgments

This paper was supported by the funds. The funds are as follows: The Science and Technology Project of PowerChina Guizhou Electric Power Engineering Co., Ltd (GZEDKJ-2024-06).

References

1. Yuan, S., Zhang, Z., Dang, T., Yang, Y., Yuan, F., and Bu, C., (2018) Comparative study of growth status of three plant measures and its ecological functions in photovoltaic power station of Mu Us sandy land, Research of soil and water conservation, 25 (2): 235–239. 10.13869/j.cnki.rswc.2018.02.034.
2. Zheng, J., Luo, Y., Chang, R., and Gao, X., (2023) Study on impact of large-scaled photovoltaic development on local climate and ecosystem. Acta energiae solaris sinica, 44 (8): 253–265. 10.19912/j.0254-0096.tynxb.2022-0530
3. Li, Y., Eugenia, K., Safa, M., Jorge, R., Fred, K., Danie, L., et al., (2018) Climate model shows large-

scale wind and solar farms in the Sahara increase rain and vegetation. Science, 361 (6406): 1019–1022. 10.1126/science.aar5629

4. Vervloesem, J., and Marcheggiani, E., Choudhury, M., and Muys, B., (2022) Effects of photovoltaic solar farms on microclimate and vegetation diversity, Sustainability, 14, (12): 7493. 10.3390/su14127493
5. Mori, K., and Tabata, T., (2020) Comprehensive evaluation of photovoltaic solar plants vs. natural ecosystems in green conflict situations, Energies, 13, (23): 6224. 10.3390/en13236224
6. Rodriguez-Pastor, D., Ildefonso-Sanchez, A., Soltero, V., Peralta, M., and Chacartegui, R., (2023) A new predictive model for the design and evaluation of bifacial photovoltaic plants under the influence of vegetation soils, Journal of cleaner production, 385: 135701. 10.1016/j.jclepro.2022.135701
7. Kannenberg, S., Sturchio, M., Venturas, M., and Knapp, A., (2023) Grassland carbon-water cycling is minimally impacted by a photovoltaic array, Communications earth & environment, 4, (1): 238. 10.1038/s43247-023-00904-4
8. Blaydes, H., Potts, S., Whyatt, J., and Armstrong, A., (2024) On-site floral resources and surrounding landscape characteristics impact pollinator biodiversity at solar parks, Ecological solutions and evidence, 5, (1): e12307. 10.1002/2688-8319.12307
9. Liu, G., Shao, Q., Fan, J., Ning, J., Rong, K., Huang, H., et al., (2022) Change trend and restoration potential of vegetation net primary productivity in China over the past 20 years, Remote sensing, 14, (7): 1634. 10.3390/rs14071634
10. Yuan, S., Zhang, Z., Dang, T., Yang, Y., Yuan, F., and Bu, C., (2018) Comparative study of growth status of three plant measures and its ecological functions in photovoltaic power station of Mu Us sandy land, Research of soil and water conservation, 25, (2): 235-239. 10.13869/j.cnki.rswc.2018.02.034
11. Kendall, M., (1975) Rank correlation methods, Griffin, London.
12. Mann, H., (1945) Nonparametric test against the trend. Econometrica, 13: 245–259. 10.2307/1907187
13. Faqusch, H., and Grossi, G., (2024) Trend analysis of precipitation, temperature and snow water equivalent in Lombardy region, northern Italy, Sustainable water resources management, 10, (1): 18. 10.1007/s40899-023-00992-2
14. Liu, L., Sun, Z., Wu, F., Xueliang, Z., and Qian, Z., (2022) Evaluation of suitability and emission reduction benefits of photovoltaic development in Beijing-Tianjin-Hebei region, Acta geographica sinica, 77, (3): 665-678. 10.11821/dlxb202203012

Research on Zero Carbon Sports Building Technology Strategy in Regions with Hot Summers and Cold Winters

Yunxia Ding[1], Xilan Qiu[2,*], Jinxing Cai[1]

[1] Shanghai Business School, Shanghai, China
[2] Shanghai Research Institute of Building Science Co., Ltd, Shanghai, China

Abstract: With the arrival of the era of all-people sports and the rapid development of sports venue construction, under the background of "carbon peaking and carbon neutrality", green and low-carbon construction of sports venues is imperative, and it is urgent to build a suitable and mature zero carbon technology system. In this thesis, a research is conducted on canonical zero carbon sports venues both domestically and internationally, summarizing the application and adaptive forms of zero carbon technology in sports buildings, as well as difficulties in implementation from actual engineering cases. Then, starting from the functional features of sports venues, the focus of energy conservation and emission reduction of sports buildings is clarified through comparative analysis, the appropriate low-carbon technology path for sports buildings in regions with hot summers and cold winters is refined, and its effectiveness is verified by the analysis on a typical actual case, by which the zero carbon strategy concept of "source reduction first, passive design priority, ontology focus, efficiency improvement as the core, and energy enhancement as the wing" is further proposed and refined into 20 key technologies and implementation points, thus constructing a zero carbon sports venue technology system with climate adaptability, and providing technical guidance for the construction of zero carbon sports venues.

1. Introduction

With the promotion of the strategy of building a country strong on sports and the arrival of the era of all-people sports, the sports industry in China has entered rapid development. The demand for high-quality and diversified sports venues is becoming increasingly prominent. In the context of the "carbon peaking and carbon neutrality" initiative where energy resources are becoming increasingly scarce, the green and low-carbon development of sports venues has become an important development direction. In this respect, it is urgent to study zero carbon technologies suitable for efficient and economical sports buildings.

Based on research on zero carbon and low-carbon sports venues at home and abroad, the characteristics of sports buildings and building energy consumption are analysed in this study, focusing on climate regions and difficulties of implementation with hot summers and cold winters, verifying the effectiveness of technologies by the analysis on a typical actual case, and proposing zero carbon technology design strategies and key implementation technology systems for sports venue buildings in these climate regions.

2. Research on Canonical Cases of Zero Carbon Sports Venues at Home and Abroad

2.1 Investigation into Canonical Cases

In the study, research and analysis were conducted on typical cases of international zero carbon sports venues, sports buildings with near zero carbon consumption in China[1], and sports venues with low energy consumption in regions with hot summers and cold winters, and the zero carbon technology strategies of each case were sorted out, as shown in Fig. 1- Fig. 3.

2.2 Summary and Difficulty Analysis of Zero Carbon Technology in Sports Venues

Through the above case studies, the common zero carbon technologies are summarized into the following classifications as shown in Table 1. Analyzed from the dimension of actual engineering application cases, the main experience of the zero-carbonization technology path of sports buildings at home and abroad is reducing the energy consumption demand of the building itself, improving the efficiency of building energy utilization, and increasing the utilization rate of non-fossil energy, which means highlighting passive design. Research[2] shows that in regions with hot summers and cold winters,

* Corresponding author: xilanqiu2005@126.com

the impact of external windows on the building envelope of zero carbon buildings is the highest, followed by roofs and external walls, and energy efficiency is maximized through high-performance equipment, while renewable energy is an essential cornerstone of the zero carbon path. Besides, the applicability of many zero carbon technologies in new and existing buildings is also analyzed, with the application difficulties and challenges of various technologies put forward, clarifying the implementation focus for the application of core low-carbon technologies.

Fig. 1. Typical Case of International Zero Carbon Sports Venues

Fig. 2. Cases of Near Zero Carbon Sports Venues in China

Fig. 3. Cases of Green and Low Energy Consumption Sports Venues in Areas with Hot Summers and Cold Winters

Table 1. Common Technology System of Zero Carbon Sports Buildings

Classification Strategy	Technical Points	Design Principles	Importance	Applicable Construction Type		Application Difficulty
				New	Existing	
Passive design of architecture noumenon	Building size factor and window-wall ratio	Adopt climate adaptive design to reduce building energy consumption.	•	√	/	Building design
	Building envelope	Improve thermal performance, and choose high-performance insulation and thermal insulation exterior walls, energy-saving doors and windows, etc.	•	√	√	Building design
	Building airtightness	Air impervious liners are continuous and there is no thermal bridge design.	•	√	/	Economical efficiency
	Natural lighting and ventilation	Design skylights, side windows, atriums, light guides, ventilating shafts, etc.	•	√	/	Building design
	External shading	Install insulation panels on the roof and provide external shadings for windows.	•	√	√	Building design, product materials
	Green low-carbon materials	Choose green and low-carbon building materials, sustainable materials, etc.	•	√	/	Product materials
Active design of building equipment	Heating and cooling system	Apply high efficiency and energy saving cold and heat source, and use high efficiency heat pump (air source, river water source,	•	√	√	Economical efficiency, equipment and products

		ground source, etc.)				
	New trend heat recovery and ventilation system	Set up new trend heat recovery system and air supplement system	•	√	/	Economical efficiency and comfort
	Energy-saving lighting	Use LED energy-saving lamps, and zonal intelligent lighting control	•	√	√	Economical efficiency
	Efficient and energy-saving equipment	Use energy saving elevators, energy-saving distribution transformers, etc.	○	√	√	Economical efficiency
	Efficient water-saving equipment	Water saving appliances, water metering systems, etc.	○	√	√	Economical efficiency
Venue area design	Rainwater reuse	Recycle and treat rainwater on roofs, outdoors, etc.	•	√	/	System design
	Roofs and outdoor greenery	Increase carbon sink	○	√	√	Structural design
	Smart monitoring and control	Establish a comprehensive monitoring and control system for energy consumption and environment	○	√	√	Meticulous management
	Sponge city	Improve the regional microenvironment and reduce the heat island effect	○	√	√	System design
Renewable energy utilization	Photovoltaic power generation	Use building photovoltaic integration, flat and vertical photovoltaic modular components, etc.	•	√	√	System design
	Energy storage power supply system	Cooperate with photovoltaic power generation for energy storage, peak shaving for valley filling	○	√	/	System design, product and equipment

Remarks: • Important; ○ General; √ Good applicability; / Poor applicability

3. Study on the Effectiveness of Zero Carbon Suitable Technologies of Sports Buildings in Regions with Hot Summers and Cold Winters

3.1 Characteristics of Sports Buildings

Sports buildings are one of the important public buildings, which have significant characteristics of their own. There are different functional positioning according to the sports competition or training content, such as ball games, water games, ice and snow games, track and field games, etc., and great differences lie in the demands for architectural space. Compared to other public buildings, sports buildings have the following characteristics:

One is that most of them are tall and spacious. Most sports venues are low-rise buildings with reinforced concrete frame structures; they have large building surface areas, high single-story heights and large spans.

Secondly, the roof area is large and the thermal performance is weak. Sports venues have large internal spans and roof areas, mostly in the form of grid structures and steel structures. Lightweight materials such as metal and film are often adopted, and their lightweight roofs have relatively poor thermal inertia, which can easily result in strong solar radiation in summer and fast heat dissipation in winter.

Thirdly, the surrounding area is mostly open space. Most sports venues are equipped with large outdoor plazas, most of which are open and paved with hard surfaces, which can easily form a heat island effect and affect the microenvironment.

3.2 Energy Consumption Characteristics and Energy Conservation Demand Trend of Sports Buildings in Regions with Hot Summers and Cold Winters

Taking Shanghai as a typical area of hot summers and cold winters, the energy consumption structure of sports buildings in Shanghai was analysed to probe into the energy consumption focus of traditional sports buildings. Meanwhile, the energy consumption structure of public buildings in regions with hot summers and warm winters was compared with it, and the low-carbon focus of sports buildings was clear.

According to the newly published reports of Public Buildings Energy Monitoring in Shanghai and Shenzhen[3], in 2023, the average energy consumption level of sports buildings in Shanghai is 72.4kW•h/m^2•a. Their monthly power consumption intensity is shown in Fig. 4 and their energy consumption from July to September in summer accounts for about 27% of the whole year. Summer refrigeration is one of the key points of building energy consumption and their energy consumption level is lower than the average energy consumption values of public buildings in Shanghai and Shenzhen, which are 98.4 kW•h/(m^2•a) and 98.1 kW•h/m^2•a, respectively.

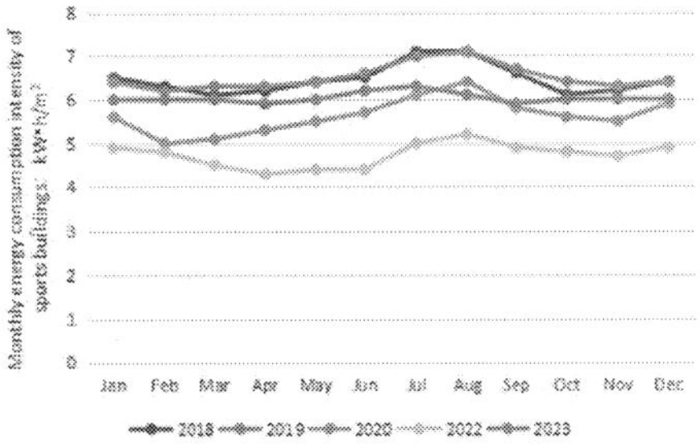

Fig. 4. Analysis of Monthly Energy Consumption Intensity of Sports Buildings in Shanghai

By comparing the energy consumption subentries of Shanghai sports buildings and the proportion of energy consumption of public buildings in Shanghai and Shenzhen as shown in Fig. 5, it presented that compared with public buildings the large category in Shanghai, the lighting socket energy consumption holds the line while power energy consumption is significantly higher, and air conditioning energy consumption is relatively less. The reason is that the lighting of sports buildings is large, and most of them are non-energy-saving lamps, leading to high energy consumption, Another cause is that sports venues are for sports activities with intense movement, and dense audience, requiring to set with sufficient new trend system. Its ventilation system is more demanding than the general public buildings, which leads to the high energy consumption of the power system (fan). Most sports venues are equipped with outdoor sites, which, being non-air conditioning areas, make the energy consumption of air conditioning low. It can be seen that lighting and power energy consumption are the focus of energy saving of sports buildings. In addition, compared with the public buildings in other climate areas, their power and special energy intensity are obviously higher, which shows that there is still great energy saving potential.

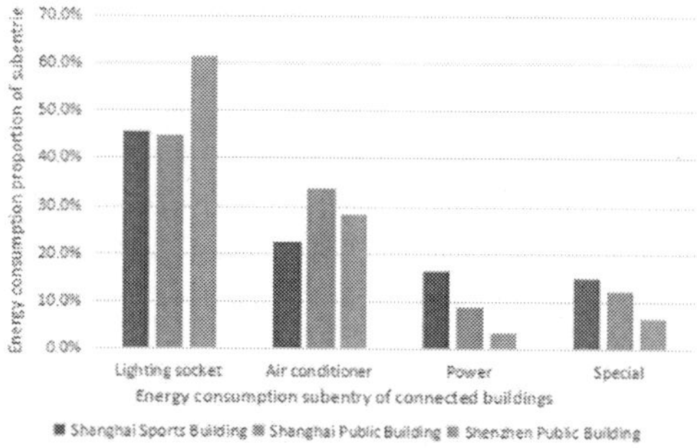

Fig. 5. Analysis Chart of Electricity Consumption between Sports Buildings and Public Buildings in Shanghai

With the development of The Times, due to the advancement and evolution of the energy consumption of sports venues, the demand for buildings' energy conservation demand has changed significantly.

For one thing, the main type of its energy is electricity, but there is a significant trend towards diversified energy demand. Traditional sports venues rely mainly on electricity for energy consumption due to their functional focus. However, with the modernization and improvement of sports venues, as well as the expansion of residential, dining, and entertainment functions, the types of gas and gasoline energy sources are also constantly expanding.

Secondly, the diversification of the demand for special energy conservation equipment for buildings is accelerating. The main energy consumption in the venue includes maintenance of sports facilities, air conditioning, lighting, power, office equipment, etc. The energy consumption structure of different sports venues varies significantly. Professional sports facilities such as ice and snow facilities, swimming and diving facilities require maintenance. This kind of special energy consumption is the focus of sports building energy consumption[5] while the main energy consumption of ordinary venues comes from air conditioning equipment.

Thirdly, the year-round operation mode has a high demand for seasonal differentiated energy-saving response. Most sports venues are open year-round, and the building itself has high requirements for climate adaptability. It maximizes the energy-saving strategy of the building body in different seasons to achieve automatic response. Strong solar radiation in summer and weak radiation in winter put higher demands on the thermal performance of the building envelope structure.

3.3 Analysis on the Suitability and Effectiveness of Low-carbon Technology for Sports Buildings in Regions with Hot Summers and Cold Winter Areas

Regions with hot summers and cold winters have climate characteristics of high temperatures, humidity, and rainfall in summer, and cold and dry winters. The zero carbon design of sports venue buildings can be analyzed for technical suitability from the following aspects.

One is to plan venue selection and design reasonably. Sports buildings have large spans and large spaces, and their building interfaces are often long and high. The building site selection and orientation will not only affect the indoor lighting and ventilation of the venue, but also affect the thermal environment around the building, thereby indirectly affecting the indoor space environment. The venue selection should be as far away from high-density areas as possible, which is conducive to the application of natural ventilation and lighting. Priority should be given to choosing areas with good natural environment and easy transportation; venue design should be tailored to local conditions, building orientations that are significantly better adapted to climate than east-west north-south layouts should be chosen[6], and micro environments need to be optimized based on regional wind environment analysis, and plan layouts.

The second is to focus on the climate adaptability design of building envelope structures. Optimize the design of external wall insulation, actively use external insulation composite inorganic material walls, and adopt light coloured outer surfaces; To address the weak areas of summer and winter insulation in building exterior windows, measures should be taken to improve sealing and insulation performance, enhance the shading coefficient of exterior window glass, balance lighting and insulation requirements, optimize exterior window shading design based on orientation design, promote the integration of exterior windows and shading, innovate shading applications, and maximize summer and winter insulation functions[7]. Combining roof materials, the thermal performance should be improved and roof greening needs to be reasonably set up.

Thirdly, endowment conditions should be made full use of to select appropriate low-carbon technologies. In response to the climate characteristics of abundant rainfall in summer, carry out landscape green design, increase regional carbon sinks, reasonably set up rainwater reuse systems, and adopt green facilities such as sponge cities; Based on natural endowments and surrounding conditions, fully utilize renewable energy, select suitable air conditioning technologies, and maximize the reduction of carbon emissions.

In the study, a typical green sports building in Shanghai was probed into and the effectiveness of the technology strategy was verified through the implementation or expected effect of key green and low-carbon technologies. The core low-carbon technology adopted in the typical case is shown in Fig. 6. Corresponding results of the implementation effect are also obtained after simulation analysis and calculation, which meet the requirements of national standard "Assessment standard for green building" GB / T50379, and winning Certificate of Green Building Design Label. Through analysis, the energy consumption per unit area of the project reaches 34 kW•h/(m^2•a), which meets the advanced value requirements of Shanghai local standard "Guideline of efficient energy conservation for large and medium-sized sport buildings", which fully explains the comprehensive effect of the above green and low-carbon technologies, and the technical strategies of passive optimal design, energy efficiency improvement of energy using equipment and utilization of renewable energy have a significant effect on energy conservation and emission reduction of sports venues.

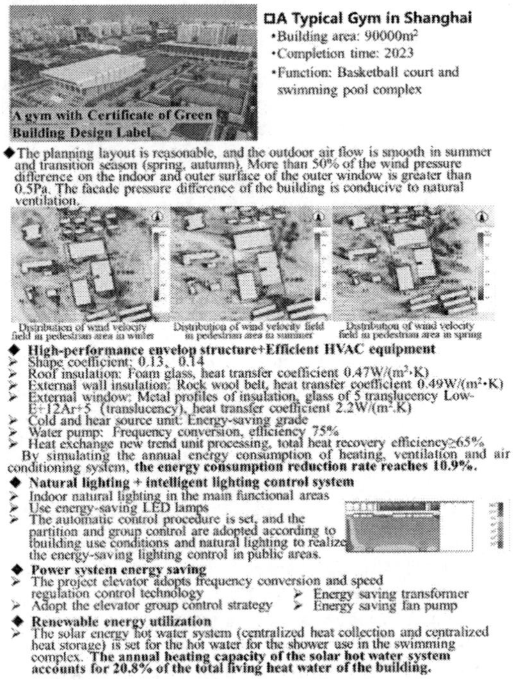

Fig. 6. Technical Strategies and Results of the Typical Green Sports Building in Shanghai

4. Zero Carbon Technology Strategy for Sports Venues in Regions with Hot Summers and Cold Winters

Based on the analysis, a technical strategy concept and key technology system were proposed for the zero carbon construction of sports venue buildings that are in regions with hot summers and cold winters.

4.1 Zero Carbon Technology Strategy Concept

Based on case summary and feature analysis, following the carbon reduction strategy of eliminating direct emissions and reducing indirect emissions, the zero carbon technology strategy for sports buildings in regions with hot summers and cold winters can follow the strategic concept of "source reduction first, passive design priority, architecture noumenon focus, efficiency improvement as the core, and energy enhancement as the wing". Source reduction first refers to coordinating the energy structure from the initial stage of the project, reducing direct and implicit carbon emissions at the source; Passive design priority refers to prioritizing the use of passive low-carbon technologies in architectural design, maximizing the utilization of natural resources such as natural lighting and ventilation; Architecture noumenon focus requires strengthening and improving the thermal performance of the building itself, and reducing energy demand; Efficiency improvement as the core refers to comprehensively enhancing energy utilization efficiency and reducing operational emissions while meeting demand; And energy enhancement as the wing refers to the construction of a technology system consisting of 20 items in 5 aspects by utilizing zero carbon energy sources such as renewable energy to reduce carbon emissions, as shown in Fig. 7.

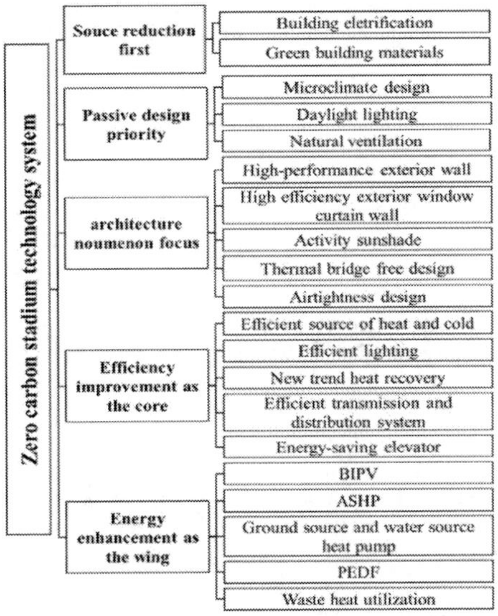

Fig. 7. Zero Carbon Sports Building Technology System in Areas with Hot Summers and Cold Winter

4.2 Key Technology System for Zero Carbon Sports Venues

Based on the above strategic concept and its core key technologies and implementation points, the key technology system of zero carbon sports venues was formed.

1. Source reduction first

One is the full electrification of building terminal energy consumption. To meet the energy needs for heating, cooking, domestic hot water, etc. in sports buildings, avoid the use of traditional fossil energy sources such as gas and oil that have direct emissions, select reasonable alternatives or choose high-efficiency electrical products to achieve full electrification of terminal energy use.

The second is green building materials. For the main structure, envelop walls and interior walls, decorative materials, building materials and appliances, research and development of new materials should be stepped up and various green and low-carbon building materials should be applied on a large scale to ensure the proportion of green building materials and reduce the hidden carbon emissions of buildings.

2. Passive design priority

One is microclimate design. In building venue planning, create a suitable microclimate, optimize the layout of buildings and landscape spaces through venue wind environment analysis, and form comfortable and reasonable regional wind corridors according to local conditions; Reasonably arrange light coloured paving materials to reduce the reflection radiation of hard floors; Through landscape design, plan outdoor activity areas and plaza spaces to be shaded by trees or structures to reduce the intensity of regional heat islands.

The second is natural lighting. Optimize the form design, apply natural lighting, implement top natural lighting methods such as diffuse reflection of skylights and fully transparent ceilings, and adopt active technology strategies such as lighting ducts and automatic chasing systems to apply sunlight.

The third is natural ventilation. Through wind environment simulation and comprehensive consideration of the dominant wind direction in the region, natural ventilation in the transition season can be optimized. By designing the entrances and exits of sports venues based on the prevailing wind direction throughout the year, and making reasonable use of the thermal pressure difference, it is possible to consider installing fans in the roof ventilation structure to jointly form a mechanical auxiliary ventilation system.

3. Architecture noumenon focus

One is high-performance exterior walls. Optimize the building form factor and enhance the thermal performance of the exterior wall enclosure structure. Based on the characteristics of hot summers and cold winters, the heat transfer coefficient K value of the exterior walls and roofs of the venue can reach 0.15-0.35W/(m^2·K), and integrated external wall insulation such as non-demoulding, sandwich insulation, and reverse heating, as well as aluminium panel curtain wall external insulation, can be used.

The second is efficient exterior windows/curtain walls. By improving the heat transfer coefficient K value of the curtain wall/external window through architectural design to 1.4 W/(m^2·K), the thermal insulation performance of the external window, i.e. the solar heat gain coefficient (SHGC), is $\leq$ 0.30 in summer and $\geq$ 0.40 in winter; Passive-design windows with strict requirements for window aluminium profiles, insulation strips, glass, and hardware can be selected; The airtightness performance of external windows should not be lower than level 8.

The third is activity shading. In response to the demand for thermal insulation and winter heating in regions with hot summers and cold winters, shading component design can be carried out in combination with the building form of the venue, which can adopt adjustable external shading, central shading, adjustable internal shading, etc.

The fourth is the design without thermal bridges. A series of thermal bridge treatments can be carried out in areas where thermal bridges are prone to form in structural systems such as exterior walls, exterior doors and windows, shading facilities, parapets, roofs, basements, and floors. Measures such as pre-embedded anchoring of thermal bridges can be used.

The fifth is airtightness design. The overall airtightness level of the building is n50 $\leq$ 1.0, ensuring the continuity of the building's airtightness. Specific measures can be taken to carry out airtightness design for enclosure structures, doors and windows, hole sealing design, structural connection sealing design, etc.

4. Efficiency improvement as the core

One is efficient cold and heat sources. Scientifically and rationally design heating and cooling plans, and adopt distributed or unit heating and cooling to meet the needs of local periods. The project can be equipped with a separate water machine+heat pump system, as well as a separate cold and heat source system, using multi split, ground source heat pump+air source heat pump/cooling tower. The energy efficiency level of the cold and heat source units or systems should reach level one.

The second is intelligent and efficient lighting. Carry out refined design for various lighting scenarios, select high-efficiency and energy-saving lamps as the lighting source, implement intelligent control systems, and achieve graded and zoning control of the lighting system to meet the adjustable lighting needs of diverse lighting areas in the scene.

The third is new trend purification and heat recovery. In the HVAC design, the unit uses a $PM_{2.5}$ filter with a filtration level of MERV14. An indoor air quality detector is installed. When the CO_2 concentration exceeds 700ppm, fresh air is activated, which is beneficial to the health of indoor personnel. Adopting fresh air heat recovery technology, that is, the unit adopts exhaust total heat recovery, with a refrigeration enthalpy efficiency of>65% and a heating enthalpy efficiency of>70%, and a transitional season fresh air bypass function.

The fourth is an efficient transmission and distribution system. Reasonably design the transmission and distribution system to avoid long-distance transportation; The transmission and distribution equipment such as fans and pumps are equipped with variable frequency speed regulation, which can achieve high-efficiency energy-saving levels.

The fifth is energy-saving elevators. As a supporting facility, the elevator system should adopt energy-saving control and drag systems, such as VVVF drive system type elevators, and have group control functions.

5. Energy enhancement as the wing

One is the integrated application of building photovoltaics. Make full use of the external surface conditions of the building, implement the integrated design of building photovoltaic, and improve the functional efficiency, beauty, safety and other performance of the photovoltaic system. Utilizing renewable energy, especially photovoltaics, is a must-have strategy for most low-carbon and green sports venues. The integrated components of facade photovoltaics, roof photovoltaics, and transparent photovoltaics railings are applied to achieve the integration of photovoltaics and intelligent shading.

The second is air source heat pump heating. In response to the typical characteristics of "partial time" and "partial space" for winter heating in sports venues in hot summer and cold winter areas, air energy distributed energy stations can be selected, which can be coupled with energy sources such as electricity and solar energy to achieve good comprehensive benefits.

The third is the development and utilization of ground source water source heat pumps. Based on the situation of regional natural resources, adapting measures to local conditions, fully utilizing shallow geothermal energy[8] and water sources to provide heating and cooling for venues, and even providing energy for sports facilities such as hot spring swimming.

The fourth is the optical storage direct flexible system. Based on the photovoltaic power generation and building energy demand, select and determine suitable and efficient energy storage equipment and capacity; Coordinate expandable energy storage facilities such as batteries, electric vehicles, and charging stations within the venue; Establish a direct current distribution equipment system for sports venues and apply direct current; Perform power load regulation, proactively respond, and solve the problem of supply-demand mismatch.

The fifth is the utilization of waste heat. Based on the industrial waste heat resource conditions around the project area and combined with venue requirements, develop an efficient, low-carbon, safe, and economical waste heat utilization plan to achieve multi-level matching applications at the regional scale.

5. Conclusion

Through investigating into canonical zero carbon sports building projects at home and abroad, the zero carbon technology application and adaptive forms and the difficulties in implementation were summarized. Then based on the characteristics of sports venues, a comparative analysis was conducted to analyze and clarify the focus of energy conservation and emission reduction of sports buildings and the trend of energy conservation demand. On the basis of analyzing the suitability and effectiveness of low-carbon technologies, the strategic concept of "source reduction first, passive design priority, architecture noumenon focus, efficiency improvement as the core, and energy enhancement as the wing" was proposed for zero carbon sports venues in regions with hot summers and cold winters. 20 key zero carbon technologies were constructed, forming a technical system with practical guidance, and providing guidance and support for the zero carbon design and construction of sports venues. With the development of innovative technologies, zero carbon new technologies and materials will promote the progress of zero carbon construction, and the innovation and iteration of the technological system will continue to advance. Sports venues will also usher in the zero carbon era.

References

1. Beijing Daily Press Group. Beijing Municipal Database Records the Xiaokang Project - Major Sports Projects Wukesong Ice Sports Center [DB/OL].(2022.10).https://xiaokang.bjd.com.cn/content/article/104801.html.
2. Xia Changhua, Yang Mengdie, Zhang Zhihong, et al. Energy saving optimization design of zero carbon building envelope structure in regions with hot summers and cold winters- taking a zero carbon exhibition hall as an example. Construction Science and Technology, 2023, (19): 62-66.
3. The Report of Shanghai Public Buildings Energy and Carbon Emission Monitoring 2023. [R]. Shanghai: Shanghai Housing and Urban-Rural Development Management Commission, 2024.
4. The Report of Shenzhen Public Buildings Energy Monitoring 2023 [R]. Shenzhen: Shenzhen Housing and Construction Bureau, 2024.
5. Zhu Fangwei, Li Jun. Energy consumption calculation and analysis of air conditioning system in a sports building in a hot summer and cold winter area [J]. Refrigeration and Air Conditioning, 2021, 21 (12): 70-74.
6. Deng Yangbo, Study on climate adaptability of high rise office buildings in regions with hot summers and cold winters [D]. Huazhong University of Science and Technology, 2004.
7. Research on the acceptable range and design strategies of sports population in national fitness centers in regions with hot summers and cold winters[D]. South China University of Technology, 2022.
8. Sokkel Ellinson, Sokol Johannsson, Karin Chatterner, et al. Integrated utilization of geothermal energy in large sports facilities - Icelandic Method [C]. Proceedings of the Beijing Geothermal International Symposium. 2002:9.

Analysis of Carbon Emission Characteristics Over the Entire Life Cycle of Indoor Substation Structures

Jingyun Wu*, Zheng Huang[a], Hao Liu[b]

Economic Research Institute, State Grid Jiangsu Electric Power Company, Nanjing 210008, China

Abstract. As a key infrastructure in the power industry, substations play an important role in achieving the strategic goal of "dual carbon", and their sustainable development concept radiates to the entire field of structural buildings and ecological buildings. Based on a systematic review and in-depth analysis of carbon emissions across the entire life cycle of substations, this paper, in combination with the characteristics of substation projects, analyzed the boundaries for carbon emission calculations and carbon emission accounting methods. It calculates the carbon emissions over the entire life cycle and compares and analyzes the carbon emissions under different structural forms. The study finds that the carbon emissions during the operation phase dominate, accounting for as much as 54% in the entire life cycle of substation. Furthermore, steel-structured substations reduce carbon emissions by approximately 10% compared to concrete structure substations. Considering energy conservation and emission reduction comprehensively, steel structure substations are superior to concrete structure substations. The research results not only provided strong data support and theoretical basis for the low-carbon design, construction and operation of substation, but also contributed to the energy conservation and emission reduction in the construction industry, civil engineering and hydraulic engineering field.

1 Introduction

In recent years, scholars have made a lot of achievements in the study of key issues related to the carbon emission management of the substation and the power industry throughout the life cycle. These achievements not only promote the green development of the power industry, but also are closely linked to the concepts of ecological architecture, sustainable development and structural construction [1-2]. In terms of life cycle management, Jin et al. [3] used sensitivity analysis to study the life cycle cost of offshore wind power projects, and their research methods and results provided useful references for the field of ecological construction and sustainable development. Zeng et al. [4], aiming at the goal of green and low-carbon construction and operation of substation, proposed to promote green and low-carbon transformation in the aspects of substation building system, auxiliary energy supply and energy-saving equipment, which not only helps to reduce the carbon emissions of traditional substation (which can be reduced by 19.3%), but also promotes the development of structural buildings in a more environmentally friendly and sustainable direction.

Based on the life-cycle cost of electrochemical energy storage, Wang et al. [5] studied the economy of energy storage participating in auxiliary service of power frequency regulation, providing a reference for the construction of frequency modulation energy storage projects and a new idea for the energy management of ecological buildings. Yuan et al. [6] Based on the LCC theory, combined with the actual situation of substation equipment, and based on the life-cycle cost of equipment operation, the decision analysis of equipment updating and upgrading was made, and the economic transformation plan of substation was proposed. In terms of carbon emission management and countermeasures to adapt to the requirements of "dual carbon", scholars' research has more deeply integrated the concept of ecological architecture and sustainable development. He et al. [7] designed the carbon emission calculation framework for the whole life cycle of the substation by using Monte Carlo simulation method, and measured the carbon emissions in the four stages of materialization, construction, operation and demolition of the substation. This framework provided a scientific basis for carbon emission management in the field of ecological building. In view of the high proportion of renewable energy equipment and power electronic equipment in the power system, Ding et al. [8] put forward technical paths to meet the needs of low-carbon development from the aspects of research and development, design, manufacturing and application of new equipment. These technical paths not only contribute to the low-carbon transformation of the power system, but also provide technical support for the green upgrade of structural buildings. Zhu et al. [9] proposed an assessment method for substation carbon emissions based on the green concept, starting from the four stages of substation materialization, transportation, construction and operation and maintenance. Wang et al. [10] identified the factors affecting carbon emissions during the whole life cycle of

* Corresponding author: *wutn06@163.com; [a]1432515418@qq.com; [b]2875893494@qq.com

substation buildings, analyzed the energy saving and emission reduction potential of each stage, and proposed corresponding low-carbon strategies. These strategies not only contributed to the low-carbon development of substation, but also provided beneficial references for energy saving and emission reduction of ecological buildings and structural buildings.

Based on the background of the implementation of the "dual-carbon" strategic goal and the requirements of The Times of low-carbon transformation of substations, this paper analyzes the carbon emissions of substations with different structural forms under the "dual-carbon" goal from the perspective of the whole life cycle of substations, and puts forward suggestions for the low-carbon development of substations in combination with the concept of ecological architecture and sustainable development. These proposals aim to promote the realization of the "dual carbon" goal of the power industry, while promoting the sustainable and collaborative development of substation structural buildings in the direction of ecological buildings.

2 Calculation method for carbon emissions of substations

2.1 Boundary conditions

The bottom-up model [11] based on carbon emission analysis is commonly used for carbon emission estimation. Lifecycle assessment (LCA) is a comprehensive environmental impact assessment technique and methodology system, known as the "cradle-to-grave" approach. The study divided the life cycle into seven stages, namely building materials production, building materials transportation, construction, operation stage and demolition stage, all of which involved resource and energy consumption, waste gas and waste water discharge and solid waste.

2.2 Calculation model

Carbon footprint is a comprehensive analysis method for accounting carbon emission sources and quantities throughout the lifecycle of the study object [12]. The carbon footprint of a substation refers to the accounting of carbon sources and carbon emissions throughout its lifecycle [13]. Emission factor method is used to construct the calculation model. The carbon emission factor data mainly refer to GB/T51366-2019 [14], and the specific calculation formula was shown in Eq. (1):

$$E = \sum Q \times EF \quad (1)$$

where E is the CO2 emissions; Q is the activity level, which quantifies the activities that cause greenhouse gas emissions; and EF is the emission factor, which is the amount of CO2 emitted per unit of activity level.

The calculation model for the full lifecycle of a building includes the material production stage, material transportation stage, construction stage, operation stage, maintenance stage, demolition stage, and waste recovery stage. The calculation formula for this is shown in Eq. (2):

$$C_{lc} = C_{mp} + C_{mt} + C_{co} + C_{op} + C_{de} \quad (2)$$

where Clc is the total carbon emissions over the full lifecycle of the building (kgCO2); Cmp is the carbon emissions during the material production stage (kgCO2); Cmt is the carbon emissions during the material transportation stage (kgCO2); Cco is the carbon emissions during the construction stage (kgCO2); Cop is the carbon emissions during the operation stage (kgCO2); and Cde is the carbon emissions during the demolition stage (kgCO2).

3 Calculation method for carbon emissions of concrete structure substation

3.1 Project overview

The 220kV Chunxi Substation is located in Nanjing City, Jiangsu Province. The substation has two floors above ground and one floor underground, with a total building area of 4209m^2 and a design service life of 50 years. The Chunxi Substation adopts an indoor layout, with main buildings including the power distribution equipment building, fire pump room, and auxiliary rooms. The power distribution equipment building has one floor above ground and is of concrete frame structure, including 110kV GIS room, 220kV GIS room, 10kV power distribution and assembly room, secondary equipment room, battery room, capacitor room, reactor room, etc. The auxiliary rooms are also of concrete frame structure, housing a guard room, fire control room, etc.

3.2 Carbon emission characteristics of concrete structure substation

3.2.1 Carbon emission in the material production and transportation stage

Carbon emissions during these stages were calculated based on the consumption of building materials, with corresponding emission factors taken from GB/T51366-2019 [14]. The consumption of building materials is set according to the substation design plan, material procurement list, etc. Carbon emissions during the transportation stage were obtained by multiplying the quantity of building materials, transportation distance, and carbon emission factor per unit distance. The main building materials used and their carbon emissions were shown in Table 1.

Based on the above consumption and corresponding production carbon emission factors, the total carbon emissions from production and transportation were approximately 8407 tCO_2, with production carbon emissions accounting for 97% of this stage and transportation carbon emissions accounting for 3%.

Table 1. Carbon Emissions from Building Material Production and Transportation.

Materials	Consumption	Unit	Carbon emissions in the Production /tCO_2	Carbon emissions in the transportation /tCO_2
Steel	1640	t	3905	27
Concrete	10389	m^3	3584	168
Cement mortar	321	t	236	6
Iron pieces	24	t	55	0.4
Rubble	785	t	4	13
Window	203	m^2	52	0.02
Bricks	80	m^3	28	3
The door	213	m^2	37	0.01
Sand	877	t	6	14
Water Pipes	22	t	181	0.4
Paint	24	t	85	0.4
Plaster	0.97	t	3	0.01
Asbestos	0.14	t	1	0.002

3.2.2 Carbon emission in the construction stage

The energy consumption method for construction processes is adopted to estimate energy consumption during this stage [15], combined with the substation construction engineering plan [16], to obtain the energy consumption for this stage. The consumption of gasoline, diesel, and electricity was 0.087t, 1.263t, and 4766kW·h, respectively, resulting in carbon emissions of 804tCO_2 for this stage.

3.2.3 Carbon emission in the operation stage

Carbon emissions during the operation stage are calculated based on electricity consumption, which includes the power used by equipment, lighting, ventilation, air conditioning, etc. Using the power of the substation to calculate the electricity consumption, the electricity consumption includes the equipment, lighting, ventilation and air condition in the substation, the average annual electricity consumption for equipment is 250300kW·h, 30000kW·h, 8700kW·h, and 102800kW·h, respectively. With a design service life of 50 years for the substation, the carbon emissions during the operation stage are approximately 12056.92tCO_2.

3.2.4 Carbon emission in the demolition stage

Carbon emissions during the demolition stage include those from demolition construction, waste transportation, and negative carbon from material recycling. Using a method similar to that of the construction stage, carbon emissions were obtained by multiplying the quantity of each material, transportation distance, and transportation factor coefficient. The carbon emissions during the demolition stage were 1085.43tCO_2.

4 Calculation method for carbon emissions of steel structure substation

4.1 Project overview

The 220kV Torch Substation is located in Suzhou City, Jiangsu Province. The substation has two floors above ground and one floor underground, with a total building area of 3230 m^2 and a design service life of 50 years. The substation adopted an indoor layout, with core buildings including the power distribution equipment building, fire pump room, and auxiliary facilities room. The power distribution equipment building has one floor above ground and adopted a steel structure frame as its structural support. It was subdivided into multiple functional areas such as 110kV and 220kV (GIS) rooms, 10kV power distribution equipment assembly room, secondary equipment monitoring room, battery storage room, capacitor room, reactor room, etc. The auxiliary facilities room was also constructed with a steel structure frame, housing key departments such as the security guard room and fire command center.

4.2 Carbon emission characteristics of steel structure substation

4.2.1 Carbon emission in the material production and transportation stage

During the calculation of carbon emissions, classification was based on the consumption of building materials, with carbon emission factors taken from GB/T51366-2019 [14]. The consumption of building materials was derived from the substation design plan, material procurement list, and other documents. For carbon emissions during the transportation stage, they were obtained by multiplying the usage of building materials, transportation distance, and carbon emission factor per unit distance. The main building materials used and their carbon emissions are shown in Table 2.

Table 2. carbon emissions from production and transportation of building materials.

Materials	Consumption	Unit	Carbon emissions in the Production /tCO_2	Carbon emissions in the transportation /tCO_2
Steel	1765	t	4202	29
Concrete	7041	m^3	2429	114
Cement mortar	991	t	728	16
Iron pieces	27	t	63	0.4
Rubble	9836	t	43	159
Window	141	m^2	36	0.02
Bricks	97	m^3	33	4
The door	189	m^2	32	0.01
Sand	4058	t	27	66
Water pipes	3	t	26	0.05
Paint	5	t	19	0.09
Plaster	0.7	t	2	0.01
Asbestos	1.3	t	3	0.02

Based on the above consumption and corresponding production carbon emission factors, the total carbon emissions from production and transportation were approximately 8032tCO_2. Among them, production carbon emissions account for 95% of this stage, while transportation carbon emissions account for 5%.

4.2.2 Carbon emission in the construction stage

The energy consumption method for construction processes was adopted to estimate energy consumption during this stage, combined with the substation construction engineering plan, to obtain the energy consumption for this stage. The consumption of gasoline, diesel, and electricity is 0.055t, 1.284t, and 3746 kW·h, respectively, resulting in carbon emissions of 919tCO_2 for this stage.

4.2.3 Carbon emission in the operation stage

The carbon emissions during the operation stage of the substation are determined by calculating its electricity consumption, which covers the energy consumption of equipment, lighting, ventilation, and air conditioning systems. According to the design plan, the average annual electricity consumption of the substation's equipment is 200200kW·h, the lighting system is 18000kW·h, the ventilation system is 8700kW·h, and the air conditioning system is 85100kW·h. Considering the design service life of the substation to be 50 years, the total carbon emissions during the operation stage can be calculated to be approximately 10086.14tCO_2.

4.2.4 Carbon emission in the demolition stage

Carbon emissions during the demolition stage include those from demolition construction, waste transportation, and negative carbon from material recycling. Using a method similar to that of the construction stage, carbon emissions are obtained by multiplying the quantity of each material, transportation distance, and transportation factor coefficient. The carbon emissions during the demolition stage were 1007.65tCO2.

5 Comparative analysis of carbon emissions between concrete and steel structure substations

After aggregating and analyzing the data, the carbon emissions for each lifecycle phase of the two structures were shown in Table 3. The total carbon emissions for the full lifecycle of the concrete structure were 22353.35t, while the total carbon emissions for the full lifecycle of the steel structure were 20044.79t, representing 10% reduction in carbon emissions compared to the concrete structure. There was a significant difference in carbon emissions between the concrete and steel structure schemes during the operation stage, reaching 1970.78t (16.3%). This was mainly due to the better thermal insulation performance of the steel structure, which reduced the heat transfer between the indoor and outdoor environments, significantly lowering the indoor air conditioning load. Additionally, there is also a relatively large difference in carbon emissions during the material production stage of the two structures. This is because the steel structure uses less concrete, and concrete production generates a large amount of CO_2.

Table 3. Carbon emission calculation results of concrete and steel structure in each stage.

Stage	Concrete structure		Steel structure	
	carbon emissions/t	carbon emissions per unit area/t	carbon emissions/t	carbon emission per unit area/t
Production	8177	1.94	7643	2.37
Transportation	230	0.05	389	0.12
Construction	804	0.19	919	0.28
Run	12056.92	2.86	10086.14	3.12
Demolition	1085.43	0.26	1007.65	0.31
Total	22353.35	5.31	20044.79	6.21

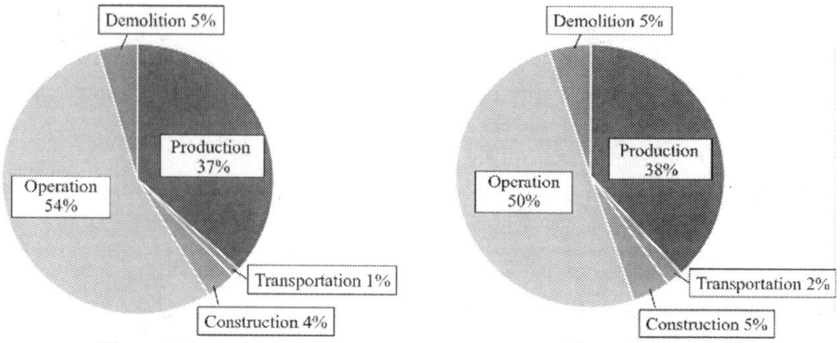

(1) concrete structure (2) steel structure

Fig. 1. Proportion of carbon emission in each stage of substation.

The carbon emission proportions for each stage of the concrete and steel structures were shown in Fig. 1.

It can be observed that among the two structural types, the carbon emissions during the building operation stage account for approximately 52%, the carbon emissions during the building materialization stage (production, transportation, and construction) account for about 43%, and the building demolition stage accounts for 5%. Among them, the operation stage was the most significant contributor to carbon emissions. Unlike civil buildings, the primary energy consumption of substation buildings comes from cabinet equipment and air conditioning for cooling. Additionally, due to the excellent thermal insulation performance of steel structures, their carbon emissions during the operation stage account for a relatively lower proportion. Furthermore, the proportions of carbon emissions during the building material production stage were almost identical for both concrete and steel structures, but the actual carbon emissions for concrete structures were slightly higher than those for steel structures.

6 Conclusion

Based on the carbon emission accounting method and the principle of sustainable development in ecological building structure, the carbon emission of substation building materials in production and transportation, construction, operation and demolition were calculated. The results indicated that the carbon emission in substation operation stage is dominant, followed by the construction stage. Therefore, the focus of carbon reduction technological retrofitting for substations should be on the construction and operation stages. Furthermore, steel structures reduce overall carbon emissions by approximately 10% compared to concrete structures. From the perspective of energy conservation and emission reduction, steel structure substations were superior to concrete structure substations. Based on the above research and findings, the following suggestions were proposed:

(1) Reasonably select the structural form of substations to reduce carbon emissions throughout their life cycle.

(2) Research and apply low-energy building technologies for substations, reducing the energy consumption of substations in terms of building material selection, envelope structure design, temperature control, and other aspects.

(3) Explore and develop integrated energy utilization technologies for substations, utilizing photovoltaic and geothermal energy for auxiliary power supply and improving the utilization rate of renewable energy.

(4) Establish economic incentive policies for carbon reduction to promote the clean and efficient development of the energy system.

References

1. Liu W B, Wang Y P, Liu Z M, et al. Discussion on the Reform of Power Grid System to Adapt to the New Power System[J]. Price: Theory & Practice, 2022, (12): 13-18+81.
2. Li Y Q. Accelerate the construction of a new energy system[J]. Chinese oil companies, 2023, (03): 39-40.
3. Jin C Y. Cost composition of whole life cycle and sensitivity analysis of offshore wind power project[J]. Solar Energy, 2022, (03): 10-16.
4. Zeng W H. The design of green and low-carbon substations is key[J]. Management of China Electric Power Enterprises, 2022, (30): 25-27.
5. Wang K. Research on the economics of frequency modulation based on the whole life cycle cost of energy storage[J]. Straits Science, 2021, (10): 28-31+35.
6. Yuan J W, Wang Y, Zhang Z, et al. Research on Transformation Equipment Technical Transformation Project Application Based on Life Cycle Cost[J]. Northeast electric power technology, 2021, 42(11):1-4+23.
7. He Y C, Hu C, Zhang H J, et al. Life-cycle carbon emission evaluation based on Monte Carlo simulation substation[C] MCC Construction Research Institute Co., Ltd, 2022: 8.
8. Ding Y F, Han W D, Wang X Y, et al. The development of new power system equipment and the clean and low-carbon transformation of energy[J]. Management of China Electric Power Enterprises, 2022, (28): 46-49.

9. Zhu X H. A carbon emission evaluation method for substations based on the concept of green construction under the "dual carbon" goal[J]. Shanghai Light Industry, 2023, (01): 93-95.
10. Wang M L, Chen M Y, Kang H, et al. Research on Carbon Emission Calculation Method of Substation Building and Low-carbon Case Practice[J]. Smart Buildings & Smart Cities, 2023, (04): 98-100.
11. Zhang Q F, Fang K, Xu M, et al. Research progress on carbon footprint based on input-output analysis[J]. Journal of Natural Resources, 2018, 33(04): 696-708.
12. Hu Y S, Zhuang D, Zhu K, et al. A Comparative Study on Lifecycle Carbon Emissions of Concrete Structure and Steel Structure Substations[J]. Building Science, 2022, 38(12): 275-282.
13. Sun Y L. Carbon Emission Accounting and Case Analysis for Substations[J]. Journal of Science and Technology, 2023, 25(16): 223-227.
14. GB/T51366-2019, Building Carbon Emission Calculation Standards[S]. Bei Jing: China Architecture & Building Press, 2019.
15. Fan Y P, Gao Y, Li Y. A Systematic Analysis of Carbon Neutrality and Enterprise Life Cycle Costs[J]. Journal of systems Science, 2023, 31(04): 79-84.
16. State Grid Jiangsu Electric Power Co., Ltd. Construction Department. State Grid Jiangsu Electric Power Co., Ltd. Power Transmission and Transformation Engineering Standardization Design Atlas Collection of 110kV Substation Modular Construction Yuan 110-A2-6 Scheme Product Construction Drawings and Typical Practices[M]. China University of Mining and Technology Press, 2019.

Photovoltaic Potential of Elevated Metro Stations: A Case Study of Shanghai, China

Hui Mao, Lina Zhang* and Gaijing Zhang

Shanghai research institute of building sciences Co., LTD, Shanghai, China

Abstract. As an essential part of urban development, the urban rail transit sector has seen rapid growth in recent years. In alignment with energy-saving and carbon reduction goals, improving energy efficiency in this sector is critical. Elevated metro stations, situated above urban roads with minimal obstructions, present an ideal opportunity for photovoltaic integration. This study investigates the PV potential of Shanghai's elevated stations through field investigation, numerical simulations and case studies. Taking Shanghai Rail Transit Line 17 as an case study, the photovoltaic application potential of the roof and facade of the elevated station is estimated, the results indicate an annual PV generation of 3.6 million kWh, potentially meeting over 30% of the stations' energy demand.

1 Introduction

In China, the transport sector plays a significant role in energy consumption, with a growing share due to expanding infrastructure and increased passenger and freight demand. Given the growing energy demands, energy efficiency and decarbonization in the transport sector are critical for reducing greenhouse gas emissions and meeting global climate goals. As an important means of energy conservation and emission reduction, photovoltaic (PV) technology has broad application prospects in urban rail transit. In 2023, the Shanghai government issued an implementation Plan, which mandates that new transportation facility buildings in Shanghai must install PV systems covering no less than 50% of rooftop areas [1]. Elevated stations, as key components of urban rail transit, offer ideal conditions for PV applications due to their large roof and façade areas and minimal surrounding obstructions [2].

Currently, the application of PV systems in urban rail transit remains in its early stages of development in China. This paper reviews global PV applications in urban rail transit, and explores the architectural and structural characteristics of elevated stations that influence PV system design and performance, including station platform layouts (side or island platforms), roof configurations (sloped, flat, or arched), and material choices (metal, membrane structures, etc.). To quickly assess the PV potential of elevated stations, simulations based on an architectural theoretical model were performed. Taking Shanghai Metro Line 17 as a case study, this paper provid an evaluation of the potential for PV integration in elevated metro stations.

2 Case studies of PV integration in urban rail transit

Globally, many cities have incorporated PV technology into urban rail transit infrastructure. In New York City, for instance, the Stillwell Avenue Station installed 7,060 m² of thin-film PV modules on its train shed roof, generating approximately 250,000 kWh of electricity annually [3]. Similarly, the New Delhi Station in India has implemented Building Attached Photovoltaic (BAPV) with an installed capacity of 500 kW, which meets the station's full electricity requirements [4]. In the United Kingdom, London's Blackfriars Station has deployed 6,000 m² of PV panels on its roof, providing 1,100 kW of capacity, accounting for roughly 50% of the station's electricity consumption [5]. Rotterdam Central Station in the Netherlands utilizes building-integrated photovoltaic (BIPV) technology, integrating 10,000 m² of PV panels into the station's architecture to generate power for operational needs [6]. Likewise, the high-speed rail infrastructure in Antwerp, Belgium features 50,000 m² of PV panels installed atop a railway tunnel, supplying energy for critical infrastructure, including signaling, lighting, and station heating [7].

In China, PV technology has also been successfully adopted in urban rail transit systems. The Zhangguozhuang Station on Beijing's Metro Line 14, for example, is an early pioneer in this domain. By installing polycrystalline silicon PV modules on the roof of the station, the system provides up to one-third of the station's daily energy needs, powering essential services such as lighting, air conditioning, and elevators [8]. Similarly, elevated stations on Shenzhen Metro Line 6 employ BIPV systems using monocrystalline silicon PV modules, supplying approximately 30% of the station's total energy consumption. During peak solar generation periods, the

* Corresponding author: zhanglina1@sribs.com

system can meet over 80% of the station's electricity demand [9]. In Shanghai, the deployment of distributed PV systems in metro vehicle depots has expanded since 2014. Through a phased implementation process, multiple depots have been outfitted with PV arrays, reaching a cumulative installed capacity of 12,400 kW by 2021 [10]. This initiative has enhanced energy efficiency and significantly reduced carbon emissions. Other examples of PV integration in Chinese rail systems include Jinan Metro Line 1 [11], Qingdao Metro Line 13 [12], and Qingdao Railway Station [13] etc,. Table 1 summarizes key case studies of PV applications in urban rail transit, both internationally and domestically. It is found that while PV installations in Shanghai's rail sector are largely concentrated in vehicle depots, the use of photovoltaics in elevated stations remains relatively limited.

Table 1. PV utilization cases in the domestic and international urban rail transit.

Location	Station/Line	Details
New York, USA [3]	Stillwell Avenue Subway Station	Operational since 2005, 7060 m² installation area. Equipped with BIPV using thin-film modules. Provides a portion of the station's power.
Gandhinagar, India [4]	Gandhinagar Railway Station	Operational since 2014, 500 kWp capacity BAPV. Provides part of the station's power demand.
Blackfriars, UK [5]	Blackfriars Station	Operational since 2012, 6000 m² installation area and 1100 kWp capacity. BAPV installation. Capable of generating around 90 million kWh annually, meeting nearly half of the station's power needs.
Utrecht, Netherlands [6]	Utrecht Central Station	Operational since 2014. BIPV system installed. Contributes to the station's power requirements while promoting sustainability.
Antwerp, Belgium [7]	Antwerp Central Railway	Operational since 2011, 50,000 m² of BAPV. The system provides energy for the station and nearby rail infrastructure.
Beijing, China [8]	Line 14	Operational since 2013, 2500 m² installation area and 60 kWp capacity. BAPV with multi-crystalline silicon modules. Offsets one-third of the station's operational power consumption.
Shenzhen, China [9]	Line 6	Operational since 2018, 1500 m² installation area and 184 kWp capacity. BIPV installation to meet 30% of the station's electricity demand.
Shanghai, China [10]	Various metro vehicle depots	Operational between 2019-2021, 12400 kWp BAPV systems installed across multiple railway stations. Provides green energy, reducing emissions and operational costs for the railways.

3 Architectural and structural characteristics of Shanghai elevated stations

3.1 Platform configurations in typical elevated stations in Shanghai

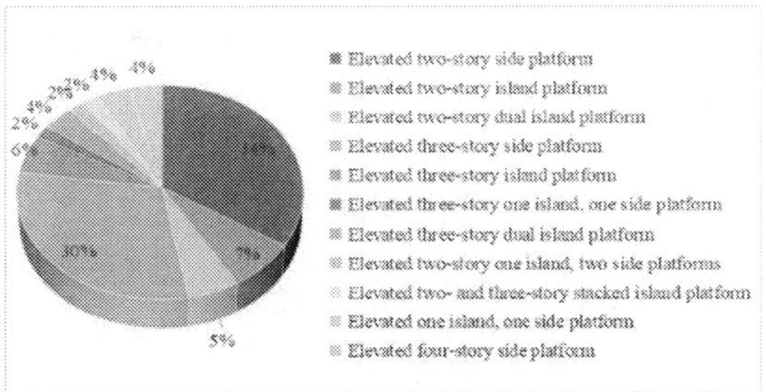

Fig. 1. Platform configurations of elevated stations in shanghai metro system (specific lines).

As of late 2023, Shanghai's metro system encompassed 20 lines, including a magnetic levitation line, with a total of 508 stations. Among these, 111 are elevated stations, representing approximately 22% of the total. There are two predominant platform configurations: side-platform and island-platform designs. In side-platform stations, tracks run through the center, flanked by platforms on either side, whereas island-platform stations have a single central platform bordered by tracks on both sides. The platform configuration plays a crucial role in determining how PV systems can be integrated into station infrastructure. A survey of over 50 elevated stations across multiple metro lines in Shanghai was conducted. As illustrated in Fig. 1, Shanghai's elevated stations predominantly utilize two-story and three-story configurations with side platforms. Compared to single-level elevated stations, these multi-level designs offer larger vertical façade areas, which provide more extensive surfaces for potential PV installations.

3.2 Roof designs in typical elevated stations in Shanghai

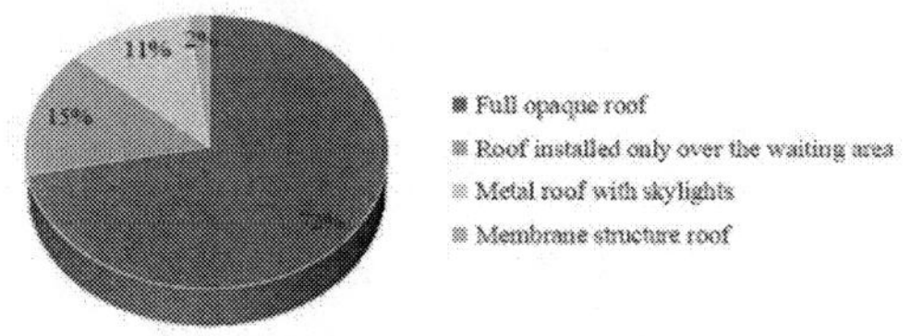

Fig. 2. Roof configurations of elevated stations in shanghai metro system (specific lines).

To estimate the PV potential of elevated stations, a survey was conducted to assess the roof designs of elevated stations across several Shanghai metro lines. The results, shown in Fig. 2, indicate that the roof structures of Shanghai's elevated stations generally fall into four categories. Approximately 72% of the stations feature a full opaque roof, while 15% have roofs that only cover the platform areas, leaving the tracks exposed. About 11% of the stations employ metal roofs with integrated skylights, while 2% utilize membrane structures. These different roof designs significantly influence the feasibility and optimization of PV system integration, with full roof offering the greatest potential for maximizing PV generation.

4 PV system design in Elevated Metro Stations

The unique structural and environmental characteristics of elevated metro stations present both opportunities and challenges for PV system integration. Positioned above ground and exposed to higher levels of vibration, wind loads, and potential urban pollutants, these stations demand PV systems designed to endure such conditions. Selecting durable materials is essential for sustaining long-term performance in an elevated environment. Additionally, reinforcing the installation structure to manage vibrations and movement due to passing trains is crucial. Given that urban pollution and debris accumulation can lower PV efficiency, establishing safe access points for regular cleaning and inspection is required. Automated or remote monitoring systems may also be suggested to quickly detect and address performance issues.

Measures can be implemented to enhance the efficiency of PV systems, such as optimizing the tilt angle, adopting high-efficiency PV modules, designing an effective layout and integrating transparent PV for skylights. PV application recommendations for different roof designs are summarized as follows:

- Full opaque roof: This type of station has a large roof area, making it well-suited for PV installations. Both crystalline silicon and thin-film solar panels can be used. Panels can be tilted to achieve maximum sunlight exposure. And adding ventilation gaps between modules can prevent overheating.
- Roof installed only over waiting area, track exposed: This type of station features a roof only above the waiting area, while the tracks remain exposed. PV modules can be installed on the roof area, with both crystalline silicon and thin-film solar panels as suitable options.
- Metal roof with skylights: This station type uses metal as the primary roof material, with skylights to enhance natural lighting and ventilation. Crystalline silicon or thin-film solar panels can be installed on the metal surface, and transparent thin-film panels can be integrated into skylights.
- Membrane structure roof: The membrane roof is made of lightweight, translucent material, offering good aesthetics and natural lighting. Thin-film PV modules are ideal due to their ability to conform to curved surfaces. Strategic placement of the modules should minimize shading while retaining the membrane's translucency to preserve natural light within the station.

5 PV power generation estimation of elevated stations

5.1 PV potential simulation based on the architectural theoretical model

Geographical factors, such as latitude and solar resources, significantly impact the feasibility of deploying PV systems. As shown in Table 2, China's solar energy resources are classified into four levels based on geographic location. Regions in Classes I to III are considered favorable for solar energy utilization while Class IV regions have relatively scarce solar resources. Shanghai, classified as a Class III region, receives an average annual solar radiation of about 4,461 MJ/(m²·a), making it viable for PV installations.

Table 2. Abundance level of solar energy resources in China.

Level	Resource Zone	Annual Total Radiation (MJ/m²)	Annual Total Radiation (kWh/m²)	Average Daily Radiation (kWh/m²)
Most Abundant	I	≥ 6300	≥ 1750	≥ 4.8
Very Abundant	II	5040 - 6300	1400 - 1750	3.8 - 4.8
Abundant	III	3780 - 5040	1050 - 1400	2.9 - 3.8
Average	IV	≤ 3780	≤ 1050	≤ 2.9

To assess the influence of orientation and tilt on PV efficiency, simulations were performed based on an architectural theoretical model using modeling software Rhino and Grasshopper. The meteorological data for Shanghai was used in the simulation. The simulations consider 17 different orientations and tilt angles. The results, presented in Fig. 3 and Table 3, indicate that a south-facing orientation with a 25° tilt (S1) yields the highest solar radiation, reaching 1287.67 kWh/m², making it the optimal orientation for PV installations. Other favourable orientations include southwest-facing with a 25° tilt (SW1), horizontal (TOP), west-facing with a 25° tilt (W1), and southeast-facing with a 25° tilt (SE1), which exhibit less than a 10% reduction in solar radiation compared to the optimal S1 orientation. The least favourable orientation is north-facing with a 90° tilt (N2), where the solar radiation is only 20% of that received by the optimal S1 orientation.

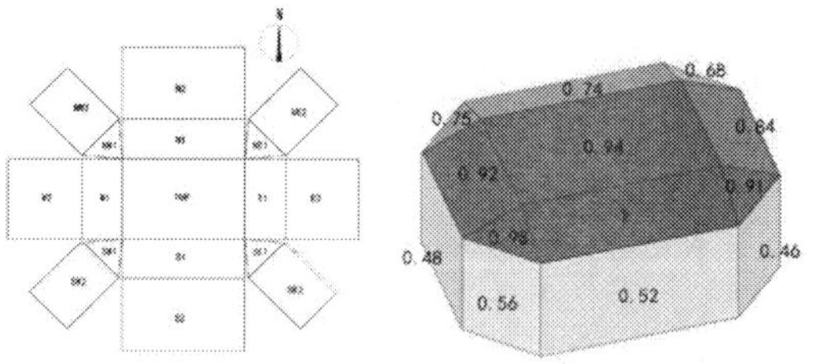

Fig. 3. Relative PV efficiency for various orientations in an architectural theoretical model.

Table 3. Solar radiation levels for various orientations in the architectural theoretical model.

ID	Orientation	Tilt Angle	Annual Solar Radiation (kWh/m²)	Relative PV Efficiency Compared to Optimal Orientation S1
TOP	Horizontal	0	1206.49	93.69%
S1	South	25°	1287.67	100.00%
S2	South	90°	665.68	51.70%
SE1	Southeast	25°	1168.22	90.72%
SE2	Southeast	90°	595.89	46.28%
SW1	Southwest	25°	1264.65	98.21%
SW2	Southwest	90°	715.33	55.55%
E1	East	25°	1079.71	83.85%
E2	East	90°	476.54	37.01%
W1	West	25°	1183.92	91.94%
W2	West	90°	613.44	47.64%
NE1	Northeast	25°	876.49	68.07%
NE2	Northeast	90°	329.34	25.58%
NW1	Northwest	25°	965.22	74.96%
NW2	Northwest	90°	403.52	31.34%
N1	North	25°	952.2	73.95%
N2	North	90°	256.72	19.94%

5.2 Estimation of PV power generation

Using the annual solar radiation results of the theoretical model in the previous section, the PV power generation for the roofs and façades of the elevated stations can be calculated using Equation (1):

$$E_p = H_A \times S \times K_1 \times K_2 \quad (1)$$

where:

- H_A is the total solar irradiation on the inclined surface (kWh/m²), based on the results for each orientation from Table 3.
- S is the total area of PV modules (m²), adjusted for usable roof space and considering only transparent sections of the façades.
- K_1 is the module conversion efficiency (%).
- K_2 is the overall system efficiency (%).

To validate the accuracy of the proposed PV generation estimation method, measured data from a PV project on a villa roof in Shanghai was used [14]. Established in 2015, this project consists of 24 polycrystalline silicon PV panels arranged on an east-west sloped roof with a tilt angle of 30°, covering an area of approximately 40 m². The efficiency of the PV module is around 17%. From August 2015 to July 2016, the project recorded an annual generation of 6,028.5 kWh. Using Equation (1), the estimated annual generation was calculated to be 5,772 kWh, yielding a 4.4% difference between recorded and estimated values, which is within a reasonable range. Therefore, the estimation method based on the theoretical model proposed in this paper has certain accuracy and reliability.

5.3 PV potential of elevated stations in Shanghai Metro Line 17

This study adopts Shanghai Metro Line 17 as a case study to quantitatively evaluate its PV power generation potential of elevated stations. The line 17 comprises 14 stations, including 7 underground stations and 7 elevated stations, along with the under-construction Xicen elevated Station. Station platform layouts and roof configurations of 7 elevated stations in Shanghai Metro Line 17 were investigated. To calculate the PV potential of Shanghai Metro Line 17, the following assumptions were made. Firstly, for opaque flat roof sections, polycrystalline silicon modules with a 22% efficiency were used, and for arched roof sections and transparent parts, thin-film modules with a 15% efficiency were applied. Secondly, Façade PV installations were considered only for transparent sections, using thin-film modules with a 15% efficiency. Thirdly, the K_2 was assumed to be 75%, accounting for various system losses.

Table 4. PV power generation estimation for elevated stations in shanghai metro line 17.

Station Name	Platform Type	Roof Type	Estimated PV Generation (10,000 kWh)			
			Rooftop	South Facade	North Facade	Total
Station A	Elevated three-story island platform	Sloped full roof	80	6	2	88
Station B	Elevated two-story island platform	Flat full roof	46	4	2	52
Station C	Elevated three-story island platform	Sloped full roof	34	1	0.5	35.5
Station D	Elevated three-story side platform	Arched metal roof with skylights	46	0	0	46
Station E	Elevated three-story side platform	Arched metal roof with skylights	46	0	0	46
Station F	Elevated two-story island platform	Flat full roof	46	2	1	49
Station G	Elevated two-story island platform	Flat full roof	42	1	0.5	43.5
Total	-	-	340	14	6	360

The estimated PV power generation for the seven elevated stations along Line 17 is summarized in Table 4. The total projected PV output is approximately 3.6 million kWh per year, with roof installations accounting for the majority (3.4 million kWh). The south-facing façades are estimated to contribute 140,000 kWh, while the north-facing façades add approximately 60,000 kWh. There are significant differences in the projected PV generation across the stations due to variations in roof and façade designs. Station A, a three-story island platform station with a large, continuous roof and extensive usable façade areas, is estimated to have the highest PV generation capacity. In contrast, Stations D and E, which feature arched roofs that significantly shade their façades, were not considered suitable for façade PV installations.

The analysis shows that the elevated stations on Line 17 have the potential to generate approximately 3.6 million kWh of electricity annually, which could meet over 30% of the total energy demand for these elevated stations. This generation capacity would result in a reduction of carbon emissions by approximately 1.5 tons annually. For the cost-benefit analysis, given the structural vibration risks associated with elevated stations, additional reinforcement is required when installing the PV system, leading to a slightly higher construction cost compared to other scenarios. Assuming a construction cost of ￥4/W, the installed PV capacity is about 3.6 MWp, the total installation cost amounts to around ￥14.4 million. Annual maintenance costs are projected at around ￥0.108 million. The electricity rate for the metro system in Shanghai is around ￥0.8/kWh. The payback period is approximately five to six years, demonstrating strong economic viability and substantial financial returns.

6 Conclusion

This paper conducted a comprehensive analysis of PV applications in elevated urban transit stations, with a focus on station platform layouts, roof types and their PV potential. A quick estimation method for PV application potential based on the simulation of building parametric design platform is proposed. The accuracy and applicability of this method have been verified. The PV potential of elevated stations in Shanghai Metro Line 17 is estimated to be 3.6 million kWh annually, meeting over 30% of the energy needs of these stations. The payback period is approximately five to six years. The PV power generation of the elevated station mainly benefits from its large roof area, which provides ample space for solar installations and therefore offers great potential for PV integration. With proper planning and design, elevated stations can become key distributed PV generation sites, offering both economic and environmental benefits. The research provides critical insights into the underexplored area of PV integration in urban transit infrastructure, offering theoretical support for the application of PV technology in new and retrofitted elevated station projects, and contributing to sustainable development in urban transit systems.

Acknowledgments

This article is one of the interim results of the Shanghai 2022 'Science and Technology Innovation Action Plan' science and technology support carbon peak and carbon neutrality special project, *Research and Demonstration of Key Technologies for Energy Saving and Carbon Reduction in Rail Transit Stations for Zero-Carbon Urban Rail (22dz1208100).*

References

1. Shanghai Municipal Transportation Commission, Shanghai Municipal Development and Reform Commission. (2023) Implementation Plan for Photovoltaic Promotion and Application in Shanghai's Transportation Sector. https://www.shanghai.gov.cn/gwk/search/content/1158498252d74a60a178254bc594a432
2. Dong, J., Sun, W. (2017) Application and Prospects of Solar Energy in Rail Transit Buildings. J. Construction Science and Technology, 14: 78–79.
3. Renewable energy world. (2005) Solar BIPV pulls into Brooklyn Train Station. https://www.renewableenergyworld.com/om/monitoring/solar-bipv-pulls-into-brooklyn-train-station-33583/#gref
4. Manju, S., Sagar, N. (2017) Progressing towards the development of sustainable energy: A critical review on the current status, applications, developmental barriers and prospects of solar photovoltaic systems in India. Renewable and Sustainable Energy Reviews, 70: 298–313.
5. Li, Z. (2012) Solar Bridge across the Thames in London. Traffic & Transportation, 28,01: 30.
6. Foyle, R. (2023) Switching tracks: Railway stations as sites for renewable energy generation in the Netherlands. Netherland: University of Groningen.
7. Amar T. (2016) Europe's first solar powered train tunnel goes live on Belgian high-speed line. https://www.engadget.com/2011-06-07-europes-first-solar-powered-train-tunnel-goes-live-on-belgian-h.html
8. Beijing Daily. (2013) China's First Subway PV Station to be Grid-Connected Next Month in Beijing. https://www.gov.cn/gzdt/2013-06/29/content_2436659.htm
9. Zhou, C. (2014) Design of Solar Photovoltaic Power Generation System for Metro Elevated Stations. J. Urban Rapid Rail Transit, 27(06): 104–108.
10. Li, J. (2022) Three Metro PV Projects in Shanghai Completed! Expected Annual Power Generation of about 11.7 Million kWh. https://baijiahao.baidu.com/s?id=1724991583123287823
11. Sun, E. (2023) Application of Photovoltaic Power Generation System in Elevated Stations of Jinan Metro Line 1. J. Solar Energy, (02): 13–19.
12. Qingdao West Coast News. (2022) Green Power! Metro Photovoltaics Jointly Draw the Dual Carbon Blueprint. https://www.xihaian.gov.cn/ywdt/zwyw/202204/t20220412_5474841.shtml
13. Luo, D., Yin, P. (2010) Design and Application of BIPV Grid-Connected Photovoltaic Power Generation System in Qingdao Railway Station. J. Advanced Materials Industry, (02): 26–33.
14. Yatan Energy. (2017) Real data tells you how much electricity a 6kW home photovoltaic power station can generate. https://www.sohu.com/a/169074513_820840

Current Situation, Trend and Peak-Reaching Path of Low-Carbon Development of Water Transportation in Zhejiang Province

Meng Lu[1], Yanhua Wei[1], Yilei Lu[2,*], Yiying Zhou[2]

[1]Jiaxing Port and Shipping Management Service Center, Jiaxing, Zhejiang, China
[2]China Waterborne Transport Research Institute, Ministry of Transport, Beijing, China

Abstract. This paper analyzes the development status and trends of the water transport industry in Zhejiang Province in detail, including port infrastructure and operation, ship transport, waterway infrastructure and operation and other aspects. Meanwhile, it summarizes the main achievements of low-carbon development in the water transport industry and discusses the situational requirements. The water transport industry in Zhejiang Province has made remarkable progress in infrastructure construction, optimization of transportation organization, and green and low-carbon transformation. However, it still faces challenges from multiple aspects such as climate change, national policies, and green transportation construction. Finally, the paper proposes the implementation path for carbon peaking in the water transport industry of Zhejiang Province, covering aspects such as upgrading transportation equipment, optimizing organizational coordination, providing infrastructure support, enhancing intelligent innovation, and activating the roles of the government and the market.

1 Introduction

Low-carbon development is a development model that takes reducing greenhouse gas emissions as one of the main goals and realizes the coordination of economic growth and environmental protection through a holistic manner including economic, technological, and policy aspects[1]. Nowadays, greenhouse gas emission reduction targets are forcing the global economy to transform in the direction of low-carbon development. "Carbon peaking", that is, carbon dioxide emissions reach a peak[2], has become a recent goal that attracts much attention in low-carbon development. As an essential part of the comprehensive transportation system, the water transport industry plays a vital role in supporting economic and social development[1]. With the increasing global concern on climate change, promoting low-carbon development in the water transport sector has become a key initiative for sustainable development[2]. As a major province for water transport in China, Zhejiang's progress in low-carbon development within this industry serves as an important example. For water transport industry in Zhejiang province, the existing research at present focuses on specific method details such as quantification of carbon emission[3-4], efficiency[5] and influencing factors[6], lacking research on macroscopic trend analysis and path formulation. This paper provides a detailed analysis of the current status and trends in Zhejiang's water transport sector, summarizes the main achievements in low-carbon development, examines the industry's requirements, and proposes pathways to reach peak carbon emissions in water transport.

2 Current Status and Trends of the Water Transport Industry in Zhejiang Province

2.1 Port Infrastructure and Operations

(1) Current Status

Zhejiang boasts abundant coastal resources, with a coastline totaling 6,646 kilometers, accounting for 21% of China's coastline (the longest among all provinces in China), which includes 1,800 kilometers of mainland coastline, 4,800 kilometers of island coastlines and 506 kilometers of planned deep-water port shoreline. Key coastal ports include Jiaxing Port, Ningbo-Zhoushan Port, Taizhou Port, and Wenzhou Port, along with nine major inland ports such as Hangzhou Port and Huzhou Port. The number of production-use berths of coastal ports in Zhejiang accounts for 18.9% of the national total, while those of inland ports account for 25.3%.

The region's unique resources and well-developed infrastructure contribute to its impressive operational results. In 2022, Zhejiang's ports achieved a total cargo throughput of 1.92 billion tons, with coastal ports handling 1.54 billion tons (ranking third nationwide) and inland ports handling 380 million tons (fourth nationwide)[7]. Ningbo-Zhoushan Port, the world's largest comprehensive seaport by cargo throughput, processed 33.35 million TEUs in 2022, ranking second among mainland ports and third globally. Fig. 1 provides an overview of berths and throughput for coastal and inland ports across various regions, showing that operational activities of coastal ports are concentrated in Ningbo and Zhoushan, while those of

* Corresponding author: luyilei@wti.ac.cn

inland ports are centered in Jiaxing, Huzhou, and Hangzhou.

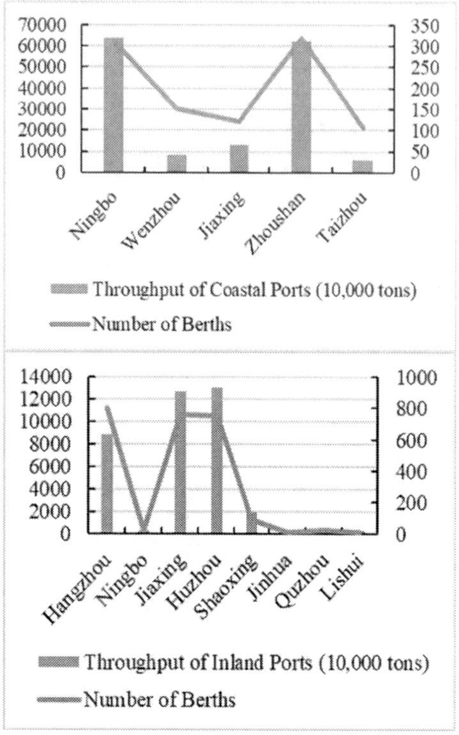

Fig. 1. Berth amounts and Throughput Statistics of Coastal and Inland Port.

(2) Development Trends

In the ten years from 2012 to 2022, the number of berths at ports in Zhejiang Province showed an overall trend of decreasing. However, port throughput achieved steady growth (inland port throughput declined in 2022 due to the impact of the COVID-19), reflecting the continuous optimization of coastal port management and operation in Zhejiang Province, as shown in Fig. 2. By reducing non-productive or low-efficiency berths, the overall operational efficiency of ports in Zhejiang Province has been improved.

In the future, Zhejiang will continue optimizing its integrated port layout. Firstly, following the "One Body, Two Wings, Multi-Link" approach, Zhejiang aims to advance Ningbo-Zhoushan Port by developing ten-million-TEU container berth clusters and hundred-million-ton bulk cargo berth clusters, strengthening its position as an international hub and promoting efficient use of port resources. This includes projects such as constructing container terminals in the Meishan Port area and creating crude oil storage bases in areas like Beilun. Secondly, Zhejiang will foster coordinated development of ports along its northern and southern wings, enhancing their infrastructure to support regional growth. This includes transforming Jiaxing Port into a river-sea intermodal hub in the Yangtze River Delta and boosting its container throughput capacity, accelerating the construction of port facilities in the Wenzhou and Taizhou regions to establish feeder container ports and a range of multipurpose terminals, thereby providing solid support for local economic and social development.

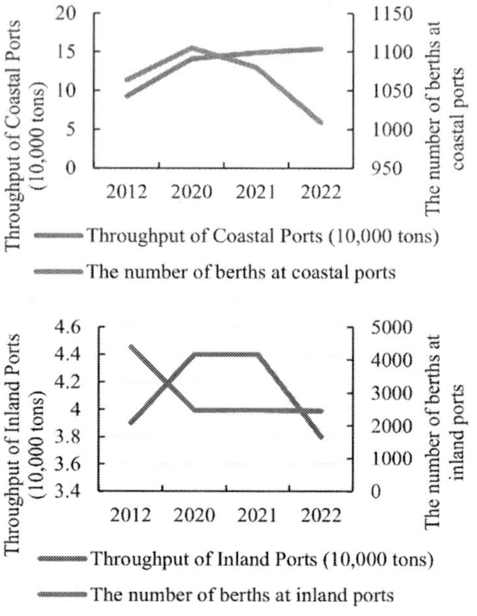

Fig. 2. Trends of Berth amounts and Throughput of Coastal and Inland Port.

Meanwhile, Zhejiang will advance large-scale, consolidated development of its inland ports, including accelerated layout and construction of large-scale terminals along inland waterways to achieve full coverage of public operation areas across inland ports. The province will strengthen the classification and integration of port resources to improve shoreline utilization efficiency. It will also develop large-scale, efficient port operation zones, enhance the container and bulk cargo transport systems, strengthen the comprehensive hub functions of ports, and promote the development of multimodal transport. Additionally, smart port construction will progress, with the establishment of automated coastal terminals and selective deployment of automated inland terminals, alongside scaled applications of "5G-enabled smart terminal" solutions, such as remote equipment control systems.

2.2 Status of Shipping Transport

(1) Current Status

As of 2022, Zhejiang had 12,510 transport vessels, with a combined net tonnage of 32.98 million tons, representing 11.1% of the national total. Of these, there were 8,705 inland cargo vessels, with a net tonnage of 5.56 million tons, while oil and specialized container vessels represented a smaller proportion[7]. For maritime transport, there were 2,432 cargo vessels with a combined net tonnage of 26.56 million tons. Passenger vessels primarily operated in Ningbo, Wenzhou, Zhoushan, and Taizhou, while the number of ocean-going vessels remained small and consisted mainly of bulk carriers.

In 2022, Zhejiang's annual passenger volume reached 24 million people, with a passenger turnover of 350 million passenger-kilometers; inland vessels accounted for 47 million passenger-kilometers, and coastal vessels for 306 million. Total cargo volume for the year was 1.102 billion tons, with a cargo turnover of 1,060.79 billion ton-kilometers. Inland vessels accounted for 39.27 billion ton-kilometers, and coastal and ocean-going vessels accounted for 1,021.52 billion ton-kilometers. Fig. 3 show the proportions of cargo and passenger volumes by city, with cargo transport concentrated in Ningbo, Zhoushan, and Taizhou, and passenger traffic focused in Zhoushan.

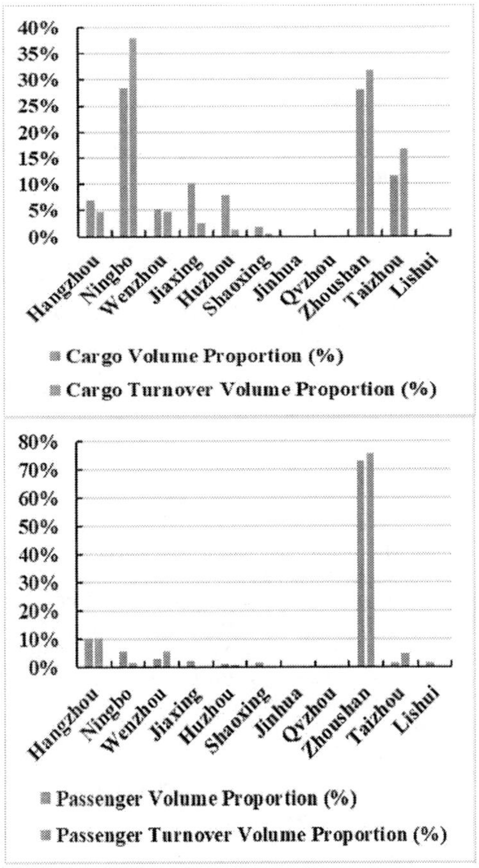

Fig. 3. Proportion of Cargo and Passenger Volume Transported by Vessels by City.

(2) Development Trends

From 2012 to 2022, the number of various types of transport ships in Zhejiang Province generally showed a downward trend amid fluctuations (the number of ocean-going transport ships has rebounded in recent years), while the cargo volume and turnover continued to grow, as shown in Fig. 4. This is because the fleet is developing towards standardization and large-scale, and the operational efficiency has also increased accordingly.

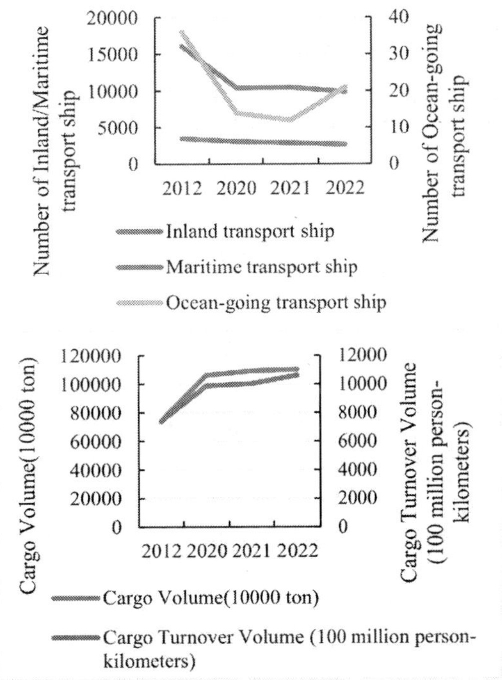

Fig. 4. Trends of Ship numbers, Cargo Volume and Turnover.

In the future, bulk cargo is expected to shift increasingly to waterway transport, with total waterway freight potentially reaching 1.4 billion tons. In response to this development trend, Zhejiang will focus on high-quality development to upgrade the modern shipping industry. Firstly, this includes updating transport equipment, encouraging early retirement of outdated vessels, and developing 500-1000-ton standard inland vessels, with a shift in inland container ships to accommodate 96 and 120 TEUs. The aim is to promote specialized ships for direct river-sea and river-river container transport. Secondly, it involves professionalizing and scaling up the shipping industry, encouraging mergers and restructuring among shipping companies, fostering large-scale enterprises, and supporting small and medium-sized shipping firms to diversify into specialized, high-value segments like LNG and automobile roll-on/roll-off transportation. It also includes developing inland container transport, actively promoting the conversion of suitable bulk cargo to containerized transport and creating high-quality inland container routes. Thirdly, it is important to upgrade and enhance traditional shipping services by promoting the intelligent development of port cargo handling, strengthening the construction of crew service systems, improving the level of international ship repair, and supporting the development of ship supply services. Fourthly, innovation in high-end shipping services should be pursued by expanding shipping financial services, cultivating port and shipping consulting research institutions, and establishing education and training bases for shipping talent along with specialized shipping service clusters.

2.3 Waterway Infrastructure and Operations

(1) Current Status

The inland waterway network in Zhejiang Province features a "Northern Network, Southern Line" layout. The northern region of Zhejiang has a well-established inland waterway network, while Hangzhou-Jiaxing-Huzhou area takes the lead, handling nearly all transportation of construction materials, non-metallic ores, coal and fertilizers. Southern Zhejiang has only three main waterways with limited navigability. In 2022, Zhejiang's navigable inland waterway length reached 9,770 kilometers, representing 7.6% of the national total, with graded waterways of 5,041 kilometers. However, only 481 kilometers are above Level III, accounting for 4.9% of the province's total length and 3.3% of the nation's total length above Level III[7]. Currently, 16 ship locks operate within the province, primarily serving cargo like sand, gravel, coal, steel, and construction materials.

(2) Development Trends

From 2012 to 2022, the navigable mileage of inland waterways in Zhejiang Province maintained stable growth. Among them, the expansion and improvement of the Class III waterway network is the most remarkable, as shown in Fig. 5.

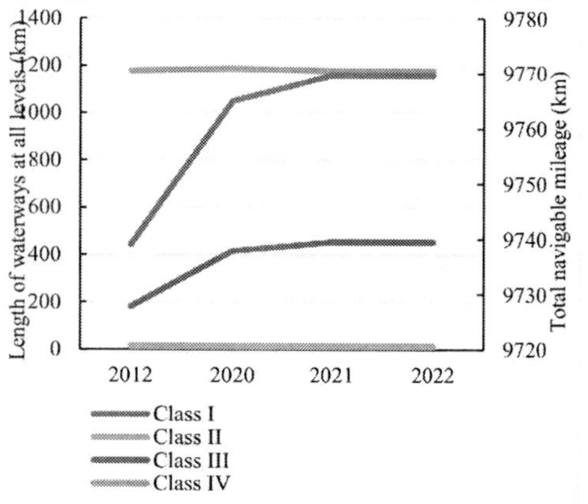

Fig. 5. Trends of waterways length at all levels.

In the future, Zhejiang will establish a Y-shaped main framework of class 1,000-ton waterways consisting of the Grand Canal, Hangzhou-Ningbo Canal, and Qiantang River, using the integration of the Yangtze River Delta as an opportunity to accelerate the improvement of inter-provincial waterway capacity. The province aims to create culturally enriched tourist waterways with Zhejiang characteristics, supporting the development of tourist routes in 26 mountainous counties. Efforts will be made to enhance the standardization of waterway maintenance and routine upkeep, construct anchorage areas along key waterways, and expand the functionality of water service areas. A smart waterway construction guide will be developed, along with the improvement of electronic navigational charts for key inland waterways, so as to build distinctive demonstration smart waterways. Additionally, the province will promote intelligent shipping, create a unified lock and inland shipping intelligent dispatch platform across the province, advance electronic navigation assistance for inland vessels, construct a 2.0 smart logistics cloud platform linking the "four ports" (Ningbo Port, Zhoushan Port, Wenzhou Port, Taizhou Port), and develop a smart service platform for container river-sea intermodal transport in northern Zhejiang.

3 Major Achievements in Low-Carbon Development of the Water Transport Industry

3.1 Specialization and High-Quality Development of Water Transport Infrastructure

Coastal port development has trended toward larger scales, increased capacity, and greater efficiency. Zhejiang has built a number of large specialized berths, strengthening its international capacity for distributing key strategic materials. Ningbo-Zhoushan Port boasts 205 deep-water berths of over 10,000 tons, including 129 super-large berths over 50,000 tons, the highest number of such berths in mainland China.

The revival plan for inland water transport is progressing smoothly, with high-level waterway networks expanding from northern Zhejiang to the entire province. All 11 cities have achieved river-sea connectivity. A sound maintenance management mechanism for waterways has been established, implementing standardized maintenance for key and important branch waterways. The structure of inland port berths has been optimized, with enhancements in scale, size, and specialization.

3.2 More Efficient Modes of Water Transport Organization

The transport structure is shifting toward "road-to-waterway" transfers. In 2022, after a decline due to external factors, waterway cargo turnover resumed growth, accounting for 78.3%, with significant increases in both cargo volume and turnover. Multi-modal connections have improved, with rapid growth in river-sea, sea-rail, and river-port transport, enhancing the overall structure of transport.

3.3 Green and Low-Carbon Transition of Ports and Operational Vessels

Operational vessels are evolving toward green, clean energy. By 2022, the province had 603 clean and new energy vessels. Notably, China's first 2,000-ton hybrid vessel, "Dongxing 100," began operations in Huzhou.

Zhejiang is building green and low-carbon ports, promoting solar, wind, and LNG applications in ports, and improving terminal handling technologies for bulk cargo like ore and coal. Ship power facilities are expanding and upgrading, setting industry standards.

Actively exploring LNG and biomass fuel bunkering. Ningbo-Zhoushan Port has become the fifth hub port worldwide capable of providing LNG bunkering services for international vessels and the first port in China to offer multiple types of new green fuel bunkering services at offshore anchorages.

Significant progress has been made in smart port and maritime construction. Ningbo-Zhoushan Port leads globally in digital port technologies, becoming the first in the nation to realize fully paperless and visually trackable container import-export operations. The inland river electronic navigational charts have been completed, and both the digital maritime comprehensive management platform and the intelligent maritime supervision platform are operational. The "Zhejiang Lock Pass" system has been launched, increasing vessel data sharing rates and key area monitoring coverage.

3.4 Continuous Improvement in Energy Efficiency and Carbon Reduction Management Capacity

Digital governance of carbon emissions in the water transport sector has advanced, with all AIS data from inland river vessels integrated into a central data hub, totaling 6,068 vessels and a verification rate of 46.8%. Vessel emission control has been effective, with intensified regulation on high-emission vehicles and vessels, and an increase in the frequency of fuel quality inspections. In 2022, the fuel compliance rate for inland waterway vessels in the province reached 99.3%.

4 Requirements for Low-Carbon Development in the Water Transport Industry

4.1 Climate Change as a Shared Challenge for Humanity Today

The urgency of global climate change is increasing, and actively addressing climate change has become a global consensus, as set out in international agreements such as the Paris Agreement, which outline goals and principles for climate action. Currently, the negative impacts of climate change are progressing more rapidly, widely, and intensely than initially anticipated. To meet the 1.5°C temperature control goal, 151 countries globally have set carbon neutrality targets as of May 2024. Facing this trend, the water transport industry urgently needs to transform to low-carbon development to adapt to the increasingly strict international carbon emission control environment and reduce the adverse impacts of increased costs such as carbon taxes on the industry development.

4.2 China's "Carbon Peak" and "Carbon Neutrality" Goals Prompt Energy Efficiency and Carbon Reduction

In the face of global challenges, China has proactively assumed responsibility for international carbon reduction. President Xi Jinping has committed to achieving a carbon peak by 2030 and carbon neutrality by 2060 and has announced relevant emission reduction targets. In this framework, China has developed documents and action plans, such as the "Carbon Peak Action Plan by 2030" and the "2024-2025 Energy Efficiency and Carbon Reduction Action Plan,"[8] which emphasize the rapid establishment of a clean energy and low-carbon transportation system to promote energy savings and emissions reductions across sectors. Facing this domestic policy environment, the water transport industry needs to formulate a carbon peak path sooner. It should strengthen its own clean and low-carbon advantages compared to other transportation methods while achieving emission reduction goals.

4.3 Low-Carbon Development as a Requirement for Green Transportation and World-Class Port Construction

The CPC Central Committee and the State Council issued the "Outline for Building China into a Strong Transportation Power" and the "Outline of the National Comprehensive Three-Dimensional Transportation Network Plan," calling for green, low-carbon development in the transportation sector and underscoring the urgency of a low-carbon development path for the water transport industry in building a strong transportation nation. The Ministry of Transport, along with eight other ministries, jointly issued "Guiding Opinions on Building a World-Class Port," emphasizing the need to promote port decarbonization, strengthen the clean, low-carbon energy system, and enhance resource conservation and recycling. The "14th Five-Year Plan for Green Transportation Development"[9] specifies low-carbon water transport strategies, including rail-water multimodal transport and the promotion of shore power facilities. The above industry policies, on the one hand, take waterway transportation as an alternative to high-emission highway transportation, providing new opportunities for industry development. On the other hand, these policies draw a blueprint for the low-carbon development of the water transport industry, indicating the direction and proposing a path for transformation and upgrading.

4.4 Zhejiang Province's Carbon Peak Deployment Requirements for Accelerated Low-Carbon Transition

As the birthplace of the "Lucid Waters and Lush Mountains Are Invaluable Assets" theory[10], Zhejiang has consistently adhered to green development principles, actively advancing the deployment of carbon peak and carbon neutrality strategies in the transportation sector to align with national goals. Zhejiang has introduced various plans, including the "14th Five-Year Plan for Water Transport Development in Zhejiang Province" and the "Implementation Plan for Building a Modern Inland Waterway Transportation Demonstration Province in Zhejiang Province," which call for high-quality development across all areas of water transport. These include strengthening shore power facilities, promoting the use of clean, low-carbon energy, advancing a modern inland waterway system, and establishing low-carbon inland water service areas. Under the guidance of a series of more detailed local policies, the water transport industry in Zhejiang Province needs to take various energy-saving and carbon reduction actions sooner to accelerate the clean transformation. While meeting policy requirements, it should make full use of the existing policy environment support.

5 Pathways to Achieving Carbon Peak in the Water Transport Industry

5.1 Accelerating Low-Carbon Upgrades of Waterway Transport Equipment

Promote the use of new and clean energy equipment, including expediting the research and application of new energy vessels, advancing clean energy use in port mobile machinery and vehicles, and applying hydrogen-powered machinery and vehicles at ports where feasible. Transition all public service vehicles to electric. Advance the iterative upgrading of transport vessels and port equipment, including scaling up and standardizing transport fleets, raising energy efficiency standards for new vessels and port projects, and accelerating the replacement or upgrading of outdated transport vessels. Orderly promote the elimination of outdated, high-energy-consuming fuel equipment at ports. Advance the application of energy-saving technologies in transportation equipment and loading machinery, including actively promoting energy-saving technologies for vessels and further advancing energy-saving technologies for port handling equipment, to comprehensively enhance the energy efficiency of common water transport infrastructure.

5.2 Optimizing Collaborative Water Transport Organization

Accelerate the construction of multimodal transport infrastructure, including expanding river-sea intermodal port hubs, enhancing river-sea channel capacity, and completing rail transport branch networks to access ports. Innovate and optimize multimodal transport business models by developing sea-rail transport train services, optimizing river-sea-river transport service systems, and improving multimodal transport connectivity. Enhance water transport operation efficiency by vigorously developing inland river container transport, increasing the operational efficiency of transport vessels, and boosting the intelligent level of port production organization and scheduling.

5.3 Strengthening Low-Carbon Support for Water Transport Infrastructure

Enhance the development level of water transport infrastructure by continuously increasing the scale and specialization of coastal ports and advancing the large-scale, intensive development of inland ports. Build a thousand-ton class inland waterway network connecting rivers and seas, and strengthen carbon emission management throughout the entire lifecycle of new water transport projects. Improve the supply of clean and low-carbon energy for water transport infrastructure, including actively enhancing the self-sufficiency of green electricity for water transport facilities, perfecting the clean energy supply infrastructure for waterway transport hubs, and planning for zero-carbon fuel supply for coastal port vessels. Promote the establishment of comprehensive energy service stations along inland waterways, and continue advancing the construction and use of shore power facilities. Enhance the ecological carbon sink capacity of water transport infrastructure by improving the carbon sink capacity of greening systems in ports and water service areas, as well as the greening level along waterways, while widely promoting the creation of Beautiful Waterways.

5.4 Improving the Intelligence Level and Innovation Capacity of Water Transport

Promote the intelligent development of equipment and facilities in the water transport industry, including accelerating the construction of smart ports, strengthening technological breakthroughs in intelligent vessels, promoting the coordinated scheduling and automatic control of locks, and advancing the demonstration construction of intelligent inland waterways. Strengthen the intelligent management of energy consumption and carbon emissions by establishing a dynamic monitoring system for energy consumption and carbon emissions in the water transport sector, enhancing the intelligent management capacity for energy consumption and carbon emissions of key enterprises. Promote the innovation and application of key technologies, including increasing the promotion of green and low-carbon technologies and actively conducting research and development on key carbon reduction technologies in the water transport industry.

5.5 Leveraging the Dual Effects of Government Guidance and Market Mechanisms

Establish a statistical accounting system for energy and carbon emissions in the industry, including developing a comprehensive system for accounting energy and carbon emissions and establishing a mechanism for the regular preparation of greenhouse gas emission inventories. Strengthen the management of industry carbon reduction targets and provide demonstration leadership by enhancing the transition from energy consumption dual control to carbon emission dual control, establishing a responsibility and assessment mechanism for carbon peak targets in the water transport sector, and actively carrying out low-carbon and zero-carbon pilot demonstration projects. Actively utilize carbon trading and carbon finance market mechanisms, including piloting carbon efficiency codes in the water transport sector, encouraging enterprises to participate in green power, green certificate and carbon emission trading, and promoting the use of new business models such as carbon finance.

6 Conclusions

The water transport industry in Zhejiang Province is characterized by the scaling and specialization of ports, continuous improvement in vessel transport capacity, and gradual enhancement of waterway infrastructure. Significant achievements have been made in low-carbon development, including the specialization and upgrading of water transport infrastructure, more efficient waterway transport organization models, green and low-carbon transformations of ports and vessels, and improved energy-saving and carbon reduction management capabilities. However, in view of climate change, the national carbon peak and neutrality targets, the promotion of green transportation, and Zhejiang Province's own carbon peak deployment requirements, the water transport industry must continue its efforts. In the future, pathways to achieve low-carbon sustainable development in the water transport sector should include accelerating low-carbon upgrades of waterway transport equipment, promoting collaborative optimization of transport organization, strengthening low-carbon support for infrastructure, enhancing intelligence levels and innovation capacity, and leveraging the dual effects of government guidance and market mechanisms, thereby providing robust support for the green transition in the transportation sector of China.

Based on the current status and situation requirements of low-carbon development in the water transport industry of Zhejiang Province, this study proposes specific implementation paths for carbon peaking from a macro level and provides a reference for the formulation of relevant policies. In the future, policy research can continue to be carried out from both quantitative and qualitative aspects. In terms of quantitative research, it is recommended to predict the possible growth of carbon emissions in the near and long term based on the emission trend in previous years and in combination with the target scale of industry development. On this basis, quantify the emission reduction benefits of various low-carbon development paths to determine the specific time nodes for carbon peaking in ports, shipping, and other industry areas. In terms of qualitative research, it is recommended to actively explore the specific implementation paths of various market-oriented carbon finance policies, in order to promote industry entities to take emission reduction actions through effective economic incentives. In addition, in terms of specific technological innovation, it is recommended that future research pay attention to key technologies that support the low-carbon development of the water transport industry, such as the promotion and application of various clean energies such as hydrogen energy in the water transport industry, the construction of electric ships and new energy ships.

Acknowledgment

This paper is completed with the support of the Science and Technology Planning Project of Zhejiang Provincial Department of Transportation, "Technical Research on Carbon Peak in Zhejiang Water Transport Industry"(Grand NO.2023023).

References

1. Chew, Tin, Lee, Haslenda, Hashim, Chin, Siong, Ho, Yee, "Sustaining the low-carbon emission development in Asia and beyond: Sustainable energy, water, transportation and low-carbon emission technology." Journal of Cleaner Production (2017).
2. D Zhou, F Huang, Q Wang, X Liu, "The role of structure change in driving CO2 emissions from China's waterway transport sector." Resources Conservation and Recycling 171.1:105627(2021).
3. M Jiang, B Wang, Y Hao, S Chen, Y Wen, Z Yang, "Quantification of CO2 emissions in transportation: An empirical analysis by modal shift from road to waterway transport in Zhejiang, China." Transport Policy 145: 177-186 (2024).
4. H Wang, Q Yue, R Liu, Y Zhang, H Wang, Q Wangl. "Regional water-energy-carbon footprint in the context of sustainable development based on EE-MRIO: cases of Jiangsu, Zhejiang and Shanghai." Environment, Development and Sustainability :1-29 (2024).
5. H Dan, P Gao, Z Sun, L Yui-Yip, "Measuring Water Transport Efficiency in the Yangtze River Economic Zone, China." Sustainability 9.12:2278(2017).
6. GU Zhang-Feng, XU Li-Hua, MA Qi-Wei, YJ Shi, LU Zhang-Wei, WU Ya-Qi. "Spatio-temporal evolution of carbon emissions in metropolitan areas and its influencing factors: A case study of Zhejiang province." Journal of Natural Resources 37.6:1524-1539(2022).
7. Zhejiang Port and Shipping Administration Center, "Zhejiang Port and Shipping Statistical Yearbook," (2022).
8. The State Council. "Energy Conservation and Carbon Reduction Action Plan for 2024-2025," May 29, 2024,https://www.gov.cn/zhengce/content/202405/content_6954322.htm.
9. Ministry of Transport. "The '14th Five-Year Plan' for the Development of Green Transportation," October 29, 2021, https://www.gov.cn/zhengce/zhengceku/2022-01/21/content_5669662.htm.
10. FU Linlin, MAO Xiaobao, MAO Xiaohong. "Practical models and path innovation of the transformation of 'lucid waters and lush mountains are invaluable assets' in Zhejiang Province". Journal of Zhejiang Agricultural Sciences, 61(12): 2469-2473+2477, (2020) (In Chinese Version). DOI: 10.16178/j.issn.0528-9017.20201209.

Renewable Energy Development in Funafuti: Photovoltaic System Design and Economic Performance

Yuan Zhuang *

School of Earth and Environment, University of Leeds, Leeds, LS2 9JT, United Kingdom

Abstract. Funafuti, the capital of Tuvalu, currently depends heavily on imported fossil fuels for its electricity generation, rendering the energy supply both expensive and environmentally unsustainable. However, the territory's abundant solar resources offer a promising pathway toward cleaner, renewable energy sources. This research investigates the design and economic evaluation of a photovoltaic (PV) energy system for Funafuti, with the aim of reducing dependence on fossil fuels and promoting renewable energy adoption. The study utilizes HelioScope for PV system design and HOMER Pro for system integration and economic analysis. The findings suggest that while a 100% renewable energy system can significantly lower emissions, it is the judicious integration of storage that is pivotal to achieving economic viability The most cost-effective system configuration substantially reduces the net present cost (NPC) and operational expenses when compared to fossil fuel-based systems, thereby emphasizing the critical role of storage in bolstering both energy security and economic efficiency. The lowest-cost system configuration significantly reduces the net present cost (NPC) and operational costs compared to fossil fuel-based systems, highlighting the critical role of storage in enhancing both energy reliability and cost-effectiveness.

1. Introduction

Funafuti, the capital of Tuvalu, is an ecologically fragile site composed of nine small atolls in the central South Pacific [1]. While the island benefits from abundant marine resources, it faces geographical isolation, especially regarding energy supply [2]. As a remote island, it depends heavily on imported diesel fuel, which is costly, environmental harmful, and vulnerable to global fuel price fluctuations. The tropical marine climate also accelerates these issues, accelerating equipment degradation and maintenance needs [3]. However, the island's consistent solar resources offer a viable pathway toward energy independence and sustainability. By 2015, five PV systems had been established on the island [4]. This amount of renewable energy systems can not enable Funafuti to move away from diesel generators entirely.

This study aims to explore and design efficient photovoltaic (PV) energy systems suitable for Funafuti conditions. Specifically, this study will achieve four main goals: firstly, access the ability of PV systems to meet local energy demand; secondly, identify and analyse the factors efficiency of PV systems; thirdly, access the impact of adopting energy storage technologies on system reliability; fourthly, conduct a holistic economic analysis of a 100% renewable energy system. This study utilises HelioScope for PV design and HOMER Pro for system integration and economic performance analysis and aims to provide a direction for Funafuti's transition to a sustainable energy future.

2. Methods

Helioscope is commonly used in studies that require accurate solar array modelling, especially in areas where shading and environmental conditions affect PV efficiency. The study highlights the effectiveness of Helioscope in capturing PV performance in actual shading situations, making it applicable to areas like Funafuti where tree and building shading may affect output in the Region [7]. The system's total capacity is 130 kW, using 480 Trina Solar 320W modules with SMA Sunny Tripower 24000TL-US inverters (see circus diagram in Figure 2). Installed on flat ground near Fetuvalu High School, the system features a 50° tilt and 243° azimuth to optimise generation. Shading from nearby structures was considered, and HelioScope's shading report indicated no significant effects.

HOMER Pro has been widely validated for hybrid system modelling, particularly in island and remote areas. This study discuss HOMER's utility in optimizing energy allocation and its effectiveness in accessing the economic performance of systems that rely on renewable energy sources [8]. This supports its application in Funafuti, which requires a balance of solar PV and energy storage to minimise the use of diesel. Version 3.14.2 was used to find optimal system configurations and run sensitivity analyses on power sources, such as limiting a 600-kW generator to 0, 300, or 600 kW. In HOMER Pro, the PV system was grid-connected, supplying excess power back to the grid. For diesel generators, capital costs were set at

* Corresponding author: ee22yz3@leeds.ac.uk

\$500,000 per 600 kW, with replacement at \$460,000, and O&M at \$428/hour, assuming a 15,000-hour lifetime. Diesel was priced at \$1.65 per litre [9]. For PV, costs were \$3,961.2 per kW capital, \$1,085 replacement, and \$150 per kW O&M annually, totalling \$375,360 capital, \$420,265 replacement, and \$13,630 O&M. The system's derating factor was 70%, with a 2,773 kW converter, costing \$115 per kW capital, \$75 per kW replacement, and \$6,047 annually for O&M.

Accurate data inputs are essential for reliable PV system modelling. Two studies both highlight the importance of localised climate and load for modelling renewable energy systems in the Pacific Islands [1][3]. This approach ensures that predictions are based on actual environmental and usage conditions, which is essential to ensure project feasibility. The photovoltaic system design and data sources for this study cover solar energy resources, Funafuti's climate, and electricity load. Solar radiation data was sourced from HOMER Pro, which uses NASA's POWER database and is adjusted with HelioScope's model as shown in Figure 1 [5]. Located around 8° South, Funafuti enjoys consistent year-round solar radiation, with an annual average of 5.25 kWh/m²/day. Climate data, including temperature, wind speed, and solar angles, were obtained from the POWER database and HelioScope, reflecting a stable tropical maritime climate with a 28°C annual average and minimal monthly fluctuations. These data informed assessments of temperature losses, power output, and system efficiency.

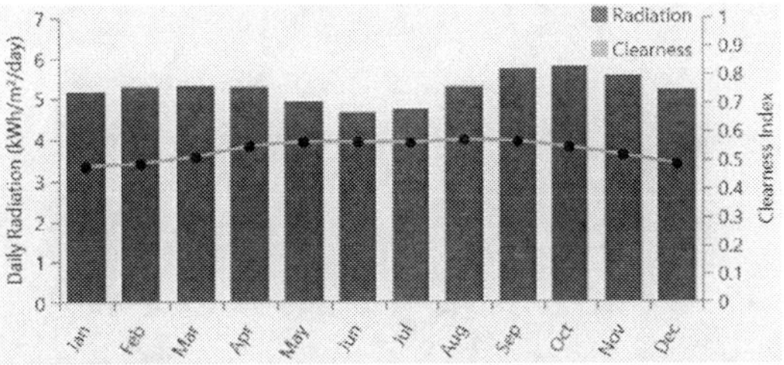

Figure 1. Monthly Average Solar Global Horizontal Irradiance (GHI) in Funafuti [5].

Electricity load data for Funafuti, taken from 2014 power plant records, were adjusted to reflect daily averages during power outages, supporting a 25-year forecast. Funafuti's load is driven primarily by government offices, with a weekday peak of around 900 kW and lower weekend demands between 400 and 500 kW, while residential demand dominates nighttime usage [6].

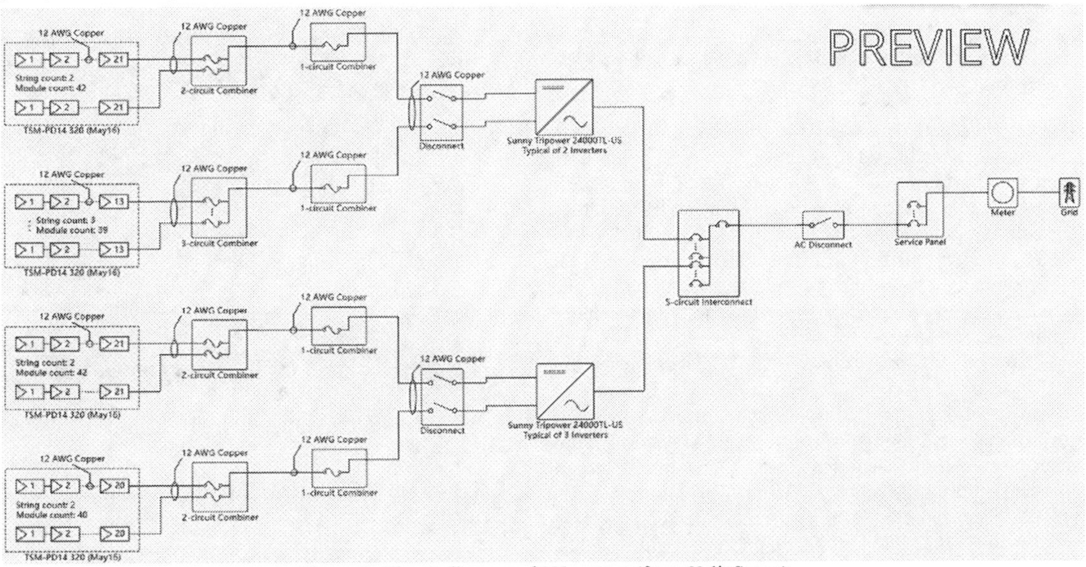

Figure 2. Circuit diagram of PV system (from HelioScope).

3. Results

3.1 Evaluating the Proposed PV System for Meeting Funafuti's Energy Needs and Reducing Fossil Fuel Dependence

The system's performance ratio (PR) is 69.3%, indicating its efficiency. The performance ratio is a key indicator of a photovoltaic (PV) system's efficiency, comparing the actual energy output under real-world conditions to the system's output in ideal conditions. A performance ratio exceeding 80% is indicative of a well-designed, efficient system. In this case, the performance ratio of 69.3% shows that there is still significant room for improvement in the system's design and performance.

According to HelioScope's simulation, the designed PV system for Funafuti has an annual energy output of 143.4 MWh. This indicates that the PV system can substantially meet the daily electricity demands of Funafuti Fetuvalu High School and also provide power to adjacent areas, especially during sunny weather. This reduces the dependence on the grid and, in turn, the reliance on diesel-powered electricity generation. Funafuti currently depends on imported diesel, which contributes to high carbon emissions and fuel costs. The integration of a PV system can partially offset these fuel-based electricity needs.

Based on HelioScope's environmental impact report, the PV system can reduce diesel consumption by 5,955 barrels of oil per year. Additionally, the system's output is equivalent to eliminating the combustion of 1,099,607 gallons of gasoline. This results in a significant reduction in greenhouse gas emissions without increasing electricity generation costs, thus having a positive environmental impact.

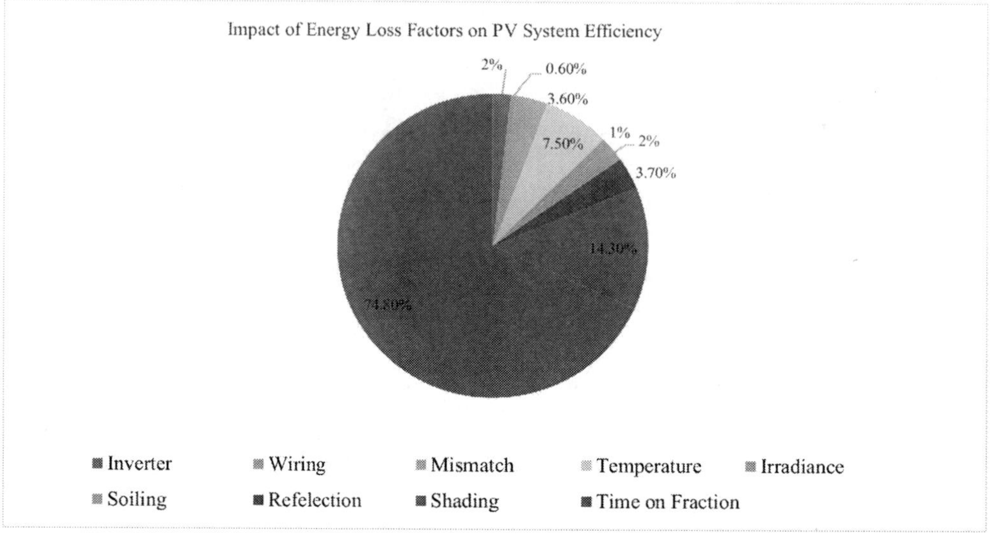

Figure 3. Energy Loss Distribution and Effective Operation of PV System

3.2 Factors Contributing to Energy Losses in PV Systems

Figure 3 presents the proportion of time the PV system operates effectively and the factors that could potentially hinder its performance. The Time on Fraction represents the percentage of the year during which the PV modules are actively producing electricity. A Time on Fraction of 74.8% shows that the PV system operates at approximately three-quarters of the year, while the remaining time is impacted by various factors that hinder optimal performance.

One of the most significant factors affecting performance is shading loss, which caused a loss of 533 kW/m^2. Although the PV system was installed on a flat surface and the design considered potential shading from nearby trees, insufficient space was allocated to account for shading from the school buildings (see Figure 4). The shading from these buildings particularly affects the row of PV array cause mutual shading between the panels, further impacting performance.

Another notable issue is temperature-related energy losses. Every PV module comes with a temperature coefficient, which describes how its performance varies with changes in temperature. In a tropical region like Funafuti, ambient temperatures are generally high, particularly around midday when solar irradiance peaks. This can raise the operating temperature of the PV modules, reducing their efficiency. HelioScope predicts an average operating temperature of 36.3°C for the PV modules. Similar to temperature, many other factors influencing performance are largely external and beyond control, resulting in unavoidable system losses.

Figure 4. Shading Heatmap

3.3 Impact of Energy Storage Addition on the Energy System

Table 1: Economic results of 3 Cases.

	Two 600kW Generators (base case)	Lowest Cost System with no Storage	Lowest Cost with a Storage
NPC	$118M	$112M	$71.7M
Initial Capital	$1.00M	$10.1M	$27.2M
O&M	$7.03M/yr	$6.13M/yr	$2.68M/yr
LCOE	$1.73/kWh	$1.65/kWh	$1.05M/kWh
IRR		8.6%	18%
ROI		5.8%	13%
Simple Payback		10 years	5.2 years

Table 1 presents a comparison of the economic performance between a system powered solely by two 600 kW diesel generators and two optimised hybrid systems.

The results indicate that the integration of energy storage significantly lowers the system's net present cost (NPC), with a reduction from $112M for the PV-diesel hybrid system to$71.7M for the system equipped with energy storage. In contrast, the NPC for the diesel-only baseline system is $118M, which underscores that the energy system with storage can achieve long-term cost savings of 39.2%.

Regarding operation and maintenance (O&M) costs, the system with energy storage incurs an annual cost of $2.68M, significantly lower than the $6.13M/year for the hybrid system without storage and the $7.03M/year for the diesel-only system. This reduction is due to the storage system lessening the frequency of diesel generator start-ups. Furthermore, by charging during off-peak hours and discharging during periods of high demand, the storage system helps to alleviate the pressure on generators during peak loads. Since solar PV generates power only during daylight hours, it is unable to provide electricity at night without the aid of storage. Therefore, an economic comparison of a PV-only system without storage is not included, as HOMER Pro does not offer this optimisation option.

The impact of energy storage on the levelised cost of energy (LCOE) is particularly notable. The LCOE of the system with storage is $1.05/kWh, compared to $1.65/kWh for the system without storage, representing a 36.4% reduction in cost. Compared to the diesel-only baseline system, the PV system with storage demonstrates a clear economic advantage. This indicates that storage not only improves power supply reliability but also significantly reduces generation costs.

In terms of the return on investment (ROI), the inclusion of energy storage increases the ROI from 5.8% to 13%. Additionally, the investment payback period is shortened from 10 years to 5.2 years, significantly accelerating the recovery of capital. This is due to the storage system's ability to increase self-consumption of generated energy, reduce reliance on external power purchases and fossil fuels, and improve overall economic efficiency.

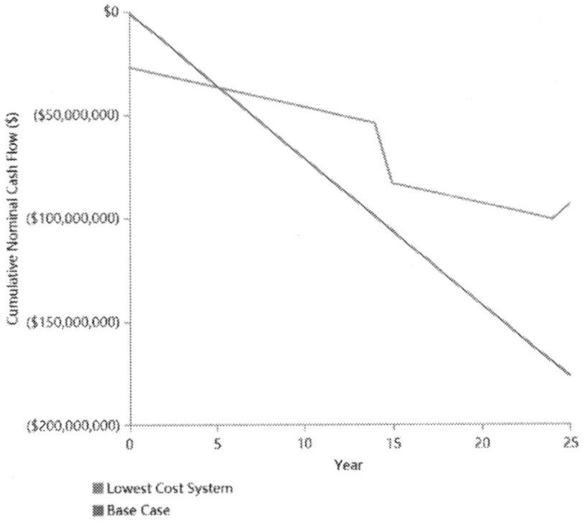

Figure 5. Cumulative Nominal Cash Flow of two energy systems in 25 years.

Furthermore, the cumulative cash flow trends over the next 25 years differ significantly between the systems (see Figure 5). With the addition of energy storage, the cash flow stabilises at a slower rate, although there are brief periods of increased costs due to maintenance and replacements. However, over time, the system returns to a steady state, and in the final year, there is even a positive cash flow, indicating that the system with storage not only pays off but potentially generates surplus revenue.

3.4 A 100% renewable energy system with zero fuel consumption enhances economic viability.

Table 2. Economic results of 2 cases

	Only renewable energy system (base case)	Lowest Cost energy system
NPC	$113M	$71.7M
Initial Capital	$57.9M	$27.2M
O&M	$3.28M/yr	$2.68M/yr
LCOE	$1.65kWh	$1.05M/kWh
IRR		0%
ROI		0%
Simple Payback		0 year

Renewable Fraction refers to the proportion of total energy generation that comes from renewable sources. In HOMER Pro, energy storage systems are also included in this calculation. "Although the most cost-effective system configuration may not depend solely on renewable energy, it nonetheless attains a Renewable Fraction of 76.7%, representing a 45% increase compared to the most economical system without energy storage.

As shown in Table 2, a comparison between two system configurations reveals that, although both systems include energy storage, the lowest-cost system remains more economically viable. This demonstrates that a 100% renewable energy system does not necessarily guarantee better economic performance. A 100% renewable system, devoid of fuel-based power generation, necessitates additional PV capacity to offset the deficit, which in turn increases the load on the converter. Moreover, since PV cannot generate power at night, the energy storage system must be used more intensively. This results in an increased number of charge-discharge cycles, which in turn raises the likelihood of maintenance and the frequency of battery replacements, effectively doubling the required storage capacity.

Additionally, Figure 6 illustrates the trends shown in Table 2. Initially, the capital costs for the two systems are different, but by year 14, the costs for PV maintenance start to rise. The system relying exclusively on PV requires more extensive maintenance, leading to a significant decrease in cash flow between years 14 and 15, further widening the economic gap between the two systems. This accounts for the absence of values in the final three rows of Table 1, as there is no convergence in the financial performance of the two systems across the 25-year duration.

Although the downward trend in cash flow appears similar for both systems in Figure 6, the PV-only system is more heavily reliant on energy storage to maintain power supply, necessitating frequent battery discharge cycles to meet energy demand.

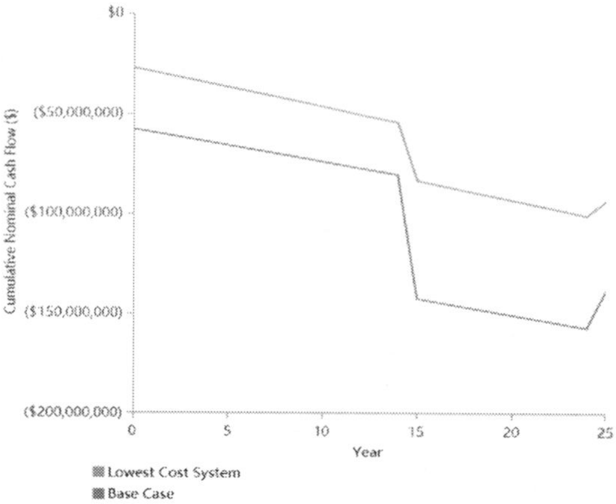

Figure 6. Cumulative Nominal Cash Flow of two energy systems in 25 years.

4. Discussion

This study analyses the design of a photovoltaic system and its energy storage configuration in Funafuti, focusing on the impact on the energy system's economic feasibility and sustainability. The results indicate that although energy storage increases the initial investment, its long-term economic benefits make the project more viable. The storage system reduces reliance on external energy during peak load periods, decreases material consumption, and cuts down on additional generation costs. This is particularly true during peak demand at night, when energy storage significantly reduces the need for fossil fuel-based power generation. Therefore, energy storage technology is not only a means of increasing renewable energy penetration but also a key component in ensuring the economic stability and efficiency of the energy system [10].

However, while the storage system significantly improves economic performance, the study also highlights the importance of optimised design. There are substantial differences in system economics depending on the configuration of energy storage and the layout of PV panels. This suggests that precise parameter selection and long-term maintenance planning will be critical to the success of future projects. Future research is needed to further assess the influence of storage technology on the levelised cost of energy (LCOE). As storage technology advances and costs decrease, it is expected that LCOE will continue to fall, making systems entirely reliant on renewable energy more economically feasible.

The current tilt of the PV panels is not optimal; it can be adjusted to approximately 8°, closer to Funafuti's latitude, to maximise solar radiation capture [11]. HelioScope simulations for Funafuti suggest an optimal tilt of 11.3°. Since Funafuti is located in the Southern Hemisphere, the PV panels should ideally face true north to maximise panel efficiency and reduce material use, as recommended by HelioScope's -1.1° orientation. Although north-facing panels may decrease overall power output due to a reduced number of panels in a given area, they provide enhanced efficiency. The original orientation of 243° faces northwest, meaning the PV system generates less electricity in the morning and relies more on afternoon sunlight, which may suit higher afternoon demand [12]. Moreover, the current PV layout is somewhat crowded, which may not appreciably enhance overall power output and could affect the efficiency of individual panels.

5. Conclusion

The introduction of energy storage systems plays a key role in enhancing the economic viability of Funafuti's energy system. As shown in the data, adding storage significantly lowers the system's net present cost (NPC), reducing total costs by approximately 40%, with notable decreases in operational costs and fuel consumption. Expanding the capacity of storage systems is recommended to further optimise the balance between electricity supply and demand, reducing dependency on diesel generation. Efficient maintenance and management are also essential; regular maintenance can prevent high repair costs and downtime due to equipment failure [13].

Increasing the proportion of renewable energy in the system is also necessary. While the renewable share has already seen considerable growth, additional solar generation capacity—particularly during peak daytime demand—can better leverage Funafuti's abundant solar resources and reduce diesel consumption. Furthermore, storing excess daytime energy for use during nighttime or lower irradiance periods via storage systems can enhance overall energy availability [14].

The results of this study have broader implications for the adoption of renewable energy in other island nations and remote areas with similar geographic and economic constraints. For these areas, where fuel imports are costly and renewable resources such as solar energy are abundant, hybrid energy system present a feasible and sustainable path forward. This study emphasises the importance of careful system configuration, particularly the integration of storage solutions, which improves energy reliability and reduces the high cost of diesel generation.

To further optimise the system, future work could focus on improving the orientation and tilt angle of the PV panels, exploring advances in storage technology to reduce costs and extend battery life, and enhancing predictive maintenance through data analytics to reduce downtime. Pilot projects to adapt these research findings to real-world conditions, adjusted to actual energy demand patterns and maintenance requirements in island environments, could also benefit real-world applications.

References

1. Islam, S.N., Reinstädtler, S., Reza, M.S., Afroze, S. and Azad, A.K., 2023. Climate change versus livelihoods, heritage and ecosystems in small Island states of the Pacific: a case study on

Tuvalu. *Environment, Development and Sustainability*, *25*(8), pp.7669-7712.

2. Lal, R. and Kumar, S., 2022. Energy security assessment of small Pacific Island Countries–Sustaining the call for renewable energy proliferation. *Energy Strategy Reviews*, *41*, p.100866.
3. Ebinger, J.O., 2011. Climate impacts on energy systems: key issues for energy sector adaptation. World Bank Publications.
4. Raturi, A., Singh, A. and Prasad, R.D., 2016. Grid-connected PV systems in the Pacific Island Countries. Renewable and Sustainable Energy Reviews, 58, pp.419-428.
5. NASA POWER. No date. V.2.4.0. [online]. [Accessed 8 October 2024]. Available from: https://power.larc.nasa.gov/data-access-viewer/
6. Funafuti Load Data in 2014. [online]. Available from: https://docs.google.com/spreadsheets/d/1V9abj4bWQi5tHKmJSNYHJaCyDXo9zNzAT6RQphDoRFs/edit?usp=sharing
7. Thulasingam, M. and Periyanayagam, A.D.V.R., 2023, December. Review of Optimization Tools Used for Design of Distributed Renewable Energy Resources. In *International Conference on Renewable Energy, Green Computing, and Sustainable Development* (pp. 160-174). Cham: Springer Nature Switzerland.
8. Babu, M.K. and Ray, P., 2023. Sensitivity analysis, optimal design, cost and energy efficiency study of a hybrid forecast model using HOMER pro. Journal of Engineering Research, 11(2), p.100033.
9. Global Petrol Prices, 2024. Tuvalu diesel prices. [online]. [Accessed 12 October 2024]. Available from: https://www.globalpetrolprices.com/Tuvalu/diesel_prices/
10. Sánchez, A., Zhang, Q., Martín, M. and Vega, P., 2022. Towards a new renewable power system using energy storage: An economic and social analysis. *Energy Conversion and Management*, *252*, p.115056.
11. Esfahani, S.K., Karrech, A., Cameron, R., Elchalakani, M., Tenorio, R. and Jerez, F., 2021. Optimizing the solar energy capture of residential roof design in the southern hemisphere through Evolutionary Algorithm. *Energy and Built Environment*, *2*(4), pp.406-424.
12. Tiwari, G.N. and Gupta, N., 2022. Photovoltaic Thermal Passive House System: Basic Principle, Modeling, Energy and Exergy Analysis. CRC Press.
13. Sarbini, N.N., Ibrahim, I.S., Abidin, N.I., Yahaya, F.M. and Azizan, N.Z.N., 2021. Review on maintenance issues toward building maintenance management best practices. *Journal of Building Engineering*, *44*, p.102985.
14. Akbari, H., Browne, M.C., Ortega, A., Huang, M.J., Hewitt, N.J., Norton, B. and McCormack, S.J., 2019. Efficient energy storage technologies for photovoltaic systems. Solar Energy, 192, pp.144-168.

Substitution Efficiency Test of $CH_3{}^{127}I$ Photoelectric Method for Traditional Radioactive $CH_3{}^{131}I$ Method

Yao Tian[1,a], Sijun Fan[2], Weiye Yin[2,*], Shuai Zhou[1], Zhaojie Li[2], Shuai Yang[2], Chen Wang[1]

[1]Yangjiang Nuclear Power Co., Ltd., Yangjiang, Guangdong, China
[2]Suzhou Nuclear Power Research Institute Co., Ltd., Suzhou, Jiangsu, China

Abstract. Iodine adsorber is a purification and adsorption functional unit for possible radioactive iodine in the ventilation system of nuclear power plant. In order to ensure its functional availability, the adsorption performance efficiency test should be conducted every year. At present, the mainstream tracer used in the test is $CH_3{}^{131}I$ and no unmanned automation has been realized, which has the risk of radioactive contamination to the test personnel and the environment. In order to realize the safety of the test in essence, this paper attempts to use non-radioactive $CH_3{}^{127}I$, which has almost the same physical and chemical properties as $CH_3{}^{131}I$, as the test tracer to replace the iodine efficiency test of the iodine adsorber. A large number of test data are compared to explore the result gap between the two tracers and the test method. Finally, the feasibility of $CH_3{}^{127}I$ photoelectric method in iodine efficiency test was verified.

1 Introduction

Nuclear power plants will produce a variety of radionuclides in daily production, which will release a variety of radioactive rays, including alpha rays, beta rays and gamma rays, etc., which have great harm to the human body, among which there is a nuclide I-131, its half-life is eight days, in the process of decaying into Xe-131 will be accompanied by the release of a large number of gamma rays. It can easily pose a risk of internal contamination to nuclear power plant workers. In 2011, a serious accident occurred in the Fukushima nuclear Power Plant in Japan due to the earthquake and tsunami, which released a large amount of radioactive iodine-131, causing great physiological health hazards to the surrounding people[1-3]. As a purification unit for this nuclide, iodine adsorbers are applied in the ventilation system of nuclear power plants[4]. Every year, nuclear power plants conduct regular efficiency tests on the functional units of iodine adsorbers, and the current test method is the radioactive I-131 method.

2 Principle and procedure of iodine efficiency test in nuclear power plant

2.1 Principle of efficiency test for radioactive $CH_3{}^{131}I$

As an important structural unit in the ventilation system of nuclear power plants, the functional unit of iodine adsorbers ensures that nuclear power plants can prevent the further diffusion and leakage of radioactive materials when risks occur. However, due to the life cycle of the iodine adsorbers of its components, it is necessary to conduct regular performance efficiency tests on them. Many substances, such as freon[5] and cyclohexane, have been used as tracers, but the results obtained by this test cannot directly determine the adsorption performance of the iodine adsorber for methyl iodine. After years of changes in test methods[6], the current iodine efficiency test scheme in nuclear power plants is as follows: $CH_3{}^{131}I$ was used as the test tracer[7]. Mixed vapor containing $CH_3{}^{131}I$ was generated by a special iodine source generator and injected into the iodine adsorbent filter unit to be tested. After sampling upstream and downstream, the samples were tested for radioactivity and then the purification efficiency of this functional unit was calculated. The traditional test method uses dimethyl sulfate $(CH_3)_2SO_4$ to react with iodine source $Na^{131}I$ to produce radioactive methyl iodine ($CH_3{}^{131}I$), and the reaction formula is as follows:

$$(CH_3)_2SO_4+2Na^{131}I=Na_2SO_4+2CH_3{}^{131}I$$

The use of dimethyl sulfate $(CH_3)_2SO_4$ is a highly toxic chemical, the effect is similar to mustard gas, acute toxicity is similar to phosgene, 15 times greater than chlorine. It has a strong irritating effect on the eyes and upper respiratory tract, and a strong corrosive effect on the skin. It can cause conjunctival congestion, edema, corneal epithelial exfoliation, partial necrosis of trachea and bronchial epithelial cells, and perforation of mediastinum or subcutaneous emphysema. In addition, it can also damage the liver, kidney and myocardium, etc., and can cause burns, blisters and deep necrosis after skin contact. The transportation, use and storage of dimethyl sulfate are highly risky and harmful, and have very strict control requirements and high management costs.

* Corresponding author: ywyhit2017@163.com, a1989724012@163.com

Due to the different operating air volume of different on-site systems[8], radioactive $CH_3{}^{131}I$ required for iodine test should be injected according to the on-site air volume, so as to avoid affecting the test results and causing errors.

After the improvement, the isotope replacement method is now used, that is, the use of non-radioactive $CH_3{}^{127}I$, ethanol and water in a certain ratio of isotope replacement liquid, compared with the dimethyl sulfate method, the substance used is less toxic and less risky. The formula is:

$$CH_3{}^{127}I+Na^{131}I=CH_3{}^{131}I\uparrow+Na^{127}I$$

2.2 Iodine generator working principle

The working structure of the iodine generator is mainly composed of two parts, respectively, the negative pressure injection part and the radioactive source part. Because the boiling point of methyl iodine at atmospheric pressure is 40.3°C, the boiling point of methyl iodine can be further reduced by using a vacuum pump to reduce pressure. In order to successfully inject radioactive methyl iodine into the system required for the test, replacement liquid (mainly composed of $CH_3{}^{127}I$, ethanol, and water) will be configured in advance during the configuration of radioactive sources, because ethanol, as a benign solvent of organic matter and its strong volatility, can inject all radioactive sources into the system more stably. Because during the work of the iodine generator, the generator box is in a state of negative pressure, such as the radioactive source leaks, because of the existence of negative pressure radioactive materials can not leak into the environment outside the box, providing a guarantee for subsequent treatment.

2.3 Adsorption principle of iodine adsorber

The iodine adsorber of nuclear power plant is filled with nuclear grade coconut shell activated carbon. This technology is widely used to remove airborne radioactive iodine, because the adsorption of activated carbon produces a large specific surface area with its various micropore structures, and thus produces a physical adsorption effect on methyl iodine gas. Due to the low physical adsorption force, there is adsorption and desorption equilibrium, and it is greatly affected by temperature. In order to ensure the normal availability of iodine adsorbers, a heater unit is set up in the ventilation system of nuclear power plants to reduce the humidity of the ventilation gas and delay the aging of iodine adsorbers. However, due to the increase in temperature, the exchange between molecules is becoming more frequent. Therefore, in order to ensure that the iodine adsorbers reduce the radioactive methyl iodine desorption caused by temperature rise, activated carbon is often chemically impregnated. The chemical impregnating agent is infiltrated into the microporous structure of activated carbon to bind the adsorbed methyl iodine molecules through chemical energy, so that it is not easy to break away. In addition, if I-127 is present in the impregnation, the isotope exchange between I-131 and I-127 will eventually reach equilibrium. It can be seen that the main adsorption methods of iodine adsorbers are: physical adsorption, chemical adsorption and isotope replacement adsorption.

3 Potential risks of $CH_3{}^{131}I$ as tracer

Iodine-131 in $CH_3{}^{131}I$ is a radioactive isotope of elemental iodine, which is an artificial radionuclide (nuclear fission product). In addition to causing a certain amount of external radiation to the test personnel during the test, the most important risk is that the tracer is a gas, which will cause serious internal radiation risk to the test personnel and serious pollution risk to the environment after leakage. If the duct of the ventilation system is damaged or the iodine adsorber fails, it will cause gas pollution in the nuclear power plant building. Especially if the ventilation system of the main control room of the nuclear power plant leaks, $CH_3{}^{131}I$ may enter the main control room, posing a serious threat to the personnel in the main control room.

It can be seen that although nuclear power plants currently have complete working procedures for iodine tests, the tracer used is still radioactive $CH_3{}^{131}I$, which cannot eliminate the risk of personnel and other public property losses caused by radioactive material leakage. Eliminating this test risk has become an urgent need for nuclear power operators. New iodine adsorber efficiency test methods also need to be explored and tested[9]. The purpose of this change is to replace the high-risk radioactive methyl iodine released in the test, and adopt more safe and reliable non-radioactive methyl iodine, which essentially eliminates the safety risks, improves the safety operation and management level of nuclear power plants, and does not reduce the safety and reliability index of equipment, and greatly reduces the difficulty of the current complicated iodine efficiency test.

4 $CH_3{}^{127}I$ non-radioactive iodine efficiency test method introduction

4.1 Introduction to the principle of non-radioactive method

Table 1. Physical and chemical properties of CH_3I.

Item	Data
Fusing point	-64 °C
Boiling point	40.3 °C
Point of flammability	7.8 °C
Density	2.28 g/cm^3
Saturated vapor pressure	53.32 kPa(25.3 °C)
Appearance	Colorless transparent liquid
Solubility	Slightly soluble in water, soluble in ethanol, ether and other organic solvents

Methyl iodine is a common halogenated hydrocarbon, the chemical formula is CH_3I, MeI, is an iodine substitution of methane, can be miscible with common organic solvents, pure product is colorless transparent liquid, exposed to sunlight will decompose iodine elemental and purple, can be removed by adding metal copper. Iodomethane is a commonly used methylation reagent in organic synthesis. Table 1 shows the basic physical and chemical properties of methyl iodine.

Because of the low boiling point of methyl iodine, the field test can be carried out by using methyl iodine-nitrogen mixed high-pressure gas cylinders. $CH_3{}^{127}I$, as a non-radioactive methyl iodine, greatly reduces the risk of transport of the original radioactive methyl iodine tracer.

Since the tracer was replaced with non-radioactive $CH_3{}^{127}I$, the original radioactive assay method for $CH_3{}^{131}I$ was no longer effective. The concentration variation of methyl iodine tracer gas after passing through the iodine adsorber is very low, generally only the nmol/mol level content. Only by developing accurate trace methyl iodine measurement technology can the concentration of the tracer downstream of the iodine adsorber be measured, and the purification capacity of the iodine adsorber can be further evaluated. It is difficult to measure nmol/mol level methyl iodine in the air by general chemical methods. In order to accurately measure the concentration of upstream and downstream non-methylated iodine in the experiment, photoelectric analysis method is adopted for non-methylated iodine.

The photoelectric sensor ionizes the tracer gas, and the amount of charge generated is proportional to the gas concentration, and the tracer concentration can be calculated through the standard concentration gas calibration[10-11]. The principle of photoelectric sensor is shown in Figure 1.

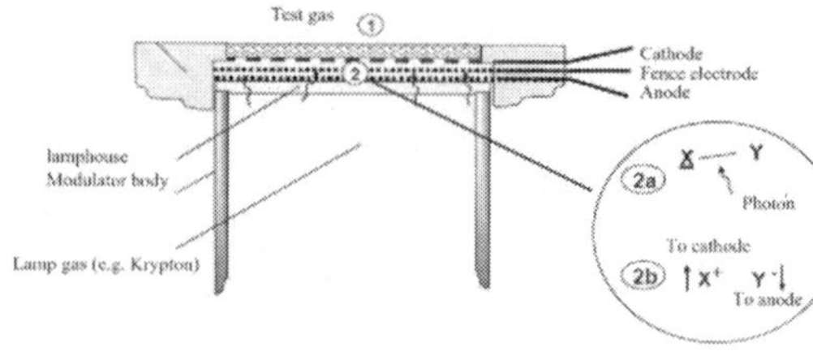

Figure 1. Principle of photoelectric sensor.

In order to test and verify the response characteristics of the selected photoelectric sensor, the test device and pipeline are set up in the laboratory, and the concentration, temperature, back pressure and flow rate are taken as variables respectively to realize the response characteristics of the sensor probe. Design a process component with small "dead space" and short response time to form a test loop (as shown in Figure 2) for testing. The diluted gas uses the air purified and filtered by activated carbon, and controls the injected mass flow through the needle regulating valve; The calibrated methyl iodine gas controls the reagent flow through the needle control valve, and the clean air is mixed into the gas of the measured concentration in the rotary mixer. Humidification and heating links can be added to the air front section to change the gas parameters. The flow rate of the pump is adjustable, and the temperature, pressure and other parameters of the gas can be measured in the photoelectric probe box. A regulating valve group, flowmeter and differential pressure meter for adjusting sampling flow rate and sampling back pressure are provided in the photoelectric probe box.

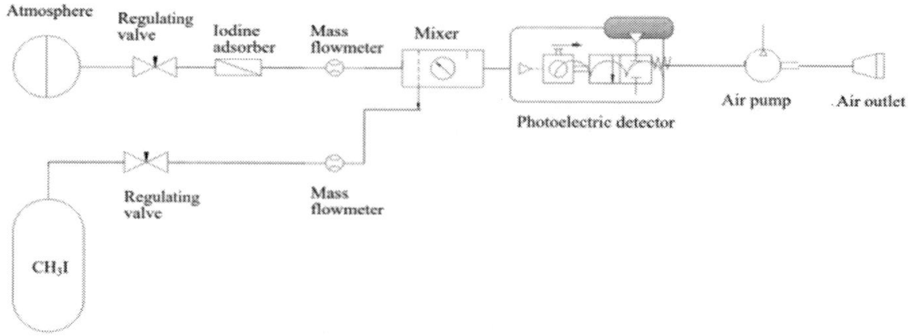

Figure 2. Process flow of sensor characteristic test.

In order to obtain different concentrations of iodomethane gas, flow ratio method was used for analysis in the laboratory. The standard concentration of iodomethane gas is mixed with the purified air. Precise flow measurement control with mass flow meters. For the standard concentration of iodomethane gas, use a mass flowmeter with a range of 0-50 ml/min, and use a mass flowmeter with a range of 0-5 l/min for purifying air. The total flow rate is controlled at about 2-4 l/min. Through flow distribution, the standard concentration of gas can be diluted 40 to 4000 times, and various concentrations of gas can be obtained according to the need.

After 1000 dilution of iodomethane with a concentration of 102 nmol/mol (2% accuracy) by flow ratio method, 0.1 nmol/mol concentration of iodomethane can be obtained. After gas is passed into the sensor, the voltage curve of the sensor changes as shown in Figure 3. By dividing the red line in the figure, it can be found that the sensor voltage has a slight rise after injection of 0.1 nmol/mol concentration, indicating that it has been able to react to 0.1 nmol/mol concentration gas.

https://doi.org/10.1051/e3sconf/202561702011

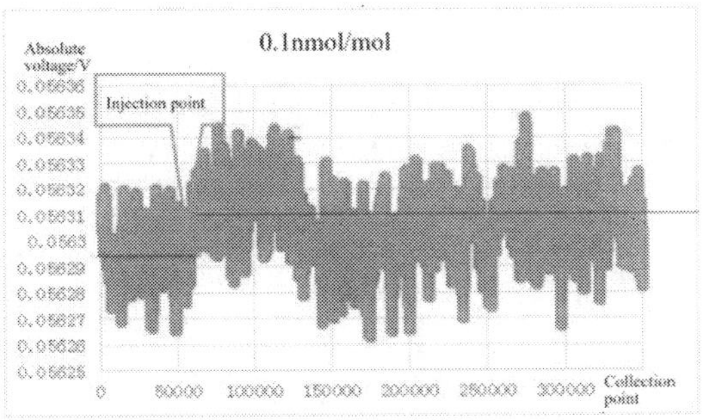

Figure 3. Voltage change caused by methyl iodine at 0.1 nmol/mol concentration.

After the concentration is increased from 0.1 to nmol/mol, the voltage change curve is already obvious, as shown in Figure 4.

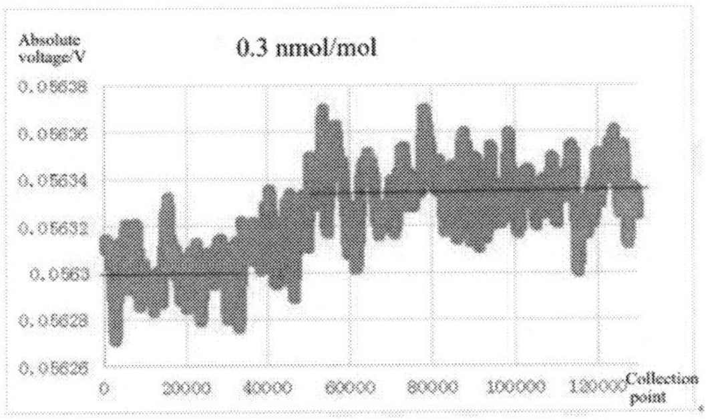

Figure 4. Voltage change caused by methyl iodine at 0.3 nmol/mol concentration.

It can be seen from the data that the sensor output voltage can also change at a concentration as low as 0.1 nmol/mol; When the concentration is increased to 0.3 nmol/mol, the sensor can effectively identify.

In addition to ensuring the sensitivity of the photoelectric sensor, it is also necessary to ensure the high-line relationship between its voltage increase and methyl iodine concentration, because the test equipment has two sets of equipment upstream and downstream equipment groups, so there are two sets of photoelectric sensing equipment available, for which the relevant experiments are also designed and the following data are obtained.

Upstream sensor measurement:

Different concentrations of gas can be obtained using the flow distribution method. The standard gas of 5.3 μmol/mol and 303 μmol/mol is diluted respectively, and finally the concentration gas of 10~5000 nmol/mol can be matched, and the specific data are shown in Table 2.

As can be seen from the data in Table 2 above, the photoelectric sensor has a good linear relationship for methyl iodine at different concentrations, and will not cause distortion of data deviation when the concentration is too large or too small. The Figure 5 curve below is drawn after data summary. As can be seen from the figure, the sensor voltage increase presents a linear change relationship with methyl iodine gas concentration. From this, the calibration coefficient between voltage and concentration can be obtained, which can be used to calculate the methyl iodine gas concentration to be measured according to the measured voltage increase.

Table 2. Upstream photoelectric sensor linear experiment data sheet.

No.	C_{CH3I} (nmol/mol)	CH_3I flow CH_3I (ml/min)	Air flow (ml/min)	△U(V)	C(nmol/mol)
1	5300	6.489	2912.3	0.001308	11.782
2	5300	12.434	2908.6	0.002318	22.561
3	5300	20.411	2894.6	0.003736	37.111
4	5300	27.521	2885.8	0.005146	50.067
5	5300	36.525	2874.0	0.007078	66.512
6	5300	42.860	2855.7	0.008592	78.371
7	5300	49.939	2844.2	0.010046	91.452
8	5300	60.315	2833.3	0.012179	110.472
9	303000	1.348	2975.4	0.016804	137.200
10	303000	2.421	2977.6	0.029951	246.183
11	303000	4.309	2989.2	0.051502	436.105
12	303000	10.257	2982.8	0.126658	1038.360
13	303000	20.016	2971.0	0.241346	2027.677
14	303000	29.465	2963.8	0.341750	2982.628
15	303000	41.210	2954.8	0.458894	4167.709
16	303000	49.902	2946.4	0.539281	5046.214

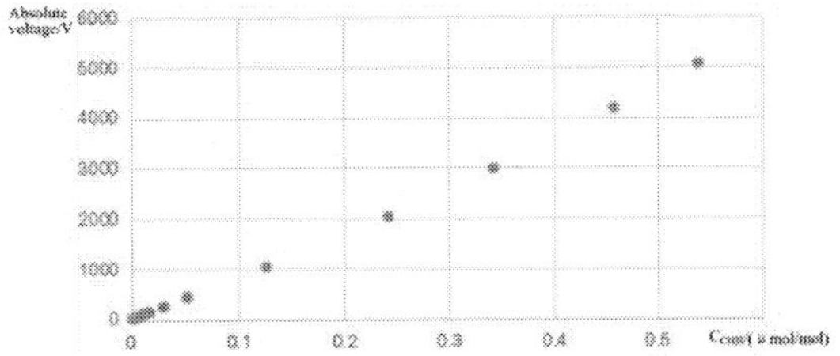

Figure 5. Upstream photoelectric sensor linear test data.

In the ventilation test, the downstream concentration data is the biggest factor affecting the test data, and the detection accuracy of very low concentration methyliodine[12] needs to be ensured.

Under the same test conditions, the downstream sensor was tested against the very low concentration (nmol/mol level) methyliodine concentration to obtain the following data.

Downstream sensor determination:

The standard gas of 102 nmol/mol, 1.1 μmol/mol, 5.3 μmol/mol is diluted, and finally the gas of 0.1~100 nmol/mol concentration can be matched, and the specific data are shown in Table 3.

The relationship between voltage increase and concentration is made into a scatter chart, as shown in Figure 6. As can be seen from the figure, there is a linear relationship between methyl iodine concentration and sensor voltage increase. In order to calculate the concentration of methyl iodine gas more accurately, the segmented linear method can be used to calculate the voltage rise of the sensor, and the segmented calibration coefficient between the voltage and the concentration can be obtained.

Table 3. Linear experimental data table of downstream photoelectric sensor.

No.	C_{CH3I} (nmol/mol)	CH_3I flow CH_3I (ml/min)	Air flow (ml/min)	△U(V)	C(nmol/mol)
1	102	2.669	2697.2	0.000009	0.101
2	102	8.433	2695.8	0.000032	0.318
3	102	20.121	2690.0	0.000079	0.757
4	102	30.968	2679.9	0.000111	1.165
5	102	40.183	2670.0	0.000139	1.512
6	102	50.331	2664.6	0.000184	1.891
7	1100	3.434	2799.1	0.000129	1.348
8	1100	6.920	2794.2	0.000232	2.717
9	1100	9.963	2794.2	0.000307	3.908
10	1100	13.699	2790.0	0.000400	5.374
11	1100	17.105	2788.8	0.000502	6.706
12	1100	21.040	2785.7	0.000601	8.246
13	5300	5.740	2791.6	0.000648	10.875
14	5300	8.815	2788.6	0.001091	16.701
15	5300	12.726	2786.4	0.001666	24.096
16	5300	15.160	2779.9	0.002094	28.746
17	5300	18.951	2777.4	0.002812	35.918
18	5300	25.169	2770.1	0.004042	47.722
19	5300	32.777	2764.9	0.005765	62.094
20	5300	37.921	2761.6	0.007212	71.792
21	5300	43.760	2757.1	0.008824	82.806

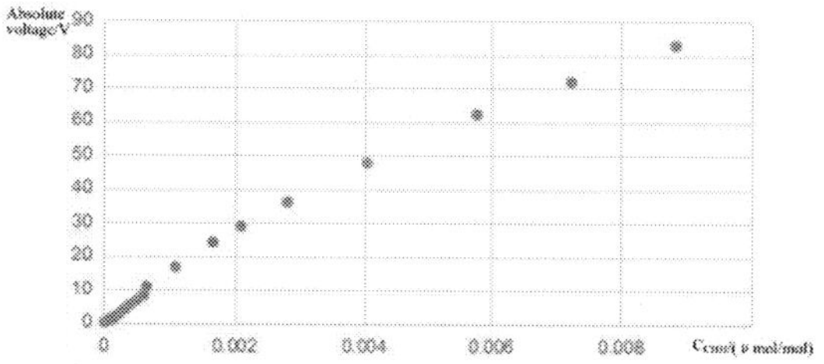

Figure 6. Downstream photoelectric sensor linear test data.

4.2 $CH_3{}^{127}I$ Non-radioactive iodine efficiency test method Field test data

In order to compare the purification efficiency test results of iodine sorbers by non-methylated iodine method and traditional radioactive method, three iodine sorbers with good, medium and poor filtration efficiency were selected on the ventilation system test bench for comparison test, and 10 groups of iodine sorbers with different systems were selected for comparison test on the power plant ventilation system.

4.2.1 Comparison test on off-line test bench

The test contents are as follows: 1) a radioactive test was performed; 2) Perform one non-radioactive test, and the injection standard is 1 μmol/mol per minute; 3) Perform two non-radioactive tests, and the injection standard is 5 μmol/mol per minute; 4) Perform two non-radioactive tests, and the injection standard is 10 μmol/mol per minute; 5) Perform a radioactive test. In the test on the off-line bench, the upstream and downstream gas pressures do not change much, and the voltage increment of the probe can be modified into the actual concentration value by limited use of the concentration coefficient. The data obtained from the non-radioactive test and the radioactivity data are shown in Table 4.

Table 4. Off-line test data.

Old iodine adsorber					
Item	E	η	≥99%	≥99.7%	≥99.9%
R-1	26.14	96.174%	×	×	×
NR-1μmol/mol	21.70	95.391%	×	×	×
NR-5μmol/mol-1	28.15	96.447%	×	×	×
NR-5μmol/mol-2	28.79	96.527%	×	×	×
NR-10μmol/mol-1	28.20	96.455%	×	×	×
NR-10μmol/mol-2	27.54	96.369%	×	×	×
R-2	22.47	95.550%	×	×	×
Normal iodine adsorber					
Item	E	η	≥99%	≥99.7%	≥99.9%
R-1	885	99.887%	√	√	×
NR-1μmol/mol	521	99.808%	√	√	×
NR-5μmol/mol-1	636	99.842%	√	√	×
NR-5μmol/mol-2	475	99.789%	√	√	×
NR-10μmol/mol-1	625	99.840%	√	√	×
NR-10μmol/mol-2	613	99.837%	√	√	×
R-2	611	99.836%	√	√	×
Fresh iodine adsorber					
Item	E	η	≥99%	≥99.7%	≥99.9%
R-1	8655	99.988%	√	√	√
NR-1μmol/mol	4622	99.978%	√	√	√
NR-5μmol/mol-1	8009	99.988%	√	√	√
NR-5μmol/mol-2	4928	99.980%	√	√	√
NR-10μmol/mol-1	7642	99.987%	√	√	√
NR-10μmol/mol-2	8243	99.988%	√	√	√
R-2	5643	99.982%	√	√	√

It can be seen from the Table 4 that the purification coefficient of the old, normal and fresh three types of typical iodine adsorbers was accepted by the non-radioactive method, and the judgment of the three types of criteria of 99%, 99.7% and 99.9% were all correct; The results of comparison of the purification efficiency of the radioactive test show that the values are similar and slightly conservative. In the application of low concentrations of tracers, it can be found that even if the new filter is tested with a concentration of 1 μmol/mol of methyl iodine for 1 minute (high purification coefficient), correct conclusions can be obtained and the purification efficiency value is close to that of the radioactive method, which is also due to the ability of the measurement probe to recognize very low limit concentrations of methyl iodine.

4.2.2 Comparative test on an online system

The online comparison test was carried out in the periodic test window of the efficiency of the iodine adsorbers in different ventilation systems. After the iodine loop was put into operation, the parameters of the air volume, temperature and humidity of the system were adjusted to reach the range required by the program. The radioactivity test is performed first, and after the radioactivity test is completed, the non-radioactive test is performed at the measuring point of the existing test. After continuously observing that the background value measured by the probe is stable and the change is less than 100 μv within 10 minutes, the data is collected for 2 minutes as the background value. After injecting 5 μmol/mol concentration of non-methylated iodine for 1 minute, the collected data were recovered to the cumulative amount of downstream methyliodine near the background value and then stabilized for 3 minutes. By substituting the probe calibration curve, the voltage increment of upstream and downstream is modified into the concentration value, and the purification coefficient is obtained. Comparative tests were carried out on the functional units of iodine adsorbers of 10 groups with different systems, among which control test 3 was carried out twice on the same functional unit of iodine adsorbers, and the comparative data were shown in Table 5.

Table 5. Summary of online test data.

No.	Comparison test 1		
	E	η	≥99.7%
R	13723	99.993%	√
NR	1752	99.943%	√
No.	Comparison test 2		
	E	η	≥99.7%
R	4370	99.977%	√
NR	4032	99.975%	√
No.	Comparison test 3-1		
	E	η	≥99.7%
R	653	99.847%	√
NR	392	99.745%	√
No.	Comparison test 3-2		
	E	η	≥99.7%
R	653	99.847%	√
NR	489	99.796%	√
No.	Comparison test 4		
	E	η	≥99.7%
R	5398	99.981%	√
NR	2826	99.965%	√
No.	Comparison test 5		
	E	η	≥99%
R	3711	99.973%	√
NR	5970	99.983%	√
No.	Comparison test 6		
	E	η	≥99%
R	279	99.642%	√
NR	153	99.347%	√
No.	Comparison test 7		
	E	η	≥99%
R	140	99.286%	√
NR	173	99.421%	√
No.	Comparison test 8		
	E	η	≥99%
R	390	99.744%	√
NR	605	99.837%	√
No.	Comparison test 9		
	E	η	≥99%
R	252	99.602%	√
NR	308	99.676%	√
No.	Comparison test 10		
	E	η	≥99%
R	1121	99.911%	√
NR	494	99.798%	√

According to Table 5, the acceptance criteria of the comparison program, the judgment results of the 11 non-radioactive tests were all consistent with the judgment results of the availability of the iodine adsorber, which was consistent with the evaluation conclusions of the radioactive method, and did not lead to a difference in the judgment results, nor did the iodine adsorber be replaced in advance due to conservative results. Except for the comparison results of tests 5, 7, 8 and 9, the results of other non-radioactive tests are more conservative than those of radioactive tests, and the difference brought by conservative tests can be accommodated for engineering applications.

5 Conclusion

Due to the differences in the physicochemical properties of $CH_3{}^{131}I$ and $CH_3{}^{127}I$, replacing radioactive tracers that are harmful to humans with non-radioactive tracers can further reduce the daily production risk of nuclear power plants. The comparison test is based on a large number of known studies on the role of methyl iodine gas in iodine adsorbers, and does not change the physicochemical properties of the iodine adsorbers for iodine purification. The purification mechanism considered in the non-radioactive test is consistent with the radioactive method, which provides a new method for the purification acceptance test of iodine adsorber. After a large number of test data comparison, it can be seen that the error of the non-radioactive method is very small compared with the radioactive method, and the final purification efficiency data are also smaller than the radioactive test method, in line with the conservative principle of nuclear power, so it can be applied to the efficiency detection test of the iodine adsorber purification unit in the ventilation system of nuclear power plant.

References

1. Shilian W, Qi L, Xinjun Z, et al. Analysis of the Dispersion Timeline and Isotope Activity Ratio Characterization of Airborne Radionuclides Released from the Fukushima Daiichi Nuclear Power Plant Accident[J]. ACS EARTH AND

SPACE CHEMISTRY, 2021, 5(10):2938-2946。

2. Yumiko Y, Makiko O, Hitomi M, et al.Risk perception regarding implementation of iodine thyroid blocking during a nuclear disaster of mothers living near a nuclear power station in Japan.[J]. Endocrine journal, 2021, 68(5):553-560.
3. Eriko S, Christopher E, J. B P. Lessons from Fukushima: Potassium Iodide After a Nuclear Disaster[J]. AJN, American Journal of Nursing, 2021, 121(2):63-67.
4. Zhang Xindgan. Nuclear ventilation and air Purification[M]. Beijing: Atomic Energy Press, 1993:22-47.
5. Chen Xiaofei, Guan Zhenggang, Feng Lifa. Study on filtration efficiency Test technology of iodine sorbent by Freon Process[J]. Plant Maintenance Engineering, 2021, (09):14-16.
6. Wang Hongkai, Chen Rongtian, Chen Jianli, et al. Optimization of iodine absorber test method for DVC System in Nuclear Power Plant[J]. Electric Engineering, 2022, (24):208-209+212.
7. Gao Linfeng, Yao Yanyan, Wang Lei, et al. Comparative study of different test methods using radioactive methyl iodine as tracer to measure iodine removal efficiency of nuclear ventilation system[J]. Radiation Protection, 2018, 38 (02):137-141.
8. Chen Jianli, Chen Rongtian, Gao Linfeng, et al. Effect of operating air volume on performance of iodine adsorber in ventilation system of nuclear power plant[J]. Heating Ventilating & Air Conditioning, 2019, 51(10):150-152+12.
9. Zhang Jirong, Wang Zhen, Zhang Chong-wen, et al. Study on comprehensive evaluation method of iodine adsorber performance in nuclear air purification system[J]. Sichuan Environment, 2022, 41 (03):186-190.
10. Zhao Gaoxin, Fan Sijun, Sun Jiyi, Chang Wei and Yin Weiye. Effect of $CH_3{}^{127}I$ tracer Method on purification efficiency of iodine adsorber in nuclear power plant[J]. Nuclear Electronics & Detection Technology, 2024, 44 (04): 738-743.
11. Wang Jinren, Duan Zhaoli. Application of photoelectric sensor in gas concentration measurement[J]. Application of IC, 2022, 39 (03):130-131.
12. Zhang Jirong, Li Yongguo, Wang Haiwei, et al. Experimental study on low concentration methyl iodine analysis method suitable for performance evaluation of iodine adsorber[J]. Radiation Protection, 2015, 35 (03):158-162.

A New Type of Cable Duct Channel and Design Method with Composite Support of Gravity Retaining Wall and Premixed Fluid Solidified Soil

Zhenhai Zhou* and Linkang Zhou

Suzhou Electric Power Design Institute Co., Ltd. Suzhou, Jiangsu, China

Abstract. Laying power cables from 10kV to 220kV in cable protection tubes is currently one of the most widely used and economical construction methods for urban transmission lines. According to the existing technical solutions, the outside of the protection tube is generally backfilled with well-graded medium-coarse sand or encapsulated with reinforced concrete, which not only wastes suitable construction materials but also causes a large amount of soil disposal, which is not conducive to environmental protection and soil and water conservation. In this paper, based on analyzing the practice of applying premixed fluid solidified soil to replace the backfill sand or reinforced concrete encapsulated structural layer outside the discharge pipe for the trench enclosed by a gravity retaining wall in soft soil, the technical scheme of using premixed fluid solidified soil to mention the backfill sand or encapsulated concrete layer is proposed. After the completion of the construction, the premixed fluid solidified soil will form a gate-type hydro mud soil protection structure with a gravity retaining wall to base on guarantee the safety of the power channel; it reduces the disposal of soil, saves building materials, and provides a new program for the green construction of the urban cable line protection pipe channel.

1 Introduction

In the construction of urban power grid projects, overhead lines cover a large area; their impact on the surrounding environment is also more significant and affects the cityscape. So, in recent years, the new urban transmission and distribution lines have been mainly laid by underground cable [1]. In the 220kV and the following voltage level cable project, a row of pipe protection is the most common and economical way of laying, and it has gradually become the main channel form of the transmission and distribution lines in China's major cities [2].

According to the different environments for laying pipe channels, at present, the pipe is protected mainly by sand filling or reinforced concrete, and the residual soil in the foundation trench is transported out to be discarded to meet the requirements of the local city for environmental protection and soil and water conservation. Using sand protection backfill will waste the river sand, a good building material. At the same time, the ability to prevent external damage is weak, easily affected by geological drilling, pile construction, and other mechanical operations, and leads to the damage of the cable protection pipe; the use of reinforced concrete encapsulation, although it has a better ability to prevent external breakage, it is a significant investment, the construction period is long, and it is not easy to dismantle the restoration of the underground space after occupying such problems. At the same time, whether it is backfilled sand protection or encapsulated reinforced concrete protection, it will produce more abandoned soil and is not conducive to urban environmental protection and soil and water conservation [3].

Premixed fluid solidified soil is a new type of backfill material consisting of soil, curing agent, and other additives mixed in proportion, with good fluidity, self-compacting, and fast curing speed [4]. This material is commonly used for backfilling trenches in construction projects, especially in small spaces or complex structures. The soil material can directly use the residual soil of the trench excavated from the cable duct. In contrast, the unconfined compressive strength of the fluid soil after curing can be greater than or equal to 0.8MPa [5], which is close to that of the hydraulic soil mixing pile. It can become a protective body with greater rigidity and strength after curing in the ground.

A new construction method for cable conduit engineering using gravity cement soil wall support for soft soil layers such as silt and silty soil is studied directly using a premixed fluid solidified soil backfilling. After solidification, the flow-solidified soil forms a gate-shaped conduit surrounding the protective wall with the gravity cement soil wall. Suggestions for parameter values for related design are proposed. Compared with the existing method, it has the advantages of less external soil waste, faster construction speed, minor occupation of

* Corresponding author: zhouzhenhaid@qq.com

https://doi.org/10.1051/e3sconf/202561702012

underground space, and higher strength of the protective wall around the pipeline.

2 Protection program for the laying of existing cable ducts

Cable laying technology using protection tubes is widely used in 10kV ~ 220kV cable line projects. Compared with cable trench, tunnel, and other laying methods, which are more widely used, the investment is also smaller.

Due to their shortcomings, Traditional high-voltage power cable conduits, such as cast-in-place concrete pipe, asbestos pipe, steel pipe, etc., have been made of new materials out of the cable conduit. Currently, a wide variety of cable conduits are on the market, but their performance varies. The material can be divided into plastic conduit, fibber cement conduit, metal material conduit, carbon bellows, and other categories, but the most widely used are CPVC, HPDE, MPP, and other plastic protection pipes.

As the pipe is laid chiefly in the urban road environment, and the need for a certain depth of burial, the protection of the pipe needs to withstand soil pressure, vehicle loads, and other large loads, but also may encounter unclear positioning and the occurrence of drilling, piling construction triggered by the external breakage problem, it is necessary to backfill the plastic protection of the pipe outside the sand or encapsulated in the plain concrete for protection, the current typical protection scheme as shown in Figures 1 and 2.

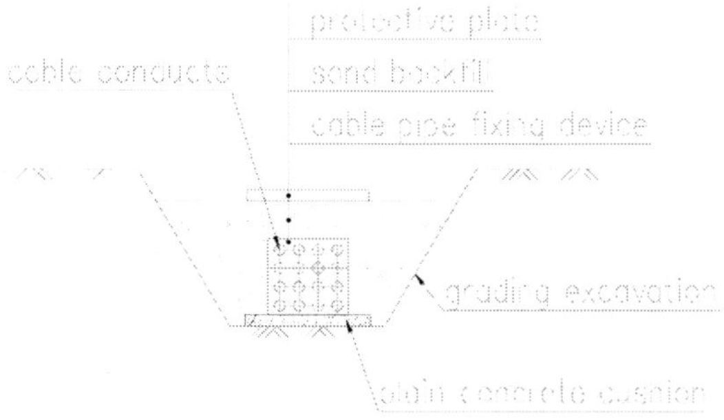

Figure 1. Cable protection pipe work protected by sand filling.

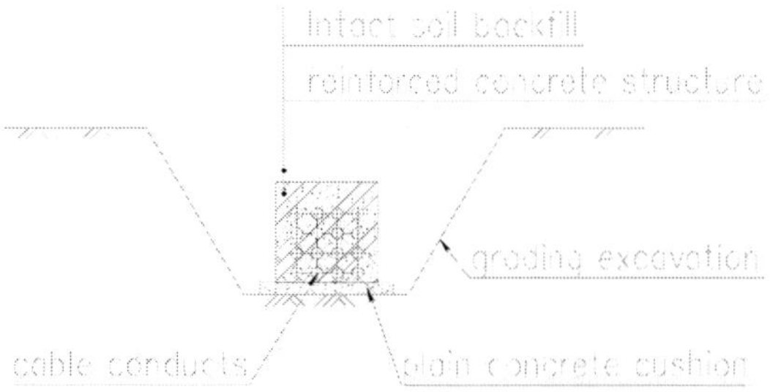

Figure 2. Cable protection pipe work protected by sand filling.

3 The cable trench is enclosed by a gravity retaining wall

According to the requirements of GB50217-2018 and other technical standards [6], when the cable discharge pipe is buried in the ground, the top of the uppermost pipe must have a cover of not less than 0.50m. The thickness of the cover should not be less than 0.50m. At the same time, considering the spacing between multiple pipes, the excavation depth of its trench is close to 2.0m. When the construction is carried out in silt and silt soil layers, it is necessary to adopt a support structure for the foundation trench.

Regarding the trench with an excavation depth of not more than 2.0m in the urban environment, the Larsen steel sheet pile is mainly used for support. Still, the mechanical equipment for driving and pulling out the steel sheet pile is significant. Its construction machinery, whether it adopts the static pressure type or vibration type, will have a particular influence on the surrounding environment, and at this time, adopting a gravity-type cement soil wall is also an option.

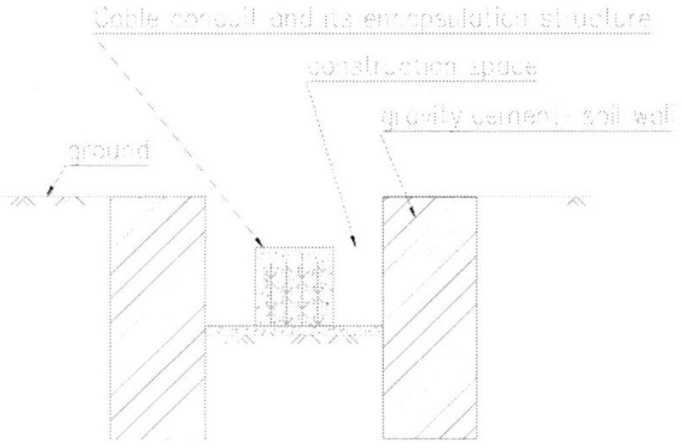

Figure 3. Conventional drainpipe construction with cement soil wall support.

In the traditional cable trench structure supported by gravity cement soil wall as shown in Figure 3, the determining factors for the size of the pit include the height of the pipe and its encapsulated structural layer, the thickness of the soil cover; the width is determined by the width of the pipe and its encapsulated structural layer, and the staffed operating space, and the staffed operating space in the pit is mostly taken to be 0.8m to 1.20m, mainly considering the installation and removal of the formwork during the construction stage of the encapsulated structural layer of the pipe. The main reason is to consider the installation and removal of formwork during the construction stage of the pipe wrapping structure layer. Usually, when the outer dimension of the pipe and its encapsulated structural layer is 1.2m wide×1.0m high, the inner clear space of the foundation trench is 2.8m wide×2.0m high.

 https://doi.org/10.1051/e3sconf/202561702012

4 Cable Drainage Channel Enclosed by Gravity Retaining Wall using premixed fluid solidified soil in Fertiliser Troughs

4.1 Premixed fluid solidified soil and its application in trench backfilling

Premixed fluidized cured soil is a new type of trench backfilling technology emerging in recent years, and there are some cases of successful application in China. Li Xiaoliang [7] introduced the principle of premixed fluid-cured soil technology and elaborated on the premixed fluid-cured soil filling process and its construction quality control points in combination with the filling characteristics of the narrow space of the foundation pit backfill of Chengdu Tianfu International Airport. The engineering application practice proves that using premixed fluid-cured soil filling technology effectively solves the filling problems in limited space and improves construction efficiency by ensuring the stability of project quality.

4.2 Construction process of cable discharge channel enclosed by gravity retaining wall using premixed fluid soil backfilling

The cable trench with a gravity retaining wall is generally excavated to a depth of not more than 5.0m (JGJ 120 stipulates that it is not more than 7.0m) [8], and the process of its construction work is mainly as follows:

(1) Trial mixing of premixed fluid-cured soil at the project site.

(2) Construction work and maintenance of gravity cement soil wall.

(3) Excavation of foundation trench, earth transport to fluidized soil processing plant, etc.

(4) Construction of power channel structures, including necessary foundation treatment, bedding construction, construction of discharge pipes or cable trenches, and necessary maintenance of reinforced concrete.

(5) Premixed fluid-cured soil backfills, 7d maintenance period.

(6) Restoration of ground for cable channels.

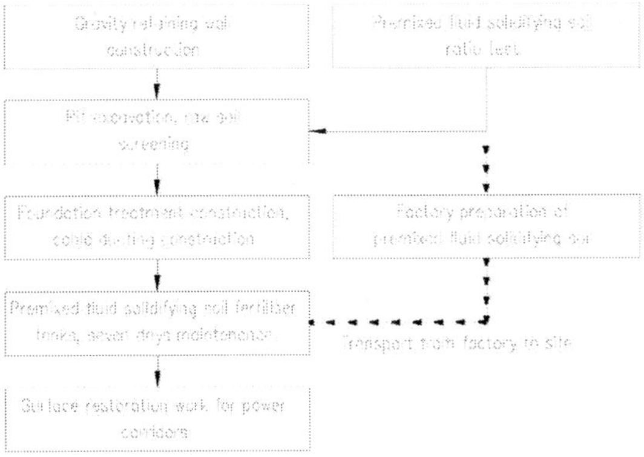

Figure 4. Construction process flow of gravity retaining wall using premixed fluid-cured soil backfilling.

The construction process flow of the cable discharge pipe and cable trench project supported by a gravity cement soil wall with a premixed fluid-cured soil backfilling is detailed in the following figure.

As Figure 4 shows, using premixed fluid-cured soil backfilling does not increase the space and time needed to apply gravity-cemented earth walls.

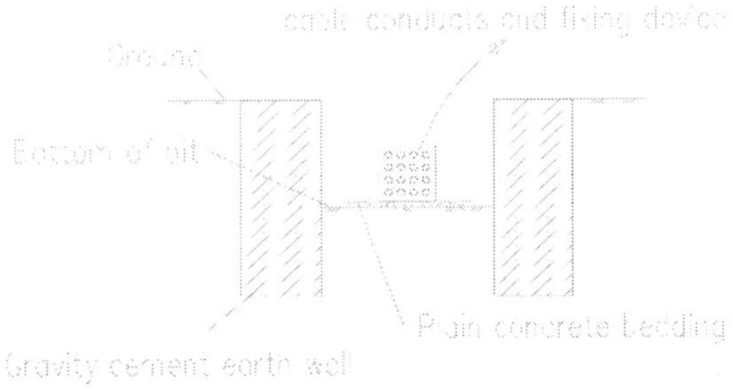

Figure 5. Diagram of discharge pipe channel with gravity retaining wall support.

Figure 5 shows a gravity retaining wall's construction process flow chart supported by a gravity retaining wall with a premixed fluidized soil backfilling. The structure of the gravity retaining wall with premixed fluidized soil backfilling and its internal discharge pipe channel is detailed in Figure 6.

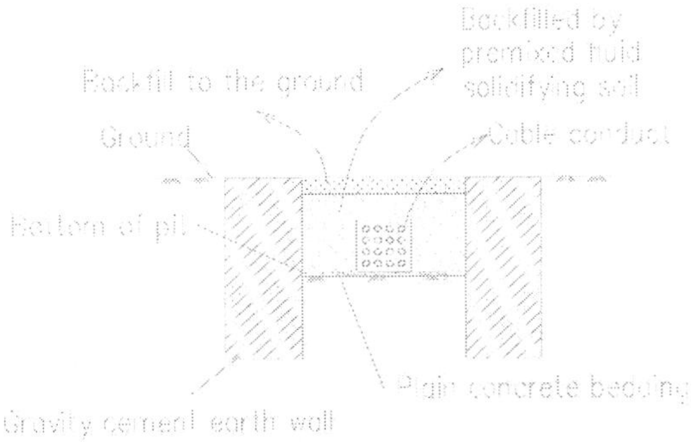

Figure 6. Diagram of discharge pipe channel using gravity retaining wall support with premixed fluid soil backfilling.

As can be seen from the schematic diagrams in Figures 5 and 6:

(1) The premixed fluidized soil backfilling has little effect on the construction process of gravity retaining walls;

(2) The use of premixed fluidized soil curing can make use of the soil excavated in the foundation trench, the soil backfilled on the ground surface, and the displacement of part of the soil by the row of pipes and their encapsulated structural layer, which can be used as the external abandonment of the soil;

(3) Premixed fluid-cured soil has little influence on the design of the cable trench calculation; the temporary support structure calculation may not take into account the impact of this part [9].

4.3 Key Design Parameters for Cable Drainage Channel Enclosed by Gravity Retaining Wall Using Premixed Fluidized Soil Backfilling

From the process point of view, the use of ready-mixed fluid-cured soil does not require the construction personnel to go down to the base trench to erect formwork, tie reinforcement, and perform other operations; they only need to go down to the base trench to construct the vegetative concrete bedding and tie the row of pipes, so its operating space can be appropriately reduced, concerning the experience of related projects, in the depth of the trench does not exceed 2.0m, you can narrow the row of pipes on one side of the understaffed operating space to 0.4m to 0.60m, at this time, the width can be reduced by 0.8m. In the height direction, compared with the reinforced concrete encapsulated pipe structure, the thickness from the ground can be calculated from the top of the highest layer of the pipe, and the depth of pit excavation can be reduced by 0.20m, further reducing the risk of construction operations.

The design of the cable discharge pipe access for the gravity retaining wall pit support using premixed fluid-cured soil f backfilling includes the support structure and the channel.

In the actual construction of the project, the size of the base trench of the discharge pipe access includes the following parts:

(1) cable protection pipe spacing composed of rows of pipe outsourcing range, which includes cable protection pipe combing, spacing composed of rows and columns, according to the different diameters of the pipe, the general horizontal spacing of 230mm ~ 280mm, the vertical spacing of 240mm ~ 280mm. Net distance between the rows of pipes is mainly used to control the pipe sleeper, and the permissible deviation of its spacing value is the same row of holes spacing ≤ 5mm, vertical spacing ≤ 20mm. Vertical spacing ≤ 20mm.

(2) The extension length of the outer protective body of the pipeline is usually not less than half of the center distance of the pipeline column.

(3) When the base trench of the row of pipes is constructed by sloping, its bottom is not less than 600mm; when the construction is carried out by supporting and blocking structure, its width is not less than 800mm.

(4) The soil cover on the top of the pipe must be at least 0.50 m thick, and it is usually set at 0.70 m to meet the requirements of the longitudinal slope.

(5) The width of the construction work surface of the row of pipes, in addition to considering the width of the row of pipes, but also need to consider the row of pipes under the side wall of the human operating space, the row of pipes in the base trench of the support type structure, the top of the row of pipes, such as the construction of the work of the enclosure.

(6) Trench edge 1.0m range is strictly prohibited within the pile of soil, equipment, materials, etc.; pile load height should not exceed 1.5m.

Diagram of optimized channel dimensions is shown in Figure 7.

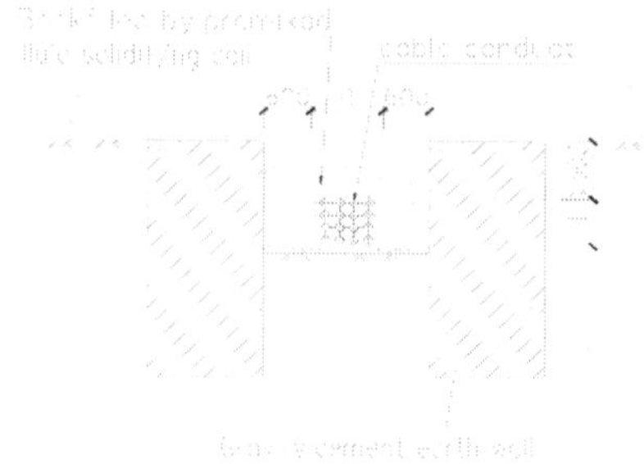

Figure 7. Diagram of optimized channel dimensions.

4.4 Analysis of Benefits

The cable discharge channel enclosed by a gravity retaining wall using a premixed fluid soil backfilling has multiple benefits, which mainly include:

(1) A gravity retaining wall used as a shallow pit support structure in silt and silty soil layers can avoid the use of large-scale mechanical construction operations such as sheet piles and grouting piles on the site while forming a final gate-type support structure together with the premixed fluidized soil to realize the combination of a cemented soil wall with permanent proximity.

(2) The premixed fluidized soil backfilling directly utilizes the original pit residual soil, reduces the external discarded earth by more than 60%, and avoids using river sand or reinforced concrete. Thus, low-carbon and environmentally friendly cable trench construction is realized.

(3) By optimizing the construction process, the width of the foundation trench is reduced by 0.8m and the height is reduced by 0.20m compared with the use of reinforced concrete wrapped protection drainage trench, which not only reduces the amount of work, but also reduces the use of underground space under the premise of ensuring safety.

(4) Each standard working section of not more than 30m can save more than 5 days of on-site construction period.

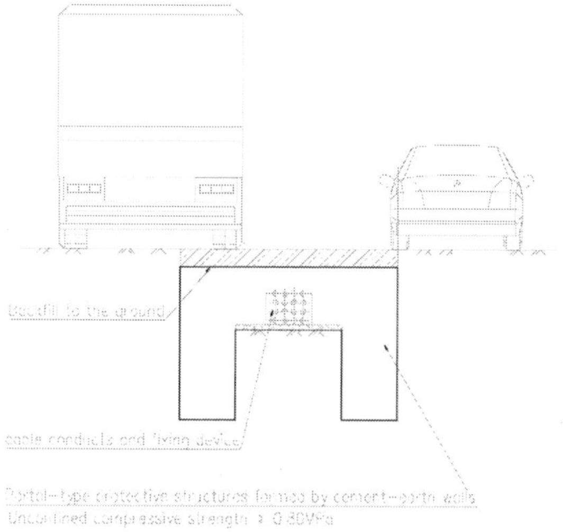

Figure 8. Portal-type protective structure formed by premixed fluidized soil and gravity-cemented earth wall.

Figure 8 shows the permanent gated drain protection channel formed by a gravity cement soil wall and premixed fluid soil.

5 Conclusion

In this paper, for the cable duct project, which is supported by cement soil wall in soft soil layers such as silt, silty soil, etc., a new type of cable duct channel is proposed which directly adopts premixed fluid cured soil to fertilize the tank, and after curing the fluid soil and gravity cement soil wall together form a gate-type protective wall around the duct. Based on analyzing the construction method, it puts forward the optimization of the relevant design parameter value, which has the advantages of less abandonment of earth, faster construction speed, less occupation of underground space, and less protection structure around the duct than existing cable duct channel construction scheme. Compared with the existing cable duct channel construction scheme, it has the advantages of less discarded earth, fast construction speed, small occupation of underground space, and high strength of pipe peripheral protection wall, making it a new cable channel form worthy of extensive promotion.

Acknowledgments

The authors would like to thank financial support provided by the Suzhou Electric Power Design Institute Co., Ltd through Grant No. JSDL24061914020002.

References

1. Cai, R., Wang, X., Cheng Vong, C., Zhao, S., & Zhang, T. (2024). Low-carbon urban development hot topics and frontier evolution: a bibliometric study from a global perspective. Frontiers in Built Environment, 10, Article 1464529.
2. Huang M, Sun T, Wang L. Application of Premixed Solidified Soil in Backfilling of Foundation Trench[J]. IOP Conference Series: Earth and Environmental Science, 2020, 510(5):052062 (6pp).
3. Xu-Guang Z. Study on Long Spiral Drilled Premixed Fluid Solidified Soil Composite Foundation in Subcenter of Beijing[J]. Building Technology Development, 2018.
4. Yang Q, Zhang N, Han L, et al. Preparation and characterization of premixed fluidized solidified soil with municipal solid waste incineration bottom ash[J]. Construction and Building Materials, 2024, 442.DOI:10.1016/j.conbuildmat.2024.137541.
5. Zhang Xin, Ji Haifeng, Yang Zuoqing. Experimental Study on Engineering Characteristics of Fluidized Solidified Soil Containing Collapsible Loess and Gravel[J]. Geotechnical Engineering Technique, 2024, 38(5):611-617.
6. Ministry of Housing and Urban-Rural Development of the People's Republic of China. Standard for design of cables of electrical power engineering[S]. Beijing: China Plan Press, 2018. (in Chinese)
7. Li Xiaoliang, Li Peng, Wu Dongyang, et al. Practice and Application of Pre-mixed Fluidized Solidified Soil Filling Construction Technology [J]. Sichuan Architecture, 2023, 43(03): 234-236. (in Chinese)
8. Ministry of Housing and Urban-Rural Development of the People's Republic of China. Technical specification for retaining and protection of building foundation excavations [S]. Beijing: China Building Industry Press, 2012. (in Chinese)
9. China Engineering Construction Standardization Association Standards. Technical standard for backfilling project by using premixed fluid solidifying soil [S]. Beijing: China Building Industry Press, 2022. (in Chinese)

Research on Implementation Pathways for Near-Zero Carbon Emissions in High-Density Urban Districts

Guoshu Yang*, Anni Guo, Gaijing Zhang

Shanghai Institute of Building Science Co., Ltd, Shanghai, China

Abstract. During the 14th Five-Year Plan period, the transition to low-carbon development in urban and rural construction has increasingly focused on achieving scalable low-carbon development and piloting near-zero carbon emission projects at the meso-scales and micro-scales. High-density urban districts in city centers serve as ideal settings for implementing near-zero carbon emission demonstration projects. Taking a high-density, high-quality urban districts in Shanghai as an example, this study proposes a pathway for implementing key technologies for near-zero carbon urban districts through a dual approach of "carbon source reduction" and "carbon sink enhancement," summarized as the "Two-Lows and Four-Aspects" strategy (low-energy buildings, low-emission transportation, resource circulation, ecological landscaping, energy maximization, and digital management). Additionally, the feasibility of achieving near-zero carbon emissions in high-density urban districts is calculated and validated, providing a practical reference for similar high-density urban projects aiming for near-zero carbon emissions.

1 Introduction

In 2021, the State Council issued the Action Plan for Peak Carbon Dioxide Emissions Before 2030, emphasizing the acceleration of green and low-carbon transformation in urban and rural development and the creation of energy-saving, low-carbon industrial parks that meet international advanced standards. The same year, the Ministry of Ecology and Environment released the Guidelines for Coordinating and Strengthening Climate Change Response and Ecological Environment Protection, advocating for regions with solid foundations to explore pilot projects for near-zero carbon emissions and carbon neutrality. According to the 2023 China Building and Urban Infrastructure Carbon Emissions Research Report, the total life-cycle carbon emissions of the national construction sector in 2021 reached 5.01 billion tons of CO_2, accounting for 47.1% of the country's energy-related carbon emissions [1]. Thus, low-carbon transformation in urban and rural construction is pivotal to achieving the national carbon peaking and carbon neutrality goals.

With the growing demand for low-carbon urban transformation and the renovation of existing communities, carbon reduction strategies and pathways in urban construction are increasingly focused on the meso-scales and micro-scales, such as districts and neighborhoods [2,3]. Based on theoretical studies on urban morphology and factors influencing carbon emissions, researchers [4-6] have proposed that efficient land development, compact spatial layouts, and polycentric urban spatial structures are more conducive to low-carbon urban development. This article integrates theories and practical case studies related to high-density urban districts to establish an implementation pathway for achieving near-zero carbon emissions in central high-density districts. Using a block-level project in Shanghai as an example, the study applies scenario analysis to calculate and validate near-zero carbon emissions, aiming to provide a research framework for similar high-density districts pursuing near-zero carbon development.

2 Characteristics of High-Density Urban Districts and Carbon Emission Calculation Methods

2.1 Key Characteristics of High-Density Urban Districts

In the context of urban and regional planning, a district is typically defined as an area enclosed by roads, comprising one or more neighborhoods, and generally covering a space ranging from several hectares to 3 km² [3]. The term "high-density urban district" is more of a conventional concept, lacking a unified national standard. However, extensive research highlights four main features of such districts: high floor area ratio (FAR), high building coverage ratio, low undeveloped space ratio, and high population density [7]. The average FAR of high-density urban districts often exceeds 3.0, with some cases reaching above 10 (e.g., Hong Kong's average FAR is as high as 12–15). Building coverage ratios can range from 30% to 50%, and population densities can reach 15,000–20,000 people per square kilometer [8-11]. Therefore, high-density urban districts exhibit a certain degree of spatial

* Corresponding author: yangguoshu@sribs.com

heterogeneity compared to traditional urban environments, forming a new type of high-density, multi-dimensional urban spatial morphology.

Shanghai, as a representative megacity with high-density urban districts, exemplifies this development. Central districts in Shanghai are characterized by dual attributes of spatial intensification and high resource consumption, marked by high building densities, intensive energy and resource use, stringent environmental quality demands, and limited development space. Their spatial form, building density, functional layout, and construction-management models all have significant impacts on carbon reduction efforts. Districts, as the fundamental units of urban space, serve as key venues for embodying "people-centered" urban warmth and promoting "green and low-carbon" healthy lifestyles. They are the primary arenas for achieving dual-carbon goals in urban construction and management. Therefore, high-density urban districts shoulder the responsibility of driving innovation and establishing best practices for high-quality near-zero carbon emissions development.

2.2 Near-Zero Carbon Emission Calculation Methods at the District Scale

Currently, there is no standardized methodology for calculating near-zero carbon emissions at the district scale. However, numerous scholars, both domestically and internationally, have explored carbon emission calculation and evaluation methods at the park and community levels. Jones C. M., et al.[12] suggested that community carbon emissions primarily include transportation, energy, waste, water, embodied carbon in buildings, food, and services. Zuda Ye, et al.[13] reorganized detailed activity categories and emission factors from national greenhouse gas emission inventories to establish a carbon emission evaluation method suitable for the operational phase of green ecological urban districts in China. Yueyang Ma, et al.[14], using a commercial office block as an example, developed a carbon emission factor database for the construction phase and proposed a calculation and evaluation method for carbon emissions at the district scale. Gaijing Zhang, et al.[15], using a low-carbon eco-park in Shanghai as an example, conducted an analysis of operational phase carbon emissions through scenario analysis, finding that actual carbon reductions could reach 51% compared to conventional construction modes. Fang Li, et al.[16] developed a community carbon emission calculation method covering buildings and public facilities, transportation, waste treatment, and carbon sinks, validating the approach with data from a new community in Shanghai.

Currently, research on meso- and micro-scale carbon emission methodologies primarily focuses on near-zero carbon parks or communities, which can be adapted to district-scale near-zero carbon emission calculations. Therefore, near-zero carbon emission calculations for high-density urban districts mainly concentrate on areas such as buildings, energy, transportation, waste, and landscape carbon sinks. The specific focus for statistical calculations of energy activity data varies across fields. The most common calculation approach involves multiplying activity data by carbon emission factors.

3 Analysis of the Development Status of Near-Zero Carbon Emission Districts

3.1 Policy Trends for Near-Zero Carbon Emissions

The concept of "near-zero carbon emission district demonstration projects" was first introduced in the national *13th Five-Year Plan* in 2015, positioning near-zero carbon emission districts as a key focus for advancing various low-carbon pilot initiatives. With the ongoing implementation of "carbon peaking and carbon neutrality" policies, the development of near-zero carbon emission districts has entered a rapid growth phase during the *14th Five-Year Plan* period. As of October 2024, several local governments have issued implementation plans and policy documents related to the construction of near-zero carbon emission districts (**Table 1**). These documents outline carbon emission targets, evaluation metrics, and construction pathways, accelerating the development of near-zero carbon demonstration districts.

Table 1. Overview of Policies and Objectives Related to Near-Zero Carbon Emissions.

Province/City	Document Number	Document Name	Near-Zero Carbon Type	Carbon Emission Goals
Hubei Province	Hubei Environmental Protection Office [2020] Document No. 39	*Implementation Plan for Near-Zero Carbon Emission District Demonstration Projects in Hubei Province*	Near-zero carbon parks, near-zero carbon communities	Steady reduction in total carbon emissions; control of per capita carbon emissions for residents
Shanghai	Shanghai Environmental Climate [2021] Document No. 182	*Shanghai Low-Carbon Demonstration Creation Work Plan*	Near-zero carbon practice zones	Carbon intensity reaching the advanced level of similar areas citywide or reduced by over 50% from the baseline period
			Near-zero carbon communities	Per capita carbon intensity reaching the city's advanced level or reduced by over 40% from the baseline period

Province/City	Document Number	Document Name	Near-Zero Carbon Type	Carbon Emission Goals
Tianjin	Tianjin Environmental Climate [2021] Document No. 82	*Guidelines for the Preparation of Low-Carbon (Near-Zero Carbon Emission) Demonstration Construction Plans in Tianjin (Trial)*	Near-zero carbon demonstration zones	Per capita carbon emissions reaching 3.5 tCO_2/person
			Near-zero carbon communities	Carbon emissions reduced by ≥10% compared to the baseline year
Shenzhen	Shenzhen Environmental Protection [2021] Document No. 212	*Shenzhen Near-Zero Carbon Emission Zone Pilot Implementation Plan*	Near-zero carbon parks, near-zero carbon communities	Total carbon emissions reduced by over 40% compared to 2020
Sichuan Province	Sichuan Environmental Letter [2022] Document No. 409	*Work Plan for Near-Zero Carbon Emission Park Pilot Projects in Sichuan Province*	Near-zero carbon parks	Total carbon emissions reduced by 5%
Chengdu	Chengdu Environmental Development [2022] Document No. 15	*Work Plan for Near-Zero Carbon Emission Zone Pilot Projects in Chengdu (Trial)*	Near-zero carbon parks	Total carbon emissions steadily reduced compared to the baseline year
Chongqing	Chongqing Environmental Protection [2022] Document No. 93	*Chongqing Near-Zero Carbon Park Pilot Implementation Plan*	Near-zero carbon parks	Carbon intensity per unit of industrial added value reduced by ≥15%
Shandong Province	Shandong Environmental Development [2023] Document No. 16	*Implementation Plan for Demonstration Projects of Near-Zero Carbon Cities, Parks, and Communities in Shandong Province*	Near-zero carbon parks, near-zero carbon communities	Total CO_2 emissions reduced compared to 2020
Jiangsu Province	Jiangsu Development and Reform Commission, Resource and Environmental Department [2024] Document No. 366	*Jiangsu (Near) Zero Carbon Industrial Park Construction Guide (Provisional)*	Near-zero carbon industrial parks	CO_2 emissions per unit of industrial added value ≤0.6 tCO_2/10,000 RMB

Based on the above analysis, the construction of near-zero carbon emission zones currently focuses mainly on industrial parks, office parks, and residential communities. At the same time, there are differences in the existing evaluation systems. The carbon emission targets include four forms: reduction in total carbon emissions, reduction in carbon emission intensity, reduction in per capita carbon emission intensity, and reduction in carbon emission intensity per unit of industrial added value. Furthermore, the implementation of near-zero carbon emission zone construction mainly involves areas such as energy, buildings, transportation, industry, and ecology.

3.2 Case Studies of Near-Zero Carbon Emission Zones

Currently, there are successful domestic and international examples of near-zero carbon emission demonstration projects at the park/community scale (**Table 2**). The Europark Zero Carbon Technology Park in Germany achieves an 80-95% renewable energy utilization rate through energy, buildings, transportation, and operations. The Hammarby EcoCity in Sweden mainly achieves the recycling of waste, water, and energy through sustainable planning and design concepts. The Shanghai Expo Park in China continues to practice low-carbon and energy-saving development concepts in areas such as buildings, energy, transportation, resources, and landscaping, with a carbon emission reduction rate exceeding 30%. The "International Sailing Center" in Qingdao, Shandong, China has achieved zero-carbon community construction in aspects such as zero-carbon heating, energy efficiency improvement, terminal optimization, and renewable energy utilization.

Based on the above analysis, the construction path for near-zero carbon emission zones should focus on six key aspects: green buildings, energy efficiency, low-carbon transportation, resource recycling, green industries, and smart operations. Through the integrated application of green and low-carbon technologies and innovative management practices, low-carbon energy saving and sustainable development can be achieved.

4 Challenges in Achieving Near-Zero Carbon Emissions in High-Density Urban Districts

There are several challenges in achieving near-zero carbon emissions in high-density urban districts. First, the high building coverage rate leads to high energy consumption in buildings, and it is difficult to implement ultra-low energy buildings or near-zero energy buildings in areas dominated by super-tall structures. Second, the low development space ratio can result in conflicts in the spatial layout of low-carbon technologies, such as competition for rooftop space between green roofing and photovoltaics. Third, the cost of utilizing energy resources is high, and the limited space in high-density urban districts makes it difficult to balance economic costs with carbon reduction outcomes when using technologies like photovoltaics and energy storage. Finally, the construction process of high-density urban districts often lacks coordinated planning for the systematic integration of low-carbon technologies and spatial matching.

Table 2. Implementation Paths for Different Types of Near-Zero Carbon Emission Projects.

Location	Project Name	Scale	Zero-Carbon Honors	Energy/Carbon Reduction Goals	Key Initiatives
Germany	EUREF-Campus	5.5 hectares	Europe's first zero-carbon smart park	80% to 95% of the park's energy comes from renewable sources; the carbon emission reduction target of 80% was achieved ahead of schedule in 2014	Buildings: All new buildings are green energy-efficient buildings, with LEED Platinum certification;
					Energy: Solar, wind, geothermal; district heating; purchased agricultural biogas with an annual power generation of 2 MWh;
					Transportation: Entirely electric vehicles, with smart EV charging stations; retired batteries are recycled to create up to 1.9 MWh of energy storage;
					Operations: Energy management integrated with the overall digital solution for the smart park, based on microgrid EMA for photovoltaic, energy storage, and charging solutions.
Sweden	Hammarby Eco-Town	2.04 million square meters	Model for sustainable urban development	About 80% of total energy consumption is from renewable sources, 29-37% reduction in GHG emissions from the 1990s	Energy: Renewable energy; 100% district heating for buildings;
					Transportation: 80% of travel is by public transport;
					Waste: Precise waste sorting system, recyclable water treatment system.
Shanghai, China	Shanghai World Expo Park	5.28 square kilometers	Ministry of Housing and Urban-Rural Development peak carbon dioxide emissions and carbon neutrality case in 2021	Carbon reduction rate of 30% compared to the baseline scenario (2017)	Energy: Three distributed energy centers with a total photovoltaic installation capacity of 4.5 MW;
					Buildings: 100% of buildings meet the Green Building Two-Star standard or higher, with 61% of buildings reaching the Three-Star standard; one near-zero-energy building demonstration;
					Transportation: High-density bus + metro network, with a 10-kilometer riverside slow-traffic system;
					Resources: Water reuse initiatives; the Shanghai World Expo Park's rainwater utilization rate exceeds 20%;
					Landscaping: Roof greening on new buildings, and the Expo Cultural Park's landscaping area is expanded by 2 square kilometers.
Qingdao, Shandong Province, China	International Sailing Center" Zero-Carbon Community	45 hectares, with a building area of 227,000 square meters	Excellent case in "Zero Carbon China" in 2020	Directly reduces carbon emissions by 8,663 tons per year; building carbon emission intensity reduced to 25 kg $CO2/m^2$	Energy: Uses clean energy technologies such as seawater-source heat pumps, solar photovoltaic and solar thermal energy, wind power, wastewater-source heat pumps, and industrial waste heat recovery;
					Buildings: Energy audits and energy-saving diagnostic analysis for 10 existing buildings;
					Transportation: Entirely electric vehicles, integrated charging stations with light storage, and smart street lighting;
					Operations: Establishment of a regional energy control platform (energy consumption monitoring, smart management).

5 Calculation and Verification of Near-Zero Carbon Emissions in High-Density Urban Districts

5.1 Overview of the Urban District Case

Fig. 1. Rendering of the urban block project.

This urban district project is an important transportation hub commercial complex in Shanghai, with a total land area of approximately 60,000 m² and a total building area of about 540,000 m². The buildings consist of six towers and podiums, with the tallest tower reaching 29 floors and a height of 100-140 meters (**Fig. 1**). The floor area ratio (FAR) of the entire block is 5.0, making it a typical high-density district block.

5.2 Implementation Path for Near-Zero Carbon Emission in the Urban District

Focusing on the goal of near-zero carbon emission in the urban district, this project proposes two core emission reduction pathways: "carbon reduction sources + carbon sinks enhancement." (**Fig. 2**). Through the "Two Low, Four Transformations" approach (low energy consumption, low emissions, circularity, ecological integration, maximization, and digitalization), six major green and low-carbon technological measures will be

 https://doi.org/10.1051/e3sconf/202561702013

implemented to reduce the carbon emission intensity of the district, leading the way in near-zero carbon emission urban district construction.

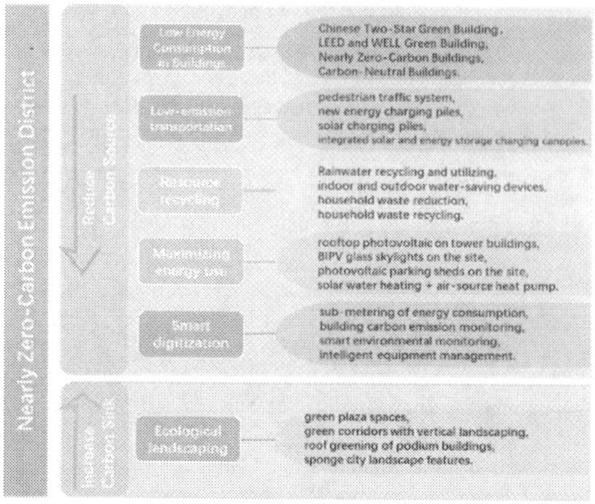

Fig. 2. Implementation Path for Near-Zero Carbon Emissions in the Urban District.

5.2.1 Low Energy Consumption in District Buildings

The buildings in the district have been constructed using an optimal technological strategy that meets the requirements of green building standards, as well as WELL and LEED Gold-level design specifications, to achieve high standards for low-carbon buildings. All new buildings fully comply with the Two-Star or higher green building standard, with office buildings T1-T6 meeting the LEED Gold-level design requirements. Additionally, the T1 commercial office building meets the WELL Gold-level requirements and serves as a demonstration for carbon-neutral buildings, while the T3-02 building is a near-zero carbon building demonstration. The new buildings integrate green, low-energy technologies through high-performance building envelopes, efficient mechanical and electrical systems, energy-saving elevators, and heat recovery for fresh air. A pilot project for high-efficiency machine rooms is also being implemented, with cooling plants achieving an EER energy efficiency ratio of ≥ 5.0, contributing to deep energy saving and carbon reduction in the buildings.

5.2.2 Low Emissions in District Transportation

The district's transportation system is organized with a focus on pedestrian networks as part of a multi-level urban traffic system. This includes a connected underground space, accessible ground-level district squares, and a two-story pedestrian corridor system, creating a multi-layered, integrated transportation network (**Fig. 3**), which allows for seamless one-stop transfers for pedestrians. Additionally, low-carbon transportation infrastructure within the district has been improved, with the proportion of new energy charging facilities reaching 15%. Two pilot projects for photovoltaic-storage-integrated parking shelters and two solar charging stations have been constructed. Furthermore, the district's public lighting system uses 100% LED energy-efficient lamps.

Fig. 3. Rendering of the District's Multi-Level Pedestrian System.

5.2.3 Resource Recycling in the District

The district implements the utilization of water resources and solid waste recycling. A rainwater reuse system has been set up to collect rainwater from rooftops, courtyards, and other areas. After treatment, the rainwater is used for landscaping irrigation and road cleaning. The rainwater collection pond has a capacity of 660 m^3, with a rainwater recovery utilization rate of 3%, achieving resource recycling of rainwater. Efforts are also made to reduce waste generation at the source and promote the recycling of household waste. All domestic waste is transported off-site for disposal, with a resource utilization rate of over 80%. In addition, both indoor and outdoor areas use high-efficiency water-saving facilities, and landscape irrigation employs water-saving methods. All indoor sanitary appliances are equipped with water-saving devices.

5.2.4 Ecological Landscaping in the District

Fig. 4. Central Park and Aerial Walkway Rendering.

The district's landscaping focuses on creating three-dimensional urban public spaces, with the layout featuring a central city park, green plazas, green roofs on podium buildings, and elevated green corridors, among other green spaces. The green coverage rate exceeds 20%, with over 30% of the podium roofs dedicated to greening, creating a district environment where greenery is visible from every window (**Fig. 4**). A resilient sponge city landscape is strategically planned, including approximately 5,800 m² of green roofs and a 60 m² rain garden, to create an ecologically sustainable and visually appealing sponge city demonstration district that is also highly people-friendly.

5.2.5 Maximization of Energy Use in the District

Energy carbon reduction is a key strategy for achieving near-zero carbon emissions in the district. The project aims to maximize energy utilization within the district. Due to significant shading from the podium buildings in this high-density district, photovoltaic panels are mainly installed on the rooftops of the towers. The photovoltaic capacity on the rooftops of T1 and T3 towers is approximately 0.12 MW, with a 70-kWh energy storage system, forming a carbon-neutral building and near-zero carbon building demonstration highlight (**Fig. 5**). The rooftop photovoltaic capacity on T2, T4, and T6 is approximately 0.18 MW. The district also includes two BIPV (Building Integrated Photovoltaic) glass skylights and two solar photovoltaic parking canopies to supply power for the underground garage lighting. The total annual green electricity generation for the project is approximately 840,000 kWh/year. Additionally, a 100 m² solar thermal panel system and air-source heat pump water heating system have been installed on the roof of the T1 podium to preheat water for the staff cafeteria.

Fig. 5. Rendering of the Photovoltaic System on the Rooftops of T1 and T3 Towers.

5.2.6 Digitalization of District Management

The district adopts smart management by establishing a digital twin platform for carbon emission monitoring. This platform integrates real-time monitoring and management of building energy consumption, carbon emissions, and district environmental data. The digital platform is linked with the building's mechanical and electrical systems for interactive control, achieving intelligent group control of operational equipment and energy-saving management. Additionally, it is planned to integrate photovoltaic systems, charging stations, landscape lighting, and other energy usage into the digital platform, enabling comprehensive smart management of the district's operations.

5.3 Carbon Emission Calculation and Verification for the District

The carbon emission accounting boundaries for this project are defined in terms of buildings, transportation, resources, carbon sinks, energy, and other related sectors. A scenario analysis method is used to establish a baseline scenario and a near-zero carbon scenario, setting different levels of low-carbon target parameters for each sector (**Table 3**). The carbon emission goals for the district are calculated and verified as follows:

Baseline Scenario: This assumes the carbon emissions of the district project under normal economic and social development trends, without any low-carbon measures being taken.

Near-Zero Carbon Scenario: This assumes the district project implements advanced carbon reduction technologies to significantly lower carbon emissions, achieving near-zero carbon levels or meeting the best practices for similar areas in the city.

Based on the research findings in Section 3.1 and Section 3.2, the minimum carbon reduction requirement for near-zero carbon emission districts in China is a carbon reduction rate of over 40%. Typical cases of near-zero carbon emissions at home and abroad can reduce carbon emissions by 30% to 80% (Different baseline scenarios).

After calculation and analysis, the carbon emissions for this district project under the baseline scenario are 36,000 tCO_2/a, and under the near-zero carbon scenario, the carbon emissions are 22,000 tCO_2/a. The total carbon reduction for the entire district is 15,000 tCO_2/a, achieving a reduction rate of 40.3% (**Table 4**), meeting the target of over 40% and reaching the level of a near-zero carbon emission district in China. Approaching the carbon emission reduction targets of typical overseas near-zero carbon emission neighborhoods.

Table 3. Low-Carbon Parameters for Different Scenarios.

Sector			Low-Carbon Parameters	
			Baseline Scenario	**Near-Zero Carbon Scenario**
Buildings	New office and commercial buildings		100% basic level	100% Two-Star Green Building certification, with T3-02 as a near-zero carbon building and T1 as a carbon-neutral building
Transportation	Transportation	Proportion of new energy charging stations	10%	Increased to 15%
	Lighting	Road lighting energy efficiency	100% energy-saving lighting	100% energy-saving lighting
Resources	Water resources	Non-traditional water source utilization rate	None	≥3%
	Solid waste treatment	Solid waste resource utilization rate	Current rate: 42%	Increased to 80%

Sector		Low-Carbon Parameters	
		Baseline Scenario	Near-Zero Carbon Scenario
Landscape	Ancillary green space	Green space ratio of 20%	Green space ratio of 20% + 30% green roof coverage on the podium buildings
Energy	Diversified photovoltaic systems	None	Photovoltaic capacity on T1 and T3 roofs is about 0.12 MW; on T2, T4, and T6 roofs about 0.18 MW; BIPV skylights, T5 + T6 photovoltaic carports

Table 4. District Carbon Emission Analysis.

Sector	Carbon Emissions		Carbon Reduction (tCO_2/a)
	Baseline Scenario Carbon Emissions (tCO_2/a)	Near-Zero Carbon Scenario Carbon Emissions (tCO_2/a)	
Buildings	29207	17186	12021
Transportation	2159	2073	86
Water Resources	213	141	72
Solid Waste	4845	2708	2137
Carbon Sinks	-17	-26	9
Energy	0	-357	357
Total	36407	21726	14682
Carbon Reduction Rate Compared to Baseline (%)			40.3%

6 Conclusion

High-density districts are a key area for achieving carbon peak and carbon neutrality goals in megacities. Urban center high-density districts are characterized by four main features: high building floor area ratio, high building coverage ratio, low development space ratio, and high population density. The near-zero carbon emission construction in these districts emphasizes the integrated application of green low-carbon technologies and innovative management mechanisms.

1. This article uses a high-density district project in Shanghai as a case study, exploring and establishing a near-zero carbon emission construction implementation path based on the "Two Low and Four Circular" approach. The study also employs scenario analysis to convert and analyze the numerical transformation of the district's near-zero carbon construction indicator requirements, further validating the feasibility of achieving a carbon emission reduction rate of over 40% in high-density districts.

2. The characteristics of different types of high-density neighborhoods vary, but there is consistency in the construction and implementation paths to achieve near-zero carbon emissions. That is, the technology path of "carbon reduction sources + carbon sinks enhancement", and the necessary market pathways for carbon offsets such as CCER carbon trading and Renewable Energy Certificate. The Market Pathway is primarily used for carbon-neutral building of small-scale volumes. District with higher average floor area ratios have greater difficulty in adopting technological measures to achieve near-zero carbon emissions, it need more efficient technological innovation practices.

3. Faced with the challenges of near-zero carbon transformation, the construction of high-density districts focuses on key measures such as low-energy buildings, low-emission transportation, resource circulation, ecological landscaping, energy maximization, and digital management. Through technical innovations such as near-zero carbon buildings, carbon-neutral buildings, integrated solar storage charging, BIPV (Building-Integrated Photovoltaics), and energy-carbon monitoring digital twin platforms, the study provides a path to lead high-density districts toward near-zero carbon emissions and green sustainable development, contributing valuable experience to the peak carbon dioxide emissions and carbon neutrality efforts in the field of urban and rural construction.

Acknowledgments

1 Shanghai 2022 "Science and Technology Innovation Action Plan" Technology Support for Carbon Peak and Carbon Neutrality Special Project: "Research on Carbon Emission Accounting Methods and Integrated Control Application Scenarios for High-Density, High-Quality Districts with Near-Zero Carbon Emissions" (22dz1207902)

2 National Key R&D Program Project: "Carbon Reduction Technology System for Building Urban Integration and Verticalized Land Use Based on Land Utilization and Spatial Compounding" (2023YFC3804104)

References

1. China Association of Building Energy Efficiency, Chongqing University. "2023 China Building and Urban Infrastructure Carbon Emission Research Report" [R]. 2023-12-27.
2. Sun Juan. "Integrated System of Urban District Carbon Reduction Planning Methods" [J]. *Urban Planning Forum*, 2022(6): 102-109.
3. Leng Hong, Zhao Yan, Yuan Qing. "Carbon Reduction Pathways and Strategies for Urban Form Regulation" [J]. *Urban Planning Forum* 2023(1): 54-61.

4. FONG W K, MATSUMOTO H, HO C S, et al. Energy Consumption and Carbon Dioxide Emission Considerations in the Urban Planning Process in Malaysia [J]. *Journal of the Malaysian Institute of Planners*, 2008(6):101 - 130.
5. Liu Zhilin, Qin Bo. "Urban Form and Low-Carbon Cities: Research Progress and Planning Strategies" [J]. *Urban Planning International*, 2013, 28(2): 4-11.
6. Lv Bin, Sun Ting. "Research on the Compactness of Urban Spatial Form from a Low-Carbon Perspective" [J]. *Journal of Geographical Research*, 2013, 06:10.
7. Chen Xinying. "Study on the Green Land Energy Saving and Carbon Reduction Evaluation and Design Optimization Strategies for High-Density Urban Districts — A Case Study of Guangzhou" [D]. *South China University of Technology*, 2023.
8. Liu Sibei, Jia Beisi. "Analysis Method of High-Density Hong Kong Urban Form Represented by Pedestrian Integrated Skyscrapers" [J]. *Architecture Worlds*, 2023, 38(04).
9. Li Sheng. "Park-Oriented City — New Strategies for Sustainable Development of High-Density Urban Districts in France and Chinese Metropolises" [J]. *Residential District*, 2014(5): 2.
10. Cui Guangxun, Zhang Yuan, Fan Yue, et al. "Research on the Generation Mechanism and Methods of High-Density District Layout Forms — A Case Study of Shenzhen" [J]. *Architecture Worlds*, 2024, 39(03).
11. Ling Changlong, Liu Zhihang. "Comparison and Insights of High-Density and Compact Development Concepts in Chinese and Foreign Cities" [J]. *Urban Planning International*, 2024, 39(2): 1-8.
12. JONES C M, KAMMEN D M. Quantifying carbon footprint reduction opportunities for U.S. Households and communities[J]. *Environmental Science & Technology*, 2011, 45(9):4088-4095.
13. Ye Zuda, Bai Wei. "Carbon Emission Measurement, Verification, and Reporting Management System for Green Ecological Urban Districts in the Operational Stage" [J]. *Urban Development Studies*, 2018(1): 1006-3862.
14. Ma Yueyang, Wang Xiaolin, Yang Yusheng. "Research on Carbon Emission Calculation and Evaluation Methods in the Construction Phase of High-Density Districts" [J]. *Building Construction*, 2024, 46(10): 1724-1727.
15. Zhang Gaiqing, Li Wan, He Xixing. "Carbon Emission Calculation and Analysis in the Operational Stage of Low-Carbon Districts" [C]. *The 15th International Conference on Green and Energy-Efficient Building and New Technologies and Products Expo*, 2024-11-17.
16. Li Fang, Song Yashan, Zhou Yibing. "Discussion on Community Carbon Emission Calculation Methods and Case Analysis" [J]. *Journal* of *Green Building*, 2022(005): 014.

Analysis and Research on the Influence of Civil Aircraft System on CO2 Emission

Kexin Sun*, Yi Yang, Zhaojie Xuan

COMAC Shanghai Aircraft Design and Research Institute, 5188 Jinke Road, Shanghai 200000, China

Abstract. Based on the international standard ICAO Annex 16 Volume III, this paper adopts the CO2 emission airworthiness calculation method proposed by ICAO to carry out sensitivity analysis and research on the influence of air intake volume of pneumatic system, power extraction of electric system and hydraulic system from engine on civil aircraft CO2 emission. The methods of reducing fuel consumption and CO2 emission of aircraft are discussed from the perspective of air intake volume and power extraction.

1 Introduction

As assessed by the United Nations Intergovernmental Panel on Climate Change in 2022, CO2 emissions from the aviation sector of 2019 accounted for 1.8% of total greenhouse gas emissions [1].

Compared to industries such as electricity and industry, the total carbon emissions from aviation are lower. However, according to ICAO's (the International Civil Aviation Council) forecasts as shown in Fig 1, if no measures are taken, CO2 emissions may increase in a rapid way, which is 3 times that of 2019 by 2050[2].

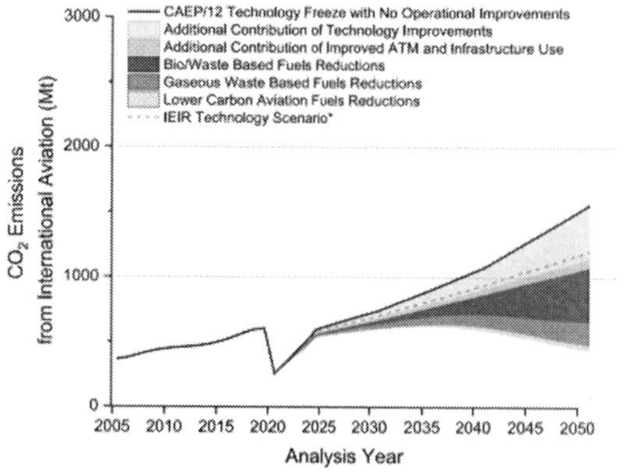

Fig. 1. ICAO's forecasts on net CO2 emissions from international aviation, 2005 to 2050.

Considering the impact of aviation carbon emissions, ICAO has established a Long-Term Ideal Goal (LTAG) in 2022 to achieve net zero aviation emissions by 2050[3, 4]. The carbon reduction methods in the aviation industry currently mainly include new aircraft technologies to improve efficiency, the use of sustainable aviation fuels, improvements in air traffic management and operations, and carbon offsetting[5, 6].

In 2017, ICAO adopted Annex 16 Environmental Protection Volume III: Aeroplane CO2 Emissions[7], which is the world's first international standard for civil aircraft CO2 emissions and the first mandatory standard for aircraft CO2 emissions. A330-900 received EASA's first aircraft CO2 certification On 26 May 2021, with Airbus becoming the first major manufacturer to receive this certification[8].

For civil aircraft manufacturers, CO2 emissions have become a necessary condition for airworthiness certification. At the same time, improving aircraft design technology, further reducing aircraft CO2 emissions based on airworthiness certification, and contributing to aviation carbon reduction have also become issues that need to be considered and studied.

2 Airworthiness calculation method of civil aircraft CO2 emissions

According to ICAO Annex 16 Volume III, the airworthiness certification of CO2 emissions from civil aircraft mainly examines the CO2 emissions of fuel consumption of the power plant during aircraft cruising. The standard defines the emission measurement value as Metric Value (MV). The calculation method for the maximum limit value of aircraft CO2 emissions varies depending on the time of applying for the type certificate, Formula (1) is for aircrafts that apply for type certificate after January 1st, 2020 and formula (2) is for aircrafts that have been certified by the existing regulations.

$$MV_{\lim} = \begin{cases} 10^{\left[-2.73780+0.681310\times \lg MTOW-0.0277861\times(\lg MTOW)^2\right]}, MTOW \le 60000 \\ 0.764, 60000 < MTOW \le 70395 \\ 10^{\left[-1.412742-0.020517\times \lg MTOW+0.0593831\times(\lg MTOW)^2\right]}, MTOW > 70395 \end{cases} \quad (1)$$

$$MV_{\lim} = \begin{cases} 10^{\left[-2.57535+0.609766\times \lg MTOW-0.0191302\times(\lg MTOW)^2\right]}, MTOW \le 60000 \\ 0.797, 60000 < MTOW \le 70395 \\ 10^{\left[-1.39353-0.020517\times \lg MTOW+0.0593831\times(\lg MTOW)^2\right]}, MTOW > 70395 \end{cases} \quad (2)$$

* Corresponding author: sunkexin@comac.cc

where:

MV_{min} is maximum permitted CO_2 emissions MV (kg/km);

MTOW is maximum take-off mass (kg).

Fig. 2 shows the CO2 emission limits for aircraft at different maximum takeoff weights.

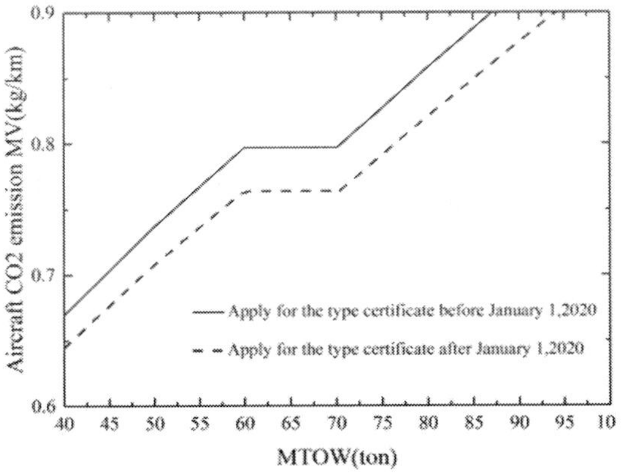

Fig. 2. Maximum permitted CO2 emissions MV with different MTOW of aircraft.

The calculation formula for the MV of CO2 emissions from civil aircraft is as follows:

$$MV=\left(\frac{1}{SAR}\right)_{AVG}\times\frac{1}{RGF^{0.24}} \tag{3}$$

where:

SAR is specific air range (kg/km) defined in formula (4);

$(1/SAR)_{AVG}$ is the average of the 1/SAR values (kg/km) for the three reference masses defined in formula (5);

RGF is reference geometric factor (non-dimensional parameter) defined used to adjust $(1/SAR)_{AVG}$;

SAR is calculated from the following formula:

$$SAR=\frac{TAS}{W_f} \tag{4}$$

where:

TAS is the true airspeed (km/h) of cruising;

W_f is total aeroplane fuel flow (kg/h) of cruising.

The standard requires SAR to take the maximum value under cruising state, that is, when the aircraft is at low, mid, and high gross mass, the maximum ratio of true airspeed to the fuel flow rate is taken as the calculation condition for SAR. This calculation value essentially represents the optimal balance between Mach number and flight altitude. After research, the optimal value generally occurs at an altitude of 35000 to 39000 feet and around the designed cruising Mach number.

Low, mid, high gross mass is defined as follows:

$$\begin{cases} M_{HGM}=0.92\cdot MTOW \\ M_{MGM}=\dfrac{M_{HGM}+M_{LGM}}{2} \\ M_{LGM}=0.45\cdot MTOW+0.63\cdot MTOW^{0.924} \end{cases} \tag{5}$$

where:

M_{HGM}, M_{MGM}, M_{LGM} is high gross mass, mid gross mass, low gross mass respectively (kg).

RGF as the reference geometric factor is a measure of the fuselage size, which is 1 square meter (m^2) divided by the defined reference geometric area to determine the non-dimensional reference geometric factor. The reference geometric area is defined as the surface area (m^2) of the maximum width of the fuselage outer model line projected from the main cabin floor to its parallel plane, including all pressurized space on the main or upper deck including aisles, assist spaces, passage ways, stairwells and areas that can accept cargo and auxiliary fuel containers. It does not include permanent integrated fuel tanks within the cabin or any unpressurized fairings, nor crew rest/work areas or cargo areas that are not on the main or upper deck (e.g. "loft" or under floor areas), and does not include the cockpit crew area. The reference geometric area has different calculation methods for single deck aircraft and aircraft with upper cabins, as shown in Fig. 3(a) and (b).

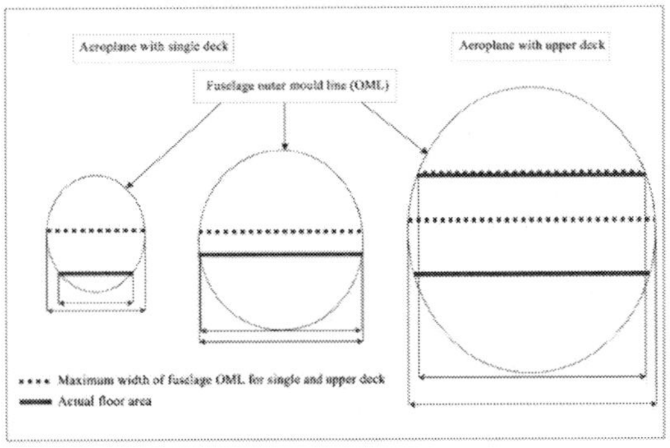

(a) Cross-sectional view

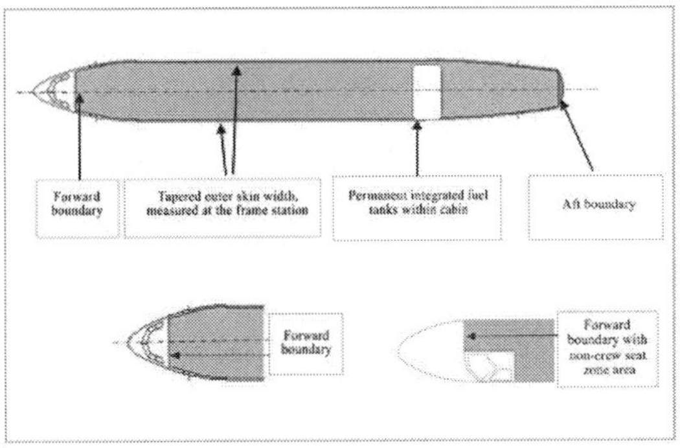

(b) Longitudinal view

Fig. 3. Diagram of reference geometric area.

https://doi.org/10.1051/e3sconf/202561702014

3 Analysis and research on the impact of pneumatic system, electric system and hydraulic system on aircraft CO2 emissions

3.1 Sensitivity analysis

According to the calculation method in Section 2, CO2 emissions can essentially be understood as "fuel consumed per passenger per flight distance". The fuel flow, drag, weight, cabin area, and other factors all have important impacts on emissions, and fuel flow is a key factor affecting emission values. From the perspective of aircraft systems, the air intake of pneumatic system from engine and the power extraction of the electric system and hydraulic system from engine affect the fuel consumption, thereby affecting the overall emission MV of the aircraft. The aircraft's pneumatic system take air from the engine, mainly providing air for cabin air conditioning, wing anti icing, engine starting, fuel tank inerting, and water tank pressurization. The engine driven pump of the hydraulic system and the generator of the electric system extract power from the engine to provide input for various systems of the aircraft.

This section focuses on the CO2 emission sensitivity analysis and research of a certain type of large aircraft using two different engine models for air intake and power extraction.

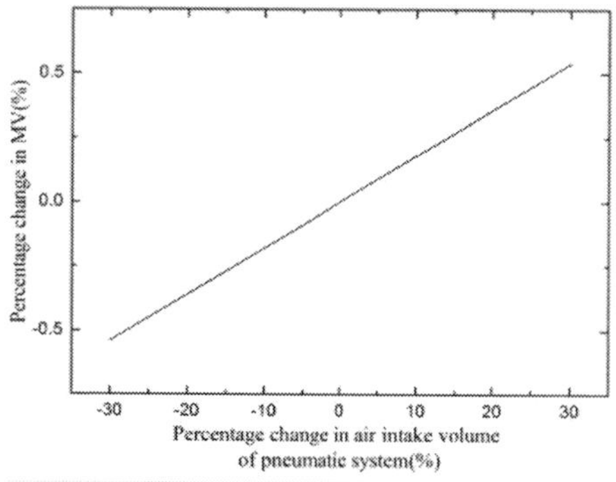

(a) Engine A

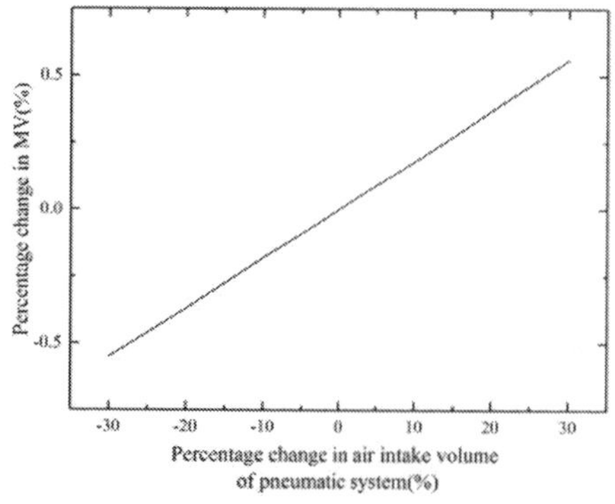

(b) Engine B

Fig. 4. Aircraft CO2 emissions vary with air intake volume of pneumatic system.

Fig. 4 shows the percentage change of CO2 emission values with respect to the air intake volume of pneumatic system from engine. As shown in the figure, the MV calculated by the two engines shows a similar linear relationship with the variation of air intake, and for every 10% increase in air intake, the MV increases by about 0.18%.

Fig. 5 shows the percentage change in CO2 emissions with power extraction of electric system and hydraulic system from engine. As the hydraulic system extracts much less power compared to the electric system, these two systems are analyzed as power extraction together. As shown in the figure, the trend of MV calculated by the two engines with power extraction is basically consistent, showing a linear relationship, and for every 10% increase in power extraction, MV increases by about 0.1%.

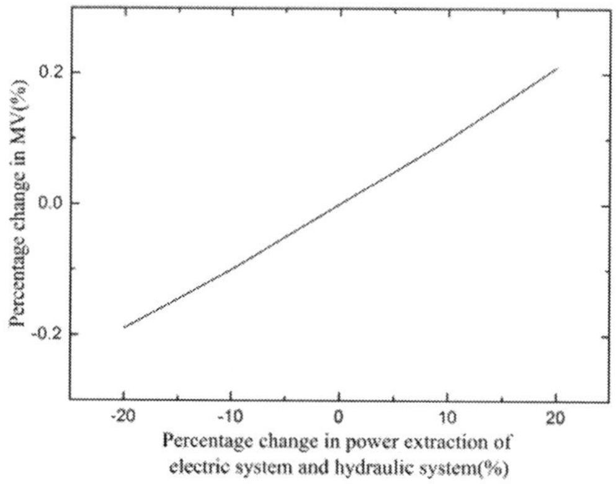

(a) Engine A

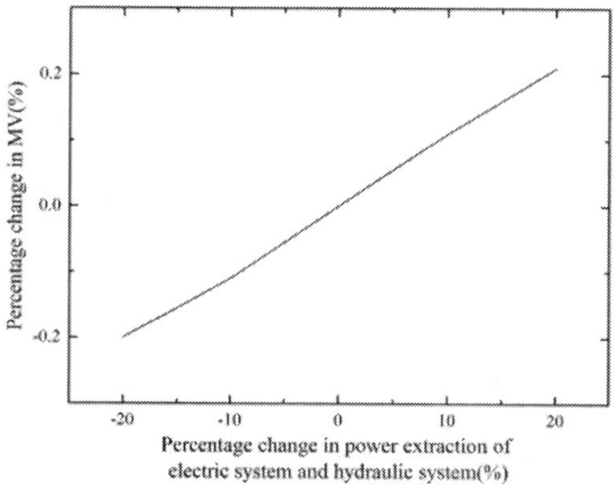

(b) Engine B

Fig. 5. Aircraft CO2 emissions vary with power extraction of electric system and hydraulic system.

According to formula (4), the fuel flow of the engine is linearly related to emissions, and the air intake and extraction power is also linearly related to fuel flow. Therefore, this result is in line with expectations. In addition, comparing Fig. 4 and Fig. 5, the benefits brought by the reduction of air intake are slightly higher than those brought by power extraction.

3.2 Emission reduction measures

The reduction of air intake volume of the environmental control system is an important research direction, which is roughly linearly related to the number of passengers[9]. The technology widely used in aircraft to reduce the air intake of environmental control systems includes: using a high-pressure water removal environment control system to lower the outlet airflow temperature of the turbine cooler, which can greatly reduce the air intake; Adopting a two-stage bleed air scheme, priority is given to bleed air from the medium pressure stage. When the medium pressure bleed air is insufficient, high-pressure bleed air is used, which can significantly reduce power loss and fuel consumption; Reasonably utilize cabin exhaust, with a portion used as recirculation air and a portion used to cool electronic equipment in the cabin[10].

Reducing the power extraction will result in less fuel consumption, thereby reducing CO2 emissions. The new adaptive power and thermal management system (APTMS) integrates the functions of emergency power system, auxiliary power system, and environmental control system in traditional airborne electromechanical systems. It uses multiple heat sink forms such as fuel and fan duct air to enhance the system's thermal management capabilities. At the same time, it adopts an adaptive combination power unit to achieve comprehensive control of engine power extraction and bleed air, greatly improving fuel economy[11].

Electric propulsion technology is a highly promising development direction, all electric propulsion is powered entirely by batteries, making it the cleanest and most efficient. However, due to limitations in battery power density, it is difficult to achieve for regional or mainline aircraft with high power levels and range. Vankan et al. studied the application of a typical parallel hybrid electric propulsion technology in multi electric aircraft. The scheme involves installing a battery motor system on the low-pressure shaft of a turbofan engine, where the low-pressure turbine and motor jointly drive the fan to provide thrust. The research results showed that the synergistic effect of hybrid electric propulsion and multi electric technology reduced fuel consumption by 12.2%[12].

4 Conclusion

Regarding the CO2 emissions from civil aircraft, from the perspective of aircraft systems, the main influencing factors are the air intake volume of pneumatic system, the power extraction of electric system and hydraulic system, which have an impact on CO2 emissions. MV shows a linear relationship with changes in air intake volume and power extraction, with an increase of about 0.18% for every 10% increase in air intake volume and with an increase of about 0.1% for every 10% increase in power extraction. Overall, the contribution of reduced air intake and power extraction to reducing MV emissions is limited. Improving the comprehensive control of power extraction and air intake can significantly improve fuel consumption and reduce emissions. Electric propulsion technology is a highly promising development direction, but it is currently not mature and requires further research.

References

1. IPCC. (2022) Climate Change 2022: Mitigation of Climate Change. https://www.ipcc.ch/report/sixth-assessment-report-working-group-3/.
2. ICAO. (2023) Environmental Report 2022. https://www.icao.int/environmental-protection/Pages/envrep2022.aspx.
3. ICAO. (2022) Long term global aspirational goal (LTAG) for international aviation. https://www.icao.int/environmental-protection/Pages/LTAG.aspx.
4. ICAO. (2022) State adopt net-zero 2050 global aspirational goal for international flight operations. https://www.icao.int/Newsroom/Pages/States-adopts-netzero-2050-aspirational-goal-for-international-flight-operations.aspx.
5. IATA. (2021) Our Commitment to Fly Net Zero by 2050.https://www.iata.org/en/programs/environment/flynetzero/.
6. ICAO. (2022) Environmental Report 2022. https://www.icao.int/environmental-protection/Pages /envrep2022.aspx.
7. ICAO. (2017) Annex 16 to the Convention on International Civil Aviation-Environmental-Protection, Volume III-Aeroplane CO2 Emissions. https://www.icao.int/environmental-protection/Pages/environment-publications.aspx.
8. Airbus. (2021) EASA certifies A330neo to ICAO's new CO2 emissions standard. https://www.airbus.com/en/newsroom/news/2021-05-easa-certifies-a330neo-to-icaos-new-co2-emissions-standard.
9. Zhao, B., Li, S, B., Zhou, S., et al. (2012) Research progress on air intake of aircraft engine air system. Progress in Aviation Engineering., 003(004): 476-485. DOI: 10.3969/j.issn.1674-8190.2012.04.017.
10. Ning, H, L., Hua, L., Liu, Y, A. (2013) Research on reducing air intake and fuel compensation loss in environmental control systems. TRAINER., (01): 50-53. DOI: CNKI:SUN:HDKO.0.2013-01-013.
11. Zhang, J, Y., Zhen, F, Y., Xie, Y, P. (2019) Analysis of the impact of integrated control of shaft power extraction and air intake on fuel economy of aircraft engines. Aeroengine., 45(6): 9-14. DOI: CNKI:SUN:HKFJ.0.2019-06-002.
12. Vankan, J., Lammen, W. (2019) Parallel Hybrid Electric PropulsionArchitecture for Single Aisle Aircraft- Powertrain Inves-tigation. MATEC Web of Conferences., 304(4): 03008.

Price Mechanisms in Green Electricity Markets and Game Analysis of Participant Behavior

Jing Liu[1], Tiantian Chen[1], Yiqing Xiao[1], and Jing Wang[2,*]

[1] State Grid Shanghai Electric Power Research Institute, Shanghai 200080, China
[2] School of Economics and Management, Wuhan University, Wuhan 430072, China

Abstract. As the global demand for carbon reduction continues to grow, promoting green electricity has become a key to achieve sustainable development. However, effectively incentivizing green electricity generators and quota subjects to participate widely in green electricity trading remains a challenge. This study constructs a bipartite network model based on the game between green electricity generators and quota subjects to analyze the impact of green electricity prices, traditional electricity prices, and green certificate market prices on decision-making behavior. The results show that an increase in green electricity prices significantly enhances market participation, while an increase in traditional electricity prices reduces the proportion of green electricity participation. This study provides a theoretical basis for formulating reasonable price mechanisms and policy incentives to promote the adoption of green electricity.

1. Introduction

As global climate change becomes increasingly severe, governments and international organizations worldwide have introduced carbon emission restrictions and environmental protection policies. The development and application of green electricity, such as wind and photovoltaic power, have become crucial strategies for advancing energy transition and achieving sustainable development. Green electricity, recognized for its renewability and environmental friendliness, has gradually gained prominence in the energy mix. However, effectively incentivizing power generation enterprises (green electricity generators) and electricity users (quota subjects) to actively participate in green electricity trading within a market-based power trading environment, while maximizing its environmental and energy value, remains a significant challenge for market design and policy-making.

In practice, the green electricity market is influenced not only by the price of electricity as a commodity but also by policy orientation, subsidy mechanisms, and the complex assessment of environmental value. Green electricity generators must weigh market returns against policy incentives when deciding whether to sell electricity to the green electricity market or the traditional electricity market. Similarly, quota subjects face a comparable trade-off when choosing between purchasing green electricity from the market or acquiring green certificates to meet policy requirements. These complex market dynamics make the decision-making behavior of both supply and demand sides unpredictable, introducing uncertainty and challenges to the efficient operation of the green electricity market. Therefore, an in-depth study of how market price mechanisms influence the decisions of green electricity generators and quota subjects is crucial for optimizing the structure of the green electricity market and enhancing market participation.

To address this issue, this study constructs a bipartite network model based on the game between green electricity generators and quota subjects, analyzing the impact of price fluctuations in the green electricity market, traditional electricity market, and green certificate market on the decision-making behavior of market participants. By simulating the strategic evolution of participants under different pricing conditions, the study highlights the critical role of market prices in regulating the supply-demand relationship for green electricity. Specifically, an increase in green electricity prices significantly enhances market participation, whereas a rise in traditional electricity prices reduces demand for green electricity. The findings provide theoretical support for policymakers in designing and managing green electricity markets, demonstrating that reasonable pricing and subsidy mechanisms can effectively guide market behavior and promote the adoption and sustainable development of green electricity.

2. Literature Review

With increasing global pressure for carbon reduction, green electricity trading has gradually become a crucial means of achieving energy transition and sustainable development. In this field, researchers have primarily focused on the pricing mechanisms, trading models, and incentive policies within the green electricity market.

First, regarding pricing mechanisms, existing literature indicates that green electricity prices and

* Corresponding author: wangjing980808@whu.edu.cn

traditional electricity prices are key factors influencing the decision-making of market participants. Higher green electricity prices can significantly enhance the participation of green electricity generators and quota subjects, while rising traditional electricity prices may suppress the consumption demand for green electricity [1,2]. Green electricity prices affect not only the production and supply decisions of generators but also the behavior of market participants. According to [3], higher green electricity prices can strongly incentivize generators to expand capacity and attract more participants to engage in green electricity trading, thereby promoting the expansion of the green electricity market. However, excessively high green electricity prices may increase the cost burden on users, potentially leading to a decline in demand and reduced market activity. Therefore, the reasonable determination of green electricity prices requires balancing the promotion of supply and the stimulation of consumption.

In addition, green certificate trading, as a complementary mechanism for advancing green electricity development, has also attracted widespread attention for its impact on market equilibrium and participant revenue [4,5]. Green certificates, as a market-based incentive tool, play a significant role in promoting green electricity and achieving carbon reduction goals. According to [6], green certificate trading markets certify renewable energy generation and enable its market-based trading, providing additional revenue for generators and a means for users to meet their green electricity consumption goals. [7] highlighted that the pricing mechanism for green certificates primarily depends on market supply-demand balance, the stringency of quota policies, and fluctuations in traditional electricity prices. When quota policy requirements are high and green certificate supply is insufficient, green certificate prices tend to rise, further stimulating green electricity production. Conversely, when certificate supply is ample but demand is weak, green certificate prices may decline, reducing economic incentives for green electricity generators.

Secondly, regarding market behavior, the decision-making behavior of different types of participants (such as green electricity generators and quota subjects) exhibits complexity in market games. Some studies have utilized game models to analyze the optimal strategies of market participants under various pricing mechanisms and policy incentives. [8] suggested that reasonable subsidy policies or price regulation mechanisms can alleviate competition between traditional electricity and green electricity, thereby promoting the overall development of the green electricity market. [9] emphasized the application of the bipartite network model in studying the supply-demand relationships and participant interactions within the green electricity market, highlighting that this modeling approach can more accurately capture the dynamic relationship between market prices and participation behaviors.

In summary, existing research has revealed the significant impact of green electricity pricing mechanisms on market behavior. However, further exploration is needed to fully consider the interconnections between green electricity, traditional electricity, and green certificate market prices, as well as the multi-participant game relationships. Based on this, this paper constructs a bipartite network game model between green electricity generators and quota subjects, focusing on the impact of pricing mechanisms on market behavior, thereby providing theoretical support for improving green electricity pricing policies.

3. Model

This study, based on bipartite networks and evolutionary game theory, constructs a model that incorporates the trading relationships between green electricity generators and quota subjects. In this model, green electricity generators must balance their participation in the green electricity market, traditional electricity market, and green certificate market to maximize their profits. Simultaneously, quota subjects need to choose between purchasing green electricity or traditional electricity and acquiring green certificates in the green certificate market to meet their energy needs and policy requirements. By characterizing the trading structure within the bipartite network, the model dynamically illustrates the behavioral evolution of participants across different markets, providing theoretical support for policymakers in designing effective and efficient market mechanisms

3.1 Market Participants

In this model, the participants are categorized into two main groups: green electricity generators (supply side) and quota subjects (demand side). Green electricity generators refer to power generation enterprises that supply green energy (e.g., wind and photovoltaic power), represented as a set $N = \{1,2,\dots,N\}$. They face two market options: they can allocate their generated electricity to the green electricity market, which combines both economic and environmental value, or to the traditional electricity market, which only provides economic value. Additionally, green electricity generators can participate in the green certificate market by generating and selling green certificates associated with their green electricity sales. Each generator must decide what proportion x_i of their total generated electricity P_i should be allocated to the green electricity market, while the remaining proportion $(1 - x_i)$ is directed to the traditional electricity market. Furthermore, generators must also decide the quantity of green certificates z_i they intend to sell in the certificate market.

Quota subjects refer to consumers or enterprises with energy demands who are also responsible for fulfilling environmental policy requirements or green electricity quotas. Represented as a set $M = \{1,2,\dots,M\}$ these subjects must meet their energy demands while complying with policy-mandated quotas to fulfill their environmental obligations. In the model, quota subjects can choose to directly purchase green electricity from the green electricity market or purchase conventional electricity from the traditional electricity market. To satisfy green electricity quota requirements, they can also buy green certificates from the certificate market. Quota subjects need to decide what proportion y_j of their total electricity demand D_j should be purchased from the green electricity market, with the remaining proportion $(1 - y_j)$ sourced from the traditional electricity market. Additionally, they must determine the quantity of green certificates g_j to purchase in order to meet their green quota obligations. These decisions are influenced by multiple factors, including electricity prices, green certificate prices, and environmental policies.

3.2 Market Structure

This model involves three primary markets: the green electricity market, the traditional electricity market, and the green certificate market, which together form the interaction structure among participants.

(1) Green Electricity Market:

In this market, electricity is traded together with its environmental value. Electricity in the green electricity market not only carries economic value but also embodies environmental attributes, namely the environmental benefits of green electricity. As a result, the price of green electricity in this market is typically higher than in the traditional electricity market, as it includes both the commodity value of electricity and the environmental value, reflecting policy support for renewable energy. The price of green electricity in this market is denoted as p_g, representing a composite of the electricity's economic and environmental value.

(2) Traditional Electricity Market:

In this market, electricity is traded solely as a commodity without any associated environmental value. Green electricity generators can allocate part of their generation to the traditional electricity market and earn revenue at the lower traditional electricity price p_e. Similarly, quota subjects may purchase electricity from the traditional electricity market to meet their basic energy needs. However, electricity purchased from this market does not satisfy environmental or green electricity quota requirements.

(3) Green Certificate Market:

The green certificate market is a separate trading platform for environmental value. Whenever green electricity generators sell a certain amount of green electricity in the green electricity market, they can generate corresponding green certificates (green credits). Quota subjects can purchase these certificates to fulfill environmental policy requirements or green electricity quotas. If quota subjects choose to purchase electricity from the traditional electricity market, they can still meet their green quota obligations by purchasing green certificates. The price of green certificates is denoted as p_z.

3.3 Bipartite Network

In this model, the trading relationships between green electricity generators and quota subjects can be represented by a bipartite network. A bipartite network refers to the connections between two different types of node sets. In this model, the node sets are green electricity generators and quota subjects, and the connections represent their transactions in various markets.

For N green electricity generators and M quota subjects, the trading structure of the bipartite network can be represented by an adjacency matrix $A \in \mathbb{R}^{N\times M}$, where each element A_{ij} represents the transaction volume between green electricity generator i and quota subject j. Based on the different market types, three distinct adjacency matrices can be defined as follows. The green electricity market adjacency matrix, denoted as $A^{green} \in \mathbb{R}^{N\times M}$, where A_{ij}^{green} represents the volume of green electricity sold by green electricity generator i to quota subject j. The traditional electricity market adjacency matrix, denoted as $A^{ele} \in \mathbb{R}^{N\times M}$, where A_{ij}^{ele} represents the volume of traditional electricity sold by green electricity generator i to quota subject j. The green certificate market adjacency matrix, denoted as $A^{cert} \in \mathbb{R}^{N\times M}$, where A_{ij}^{cert} represents the number of green certificates sold by green electricity generator i to quota subject j. The total transaction volume can be represented as the sum of the elements across these matrices.

3.4 Revenue Function of Green Electricity Generators

The revenue of green electricity generators comes from transactions in the green electricity market, traditional electricity market, and green certificate market. Their revenue function can be expressed by the following formula:

$$\pi_i = \sum_{j=1}^{M}\left(x_i A_{ij}^{green} p_g + (1 - x_i) A_{ij}^{ele} p_e + z_i A_{ij}^{cert} p_z\right) - C_i(P_i) \quad (1)$$

where, P_i represents the total electricity generation of green electricity generator i, x_i is the proportion of electricity allocated to the green electricity market by the generator, and $(1 - x_i)$ is the proportion allocated to the traditional electricity market. p_g is the price of electricity in the green electricity market, and p_e is the price in the traditional electricity market. $z_i = x_i P_i$ denotes the quantity of green certificates generated by the generator, which are sold at a price p_z. $C_i(P_i)$ is the total electricity generation cost of generator, expressed as $C_i(P_i) = c_i P_i$, where c_i is the unit cost of electricity generation. Green electricity generators aim to maximize their revenue by deciding how to allocate their electricity generation across different markets and whether to sell green certificates.

3.5 Utility Function of Quota Subjects

The utility function of quota subjects depends not only on their transactions with green electricity generators but also on their strategy y_j, which determines how much electricity to purchase from the green electricity market. By introducing the decision variable y_j, the utility function of quota subjects is expressed as follows:

$$U_j = \sum_{i=1}^{N}\left(y_j A_{ij}^{green}(u_g - p_g) + (1 - y_j) A_{ij}^{ele}(u_e - p_e)\right) - \sum_{i=1}^{N} A_{ij}^{cert} p_z - C_j(g_j) \quad (2)$$

where, $(u_g - p_g)$ represents the net utility of purchasing electricity from the green electricity market. When considering buying electricity from the green electricity market, quota subjects compare the utility per unit u_g with the price p_g . Higher prices p_g reduce the attractiveness of purchasing green electricity. Similarly, $(u_e - p_e)$ represents the net utility of purchasing electricity from the traditional electricity market. As the price p_e increases, the net utility derived from the traditional electricity market decreases. Quota subjects adjust the value of y_j to determine the proportion of electricity purchased from the green electricity market. A larger y_j indicates a higher proportion of electricity purchased from the green electricity market, which requires paying a higher green electricity price p_g . Conversely, $(1 - y_j)$ determines the proportion of electricity purchased from the traditional electricity market, where the utility per unit is u_e, and the price is p_e.

3.6 Utility Function of Quota Subjects

The strategy adjustments of participants (green electricity generators and quota subjects) follow a gradual best response dynamic. This means that in each round of the game, participants update their strategies based on the revenue or utility outcomes from the previous round, gradually approaching a local maximum of their revenue or utility. This dynamic process simulates how market participants, under conditions of limited information, adjust their strategies based on historical experience, progressively moving toward optimal behavior.

(1) Strategy Dynamics of Green Electricity Generators

For green electricity generators, the strategic variable is the proportion of electricity x_i they decide to allocate to the green electricity market, with the remaining proportion $(1 - x_i)$ allocated to the traditional electricity market. Green electricity generators adjust their strategies dynamically based on their market revenue $\pi_i(t)$ from the previous round and the behavior of other participants. This adjustment ensures that their strategy $x_i(t+1)$ in the next round is closer to the current optimal strategy. This update process can be represented using a gradual best response mechanism:

$$x_i(t+1) = x_i(t) + \beta \frac{\partial \Pi_i(t)}{\partial x_i} \quad (3)$$

where $\pi_i(t)$ represents the revenue of green electricity generator i in the t -th round of the game; $\frac{\partial \Pi_i(t)}{\partial x_i}$ is the gradient of the revenue function with respect to the strategy x_i , indicating the marginal change in revenue; and β is the adjustment rate parameter, which controls the step size of the strategy update. In each round of the game, green electricity generators evaluate the marginal revenue (the derivative of the revenue function with respect to the strategy variable x_i) and gradually adjust their participation proportions in the green electricity market and the traditional electricity market.

(2) Strategy Dynamics of Quota Subjects

The strategic variables for quota subjects are the proportion of electricity y_j they decide to purchase from the green electricity market and the quantity of green certificates g_j they purchase from the green certificate market. Quota subjects also follow a gradual best response dynamic, updating their strategies based on the marginal utility of their utility function. They adjust the proportion y_j of electricity purchased from the green electricity market and the traditional electricity market based on their utility $U_j(t)$ from the previous round and the utility differences between purchasing green electricity and traditional electricity:

$$y_j(t+1) = y_j(t) + \gamma\left(\frac{\partial U_j(t)}{\partial y_j}\right) \quad (4)$$

where $U_j(t)$ represents the utility of quota subject j in the t -th round of the game; $\frac{\partial U_j(t)}{\partial y_j}$ is the gradient of the utility function with respect to the strategy y_j , indicating the marginal change in utility; and γ is the adjustment rate parameter, controlling the step size of the strategy update. Quota subjects compare the net utility (the difference between utility and price) derived from the green electricity market and the traditional electricity market, gradually adjusting the proportion of electricity purchased. Similarly, quota subjects iteratively adjust their strategies during the game to maximize their utility.

4. Simulation Analysis

4.1 Parameter Setting

To construct the model and perform reasonable simulations, it is essential to assign values to the exogenous parameters within the model. Firstly, the price of green electricity (p_g) is generally higher than the price of traditional electricity, as it reflects not only the commodity value of electricity but also the environmental benefits and policy support. According to China's *Green Electricity Certificate Trading and Price Monitoring Data* and the *China Electricity Council*, the price of green electricity in China's green electricity market in 2023 was approximately 450 RMB per megawatt-hour (around \$65). Second, the price of traditional electricity (p_e) is relatively lower, reflecting only the basic supply-demand conditions of the market. Based on electricity pricing policies from China's National Development and Reform Commission (NDRC), the average price of coal-fired electricity nationwide in 2023 was about 380 RMB per megawatt-hour (around \$55). This data is sourced from the NDRC's *2023 National Electricity Market Price Adjustment Report*. In addition, the price of green certificates (p_z) reflects the environmental contribution value of green electricity. According to data from the China Green Electricity Certificate Trading Platform, the average trading price of green certificates in 2023 was approximately 50 RMB (around \$7.5) per certificate, as reported in the *Green Electricity Certificate Market Analysis Report* by the China Energy Research Institute. For the unit electricity generation cost (c_i) of green electricity generators, the *2023 China Electricity Cost Analysis Report* by the International Energy Agency (IEA) indicates that the cost of renewable energy (such as wind and solar power) generation in China is approximately 250 RMB per megawatt-hour, reflecting the specific cost structure within the domestic market. Finally, the adjustment rate parameters β and γ, which represent the sensitivity of participants to changes in past revenue or utility, are set within a reasonable range of 0.1 to 0.3. These parameters are typically selected based on the convergence properties of the simulation. For this study, β and γ are both set to 0.2. By assigning these values, the exogenous parameters in the model are grounded in the latest market data and academic research, ensuring the accuracy and practical applicability of the model.

4.2 Baseline Model Simulation Analysis

Figure 1 illustrates the strategy changes of green electricity generators and quota subjects across different evolutionary rounds, specifically showing the average green electricity sales proportion of generators and the average green electricity purchase proportion of quota subjects. By observing the trends in these two strategies, we can understand the decision-making dynamics of market participants in the green electricity market, where economic and environmental values intersect, as well as the impact of changing market conditions on both supply and demand sides.

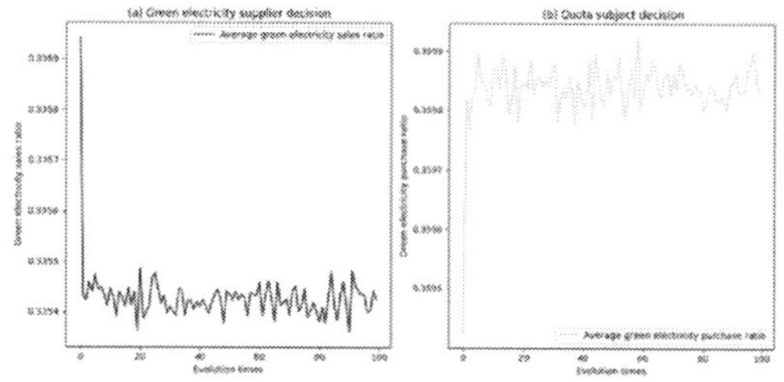

Figure 1 Baseline Model Evolutionary Analysis

In Figure 1, the left panel shows that the average green electricity sales proportion of generators rapidly decreases to approximately 0.3355 in the initial stage and then gradually stabilizes after fluctuations, ultimately remaining between 0.3354 and 0.3355. The right panel indicates that the green electricity purchase proportion of quota subjects initially rises rapidly to around 0.3596 and then exhibits relatively small fluctuations, eventually stabilizing between 0.3596 and 0.3599. For green electricity generators, the stable green electricity sales proportion suggests that, after multiple rounds of decision-making, they have gradually identified a sales strategy that aligns with market supply and demand balance, while also being influenced by market conditions and unit generation costs. For quota subjects, the relatively high proportion of green electricity purchases reflects strong demand in the green electricity market, likely driven by policy incentives or the green certificate mechanism. This highlights that policymakers can enhance green electricity consumption through well-designed market mechanisms and policy tools, achieving a balance between environmental goals and economic benefits.

4.3 Sensitivity Analysis

4.3.1 Sensitivity Analysis of Green Electricity Market Price

Figure 2 illustrates the trends in the average green electricity sales proportion of generators and the average green electricity purchase proportion of quota subjects under different green electricity market prices (p_g). By observing the sales proportion in the left panel and the purchase proportion in the right panel, the impact of market price changes on the decision-making of both parties can be revealed, helping us understand how the pricing mechanism regulates the supply-demand relationship in the green electricity market.

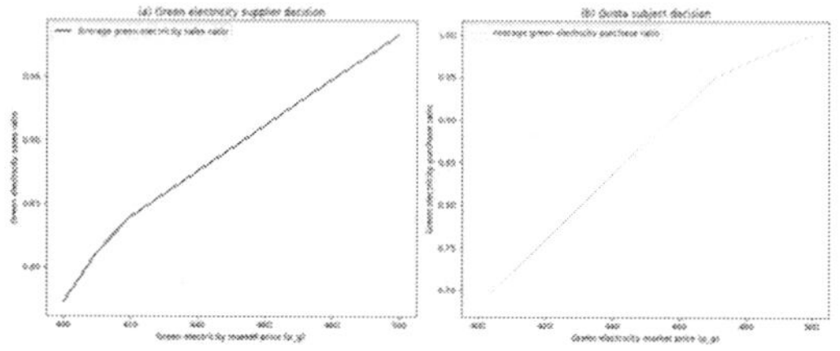

Figure 2 Sensitivity Analysis of Green Electricity Market Price

From the left figure, as the green electricity market price increases from 400 RMB/MWh to 500 RMB/MWh, the green electricity sales proportion of generators steadily rises from approximately 0.80 to nearly 1.00, indicating that higher prices encourage more generators to sell their electricity to the green electricity market. Specifically, the sales proportion reaches about 0.90 at 450 RMB/MWh and approaches 0.97 at 500 RMB/MWh. This trend demonstrates that rising market prices significantly enhance the participation of green electricity generators in the green electricity market. The right figure shows that the green electricity purchase proportion of quota subjects also increases gradually with the price, rising from around 0.70 to nearly 1.00, particularly accelerating after 460 RMB/MWh. This suggests that alongside the price increase, policy incentives and market demand drive quota subjects to favor purchasing green electricity to meet environmental and energy policy requirements. These findings highlight the critical role of pricing mechanisms in regulating market behavior. As green electricity prices rise, the supply proportion from green electricity generators increases, reflecting the incentive effect of price signals on green electricity supply. Simultaneously, quota subjects exhibit a higher proportion of green electricity purchases, indicating that reasonable pricing and policy support can significantly boost market demand and facilitate the widespread adoption of green electricity. This trend provides valuable insights for governments and energy market regulators: by adjusting prices and utilizing policy tools, it is possible to effectively guide the decisions of market participants, promote the development of green electricity markets, accelerate the adoption of clean energy, and achieve environmental goals.

https://doi.org/10.1051/e3sconf/202561702015

4.3.2 Sensitivity Analysis of Green Electricity Market Price

Figure 3 illustrates the trends in the average green electricity sales proportion of generators and the average green electricity purchase proportion of quota subjects under different traditional electricity market prices (p_e). By analyzing the decisions of green electricity generators in the left panel and quota subjects in the right panel, we can observe how changes in traditional electricity market prices affect the supply-demand relationship of green electricity in the market.

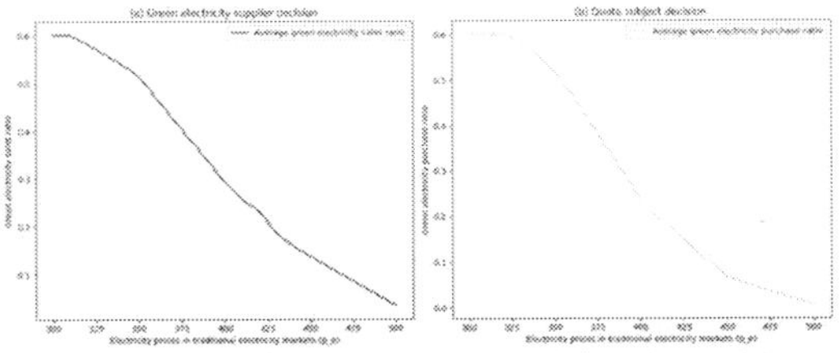

Figure 3 Sensitivity Analysis of Electricity Market Price

From the left figure, it can be observed that as the traditional electricity market price increases from 300 RMB/MWh to 500 RMB/MWh, the green electricity sales proportion of generators gradually declines from nearly 0.60 to around 0.05, with a particularly sharp drop after 400 RMB/MWh. In the right figure, a similar downward trend is evident in the green electricity purchase proportion of quota subjects, which decreases from approximately 0.60 to near zero. Specifically, when the price reaches 450 RMB/MWh, the green electricity purchase proportion falls below 0.10. This indicates that as traditional electricity market prices rise, both green electricity generators and quota subjects gradually reduce their participation in and purchase of green electricity. This trend is likely because the high traditional electricity prices encourage generators to allocate more electricity to the traditional market to achieve higher profits, while quota subjects may find the traditional electricity market more attractive due to its lower cost compared to green electricity. These findings highlight the significant impact of traditional electricity prices on the green electricity market. When traditional electricity prices rise, green electricity generators may prefer the traditional market for better profitability, and quota subjects may reduce their green electricity purchases due to cost considerations. This provides an important policy signal for energy market regulators: to stabilize or increase the supply and demand in the green electricity market, it is essential to balance the pricing mechanism of the traditional electricity market to ensure the economic appeal of green electricity. Furthermore, policymakers can use subsidies, tax incentives, or similar measures to maintain the competitiveness of green electricity and promote the broader adoption of clean energy.

5. Conclusion

This study develops a network evolutionary game model between green electricity generators and quota subjects to study their decision-making evolution in response to different market options for electricity value and environmental value. By incorporating the green electricity market, traditional electricity market, and green certificate market, the model analyzes the strategic changes of green electricity generators and quota subjects under various pricing conditions, with a particular focus on how market prices influence the behavior of supply and demand participants. Using evolutionary game methods, the model provides a detailed depiction of the sales decisions of green electricity generators and the purchasing decisions of quota subjects, simulating their response trends under varying market price scenarios.

The simulation results lead to the following conclusions: as the green electricity market price increases, the willingness of green electricity generators and quota subjects to participate significantly improves, with the green electricity sales proportion and purchase proportion both approaching 1.00. Conversely, when the traditional electricity market price rises, green electricity generators gradually reduce their green electricity sales proportion, shifting toward the traditional electricity market, while quota subjects also decrease their demand for green electricity. Notably, when the traditional electricity price exceeds 450 RMB/MWh, the participation of both green electricity generators and quota subjects in the green electricity market becomes nearly zero. This indicates that both the supply and demand sides of the green electricity market are highly sensitive to price signals, demonstrating the significant regulatory role of the pricing mechanism within the market.

Overall, governments and market regulators need to carefully balance the price differences between the green electricity market and the traditional electricity market in market design to avoid the adverse impact of rising traditional electricity prices on the competitiveness of the green electricity market. To promote the development of green electricity, policymakers can ensure its economic appeal to both supply and demand sides through subsidies, tax incentives, or green certificate trading mechanisms. Additionally, regulators can dynamically adjust the price leverage between green electricity and traditional electricity to guide the market in better aligning with environmental policies, thereby driving the adoption of clean energy and achieving sustainable development.

References

1. Sundt, S., Rehdanz, K., 2015. Consumers' willingness to pay for green electricity: A meta-analysis of the literature. Energy Economics 51, 1–8. https://doi.org/10.1016/j.eneco.2015.06.005
2. Choi, W., Kang, S., 2023. Greenhouse gas reduction and economic cost of technologies using green hydrogen in the steel industry. Journal of Environmental Management 335, 117569. https://doi.org/10.1016/j.jenvman.2023.117569
3. Yu, X., Dong, Z., Zhou, D., Sang, X., Chang, C.-T., Huang, X., 2021. Integration of tradable green certificates trading and carbon emissions trading:

How will Chinese power industry do? Journal of Cleaner Production 279, 123485. https://doi.org/10.1016/j.jclepro.2020.123485

4. Hui, W., Xin-gang, Z., Ling-zhi, R., Fan, L., 2021. An agent-based modeling approach for analyzing the influence of market participants' strategic behavior on green certificate trading. Energy 218, 119463. https://doi.org/10.1016/j.energy.2020.119463
5. Xu, S., Xu, Q., 2023. Optimal pricing decision of tradable green certificate for renewable energy power based on carbon-electricity coupling. Journal of Cleaner Production 410, 137111. https://doi.org/10.1016/j.jclepro.2023.137111
6. Yu, X., Ge, S., Zhou, D., Wang, Q., Chang, C.-T., Sang, X., 2022. Whether feed-in tariff can be effectively replaced or not? An integrated analysis of renewable portfolio standards and green certificate trading. Energy 245, 123241. https://doi.org/10.1016/j.energy.2022.123241
7. Liu, D., Luo, Z., Qin, J., Wang, H., Wang, G., Li, Z., Zhao, W., Shen, X., 2023. Low-carbon dispatch of multi-district integrated energy systems considering carbon emission trading and green certificate trading. Renewable Energy 218, 119312. https://doi.org/10.1016/j.renene.2023.119312
8. Meng, Q., Li, M., Liu, W., Li, Z., Zhang, J., 2021. Pricing policies of dual-channel green supply chain: Considering government subsidies and consumers' dual preferences. Sustainable Production and Consumption 26, 1021–1030. https://doi.org/10.1016/j.spc.2021.01.012
9. Sun, X., Fang, W., Gao, X., An, H., Si, J., Wei, H., 2024. Dynamic interactions among new energy metals and price adjustment strategies: A cross-industry chain perspective. Energy 303, 131923. https://doi.org/10.1016/j.energy.2024.131923

Evaluation of the Application and Development Potential of Distributed PV in Industrial Parks–Take the 180MW Distributed Photovoltaic Project as an Example

Yetian Long[1,a], Shuming Lu[1,b], Shuaiming Chen[1,c*], Chen Chen[1,d], Guanghua Chen[1,e], Xianjin Shan[1,f], Yuanyuan Hu[2,g]

[1]Southwest Branch of China Construction Third Bureau Group Co., Ltd Southwest Branch of China Construction Third Bureau Group Co., Ltd, Chengdu, Sichuan, China
[2]State Grid Sichuan Electric Power Company Tianfu New District Power Supply Company, Chengdu, Sichuan, China

Abstract. As one of the main sources of global greenhouse gas emissions, realizing the goal of "carbon peak, carbon neutrality" has become a major problem to be solved urgently. The application of distributed photovoltaic technology, as one of the effective ways to promote the realization of carbon peak and carbon neutrality in industrial parks, is particularly critical. This paper analyzes the application status of distributed photovoltaic in industrial parks in depth, and focuses on the application scenarios and technical standards of related technologies. Taking the 180MW distributed photovoltaic project as an example, the engineering benefits and development potential of the project are analyzed in detail, and the problems of distributed photovoltaic in the industrial park are pointed out. Through this analysis, we aim to realize the green aluminum industry and clean energy industry integration collaborative win-win development, power cloud aluminum construction "national environment friendly enterprise", for more similar to aluminum industrial park in promoting distributed photovoltaic applications provide useful reference and reference, in order to promote the development of industrial park green low carbon.

1 Introduction

According to the National Energy Administration in the first half of 2024, the national provincial photovoltaic construction in the first half of 202 4 Yunnan photovoltaic new capacity reached 6.663 million KW, ranked the fourth: distributed photovoltaic new installed scale reached 680000 KW, including industrial and commercial photovoltaic new capacity 606000 KW, distributed photovoltaic new installed proportion as high as 89.12%, industrial and commercial photovoltaic in distributed photovoltaic development momentum.

This data reflects the rapid development of the photovoltaic industry in Yunnan province. As a major green energy province, Yunnan province has rich solar energy resources. In recent years, it has increased the layout of new energy, planned and built a number of photovoltaic power generation bases, and the output value of the whole photovoltaic industry chain has jumped to a 100 billion industry [1]. In addition, industrial and commercial distributed photovoltaic occupies an important position in the distributed photovoltaic market due to its high space utilization efficiency and high project flexibility. This trend has also been supported by national policies, such as "promoting the development mode" and other policies, which further promote the development of distributed industrial and commercial photovoltaic [2]

In terms of green certificate trading mechanism, green power certificate (green certificate) is an important market-oriented means to promote the development of renewable energy industry. Through the green certificate trading, enterprises can purchase the certified carbon emission reduction to offset the carbon dioxide emissions generated by the park, so as to achieve the carbon neutrality target [3]. In addition, distributed photovoltaic power generation projects have a low cost in the life cycle, and can bring significant economic and social benefits

Considering the application of distributed photovoltaic in industrial parks, it can not only help reduce the energy consumption and carbon emissions of enterprises, but also maximize the economic benefits through the green certificate trading mechanism [4]. Therefore, the exploration and evaluation of the application potential of distributed PV in industrial parks is of vital significance for promoting the sustainable development of industrial parks and improving the efficiency of energy utilization.

2 The application status of distributed photovoltaic industry in industrial parks

On September 27, 2021, energy bureau of Yunnan province officially released the whole county (city, area) roof distributed photovoltaic pilot work propulsion plan, clearly put forward the implementation of roof distributed photovoltaic development of three key projects, namely urban roof photovoltaic development project, park roof photovoltaic development project and rural roof

c* Corresponding author: c*279103361@qq.com, a907002111@qq.com, b370321845@qq.com, d908362269@qq.com, e3491917907@qq.com, f515567780@qq.com, g381093883@qq.com

photovoltaic development project. Among them, for the rooftop photovoltaic development project in the park, the plan emphasizes the comprehensive implementation of rooftop distributed photovoltaic power generation projects in economic development zones, industrial parks, industrial and commercial factories and other areas, and stipulates that under the premise of meeting the conditions of construction projects, the proportion of industrial and commercial rooftop photovoltaic installation in the pilot county should not be less than 30%.

By the end of June 2024, the total grid-connected photovoltaic power generation capacity in Yunnan province has reached 27.481 million kilowatts. Among them, the cumulative grid-connected capacity of centralized photovoltaic is 25.461 million kilowatts, and the cumulative grid-connected capacity of distributed photovoltaic is 2.02 million kilowatts. In distributed photovoltaic, the cumulative grid-connected capacity of household photovoltaic is 290,000 kilowatts, and the cumulative grid-connected capacity of industrial and commercial photovoltaic is 1.73 million kilowatts. As can be seen from the above data, the cumulative grid connected amount of distributed photovoltaic in Yunnan province only accounts for 7.93% of the total centralized photovoltaic grid connected amount, and the total amount of industrial and commercial photovoltaic only accounts for 6.30% of the total photovoltaic grid connected amount in Yunnan Province(Figure 1), showing that there is still a large amount of development space for the promotion and application of industrial and commercial rooftop photovoltaic power stations in Yunnan Province.

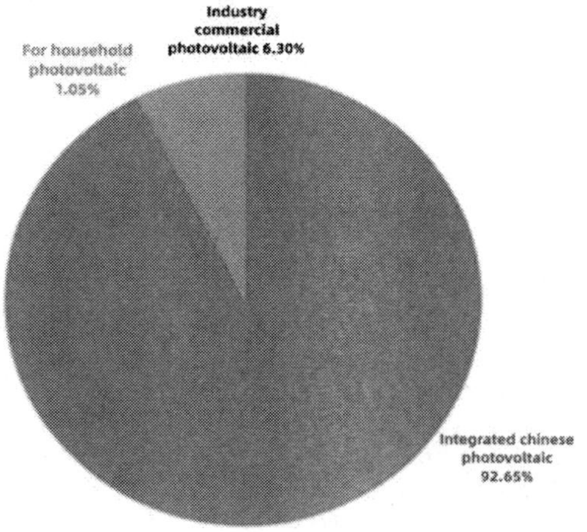

Figure 1. The cumulative proportion of all kinds of pv in Yunnan Province.

2.1 Application scenarios of distributed PV in industrial parks

With its rich roof and facade space resources, industrial parks provide ideal conditions for the deployment of distributed PV. These Spaces not only provide sufficient space for the installation of photovoltaic panels, but also the stable high power demand in the industrial park, providing a favorable market environment for the grid-connected operation of distributed photovoltaic. With the economic advantages of distributed photovoltaic gradually emerging, its application in industrial parks is also becoming increasingly popular.

We can make full use of the idle resources in the industrial park, such as the roof and facade of sewage treatment stations, workshops, warehouses, office buildings, to carry out reasonable planning and layout of distributed photovoltaic power generation facilities. This can not only meet the electricity demand in the park, but also realize the diversification of power generation modes.

According to the guidance of The State Council of The State Council of the Peoples Republic of China on Promoting the Healthy Development of photovoltaic Industry, the promotion of distributed PHOTOVOLTAIC power generation system in industrial parks has important strategic significance [5]. First, this policy is in line with the overall goal of national low-carbon development, and helps to promote the optimization of the energy structure and the transformation of green energy. Secondly, industrial parks usually have high electricity price and large power demand. Through the construction of distributed photovoltaic power generation system, it can not only reduce the operating costs of enterprises, but also reduce the dependence on traditional energy, so as to improve the efficiency and sustainability of energy use.

In addition, the promotion of distributed photovoltaic power generation system can also bring direct economic benefits. The research shows that the internal rate of return of distributed photovoltaic power generation projects in most areas can reach more than 8%, and the static investment payback period is 5~11 years, showing good economy and high investment value. At the same time, the governments support policies for distributed photovoltaic power generation projects, such as electricity price subsidies and tax incentives, will further enhance the attractiveness of the projects [6].

2.2 Current status of technical standards of distributed PV in industrial parks

In the context of the global energy structure transformation and the rapid development of renewable energy, distributed photovoltaic, as an efficient and environmentally friendly energy solution, is becoming more and more common in industrial parks. However, the complexity and diversity of this technology also bring a series of challenges to the development and improvement of technical standards.

First of all, from the perspective of technical standards, the application of distributed PV in industrial parks covers many fields such as photovoltaic modules, inverters, energy storage systems and grid-connected technology. The technical standards in these fields have different focuses, including equipment performance, safety, reliability, economy and other factors. In view of the continuous innovation and progress of photovoltaic technology, the relevant standards are also constantly updated and improved. Therefore, for industrial parks, timely understanding and mastery of the latest technical standards is the key to ensure the smooth implementation of distributed photovoltaic projects.

Secondly, the application of distributed PV in industrial parks is affected by local policies, regulations, market environment and other factors [7]. The promotion and support of distributed PV vary different regions, which affects the formulation and implementation of technical standards to a certain extent. At the same time, the intensification of market competition also promotes photovoltaic enterprises to continuously launch new products and new technologies, which brings new challenges to the formulation of technical standards.

In addition, the application of distributed photovoltaic in industrial parks also needs to consider the coordination and integration with the existing power grid. Because the distributed photovoltaic grid-connected network is more dispersed and random than the centralized one, its grid-connected operation puts forward higher requirements for the stability and security of the power grid. Therefore, in the process of formulating technical standards, the coordinated operation of distributed photovoltaic and the power grid should be fully considered to ensure the safe and stable operation of the power system [8].

To sum up, the wide application of distributed photovoltaic technology in industrial parks requires multiple challenges such as the diversity of technical standards, the impact of policies and market environment, and the coordination and integration with the power grid. This requires the industrial park to not only pay attention to technological innovation, but also pay attention to close attention to policy trends, strengthen communication and cooperation with the power grid, in order to realize the healthy development and effective application of distributed photovoltaic technology.

3 Assessment of the development potential of distributed PV in industrial parks

Taking the 180MW distributed PV project of Yunaluminum as an example, the project benefits of the project were evaluated and the development potential of distributed PV in the industrial park of Yunnan province was discussed.

3.1 Project Overview

The planned total installed capacity of Yunal 180MW distributed photovoltaic project is 180.298MW, The project owners are 6 subordinate enterprises of Yunnan Aluminum Co., Ltd., The main contents include the investment, construction and operation services of distributed photovoltaic projects, In accordance with the principle of "spontaneous self-use and full consumption", Planning to build 6 enterprise distributed photovoltaic power stations, All the electricity generated capacity is directed to the nearby consumption.

The project construction sites are located in four prefectures in Yunnan Province, including two Yangjie Industrial Parks in Jianshui County, Honghe Hani and Yi Autonomous Prefecture, one Datun Town in Gejiu City, one provincial industrial park in Zhaoyang District of Zhaotong City, one Matang Industrial Park in Wenshan Miao and Zhuang Autonomous Prefecture, one Chengbei Industrial Park in Fuyuan County of Qujing City and one Baishui Town in Zhanyi District of Qujing City.

Project using roofing, parking lot, idle land location (Figure 2), the construction of distributed photovoltaic + wisdom facilities + day 1 wisdom control and service platform of integrated energy integration, realize "source-net-load-storage-control-with" energy supply link between production coordination, control coordination, demand coordination and interaction between production and consumption, build distributed green energy station, combining the factory actual need to expand other comprehensive wisdom energy energy elements. Project to "reduce the energy cost, improve the efficiency of comprehensive energy use, provide green low carbon can environment, wisdom energy technology application and green aluminum industry integration" as the goal, create environment comprehensive perception, intelligent, digital, networking, collaborative integration of green wisdom energy system, realize green aluminum industry and clean energy industry integration collaborative win-win development, power cloud aluminum construction "national environment friendly enterprise".

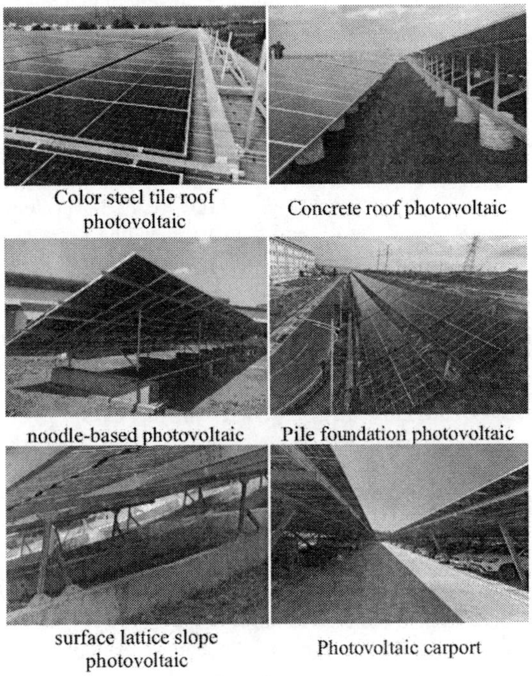

Figure 2. Photovoltaic Layout of Yunaluminum 180MW distributed PV project.

3.2 Project benefit evaluation

The benefit evaluation of this project can be discussed from three aspects: economy, environment and society, as shown in Figure 3.

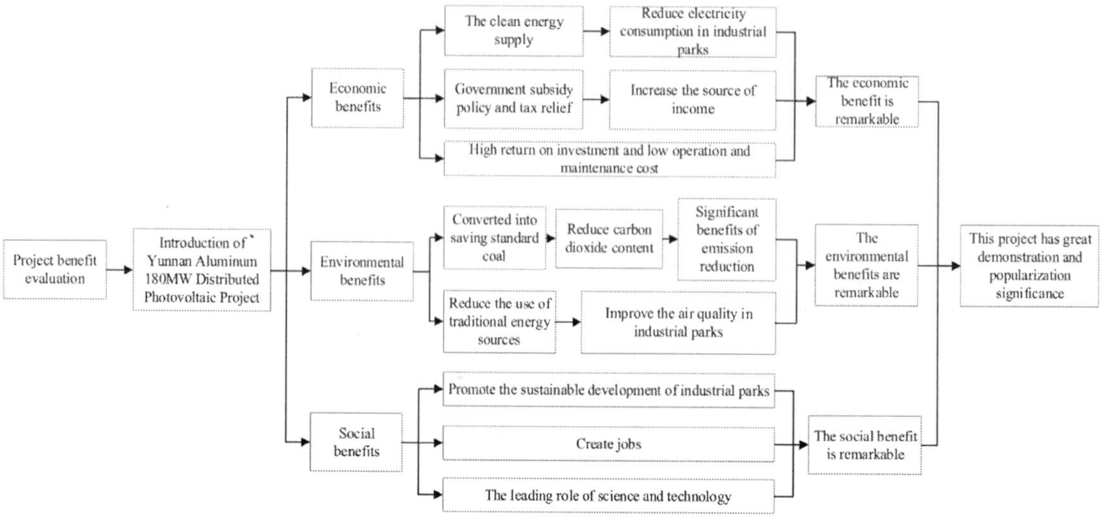

Figure 3. Project benefit evaluation model Economic benefits.

3.2.1 Economic Benefits

a) Lower electricity cost

The power consumption area of aluminum plant is mainly concentrated in the production process of electrolytic aluminum, which requires a large amount of electricity to drive the electrolytic reaction. The installed capacity of the distributed solar photovoltaic power generation system of the 180MW distributed photovoltaic project is 180.298MW, and the annual generating capacity is 213,244,300 kw.h. The upper limit of the current electricity price is 0.516 yuan / (kw.h), the lower limit is 0.468 yuan / (kw.h), and the corresponding peak-valley price difference is 0.048 yuan / (kw.h). By the arithmetic average calculation, the average price is 0.492 yuan / (kw.h), saving the annual energy cost of 104,916,200 yuan. The application of distributed solar photovoltaic power generation system will not only help to reduce the dependence of aluminum plants on the national grid, but also reduce the energy cost of aluminum plants.

b) State subsidies and tax relief

The state has issued a series of supporting policies to promote the application of distributed photovoltaic in industrial parks. For example, in terms of financial subsidies, the state has provided a certain amount of financial subsidies for distributed photovoltaic projects, which reduces the investment cost of 180MW distributed photovoltaic projects. In terms of tax incentives, the state provides tax relief policies for Yunnan Aluminum Co., Ltd., which further reduces the burden of enterprises.

c) High return on investment

From the perspective of investment, based on the national policy support and power station 25 years operation period, cloud aluminum 180MW distributed photovoltaic project in 12 years to realize cost recovery, at the same time, the operation cost of photovoltaic system is relatively low, only need to clean photovoltaic panels and check the operation of the equipment regularly, no need to invest a lot of manpower and material resources.

Project return on investment (Return on Investment, ROI) is one of the indicators to measure the effect of investment project income, and its calculation formula is as follows:

$$ROI=(NP\div C)\times 100\% \quad (1)$$

The calculation formula of annual net profit is:

$$NP=R-G-H \quad (2)$$

The calculation formula of the depreciation rate is:

$$D=(1-S)\div N\times 100\% \quad (3)$$

The calculation formula of the depreciation amount is:

$$G=C\times D \quad (4)$$

where: R is the total income, which is all the total income of the investment project within a certain period; C is the total investment, the total cost of the investment project; D is the depreciation rate; S is the residual value rate; N is the depreciation life; G is the depreciation amount; and H is the cost. The calculation result of the ROI can be expressed as a percentage, reflecting the return effect of the investment project.

The Project payback period (Payback Period, PBP) refers to the time required from the start of the project investment to the recovery of the total investment. The calculation formula is:

$$PBP=I\div CF \quad (5)$$

where: I is the investment amount, namely the total investment amount of the project; CF is the annual net cash flow, that is, the net cash flow generated by the project each year, and the total income of the project minus the total expenditure. The payback period can be used to assess the risks and benefits of the investment project. Usually, the shorter the payback period, the smaller the risk and the higher the benefits.

The total investment of the project is 271.35 million yuan, the annual cost is 1.5 million yuan, and the annual total income is 55 million yuan. The depreciation of fixed assets is calculated by the straight-line depreciation method, the residual value rate is 5%, and the depreciation life of equipment is 20 years. According to the above data, the return on investment is 14.9%. In this project, the annual net cash flow is 20.79 million yuan, and the payback period of the project is about 13 years, which means that the project can recover all the investment after 13 years of operation.

3.2.2 Environmental benefits

a) Help to reduce carbon emissions in industrial parks

After the completion of the photovoltaic power station project, it can provide about 213.2443 million kWh for Yunnan Aluminum Co., LTD every year. Compared with coal-fired power plants, it can save about 65,500 tons of standard coal consumption per year, and correspondingly reduce the emission of various air pollutants every year, including carbon dioxide (CO_2) About 212,500 tons, sulfur dioxide (SO_2) emissions of about 6,396 tons, nitrogen oxide (with NO_2) About 3,198.318 tons, but also can reduce the dust emissions by about 58,000 tons.

b) Improve the air quality in industrial parks

The installation of distributed PV can reduce the dependence of industrial parks on fossil energy and reduce the risk of air pollution. Photovoltaic power generation is a clean and renewable energy, which has good environmental benefits, can reduce environmental pollution to a great extent, and practice the development concept of "clear water and green mountains are gold and silver mountains" with practical actions.

3.2.3 Social benefits

a) Promote the sustainable development of industrial parks

Distributed photovoltaic can provide stable clean energy for industrial parks, reduce energy consumption and pollution, balance the single supply of energy in industrial parks, realize the integration and win-win development of green aluminum industry and clean energy industry, and help cloud aluminum to build a "national environment-friendly enterprise". Through the promotion and application of distributed photovoltaic technology, industrial parks can gradually reduce their dependence on fossil energy, improve energy efficiency, and achieve green and low-carbon development. This will not only help to improve the competitiveness of enterprises, but also to lay a solid foundation for the long-term development of the industrial park.

b) Creating job opportunities

The construction, operation and operation of distributed photovoltaic provide rich employment opportunities for many positions such as technical personnel and operation management personnel.

In the construction stage of distributed PV, a large number of technical personnel with professional knowledge and skills are required to participate, including designers, electrical engineers, structural engineers, etc. They are responsible for the design, construction and installation of the photovoltaic power station.

In the distributed photovoltaic operation stage, the participation of operation and management personnel is required. They carry out the operation management of distributed photovoltaic, formulate the operation strategy of the power station, and optimize the operation efficiency of the power station. At the same time, we will communicate and coordinate with government departments, power companies and users to ensure the normal operation and revenue of the power station.

The distributed photovoltaic industry also drives the development of related industrial chains, such as photovoltaic equipment manufacturing, material supply, intelligent monitoring and other fields, which also provide a lot of opportunities for the job market.

c) The leading role of science and technology

Cloud aluminum 180MW distributed photovoltaic project research using photovoltaic dc grid technology, the technology breakthrough, reduce the inverter, booster, rectification in the process of power loss, and energy utilization efficiency is higher, for large-scale green electricity direct electrolytic aluminum industry to provide technical support, further improve the electrolytic aluminum industry utilization level, renewable energy efficiency level, also for the domestic green electricity conversion and new power system provide experience and demonstration.

To sum up, the application of distributed photovoltaic technology in industrial parks has multiple advantages, including reducing the electricity cost of the plant, improving the energy self-sufficiency rate, reducing carbon emissions, high return on investment and low operation and maintenance cost. These advantages make the distributed photovoltaic technology become an ideal choice for the green development of industrial parks, and help to promote the sustainable development process of the whole society.

3.3 Evaluation of the development potential of distributed PV in industrial parks

As a clean energy vigorously promoted by the country, distributed photovoltaic has been widely concerned by the society. In particular, the development potential of distributed photovoltaic in industrial parks is great, which is of great significance for promoting the green development and energy utilization in the parks. The development potential of distributed photovoltaic in industrial parks is mainly reflected in the following aspects:

First of all, industrial parks usually have a large number of roof, facade space and idle ground, which provides good conditions for the installation of distributed photovoltaic. By making rational use of these space, the photovoltaic power generation capacity of the park can be greatly improved and the dependence on traditional energy can be reduced.

Secondly, the demand for electricity in industrial parks is large and stable, which provides a good environment for the grid-connected operation of distributed photovoltaic.

Through grid-connected operation, the distributed photovoltaic system can transfer the excess electricity to the grid, realizing energy sharing and optimal allocation.

In addition, with the progress of photovoltaic technology and the reduction of the cost, the investment cost of distributed photovoltaic in industrial parks is getting lower and lower, and the economy is becoming increasingly prominent. Through reasonable investment planning and scientific operation and management, distributed photovoltaic projects can achieve good economic and social benefits in industrial parks.

4 Analysis of the problems existing in the application of distributed photovoltaic in industrial parks

Although the promotion of distributed photovoltaic as an important means of green energy utilization in industrial parks has broad application prospects and huge development potential, it still faces a series of challenges and problems in the actual operation process.

4.1 Policy support is still insufficient

Although the national level has issued a series of policies to support the development of the photovoltaic industry, there are still some problems in the specific implementation process, such as the insufficient policy implementation and the incomplete policy coverage. In addition, there are great differences in policies between different regions, which leads to the difficulties of industrial parks facing difficulties such as poor policy convergence and policy gaps when promoting distributed photovoltaic [9].

4.2 Lack of unified technical standards

The installation, operation and maintenance of distributed photovoltaic system involve technical standards in many fields, but no unified technical specifications have been formed at present. This not only increases the difficulty and risk of selecting photovoltaic equipment in industrial parks, but also affects the performance and reliability of distributed photovoltaic systems. The lack of uniform technical standards may lead to a waste of resources and repeated investment.

4.3 Financing problems need to be solved urgently

Due to the large investment scale and long return cycle of distributed photovoltaic projects, industrial parks often face the problem of capital shortage in the process of promotion. At the same time, due to the particularity of the photovoltaic industry, the traditional financing channels may not be fully applicable to distributed photovoltaic projects. Therefore, how to innovate financing methods and reduce financing costs has become a key problem to be solved in promoting distributed photovoltaic in industrial parks.

4.4 Uneven number and distribution of industrial parks

As an important carrier of industrial activities, the number and distribution of industrial parks directly affect the potential market scale of distributed photovoltaic development.

5 Study Conclusions and Development Recommendations

(1) The application of distributed photovoltaic in the industrial park has multiple advantages. It can effectively use idle resources, improve the efficiency of resource utilization, and then improve the economic and social benefits of the industrial park. At the same time, reducing carbon emissions and environmental pollution will help improve air quality and enhance environmental benefits. For example, for every kilowatt hour saved, 0.4kg of standard coal is saved, and pollution emission is reduced by 0.272kg of carbon dust, 0.97kg of carbon dioxide, 0.03kg of sulfur dioxide and 0.015mg of nitrogen oxides.

(2) In the design, construction and operation process of distributed Pv, new materials, new technologies and new processes should be actively explored and developed to improve the output efficiency and stability of the system. By continuously optimizing the technical scheme and improving the management level, the operation and maintenance cost can be further reduced and the overall efficiency of the project can be improved. Future research will continue to focus on improving photoelectric conversion efficiency, reducing production costs and enhancing the stability and service life of materials.

(3) The government should strengthen the policy guidance and support for the application of distributed photovoltaic in industrial parks, improve the relevant policy system, and ensure the effective implementation and implementation of the policies. At the same time, a cross-departmental coordination mechanism should be established to simplify the project filing and approval procedures and improve the efficiency of project development and management. For example, the National Energy Group plans to increase the installed capacity of renewable energy to 40% during the Tenth Five-Year Plan period, which provides policy guarantee for the development of photovoltaic industry.

(4) Strengthen technological research and innovation, and promote the unification and standardization of technical standards. For the combined application of photovoltaic and industrial plants, a standardized system suitable for industrial parks is established to promote technological innovation and achievement transformation. New materials, such as graphene, have become a research hotspot because of their excellent conductivity and flexibility, which is expected to supplement the shortage of silicon materials.

(5) Improve the economy of distributed photovoltaic projects in industrial parks. By optimizing project design, improving power generation efficiency, and reducing operation and maintenance costs, the revenue capacity of the project is enhanced. Using efficient and low-cost

photovoltaic modules and intelligent control technology can significantly improve the power generation efficiency and economy of the system. For example, the power generation efficiency of crystalline silicon photovoltaic modules has increased from 15% to about 19%. By increasing the area of single silicon wafer and optimizing the stacking mode, the LCOE (leveled power generation cost) of the project can be further improved. In addition, reasonable selection of DC voltage level of photovoltaic system and optimization of inverter capacity ratio are also helpful to reduce LCOE.

To sum up, the application of distributed photovoltaic in industrial parks has great potential and broad prospects. By constantly overcoming challenges and problems, strengthening policy support and technological innovation, we will promote the wide application of distributed photovoltaic in industrial parks.

References

1. This print is comprehensive. The "acceleration" of photovoltaic industry development in Yunnan Province [J]. Times Finance, 2023, (09):11-11.
2. Chen Yong, Guo Yili, hd. Industrial and commercial distributed photovoltaic "1 + 4" full life cycle risk prevention and control strategy [J]. China Electric Power Enterprise Management, 2024 (9): 76-78.
3. Zhu Bingcheng, Ou Jiaxin. The standard construction of green power certificate trading system in China under the goal of "double carbon" [J]. Journal of Jiangsu University (Social Science Edition), 2023, 25 (5): 29-45.
4. Gemdiway New Industry Research Institute. Park photovoltaic is one of the best ways to practice carbon neutrality in industrial parks [J]. Residential and Real Estate, 2024.
5. Zhou Dan. The State Council issued several Opinions on Promoting the Healthy Development of the photovoltaic Industry [J]. Chinese and Foreign Energy, 2013 (8): 90.
6. He went slowly. Economic analysis and development policy research of distributed photovoltaic power generation project [D]. North China Electric Power University, 2019.
7. Ge Shanqi, Liu Hong. Problems and solutions existing in the development of distributed photovoltaic power generation [J]. Technology and Markets, 2017, 24 (9): 286.
8. Zhang Di. Research on the coordinated optimization operation technology of photovoltaic power generation planning and power grid under the market environment of distributed generation [D]. Hubei: Huazhong University of Science and Technology, 2021.
9. Moran. Research on the implementation deviation of local government photovoltaic power generation policy [D]. Chongqing: Southwest University, 2016.

Public Transportation Commuting: Inequity and Polycentric Urban Structure

Wantong Fu *

Dalian University of Technology, Dalian 116081, China

Abstract. The polycentric urban structure is a way to solve many problems faced by a single big city, such as traffic congestion, environmental pollution, resource shortage and so on. Meanwhile, workers who take public transportation are importantly representative of the whole population of a city. Currently, there is limited research estimating and quantitatively analyzing the impact of polycentric urban structure on morning commute. This research fills this blank by analyzing the current utilization of morning bus among workers in the multi-center cluster city of Beijing. The study uses multiple linear regression to check the fairness of commuting. The result shows that bus commute is relatively equitable in terms of commuting time and distance, but not in terms of bus station density. However, it's interested that commuting time is less equitable than commuting distance.

1 Introduction

The multi-center cluster city model represents a form of urban development that differs from the traditional single-centric city. It involves the emergence of several relatively independent central regions within the city, each possessing a certain degree of economic, cultural, and social autonomy. In the process of modern urban development, the traditional single-center city, after reaching a certain scale, begins to reflect the structural agglomeration diseconomy.[18] Because the population as well as industrial and commercial activities are highly concentrated in the small area the single center city, it is inevitable that the population density of the downtown area is too high, and industrial and residential areas are intertwined.[10] As large and rapidly growing first-tier cities in China, such as Beijing, Shenzhen, and Shanghai continue to experience "urban diseases" like traffic congestion, housing shortages, and employment difficulties due to annual population influxes, there is a need for these cities to establish internal polycentric urban structures at an urban scale. Polycentricity is supposed to reduces the negative agglomeration effects, and it facilitates cities in regaining positive momentum when employment reclusters in subcenters.[12] This will help alleviate population pressure, optimize resource allocation, address regional development imbalances and agglomeration diseconomies while also facilitating sustained economic growth by attracting more talent and investment. It is believed that urban dwellers in polycentric cities might benefit from congestion relief and from increased accessibility to jobs and services.[3]

Compared with single-center city, multi-center city with higher work-residence balance within each group has greater advantages in reducing residents' commuting distance and easing the traffic pressure in the central area.[7] There are many sub-centers of employment in Megacity like Beijing metropolitan area, and the distribution of employment presents a polycentric spatial structure. But employment still presents a trend of centripetal agglomeration.[13] The employment accessibility of streets decreases as the distance from the city center increases.[9]

It was found that commuting time and cost play a significant mediating role in the effect of commuting mode on subjective well-being.[4] The longitudinal findings indicate that compared with the car, public transport as the primary commuting mode was associated with better psychological wellbeing because of the physical activity and social contact involved.[14] According to 2021 data from the Ministry of Transportation, the number of public transport passengers in 36 major cities is projected to reach nearly 53 billion, with approximately 46% of residents opting for public transit as their primary mode of commuting. As subway and bus systems continue to improve, an increasing number of individuals are expected to choose public transportation for their travel needs. However, there are more studies on commuting in big cities based on macro level, and fewer studies on micro individual differences based on different social attributes. [8]

The construction of polycentric urban structure can effectively alleviate the traffic congestion of single-center city and reduce the commuting distance of residents.[11] The polycentric urban structure is superior to single-center structure in alleviating urban congestion and improving travel efficiency. When the city grows to a certain scale, the spatial structure is the key factor to determine the efficiency of the city.[15] For high school earners who can have high mobility, low and middle income earners are

* Corresponding author: fuwantong_0402@163.com

increasingly dependent on commuting public transport in the city. As urban space expands, jobs become more dispersed and low-and middle-income people migrate in the case of weaker capacity, the location selection behavior of residential and employment location, commuting and other activities are more and more the passive and weak condition.[20] Thus, the evaluation method of public transport employment accessibility considering job-housing matching and competition, diversity and accessibility of public transport is proposed.[16] In previous studies, it has been proved that Polycentric centers are assisting in the reduction of commuting travel times and maintaining the share of public transport journeys. [1]

There are plenty of studies have demonstrated the significant benefits of a polycentric urban structure in addressing traffic congestion. One analysis showed that increasing proximity to the nearest employment center reduces commuting distance, even after controlling for the job attributes located in the neighborhood of workers.[6] However, these studies didn't put their focus on the impact on public transport commuters. Therefore, this research examines the average commute time and distance of commuter, as well as the distance of the bus station density to the nearest CBD.

2 Case study and data description

To select an ideal polycentric urban structure, this study compared four megacities in China, which are Beijing, Shenzhen, Shanghai and Tianjin. Finally, this study chose Beijing as my target. Considered that it is a typical ring city, with a population of 21.8 million and an area of 16,400 square kilometers. The range of study is inside the fifth Ring Road of Beijing, which is about 1,537 square kilometers, dividing in to 104 subdistrict offices. Above all, there are 5 CBD in total being the economic and trade centers in this area. The study area is showed in Figure 1.

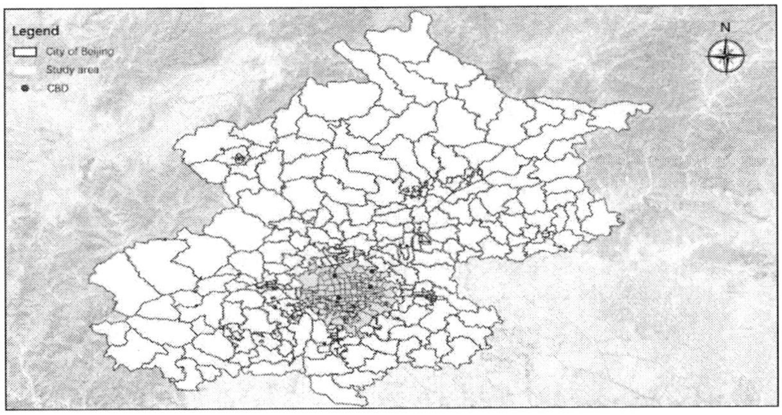

Figure 1. Study area. (Source: Self-made)

This study included information from Beijing Traffic Management Bureau, acquiring the data of about 7800,000 bus passengers in Beijing in December 18, 2021.

The research inquired complementary data in addition to bus passengers data. The followings are the data sources, while the processing and methods will provided in the next part.

The study gained lots of information from the National geographic information resources directory service system. Including longitude and latitude of places, the area of CBD, geographical information within the Fifth Ring Road of Beijing, and the places of 104 subdistrict offices. Considering that the residential opportunities, employment opportunities and public service facilities in Beijing are mostly distributed by the street, the street is chosen as the spatial unit to measure the relationship between employment and housing. National Geograohic Information Center provides some basic geographic data on the Internet and allowed users to download free of charge.

Meanwhile, The study gained information of the population in 2021 from the Demographic. The Demographic conducts the survery every ten years.

At last, GDP of each district can be seen at the National Bureau of Statistics. The study used some data from 2021. The data for all models are listed in Table 1.

Table 1. Summary of key variables considered in the analyses.

Variable	Description	Source
Measure Objects		
Average commuting time	Average commuting time of all the workers in every subdistrict office	Beijing Traffic Management Bureau 2021
Average commuting distance	Average commuting distance of all the workers in every subdistrict office	Beijing Traffic Management Bureau 2021
Bus station density	The number of bus station in unit area	National geographic information resources directory service system
Variables		
Distance to the center	Average distance from bus station to the Beijing center in the subdistrict office	National geographic information resources directory service system

Distance to the nearest CBD	Average distance from bus station to the nearest CBD in the subdistrict office	National geographic information resources directory service system
Other Influence Factors		
Population	Total number of population in the subdistrict office	Demographic 2021
Population Density	The number of people living in each unit area	National geographic information resources directory service system
Area	The district to which the sub-district office belongs	National geographic information resources directory service system
Average station passage flow	Average sample data in each subdistrict office	National geographic information resources directory service system
Sample data	Total number of all the sample data in every bus station	Beijing Traffic Management Bureau 2021
Subdistrict office area	Area of sub-district office	National geographic information resources directory service system

3 Methodology

To accurately estimate commute time and distance, the study uses the real data from the morning bus commute on December 18, 2021, as well as the demographic and spatiotemporal data for every subdistrict office that mass point is in the study area. Firstly, This study selected and matched all the travel records. Then, calculated the average time and distance. After assembling the required data, this study made the Multiple linear regression model and has three measure objects and two variables.

3.1 Identification of people

Using on bus and off bus data about 7,000,000 of all bus riders in Beijing on December 18, 2021. After excluding the elderly and bus employees, using programming software to select the time between 7.am and 9.am (includeing the first time getting on the bus in this time period and the last time getting on the bus in this time period), and to match the data of each person's first and last boarding. Finally this study obtained data on all commuters who got on the bus in the morning rush hour, about 220,000 in total. Combining the latitude and longitude coordinates of the pick-up and drop-off bus stations, the study gained the commuting time and commuting straight-line distance of all the communicators. According to the first pick-up bus station, the study allocated every person's information to 104 subdistrict office, and obtained the average commuting time and average commuting straight-line distance.

3.2 Distance to the center

The study divides "center" into two concepts. The first concept is the center of the ring structure in the Beijing built-up area, namely Tiananmen Square (116.3912 ,39.9062). This study defines the second concept of "center" as the central business district that serves as the nucleus of the city within a polycentric urban structure. The study specifically focuses on five CBD centers along Beijing's Fifth Expressway, namely Financial Street, Beijing CBD, Zhongguancun, Lize and Wangjing. The study obtained the footprint of each CBD and determined the latitude and longitude of their centers. Using the same method, the study obtained the latitude and longitude of all the bus stops needed for the study in this open platform and calculated the distance between each bus stop and their nearest CBD.

3.3 Administrative district

Subdistrict offices are township level administrative districts in China, consisting of multiple communities. The study took the subdistrict office as the basic unit, and studied the average commuting time and average commuting distance within the unit, as well as the density level of bus stops within the basic unit.

There are a total of 104 subdistrict offices within the Fifth Ring Road of Beijing, which is used for the analysis of Multilevel linear regression model.

3.4 Multiple linear regression model

The concept of multiple linear regression model refers to a linear regression model containing multiple explanatory variables, which is used to explain the linear relationship between the explained variable and several other explanatory variables. It is more effective and practical to predict or estimate dependent variables by the optimal combination of several independent variables than to predict or estimate only one independent variable. To use a multiple linear regression model, some assumptions need to be satisfied. Firstly, this study tested the linear relationship between independent variables and dependent variables by drawing scatter plots. Then, performed linear regression analysis to verified the normality, independence, variance homogeneity of the data and presented reports. After meeting the four basic preconditions of linear regression analysis, the model is made to test the relationship between the independent variable and the dependent variable. The fomulation is:

$$\begin{aligned} Y &= f(X_1, X_2, \ldots, X_P) + \varepsilon \\ &= \beta_0 + \beta_1 X_1 + \beta_2 X_2 + \ldots + \beta_P + \varepsilon \end{aligned} \quad (1)$$

Y is called the dependent variable, X is called the independent variable, f(x) is called the regression function, and Ɛis a random error with a distribution independent of the independent variable, which usually assumed to be a positive variable with a mean of 0.

4 Empirical results

This section presents the results of the empirical analyses in Beijing following the described methodology. First, the spatial distribution of commuting condition in range of study was mapped. Then, this study conducted the Multiple linear regression model. Finally, the study analyzed bus commuting fairness in polycentric urban structure.

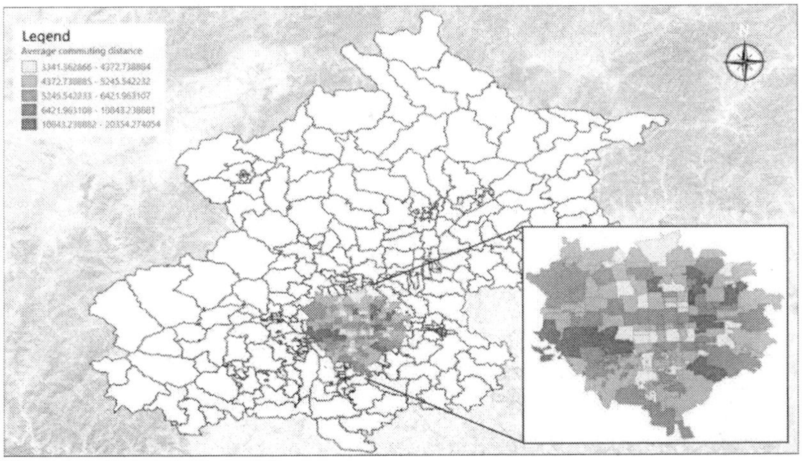

Figure 2. Average commuting distance. (Source: Self-made)

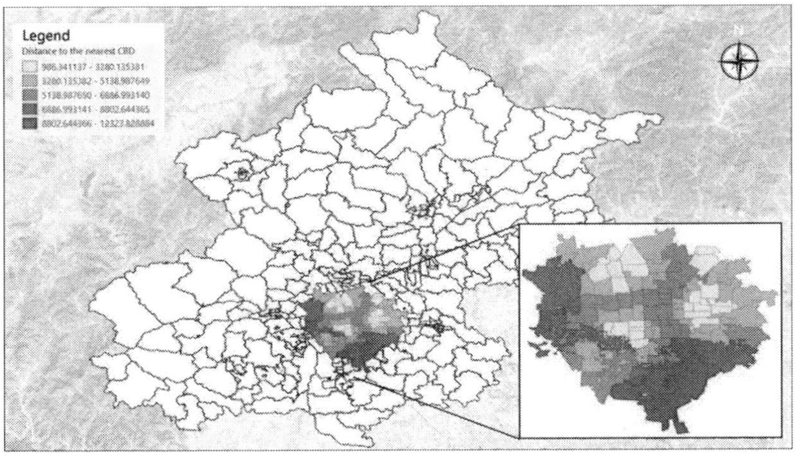

Figure 3. Distance to the nearest CBD. (Source: Self-made)

4.1 Commuting condition

A total of 104 subdistrict offices were included in the study, all of these areas are in the Beijing built-up area. By sifting through 7,000,000 raw data in December 18, 2021, a total of 52,327 commuters riding a bus on 7.am to 9.am in study area were obtained. Figure 2 shows that the maximum Average commuting distance reached 20,354 meters and the minimum reached 3,341 meters. Figure 3 shows that CBDs are mainly concentrated in the central and northeast regions. The region with the longest distance is in the south.

4.2 Multiple linear regression model

In Table 2 The Model summary[b] is an important index to judge the linear relationship between the independent variables and dependent variables, and also reflects the degree of regression fitting. For the resulting output, Durbin-Watson value are 1.637, 1.574 and 2.099. The value of this statistic ranges from 0 to 4, and if the residuals are independent of each other, the value is ≈ 2. If the Durbin-Watson value deviates from 2 and approaches in the direction of 0 or 4, it indicates that the residual is not independent. In this example, the Bus station density value is 2.099, which is very close to 2, indicating that there is no obvious correlation between the residuals, that is, the residuals are independent. And for Average commuting time and distance value are also close to 2 too. It means that the residuals are independent too.

In general, the value of Adjusted R square is relatively unaffected by the number of independent variables, and the result is more reliable. This example includes multiple arguments, the results means that all independent variables can separately interpret 7.4% of Average commuting time, 9.3% of Average commuting distance and 57.4% of Bus station density. This indicates that in this case, the distance to the nearest CBD and the center can best explaint Bus station density, and the distance to the nearest CBD and the center has a high influence on Bus station density. However, for the Average commuting time and distance, the model has a poor fit.

Table 2. Model summary[b].

Model	R	R Square	Adjusted R Square	Std. Error of the Estimate	Durbin-Watson
Average commuting time	.333[a]	.111	.074	.0014161818	1.637
Average commuting distance	.359[a]	.129	.093	1218.63157161064120	1.574
Bus station density	.769[a]	.591	.574	1.747900111	2.099

Table 3. Coefficients[a].

Model		Unstandardized		Standardized Coefficients	t	Sig.	Correlations			Collinearity Statistics	
		B	Std. Error	Beta			Zero-order	Partial	Part	Tolerance	VIF
Average commuting time	(Constant)	.014	.001		22.266	.000					
	Distance to the center	-9.991E-8	.000	-.251	-2.289	.024	-.069	-.225	-.218	.752	1.330
	Distance to the nearest CBD	1.194E-7	.000	.191	1.892	.061	.180	.188	.180	.888	1.126
	Population Density	.000	.000	-.183	-1.745	.084	-.144	-.174	-.166	.827	1.209
	Average station passage flow	3.027E-5	.000	.198	2.007	.048	.164	.199	.191	.928	1.077
Average commuting distance	(Constant)	4511.234	525.974		8.577	.000					
	Distance to the center	.002	.038	.007	.066	.947	.168	.007	.006	.752	1.330
	Distance to the nearest CBD	.086	.054	.158	1.582	.117	.215	.158	.149	.888	1.126
	Population Density	-154.957	142.885	-.112	-1.084	.281	-.171	-.109	-.102	.827	1.209
	Average station passage flow	34.526	12.976	.260	2.661	.009	.286	.260	.251	.928	1.077
Bus station density	(Constant)	7.821	.754		10.367	.000					
	Distance to the center	-.000268	.000	-.371	-4.982	.000	-.618	-.450	-.322	.752	1.330
	Distance to the nearest CBD	-8.140E-5	.000	-.072	-1.045	.299	-.306	-.105	-.068	.888	1.126
	Population Density	1.034	.205	.358	5.045	.000	.543	.454	.326	.827	1.209
	Average station passage flow	-.089	.019	-.321	-4.792	.000	-.451	-.436	-.310	.928	1.077

Meanwhile, the Table 3 Coefficients[a] of the result out put shows two statistics of clooinearity diagnosis, namely Tolerance (tolerance) and VIF (variance inflation factor). It is generally believed that if the Tolerance < 0.2 or VIF > 10, it suggests that the problem of multicollinearity between independent variables should be considered. Collinearity refers to the strong correlation between two or more predictors in the process of fitting linear regression model. In three tables, Tolerance of each variable > 0.2, and VIF < 10, indicating that there is no collinearity between independent variables in the three models.

Through the analysis of the above results, it is known that the research data in this example conforms to the applicable conditions of multiple linear regression, and the results of the regression model will be formally interpreted below.

4.2.1 Evaluate the pros and cons of regression model

In the Table 2 of the result output shows the fitting of the model, in which the complex correlation coefficient R are 0.333, 0.359 and 0.769, reflecting the degree of linear correlation between all independent variables x and dependent variable y.

The larger the value, the closer the linear correlation is. The coefficient of determination R Square (R2) are 0.111, 0.129 and 0.591, indicating that 11.1%, 12.9% and 59.1% of the variation in VO2 max could be explained by Average station passage flow, Population Density, Distance to the nearest CBD and Distance to the center.

The closer R2 is to 1, the better the model fits the data. Adjusted R Square (R2) are 0.074, 0.093 and 0.574, which is similar to R2 and is also an important indicator to measure the model. The larger the value, the better the model fitting effect.

Combining three kinds of data shows, it means that Bus station density is most affected by the dependent variables.

4.2.2 Test of global regression effect

The Table 4 with the result output illustrates the ANOVA results to test the overall significance of the regression model, where the F statistic are 3.052, 3.629, 35.383. However, only P-value of Bus station density is less than 0.01. So the best model is the Third one.

Table 4. ANOVA[a].

Model		Sum of Squares	df	Mean Square	F	Sig.
Average commuting time	Regression	.000	4	.000	3.052	.020[b]
	Residual	.000	98	.000		
	Total	.000	102			
Average commuting distance	Regression	21554977.898	4	5388744.475	3.629	.008[b]
	Residual	145536164.918	98	1485062.907		
	total	167091142.816	102			
Bus station density	Regression	432.402	4	108.101	35.383	.000[b]
	Residual	299.405	98	3.055		
	Total	731.808	102			

4.2.3 Test of partial regression coefficient

In Table 3 (Coefficents[a]) shows the Coefficients of the regression model (B), the T-statistic of the regression coefficient test and its P-value (Sig.)

If the B-value is positive, it indicates a positive correlation between the dependent variable and the independent variable; if the B-value is negative, it suggests a negative correlation between the dependent variable and the independent variable. In terms of Average commuting time, B-value of Distance to the center is negative but Distance to the nearst CBD is positive. It implies that the further one is from the center, or the closer one is to the nearest CBD, the Average commuting time is significantly reduced. But in terms of Average commuting distance, B-value of Distance to the center and Distance to the nearst CBD are both positive. The further from the center or CBD, the Average commuting distance is further. Finally, in terms of Bus station density, B-value of Distance to the center and Distance to the nearst CBD are both negative. When it closer to the nearst CBD or center, the number of bus station improves.

value, also known as significance value or Sig. Value, is used to describe the probability of a certain event. Its value ranges from 0 to 1. Generally, there are three criteria for judging: 0.01, 0.05 and 0.1.In the result output of Table 3, only in the impact of Distance to the center on Bus station density is the P-value< 0.01. Meanwhile, for the impact of Distance to the center on Average commuting time, the P-value < 0.05, and for the impact of Distance to the nearest CBD on Average commuting time, the P-value < 0.1.

At the test level of α=0.05, it can be considered that their partial regression coefficients are not 0 and have statistical significance, and can be included in the final regression model.

5 Discussion and policy implications

5.1 Average commuting time and distance are relatively equitable but Bus station is not in polycentric urban structure

First, the study noticed that polycentric urban structure do make public transport commuting fairer. Public commuting distance is the most equitable, because the impact of neither distance from bus station to the center or nearest CBD inapparent. It is the same as previous studies, there is no significant relationship between the employment accessibility values of each street and district and their distance from the CBD.[17] Public commuting time is the second-fair that the distance from bus station to the center has a greater impact on him, and the distance to CBD has a lesser impact on him, but both have an evident impact. Commuting time is positively correlated with the distance to the CBD, and negatively correlated with the distance to the center. Bus station density is the most unfair that distance to the center greatly affects station density. The closer a location is to the center, the higher the concentration of bus stations. This may be related to the economy and infrastructure development level.

5.2 Insights for developing public Commuting Equity in polycentric urban structure

There may be many reasons why the commuting distance may show little variation across different sub-districts, yet the commuting time can differ significantly. Firstly, It could be that the closer the area is to the center, the more heavy the traffic. The government can shorten the commuting time by creating bus roads. Separate bus travel from car travel to make bus travel smoother. Or the bus station is riddled with numerous detours. When the travel OD data of residents in a traffic community is known, the optimization model of the receiving bus line can be established based on the objective function of the maximum passenger turnover of the MRT line and the rail transit on the unit length.[5] In the case of having a subway for medium and long distance commutes, customised traffic can be provided for heavily used parts of the route.

After the city changes from single-center structure to polycentric structure, the traditional bus stop layout should be adjusted. The traditional bus stop layout principle is based on the layout of traffic flow and capacity, and multiple bus stops are set in the sections with large traffic flow to improve the operating efficiency of the bus system and avoid congestion and inconvenience caused by too many passengers. When the urban pattern becomes polycentric, the traffic flow in each area gradually equalizes. Therefore, the number of bus stops on many roads should be increased or reduced according to the theory of demand. Using the distance from the center as a benchmark, the quantity of bus stops is adjusted based on the land use type. For instance, a greater number of bus stops may be established in the vicinity of residential areas.

5.3 Possible future developments in Bus Commuting

The formation of the polycentric urban structure is not the product of the development of the market economy itself, which needs the active intervention and organization of the government.[11] Even if the compactness of the spatial structure decreases with the growth of the city, the spatial organization efficiency can still be better if the balanced allocation of public service facilities at all levels is emphasized.[19] The government can slightly intervene in the construction of CBD at a distance from the center to ease traffic pressure, or build more bus stops to balance the allocation of resources. As the previous study said, prescriptions of higher degrees of self-sufficiency in new peripheral job centres are more likely to reduce commuting duration and distance.[2] When the polycentric urban structure develops into a decentralized structure, it is expected that public transport will become more equitable.

5.4 Trade-offs between polycentricity and other forms of urban planning in terms of Commuting Equity

Mixed-use development refers to the integration of different functional uses in urban planning, so that the same area can have residential, commercial, office, entertainment and other functions at the same time. This kind of planning strategy is actually the presentation of a decentralized concept on small plots. In mixed-use development, because the needs of residents can be met in a shorter distance, the use of short-distance buses is encouraged in such areas, which can reduce the waiting time and travel time of residents.

Hotspot landmarks (landmarks of a city, generally of historical significance), which cannot be replaced even in a polycentric structure, naturally form tourist centers. Therefore, in the multi-center structure, it is possible to build a tour bus specifically for it, connecting the train station and the subway station.

5.5 Alternative Transportation modes in Polycentric urban structures

First of all, to encourage shared bike travel, you can set up shared bike stops near residential areas and bus stops to shorten the travel time between the two. Secondly, the government should increase investment in the construction of non-motorized lanes and sidewalks to improve the safety and comfort of these modes of transportation, so that people are more willing to choose bicycles or walking for short trips.

As driverless car technology matures and urban decentralization becomes more common, shared cars will become more popular in the future. Since autonomous vehicles use electricity for energy consumption, it is important to ensure that the charging pile is multicentre and equitable.

6 Conclusions

This study conducted this research to probe public transportation commuting equity in polycentric urban structure. First, This study estimate all the average commuting time and distance as well as bus station density of each sub-district. Most importantly, Multiple linear regression model shows that the polycentric urban structure has some fairness in terms of public transportation and behaves differently in commuting distance and time as well as bus station density.

In analyzing the three dependent variables, two key findings have emerged. Firstly, commuting distance is deemed the most equitable, followed by commuting time as the second most equitable. Commuting time is positively correlated with the distance to the CBD, and negatively correlated with the distance to the center. Conversely, bus station density is considered to be quite inequitable. As one moves further from the center, the density of bus stops decreases.

Then the study discuss what policies should governments implement to make public transport fairer. From the perspective of CBD distribution, the government may implement minor interventions in the construction of Central Business Districts (CBDs) located away from the city center to alleviate traffic congestion, or establish additional bus stops to better distribute resources. From the perspective of bus subdivision, in the case of having a subway for medium and long distance commutes, customised traffic can be provided for heavily used parts of the route. As the urban structure evolves from a polycentric to a decentralized model, it is anticipated that public transportation will become more equitable.

Finally, the experiment still leaves much to be desired. Only one day of bus records data were utilized, which means that the results cannot fully represent the comprehensive characteristics of every weekday. And the station at which the passenger alights is not necessarily their place of employment. In addition, other factors that could affect commuting time and distance, such as socio-economic status or work hours, were not considered. Therefore, this research could be extended by applying a longer period of bus commuting, and considering more influencing factors, to make the research more rigorous.

References

1. ALPKOKIN P, HAYASHI Y, BLACK J, et al. Polycentric employment growth and impacts on urban commuting patterns: Case study of Istanbul[J]. Journal of the Eastern Asia Society for Transportation Studies, 2005, 6: 3835-3850.
2. Asikhia M O, Nkeki N F. Polycentric employment growth and the commuting behaviour in Benin Metropolitan Region, Nigeria[J]. Journal of Geography and Geology, 2013, 5(2): 1-17.
3. Cabrera-Arnau C, Zhong C, Batty M, et al. Inferring urban polycentricity from the variability in human mobility patterns[J]. Scientific Reports, 2023, 13(1): 5751.
4. Deng Y, Wang L, Chen M. Commuting and its spillover effects on subjective well-being: Evidence from China[J]. Transportation Research Part D: Transport and Environment, 2024, 126: 104001.
5. Guan Shuwen. Research on bus route optimization based on the urban rail transit[D]. Dalian Jiaotong University, 2014.
6. Ha J, Lee S, Kim J H, et al. Do employment centers matter? Consequences for commuting distance in the Los Angeles region, 2002–2019[J]. Cities, 2024, 145: 104669.
7. Liu Dinghui. A Study on Influence of Urban Spatial Structure on Residents' Commute Travel: The Case of Chengdu and Lanzhou[J]. World Regional Studies, 2015, 24(04):78-84+93.
8. Liu Dinghui, Zhu Chaohong, Yang Yongchun. Teview on Urban Commuting Study of China in the Perspective of Geography[J]. Urban Development Studies, 2012, 19(10):55-59.
9. Liu Zhilin, Wang Maojun. Job Accessibility and Its Impacts on Commuting Time of Urban Residents in Beijing: From a Spatial Mismatch Perspective [J]. Acta Geographica Sinica, 2011, 66(04):457-467.
10. Shi Yishao. From A Single Center City to A Multi-centers City: Spatial Organization Pattern of Netriopolitan Development in China [J]. Urban Planning Forum, 1999, (03):36-39+26-80.
11. Sun Bindong, Shi Wei, Ning Yuemin. An Empirical Study on the Polycentric Urban Structure of Shanghai Strategies in Future[J]. Urban Planning Forum, 2010, (01):58-63.
12. Sun B, Zhang T, Li W, et al. Effects of polycentricity on economic performance and its dependence on city size: The case of China[J]. Land, 2022, 11(9): 1546.
13. Sun Tieshan, Wang Lanlan, Li Guoping. Distributions of Population and Employment and Evolution of Spatial Structures in the Beijing Metropolitan Area[J]. Acta Geographica Sinica, 2012, 67(06):829-840.
14. Tao Y, van Ham M, Petrović A. Changes in commuting mode and the relationship with psychological stress: A quasi-longitudinal analysis in urbanizing China[J]. Travel behaviour and society, 2024, 34: 100667.
15. Tu Ting, Sun Bindong. Problems of Shanghai's Urban Traffic in Monocentric and Polycentric Perspective and Their Improvement Tactics. Communication & Shipping2009, 23 (3) : 1-4 + 55, DOI: 10.16487 / j.carol carroll nki issn2095-7491.2009.03.003.
16. Wu Lingling. Evaluation of Transit Equality based on Job Accessibility by Transit[D]. Wuhan University, 2019.
17. Wu Lingling, Huang Zhengdong, Jiang Haiyan, Cai Yunnan. The Study of Public Transport Job Accessibility of the Middle-Low-Wage Workers. [J]. Urban Development Studies, 2018, 25 (08) :117-124.
18. Wu Yizhou, Lai Shigang, Wu Cifan. Research on polycentricity: concept, connotation, and spatialfeatures[J]. City Planning Review, 2016, 40(06):23-31.
19. Zhang L, Zhang L, Liu X. Evaluation of urban spatial growth performance from the perspective of a polycentric city: A case study of Hangzhou[J]. Land, 2022, 11(8): 1173.
20. Zheng Siqi, Zhang Xiaonan, Xu Yangfei, Xu Junyan. Urban Spatial Mismatch and Traffic Congestion--Empirical Study on Jobs-housing Unbalance and Over-concentration of Public Service in Beijing[J]. Reform of Economic System, 2016, (03):50-55.

Analysis and Thinking on the Difficulties of Substation Location in Guizhou under the Background of New Energy

Linlin Li[a], Jiaxing Li[b], Shujuan Bai[c], Huafeng Chen[d]

China Power Engineering Consulting Group Guizhou Electric Power Design and Research Institute Co., Ltd., Guiyang, Guizhou, China

Abstract: This paper takes the site selection of power supply side substation in Guizhou as an example, analyzes and discusses the characteristics of this kind of substation site selection process, and provides a new idea and way for the site selection of power supply side substation under the background of new energy.

1. Problems faced by substation site selection in Guizhou mountainous area

Under the background of new energy, there are the following problems in the site selection of power supply side substations in Guizhou mountainous areas: ① Power stations are scattered, and the principle of proximity (adjacent to the site) cannot be simply adopted for the site selection of supporting substations. ② Whether the tower foundation position of high-voltage line (especially the terminal tower) can meet the access direction and the feasibility of natural conditions becomes the decisive factor. ③ The traffic conditions in mountainous areas are poor, and the transportation of large pieces of substations is easily restricted by road slopes, bridge opening and tunnels. ④ The water supply and drainage conditions in mountainous areas are complex, and the station site needs not only sufficient water for fire fighting and operation and maintenance, but also smooth drainage to prevent waterlogging. ⑤ Guizhou is a typical karst landform. This geological environment is accompanied by geological disasters such as mudslides, caves, dangerous rocks and collapses, which will cause hidden dangers to the safety of the station site. ⑥ Because of the steep terrain in Guizhou mountainous area, it is impossible to directly choose the flat land as the station site.

2. Analysis on the Difficulties of Substation Location in Guizhou Mountainous Area

Aiming at the problems in the previous chapter, this chapter analyzes the difficulties of substation (mains side) location in Guizhou mountainous areas through practical cases and engineering experience, and summarizes seven solving factors.

2.1 Location of Substation and Power Generation Project

The photovoltaic power station in a mountainous area of Guizhou (relative to the sea or hills) is characterized by scattered plots, small plots and large span (Figure 1). Taking a photovoltaic project as an example, the total installed capacity of the project is 120MW, covering an area of more than 1, 750 mu, with a span of 5.4km from north to south and 3.1km from east to west, involving 36 plots and 46 plots. Therefore, the principle of proximity cannot be adopted in site selection. After comparison and selection of several stations, the substation is finally located in the middle of the station (site1), which saves about 15km of collecting lines compared with the south side of the outgoing line (site 2). Figure 2 shows the wind power station of Guizhou wind power plant. The maximum span between generators of the power station is 11.9km. At first, it was assumed that the power station was located at the southwest outgoing line side (site2) and the length of gathering and transmission line was 32.5km, which was later revised to the middle of the power station (site1). The length of gathering and transmission line is reduced from 32.5km to about 25km, and the power consumption brought by centralized high-voltage transmission is reduced by about 3%.

Therefore, in mountainous areas, the power supply side substation is located in the center of the station area, which increases the length of a small part of high-voltage transmission lines, but saves a lot of collector line length and power line loss, which is more economical.

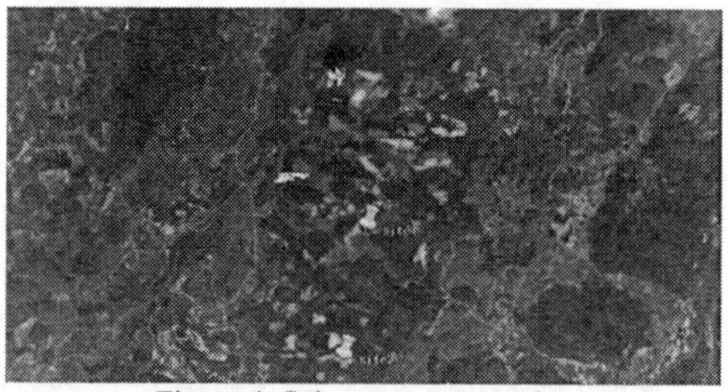

Figure 1. Solar power generation

[a] 1259444711@qq.com, [b] 2039185515@qq.com, [c] baishujuan-gzy@powerchina.cn, [d] chenhuafeng-gzy@powerchina.cn

Figure 2. Wind power

2.2 External conditions of terminal tower in substation

Table I below is a list of the external conditions of station A and station B. By comparing a series of conditions, the geological features, lood control and drainage, bad geology, external traffic, construction water supply, power supply, environmental protection and town planning of the two stations are basically the same. Although the site of Scheme B is smaller than that of Scheme A, the terminal tower is located in the geological subsidence area, so Scheme A was finally chosen.

Therefore. The good external conditions of substation terminal tower have become the decisive factor for substation site selection. [1]

TABLE I. TERMINAL TOWER CASE COMPARISON TABLE

Table Head	*case*		
	A	B	contrast
geographical position	Dakang Village	Baiyudong Village	equal
Flood control and drainage	The site is not threatened by the flood level with a return period of 100 years.	The site is not threatened by the flood level with a return period of 100 years.	equal
Incoming and outgoing line situation	The venue is open, and the outgoing line is on the southwest side, which is convenient.	The venue is narrow, and the outgoing line is northbound, sinkzone	A
External traffic	The entrance road is connected from the country road, with a length of about 80 meters.	The entrance road is connected from the country road, with a length of about 51meters.	equal
Construction water supply and power supply	Construction and domestic water is led to tap water in nearby villages, with a length of about 300 meters; Construction electricity is connected by 10kV nearby.	Construction and domestic water is led to tap water in nearby villages, with a length of about 51 meters; Construction electricity is connected by 10kV nearby.	equal
environmental protection	The site itself has no pollution to the environment, so there is no need to take environmental protection measures; There are no industrial and mining enterprises near the site, and there is no phase.	The site itself has no pollution to the environment, so there is no need to take environmental protection measures; There are no industrial and mining enterprises near the site, and there is no phase.	equal
floor area	16.75mu	13.71mu	B
Construction, operation and living conditions	Close to rural roads, convenient transportation, convenient construction, and convenient life and management of employees.	Close to rural roads, convenient transportation, convenient construction, and convenient life and management of employees.	equal
expropriate	The site is non-basic farmland, and outside the urban planning area, land acquisition is relatively easy.	The site is non-basic farmland, and outside the urban planning area, land acquisition is relatively easy.	equal

2.3 Geological and Geomorphological Conditions of Substation

Guizhou is a typical karst landform, which has caused geological risks and disasters in special areas of Guizhou. This geological disaster will cause potential safety hazards to the station site. This section summarizes the current situation of common unfavorable geological conditions in Guizhou site selection in Table II, which can be used as a reference for site exploration, and try to avoid site selection in this area if conditions permit. [2]

TABLE II. GEOLOGICAL CONDITIONS TABLE

	situation
Landslide	The landslide body in loose layer is mostly composed of alluvium, diluvium and colluvium. The sliding body is only a few meters small, and the big one can reach hundreds of meters. The sliding structural plane is mostly the contact surface between loose layer and underlying bedrock. A small number of sliding surfaces are located in loose layers, but the scale of such landslides is relatively small. The main deformation and failure mechanisms of landslides are traction, creep-cracking, and are affected by atmospheric precipitation, and generally slide after concentrated continuous rainfall.

	Landslides between bedrock are mainly medium and small landslides, which mainly occur in clastic rocks such as sandstone and mudstone. The hazard of bedding landslide in bedrock is more serious, and the landslide is often distributed in the lower part of the slope, and the gullies are developed.

Dangerous rock and collapse	In the development of structural erosion, natural near-vertical cliffs are often formed in deep rivers, and the lithology is mostly hard rock. These areas where structural cracks and unloading cracks are extremely developed provide material sources and internal and external dynamic conditions for the formation of collapse, and it is easy to form medium and small bedrock collapse and loose layer collapse. 
Gully and debris flow	Debris flow can be divided into two types according to the landform: gully debris flow and slope debris flow, of which gully debris flow is the largest and most destructive. Gully-type debris flows mostly occur in V-shaped gullies with slope gradient over 20%, rich loose solid deposits in gullies and gullies, and poor vegetation development. It has obvious formation area, circulation area and accumulation area. The accumulation fan is mainly composed of sub-angular-sub-circular gravel, sand and clayey silt. 
karst	Fissures, troughs, stone buds and caves are mainly developed in carbonate strata such as limestone and dolomite, and their development degree is mainly controlled by the solubility of rocks, geological structures and geological conditions such as surface water and groundwater, and the development scale varies. 
	Goaf and ground subsidence
Goaf and ground subsidence	

2.4 Land nature in land spatial planning

Under the control of land and space use, the location of substation should also be planned reasonably in combination with the nature of land to ensure the availability of land.

In China, If you want to get the land-related documents approved by the government, you must avoid choosing the land types that cannot be used: three areas and three lines, natural forest protection, engineering afforestation, mining area, returning farmland to forests, drinking water source protection forest data, nature reserve data, etc.

2.5 Transport condition

First of all, the inbound roads need to meet the transportation conditions of large pieces, and the transportation path should avoid villages, tunnels and wall holes as far as possible. Considering whether the bearing capacity of the bridge is sufficient, it is recommended to recommend road locations near provinces and counties, which can not only reduce road reconstruction, but also facilitate the travel of operation and maintenance personnel; Secondly, the slope and turning radius of road design should also meet the fire protection requirements.

2.6 Determination of hydrological and meteorological conditions

Substation site is a depression, which easily leads to water accumulation in the substation, and the lack of water will affect the construction and operation and maintenance. Therefore, in the process of site selection, it is necessary to judge whether the substation site is a depression, the source of flood and drainage, and the direction of rainwater and wastewater discharge to judge the drainage conditions. Secondly, judge whether there are villages around that can provide domestic water and drainage access points, and determine their water supply conditions.

2.7 Earthwork balance

TABLE III. Comparison Table of Earthwork Balance Cases

name	*unit*	*CASE1*	*CASE2*	*CASE3*
spoil	m^2	650	5600	2555
Fill retaining wall	m^3	2618	553	1300
Site elevation	m	1558.1	1558.6	1560.72-1557.82
Highest retaining wall	m	10.5	8.60	7.89
Site elevation difference	m	0	0	2.9
Coverage area of fence	h m^2	0.581	0.493	0.493

In engineering practice, in order to reduce the procurement and transportation of earthwork, we should try our best to achieve the balance of excavation and filling of earthwork [3]. At the initial stage of design, three schemes will be compared and selected: ① Completelyinclined (cass1). ② Slope+retainingwall (cass2). ③ Adjust the slope of the site and lower the high retaining wall (cass3). Table III shows the comparison and selection parameters of three design schemes of a certain site. The relevant parameters are:

occupied area, fence area, maximum height of retaining wall and site height difference.

These parameters can determine whether the site selection of substation is feasible or not, and whether it is necessary to relocate to a new site.

3. Approach to Substation Location in Guizhou Mountainous Area

Steps of substation site selection under the background of new energy:

As a whole:

Combined with the new energy power generation project, the supporting substation is arranged in the middle of the station site, and compared with the substation site in the access direction, the approximate geographical scope is determined macroscopically.

In detail:

- Select multiple sites according to terrain or contour lines. [4]
- Determine the nature of the land available according to government documents.
- Make sure there is no adverse geological environment.
- Determine the recommended site according to the conditions of outgoing line and access system.
- Make sure that there is a water supply source and the drainage of the station site is smooth.
- Make sure that the transportation conditions are good, in particular, it can meet the needs of large-scale transportation. [5]
- After selecting the site, it is determined that the location of the terminal tower is reasonable and has good conditions for sending out high-voltage lines.
- Conduct site design to determine its feasibility.

4. Conclusions

Based on practical engineering experience and case study, this paper firstly summarizes the problems and difficulties encountered in the site selection of substation (power supply side) in Guizhou mountainous area, secondly, analyzes the ideas and methods to solve these problems through practical cases and theories, and finally summarizes the site selection steps of substation (power supply side) in Guizhou mountainous area, trying to explore a way of preliminary site selection of substation (power supply side) in mountainous area is the significance of this paper.

References

1. Jun Peng, Key Factors for Selecting Substation Sites in the Karst Mountainous Areas of Guizhou Province [J]. LOW CARBON WORLD 2019, 10
2. LI Xiaoquan, ZHANG Huilong.Study on Substation Planning and Design Optimization in Power System Construction [J]. Transactions of China Electrotechnical Society 2023, 12
3. LIANG Yongquan, Design and Reflection of site leveling [J]. Building Technology Development, 2022, 8
4. LUO Bing, DAI Changdong, NI Chuan, ZHAO Guoliang, LIU Ling Analysis of Site Selection and Installation Scale of Mountain Area Massive Centralized Photovoltaics [J]. Electric Engineering, 2024, 03
5. Liu Yuntian, Researeh on intelligent site selection and line selectionplanning management of substaticn engineering under thebackground of digitalization [D]. 2022, 5

E3S Web of Conferences **617**, 03001 (2025) https://doi.org/10.1051/e3sconf/202561703001
EEUPD 2024

Study on Post-assessment System of Countryside with Distinctive Characteristics in Jiangsu Province

Xiu Li[1,2,*], Yanlin Liu[2] and Lan Yu[2]

[1]Shinawatra University, 12160 Pathunm Thanni, Thailand
[2]Jinken College of Technology, 210000 Nanjing, China

Abstract: To further promote rural revitalization in all respects, Jiangsu Province focuses on advancing the construction of countryside with distinctive characteristics based on local realities. While achieving phasic gains, addressing changes in various data after countryside construction and consolidating the achievements through a sustained and effective management mechanism remain urgent issues. Therefore, establishing a post-assessment system of countryside with distinctive characteristics is necessary. This mechanism is analysed and discussed from conceptual innovation to method innovation in this paper.

1. Introduction

The 19th National Congress of the Communist Party of China has made significant decisions to implement rural revitalization strategy. This strategy has been widely responded to across the country, with many provinces devising action plans based on local conditions [1]. Jiangsu Province took the lead in practical actions to promote rural revitalization. After long-term unremitting efforts, Jiangsu Province's initiatives in construction of countryside with distinctive characteristics have achieved notable results. A large number of named countryside with distinctive characteristics have demonstrated the diverse values of new rural areas in new era and have been fully recognized by the government, scholars, and masses [2].

To consolidate the achievement of rural construction, it is urgent to establish a sustained and effective management mechanism. An effective post-assessment system should be developed based on different regions and economic conditions to monitor the changes of basic data and rural development when faced the unique rural countryside in Jiangsu Province. Hence, establishing this system plays a crucial role in boosting the high quality of development and ensuring the sustainability of countryside with distinctive characteristics in China.

2. Conceptual Innovation

2.1 Retrospect

The construction of distinctive characteristic countryside in Jiangsu Province revolves around the keywords "characteristics", "pastoral" and "countryside", embodying the spirit of people-centered. According to the department, the amount of countryside with distinctive characteristics has reached of 752, with 159 of them receiving recognition in 2023. The distribution of these countryside is illustrated in Fig. 1.

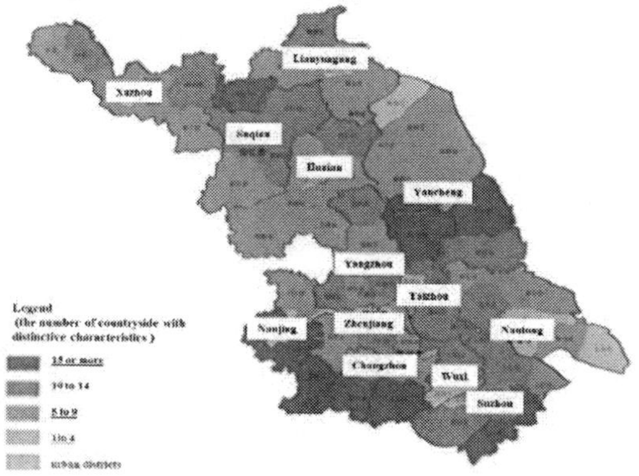

(a) The distribution in different regions

(b) Recognized countryside

Fig. 1. The distribution of countryside with distinctive characteristics.

* Corresponding author: eli851929@gmail.com

With the deep advance of the construction of countryside with distinctive characteristics, several problems and challenges have also been exposed, such as the unbalance of development of industries, the deficiency of web infrastructure, and the imperfect governance system. Fin response, the provincial government of Jiangsu is constantly improving policies and promulgated a series of the policy documents, providing guidance and requirements for the construction of countryside with distinctive characteristics. Meanwhile, a government-led program has been initiated to review and assess the outcomes of the distinctive characteristic countryside construction efforts. Therefore, how to establish a post-assessment system and how to develop quantitative and qualitative criteria are the key to well carry out the construction of countryside with distinctive characteristics.

2.2 Conceptual Innovation

The post-assessment mechanism is a concept proposed for the retrospective evaluation of recognized countryside with distinctive characteristics. It does not simply replicate existing policy documents but instead focuses more on aspects such as countryside industrial development, vitality, and sustainable governance. Taking into account variations in geographical location and basic economic conditions of countryside, it specifically assesses changes in fundamental data post-construction to make targeted judgments on countryside development. This assessment mechanism converts the effectiveness of efforts into an enduring force for countryside economic development and improved livelihoods, providing strong guarantees for the practice of rural revitalization. Besides, this post-assessment mechanism has conducted a detailed acceptance and evaluation of the construction achievements of distinctive characteristic countryside, which is the first unified implementation in Jiangsu Province and the first in China, and even in the world, offering significant insights and useful references for rural management practices in other regions.

3. Methodological Innovation

In addition to conceptual innovation, in order to better maintain, consolidate, and develop the achievements of distinctive characteristic countryside construction, as well as continually enhancing their quality and charm, there are also corresponding innovations in working methods.

The provincial review and assessment of distinctive characteristic countryside construction, known as "Walking into the countryside and looking back at the countryside" special action, was first implemented in Nanjing, Jiangsu Province in August 2023, led by the Jiangsu Provincial Department of Housing and Urban-Rural Development. This action established a joint work camp consisting of 8 expert teams, employing methods such as questionnaire surveys, discussion and exchange, and on-site visits to comprehensively and accurately understand countryside industrial development, basic infrastructure, supporting facilities, long-term management, and countryside governance, etc., and provided comprehensive and accurate evaluations.

By integrating an expert scoring system and employing Analytic Hierarchy Process (AHP) and Fuzzy Comprehensive Evaluation (FCE) methods, a multidimensional evaluation system was constructed. Through field research data, reasonable evaluation indicators were set to ensure coverage of both qualitative and quantitative criteria, enhancing the objectivity and comprehensiveness of the evaluation system. Ultimately, these evaluation criteria rely not only on objective data analysis but also incorporate judgments from experts, making the evaluation results more convincing.

3.1 Key points of post-assessment

Based on the three major connotations of "characteristics", "pastoral" and "countryside" in distinctive characteristic countryside, as well as the five major rural revitalization requirements of "industry, talent, culture, ecology, and organization", the four aspects of industrial development, long-term management, countryside governance, and work organization were ultimately determined as primary indicators, covering a total of 23 secondary indicators.

3.2 Explanation of assessment elements

Based on the guidance requirements in the document issued by Jiangsu Provincial joint conference on the construction of distinctive characteristic countryside, this paper summarized and clarified the definitions, explanations, and coverage of the above indicators mentioned in Table 1 [3].

3.2.1 Industrial development

This element focuses on assessing the overall development status of distinctive industries, agricultural production, and rural tourism in each countryside. It aims to understand the current development situation across villages, providing a basis for future development planning and policy-making. Key elements include evaluating the diversity and growth of local industries, agricultural productivity, and the attractiveness of rural tourism [4].

3.2.2 Long-term management

Long-term management is the key to ensuring the sustainable development of distinctive characteristic countryside. The evaluation work should focus on whether the village has established a long-term management mechanism and a stable source of funds. In addition, consideration should be given to rural resource management, environmental protection, infrastructure construction, and continuous improvement of the ecological environment. By evaluating these aspects, deficiencies in management systems and mechanisms can be identified, providing suggestions and improvement directions for building a robust management system [5].

Table 1. Key factors of post-assessment system.

Primary indicators (Assessment elements)	Secondary indicators (Assessment factors)
Industrial development	Characteristics of leading industries
	Integration of agricultural industry chains
	Brand building of agricultural products
	Economic income of countryside collective
	Changes in the permanent population of countryside
	Changes in the sources of income for villagers
	Annual changes in tourism revenue
Long-term management	Management funds and management models
	Participation of diverse entities
	Ecological environment conservation
	Infrastructure support
	Waste management
	Greening management
	Public service facilities support
Countryside governance	Leadership of village cadres
	Civilization practice station in new era
	Rural sports activities
	Self-governance activities
	Local elites
Work organization	Innovation and reform of work systems
	Dynamic update of data
	Publicity and promotion
	Commendation and honors

3.2.3 Countryside governance

Countryside governance is the core content of the construction of distinctive characteristic countryside. This element focuses on the construction of countryside level organizations, the implementation of countryside self-governance, and effectiveness of grassroots governance. Evaluating rural governance effectiveness helps identify shortcomings in governance systems, leading to specific improvement measures to enhance governance efficiency [6].

3.2.4 Work organization

Work organization is a key guarantee for the construction of distinctive characteristic countryside. The evaluation should closely monitor whether the recognized countryside continues to undergo dynamic updates, and whether the updated construction projects truly meet the development needs of the countryside and the actual needs of the villagers. By conducting a detailed evaluation of work organization and gaining insight into the current status of organizational construction in various rural areas, targeted suggestions and improvement directions are proposed to improve work efficiency and promote further development of rural construction.

3.3 Quantization

Based on the above assessment elements and expert field research, a scientific and reasonable evaluation standard has been established, and each factor has been quantified. Each factor adopts a five-point standardized index, divided into three levels: 0-1, 2-3, and 4-5, ensuring that each level matches the evaluation criteria. The specific scoring criteria are shown in Table 2.

3.4 Classification

Referring to FCE method, four levels for assessing the effectiveness of construction of distinctive characteristic countryside have been determined, as depicted in Table 3. Referring to existing standards [7, 8], the assessment results intuitively demonstrate the construction status of distinctive characteristic countryside, which can contribute to facilitate horizontal comparisons among villages in similar locations and objectively analyze their strengths and weaknesses.

Table 3. Classification based on FCE method.

Level	Total score
Excellent	80-100
Great	70-79
Good	60-69
Bad	0-59

Table 2. Quantitative criteria for assessment factors.

Secondary indicators (Assessment factors)	Scoring		
	0-1 Points	2-3 Points	4-5 Points
Characteristics of leading industries	No characteristic industries	Has certain characteristics that can be enhanced to become attractive points	Distinctive industrial characteristics, forming 1-2 flagship industries with unique features.
Integration of agricultural industry chains	No integrated development	A certain degree of integration	Deep integration of agriculture with secondary and tertiary industries
Brand building of agricultural products	No distinctive brands	Agricultural product brands with average popularity	Well-known agricultural brand with high recognition
Economic income of countryside collective	Below the level of this county (city, district)	At the same level as the county (city, district)	High than the level of the county (city, district)
Changes in the permanent population of countryside	No change	Few returnees for employment or entrepreneurship	Increase in the number of returnees for employment and entrepreneurship
Changes in the sources of income for villagers	No change or decrease	No significant increase in disposable income	Rising disposable income

Annual changes in tourism revenue	No increase or decrease	Some increase, but low reception volume	An increase in the number of tourists and tourism revenue
Management funds and management models	No stable funding source, no innovative management mode	Generally average	Established long-term management mechanisms, stable funding for conservation, and innovative management models.
Participation of diverse entities	No multiple participation	Moderate participation from diverse entities, but low involvement	Active participation from villagers, designers, local elites, and social capital.
Ecological environment conservation	Lack of ecological resources, severe damage, and lack of distinctive features	No prominent ecological features	Abundant ecological resources, characterized by pastoral landscapes.
Infrastructure support	Inadequate garbage classification facilities and lack of ecological water treatment facilities	Good completion of garbage classification, rainwater and sewage network renovations, with average water purification facilities.	Effective garbage classification facilities, completed rain and sewage network renovations, and sound water purification facilities
Waste management	No public toilets available	Has a countryside history museum (village cultural center) and one 2A-rated public toilet	countryside history museum and 2 or more 2A-rated public toilets.
Greening management	Greening without management and maintenance	Greenery is naturally beautiful, but management effect is average	Greening is natural and beautiful, and management are implemented
Public service facilities support	Poor use and maintenance of various industrial service facilities and public facilities	The use and maintenance level are general	Well utilized and maintained
Leadership of village cadres	Limited role	Leaders have some leadership	Effective leadership
Civilization practice station in new era	No or few activities	Has some related activity projects	Well-organized events
Rural sports activities	No or few activities	Has some related activity projects	Rich cultural and sports activities.
Self-governance activities	No or few activities	Has some related activity projects	Sound autonomous organizations
Local elites	No outstanding person	Not playing a demonstrative and leading role	Leading by outstanding local elites
Innovation and reform of work systems	No innovative breakthroughs in the work system	Some innovation and reform in work systems.	Sustainable development of countryside
Dynamic update of data	No updates or construction projects	Some dynamic updates and construction projects	Dynamic updates
Publicity and promotion	No publicity	Less publicity	Promotion in provincial-level and higher media.
Commendation and honors	No honor after recognizing	Few honors received.	Received provincial-level, or higher honors

3.5 Data and results

20 countryside projects in the first and second batches of pilot projects announced by Jiangsu Province, which have been officially checked and recognized, were selected as samples in this paper. These samples cover basic types of characteristic countryside across all regions of the province, showing high typicality and representativeness in industrial development model and construction effectiveness.

The post-assessment results are given in Table 4. The results show that there are 9 countryside with excellent performance evaluation, accounting for 45%, 8 countryside with great performance evaluation, accounting for 40%, and 3 countryside with good performance evaluation, accounting for 15%. Overall, relative well performance in the construction of distinctive characteristic countryside has gained in the province.

Table 4. Post-assessment results of 20 typical characteristic countryside.

City	County or district	Name of countryside	Industrial development	Long-term management	Countryside governance	Work organization	Total score	Assessment result
Nanjing (2)	Jiangning	Shecun	33	28	20	16	97	Excellent
	Gaochun	Qingshancun	21	35	18	10	79	Great
Wuxi (2)	Huishan	Taoyuancun	28	30	20	7	85	Excellent
	Huishan	Taoyuancun	21	21	20	10	75	Great
Suzhou (2)	Kunshan	Jinhuacun	28	33	20	16	97	Excellent
	Wujiang	Anqiaocun	33	29	19	16	97	Excellent
Changzhou (2)	Liyang	Yuxiangcun	33	29	20	17	99	Excellent
	Liyang	Dainancun	21	21	20	16	78	Great

Zhenjiang (2)	Dantu	Shiyecun	21	21	20	12	74	Great
	Dantu	Xianfengcun	28	28	20	16	92	Excellent
Nantong(2)	Rugao	Gujiazhuang	21	28	15	12	76	Great
	Rugao	Qianchangcun	28	28	20	16	92	Excellent
Taizhou (2)	Xinghua	Tangzhuangcun	21	21	15	12	69	Good
	Taixing	Qixiangcun	28	35	20	16	99	Excellent
Xuzhou (1)	Tongshan	Niyuancun	28	28	20	16	92	Excellent
Lianyungang (1)	Ganyu	Lushancun	21	21	15	13	70	Great
Huaian (1)	Jinhu	Gaoqiaocun	21	22	15	12	70	Great
Suqian (2)	Siyang	Denglonghu	21	21	15	12	69	Good
	Siyang	Babao	21	28	15	12	76	Great
Yancheng (1)	Dongtai	Guanju	21	14	15	12	62	Good

4. Conclusions

Against the backdrop of China's rural revitalization strategy and Jiangsu Province's distinctive characteristic countryside construction action, this paper discussed a post-assessment mechanism that covers conceptual innovation to methodological innovation, and conducts on-site sample evaluation and application. The specific results are as follow.

(1) The post-assessment mechanism, as an innovative measure, is important and practical in its ability to translate the effects and impacts of policy implementation into a driving force for promoting sustainable countryside development. This post-assessment mechanism not only effectively monitors and evaluates the current status and trends of rural development but also allows for more precise and targeted policy recommendations based on specific needs and development conditions of different villages. Meanwhile, the mechanism basically covers the main indicators of rural development and is applicable to other areas across the country.

(2) Practice has shown that the post-assessment mechanism has demonstrated initial effectiveness in promoting optimization and upgrading of village industries, enhancing village vitality, and achieving long-term management. This mechanism has enhanced the transparency of policy implementation and public participation, providing solid data support and decision-making basis for adjustments and optimizations of China's rural revitalization strategies.

(3) The mechanism has challenges related to timeliness and accuracy of the data used, which could potentially affect the accuracy and reliability of assessment results. Therefore, in future practices, there should be further emphasis on enhancing the dynamism, openness, and diversity of the mechanism, continuously improving and refining the post-assessment process.

References

1. C.R. Liao, M.Q. Chen, *Journal of Agro-Forestry Economics and Management*. The theoretical logic, scientific connotation and achieving methods of rural revitalization strategy **16**, 6 (2017) (in Chinese)
2. X.R. Ge, S. Liu, *J. Shanxi Agric. Univ. (Social Science Edition)*. Literature review of rural revitalization research since 19th CPC National Congress **18**, 1 (2019) (in Chinese)
3. Y.S. Liu, Y.H. Li, *NATURE*. Revitalize the world's countryside **548** (2017)
4. O.K. Ayobami, H.N.B. Ismail, *Australian Journal of Basic & Applied Sciences*. Host's supports for voluntourism: a pragmatic approach to rural revitalization **7**, 4 (2013)
5. A. Nonaka, H. Ono, *Japan Agricultural Research Quarterly*. Revitalization of rural economies though the restructuring the self-sufficient realm **49**, 4 (2015)
6. T.G. Johnson, *American Journal of Agricultural Economics*. Entrepreneurship and development finance: Keys to rural revitalization **71**, 5 (1989)
7. Jiangsu Provincial Market Supervision and Administration Bureau, Standard for construction of characteristic pastoral countryside: DB32/T4417-2022 (2022) (in Chinese)
8. Urban Planning Society of China, Guidelines for the construction of characteristic idyllic countryside: T/UPSC 0004-2021 (2021) (in Chinese)

An Empirical Study on Win-win Strategies for the Theme Park Sustainable Development and Tourist Satisfaction Based on IPA

Qifeng Yin[1], Xingmei Yang[2] and Yanhong Liu[3,*]

[1] Faculty of Economics and Business Administration, Yibin University, Yibin 644000, China
[2] College of Information and Intelligent Manufacturing, Sichuan Winshare Vocational College, Chengdu 611300, China
[3] School of Physical Education, Shenzhen University, Shenzhen 518060, China

Abstract. As China's tourism industry enters a new era of high-quality development, the immersive experience becomes a business form full of challenge to keep theme parks' sustainable development. In order to promote win-win strategies for theme parks and tourists, this paper takes Chengdu Happy Valley as an example to focus on the tourist satisfaction and constructs the evaluation system. Questionnaire survey and IPA analysis method have been used to obtain the immersive experience quality data. Based on SPSS27.0 software, IPA quadrants of Tourists' satisfaction have been analyzed. The results show that four quadrants are formed as relative advantage area, maintenance management area, key improvement area and priority improvement area. Specific suggestions have been put forward to enhance the theme park core competitiveness and tourists' satisfaction, so as to provide references for the sustainable development of theme parks.

1 Introduction

Judging from the volume base of domestic cultural tourism and holiday consumption market, as well as the growth trend in the next 10 to 20 years, the theme park market will continue to be active [1]. The Global Theme Park and Museum 2023 Report, jointly released by AECOM and TEA, notes that despite challenges such as severe weather, supply chain issues, inflation, and more, the global theme park industry is demonstrating strong resilience and resilience, with nearly every region approaching or exceeding 2019 attendance levels. The world's top 10 theme park groups saw 512 million visitors in 2023, up 30% from 393 million in 2022. TOP25 visitors totaled 244.6 million, up 23% from 2022. In order to stand out among the competitors, theme parks take tourists as the core, improve the experience effect and satisfaction of tourists.

The researchers have given different definitions of tourist satisfaction. Pitzam (1978) defined tourist satisfaction as the result of comparison between the expectation of tourists before going to a tourist destination and the actual experience of reaching the destination [2]. Hughes (1991) believed that satisfaction is relative. Although the tourists' experience in the tourist destination does not reach the initial expectation, tourists are still satisfied with the tourism [3]. Sun and Chen (2015) regarded tourist satisfaction as the personal experience of the integration of mind and body formed by comparing the perceived experience effect and expectation of tourism product or service in a certain tourist destination [4].

Scholars have analyzed the tourist satisfaction and influencing factors of the experience quality in theme parks from different perspectives. Liang and Li (2021) pointed out that a theme park is a dedicated space featuring five main characteristics as thematic identity, closed space with controlled access, hybrid consumption, performative labor and merchandising [5]. Chan et al. (2019) deemed that service quality plays a pivotal role in affecting the performance of and visitor satisfaction with theme parks [6]. Chang et al.(2021) considered that the recreational facilities, theme characteristics and fast pass drawing are the most valued theme park attributes for tourists [7]. Lee(2019) expected new market expansion by continuously creating new demands from consumers with adopting innovative technologies for more convenient facilities and more thrilling rides [8]. Zhang et al.(2017) indicated that visitor movement is influenced by attraction attributes and spatial layout attributes [9].

Evaluation system began to enter the field of view of researchers. Zhou et al.(2017) constructed the evaluation index system of tourist satisfaction in theme parks from five dimensions, including emotional expression, tourism function, place environment, theme value and negative evaluation, and discussed the multidimensional influencing factors of tourist satisfaction [10]. Fu et al. (2022) revealed that the immersion through theme park storytelling involves four stages, namely spatial-temporal immersion, sensory immersion, conceptual-imaginative immersion and emotional immersion [11].

In general, progress has been made in academic research on tourist satisfaction and theme parks, but there are also some evaluation models that focus too much on

* Corresponding author: lucyliu@szu.edu.cn

the psychological dimension and ignore the technological dimension of management. Theme park sustainable development should include not only tourist psychology but also the coordinated development of technological and environmental factors. Therefore, in order to promote win-win strategies for theme parks and tourists, this paper takes Chengdu Happy Valley as an example. Happy Valley chain brand is China's first chain theme park brand. Founded on October 1, 1998, it is the core products of Overseas Chinese Town (OCT), one of the first national 5A-level tourist attractions. In 2023, Happy Valley Group ranked in sixth place with 35.71 million visitors of the world's top 10 theme park groups, a year-on-year increase of 57%. Chengdu Happy Valley is the third stop of Happy Valley chain brand in China after Shenzhen and Beijing. As a large amusement theme park, it mainly attracts college students who are representative group of young people. This selection reflects the satisfaction of college students and influencing factors of the experience quality, which are included in the win-win strategies for theme parks and tourists.

2 Materials and methods

2.1 Research methods

Questionnaire survey is adopted to obtain the experience quality data of college students in Chengdu Happy Valley in this study. Based on IPA analysis method, SPSS27.0 software are used to conduct reliability test, validity test and IPA quadrant analysis on the questionnaire data. The questionnaire is designed and compiled based on the prior research on the tourists' satisfaction of three parts: the first part is the evaluation of expectation and satisfaction, the second part is the tourism preference, and the third part is the basic information of tourists [12]. It contains 8 first-level evaluation factors and 39 second-level evaluation factors, with Likert Scale 5 being used.

2.2 Data source

The questionnaire has been distributed from July 1, 2022 to July 5, 2022 both online and offline with 20 questionnaires pretested on Wechat, Moments, QQ and Qzone, among which 18 being valid with a recovery rate of 90%. The language expression in the questionnaire has been fine-tuned according to the comments from the pretest feedback. Then the questionnaires were officially distributed, and 158 questionnaires were collected online by July 5. Considering the limitation and false answers, the questionnaires that take less than 90 seconds to complete are excluded. Finally, 123 valid online questionnaires were collected, with an effective rate of 77.85%. 100 questionnaires were distributed, and 100 valid questionnaires were collected in Chengdu Happy Valley. Totally, 258 questionnaires were collected by the two methods, of which 223 were valid, with an effective rate of 86.43%.

As shown in Table 1, female tourists account for 62.78%, juniors and seniors account for 38.12% and 26.01% respectively, 1,000-1,500 RMB incomers account for the most proportion with 45.74%, less than 50% of the tourists have been there twice or more times, 52.02% of the tourists spend less than 500 RMB each time in Chengdu Happy Valley, most tourists travel with their friends, accounting for 66.37%, half day to one day tour accounted for 69.06%, 91.48% of the tourists are willing to recommend Chengdu Happy Valley to other potential tourists.

Table 1. Tourist information statistics table.

Items	Characteristics	Number	Scale (%)
Gender	Male	83	37.22
	Female	140	62.78
Grade	Freshman	32	14.35
	Sophomore	48	21.52
	Junior	85	38.12
	Senior	58	26.01
Monthly living Expenses	Under 1,000 RMB	24	10.76
	1000-1500 RMB	102	45.74
	1500-2000 RMB	59	26.46
	Above 2,000 RMB	38	17.04
Times	Once	118	52.91
	Twice or more times	105	47.09
Expenditure	Within 500 RMB	116	52.02
	500-1,000 RMB	77	34.53
	Above 1,000 RMB	30	13.45
Companion	Alone	19	8.52
	Friend	148	66.37
	Family	35	15.7
	Others	21	9.41
Time	Less than half day	32	14.35
	Half day to one day	154	69.06
	Above one day	37	16.59
Recommended intention	Yes	204	91.48
	No	19	8.52

3 Analyses and results

3.1 Reliability and validity analysis

The reliability test of expectation (39 items) and satisfaction (39 items) in the questionnaire have been conducted by SPSS27.0 software, and the theta coefficient is 0.950 and 0.913 respectively, indicating that the reliability quality of the research data is high and can be further analyzed. The validity test of the questionnaire data shows that the KMO value is 0.861, Bart sphere value is 9791.837, DF value is 3003. When the KMO value is greater than 0.8, the research data are very suitable for information extraction, and the validity can be reflected from the side.

3.2 Importance and satisfaction analysis

3.2.1 T-test analysis

The expectation of tourists on the eight elements of tasting, residing, traveling, sightseeing, purchasing, entertaining, hardware facilities and software service in Chengdu Happy Valley all ranges from 3.390 to 3.839, as shown in

Table 2, indicating that the expectation of tourists is moderately high.

Tourists have the lowest expectation on the distribution and accuracy of signposts in the park, and higher expectation on the rest and storage facilities. The lower limit of standard deviation of importance is 0.881, and the upper limit is 1.311. Lower standard deviation values mean higher t-values.

The average satisfaction of tourists with the six tourism elements, hardware facilities and software service of Chengdu Happy Valley range from 3.099 to 3.883, as shown in Table 3. Tourists have the lowest satisfaction with the commodity price and the highest satisfaction with the excitability of amusement projects. The lower limit of standard deviation of satisfaction is 0.988, and the upper limit is 1.229. Lower standard deviation values mean higher t-values. The mean value of satisfaction is slightly lower than the mean value of expectation, indicating that the tourists' sense of gain is relatively low.

Table 2. Results of the expectation T-test analysis.

Metric	Sample capacity	Least value	Crest value	Average value	Standard error	t	p
Catering price	223	1.000	5.000	3.502	1.123	46.590	0.000**
Catering selection	223	1.000	5.000	3.704	1.132	48.864	0.000**
Catering taste	223	1.000	5.000	3.475	1.311	39.583	0.000**
Catering service	223	1.000	5.000	3.664	1.170	46.770	0.000**
Catering hygiene	223	1.000	5.000	3.762	1.179	47.668	0.000**
Accommodation accessibility	223	1.000	5.000	3.480	1.207	43.037	0.000**
Accommodation price	223	1.000	5.000	3.583	1.197	44.690	0.000**
Accommodation hygiene	223	1.000	5.000	3.668	1.247	43.913	0.000**
Accommodation feature	223	1.000	5.000	3.444	1.180	43.588	0.000**
Convenient transportation to the park	223	1.000	5.000	3.709	1.107	50.027	0.000**
Cost of transportation to the park	223	1.000	5.000	3.498	1.142	45.720	0.000**
Duration to the park	223	1.000	5.000	3.735	1.102	50.638	0.000**
Distribution and accuracy of signposts	223	1.000	5.000	3.390	1.250	40.490	0.000**
Tour route design	223	1.000	5.000	3.821	0.961	59.398	0.000**
Ticket price	223	1.000	5.000	3.655	1.256	43.440	0.000**
Ticket promotion	223	1.000	5.000	3.511	1.102	47.569	0.000**
Tour environment	223	1.000	5.000	3.740	1.199	46.596	0.000**
Enjoyment of performance	223	1.000	5.000	3.659	1.167	46.842	0.000**
Interactivity of performance	223	1.000	5.000	3.453	1.145	45.017	0.000**
Commodity feature	223	1.000	5.000	3.583	1.193	44.831	0.000**
Commodity price	223	1.000	5.000	3.511	1.294	40.521	0.000**
Commodity quality	223	1.000	5.000	3.610	1.176	45.836	0.000**
Convenient portable commodity	223	1.000	5.000	3.439	1.187	43.261	0.000**
Novelty of amusement projects	223	1.000	5.000	3.614	1.096	49.231	0.000**
Excitability of amusement project	223	1.000	5.000	3.596	1.166	46.071	0.000**
Richness of amusement project	223	1.000	5.000	3.673	1.145	47.894	0.000**
Waiting time for amusement project	223	1.000	5.000	3.655	1.227	44.467	0.000**
Waiting mode for amusement project	223	1.000	5.000	3.520	1.192	44.084	0.000**
Safety of amusement projects	223	1.000	5.000	3.664	1.248	43.838	0.000**
Rest and storage facilities	223	1.000	5.000	3.839	0.881	65.087	0.000**
Reasonable layout of toilets	223	1.000	5.000	3.570	1.285	41.476	0.000**
Toilet hygiene	223	1.000	5.000	3.758	1.195	46.967	0.000**
Wireless network coverage and signal degree	223	1.000	5.000	3.592	1.197	44.827	0.000**
Tourist Service Center	223	1.000	5.000	3.619	1.175	45.984	0.000**
Safety facilities	223	1.000	5.000	3.650	1.268	43.002	0.000**
Service attitude	223	1.000	5.000	3.744	1.132	49.399	0.000**
Response and handling of tourist complaints	223	1.000	5.000	3.650	1.242	43.871	0.000**
Consulting service	223	1.000	5.000	3.713	1.106	50.126	0.000**
Professionalism of service personnel	223	1.000	5.000	3.717	1.161	47.819	0.000**

$^{*}p<0.05$ $^{**}p<0.01$

Table 3. Results of the satisfaction with the T-test analysis.

Metric	Sample capacity	Least value	Crest value	Average value	Standard error	t	p
Catering price	223	1.000	5.000	3.260	1.109	43.908	0.000**
Catering selection	223	1.000	5.000	3.278	1.202	40.724	0.000**
Catering taste	223	1.000	5.000	3.395	1.142	44.408	0.000**
Catering service	223	1.000	5.000	3.462	1.138	45.434	0.000**
Catering hygiene	223	1.000	5.000	3.695	1.072	51.458	0.000**
Accommodation accessibility	223	1.000	5.000	3.390	1.133	44.676	0.000**
Accommodation price	223	1.000	5.000	3.193	1.172	40.697	0.000**
Accommodation hygiene	223	1.000	5.000	3.507	1.017	51.477	0.000**
Accommodation feature	223	1.000	5.000	3.507	1.114	46.987	0.000**
Convenient transportation to the park	223	1.000	5.000	3.646	1.080	50.403	0.000**
Cost of transportation to the park	223	1.000	5.000	3.511	1.056	49.637	0.000**
Duration to the park	223	1.000	5.000	3.664	1.004	54.493	0.000**
Distribution and accuracy of signposts	223	1.000	5.000	3.507	1.026	51.035	0.000**

Metric	Sample capacity	Least value	Crest value	Average value	Standard error	t	p
Tour route design	223	1.000	5.000	3.798	1.074	52.827	0.000**
Ticket price	223	1.000	5.000	3.265	1.110	43.930	0.000**
Ticket promotion	223	1.000	5.000	3.475	1.189	43.664	0.000**
Tour environment	223	1.000	5.000	3.762	1.148	48.955	0.000**
Enjoyment of performance	223	1.000	5.000	3.700	1.101	50.200	0.000**
Interactivity of the performance	223	1.000	5.000	3.673	1.161	47.250	0.000**
Commodity feature	223	1.000	5.000	3.574	1.179	45.272	0.000**
Commodity price	223	1.000	5.000	3.099	1.162	39.817	0.000**
Commodity quality	223	1.000	5.000	3.377	1.120	45.032	0.000**
Convenient portable commodity	223	1.000	5.000	3.426	1.152	44.417	0.000**
Novelty of amusement project	223	1.000	5.000	3.632	1.061	51.129	0.000**
Excitability of amusement project	223	1.000	5.000	3.883	1.125	51.548	0.000**
Richness of amusement project	223	1.000	5.000	3.623	1.182	45.762	0.000**
Waiting time for amusement project	223	1.000	5.000	3.166	1.228	38.494	0.000**
Waiting mode for amusement project	223	1.000	5.000	3.211	1.229	39.026	0.000**
Safety of amusement project	223	1.000	5.000	3.700	0.988	55.896	0.000**
Rest and storage facilities	223	1.000	5.000	3.592	1.103	48.650	0.000**
Reasonable layout of toilets	223	1.000	5.000	3.498	1.069	48.856	0.000**
Toilet hygiene	223	1.000	5.000	3.574	1.100	48.527	0.000**
Wireless network coverage and signal degree	223	1.000	5.000	3.592	1.078	49.769	0.000**
Tourist Service Center	223	1.000	5.000	3.632	1.123	48.311	0.000**
Safety facilities	223	1.000	5.000	3.655	1.087	50.198	0.000**
Service attitude	223	1.000	5.000	3.673	1.059	51.775	0.000**
Response and handling of tourist complaints	223	1.000	5.000	3.462	1.073	48.196	0.000**
Consulting service	223	1.000	5.000	3.776	1.041	54.140	0.000**
Professionalism of service personnel	223	1.000	5.000	3.601	0.990	54.329	0.000**

*p<0.05 **p<0.01

3.2.2 Weight analysis of priority graph

The weight analysis method of priority graph is to give the importance order and priority order by comparing several indexes or objectives. The lower limit value of TTL of tourists' expected value is 0.500, and the upper limit value is 38.500. The larger the average value is, the larger the TTL value is and the higher the weight is. If not, the lower it is. As shown in Table 4, tourists have great expectation on the catering hygiene, duration to the park, tour route design in the park, tour environment, rest and storage facilities, toilet sanitation, service attitude, consulting service and professionalism of service personnel in Chengdu Happy Valley. It has little expectation on the distribution and accuracy of signposts in the park, convenient portable commodity, catering taste, accommodation features and interactivity of the performance.

Table 4. Results of weight calculation.

Metric	Average value	TTL (Index Score)	Weighted value
Catering price	3.502	7.500	0.986%
Catering selection	3.704	28.500	3.748%
Catering taste	3.475	4.500	0.592%
Catering service	3.664	25.000	3.287%
Catering hygiene	3.762	36.500	4.799%
Accommodation accessibility	3.480	5.500	0.723%
Accommodation price	3.583	13.000	1.709%
Accommodation hygiene	3.668	26.500	3.485%
Accommodation features	3.444	2.500	0.329%
Convenient transportation to the park	3.709	29.500	3.879%
Cost of transportation to the park	3.498	6.500	0.855%
Duration to the park	3.735	32.500	4.274%
Distribution and accuracy of signposts	3.390	0.500	0.066%
Tour route design	3.821	37.500	4.931%
Ticket price	3.655	22.000	2.893%
Ticket promotion	3.511	9.000	1.183%
Tour environment	3.740	33.500	4.405%
Enjoyment of performance	3.659	23.500	3.090%
Interactivity of performance	3.453	3.500	0.460%
Commodity feature	3.583	13.000	1.709%
Commodity price	3.511	9.000	1.183%
Commodity quality	3.610	16.500	2.170%
Convenient portable commodity	3.439	1.500	0.197%
Novelty of amusement project	3.614	17.500	2.301%
Excitability of amusement project	3.596	15.500	2.038%
Richness of amusement project	3.673	27.500	3.616%
Waiting time for amusement project	3.655	22.000	2.893%
Waiting mode for amusement project	3.520	10.500	1.381%
Safety of amusement project	3.664	25.000	3.287%
Rest and storage facilities	3.839	38.500	5.062%
Reasonable layout of toilets	3.570	11.500	1.512%
Toilet hygiene	3.758	35.500	4.668%
Wireless network coverage and signal degree	3.592	14.500	1.907%
Tourist Service Center	3.619	18.500	2.433%
Safety facilities	3.650	20.000	2.630%
Service attitude	3.744	34.500	4.536%
Response and handling of tourist complaints	3.650	20.000	2.630%
Consulting service	3.713	30.500	4.011%
Professionalism of service personnel	3.717	31.500	4.142%

The lower limit value of TTL of tourists' expected value is 0.500, and the upper limit value is 38.500. The larger the average value is, the larger the TTL value is and the higher the weight is. If not, the lower it is. As shown in Table 5, tourists are more satisfied with the catering hygiene, tour route design, tour environment, enjoyment and interactivity of performance, excitability and, safety of amusement projects, safety facilities and consulting service in the park. Tourists are less satisfied with the catering price, accommodation price, commodity price, waiting time and method of amusement projects in Chengdu Happy Valley.

Table 5. Results of Satisfaction Optimization Chart Weights.

Metric	Average value	TTL (Index Score)	Weighted value
Catering price	3.260	4.500	0.592%
Catering selection	3.278	6.500	0.855%
Catering taste	3.395	9.500	1.249%
Catering service	3.462	12.000	1.578%
Catering hygiene	3.695	32.500	4.274%
Accommodation accessibility	3.390	8.500	1.118%
Accommodation price	3.193	2.500	0.329%
Accommodation hygiene	3.507	16.500	2.170%
Accommodation feature	3.507	16.500	2.170%
Convenient transportation to the park	3.646	27.500	3.616%
Cost of transportation to the park	3.511	18.500	2.433%
Duration to the park	3.664	29.500	3.879%
Distribution and accuracy of signposts	3.507	16.500	2.170%
Tour route design	3.798	37.500	4.931%
Ticket price	3.265	5.500	0.723%
Ticket promotion	3.475	13.500	1.775%
Tour environment	3.762	35.500	4.668%
Enjoyment of performance	3.700	34.000	4.471%
Interactivity of performance	3.673	31.000	4.076%
Commodity features	3.574	20.000	2.630%
Commodity price	3.099	0.500	0.066%
Commodity quality	3.377	7.500	0.986%
Convenient portable commodity	3.426	10.500	1.381%
Novelty of amusement projects	3.632	26.000	3.419%
Excitability of amusement project	3.883	38.500	5.062%
Richness of amusement project	3.623	24.500	3.222%
Waiting time for amusement project	3.166	1.500	0.197%
Waiting mode for amusement project	3.211	3.500	0.460%
Safety of amusement project	3.700	34.000	4.471%
Rest and storage facilities	3.592	22.000	2.893%
Reasonable layout of toilets	3.498	14.500	1.907%
Toilet hygiene	3.574	20.000	2.630%
Wireless network coverage and signal degree	3.592	22.000	2.893%
Tourist Service Center	3.632	26.000	3.419%
Safety facilities	3.673	31.000	4.076%
Service attitude	3.655	28.500	3.748%
Response and handling of tourist complaints	3.462	12.000	1.578%
Consulting service	3.776	36.500	4.799%
Professionalism of service personnel	3.601	23.500	3.090%

3.3 IPA quadrant analysis

The expectation variable is X axis, and the satisfaction variable is Y axis. The mean values of expectation and satisfaction are the corresponding reference axes of X and Y axis, which are divided into four quadrants: relative advantage area, maintenance management area, key improvement area and priority improvement area. All data are reflected in the four quadrants, as shown in Fig 1.

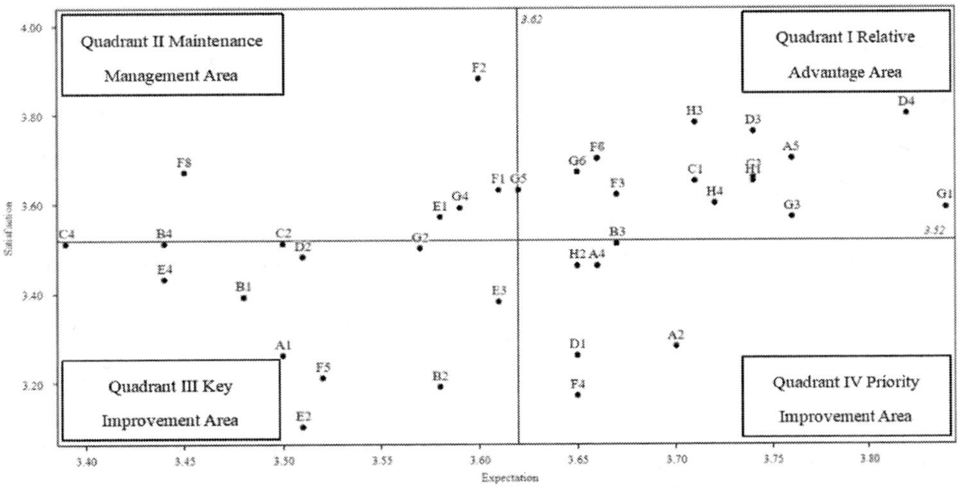

Fig 1. Quadrantal diagram of expectation and satisfaction.

A1 Catering price, A2 Variety of catering selection, A3 Catering taste, A4 Catering service, A5 Catering hygiene, B1 Accommodation accessibility, B2 Accommodation price, B3 Accommodation hygiene, B4 Accommodation feature, C1 Convenient transportation to the park, C2 Cost of transportation to the park, C3 Duration to the park, C4 Distribution and accuracy of signposts, D1 Ticket price, D2 Ticket promotion, D3 Tour environment, D4 Tour route design, E1 Commodity feature, E2 Commodity price, E3 Commodity quality, E4 Convenient portable commodity, F1 Novelty of amusement project, F2 Excitability of amusement project, F3 Richness of amusement project, F4 Waiting time for amusement project, F5 Waiting mode for amusement project, F6 Safety of amusement project, F7 Enjoyment of performance, F8 Interactivity of performance, G1 Rest and storage facilities, G2 Reasonable layout of toilets, G3 Toilet hygiene, G4 Wireless network coverage and signal degree, G5 Tourist Service Center, G6 Safety facilities, H1 Service attitude, H2 Response and handling of tourist complaints, H3 Consulting service, H4 Professionalism of service personnel.

3.3.1 Quadrant I Relative Advantage Area (high expectation, high satisfaction)

This area relates to 14 elements under 6 first-level indicators. It includes the catering hygiene in the element of "tasting". The "traveling" elements are the convenient transportation and duration to the park. The "sightseeing" element is the tour environment and tour route design. The "entertaining" elements are the enjoyment of the performance, the richness and safety of amusement projects. Hardware Facilities of the park include rest and storage facilities, toilet hygiene and safety facilities. The elements of software service include service attitude, professionalism of service personnel and consulting service.

From the first quadrant, it can be concluded that college students pay more attention to the above 14 aspects and have relatively high satisfaction during their visit to the theme park. The transportation is developed and convenient. The official website provides different tour routes for tourists, so tourists are satisfied with the external and internal traffic. It has more than 130 experience items, including 43 entertainment facilities and equipment, 58 cultural and ecological landscapes, 10 art performances, 20 themed games, which well meet the students' demand for enjoyment of the performance and richness of amusement projects. The park regularly inspects and maintains the amusement equipment and the detailed safety inspection of tourists by the staff on site, which makes the tourists have high satisfaction degree with the safety of Happy Valley. The park continues to care about the needs of employees, and encourages them to receive every tourist confidently and enthusiastically. The employees can quickly respond to the needs of the tourists, so the tourists have high satisfaction on the service attitude, consulting service and professionalism of service personnel. The park should maintain and appropriately enhance its advantages in 14 elements, so as to enhance its core competitiveness and provide tourists with better sense of immersive experience.

3.3.2 Quadrant II Maintenance Management Area (low expectation, high satisfaction)

This area relates to 6 elements under 3 first-level indicators. In terms of the six elements of tourism, it involves the interactivity of performance in the element of "traveling", the commodity features in the element of "purchasing", the novelty and excitability of amusement projects in the element of "entertaining". As far as the hardware facilities of the park are concerned, they mainly include Tourist Service Center and wireless network coverage and signal degree.

From the second quadrant, it can be concluded that college students pay less attention to the above 6 aspects during the visit of Chengdu Happy Valley, but the actual experience satisfaction is relatively high. "Grasp the strong men", "Happy parade" and other wonderful performances can better meet the needs of tourists for interactive performance projects. Roller coaster projects such as "flying over the Mediterranean", "Sky coaster" and "Desert Rally" can well meet tourists' requirements for the novelty and excitement of entertainment projects. Therefore, tourists have relatively high satisfaction with the interactivity of performance, the novelty and excitability of amusement projects. Tourist Service Center, wireless network coverage and signal degree can better have tourists to provide convenient and quick service. Due to the low expectation of tourists on commodity features, the satisfaction degree is relatively high. The park needs to maintain high level of tourist satisfaction in 6 elements.

3.3.3 Quadrant III Key Improvement Area (low expectation, low satisfaction)

This area relates to 13 elements under 7 first-level indicators. In terms of the six elements of tourism, it mainly includes the price and taste of catering in the element of "tasting", the accessibility, price and characteristics of accommodation in the element of "residing", the cost of transportation to the park in the element of "traveling", the distribution and accuracy of signposts, and the means of ticket promotion in the element of "sightseeing". The quality, price and portability of commodities in the "purchasing", and the waiting mode of amusement projects in the "entertaining" are also included in it. As far as the "hardware facilities" of the park are concerned, the toilet layout is reasonable.

From the third quadrant, it can be concluded that college students pay low attention to the above 13 aspects during their visit to Chengdu Happy Valley. However, from the specific value, the attention degree is significantly higher than the satisfaction degree, indicating that the satisfaction of tourists is far lower than the expectation. Due to the high price of goods, catering and surrounding accommodation, and the limited economic conditions of college students, they generally do not choose the commodities, food, and the surrounding accommodation facilities. Therefore, the expectation and satisfaction of college students on the above three aspects are relatively low. Due to the need to wait in line during holidays and weekends, low expectation and satisfaction for waiting mode of amusement projects are shown.

3.3.4 Quadrant IV Priority Improvement Area (high expectation, low satisfaction)

This area relates to 6 elements under 5 first-level indicators. In terms of the six elements of tourism, it mainly includes the service and selection of catering in the element of "tasting", the accommodation hygiene in the element of "residing", the ticket price in the element of "traveling", and the waiting time of amusement projects in the element of "entertaining". As far as "software service" is concerned, it is mainly the response and handling of tourists' complaints.

From the fourth quadrant, it can be concluded that college students pay high attention to the above 6 aspects but have low satisfaction during their visit. Chengdu Happy Valley Park is dominated by fast food and snacks, with few choices and no local characteristics. Restaurants provide fewer locations and fewer service personnel. When encountering meal points or peak periods, catering service personnel are too busy and the service is not in place, which leads to low satisfaction of tourists in terms of catering selection and the concerned service. Although there is a discount for college students, ticket price is still relatively more expensive for them without economic resources, so tourists have low satisfaction with the ticket price. Tourists wait in line for a long time on some popular amusement projects, such as "Roller coaster", "Big pendulum hammer" and "Sky wheel", so their satisfaction are relatively low.

3.4 Improvement strategy

Based on the quadrant diagram of tourist expectation and satisfaction analyzed by IPA, the corresponding countermeasures are put forward.

3.4.1 Relative Advantage Area

Tourists' expectation and satisfaction with the elements of the area are relatively high. These elements belong to the advantages of Chengdu Happy Valley resources, the park should give full play to the existing advantages of resources, strengthen the core competitiveness. Firstly, with a variety of large-scale amusement equipment, enhance the theme story line scene, maximize the restoration of the characters and scenes in each story. Secondly, improve the tour line design, meet the needs of different groups, and enhance the sense of experience and gain of tourists. Finally, strengthen the management of hygiene and safety in the park to ensure food hygiene and environmental hygiene. Meanwhile, regular inspection and maintenance of the safety facilities and equipment in the park, to ensure the safety of tourists. In short, the relative advantage area can provide tourists with high-quality tourism experience, improve the revisit rate, and create more benefits, so as to promote the sustainable development of Chengdu Happy Valley.

3.4.2 Maintain Management Area

Tourists have relatively low expectation of the elements in the area, but their satisfaction rate is relatively high. The park needs to maintain high satisfaction by creating more cultural and creative products, providing tourists with more rich and diverse choices of commodities, so as to improve the income and popularity of the park, and enhance the tourists' satisfaction. Integrate Sichuan Opera or giant panda elements to create a local IP image of Chengdu Happy Valley, improve the publicity effect, and capture the hearts of consumers. Meanwhile, with the help of a series of new media platforms such as Xiaohongshu, TikTok, Bilibili and Kuaishou, the park will promote special commodities and various play projects, increase publicity efforts, attract the attention of tourists, and strive for more tourists. VR or AR technology is applied to play projects. When experiencing "Roller coaster", tourists will put on the equipment as if being on the scene, which will enhance the excitement of playing and improve tourists' immersive experience.

3.4.3 Key Improvement Area

Tourists' expectation and satisfaction of the elements of this area are relatively low, but from the specific value, the expectation value is higher than the satisfaction, which needs to be improved. Commodity price and food price in the park can be offered as "activity price". Tourists can enjoy discounts by taking photos, posting on Wechat moments or TikTok, while businesses can increase their profits through publicity. The park can cooperate with the surrounding hotels to launch theme rooms, and tourists can enjoy 10% or 20% discount with ticket records. Tourists who check in the hotel can also enjoy ticket discounts. Benefit sharing can not only increase the income of the park, but also promote the business of the surrounding hotels. The park can adopt dynamic pricing system to raise or lower the ticket price according to different time periods and guide tourists to visit the park at different peaks. Chengdu Happy Valley can make use of the Internet to create smart play, directly book the amusement projects through the park software, and enter the play area at the agreed time through the express channel. On the premise of maintaining healthy social distancing, it will provide more pleasant experience and improve tourists' satisfaction and sense of gain.

3.4.4 Priority Improvement Area

Tourists have high expectation of elements in this area, but the satisfaction degree is relatively low. Therefore, the park should give priority to improve its concerned elements. According to the needs of tourists, the park should add Chengdu flavor snacks in terms of catering types, introduce Shandong cuisine, Guangdong cuisine, Huaiyang cuisine, Sichuan cuisine and other local dishes, and create healthy dishes in combination with healthy diet. As far as catering service is concerned, flexible staff deployment should be implemented to ensure adequate staffing during peak hours. Meanwhile, strengthen service training, improve service system, so as to provide satisfactory service for each tourist.

Actively improve the supervision system of the park, optimize the online and offline complaint response channels, timely reply to tourist' questions, and adopt reasonable suggestions from tourists, so as to create cozy tour environment for tourists.

4 Conclusions

This study analyzes the expectation and satisfaction of tourists based on the IPA method, so as to obtain the evaluation of college students as representative of young tourists on the tourism experience quality of Chengdu Happy Valley. The expectation and satisfaction of college tourists are above the middle level in Chengdu Happy Valley. It indicates that the actual experience effect of college tourists in Chengdu Happy Valley is almost the same as expected, and the park should be improved on the basis of continuing maintenance, so as to improve tourists' satisfaction. In terms of the six elements of tourism, hardware facilities and software service, the expectation degree of college students on the first-level indicators are as follows: hardware facilities, entertainment, software service, travelling, tasting, sightseeing, residing and purchasing, indicating that college students pay more attention to the hardware facilities of the park, but have relatively low expectation on purchasing. Meanwhile, the ranking of satisfaction weight by college student tourists is hardware facilities, transportation, entertainment, sightseeing, software service, tasting, accommodation, purchasing, indicating that college students have the highest satisfaction with hardware facilities, while the lowest satisfaction with purchasing. Due to the deviation of expectation and satisfaction, the facilities and service of the park should be improved according to the expectation and satisfaction of tourists, so as to continuously improve the sense of tourist experience. Based on the quadrantal diagram of expectation and satisfaction, this study puts forward corresponding countermeasures to provide reference for the sustainable development of Chengdu Happy Valley.

Questionnaire survey has been adopted to obtain the data of tourists' experience quality of Chengdu Happy Valley. However, the questionnaire has been limited to college students, and the number of questionnaires has been far from adequate. In future research, the evaluation system of questionnaire index can be improved, the number of questionnaires can be increased, and the tourist group can be expanded. In the IPA analysis, only simple analysis has been made, without comparing the differences among various elements, nor in-depth analysis of the overall expectation and satisfaction. More research methods can be used to conduct in-depth research on tourist satisfaction so as to improve the quality of tourism experience and promote the sustainable development of theme parks.

Acknowledgments

This research was funded by Yibin Federation of Social Sciences Project “Research on Tourism Upgrading in Yibin from the Perspective of Tourism Experience”, Sichuan Provincial Department of Education Project “Research on Ecotourism Certification in the Southern Sichuan Bamboo Sea” (15SA0087) and Shenzhen University Higher Education Teaching Reform Key Project(2024JG008).

References

1. Song, R. China's Tourism Development and Analysis in 2021-2022. Beijing: Social Sciences Academic Press, 210-212(2022)
2. Pizam, A. Tourism’s Impacts: The Social Costs to the Destination Community as Perceived by Its Residents. Journal of Travel Research, 16, pp.8-12(1978)
3. Hughes K. Tourist Satisfaction: A Guided Cultural Tour In North Queensland. Australian Psychologist, 26(3), pp.166-171(1991)
4. Sun, J.X.; Chen, G.H. Tourism Consumer Behavior. Dalian: Dongbei University of Finance and Economics Press, 217(2015)
5. Liang, Z.X.; .Li,. What is a Theme Park? A Synthesis and Research Framework. Journal of Hospitality & Tourism Research, pp.1-28(2021)
6. Chan, C.S.; Liu, Y.M.; Li, C.Y. Expectation-Perception Evaluation of Theme Park Service Quality in Zhengzhou Fantawild, China.Journal of Park and Recreation Administration,37(2), pp.99-117(2021)
7. Chang, M.Y.; Hsu, Y.S.; Chen, H.S. Choice Experiment Method for Sustainable Tourism in Theme Parks.Sustainability, 13(13), pp.1-28(2021)
8. Lee, S. Systematic Approach to Theme Park. in 3rd International Conference on Management Engineering, Software Engineering and Service Sciences, pp.190-197(2019)
9. Zhang, Y.S.; Li, X.; Su, Q. Does Spatial Layout Matter to Theme Park Tourism Carrying Capacity? Tourism Management, 61, pp.82-95(2017)
10. Zhou, X.J.; Zhong, S.N.; Xu, W.Y.; Ren, X.L. Research on Multidimensional Influencing Factors of Tourist Satisfaction in Theme Parks: A Case Study of China Dinosaur Park. Geography and Geographic Information Science, 33 (06), pp.118-124(2017)
11. Fu, X.X.; Baker, C.; Zhang, W.; Zhang, R.Y. Theme Park Storytelling: Deconstructing Immersion in Chinese Theme Parks.Journal of Travel Research, pp.60-67(2022)
12. Belarmino, A.; Whalen, E.; Koh, Y.; Bowen, J.T. Comparing Guests’ Key Attributes of Peer-to-peer Accommodations and Hotels: Mixed-methods Approach. Current Issues in Tourism, pp.1-7(2017)

Simulation of Spatial and Temporal Dynamics in Habitat Quality amid Rapid Urbanization based on Random Forest Algorithm: A Lingui District, Guilin City Case Study

Xiaoyan Zhang*, Hanxin Liu, Xin Yang, Gang Wu, Junjie Tan

College of Tourism and Landscape Architecture, Guilin University of Technology, Guilin 541000, China

Abstract. Territorial space planning policies profoundly influence land use alterations, subsequently impacting regional habitat quality. Habitat quality epitomizes the capacity to sustain biodiversity. The acceleration of human activities and urban construction land expansion foregrounds habitat quality as a pivotal aspect of ecological diversity conservation and sustainable development scholarship. Analyzing the spatial and temporal fluctuations of regional habitat quality furnishes an ecological, scientific foundation for territorial space strategy. This study, centering on the rapidly urbanizing Lingui District, assesses and forecasts land use conversion and habitat quality evolution from 2000 to 2030, using the random forest algorithm, employing land use data from 2000, 2010, and 2020 through an integrated application of the PLUS-InVEST model. Furthermore, it explores spatial differentiation, spatiotemporal pattern characteristics, and determinants in Lingui District by amalgamating spatial statistical and spatial autocorrelation methodologies.

1 Introduction

As human interventions in the ecological environment escalate, the challenge of biodiversity conservation becomes increasingly acute. The imperative to harmonize development and conservation efforts, propel high-quality socio-economic advancement, improve habitat quality, and safeguard species diversity has ascended as a paramount concern for nations and nongovernmental entities worldwide. Consequently, the forecasting of land use configurations has garnered international focus [1, 2]. Habitat quality, delineating the living conditions and cornerstone for the sustainable proliferation of regional ecosystem entities, is identified as a crucial metric for assessing ecosystem vitality and biodiversity health [3-5]. Current methodologies for habitat quality evaluation prominently include the utilization of models such as the SolVES [6], MaxEnt [7], GUMBO [8], and MIMES [9] models, with the latter being widely recognized for its operational adaptability and robust theoretical structure [10, 11]. In endeavoring to anticipate habitat quality trajectories, some researchers have amalgamated habitat quality evaluation models with land use simulation and forecasting frameworks, like the CLUE-S [12], FLUS [13], CA-Markov [14], and PLUS [15] models, with the PLUS model demonstrating efficacy in comprehensive simulations across vast scales and diverse categories based on random forest. Notably, studies, for instance, those conducted by Zhou Junxin and colleagues [16], have assessed the prevailing land use habitat quality via the InVEST model and forecasted future habitat quality shifts by integrating the PLUS model, thus delineating the prospective habitat quality evolution in regions such as Jiangxi Province. While numerous studies have focused on merging these models for analytical purposes, research from a municipal viewpoint that scrutinizes habitat quality via land use modifications remains comparatively limited, with a dearth of investigations into municipal habitat quality variations under different future scenarios.

In recent epochs, Guilin City has zealously pursued the "Four Major Positions," endeavoring to ascend as a globally distinguished tourist destination and marking a noteworthy commencement. Hence, research predicated on land use data spanning 2000 to 2020, examining the spatiotemporal dynamics of habitat quality in Lingui District, Guilin City, and forecasting habitat quality alterations under three prospective development scenarios by 2030, can furnish a scientific basis and insights for the ecological civilization evolution, as well as territorial space planning and development in Lingui District, Guilin City.

2 Materials and Methods

2.1 Study Area

Lingui District is under the jurisdiction of Guilin City, Guangxi Zhuang Autonomous Region. The study region is situated as follows (Figure 1).

2.2 Data Processing

The origins of the experimental data are as follows (Table 1). The land use data are based on Landsat remote sensing

* Corresponding author: 1632808228@qq.com

information sources, combined with fieldwork verification, and the accuracy has been tested to be at 90%.

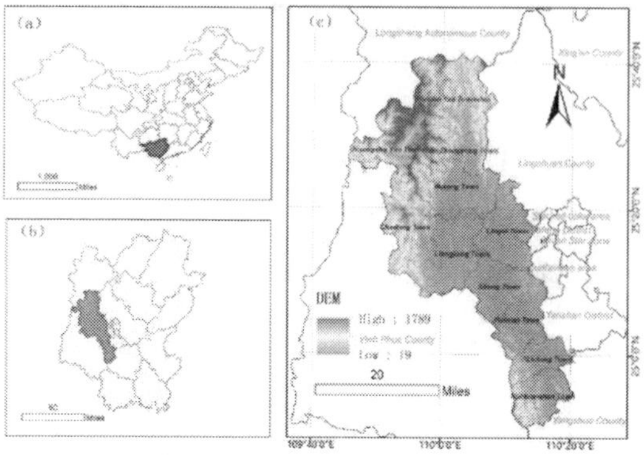

Fig. 1. Location Map of the Study Area.

Table 1. List of Research Data.

Data Type	Data Content	Data Source
Natural Data	Land Use Status	Institute of Geographic Sciences and Natural Resources Research, CAS
	Temperature	
	Precipitation	
	Vegetation Cover	National Ecosystem Science Data Center
	Terrain	Geospatial Data Cloud
Socio-economic Data	Population	WordPop
	GDP	National Ecosystem Science Data Center
	Nighttime Lights	Chinese Academy of Sciences (Fujian)
	Road Data	Open Street Map

2.3 Research Approach

This paper's study strategy, as illustrated in Figure 2, uses coupling models to determine the spatiotemporal aspects of habitat quality [17-18].

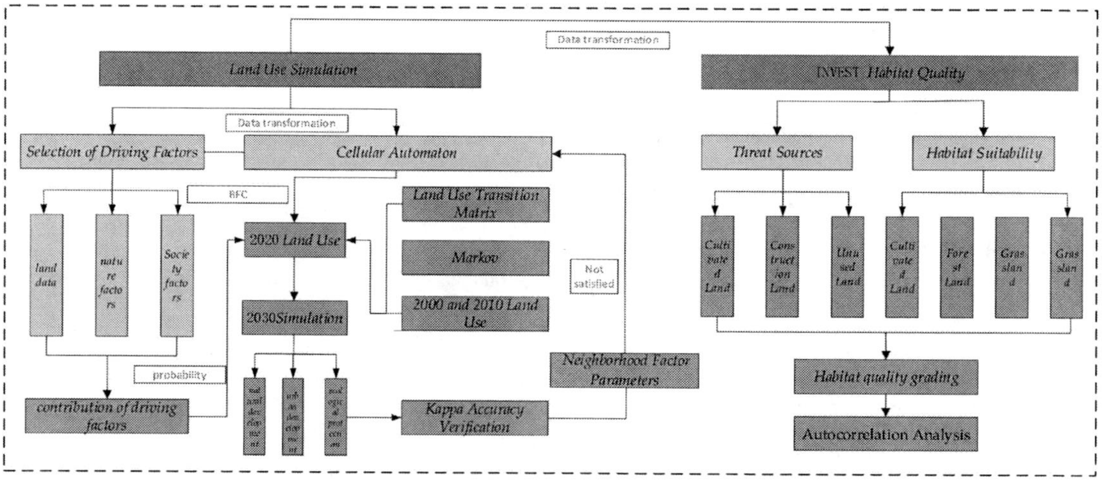

Fig. 2. Research Approach.

2.4 Habitat Quality Analysis

The InVEST model Habitat Quality module quantitatively assesses the habitat quality of the study area by considering the sensitivity of various land cover types to threat factors. The strength of ecosystem functions is measured based on the habitat quality score [20,21]. This model is characterized by requiring a small amount of input data while generating a large volume of output data. In light of the prevailing circumstances in the Lingui District, where human activities significantly alter the state of cultivated, construction, and idle lands, these categories have been designated as key threat factors. Utilizing prior scholarly work [19], parameters such as the maximum stress distance, relative weight, and the type of spatial decay for these factors have been delineated (refer to Table 2). Moreover, the study appraises the habitat suitability and the level of sensitivity to these threats across various land use classifications (refer to Table 3).

Table 2. Threat Sources.

Threat Factor Parameters	Maximum Stress Distance	Relative Weight	Spatial Decay Type
Cultivated Land	2.6	0.26	Linear
Construction Land	5.8	0.73	Exponential
Unused Land	2	0.25	Linear

Table 3. Stress Factors and Habitat Suitability.

Land Use Type	Habitat Suitability	Stress Factors
Cultivated Land	0.3	0
Forest Land	0.9	0.6
Grassland	0.7	0.8
Water Bodies	0.85	0.5
Construction Land	0	0
Unused Land	0	0

https://doi.org/10.1051/e3sconf/202561703003

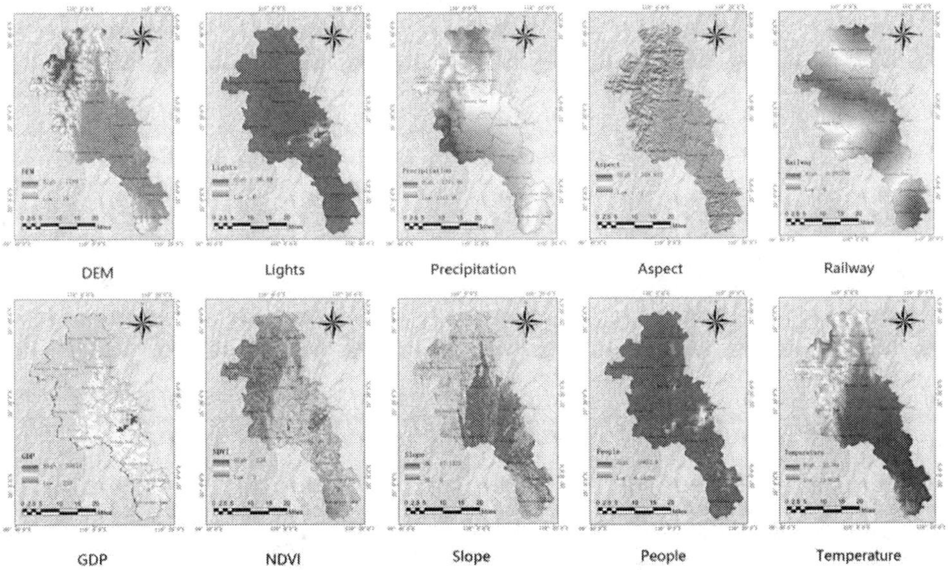

Fig. 3. Selection of Driving Factors.

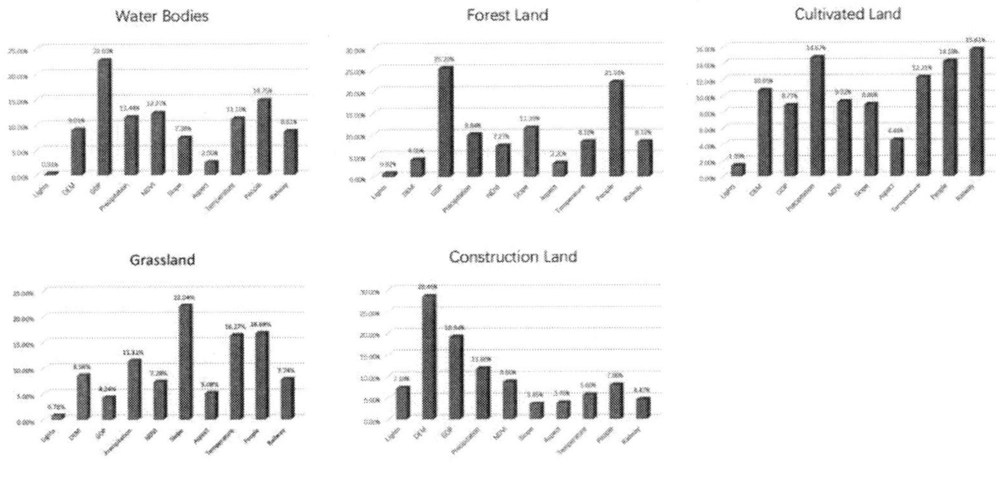

Fig. 4. Proportion of Contribution by Each Factor.

2.5 Land Use Simulation

2.5.1 Selection of Driving Factors

The PLUS model extracts land expansion for analysis in the strategy module, selecting land expansion impact factors suitable for Lingui District., as shown in Figure 3.

2.5.2 Neighborhood Factor Parameters

Neighborhood factors refer to the interaction relationships between different land use types and among different land use units within a neighborhood range. The parameters for each factor in the land use domain are set as shown in Table 4 [22-24].

Table 4. Neighborhood Factor Parameter Settings.

Land Use Type	Cultivated Land	Forest Land	Grassland	Water Bodies	Construction Land	Unused Land
Natural Development	0.23	0.63	0.08	0.01	0.05	0
Ecological Protection	0.23	1	0.07	0.02	0.27	0
Urban Development	0.23	0.63	0.08	0.01	0.3	0

2.5.3 Accuracy Verification

The Kappa coefficient is used for cross-comparison between the simulated map of 2020 and the actual status map of 2020 to verify the final results [25-26].

2.5.4 Neighborhood Factor Parameters

To explore the potential spatial dependence, association, and variation among habitat quality in Lingui District, methods combining R programming and Geoda software for spatial information analysis are employed. The Moran's I index, which includes both global and local Moran's I indices, is used for this analysis [26].

3 Research Results

3.1 Land Expansion Analysis Strategy

Utilizing a tripartite temporal framework of land use data, the datasets underwent analysis through remote sensing techniques. Leveraging a dataset encompassing 10 influential variables, a sampling-based computational approach was implemented. The ensuing probability maps delineating the potential development for various land types, along with the hierarchical impact of contributing factors on land use dynamics, are depicted in Figure 4. Demographic trends, and terrain characteristics are the overarching forces that shape land use transformations.

3.2 2030 Land Use Prediction Results

3.2.1 Neighborhood Factor Parameters

Table 5. Land Use Area in Lingui District, Guilin City, 2000-2030.

Land Use	2000	2010	2020	2030 (Natural Development)	2030 (Ecological Protection)	2030 (Urban Development)
Cultivated Land	552.25	550.98	532.98	532.4	551.52	515.69
Forest Land	1436.82	1437.04	1424.93	1423.97	1434.28	1412.01
Grassland	188.78	184.63	180.11	179.96	175.72	192.09
Water Bodies	16.35	17.47	18.12	18.32	15.04	18.75
Construction Land	43.55	47.65	81.63	81.59	59.73	97.78
Unused Land	0.09	0.09	0.09	0.09	0.04	0.09
Total	2238.85	2238.85	2238.85	2237.32	2237.32	2237.41

To achieve the carbon peak by 2030, simulations of land use changes in Lingui District were conducted (refer to Table 5). Projections within the context of natural progression anticipate a decline in the expanses of arable, respectively, by the year 2030. Concurrently, the expanse allocated to built-up areas is anticipated to augment to 81.59 square kilometers, constituting 3.65% of the entire expanse. In the urbanization trajectory, the prospective dimensions for arable, meadow, forested, and built-up zones are delineated as 515.69 square kilometers, 1412.01 square kilometers, 192.09 square kilometers, and 97.78 square kilometers, respectively. Conversely, the ecological conservation trajectory foresees an upsurge in the dimensions of arable and forested zones to 551.52 square kilometers and 1434.28 square kilometers, respectively.

3.2.2 Spatial Distribution Characteristics

The land use characteristics' spatial distribution from 2000 to 2030 (Figure 5) varies according to various scenarios, with notable shifts in the amount of land used for construction, farming, and forests. The natural development forecast exhibits a discernible decline in arable land and a gradual rise in construction land, with a noteworthy surge in construction land expansion between 2020 and 2030. Under the ecological protection scenario, the expansion of construction land is somewhat restricted, preventing excessive conversion of cultivated and forest land into construction land. A balanced consideration of urban development and ecological protection scenarios can better promote the sustainable economic development of Lingui District.

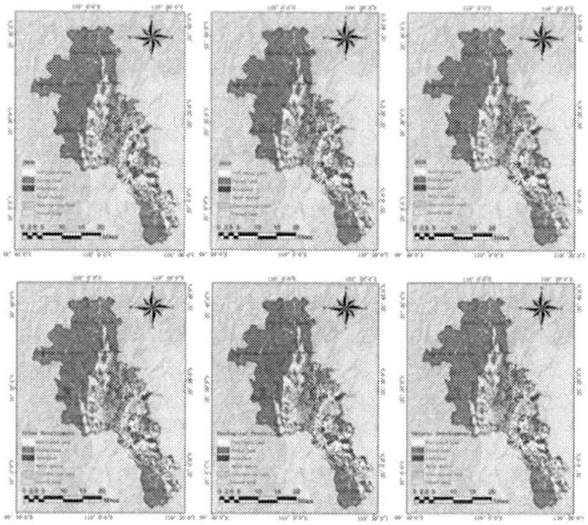

Fig. 5. Spatial Distribution of Land Use from 2000 to 2030.

3.2.3 Accuracy Verification Results

The Kappa coefficient was used for an overall consistency check between the simulation results and observational data. This study selected the year 2010 as the baseline data and applied the aforementioned method to simulate the land use of 2020. It was found that the Kappa coefficient reached as high as 0.912. Based on this, multiple sets of data were tested, indicating that the PLUS model is highly applicable in simulating future changes in the spatial pattern of land use in Lingui District.

3.3 Habitat Quality Characteristics Analysis

3.3.1 Temporal Change Characteristics

Over time, as shown in Table 6, Predictions for 2030 indicate that "poor" habitat quality will continue to increase by 0.92%, "moderate" habitat quality will continue to decline by 0.99%, and "high" habitat quality will continue to rise. In contrast, urban development saw no significant change in "moderate" habitat quality relative to 2020, whereas natural protection and ecological protection scenarios saw increases of 4.81% and 4.42%, respectively. Thus, habitat quality is linked to changes in the distribution of agricultural and forest land.

Table 6. Land Use Area in Lingui District, Guilin City, 2000-2030.

Habitat Quality	2000	2010	2020	Natural Development	Urban Development	Ecological Protection
Low	721.42	719.96	725.65	705.66	720.03	694.81
Lower	553.54	555.08	549.53	565.03	553.65	576.07
Moderate	247.41	247.4	245.3	250.11	245.62	249.72
Higher	184.37	184.3	184.87	186.5	184.7	187.04
High	531.11	531.11	532.5	529.02	532.32	528.68
Total	2237.85	2237.85	2237.85	2236.32	2236.32	2236.32

3.3.2 Spatial Change Characteristics

The study region's overall habitat quality, which accounts for 84% of its total area, is primarily classified as high, low, and lower from the standpoint of spatial pattern (Figure 6). The minority of the area is categorized within higher and moderate habitat quality ranges. This indicates a negative correlation between human activities and habitat quality development.

3.4 Spatial Autocorrelation Analysis

3.4.1 Global Autocorrelation

Global spatial autocorrelation metrics (as detailed in Table 7) reveal that the Moran's I index for the predicted habitat quality in the 2030 regional simulation surpasses 0.8. This is evidenced by Z-scores significantly exceeding 2.58 and p-values being notably less than 0.01, which underscores a pronounced spatial pattern of habitat quality clustering within Lingui District, Guilin City.

Table 7. Land Use Area in Lingui District, Guilin City, 2000-2030.

Scenario	Global Moran's I	Z-score	P-value	Threshold
Natural Development	0.863832	30.3375	<0.00001	2000.2
Urban Development	0.874247	30.71471	<0.00001	2000.2
Ecological Protection	0.874008	30.70638	<0.00001	2000.2

Combining with the univariate scatter plots, it can be observed that areas of high habitat quality tend to disperse, with a lesser degree of adjacency under the natural development scenario. Areas of low habitat quality tend to concentrate, with a higher degree of concentration relative to the other two scenarios. Ecological protection, compared to urban development, shows a higher concentration (Figure 7), demonstrating good applicability.

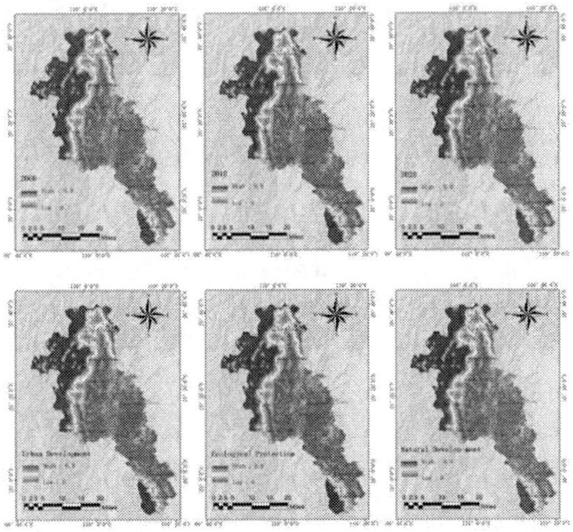

Fig. 6. Spatial Distribution Characteristics of Habitat Quality.

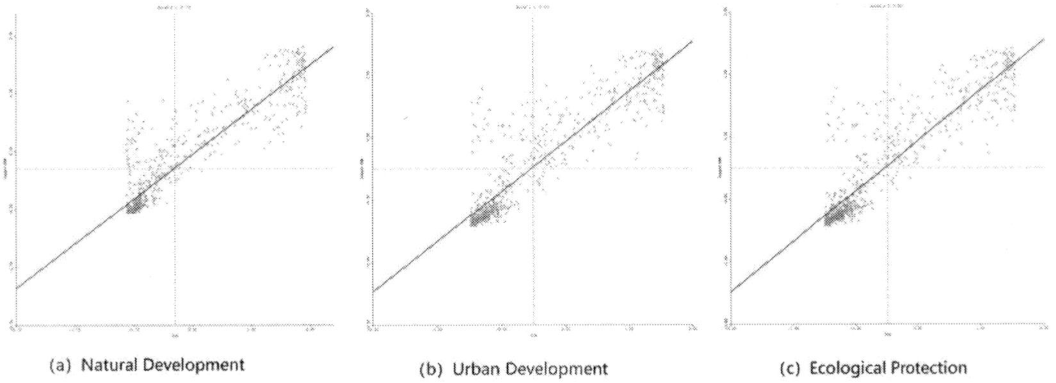

Fig. 7. Land Use Area in Lingui District, Guilin City, 2000-2030.

3.4.2 Local Autocorrelation

The local autocorrelation cluster map (Figure 8) shows that areas of "high-high" habitat quality clustering are primarily located in the western part of Wantian Yao Township, Huangsha Yao Township, the western part of Chadong Township, and the southwestern part of Nanbian Mountain Township. The "high-high" clustering areas are mainly distributed around Lingui Town, Liangjiang Town, Sitang Town, Huixian Town, and Liutang Town. The primary reason for this distribution is the continuous development of Lingui Town, resulting in diminished habitat integrity. Differences in habitat quality between the central and western parts and the delineation of urban development boundaries are related to this pattern.

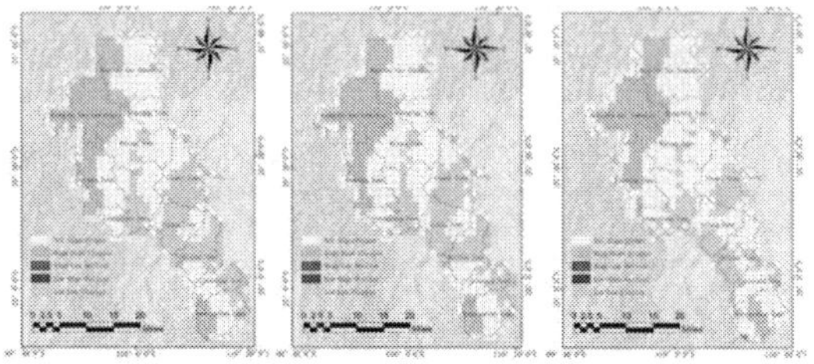

Fig. 8. Local Spatial Autocorrelation of Habitat Quality.

4 Conclusions and Recommendations

4.1 Conclusions

Dynamics Dynamics of Habitat Quality (2000-2020): The caliber of habitat quality within the area of study is significantly compromised by the encroachment of urban construction expanses, delineated by a dichotomy of "high" and "low" quality habitats, indicative of a polarized distribution. The interval from 2000 to 2020 delineated a pattern where habitat quality initially experienced an upturn, only to subsequently decline. Looking towards the year 2030, projections for both natural and urban development indicate a potential decline in habitat quality, whereas the ecological protection scenario anticipates an enhancement. Zones of superior habitat quality are predominantly situated in the autonomous townships on the western periphery of Lingui District, whereas areas of inferior quality are largely concentrated around Lingui Town and its environs, areas characterized by heightened human activity. The scenario-based delineation of habitat quality clusters underscores the potential for targeted developmental and conservation efforts tailored to habitat quality enhancement.

4.2 Recommendations

In the context of swiftly growing urban areas, spatial planning should give precedence to safeguarding pristine natural features such as mountain chains, streams, woodlands, cultivated fields, freshwater bodies, prairies, and arid zones, concurrently meeting the developmental requirements of burgeoning cities. By maintaining the construction land ratio within a 2.67%-4.37% bracket, Lingui District can achieve rapid urbanization while concurrently ensuring ecological integrity. This methodology not only averts a decline in habitat quality but also curtails the ongoing erosion of ecological demarcations to a certain degree. Furnishing empirical data for the ecological framework's spatial planning in Lingui District will further endorse Guilin's trajectory towards sustainable development.

From the point of view of land space development and protection planning, the delineation of the production and living and urban development boundaries, in addition to the limitations of the annual land supply plan, should give more consideration to its impact on the surrounding ecological environment, quantify the ecological value and economic value of each piece of land on the basis of land use simulation prediction and the InVEST model, comprehensively assess the ultimate benefits of urban expansion, and accurately determine the direction of future urban development and formulate corresponding plans, making the planning more scientific and forward-looking. The quantitative assessment of construction land expansion can accurately determine the future direction of urban development and formulate corresponding plans, making the planning more scientific and forward-looking.

References

1. GASCON C, BROOKS T M, CONTRERAS-MACBEATH T, et al. The importance and benefits of speciesL [J]. Current Biology, 2015, 25(10):R431-R438.
2. Fllman J B, Hod E, Dryer w, Pyare S. Stream physical characleristics impact habitat quality for pacific salmon in two temperale coastalwatersheds.PLoS One, 2015, 10(7) : e0132652.
3. Hilard E M, Nielen CK, Crowinger J W.Swamp rabbis as indicatos of wilife habitat quality in boltomlland hardwood forest ecosystems.Ecological Indicators, 2017, 79:47-53.
4. Zhang X R, Song W, Lang Y Q, Feng X M, Yuan Q Z, Wang J 'T.Land use changes in the coastal zone ol China's Hebei Povince and thecorresponding impacts on habitat quality. Land Use Policy , 2020, 99: 104957.
5. DIODATO N, BELLOCCHI G. MedREM, a rainfall erosivitymodel for the Mediterrancan region . Journal of Hydrology, 2010, 387(1): 119-127.
6. Merow, C., Smith, M.J. and Silander, J.A., Jr (2013), A practical guide to MaxEnt for modeling species' distributions: what it does, and why inputs and settings matter. Ecography, 36: 1058-1069.
7. Boumans, Roelof, Joe Roman, Irit Altman and Les Kaufman. "The Multiscale Integrated Model of Ecosystem Services (MIMES): Simulating the interactions of coupled human and natural systems." Ecosystem services 12 (2015): 30-41.
8. Xie Yiru, Gao Peichao, Wang Xiangyu, et al.Exploring thetrade-offs between grain yield and

ecological benefits in aneconomic development context: Land-use optimization ofHeilongjiang Province[J]. Journal of Beijing NormalUniversity (Natural Science), 2020, 56(6):873-881. (inChinese with English abstract)

9. Billionnet A.Mathematical optimization ideas for biodiversity conservation. European Joumal of Operational Research , 2013, 231(3):514-534.
10. Reeder W J, Gariglio F, Carnie R, Tang C L, Isaak D, Chen Q W, Yu Z B, McKean JA, Tonina D.Some (fish might) like it hot: habitatquality and fish growth from past to future climates. Science of the Total Environment, 2021, 787: 147532.
11. Torgersen, C. E., Le Pichon, C., Fullerton, A. H., Dugdale, S. J., Duda, J. J., Giovannini, F., et al. (2021). Riverscape approaches in practice: Perspectives and applications. Biological Reviews, 97(2), 481–504.
12. Gu Hanlong, Ma Tianjun, Qian Fengkui, et al. County land use scenario simulation and carbon emission effect analysis using CLUE-S model[J]. Transactions of the Chinese Society of Agricultural Engineering (Transactions of the CSAE), 2022, 38(9);288-296.
13. We D, Wang X, Liul, etal.Maintaning key ecosystem services under multiple development scenarios: A casestudy in Guangdong-Hong Kong-Macaogrecatcr bay Area, China[J].Ecological Indicators, 2023, 154: 110691.
14. YANG J, XIE P, XI J C, et al. LUCC simulation based on thecellular aulomata simulation: a case study of Dalian economicand technological development zone . Acta Geographica Sinica, 2015, 70(03) : 461-475.
15. Li, H., Lin, J., Lei, X. and Wei, T.X. (2022) Compressive Strength Prediction of Basalt Fiber Reinforced Concrete via Random Forest Algorithm. Materials Today Communications, 30, Article ID: 103117.
16. Sijia, J., Tianjun, W., Jiancheng, L. et al. Spatial prediction using random forest spatial interpolation with sample augmentation: a case study for precipitation mapping. Earth Sci Inform 16, 863–875 (2023).
17. Wang, D., Hao, H., Liu, H., Sun, L., & Li, Y. (2024). Spatial–temporal changes of landscape and habitat quality in typical ecologically fragile areas of western China over the past 40 years: A case study of the Ningxia Hui Autonomous Region. Ecology and Evolution, 14, e10847.
18. Cheng, Y.; Zhang, X.; Song, W. Ecological Risk Assessment of Land Use Change in the Tarim River Basin, Xinjiang, China. Land 2024, 13, 561.
19. multiple land use scenarios by coupling human and natural effects. Landscape and Urban Planning, 2017, 168;94-116.
20. Liang X, Liu X P, Li D, Zhao H, Chen G Z. Uirban growth sinulation by incorporating planming policies inlo a CA-based future land-usesimulation model. International Journal of Geographical Information Seience, 2018, 32(11) : 2294-2316.
21. li C, wa Y M, Cao B P , Zheng K J, WuY, liC.Muli-scenario simulation of ecosystem service value for optimization of land use in the SichuanYunnan ecological barrier, China. Ecological Indicators, 2021, 132:108328.
22. SHI M J, WU HQ, FANX, et al.Trade - olls and synermies of muliple cosysem services fordifrent land use seenarios in the Yii river valley , China. Sustainability , 2021 , 13(3) : 1577.
23. Xie L, Wang H W, Liu S L. The ecosystem service values simulation and driving force analysis based on land use/land cover: A case study in inland rivers in arid areas of the Aksu River Basin, China[J]. Ecological Indicators, 2022, 138.
24. Ting Zhao, Jinghu Pan.Ecosystem service trade-offs and spatial non-stationary responses to influencing factors in the Loess hilly-gully region: Lanzhou City, China[J]. Science of The Total Environment, 2022, 157422.
25. Yang Q, Zhang P, Qiu X, Xu G, Chi J. Spatial-Temporal Variations and Trade-Offs of Ecosystem Services in Anhui Province, China. Int J Environ Res Public Health. 2023 Jan 3; 20(1): 855. doi: 10.3390/ijerph20010855.
26. LIN J, HE P, YANG L. Predicting future urban waterlogging-prone areas by coupling the maximum entropy and FLUS model[J]. Sustainable Cities and Society, 2022, 80(2022):103812.

Evaluating Urban Ecological Livability Based on the Coupling Coordination Degree Model: A Case Study of Guangzhou

Shan Huang[1,a], Baoqiang Wang[2,b*]

[1] Faculty of Innovation and Design, City University of Macau, Macau, China
[2] School of Architecture and Urban Planning, Suzhou University of Science and Technology, Suzhou, China

Abstract. The construction of an ecologically livable city has emerged as a crucial objective in Guangzhou's "14th Five-Year Plan" and the vision for 2035. This paper establishes an evaluation index system for urban ecological livability, encompassing five dimensions: economy, society, technology, environment, and nature. It employs the TOPSIS analysis method and a coupling coordination model to conduct a trend analysis of the evaluation values and coupling coordination degrees of these five systems in Guangzhou from 2014 to 2023. The findings indicate: (1) the overall ecological livability of Guangzhou shows a positive trend, with a relatively favorable development level; however, the progress of each system is constrained by various factors. (2) Since 2015, the coupling degree of the five systems has gradually increased, maintaining a good state since 2017. The calculations of coordination degree reveal a steady rise in the overall system's coupling coordination. This study enhances the index system for ecologically livable cities and offers policy recommendations to strengthen urban ecological livability, thereby providing valuable guidance for promoting sustainable urban development and the construction of ecologically livable cities.

1 Introduction

Since the onset of the Industrial Revolution in the mid-18th century, urbanization has rapidly accelerated globally, marking a swift transition from agrarian to industrial civilization [1]. This urban growth has led to a significant migration from rural areas to cities, where increased productivity has enabled residents to attain higher economic incomes and enhanced living standards. However, unchecked land development during this urbanization process has resulted in a series of developmental challenges. The balance between economic growth and environmental protection has often been overlooked, leading to issues such as environmental pollution, housing shortages, traffic congestion, and uneven regional development, thereby severely compromising urban ecological environments. In response to these urban challenges and to achieve sustainable development, academia has proposed various future urban development concepts, including ideal societies from a humanistic perspective, utopian cities, solar cities, garden cities, satellite cities, landscape cities, and low-carbon cities [2]. The evolution of these models has given rise to a new urban development paradigm: the eco-friendly city.

The concept of the "eco-city" was first introduced by UNESCO in 1971 during the "Man and the Biosphere" research program [3]. In 1996, the United Nations Conference on Human Settlements first introduced the notion of "livable communities [4]," which the eco-friendly city integrates. Current literature indicates that scholars worldwide focus on four main aspects of urban eco-friendliness: definitions, evaluation indicators, evaluation methods, and development strategies. Most researchers define eco-friendly cities from a human-centered perspective, arguing that these cities should provide suitable living environments without exceeding ecological carrying capacities [5]. Urban livability is influenced not only by objective conditions but also encompasses residents' subjective perceptions of social welfare and economic development [6]. While there is no consensus on the definition of eco-friendly cities, scholars generally agree that they should provide a favorable ecological environment and adequate living space within the limits of environmental carrying capacity while balancing external environmental factors and internal economic development. Regarding evaluation indicators, experts widely acknowledge that social development levels, economic strength, and climate conditions are critical factors influencing urban eco-friendliness. Pacione has established an objective evaluation model for urban eco-friendliness based on social, demographic, and economic dimensions [7]. As research progresses, some scholars have pointed out that residents' subjective perceptions of eco-friendliness are also pivotal; for instance, Benjamin L. Saitluanga developed an evaluation framework incorporating both objective indicators—such as family, social, economic, and accessibility factors—and subjective indicators related to residents' satisfaction with economic and social aspects, as well as their perceptions of the natural environment and infrastructure development [8]. Contemporary evaluation methods increasingly

* Corresponding author: b* wbq318@163.com
a hs3225132134@163.com

employ advanced technologies, including non-parametric kernel density estimation [9], machine learning techniques, and remote sensing technology. Some researchers have combined analytic hierarchy processes with cognitive mapping [10], remote sensing with TOPSIS methods [11], and remote sensing with machine learning to offer new perspectives for establishing a scientific evaluation framework for eco-friendly cities [12]. Further analysis by scholars exploring the determinants of urban livability has identified pathways for developing livable cities. For example, Ghasemi et al. suggest that urban elements and their functions significantly impact eco-friendliness [13], while Parker and Zingoni de Baro argue that high-quality public green infrastructure enhances residents' perceptions of urban livability [14]. These research outcomes greatly enrich the study of eco-friendly cities and contribute to efforts in creating them.

Current research primarily assesses eco-friendliness by evaluating the economic, social, and environmental development levels of cities, yet analysis of the interrelationships among these systems remains insufficiently explored. Addressing this gap, this paper adopts a perspective encompassing five major systems: economic, technological, social, environmental, and natural. By considering the development levels of each system and analyzing urban eco-friendliness from a coupled coordination perspective, this study aims to explore the interrelationships among various systems and assess historical evaluation values of these systems and their sub-indicators, thereby providing references and recommendations for constructing eco-friendly cities.

2 Study area and data sources

2.1 Study area

Guangzhou, located in South China, spans geographical coordinates of 22°26′ to 23°56′ N and 112°57′ to 114°03′ E, covering an area of approximately 7,434.40 square kilometers. As the economic, political, and cultural center of Guangdong Province, Guangzhou is recognized as a national central city, a megacity, and an international trade hub. However, a 2016 study by the Chinese Academy of Sciences ranked Guangzhou second to last in livability among 40 cities, with a livability score of 56.78 [15]. This highlights the complex relationship between urban economic development and ecological livability, which is not merely positive. Recently, the city's 14th Five-Year Plan explicitly aims to enhance the ecological environment and improve residents' well-being. Therefore, evaluating and analyzing Guangzhou's ecological livability is crucial not only for elevating residents' living standards but also for the effective implementation of the 14th Five-Year Plan. This study focuses on the relationships and collaborative levels among economic, technological, social, environmental, and natural factors in relation to urban ecological livability. Considering data availability, Guangzhou serves as the case study to explore pathways for constructing eco-friendly cities, offering valuable insights for the planning and development of similar urban areas.

2.2 Data sources

This study uses Guangzhou as a case example and compiles development indicator data for the city from 2014 to 2023. The data is sourced from *Guangzhou Statistical Yearbook*, which are publicly released by the government.

3 Methodology

The content of this study is as follows: (1) Establish a scientifically applicable evaluation index system for urban ecological livability, considering the development levels of five major systems: economic, technological, social, environmental, and natural. The entropy weight method will be used to assign weights to each indicator. (2) Employ the TOPSIS method to calculate the annual evaluation values for each indicator and system, examining the changes in urban ecological livability from 2014 to 2023 and the reasons behind these changes. (3) Conduct a comprehensive analysis of regional coupling and coordination, along with the annual evaluation values of each system and indicator, and propose policy recommendations aimed at enhancing urban ecological livability, thereby providing a reference for the construction of ecologically livable cities.

3.1 Construction of the Indicator System

3.1.1 Determination of Evaluation Indicators

The evaluation indicators for the ecological livability of Guangzhou primarily reflect the quality of urban development, guided by the aim of enhancing the city's overall quality. These indicators focus on the efficiency of economic development, livability of the city, level of technological advancement, environmental quality, and the integrity of natural systems. Therefore, the selection of indicators is tailored to the specific characteristics of urban development in Guangzhou. At the indicator level, the criteria for economic efficiency emphasize the city's efforts in promoting a low-carbon and circular economy, along with enhancing resource and energy utilization efficiency. These factors serve as prerequisites for constructing an ecological and livable city. Given the continuous upgrading of the city's industrial structure and the rapid development of the tertiary industry, indicators that reflect efficiency have been selected, such as energy consumption per unit of GDP, water consumption per unit of GDP, and carbon emissions per unit of GDP. The social system primarily focuses on the livability of urban development, emphasizing residents' sense of well-being and paying particular attention to the standard of living in a city like Guangzhou, where housing prices are high. Housing affordability has become a significant factor affecting citizens' happiness. Therefore, indicators such as the housing price-to-income ratio, urban registered unemployment rate, and Engel's coefficient have been chosen. The technological system aligns with Guangzhou's "14th Five-Year Plan" for technological innovation, focusing on the intensity of research and development (R&D) investment and educational conditions. The education sector plays a crucial role in supplying skilled personnel for research, while technological advancements facilitate industrial transformation, reduce resource consumption, and decrease pollution emissions, thereby accelerating the development of an ecological and livable city. Consequently, indicators like the number of students in higher education institutions, the number of R&D personnel, and the proportion of R&D expenditure to GDP have been selected. The environmental system emphasizes the protection of the environment during urban development and the extent of pollution management. Considering the actual situation in Guangzhou, particularly regarding water pollution, indicators such as sewage treatment rate, harmless treatment rate of municipal waste, and utilization rate of industrial solid waste have been integrated, in line with national pollutant reduction targets. Finally, the natural system highlights the importance of protecting natural conditions. In light of Guangzhou's efforts to strengthen ecological protection lines and promote ecological restoration, indicators such as forest coverage rate, proportion of protected land area, and per capita carbon sink capacity have been selected.

In the table, positive indicators are those that have a beneficial impact on ecological livability as data increases, marked with a "+" sign; negative indicators refer to those that negatively affect ecological livability as data increases, marked with a "-" sign. The selection of ecological livability evaluation indicators adheres to the principles of scientific rigor, sensitivity, and feasibility, resulting in the final version of Table 1.

Table 1. Indicator System for Ecological Livable Cities.

Objective Layer	System Layer	Criterion Layer	Indicator Layer (Unit)	Indicator Evaluation Standards	Sources of Indicator Evaluation Standards	Attribute
Ecological Livable City A	Economic System B1	Economic Efficiency C1	Per Capita GDP (Yuan) D1	210957	Reference Value: Singapore	+
			Per Capita Disposable Income (Yuan) D2	150788	Reference Value: Singapore	+
			Energy Consumption per Unit of GDP (t Coal/10,000 Yuan) D3	0.2	Reference Value: Hong Kong	+
			Water Consumption per Unit of GDP (m^3/10,000 Yuan) D4	18	Reference Value: Hong Kong	+
			Carbon Emissions per Unit of GDP (t CO_2/10,000 Yuan) D5	0.4	Reference Value: Hong Kong	+
			Tertiary Industry Ratio (%) D6	75.2	Reference Value: Hong Kong	+
	Social System B2	Social Services C2	Housing Price-to-Income Ratio (count) D7	4	Reference Value: Average Level of Developed Countries	+
			Number of Hospital Beds per 10,000 People (count) D8	120	Reference Value: Average Level of Developed Countries	+
			Urban Registered Unemployment Rate (%) D9	4	Reference Value: Average Level of Developed Countries	-
			Number of Schools (count) D10	1830	Reference Value: Average Level of Developed Countries	+
			Criminal Case Clearance Rate (%) D11	70%	Reference Value: Average Level of Developed Countries	+
			Engel's Coefficient (%) D12	30%	Reference Value: Average Level of Developed Countries	-
		Social Equity C3	Gini Coefficient D13	0.4	Reference Value: Average Level of Developed Countries	-
			Comparison of Urban and Rural Consumption Levels D14	2	Reference Value: Average Level of Developed Countries	-
	Technological System B3	Technological Development C4	Number of Students in Higher Education Institutions (count) D15	188270	Guangzhou's 14th Five-Year Plan	+
			Number of Granted Patents (count) D16	18827	Guangzhou's 14th Five-Year Plan	+
			Number of R&D Personnel (/per 10000 people) D17	28.24	Guangzhou's 14th Five-Year Plan	+
			Proportion of R&D Expenditure to GDP (%) D18	3.5	Guangzhou's 14th Five-Year Plan	+
	Environmental System B4	Environmental Protection C5	Green Coverage Rate in Built-up Areas (%) D19	30	International City Standards	+
			Per Capita Park Green Space Area (m^2) D20	12	U.S. Garden City Standards	+
			Average Noise Level in the Region (dB) D21	55	National Standards	-
		Pollution Control C6	Wastewater Treatment Rate (%) D22	95	Guangzhou's Environmental Planning Indicators	+
			Rate of Harmless Treatment of Municipal Waste (%) D23	100	Binding Indicators	+
			Utilization Rate of Industrial Solid Waste (%) D24	80	National Ecological City Standards	+
	Natural System B5	Natural Conditions C7	Forest Coverage Rate (%) D25	40	National Ecological City Standards	+
			Per Capita Water Resources (m^3) D26	1000	Reference Value: Average Level of Developed Countries	+
			Area of Nature Reserves ($10000hm^2$) D27	1.2	National Ecological City Standards	+
			Per Capita Carbon Sink Volume (t CO_2/per people) D28	0.46	Per Capita Carbon Sink in China	+

3.1.2 Establishment of Evaluation Indicator Standards

To mitigate biases arising from data selection limitations, the chosen indicators will be compared with ecological city standards from both domestic and international contexts, striving for an objective evaluation of Guangzhou's ecological livability. The criteria for evaluating these indicators consider several aspects: 1. Whenever national or international standards exist, the prescribed standard values should be adopted as closely as possible. 2. The current values of internationally recognized cities with commendable characteristics will serve as reference standards. 3. The current values of domestic cities, particularly those with higher standards, will be referenced to determine the standard values. 4. Existing theories on the coordinated development of environmental, social, and economic factors will be utilized to aim for quantifiable standard values. Based on these principles, the ecological evaluation standards for Guangzhou have been formulated. Some indicators, such as energy consumption per unit of GDP and carbon emissions per unit of GDP, reference the average values of Asian cities from the "Asian Green City Index" environmental performance assessment. Other indicators, like forest coverage rate, draw upon the standard values established by the Ministry of Ecology and Environment for ecological city assessments. Additionally, some indicators are based on relevant values from Hong Kong (Table 1).

3.2 Weight Assignment for Indicators

3.2.1 Standardized Data

Data standardization aims to eliminate differences in properties, dimensions, and magnitudes among various variables, enabling effective synthesis and comparison of indicators with different units or scales. For positive indicators, standardization is performed using Equation (1), while for negative indicators, Equation (2) is employed for standardization.

$$r_{ij} = \frac{r_{ij}' - \min(r_{ij}')}{\max(r_{ij}') - \min(r_{ij}')} \quad (1)$$

$$r_{ij} = \frac{\min(r_{ij}') - r_{ij}'}{\max(r_{ij}') - \min(r_{ij}')} \quad (2)$$

r_{ij} represents the original value of indicator i in year j, while r_{ij}' denotes the standardized value of the indicator data.

3.2.2 Determining Indicator Weights Using Entropy Weight Method

The entropy weight method determines weights based on the variability of each indicator, serving as an objective weighting approach that mitigates biases from human factors. When calculating the system weights, the weights of the indicators within each system are summed to derive the overall system weight. For annual evaluations of each system, each system is treated as an independent entity, with its weight standardized to 1. This process involves normalizing the indicator weights.

The original data consists of m evaluation indicators and n evaluation objects, represented by the original data matrix $R = (r_{ij})_{m\times n}$. The entropy weight coefficient method is employed for calculations.

$$H_i = -k\sum_{j=1}^{n} f_{ij} \ln f_{ij} (i=1,2,\cdots, m; \ j=1,2,\cdots, n) \quad (3)$$

$f_{ij} = \frac{r_{ij}}{\sum_{j=1}^{n} r_{ij}}$ and $k = \frac{1}{\ln m}$, where f_{ij} represents the proportion of the j evaluation object for the i indicator. n denotes the number of evaluation objects, and H_i is the entropy of the i indicator

$$w_i = \frac{1-H_i}{\sum_{i=1}^{m}(1-H_i)} \quad (4)$$

In the equation, $0 \le w_i \le 1$ and $\sum_{i=1}^{m} = 1$, where H_i represents the entropy of the i indicator; m denotes the number of evaluation indicators; and w_i is the weight of the i indicator.

3.3 Calculating Urban Evaluation Values Using the TOPSIS Method

The TOPSIS (Technique for Order Preference by Similarity to an Ideal Solution) method is a ranking approach that approximates the ideal solution and can be used to evaluate systems with multiple scenarios and decision indicators [16]. It ranks evaluation objects by assessing their distances from the ideal values; an evaluation object is considered the best if it is closest to the positive ideal value while being farthest from the negative ideal value. The positive ideal value is composed of the optimal values of each indicator across the years, while the negative ideal value consists of the worst values for each indicator in the same time frame. In addition to evaluating the optimal values of various indicators from Guangzhou over the years, a comparative analysis will be conducted between Guangzhou's comprehensive score for 2023 and the selected domestic and international ecological city indicator standards, as outlined in Table 2. This table is derived from the research conducted by Li et al [17].

$$V^{+} = \{\max r_{ij} \mid i = 1,2,\cdots, m\} \quad (5)$$

$$V^{-} = \{\min r_{ij} \mid i = 1,2,\cdots, m\} \quad (6)$$

V^+ represents the positive ideal value, while V^- denotes the negative ideal value. r_{ij} refers to the standardized value of the i indicator in year j.

$$D_j^+ = \sqrt{\sum_{i=1}^{m} w_i (r_i^+ - r_{ij})^2} \tag{7}$$

$$D_j^- = \sqrt{\sum_{i=1}^{m} w_i (r_i^- - r_{ij})^2} \tag{8}$$

r_{ij} represents the standardized data of indicator i in year j, r_i^+ denotes the optimal value of indicator i over the years, and r_i^- represents the worst value of indicator i across the same period. w_i is the weight of indicator iii. D_j^+ indicates the distance from the evaluation value in year j to the positive ideal value, while D_j^- denotes the distance to the negative ideal value.

$$C_j = \frac{D_j^-}{D_j^+ + D_j^-} \tag{9}$$

In the equation, C_j represents the city evaluation value for year j; a higher value of C_j indicates better urban ecological livability for that year.

Table 2. Eco-city grade of comprehensive index.

Grade	Comprehensive Score	Level of Ecological Livability
I	> 85	Excellent
II	70~85	Good
III	60~70	Average
IV	50~60	Poor
V	< 50	Very Poor

3.4 Coupling Coordination Degree Model Analysis

The coupling coordination degree model is used to analyze the level of coordinated development among entities. Coupling degree refers to the extent of interaction between two or more systems, also known as the degree of dependence; a higher coupling degree implies that changes in one system can significantly impact another. Coordination degree indicates the magnitude of positive coupling in interactions, reflecting the quality of coordination.

$$C = \left[\frac{\prod_{i=1}^{n} U_i}{(\frac{1}{n}\sum_{i=1}^{n} U_i)^n} \right]^{\frac{1}{n}} \tag{10}$$

In the equation, n represents the number of systems; U_i denotes the system evaluation value, and C is the coupling degree of the system. A higher value of C indicates a greater degree of interdependence among the systems.

$$T = \sum_{i=1}^{n} a_i U_i, \quad \sum_{i=1}^{n} a_i = 1 \tag{11}$$

U_i represents the standardized value of the i system; a_i denotes the weight of the i system, and T is the system coordination index.

$$D = \sqrt{C \times T} \tag{12}$$

D represents the system coordination degree; a higher value indicates better coordination among the systems.

Based on the classification criteria by Liao et al [18], the classification standards for the coupling coordination degree D in this study are outlined as follows (Table 3).

Table 3. Classification Criteria for Coupling Degree (C) and Coupling Coordination Degree (D).

Value Range	C	D
[0,0.1)	Extreme Decoupling	Extreme Discoordination
[0.1,0.2)	Severe Decoupling	Severe Discoordination
[0.2,0.3)	Moderate Decoupling	Moderate Discoordination
[0.3,0.4)	Slight Decoupling	Slight Discoordination
[0.4,0.5)	Near Decoupling	Near Discoordination
[0.5,0.6)	Near Coupling	Near Coordination
[0.6,0.7)	Slight Coupling	Slight Coordination
[0.7,0.8)	Moderate Coupling	Moderate Coordination
[0.8,0.9)	Good Coupling	Good Coordination
[0.9,1)	Excellent Coupling	Excellent Coordination

4 Result

4.1 Indicator Weights at Various Levels

The indicator weights are calculated using equations (1)~(4), with the results for each level's indicator weights presented in Table 4. Indicator weights reflect the contribution of a specific indicator within its system, used to calculate the comprehensive evaluation value of each system, where the sum of weights for all indicators equals 1. The weights at the system level represent the contributions of the five systems to the evaluation of urban ecological livability, which are used to compute the overall ecological livability evaluation value and the coupling coordination degrees of the five systems.

4.2 Trend of Comprehensive Evaluation Values of Five Major Systems

Based on the calculations from equations (5) to (9), we derived the evaluation values for the system across various years, as illustrated in Figure 1. The overall assessment results of the five major systems reveal two primary stages of development. The first stage, termed the "natural system lag phase" (2014-2020), experienced disruptions due to accelerated urbanization, adversely affecting the natural system. The second stage, referred to as the "tortuous development phase" (2020-2023), saw a slight lag in the natural system, while other systems exhibited relatively high scores. Notably, both the social and economic systems showed a general trend of linear growth. The environmental system experienced a significant decline in 2020 but gradually rebounded thereafter. Conversely, the technological system's evaluation steadily increased from 2014 to 2022, peaking in 2022 before beginning to decline, illustrating an initial upward and subsequent downward trend. The evaluation value of the natural system fluctuated downward from 2014 to 2022 but has shown an upward trajectory since 2022. Furthermore, by comparing the selected evaluation indicators, the comprehensive score for Guangzhou's ecological livability in 2023 was calculated to be 88, indicating an excellent level. This suggests that while Guangzhou's ecological livability is commendable, there remains significant room for further development.

Table 4. Weights of Indicators for Ecological Livable Cities.

Objective Layer	System Layer	Criterion Layer	Indicator Layer	Indicator Weights	Standardized Weights
Ecological Livable City A	Economic System B1(0.1338)	Economic Efficiency C1 (0.1338)	Per Capita GDP D1	0.0067	0.0501
			Per Capita Disposable Income D2	0.043	0.3214
			Energy Consumption per Unit of GDP D3	0.0226	0.1689
			Water Consumption per Unit of GDP D4	0.0427	0.3191
			Carbon Emissions per Unit of GDP D5	0.0166	0.1241
			Tertiary Industry Ratio D6	0.0022	0.0164
	Social System B2(0.2673)	Social Services C2(0.2462)	Housing Price-to-Income Ratio D7	0.0283	0.1059
			Number of Hospital Beds per 10,000 People D8	0.0533	0.1994
			Urban Registered Unemployment Rate D9	0.0139	0.0520
			Number of Schools D10	0.0004	0.0015
			Criminal Case Clearance Rate D11	0.1499	0.5608
			Engel's Coefficient D12	0.0004	0.0015
		Social Equity C3 (0.0211)	Gini Coefficient D13	0.0001	0.0004
			Comparison of Urban and Rural Consumption Levels D14	0.021	0.0786
	Technological System B3(0.2075)	Technological Development C4(0.4918)	Number of Students in Higher Education Institutions D15	0.0204	0.0415
			Number of Granted Patents D16	0.3907	0.7944
			Number of R&D Personnel D17	0.044	0.0895
			Proportion of R&D Expenditure to GDP D18	0.0367	0.0746
	Environmental System B4(0.0078)	Environmental Protection C5(0.0037)	Green Coverage Rate in Built-up Areas D19	0.0015	0.1923
			Per Capita Park Green Space Area D20	0.0012	0.1538
			Average Noise Level in the Region D21	0.001	0.1282
		Pollution Control C6(0.0041)	Wastewater Treatment Rate D22	0.0007	0.0897
			Rate of Harmless Treatment of Municipal Waste D23	0.0012	0.1538
			Utilization Rate of Industrial Solid Waste D24	0.0022	0.2821
	Natural System B5(0.109)	Natural Conditions C7(0.0994)	Forest Coverage Rate D25	0.0018	0.0181
			Per Capita Water Resources D26	0.0821	0.8260
			Area of Nature Reserves D27	0.0002	0.0020
			Per Capita Carbon Sink Volume D28	0.0153	0.1539

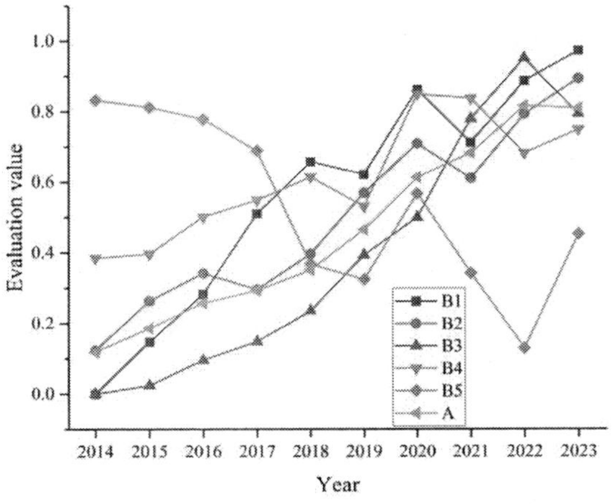

Figure 1. Trends in Evaluation Values of the Five Major Systems Over the Years.

4.3 Coupling and Coordinated Development Trends of the Five Major Systems

Based on equations (10) ~ (12), the coupling coordination degree of the five major systems in Guangzhou was calculated and classified according to the standards in Table 2, with results presented in Table 5. The coupling degree among the five systems significantly increased from 2014 to 2016, reaching the threshold for a good coupling state in 2016. From 2016 to 2023, the coupling degree displayed a gradual upward trend. Between 2014 and 2018, the coupling degree improved from a state of extreme imbalance to a slight coordination state. Although the coupling among the five systems was relatively strong during this period, their mutual coordination was insufficient for effective collaborative development. Since 2018, the coupling coordination degree has significantly increased and stabilized at a high level, indicating that the coupling mechanisms among the five systems have strengthened, fostering robust internal connections and interactions, thereby achieving mutual promotion at an elevated level.

Table 5. Coupling Degree and Coupling Coordination Degree of the Five Major Systems.

Year	C	Classification Standards	D	Classification Standards
2014	0	Extreme De-coupling	0	Extreme Discoordination
2015	0.773	Moderate Coupling	0.428	Near Discoordination
2016	0.892	Good Coupling	0.540	Near Coordination
2017	0.933	Excellent Coupling	0.606	Slight Coordination
2018	0.962	Excellent Coupling	0.694	Slight Coordination
2019	0.976	Excellent Coupling	0.740	Moderate Coordination
2020	0.977	Excellent Coupling	0.833	Good Coordination
2021	0.999	Excellent Coupling	0.848	Good Coordination
2022	0.986	Excellent Coupling	0.889	Excellent Coordination
2023	0.992	Excellent Coupling	0.922	Excellent Coordination

5 Discussions

5.1 Analysis of Trends in Evaluation Values of the Five Major Systems

5.1.1 Trends in Evaluation Values of the Economic System

Between 2014 and 2023, the development level of Guangzhou's economic system, despite experiencing some fluctuations, exhibited an overall growth trend (Figure 2). In 2014, Guangzhou began to establish a high-level economic development model, with most key economic indicators showing an upward trajectory. However, per capita GDP and disposable income witnessed significant declines from 2019 to 2021, primarily due to the outbreak of COVID-19 and the stringent pandemic control measures that restricted daily life and work, adversely affecting these indicators. Additionally, during the study period, metrics such as energy consumption per unit of GDP, water consumption per unit of GDP, and carbon emissions per unit of GDP consistently increased, indicating that Guangzhou adopted a relatively greener economic development approach, resulting in decreasing resource consumption per unit of GDP. Moving forward, sustainable technological innovation in Guangzhou can continue to facilitate green economic growth while balancing economic development and environmental protection. The proportion of the tertiary industry remained relatively stable from 2015 to 2023, as the city's industrial structure was in a preparatory phase transitioning from primary and secondary industries to the tertiary sector, resulting in minimal changes in its share. Overall, the evaluation of the economic system was influenced by per capita GDP and disposable income, experiencing a decline from 2020 to 2021, but demonstrating an upward trend in other periods.

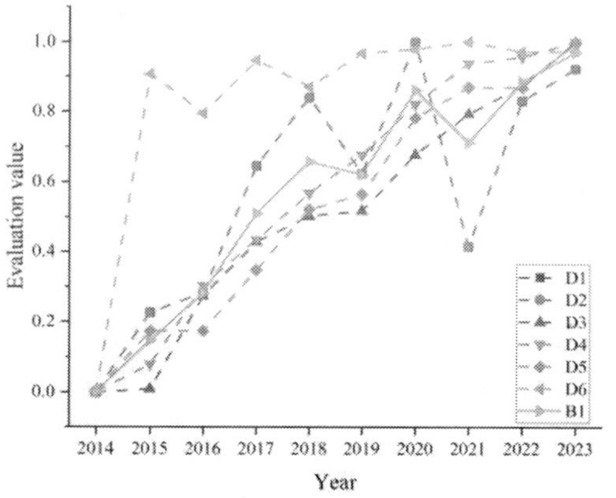

Figure 2. Trend Analysis of Evaluation Values for Various Indicators in the Economic System.

5.1.2 Trends in Evaluation Values of the Social System

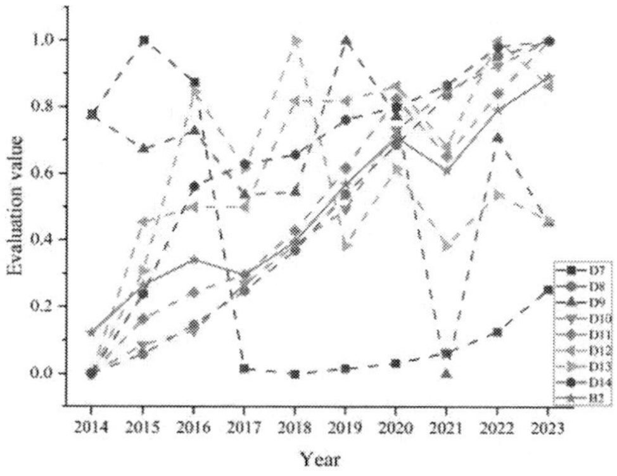

Figure 3. Trend Analysis of Evaluation Values for Various Indicators in the Social System.

Over the years, the development level of Guangzhou's social system has shown a generally upward but fluctuating trend. Key negative indicators within the social system, including the housing price-to-income ratio, urban registered unemployment rate, Engel's coefficient, Gini coefficient, and urban-rural consumption levels, have been positively processed and illustrated in Figure 3. Notably, from 2020 to 2022, fluctuations in the evaluation value of Guangzhou's social system may be related to changes in the urban registered unemployment rate, clearance rates for criminal cases, and the Engel coefficient. During this period, the widespread transmission of COVID-19 in China led to the implementation of regional lockdown measures in various parts of Guangzhou, restricting residents' work and daily life, and resulting in an increase in the urban registered unemployment rate, which undermined social harmony and stability. Additionally, the rising Gini coefficient indicates that social equity issues have become more pronounced, with increased social instability and widening wealth disparities potentially impacting residents' satisfaction with ecological livability in the city.

5.1.3 Trends in Evaluation Values of the Technological System

The development of Guangzhou's technology system continues to show an upward trend (Figure 4). This is evidenced by a linear increase in the number of university students, the volume of patent grants, and the proportion of R&D expenditures relative to GDP. These indicators suggest that the establishment of a technology-driven city has become a crucial strategic direction for promoting high-quality urban development. The intensity of efforts to advance this initiative largely determines the city's developmental stature. As one of the provincial centers for technological innovation, Guangzhou can maintain its leadership in the technology sector through strategies that encourage corporate innovation, thereby steadfastly pursuing the path of high-quality development as a technology-driven city.

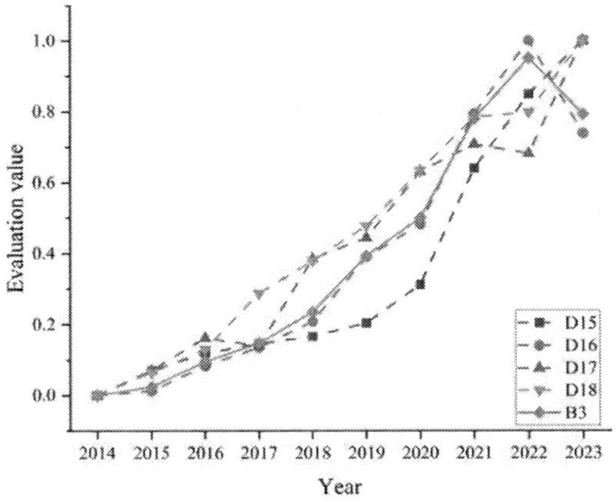

Figure 4. Trend Analysis of Evaluation Values for Various Indicators in the Technological System.

5.1.4 Trends in Evaluation Values of the Environmental System

In the environmental system, the average regional noise level serves as a negative indicator and is depicted in Figure 5 after positive processing. As shown, the sewage treatment rate and the harmless disposal rate of municipal waste exhibit a linear upward trend, while the average regional noise level has declined annually, indicating a gradual increase in urban noise levels. Concurrently, fluctuations in per capita green park area and the utilization rate of industrial solid waste are notable. In 2020, the evaluation value of Guangzhou's environmental system peaked before entering a downward trend. This change is primarily attributed to significant declines in the evaluation values of greening coverage in built-up areas, per capita green park area, average regional noise levels, and the utilization rate of industrial solid waste.

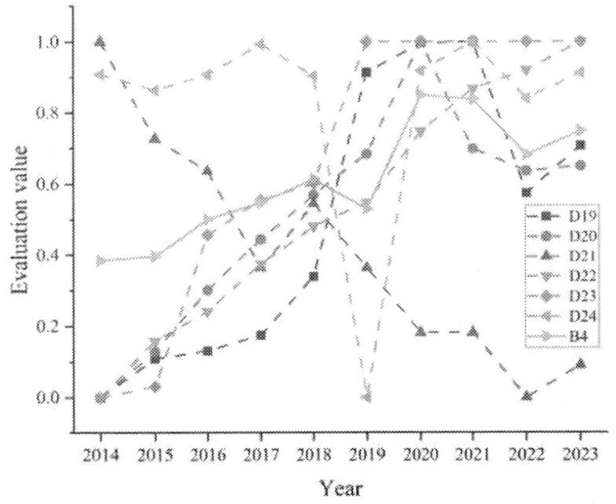

Figure 5. Trend Analysis of Evaluation Values for Various Indicators in the Environmental System.

5.1.5 Trends in Evaluation Values of the Natural System

As illustrated in Figure 6, the evaluation value of Guangzhou's natural system fluctuated downward from 2014 to 2022, likely due to a decline in per capita water resources and low forest coverage. However, the area of protected land and per capita carbon sink have increased annually, which may be attributed to the city's defined ecological regional planning and delineation of ecological protection red lines. As the capital of Guangdong Province, Guangzhou actively responds to environmental policies, accelerating ecological civilization construction, optimizing urban development layouts, and enhancing development quality. These measures have continuously improved residents' perceptions of the ecological environment and their overall happiness, thereby advancing the construction of an ecologically livable city.

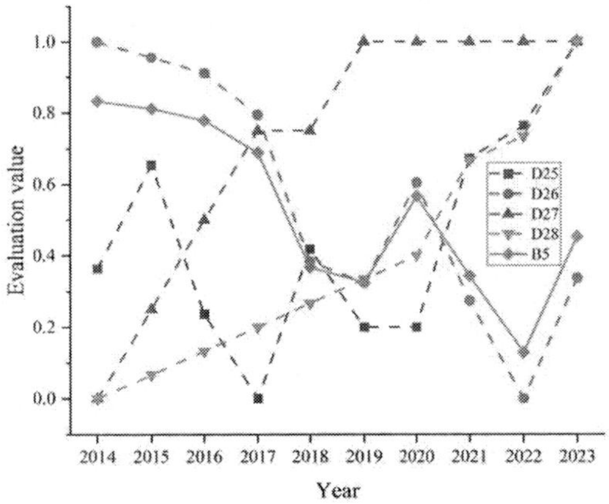

Figure 6. Trend Analysis of Evaluation Values for Various Indicators in the Natural System.

5.2 Limitations and Future Research Developments

(1) The indicator system needs improvement. Although the evaluation indicators for the ecologically livable city in this study were selected based on existing research, some indicators could not be included due to data limitations, which may lead to biased evaluation results.

(2) In analyzing the coordination levels of the five major systems, this study focused on a system-level perspective. Future research could delve deeper into specific systems or critical indicators for more comprehensive insights.

6 Conclusions

Constructing eco-friendly and livable cities is an effective strategy to address the "urban diseases" arising from urban development, and it represents a mainstream trend in future urban growth. This study employs the entropy method and TOPSIS to evaluate the ecological livability of Guangzhou, analyzing the coupling and coordination among five major systems within eco-friendly cities. The key findings indicate that the technological development level in Guangzhou shows a linear growth trend, while the economic, social, environmental, and natural systems exhibit fluctuations but overall demonstrate an upward trajectory. This suggests that Guangzhou should place greater emphasis on social equity to reduce the wealth gap in its societal development. Currently, the development of the environmental system is constrained by per capita park green space, average regional noise levels, and industrial solid waste utilization rates, whereas the natural system is primarily influenced by forest coverage and per capita water resources. Utilizing a coupling coordination model, it was found that the coupling degree among Guangzhou's economic, social, technological, environmental, and natural systems has gradually increased since 2015, maintaining an excellent state of coupling since 2017. The calculated coordination degree reflects a trend of gradual increase across the entire system. This signifies a tendency toward synchronous development among the five ecological systems, demonstrating a more coordinated approach and an ongoing enhancement of overall ecological livability.

In the development pathway of eco-friendly cities, the government should take the lead by promoting interdisciplinary collaboration and facilitating the development of the five major systems to ensure their coordinated integration. Amid rapid urbanization, the government should focus on addressing four key challenges: Firstly, resource competition poses a conflict between ecological protection and economic development due to limited land and water resources. The government should implement efficient land use planning, prioritize the protection of ecological red lines, and rationally allocate resources. Secondly, there is insufficient coordination among various government levels in implementing ecological policies, which hampers effective policy execution. Strengthening inter-departmental cooperation and establishing an ecological governance linkage mechanism can enhance policy uniformity. Thirdly, there is a lack of public awareness and participation in ecological planning. The government should enhance public outreach and education, encouraging residents to engage in urban planning processes and fostering their understanding and support for ecological initiatives. Lastly, the construction of eco-friendly cities often requires advanced technologies and funding, which many cities currently lack. Attracting social capital and innovative technologies, while promoting collaboration between enterprises and research institutions, is essential for advancing ecological projects. Through comprehensive measures, these challenges can be progressively alleviated, thereby enhancing the ecological livability of cities.

References

1. Sun, L., Chen, J., Li, Q., & Huang, D. (2020). Dramatic uneven urbanization of large cities throughout the world in recent decades. *Nature communications*, *11*(1), 5366.
2. Gong, F., & Chen, X. (2020). Study on the Evaluation of Ecological Livable City in Anhui Based on Intuitionistic Fuzzy Theory. In *IOP Conference Series: Earth and Environmental Science* (Vol. 555, No. 1, p. 012089). IOP Publishing.
3. Clüsener-Godt, M., Köck, G., & Möller, L. (2022). Editorial It is about life: 50 years of UNESCO's Man and the Biosphere Programme. *International Journal of Environment and Sustainable Development*, *21*(4), 377-387.
4. Tan, S. (2022). Research on evaluation of ecological livable city based on residents' satisfaction—Take Zhuzhou as an example. *Eco Cities*, *3*(2).
5. Kutty, A. A., Wakjira, T. G., Kucukvar, M., Abdella, G. M., & Onat, N. C. (2022). Urban resilience and livability performance of European smart cities: A novel machine learning approach. *Journal of Cleaner Production*, *378*, 134203.
6. del Mar Martínez-Bravo, M., Martínez-del-Río, J., & Antolín-López, R. (2019). Trade-offs among urban sustainability, pollution and livability in European cities. *Journal of cleaner production*, *224*, 651-660.
7. PACIONE, M. (1995). The geography of multiple deprivation in the Clydeside conurbation. *Tijdschrift voor economische en sociale geografie*, *86*(5), 407-425.
8. Saitluanga, B. L. (2014). Spatial pattern of urban livability in Himalayan Region: A case of Aizawl City, India. *Social indicators research*, *117*, 541-559.
9. Liu, Q. F., & Song, J. P., (2022). Spatio-Temporal Evolution and Prediction Simulation of Urban Livable Level in Hunan Province. *Journal of Natural Science of Hunan Normal University* (05), 1-10.
10. Reis, I. F., Ferreira, F. A., Meidutė-Kavaliauskienė, I., Govindan, K., Fang, W., & Falcão, P. F. (2019). An evaluation thermometer for assessing city sustainability and livability. *Sustainable Cities and Society*, *47*,

101449.

11. Li, Y., Wu, L., Han, Q., Wang, X., Zou, T., & Fan, C. (2021). Estimation of remote sensing based ecological index along the Grand Canal based on PCA-AHP-TOPSIS methodology. *Ecological Indicators, 122*, 107214.
12. Dong, X. Y., Hu, Z. W., Wu, J. J., Wang, J. Z., Yang, C., Zhang, J., Xia, J.Z., & Wu, G. F. (2022). Evaluation of ecological livability of dwelling area based on multi-source data: A case study of Shenzhen City. *Acta Ecol. Sin, 42*, 1-13.
13. Ghasemi, K., Hamzenejad, M., & Meshkini, A. (2019). The livability of Iranian and Islamic cities considering the nature of traditional land uses in the city and the rules of their settlement. *Habitat International, 90*, 102006.
14. Parker, J., & Zingoni de Baro, M. E. (2019). Green infrastructure in the urban environment: A systematic quantitative review. *Sustainability, 11*(11), 3182.
15. Zhang, W.Z.; Yu, J.H. *China Livable Cities Research Report*; United Nations Development Programme (UNDP) China: Guangzhou, China, 2016; p. 279.
16. Fang, X., Zou, J., Wu, Y., Zhang, Y., Zhao, Y., & Zhang, H. (2021). Evaluation of the sustainable development of an island "Blue Economy": A case study of Hainan, China. *Sustainable Cities and Society, 66*, 102662.
17. Li, B., & Wang, H. (2023). Comprehensive evaluation of urban high-quality development: A case study of Liaoning Province. *Environment, Development and Sustainability, 25*(2), 1809-1831.
18. Li, J., Sun, W., Li, M., & Meng, L. (2021). Coupling coordination degree of production, living and ecological spaces and its influencing factors in the Yellow River Basin. *Journal of cleaner production, 298*, 126803.

Research on the Guidelines for Green Ecological Planning and Construction of New Cities in Shanghai

Hongyan Pan[1,*], Guoshu Yang[1], Xilan Qiu[1]

[1]Shanghai Institute of Building Science Co., Ltd, Shanghai, China

Abstract: During the 14th Five-Year Plan period, Shanghai's five new cities (Jiading, Qingpu, Songjiang, Fengxian, and Nanhui) are designated as key development areas, with the goal of becoming future cities that lead high-quality living standards and the most ecologically advanced urban areas. In response to the evolving demands for the development of green ecological urban areas in Shanghai, the standards for urban evaluation have been revised. To fully implement the planning and construction requirements of green ecological urban areas in the new cities, a targeted green ecological planning study was conducted at the overall level for these five new cities. Following the approach of "separate cities, separate regions, and distinct focuses," target units for green ecological urban areas were defined, and a tailored indicator system ("one city, one policy") along with a list of key regional projects was developed.

1. Introduction

On February 23, 2021, the Shanghai Municipal People's Government issued the Implementation Opinions on Accelerating the Planning and Construction of New Cities during the 14th Five-Year Plan Period [1]. The document proposes that the five new cities—Jiading, Qingpu, Songjiang, Fengxian, and Nanhui—be cultivated into comprehensive node cities with a radiative and driving role within the Yangtze River Delta urban agglomeration (Figure 1). These cities are envisioned as strategic spaces forming a multi-center, multi-level, and multi-node networked urban cluster structure. High-level planning and construction of new cities have been established as a strategic proposition, accompanied by the release of the "1+6" policy documents for new cities. Among these, the 14th Five-Year Plan Special Program for Environmental Quality and New Infrastructure in New cities emphasizes green and low-carbon development, led by green ecology, defining the requirements for new city construction throughout the entire process of planning, construction, and management. The plan mandates that new cities be fully developed according to the goals of green ecological urban areas. In September 2023, the Shanghai Municipal Housing and Urban-Rural Construction Management Committee, in collaboration with four other departments, issued the Guiding Opinions on Promoting the Construction of Green Ecological Urban Areas in the City, clearly stating that new and updated urban areas within the five new cities must fully comply with the standards of green ecological urban areas and must prepare specialized green ecological plans.

The Standards for the Evaluation of Green Ecological Urban Areas in Shanghai (DG/TJ08-2253-2018) [2] have been simultaneously initiated for revision. This revision aims to align with new development concepts such as the "Dual Carbon" strategic goals [3], "People's City"[4], "Resilient City" [5, 6], "Healthy City"[7, 8], and "Digital City" [9], while meeting the fundamental requirements for green ecological construction across the five new cities.

Given the context of high-quality construction and new development needs of the aforementioned new cities, there is an urgent need to develop a special green ecological plan at the overall level for the five new cities. This planning work should be closely integrated with the revision of standards to ensure the comprehensive promotion of green ecological urban area objectives and to guide green and low-carbon development in the new cities.

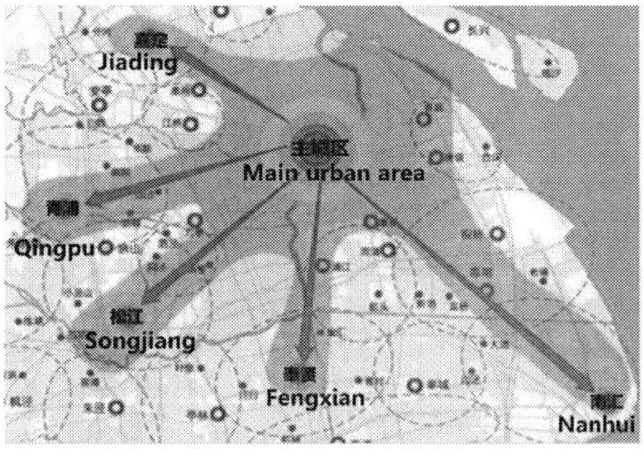

Fig. 1. New city and main city distribution relationship

2. The objectives and concepts of the research

During the 14th Five-Year Plan period, in order to achieve the comprehensive implementation of green ecological

* Corresponding author: panhongyan@sribs.com

urban area construction goals in the new cities and to promote green and low-carbon initiatives, the research aims to explore the development potential of green ecological urban areas in the new cities. Following the approach of "separate cities, separate regions, and distinct focuses," the plan involves defining target units for green ecological urban areas, clarifying overall goals and indicator systems for green and low-carbon construction, and creating a reserve list of green ecological urban areas. From the perspectives of the feasibility of implementing green ecological urban area indicators and manageability, technical documents for special green ecological planning for the five new cities will be prepared. The research is centered on the following aspects. The research ideas are outlined as follows (Figure 2):

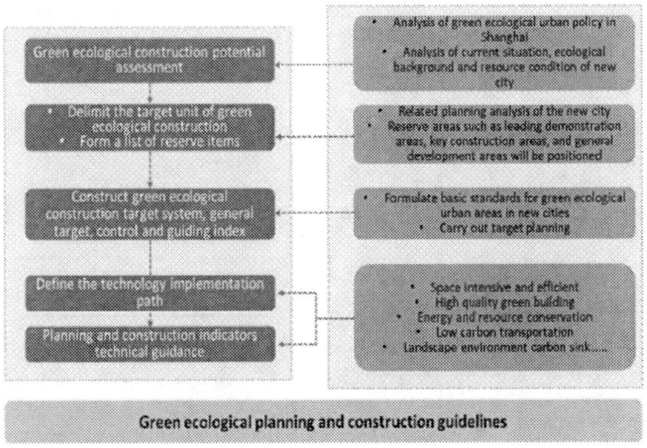

Fig. 2. The research concept

3. Core Content of the Guideline

The Guidelines serves as a green ecological guidance document at the overall planning level of the new cities, supporting and aligning with the overall and unit-level planning implementation of the new cities. It aims to explore the development potential of green ecological urban areas in the five major new cities and to reserve a number of key construction projects. The core research content is as follows:

Fig. 3. New city green ecological planning and construction guidelines

The Guidelines are divided into seven chapters and follow the approach of "separating new cities, regions, and key focuses." It delineates the target units for green ecological urban districts within new cities, formulating a distinctive indicator system based on the principle of "one city, one policy." The guidelines primarily aim to guide the planning and construction of new cities in five key areas: resilient and safe infrastructure and sponge city systems, healthy and dynamic spaces and environments, low-carbon and green buildings and transportation, efficient and sustainable energy and resources, and smart and innovative management and human-centered design (Figure 3).

3.1 Green Ecological Characteristics of New Cities Based on the Foundation

The Guidelines, grounded on the current status of the five new cities, identify their unique characteristics and differences. Based on the green ecological development conditions of each city, the guidelines establish specific green ecological features for each. For Jiading New city, the focus is on smart transportation and low-carbon buildings, aiming to create a green, human-centered ecological urban district. Qingpu New City emphasizes its natural resources and Jiangnan cultural heritage, with the goal of building a modern, lakeside green ecological urban district. Songjiang New City focuses on integrated station-city development and a balance between living and working, striving to create a convenient and livable green ecological urban district. Fengxian New City centers on vitality, livability, and age-friendly designs to establish an innovative paradigm for green ecological urban districts. Nanhui New City emphasizes low-carbon, smart, and resilient urban design, aiming to build a high-energy, high-standard coastal low-carbon city.

3.2 Comprehensive Response: Establishing the Overall Green Ecological Goals for New Cities

The Guidelines specify five overall green ecological goals for the new cities. These include ensuring that 100% of newly developed urban areas are built as green ecological districts, meeting the overarching control requirements for urban landmark revisions, planning and constructing at least one concentrated demonstration zone for ultra-low energy consumption buildings, Ensuring that the proportion of newly constructed residential buildings meeting basic or higher green building standards accounts for 100% of the total building area, setting a target that no less than 50% of the rooftop area of newly constructed government buildings, schools, and industrial facilities is equipped with photovoltaic (PV) systems, while for other types of public buildings, the proportion of rooftop area with PV installations should be no less than 30%.

3.3 Gradual Advancement: Defining Zonal and Targeted Focused Construction

Based on the green and low-carbon development potential of the five new cities, the Guidelines delineate green ecological construction target units, classifying them into three types: Leading Demonstration Zones, Key Construction Zones, and General Development Zones. Different green ecological construction goals are set for each type of zone. Leading Demonstration Zones are to be planned and built according to two-star or higher green ecological district standards. Key Construction Zones and General Development Zones are to follow one-star or higher and basic level or higher standards, respectively. Upon review, the five new cities collectively have 10 Leading Demonstration Zones, 20 Key Construction Zones (Figure 4).

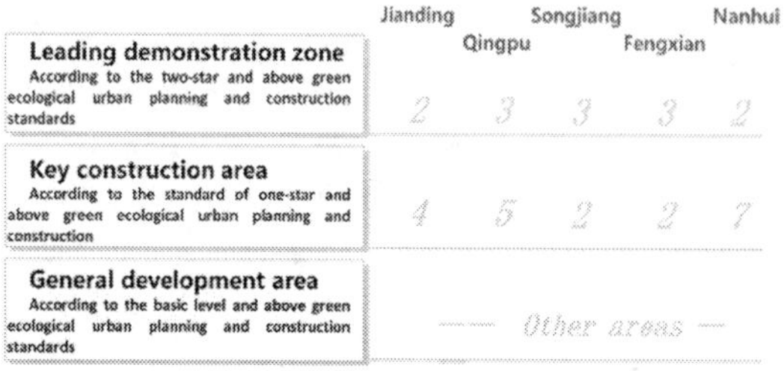

Fig. 4. New city green ecological development core area

3.4 One City, One Policy: High Standards Leading in Five Dimensions

The Guidelines focus on five key dimensions: resilient and safe infrastructure and sponge city systems, healthy and dynamic spaces and environments, low-carbon and green buildings and transportation, efficient and sustainable energy and resources, and smart, innovative management and human-centered design. These dimensions form the basis for developing a distinctive indicator system for each new city, referred to as "one city, one policy." Fully combined with international green sustainable indicators and Shanghai's green and low-carbon policy standards, a total of 91 indicators were proposed (Table 1). The indicator system applies differentiated values to Leading Demonstration Zones, Key Construction Zones, and General Development Zones, dividing the indicators into both specific and general categories. Multidimensional green and low-carbon requirements are outlined for each zone.

Table 1. Green Ecological Planning and Construction Indicator System - Example of Jiading New City (Partial)

Category	Indicator Name	Indicator Value		
		Leading Demonstration Zone	Key Development Area	General Development Area
Resilient and Safe Facilities & Sponge Infrastructure	Per capita emergency shelter area	≥3m²	≥3m²	≥3m²
	Annual runoff control rate in newly built urban areas	≥74%	≥70%	≥70%
Healthy and Vibrant Spaces & Environment	Service facility coverage rate in 15-minute community life circles	100%	100%	85%
	Continuity rate of public spaces along main rivers	100%	100%	100%
	5-minute walking accessibility rate to public open spaces	100%	100%	100%
Low-Carbon Green Buildings & Transportation	Green building star rating for newly constructed large public buildings	Three Stars	Three Stars	Three Stars
	Near-zero energy/zero carbon building demonstration projects	≥2 projects	≥1 project	≥1 project
	Proportion of parking spaces equipped with charging facilities in newly built parking lots (fast charging)	≥15% (30%)	≥15% (30%)	≥15% (30%)
Efficient and Resource-Saving Energy & Resources	Renewable energy substitution rate in buildings	≥15%	≥15%	≥15%
	Recycling rate of construction waste, such as waste concrete, renovation waste, and demolition waste	100%	≥93%	≥93%
Smart and Innovative Management & Culture	Demonstration of intelligent connected vehicle applications	≥1 project	≥1 project	≥1 project
	Green community creation compliance rate	≥65%	≥65%	≥65%

3.5 Key Demonstration: Establishing a Benchmark for Green Ecological Urban Areas in New Cities

Based on the comprehensive promotion of the entire area and in alignment with both the short- and long-term planning and development schedules for key areas within the new cities, the Guidelines have compiled and reserved a list of key projects. This includes a total of 28 green ecological urban area reserve projects, five clusters of ultra-low energy building demonstration zones, and a large-scale demonstration benchmark of ultra-low energy buildings within the core demonstration area of Dishui Lake in Nanhui New City (Figure 5). These initiatives serve as a strong foundation for future large-scale promotion and management of green ecological urban areas in new cities. According to the requirements of the Guidelines, at present, Jiading New City Yuanxiang Lake Central Active area and Jiabao Smart Bay have completed the green ecological urban planning and the creation of pilot urban areas, and Nanhui New City top scientist community, Finance Bay and other projects have also completed the relevant green ecological planning and the creation of pilot urban areas.

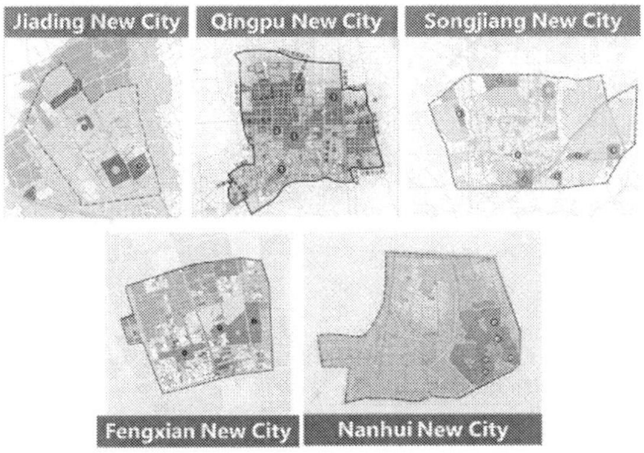

Fig. 5. Green ecological urban potential planning and construction area

3.6 Strengthen management and control, supporting relevant policies to promote the implementation of objectives

In order to promote the implementation of green ecological indicators and planning strategies for new cities, the Guidelines were officially issued by the Shanghai Municipal Commission of Housing and Urban-Rural Development in January 2024, clearly requiring each new city to implement the guidelines in combination with the requirements of the Guidelines, and taking the indicators in the Guidelines as an annual assessment task for new cities. In order to cooperate with the implementation of the guidelines, each new city has also introduced relevant control policies and measures, such as Jiading District issued the "Jiading District Green Ecological Urban Construction Management Measures (Trial)", it is clear that the construction unit should prepare the design plan and construction drawing design documents including green and low carbon sections in accordance with the planning and design conditions and green and low carbon index requirements. Fengxian New City is drafting the "Fengxian District Green ecological urban construction management Measures".

4. Conclusion

During the "14th Five-Year Plan" period, the five new cities will assume the core functions of Shanghai and comprehensively promote green and low-carbon initiatives in accordance with the requirements of new construction. This includes enhancing ecological livability, improving transportation convenience, and unlocking the development potential of green ecological urban areas within these new cities. This paper focuses on specialized planning and research for green ecology, emphasizing a 100% implementation goal for green ecological urban areas. It presents detailed implementation paths and technical strategies while studying and determining the orientation, index system, and target development areas for each new city's green ecological development. The research results provide technical support for key aspects such as subsequent planning, construction, and operation of these new cities to ensure that distinctive goals are achieved while guiding future green and low-carbon development.

Acknowledgments

Close to zero carbon emission operation control and digital platform for high density and high quality blocks 22DZ1207905.

References

1. Anonymous. Shanghai Clarifies the "14th Five-Year Plan" to Accelerate the Promotion of New City Planning and Construction Work [J]. 2021. DOI: 10.12334/j.issn.1002-8536.2021.12.008.
2. Evaluation Standards for Green and Ecological Urban Areas DG/TJ08-2253-2024.
3. Research Group of the Development Research Center of the State Council. Promoting Carbon Peak and Carbon Neutrality through Green and Low-Carbon Transition [J]. China Economic Report, 2023(6):89-100.
4. Wu Zhiqiang. Exploring New Paths for Fine Management of the "Five Major Spaces" and Creating the "Shanghai Model" of People's City [J]. Future City Design and Operation, 2022(1):45-46.
5. Yan Wentao, Ren Jie, Zhang Shangwu, et al. Resilient Urban Planning in Shanghai: Key Issues, General Framework, and Planning Strategies [J]. Urban Planning Forum, 2022(3):10. DOI: 10.16361/j.upf.202203003.
6. Weichselgartner J., Kelman I. Challenges and Opportunities for Building Urban Resilience [J]. A|Z ITU Journal of Faculty of Architecture, 2014, 11(1): 20-35.
7. Li Juan. Beyond Health: A Policy Process Study on the Construction of a Healthy City in Shanghai [M]. China Social Sciences Press, 2020.
8. Luo J., Chan E. H. W., Du J., et al. Developing a Health-Spatial Indicator System for a Healthy City in Small and Midsized Cities [J]. International Journal of Environmental Research and Public Health (IJERPH), 2022, 19. DOI: 10.3390/ijerph19063294.
9. Song Xin, Hao Dongdong, Yan Longping. Research and Exploration of New Smart City Construction in Shanghai [J]. Economic Research Guide, 2022(9):4.

Constructing the Boundary-Domain Theory: A Spatial and Social Framework

Hua Zhu*, Zhangpan Song, Xinghua Lai and Haiyang Li

School of Art and Design, Changchun University of Technology, Changchun, China

Abstract. This paper examines the concept of power spaces across various disciplines, including architectural design and urban development, to understand how spatial environments influence power dynamics and social interactions. It introduces the "Boundary-Domain" theory as a key framework for analyzing the role of physical and social boundaries in shaping specific domains. The study highlights how the interplay between boundaries and domains can either foster segregation or enhance connectivity, providing insights for addressing social and spatial challenges in modern urban contexts.

1 Introduction

In Michel Foucault's theory of power, power spaces are closely linked to architectural design. Power, in this context, goes beyond the authority of the ruling class and encompasses the strategies, techniques, economic forces, and knowledge that discipline individuals. In Discipline and Punish, Foucault discusses the modern power-knowledge system, wherein knowledge is used to unconsciously regulate and control others[1]. A central element of this theory is the principle of surveillance, which makes "disciplinary power" effective by isolating and confining individuals, subjecting them to behavioral norms, and using knowledge to enforce their conformity. This power-knowledge system turns space into a stronghold of power, where spatial hierarchies are created not only to accommodate people but also to regulate and control their actions. Thus, it exerts influence over inhabitants and enforces power through spatial organization.

We need to identify more practical, context-specific theories within the broader framework of power space to address the issues it creates in contemporary society. While the overarching theory of power space provides a valuable lens for understanding the relationship between power, space, and social dynamics, it often remains too abstract to tackle real-world challenges. By focusing on theories that are more closely aligned with everyday experiences and societal structures, we can better analyze and mitigate the impact of power dynamics in urban design, community development, and the built environment.

2 Theoretical research on power spaces

Research on power space in the field of architectural design can be categorized into three main types: theoretical research, urban development studies, and investigations of real-world issues.

In theoretical research, Jianfei Zhu compared the disciplinary institutions designed by Jeremy Bentham with the traditional power structures of Ming and Qing Beijing under the influence of Han Feizi, analyzing the general construction of power spaces[2]. Wenxin Feng also used Foucault's theory to examine the evolution of hospital spaces[3].

Urban development studies include Zhang Yang provides a new perspective on understanding planning heritage from the angle of power spaces and historic urban landscapes[4]. Jing Chen examined Beijing's central axis from a power-space perspective[5], while Yong Wang analyzed how capital and power drive urban development[6]. Zhiming Sheng and Qing Zhou propose the analytical concept of "power spaces", concluding that such spaces primarily influence community governance performance through mechanisms of public resource allocation and performance signaling[7].

In research on real-world issues, Xuan Xiao and Zhimin Chen focus on the social role of museums, exploring their evolution from symbols of power in the past to public spaces today[8]. Junyan Mei examined women's spaces[9], Nan Zhang introduced the concept of cultural power space[10], and Hui Wang demonstrated through three case studies how designers should actively communicate with developers to achieve win-win outcomes for both the public and developer[11].

The study of power spaces in fields like architectural design, urban development, and social theory reveals the complex relationship between physical environments and social hierarchies. It illustrates how power dynamics shape spaces, influencing social interactions and community structures. In this context, the "Boundary-Domain" theory is essential for understanding how physical and social boundaries define functional domains. This framework helps examine how both visible and

* Corresponding author: zhuhua@ccut.edu.cn

invisible boundaries organize space and affect experiences. By analyzing their interplay, it offers insights into how spaces can either foster segregation or enhance connectivity, addressing challenges related to power spaces.

3 Overview of Boundary-Domain Theory

3.1 Boundary

Tangible boundaries are visible structures like walls, cameras, and access control systems, commonly used in former leaders' residential areas to isolate disturbances and maintain security. These boundaries not only restrict behavior and control movement but also preserve privacy.

In contrast, intangible boundaries are shaped by policies, social rules, and economic conditions. The unique status of leaders' residential areas created intangible boundaries, including government support and social hierarchy symbols, exerting invisible control over daily life.

The historical evolution of the Xinghua Village community illustrates how intangible boundaries influence space use through social rules and cultural factors. As a prominent "high-ranking official area" in Changchun, Xinghua Village became a power symbol during the Japanese occupation, leading to strict space usage rules primarily benefiting the elite, resulting in social isolation.

After the 1948 liberation, old social rules persisted, with high-ranking officials remaining in the area, reinforcing its status as a power symbol. Over time, these cultural rules evolved, reflecting society's understanding of power and forming a distinct community culture.

In the new century, policy changes led to original residents moving away, challenging existing rules and causing spatial function degradation. The resulting vacancies and neglect reveal changes in intangible boundaries—previous symbols of power became obsolete, diminishing community culture and interactions. With new policies, the Xinghua Village community faced disconnection from a thriving urban environment, becoming a "blank zone." This transition shows that evolving intangible boundaries directly affect space usage and its relationship with the urban environment, highlighting the impact of historical and social changes on spatial forms.

3.2 Domain

By using power to enclose space and form domains, Friedrich Ratzel proposed the concept of "Lebensraum" (living space) in his Political Geography at the end of the 19th century. He argued that "the state is a living organism, and as it grows stronger, it needs to expand its territory at the expense of other states." In the 1960s and 1970s, some scholars claimed that territorial behavior was innate, a perspective known as the "biogenetic school". In the 1970s, the idea of human territoriality emerged, evolving from animal territoriality. Sociobiologist Richard Dawkins, in his book The Selfish Gene, argued that human genes lead individuals to protect those similar to them within their domain while rejecting those who are different[12]. The above concepts focus on the physical aspects of space. After the 1970s, exploration of the social aspects of territoriality increased. For example, Jean Gottmann pointed out that domains are demarcated by humans and used for specific purposes, and should be viewed within the context of political processes. By examining the evolution of state territoriality, he revealed the functional shifts in domains as society develops. In the 1980s, the emphasis shifted from the physical functions of space that meet human needs to the idea that human territoriality involves conscious strategies of spatial control within social and cultural phenomena. Thus, territoriality has shifted from a focus on physical space to social, cultural, and political dimensions.

Tangible domains refer to the concrete spatial forms and functional arrangements created within boundaries. These domains are often manifested through the physical layout of buildings, public spaces, and roads. For instance, in leaders' residential communities, the housing and public facilities form an enclosed space centered on privacy and security. The physical characteristics of these spaces significantly influence the community's daily lifestyle.

Intangible domains, on the other hand, involve aspects such as social culture, power relations, and identity within the space. Inside a leaders' residential community, residents not only share a physical space but also a sense of special identity and social circles. Within this unique social network, invisible power structures and cultural rules shape the patterns of social activities and regulate the ways in which space is used.

3.3 Boundary-Domain Theory

The Boundary-Domain Theory offers a comprehensive framework for analyzing how physical and social boundaries shape spaces and their interactions with surrounding environments. "Boundaries" encompass not only physical barriers but also social, cultural, and conceptual limits that define distinct areas or "domains." These boundaries regulate access, movement, and interaction, creating zones with unique characteristics and social dynamics. Conversely, "domains" refer to the internal spaces within these boundaries, shaped by specific functions and social interactions. By examining how boundaries influence the flow of people, goods, and ideas, the theory enhances our understanding of segregation or integration within urban or social settings.

As a micro-level analytical tool for power spaces, the Boundary-Domain Theory specifically explores how power is expressed and operates within space, revealing the social relationships and power structures hidden behind spatial forms. It emphasizes the impact of both tangible and intangible boundaries on spatial functions and social interactions, indicating that spatial division is not only physical but also social and cultural. In contrast, existing spatial organization theories, such as space syntax or urban design theories, typically focus on the structural and morphological characteristics of spatial configurations, examining how space influences movement and behavior patterns. While these theories explore spatial efficiency and functionality at a macro level, the Boundary-Domain Theory delves into the micro level, analyzing how power shapes the relationship between individuals and space. Integrating these two approaches can provide a more comprehensive perspective for spatial design and social practice, addressing both the physical characteristics of space and the underlying socio-cultural context, ultimately achieving deeper and more meaningful spatial design.

4 The interactive Relationship between Power and Space

4.1 The dual role of borders

Boundaries serve not only as tools for separation but also as guides for organizing and using space. Traditionally, they delineate different functions—residential, commercial, and public—helping maintain order by clearly defining areas, such as distinguishing "inside" from "outside" in leaders' residential communities for privacy and security.

However, boundaries can also foster openness and connection. Their form and function can be adjusted to reconnect previously isolated areas with the outside. In declining communities, softening boundaries and increasing permeability can enhance interaction and revitalize the area. For example, removing walls or incorporating semi-open designs can create more flexible, multifunctional spaces that attract external attention and use.

4.2 The gradual change of the boundary

Boundaries are not fixed but transitional zones that can shift from enclosed barriers to open public spaces, creating varied boundary experiences. For instance, walls of a leaders' residential community may have initially served as clear separators, but with changes in leadership and policy, they can transform into symbolic boundaries. Designing gradient boundaries allows different functional areas to integrate more seamlessly into the urban environment, reducing resistance to external interactions. A semi-private entrance plaza, for example, can guide visitors into the community, enhancing openness and fluidity. This approach fosters social interaction while preserving order and community identity.

4.3 Multiple dimensions of the field

A domain is not only a physical space but also a carrier of multiple social, cultural, and identity elements. The formation of a domain relies not just on its spatial form but also on the social relationships, economic activities, and cultural atmosphere within it. In leaders' residential communities, the domain is reflected not only in the enclosed living spaces but also in the symbols of power, social hierarchies, and invisible social rules.

As the leaders moved out, the community's domain began to weaken, with the previous power and identity gradually fading, turning the domain into an ordinary urban space, losing its original distinctiveness. However, this transformation also presents an opportunity for the community's reconstruction. By redefining the space's purpose and function, new cultural and social activities can be introduced into the domain, creating a new space adapted to contemporary needs.

4.4 Dynamic balance of boundary and domain

The relationship between boundaries and domains is dynamic; they influence each other. Strengthening boundaries can stabilize a domain's function, as seen with walls and access controls that enhance community security and privacy. Conversely, opening boundaries can promote innovation and transformation, allowing areas to diversify functions and attract external resources.

Community decline often results from an imbalance between boundaries and domains: closed boundaries lose their protective roles, while the domain suffers from a lack of external interaction. Renovating old buildings aims to readjust boundaries to reactivate domain functions, helping the community reintegrate into the urban fabric. This process involves reconstructing physical space while also addressing intangible social and cultural boundaries to meet modern needs and residents' lifestyles.

5 Strategies for the reconstruction of boundaries and domains

5.1 Permeable Design of Boundaries

By introducing semi-open spaces, public transition areas, and shared functional zones, the sense of enclosure within the space can be effectively reduced, thereby enhancing the community's inclusivity. Semi-open spaces such as sunlit courtyards and open corridors provide privacy while encouraging interaction among residents, allowing people to move freely and enjoy public life. At the same time, public transition areas like green walkways, recreational seating, and gathering squares can serve as buffers between residents and the outside world, creating a friendly social environment that encourages communication, gatherings, and community events. This design strategy reduces physical and psychological barriers, making the community more open and welcoming to different social groups, thereby enhancing community cohesion and vitality.

5.2 Multifunctional Domains

Redefining community functions to meet modern living needs is essential. Introducing community activity centers, cultural exchange platforms, and commercial facilities can enrich the domain's usage patterns and enhance overall functionality. For example, activity centers can host gatherings, lectures, and interest groups, fostering neighborhood relationships and engagement. Cultural exchange platforms can organize art exhibitions and events, providing diverse experiences and strengthening cultural identity. Additionally, adding commercial facilities like coffee shops and convenience stores not only offers essential services but also facilitates social and economic interactions, promoting the community's economic vitality and sustainable development.

5.3 Interactive Design of Boundaries and Domains

Designing the relationship between boundaries and domains as fluid and complementary allows community activities to integrate naturally with the urban environment. This approach breaks down traditional spatial divisions, creating a connected network. For instance, permeable building facades, adjustable layouts, and flexible functional divisions enable community activities to link seamlessly with public spaces, enhancing residents' quality of life. Consequently, boundaries transform from isolating limits into active interfaces that embrace external cultures and resources, increasing the community's openness and attractiveness. This design philosophy not only revitalizes the community but also fosters positive interactions with the city, enriching residents' lifestyles.

6 Conclusion

Reconstruction strategies for boundaries and domains are vital in renovating old buildings, requiring a comprehensive understanding of the community's history, current state, and future vision. In historically rich communities, designers must preserve cultural heritage while using innovative designs to break down closed boundaries, transforming them into channels for communication and interaction. This process reshapes not only physical space but also the social fabric, revitalizing the community and fostering stronger connections with the outside world.

Opening boundaries weakens physical and social barriers, enabling better integration with the urban environment through public spaces, open designs, and a permeable transportation network. Domain reconstruction focuses on multifunctionality and flexibility, allowing spaces to serve diverse populations and lifestyles. These strategies not only reintegrate the community into urban life but also enhance residents' experiences by offering diverse social, cultural, and economic opportunities. Ultimately, the community evolves from a traditional residential area into a vibrant, multifunctional public space, driving its ongoing development and modernization

References

1. Michel Foucault. Discipline and Punish: The Birth of the Prison. Translated by Beicheng Liu and Yuanying Yang. Beijing: SDX Joint Publishing Company, 2003.
2. Jianfei Zhu, "A Cross-Cultural Discussion of Power Spaces: Bentham, Foucault, Han Fei, and Beijing in the Ming and Qing Dynasties," Time + Architecture, 2003(02): 104-109.
3. Wenxin Feng and Hanfei Zhu, "Hospital Space Evolution from the Perspective of Foucault's Power Space Theory," Architecture & Culture, 2022(12): 219-221.
4. Hui Wang, "Political Patterns and Changes in Urban Space During Medieval Japan: A Case Study of Kyoto," Urban Planning International, 2008(02): 3-9.
5. Yang Zhang. Planning Heritage: Historical Urban Landscapes Guided by Power Spaces [J]. New Architecture, 2023, (06): 15-20.
6. Yong Wang, Guangbin Li, and Wen Shi, "Characteristics and Mechanisms of Urban Space Production in Suzhou: A Study on the Evolution of Suzhou's Urban Spatial Structure," Modern Urban Research, 2015(11): 125-130.
7. Zhiming Sheng, Qing Zhou. Power Spaces and Governance Performance: An Analysis Based on the Shanghai Urban Community Survey [J]. Society, 2021, 41(05):1-30.DOI: 10.15992/j.cnki.31-1123/c.2021.05.001.
8. Xuan Xiao, Zhiming Chen. Spatial Metaphors and Power Spaces in the Cultural Landscape of Jingdezhen Ceramics [J]. Ceramic Research, 2023, 38(05):46.DOI:10.16649/j.cnki.361136/tq.2023.05.069
9. Junyan Mei and Ye Jiang, "Philosophical Reflection on Women's Spaces Behind 'Female-Only' Spaces," Decoration, 2019(11): 142-143.
10. Nan Zhang, "Ethical Study of Street Space Design from the Perspective of Cultural Power Space," Furniture and Interior Design, 2022, 29(02): 107-111.
11. Hui Wang, "Whose Urban Renewal?" The Architect, 2023(01): 29-36.
12. Dawkins R. The Selfish Gene[M]. Oxford: Oxford University Press, 2006:384.

Strategy and Practice of Artistic Transformation of Discarded Teaching Facilities in the University

Mingqi Shi, Jiayi Shi, Wei Cheng*

School of Urban Construction, Wuhan University of Science and Technology, Wuhan, China

Abstract: The university is not only crucial for talent cultivation but also plays a significant role in resource recycling. Recycling waste resources in universities has a positive effect on building a sustainable campus and promotes educational activities' development in a more sustainable and innovative direction. Against this background, our case study is based on renovating the public space in the teaching building of Wuhan University of Science and Technology, China. Firstly, we investigate the current situation of used teaching facilities in the university and carry out sorting statistics. Secondly, based on the theory of material aesthetics and value regeneration, we explore the possibility of recycling waste materials in college and achieve the value of recycling. Finally, the project is a space design study in the form of artistic transformation and proposes a strategy for redesigning used teaching facilities in the university.

1. Introduction

University students are quick to adopt and effective promoters of advanced culture and concepts, thus playing a role in driving the development of social waste resource recycling and enhancing public awareness of environmental protection. The recycling of discarded resources within universities mirrors the social circular economy. According to statistics, China has 3,117 higher education institutions[1], contributing to green development and transformation of students' living styles through the model of "resource-product-waste-reuse."

Our investigation found that with the continuous expansion of enrollment in higher education institutions and the increasing number of students, there has been a constant expansion and replacement of teaching facilities, resulting in a large number of abandoned teaching facilities and resource waste. Items like lecture desks, podium tables, chairs, blackboards, easels, teaching aids, and lighting fixtures, curtain instruments, etc. Although a large number of these teaching equipment are eliminated annually, it does not mean that they have completely lost their functionality. Given their large quantity, modularity, and material diversity, these discarded items still hold considerable potential for reuse and artistic transformation.

Therefore, the design method of "designing and recycling" of waste and discarded items can be used to deal with these teaching waste materials. On the one hand, it saves, recycles, and utilizes ordinary and non-renewable energy, and through various artistic means and methods, the discarded "garbage" can be reused by people. On the other hand, by the principle of minimizing the consumption and use of renewable resources, conventional consumables are used for recreation and then returned to people's lives to extend their service life[2]. By artistically processing the parts or components of these discarded teaching aids, not only can new aesthetics be interpreted and extend the lifespan of renewable resources but also reinforce the principles of sustainable and green living.

2. Literature Review

Most universities recycle old teaching materials through regeneration, which refers to transforming, repairing, or reprocessing waste into useful products, such as transforming old clothes into new clothing or repairing old furniture for reuse. Alternatively, by organizing school recycling programs or sharing platforms with social organizations, old desks, chairs, blackboards, and bookshelves can be refurbished and reused or donated to schools and institutions in need to continue their use. Although this measure is beneficial for the reuse of waste teaching aids, it requires a large amount of human, material, and technological support from social organizations and institutions, which is costly and not economical enough for universities with large replacement volumes, fast updating rhythms, and limited resources. [3] This method of resource utilization is mainly based on the establishment of various industrial recycling systems, which cannot be widely promoted and popularized. There is not much exploration in the artistic reshaping and utilization of universities, and only a few universities in China have implemented measures for the artistic transformation of old teaching aids recycling.

In developed regions of Europe and the United States, the reuse of used objects has become a way for residents to meet the needs of their daily lives, and it has gradually become popular all over the world, which has made more and more people pay attention to the study of the artistic

* Corresponding author: smq0601@wust.edu.cn

transformation of old objects.[4] For instance, the Walking Chair in Vienna is a studio that uses waste materials from daily life as its design theme, which transforms seemingly useless waste into artistic processing[5], while Manoteca in Italy, a studio that takes in discarded materials, gives new life to discarded materials through creative reintegration by designer Elisa Cavani[6]. In the UK, students from Brighton University, in collaboration with designer Duncan Baker-Brown, built the "Waste House," the first permanent structure made entirely from waste and recycled materials[7], demonstrating the possibilities of waste reuse and contributing to public art exhibitions. In Singapore, Professor Christine Yogiaman and her students repurposed retired subway seats into unique seating areas for urban landscapes, effectively reducing resource waste[8].

These international examples highlight how artistic repurposing can endow discarded educational materials with new artistic forms and functions, emphasizing waste reuse and resource recycling. Such practices raise awareness about environmental protection and sustainability among students and faculty. Through showcasing and promoting these works, more people are encouraged to engage in environmental issues, fostering broader environmental protection actions.

This study uses Wuhan University of Science and Technology as a case example to explore the feasibility of integrating artistic repurposing theories with resource recycling in educational settings, demonstrating a creative and practical approach to waste utilization in higher education.

3. Methodology

3.1 Survey on the current situation of waste materials--Taking Wuhan University of Science and Technology as an example

Wuhan University of Science and Technology (WUST) is a higher education institution specializing in engineering, especially in the fields of materials science and engineering, mechanical engineering, electronic information engineering, and other fields with strong research strength, with several special laboratories and research institutes, so the teaching area and experimental area is the main area of learning activities in the university, where the waste generated by the waste materials are mainly teaching and experimental wastes. Compared with the randomness and complexity of domestic waste, this kind of resource has a stable recycling cycle and uniformity and also has a higher recycling value.

By sorting and assessing discarded teaching facilities in instructional and laboratory areas, the utilization value of each resource category is evaluated. As shown in Table 1 below, a breakdown of the materials used in the abandoned teaching facilities is presented:

Select four common materials from university waste and assess their repurposing value based on the following five dimensions.

Evaluate material plasticity: Assess ease of processing and potential for repurposing, with some metals, wood, or plastics being more adaptable for reuse.

Table 1. Categories of materials for discarded teaching facilities

Material classification	Specific teaching facilities
Mixed paper	Carbon paper, pressure-sensitive paper, books, corrugated paper products
Metal	Alloy, iron sheet, aluminum
Wood	Packaging wooden crates, easels, desks and chairs, disposable tableware
Plexiglass	Glass plates, acrylic sheets, glassware, and other laboratory equipment
Plastic	Plastic bags, plastic bottles
Rubber products	Rubber gloves, shared bicycle tires
Textile	Curtains, military training belts, laboratory cleaning cloths

Assess design flexibility: Consider how materials can be used in design to create high-value products, such as furniture, art, or architectural elements.

Assess conversion costs: Calculate expenses for processing, labor, and any specialized equipment or tools needed to transform materials into new products.

Assess lifecycle impact: Evaluate environmental and economic effects from production to disposal.

Life Cycle Assessment: Evaluating the environmental and economic impacts of a product throughout its entire lifecycle.

Sustainability Assessment: Evaluating whether a transformation is long-term sustainable and promotes circular economy and resource efficiency.

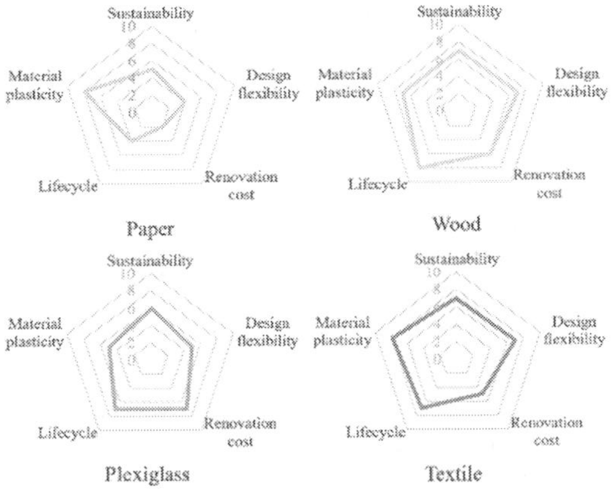

Fig. 1. Material Analysis charts

The results of the material evaluation shown in Fig. 1 give us an idea of the processing characteristics and suitability of different types of materials. For instance, paper is highly malleable, allowing it to be easily folded into various shapes with low processing costs and high adaptability. Wood is suitable for supporting structures due to its high-quality texture and hardness, and is easy and highly sustainable to process. Glass, due to its transparency and light transmission, has great potential in artistic transformations but requires specialized tools and is fragile with high maintenance costs. Conversely, textiles have excellent ductility and wrapping properties, are easy to work with, and have a long lifespan.

In the process of transforming old objects, the

characteristics of various materials should be fully utilized for artistic treatment. This approach ensures that the transformed items meet safety, durability, and aesthetic principles and maximizes each material's advantages. Therefore, selecting appropriate materials and creating based on their characteristics can fully realize the potential of the transformed pieces, enhancing both their aesthetic and functional qualities.

3.2 Material Characteristics and Design Research

In remodeling design, materials are fundamental to realizing concepts. To a certain extent, materials are the medium through which design ideas become tangible, influencing form, color, texture, strength, and processability. With the transformation of design from simple formalization to connotation, the material will not only stay at the traditional decorative level, it is more attached to the space, serving the expression of the space theme, with the double task of function and artistic expression[9]. The following analysis explores three common waste materials from higher educational institutions as cases for characterization and transformation potential.

3.2.1 Cardboard

Cardboard is a common waste material in universities, which looks fragile but has a degree of sturdiness that is not inferior to other materials. The characteristics of the cardboard material determine that it can be processed and treated with simple tools, and the most important thing is that cardboard can be biodegradable, which reduces the pollution to the environment after treatment. In 2014, students from Changsha University's Art and Design Department repurposed cardboard into tables and chairs for reading in the lobby of the library[10], exploring the artistic potential of cardboard as an art medium and practicing eco-friendly design.

3.2.2 Wood

Wood waste in universities often comes from old desks, art easels, and transport crates. Due to the modeling differences of old products, wood is in the form of strips and boards, which have better mechanical properties and aesthetic value. Wood is easy to work with and can be carved, sanded, cut, etc., using traditional woodworking tools and techniques, which allows for greater flexibility in design and adjustment. Students from the Hebei University of Technology went into the Chengde paddocks, with creative roots of Mulan paddocks, waste wood, pine tower shells, cleverly designed and made into Christmas trees, wooden baskets, wooden paintings, and other wooden handicrafts, turning waste into treasure and green fashion! [11]

3.2.3 Textiles

Textiles, with their softness and malleability, are ideal for covering surfaces, padding, and making movable decorations. Not only are they lightweight and easy to handle and install, but they are also relatively simple to process and can be adapted to different design needs, injecting unique charm and practicality into the work. In 2023, Associate Professor Lu Zhiwen[12] at Taiyuan University of Technology led a team to adopt the principle of 'addition and subtraction' to repurposed discarded textiles using cutting, stitching, and reshaping techniques to create innovative fashion collections, showcasing their potential in a fashion show.

3.3 The significance of composition theory for the artistic transformation of old items

"Points, lines, and planes" are the core elements of design compositions and modeling activities. Piet Mondrian's work attaches great importance to the balance of 'points, lines, and surfaces,' thinking about the horizontal 'lines' and vertical 'lines' in the combination of interspersed and divided in the process of the formation of surfaces, as well as various types of form, position, and proportion of the constituent units reflect the overall aesthetics of the design.

In spatial design, points have a significant visual impact[13]. According to the size of the points and the characteristics of the gathering and dispersing, it can be used to highlight the key elements in educational scrap materials. By retaining the unique parts of the scrap teaching facilities and as decorative embellishments, such as the original button handle, connectors, etc., you can create a visual focus to enhance the attractiveness of the transformed works of creativity.

The length, thickness, straightness, softness, and hardness of the line elements, as well as the representation of reality and falsehood in the interface, can be used to define shapes and structures. By rearranging and integrating the lines of used teaching facilities, their sense of composition and visual effect will be enhanced. Of course, the combination of lines is not only within a single interface but also within a three-dimensional space, which means that the design of the lines can create support frames or other structural elements, making them more functional and stable.

Surfaces are defined by their outer contours[14]. Their shape, size, direction, and whether they are flat or curved create the object's skin and 3D form. Generally, both regular and freeform surfaces can be varied through arrangements. Arranging and combining the surfaces of different teaching facilities by separating, joining, intersecting, superimposing, subtracting, folding, etc., can create objects with unique artistic styles.

3.4 Design Practice: Public Space Renovation at Wuhan University of Science and Technology

Site assessment: the site is located in the corner space of the corridor on the first floor of the No. 5 academic building, the current status of which can be seen in Fig. 2, near the graduate students' offices and laboratories, where exhibitions of architectural work are often held, but some of the guiding facilities for relaxation or stopping are lacking, and the corridor is backlit from the west at the end of the corridor, which makes the overall space dim and monotonous and lacks interest.

Fig. 2. Original Site

3.4.1 Material Selection

Selection of waste materials based on the current status of waste teaching facilities materials summarized in the previous section, and according to the design requirements, after balancing the weights of processing difficulty, load-bearing strength, and appearance effect, we finally chose to use the discarded teaching easel as the main material, the acrylic board as the decorative embellishment, and the hardwood packing box to ensure the structural strength. The selected materials are:

Waste Teaching Easels (63%): Main material

Acrylic Panels (13%): Decorative accents

Hardwood Crates (24%): Structural support

This combination ensures optimal functionality and appearance.

3.4.2 The design employs the principles of point, line, and plane composition and utilizes Mondrian's primary color scheme.

(1) Point, Line, Plane Composition Theory

Line Reconstruction: With the window as the main visual center, the design makes use of black horizontal and vertical lines of varying thicknesses to form the basic framework of the image, and a grid structure of right angles is formed[15]. The strict arrangement of straight lines conveys a sense of order and stability. Through the length and staggered position of the lines to create a dynamic balance, the asymmetry increases the visual tension and dynamism of the space as a whole.

Plane Integration: The screen is divided into rectangular or square areas of different sizes, which are either filled with color or kept white.

(2) Mondrian's Primary Color Scheme

Color Coordination: The primary colors of red, yellow, and blue, which symbolize the purity of the original, are used as the main spatial colors and are arranged in a rhythmic combination. In addition to this, black lines divide up the area, and white or grey space provides a background for the colors, creating a strong contrast and a rhythmic visual experience.

Surface Material Contrast: Paint covers the wood's original color but retains its texture, while colorful acrylic panels on the walls create a contrast between solid and translucent areas when illuminated by sunlight.

Please refer to Fig. 3 for an axonometric view of the design details and overall layout, which shows the specific implementation and overall visualization of the design in a 3D view and helps to better understand the spatial structure and layout of the design.

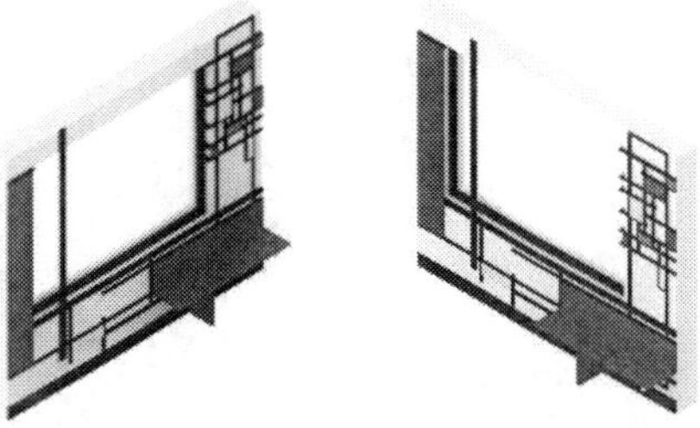

Fig. 3. Axonometric Drawing

3.4.3 Design Process: The work is divided into three parts

(1) Wooden Framework: Black wooden strips were dismantled and cut from discarded easels, painted and joined using hidden mortise and tenon joints. For detailed steps, please see please see Fig. 4.

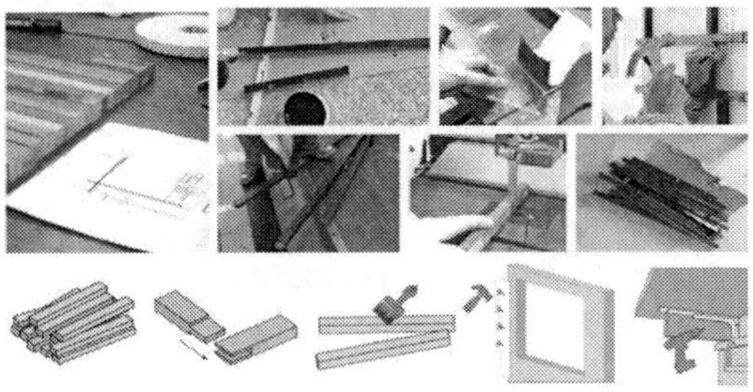

Fig. 4. Production Process

(2) Acrylic panels: Laboratory discarded acrylics are cut to size, covered with colored film and skillfully fitted into the grooves of the wooden frame. See Fig. 5 for a preview of the project's design plan and visuals.

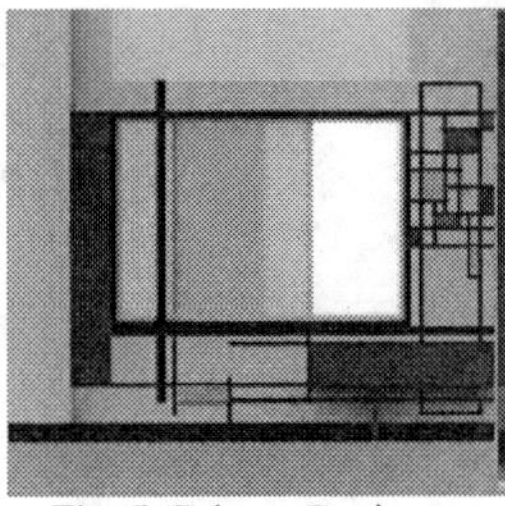

Fig. 5. Scheme Design **Fig. 6.** Project Completion

(3) Seat Support: The seat is built from plywood taken from a sturdy shipping crate and securely fastened to the wall and base using expansion bolts to enhance its stability. See Fig. 6 for the final completion of the project and a live demonstration. see Fig. 7 for the exact process and details of the construction.

Fig. 7. Construction Process

3.4.4 Design Results

Cost: Mainly from invoice data.

Residual Value: 95% (Renewable energy use/total waste x 100%)

After the artistic renovation of the corner corridor, observations made every three days at noon revealed a significant increase in foot traffic and usage. Both students and faculty frequently use the space for waiting, resting, and socializing.

3.5 Challenges and Solutions

In the art renovation project, we encountered several challenges, including the limited variety of teaching aids available, as well as issues related to the processing, durability, and aesthetics of the materials. In response, we implemented three key measures:

(1) Mobilizing resources within the school to promote data collection through teacher-student collaboration, as shown in Fig. 8, which illustrates active teacher-student exchanges;

(2) Employing modern design concepts and techniques to improve both the artistry and functionality of the work;

(3) Strengthening inter-school cooperation to promote the recycling of used teaching aids.

These initiatives not only addressed the challenges we faced but also enhanced the scalability and overall impact of the project.

Fig. 8. Teacher-student Exchange

4. Discussion

4.1 Design Principles for Repurposing Educational Waste Materials

In this study, the artistic transformation of educational waste materials not only utilized the principles of point, line, and plane composition and Mondrian's "Red, Yellow, Blue" but also incorporated architectural space design. This approach differs from traditional design, as it not only adheres to basic design principles but also meets architectural space design standards of "strength, functionality, and aesthetics." With evolving societal needs and lifestyles, these principles must adapt, aiming for a unified design that integrates function, structure, and materials. The fundamental principles for repurposing materials in space design are summarized as follows:

(1) Principle of Coherence: Ensure that all design elements—function, decoration, and materials—are harmonized through contrast, transition, emphasis, repetition, and order to achieve a unified and harmonious space.

(2) Principle of Compatibility: Select and process collected waste materials to match their form, structure, and color with the design. Ensure that the final product meets functional and decorative requirements.

(3) Principle of Usability: Ensure that repurposed products are functional and structurally sound while also addressing comfort and safety requirements for users.

(4) Principle of Environmental Sustainability: Focus on enhancing product durability with efficient structural designs and using eco-friendly adhesives to minimize environmental impact. Optimize and simplify the product's design to align with green design principles4.

4.2 Feasibility Summary of Artistic Transformation of Educational Waste Materials

By transforming waste teaching facilities into artistic devices through creative design, these resources can be effectively utilized to reduce the generation of wastes. The new products, after artistic transformation, can add beauty and creativity to the teaching public space, making the environment more attractive and energetic and enhancing the students' learning experience and comfort, which is fully in line with the concept of sustainable development. At the same time, it can become an interactive platform on campus. Through the artistic transformation of waste teaching facilities, students and staff can participate in the creative process, thus stimulating their creativity and hands-on ability, as well as promoting communication and cooperation between teachers and students. In addition, we compared the costs associated with art remodeling to those of traditional recycling. Traditional recycling depends on industrialized systems, which can be expensive. Conversely, creative design in art transformation repurposes used teaching aids into art installations, thereby reducing waste and enhancing the campus environment. Although the initial design and production costs may be significant, the long-term environmental benefits and overall value greatly surpass the initial investment.

4.3 Details on long-term sustainability and maintenance

Artistically transformed waste teaching aids significantly enhance both aesthetics and practicality; however, long-term sustainability and maintenance costs require further consideration. Taking the recreational facilities on the fifth floor of Wuhan University of Science and Technology as a case study, we utilized materials such as discarded easels, acrylic boards, and cardboard boxes to improve the functionality and visual appeal of the space through artistic processing. To promote sustainability, we selected biodegradable materials, including wood and cardboard, to mitigate environmental pollution, while the design prioritizes structural stability and durability. For maintenance, we implemented easily removable snap-on acrylic panels and invisible interface mortise and tenon timber frames to minimize both the cost and complexity of upkeep.

5. Conclusion

In the context of the circular economy, the transformation of old objects and resources has always been a significant concern. By artistically repurposing teaching waste materials on campus, we aim not only to achieve resource reuse but also to highlight the comprehensive value of innovative education, cultural heritage, and social responsibility. For architecture undergraduates, such practical activities offer valuable learning and practice opportunities, fostering their innovative spirit and environmental awareness. Additionally, these initiatives enrich campus culture, showcase the institution's innovative capabilities and social responsibility, enhance public environmental awareness and aesthetic appreciation, and promote sustainable development alongside the growth of cultural and creative industries. To integrate design into daily life, we are dedicated to making the redesign of old items an enjoyable and meaningful endeavor, allowing unused objects to regain their luster.

To further enhance the engagement of teachers and students in sustainable practices, we will explore innovative methods of participation. For instance, we will organize regular competitions and exhibitions focused on the transformation of used teaching aids to stimulate students' creativity and practical skills. Additionally, we will invite professional designers and artists to provide guidance and training, thereby enhancing the design capabilities and aesthetic sensibilities of both teachers and students. Through the implementation of these initiatives, we aim to promote the artistic transformation and recycling of used teaching aids more effectively, contributing significantly to environmental protection on campus.

References

1. Ministry of Education of the People's Republic of China. (2024) List of National Higher Educati on Institutions. (http://www.moe.gov.cn/jyb_xxgk/s5743/s5744/A03/202406/t20240621_1136990.html)
2. Wu Xiaoyan. Exploration and Research on the "Design Reincarnation" of Waste and Abandoned Items: A Case Study on Designing Lighting Fix tures from Waste Items. Industrial Design, 2018, (12): 125-126.
3. Li Xu, Zhang Yuanlu, Yu Lingjie, et al. Explora tion of the status quo and problems in the use of old things in colleges and universities[J]. Mod ern Business Industry, 2022, 43 (24): 149-150.D OI:10.19311/j.cnki.1672-3198.2022.24.064.
4. H. Li. Research on the reuse of old things base d on green design [D]. Xinjiang Normal Univers ity, 2016.
5. Yang Xi. Creative Design of Waste Items in Inte rior Decoration: A Review of Study and Trainin g in Austria. Journal of Suzhou Craft Art Vocati onal and Technical College, 2014, (01): 77-80.
6. Bey & MS, Manoteca, Kang S, et al. From Old to New: "Recreation" of Waste Furniture. Moder n Decoration (Home), 2014, (01): 114-121.
7. The Encounter of Waste, Environmental Protectio n, and Recycled Design. Daily Headlines. (2019) (https://kknews.cc/home/rllm69n.html)
8. New Life for Old Items: SUTD Design Innovati on, Practicing Sustainable Concepts, Energizing P ublic Spaces. (https://mp.weixin.qq.com/s/KnpZ6YRiWDMZHN70Fng2gA)
9. Zhou Min. Application of Decorative Materials a nd Spatial Expression of Composition Imagery. S outhwest Jiaotong University, 2014.
10. China Education News Network: DIY Waste Pap erboard Transformed into Tables and Chairs by Students of the Art and Design Department, Cha ngsha College. (2014) (http://www.ccsu.edu.cn/info/1323/18065.htm)
11. Hebei Provincial Government Website – Hebei U niversity of Technology: Creativity Boosts the U pgrade of "Mulan Woodwork". (2023) (http://www.hebei.gov.cn/columns/580d0301-2e0b-4152-9dd1-7d7f4e0f4980/202305/30/9b8995a5-4d53-11ee-985b-6018954d7f6f.html)
12. CCTV "Innovation in Progress" Program Reports on Waste Textile Recycling by the School of L ight Textile Engineering, Taiyuan University of T echnology. Student Affairs Department. (2023) (http://student.tyut.edu.cn/info/1187/6507.htm)
13. Zhou Jia. Application of Point, Line, and Plane in Interior Space Composition. Journal of Chifen g University (Natural Science Edition), 2017, 33 (06): 49-50. DOI: 10.13398/j.cnki.issn1673-260x.2017.06.019
14. Tang Jiao, Xu Hui. Application of Visual Eleme nts of Point, Line, and Plane in Three-Dimensio nal Composition in Interior Space. Drama Home, 2017, (21): 144.
15. Chen Mobao. Exploration of the Application of Mondrian's "Red, Yellow, Blue" Design Languag e. China Packaging Industry, 2014, (06): 44-45. DOI: 10.14047/j.cnki.cpi.2014.06.039

Design Method and Construction Process of Slope Greening in A Basalt Mine

Xiaoming Chu[1], Yu Yang[2,*], and Yunxiao Yu[1]

[1] Zhejiang Communications Investment Group of Shengxing Mining Industry Co., LTD., Shengzhou, Zhejiang, China
[2] Zhejiang Communications Investment Group of Lixing Mining Industry Co., LTD., Lishui, Zhejiang, China

Abstract: Through the greening management example of a basalt mine slope in Shengzhou City, Zhejiang Province, China, a variety of greening technology methods, such as local anchors and local active nets, were used. The ecological environment of the quarry slope was improved, the overall level of slope management was upgraded, and the management results were safe and reliable.

1. Project Overview

The upper part of a basalt mine in Shengzhou City is a skip-shaped face and bottom formed by historical mining with the mouth of the mine to the south, and the highest elevation of the mine is 131.4m. The highest elevation of the mine is 131.4m, the natural topography of the hillside is generally 10-25°, and the elevation of the original bottom is about 91-99 m. Four mining steps of +120m, +105m, +90m and +75m are formed at the end of the mining. The final slope angle is less than 51°, and the final platform slope angle is 60°. The final slope consists of four steps, and the minimum width of the safety platform is 4 m. The height of the steps is 15 m, and the elevation of the bottom of the mine is 60 m. The maximum height of the slope at the end of the mine is about 71.4 m. According to the height of the slope of the open-pit mine, it is a low slope (H is less than 100 m), and according to the hazard class of the slope, it is classified as grade III (the lowest level, without serious hazards and may cause casualties), and according to the slope engineering safety level, the slope is classified as Grade III[1].

2. Geological environment conditions

The mine adopts open-pit mining, and the cover layer of the ore body is basalt weathering layer, with a thickness of 3-8m in general, which consists of gravel-containing pulverized clay, pulverized clay, and crushed stones. The ore body consists of fresh basalt layer with dense structure, which is hard rock type, the thickness of the ore area varies from 5 to 50m, and the development of columnar joints, the production shape is 172°∠72°, 153°∠74°. The bottom dangling ballast body consists of porous basalt layer, porous and light, the compressive strength is obviously reduced. The deep part of the mine area is weathering layer of crystal glass chip fused tuff layer, the bottom plate of porous basalt layer, the thickness varies from 0 to 9m, and it is composed of crushed stone powdery clay, crushed stone, kaolin and so on. The overall contact surface is relatively gentle, with the following characteristics: inclination 110~170°, dip 5~10°. Through on-site investigation and combined with historical ground investigation data, it can be seen that the geological conditions of the slope at this place are relatively complicated, and the nature of the underlying tuff all-strong weathering layer is poor, which is very easy to destabilize and damage in the excavation process.

3. Greening design principles and requirements

(1) Seriously implement the laws, regulations and policy requirements of land reclamation, ensure the implementation of land reclamation obligations, rational land use, protection of arable land, prevention of soil erosion, restoration of ecological environment and protection of biodiversity[2-4].

(2) Determine the management plan according to local conditions, and try to make the management engineering measures as simple and easy as possible, and closely combine the engineering geological conditions of the slopes with the engineering measures; select the localization of the greening species, integrate with the surrounding vegetation as much as possible, and save the investment in the project.

(3) Environment-friendly. Considering the greening of the project, adopt measures combining protection and greening landscape to beautify the slope environment and create a harmonious slope ecosystem[5].

(4) Dynamic design. Carry out the principle of dynamic design in the whole process, strengthen the monitoring and supervision of the construction process, and adjust or optimize the program in time once the difference between the engineering geology and the actual is large, and the broken rock and soil area should be appropriately added with anchors.

(5) Safe and reliable, comprehensive disposal. Tailings

* Corresponding author: ethany96@163.com

slope ecological restoration project design using stabilization-based, reinforcement-based, drainage, protection and comprehensive treatment measures[6].

4. Slope greening design and process

4.1 Slope greening design

According to the geological and structural characteristics of the slope, the implementation of classification greening, respectively, the use of local anchor + spray planting grass, local active network + local crack closure anchor spraying + spray planting grass and local active network + spray planting grass and other methods of greening the slope, slope ecological restoration of the construction steps are as follows: slope excavation → flexible protection network construction → anchor construction → spray planting grass construction → reclamation[7-8].

(1) The first level: the design height of this level of slope is 15m, and the slope is 60°. It can be seen from the borehole exposure that soil-yellow and gray-white gravel-containing pulverized clay crystals glassy fused tuff is exposed at this level. Therefore, two rows of long anchors are arranged at the poor geological conditions of this level to play the role of stabilizing the loose rock body of the slope, and spray seeding grass re-greening is carried out. The treatment program for this level is: local anchors + grass spraying.

(2) The second and third level: each level of slope is 15m high with 60° slope. The upper part is strong weathering layer, containing different sizes of broken stones; the lower part is spherical weathering layer, constituting a complete weathering shell, which seems to be laminar structure, and the rock structure is broken. Considering the high height and broken rocks of these two slopes, two rows of long anchors are arranged at the poor geological conditions of these two slopes, and then the slopes are re-greened by spraying and planting grass. The treatment program for this level of slope is: local anchor + grass spraying.

(3) The fourth and fifth level: the slope height of the fourth level is 15m, the fifth level is 10m high, the upper part is basalt fully weathered layer, clay-like, thickness of about 1-2m, and the lower part is strongly weathered loose block gravel. The slope surface is mainly spraying grass re-greening, and active net protection is adopted in the fourth and fifth level platforms, and the active net adopts GPS-type flexible protection system or the same type of flexible protection system, and grouting is adopted to close the cracks and anchor spraying support at the fourth level where there are more cracks. Two rows of ordinary anchors are arranged to reinforce the first, second and third level slopes where the geological conditions are poor. The treatment program for the fourth level slope is: local active net + local crack closure anchor spraying + grass spraying, and the treatment program for the fifth level slope is: local active net + grass spraying.

4.2 Slope greening process

4.2.1 Dangerous stone cleaning

(1) When clearing the protruding blocks of rocks on the slope and refurbishing the slope, it should be carried out sequentially from the top to the bottom, and the loose soil and rocks on the slope surface must be removed in time. It is strictly prohibited to work, rest and store machinery under the dangerous rocks. Clearing the stone working surface should be staggered with the loading and transportation working surface, and it is strictly prohibited to work up and down cross [9].

(2) If signs of landslide or collapse are found during construction that jeopardize the safety of construction, the construction shall be stopped immediately, personnel and machinery shall be withdrawn, and the matter shall be reported to the relevant departments.

(3) The main entrance and exit of the construction production area shall be equipped with obvious construction warning signs and safety and civilized production regulations and prohibitions. Personnel and equipment not related to construction shall not enter the construction area.

(4) In case of rainy days, foggy days, are stopped cleanup operations, cleanup operations throughout the process, the lower part of the foundation construction suspension.

(5) The whole process of cleaning work shall be carried out under the supervision of the safety officer.

(6) In the process of cleaning, specialized technicians should be configured to command and guard the excavation to avoid the rock and soil body rolling down to the roads and villages below.

4.2.2 Construction of flexible protection network

(1) Construction process

The installation process of flexible protection net system is as follows[10]: clearing the slope → lining → review → drilling → cleaning → installation of steel rope anchors (local hollow anchors) → grouting → installation of longitudinal and transversal support ropes → acceptance → stitching → hanging the net → tensioning[11].

1) Remove floating soil and pumice that threaten the construction safety in the slope protection area, and make appropriate modifications to the local terrain (local piles and raised bodies, etc.) that are unfavorable to the construction and installation and affect the normal functioning of the system after installation. As shown in Figure 1.

Fig. 1. Clearing of floating soil and pumice

2) Place lines and measurements to determine the position of steel rope anchor holes (according to the terrain conditions, the hole spacing allows 0.3m of adjustment). Within the adjustment amount allowed for the hole spacing, the steel rope anchor hole position is selected at the low concave; the general continuous overhanging area shall not be greater than $5m^2$, otherwise it is appropriate to add a local anchor with a length of not less than 0.5m. Should be at each hole position to cut a depth of not less than the length of the exposed ring sleeve of steel rope anchors pit, general caliber 20cm, 20cm deep.

3) Drill and clear the holes for steel rope anchors according to the designed position and depth, the depth of the holes should be greater than the designed length of 5cm~10cm, and the diameter of the holes should be φ55. The steel rope anchors should be perpendicular to the slope surface as far as possible, and the angle with the horizontal surface should not be less than 15°. When the local holes on the slope surface can not be formed due to loose or broken strata, C15 concrete foundation with section size not less than 0.4×0.4m can be used to replace the rock and soil section that can not be formed.

4) Grouting and inserting steel rope anchors with cement slurry of grade not less than M25, pure cement slurry with gray-to-sand ratio of 1:1 to 1:2 and water-to-cement ratio of 0.45, and cement of 32.5 ordinary silicate cement. The construction should ensure that the slurry is full, and the maintenance of the grouted body before carrying out the next process should not be less than three days.

5) Installation of longitudinal and transverse support ropes, tensioned tightly at each end with 2 to 4 (2 when the length of the support rope is less than 15m, 4 when it is greater than 30m, and 3 in between) rope card and the exposed ring sleeve of the steel rope anchor fixed connection.

6) from the top down to hang the grille net, grille net overlap width ≥ 5cm, between two grille net as well as necessary grille net and support ropes with φ1.5 wire between the knot, when the slope is less than 45 °, the knot spacing is generally <2m; when the slope is greater than 45 °, the knot spacing is generally not greater than 1m (when possible, this process can be completed in the previous process is to be placed in the grille net under the support rope).).

7) Lay the steel rope mesh from top to bottom and sew it together, the sewing rope is φ8 steel rope, each piece of steel rope mesh is sewed together with the surrounding support rope with a sewing rope about 33m in length and pre-tensioned, and the two ends of the sewing rope are fixed with two rope cards and fixed connection with the mesh rope. The laying site diagram is shown in Figure 2[12].

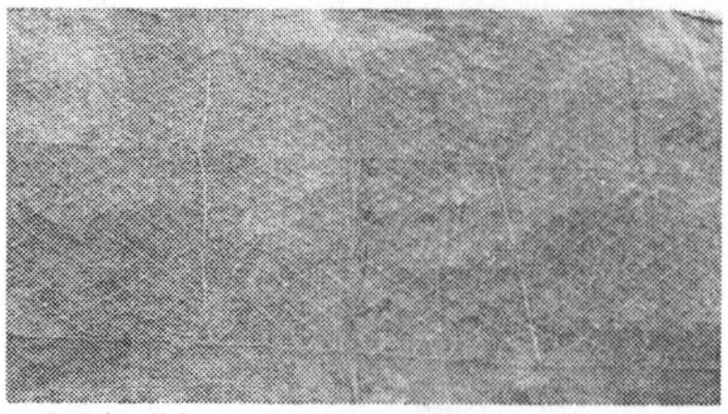

Fig. 2. Flexible protection net laying effect diagram

4.2.3 Quality inspection requirements

The quality of flexible protection net material should meet the design requirements, satisfy the requirements of impact resistance and durability, no aging, no breakage in appearance and no pollution. Detailed parameters were shown in Table 1 [12].

Table 1. Quality inspection requirements of flexible protection network.

serial umber	Inspection items	Specified value or allowable deviation
1	Flatness and degree of supply of the lower bearing surface	≤10cm
2	Overlap width	≥5cm
3	Tie spacing	≤5cm
4	Anchor spacing	450cm, except side
5	Anchor insertion length	±5cm
6	Cement mortar strength	Conform to the design requirements

The material should be close to the lower bearing surface, laid, tensioned and fixed according to the design requirements. Material joints lap, length should meet the design requirements, the upper and lower material lap joints should be alternately staggered.

4.3 Ordinary steel anchor

(1) Anchor material requirements: anchor rod adopts full-length bonded anchor rod, HRB400 steel bar, diameter of 22mm, anchor pull-out force of not less than 100kN, grouting body adopts M25 cement slurry. The quality standard of anchor material, slurry material and anchor parts should meet the current national standards.

(2) Basic test of anchor: basic test shall be carried out before the construction of anchor, and the requirements of basic test shall be implemented in accordance with the requirements of basic test in section 12.1 of Technical Specification for Geotechnical Anchor and Shotcrete Support Engineering (GB50086-2015).

(3) Anchor construction should be accurately positioned, hole position deviation ≤ 3cm, cut excess steel, head steel exposed 200mm, can be mechanically removed. Each hole of anchor should be numbered during construction to facilitate construction and quality inspection.

(4) Anchor drilling requirements: when drilling, anchor drilling requires dry drilling, water drilling shall not be used to ensure that the anchor construction does not deteriorate the engineering geological conditions of the slope rock and ensure the bonding performance of the hole wall, broken rock can be used to protect the wall of the casing. After drilling and before the construction of anchor under material, high pressure wind should be carried out to clear the hole, and after clearing the hole, the anchor should be constructed as early as possible to reduce the construction impact of collapsed hole. Drilling depth is generally required to overdrill 200-300mm, along the length of the requirement to set the centering bracket, spacing 2-3m, to ensure that the rebar is centered in the hole.

(5) Anchor grouting requirements: anchor requires the

use of M25 pure cement paste for grouting. Pure cement slurry using water-cement ratio of 0.38 ~ 0.42 (water: cement = 0.38 ~ 0.42: 1) of pure cement slurry, and mixed with cement dosage of 0.6% JN water reducer, the optimal proportion of its slurry formulations need to be carried out tensile and pull-out test, and no shrinkage, should be made according to the results of the experiments to make appropriate adjustments. The slurry should be stirred evenly and used as it is stirred, and the slurry should be used up before the initial setting. The grouting pressure should be low pressure 0.1～0.6MPa, if the rock quality is broken and lead to the leakage of grout, the grout can be injected in several times, until the hole returns to the grout. When placing the anchor, it should be well positioned, and put the grouting pipe directly into the hole, the distance from the end of the grouting pipe to the bottom of the hole is 100~200mm, and the slurry should be continuously grouted from bottom to top, without interruption.

(6) Anchor welding and tying: This anchor adopts full-length bonded anchor, and after the design strength of the pad pier reaches more than 80%, the exposed 200mm steel bar is bent and welded and tied.

(7) anchor acceptance test: anchor project should be 5% of the number of anchor acceptance test. Acceptance test should be strictly in accordance with the national standard “geotechnical anchor and shotcrete support engineering technical specifications” (GB50086-2015) 12.1 section on the acceptance test requirements.

(8) Anchor head sealing and anchoring: the anchor head should be replenished and sealed in time after the completion of anchor construction, and the outer anchor head should be sealed with concrete of the same grade as that of the anchor beam, so as to prevent corrosion damage.

4.4 Spraying grass greening

Spraying grass planting on the slope of the average thickness of spraying substrate layer is not less than 10cm. substrate layer can be done as thick as possible, so that the plant growth is good, more responsive to the slope water shortage and high temperature drought, to facilitate the maintenance of late management of the cost savings[13]. The construction process is as follows:

(1) Wet the slope;

(2) Spraying substrate mixture;

(3) will meet the requirements of the base material mixture through the air power with jetting machinery mixed spray attached to the slope surface, the thickness is greater than 10cm. guest soil and seed nutrient solution blowing attached spraying, must be left and right back and forth, subdivided, layered step by step, and to maintain a certain range and angle, hardness shall not spray too loose or too hard[14];

Fig. 3. Effect of non-woven fabric covering

(4) In the completion of the substrate surface layer spraying, covered with non-woven fabric (18g/m^2), in order to create an environment for rapid seed germination and to prevent rain washout. The covering effect is shown in Figure 3;

(5) maintenance: according to the weather conditions and seedlings watering, not too wet, too dry, after spraying if it does not rain every day watering to keep the soil moist, water spraying equipment should be used spray nozzle mobile spraying, eliminate the use of high-pressure nozzle direct sprinkler irrigation, and a certain amount of drought, so that the plant root system to the depths of the growth of the plant, conducive to over-summering. During the maintenance period should always observe the lawn's water and fertilizer, moisture mainly depends on the degree of wetness of the root system soil. Pest control must be set up plant protection patrol, adhere to the policy of prevention[15].

The effect after 2 months of growth is shown in Figure 4.

Fig. 4. Greening effect

Maintenance requirements: should ensure that the plant survival rate is greater than 95%. Thereafter, basically do not need artificial maintenance, it could grow naturally.

5. Precautions

(1) The slope construction should fully consider the impact of local seasonal climate on the construction process, and should try to choose the dry season construction to reduce the impact of rainy weather on the slope construction safety.

(2) In the construction process, should do a good job of interception, drainage measures, clean up the platform drainage, prevent the platform drainage ditch surface water flow to the first grade slope; at the same time, the first grade slope groundwater seepage location, if the amount of water is large, should be added to introduce drainage holes into the drainage system.

(3) During the construction process, if unstable wedges or loose blocks are found on the slope, they can be removed or reinforced with additional anchors.

(4) In the process of construction, if the actual geological situation is not consistent with the original design, the design should be promptly notified to adjust the program.

6. Conclusion

The slope of a basalt mine in Shengzhou City, Zhejiang Province, China, was greened and treated by adopting various greening techniques, such as local anchors and local active nets. The results showed that the greening technology was safe and reliable, it coould maintain the survival state of the slope green plants for a long time, which effectively improved the stability of the slope and achieved the effect of mine slope management.

References

1. Chinese Society of Rock Mechanics and Engineering. (2021) Design Code for Rocky Slope Engineering in Open Pit Mines[S]. T/CSRME009-2021.
2. Luo Y.M., Lei C.D., Zhou D.P., Yang T. (2006) Discussion on Vegetation Methods and Stability of Slopes Reinforced by Sns Active Protection System. Chinese Journal of Rock Mechanics & Engineering, 25(02), 235-240.
3. Vjekoslav Budimir. (2019) Technology of greening slopes stabilized by a supporting net. Journal of Road and Traffic Engineering, 65(3). DOI: https://doi.org /10.31075/PIS.65.03.08
4. Chen X.D., Zhao T.Q., Nie X.J., Guo X.M., Li P.B. (2024)Soil Substrate Characteristics for Planting Hole Greening Technology for High, Steep, Rocky Slope Vegetation in Semi-Arid Areas. Land. 13(3). DOI: 10.3390/LAND13030287.
5. Wang L.J., Tang X.G., Liu X., Zhang J.C. (2023) Active permanent greening – a new slope greening technology based on mineral solubilizing microorganisms. Frontiers in Plant Science. 14, 1219139-1219139. DOI: 10.3389/FPLS. 2023. 1219139.
6. Hattori Hiroyuki, Konoshima Masashi, Surová Diana. (2022) Visual Preferences for Slope Greening and Stabilization Techniques: the Case Study of Northern Part of Okinawa Island, Japan. FORMATH. 21(0). DOI: 10.15684/FORMATH. 21.003.
7. Liu J.R., Zhao C.X., Li C.N., Lei L.G., Ta F. (2024) Mixed planting mode is the best measure to restore soil quality in alpine mines. Soil & Tillage Research. 244, 106209-106209. DOI: 10.1016/J. STILL. 2024.106209.
8. Metallurgical Construction Association of China (MCCA). (2014) Technical Specification For The Design And Construction Of Landslide Control Works[S]. GB51016-2014.
9. Ministry of Land and Resources of the People's Republic of China. (2006) Technical Specification For The Design And Construction Of Landslide Control Works[S]. DZ/T0219-2006.
10. State Council of China (PRC). (2003) Regulations on Prevention and Control of Geological Disasters[S]. State Council Decree No. 394.
11. Provincial People's Congress Announcement. (2010) Regulations on Prevention and Control of Geological Disasters in Zhejiang Province. China. No. 18.
12. National Technical Committee for Standardization of Traffic Engineering Facilities (Highway). (2020) Flexible slope protection network system[S]. JT/T1328-2020.
13. Technical Specification For Safety Monitoring Of High And Steep Slopes In Metal And Nonmetal Open Pit Mines[S]. AQ/T2063-2018.
14. Xie J.Y., Liu Y.X., Xie M.M., Xia L. Yang R.J., Li J.G. (2024) Exploring the restoration stability of abandoned open-pit mines by vegetation resilience indicator based on the LandTrendr algorithm. Ecological Indicators, 166, 112392-112392. DOI: 10.1016/J.ECOLIND.2024.112392.
15. Zhang C.Y., Li F.Y., Li J., Zhang K., Ran W.Y. (2023) Assessing the effect, attribution, and potential of vegetation restoration in open-pit coal mines' dumping sites during 2003–2020 utilizing remote sensing. Ecological Indicators, 155. DOI: 10.1016/J.ECOLIND.2023.111003.

https://doi.org/10.1051/e3sconf/202561703009

A Study on the Morphology and Efficiency of Urban Blocks Around the Radial Plaza: Based on a Comparison of Dalian and Paris

Xinning Guo[1,*], Dongfengt Yang [2]

[1] Dalian University of Technology, Dalian 116024, China
[2] Professor, PhD Supervisor, Dalian University of Technology, Dalian 116024, China

Abstract: This study explores how ring-radial layouts in Dalian, China, and Paris, France, shape block morphology and performance. Analyzing the Zhongshan Square and Arc de Triomphe areas, the research identifies block types and compares their efficiency. Significant differences between the cities' morphologies affect block functionality, emphasizing the role of dense, multifunctional, and connected urban forms in enhancing spatial quality and performance.recommendations.

1. Introduction

The ring-radial spatial pattern has a long history in urban planning, particularly during the Baroque period, where it was widely used to symbolize power. This structure became popular across Europe, especially in Paris, and influenced the planning of cities like Saint Petersburg, Washington D.C., and Dalian, as well as Den-en-chofu in Japan [1]. Research shows this urban form creates attractive city centers but can overburden the central area [2].

With urban development shifting from sprawl to compact forms, increasing focus has been placed on the morphology of urban centers. In A City is Not a Tree, Alexander advocated for the "semi-lattice" city model, suggesting that cities should not only contain a variety of functions but also the complex relationships between them [3]. From a macro perspective, Yang Junyan developed a model to evaluate land use efficiency in urban centers based on spatial form and function [4]. Scholars like Song Yacheng introduced the "access structure" method to quantify urban block morphology [5][6]. Wang Xiaohui and others suggested that good street accessibility, appropriate building intensity, and diverse functionality are key to sustainable spatial morphology [7]. Consequently, research supports adopting high-density, multifunctional, well-connected, and high-quality urban forms to enhance efficiency while maintaining spatial quality.

This study examines the Zhongshan Square area in Dalian and the Arc de Triomphe area in Paris to quantify the burden capacity of urban blocks shaped by radial street networks. It evaluates block connectivity, compactness, and building efficiency, exploring how different block types perform in subsequent development stages to improve overall spatial quality and functional efficiency in central urban districts.

2. Study Area Overview

This study focuses on the areas surrounding the radial squares in Paris and Dalian. Due to the fact that, during the planning of Dalian in 1900, the Russian Empire made reference to Paris's Haussmann renovation, both the downtowns of Dalian and Paris are typical examples of Baroque urban composition and share similar structures. The study areas selected include 3.0 square kilometers surrounding Zhongshan Square in Dalian and 3.0 square kilometers around the Arc de Triomphe in Paris (Figure 1).

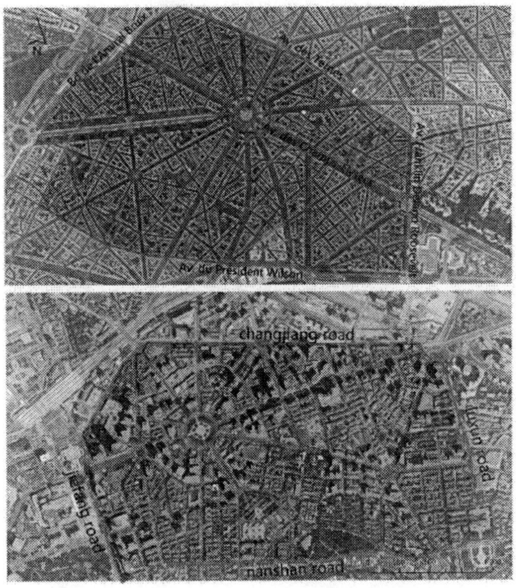

Fig. 1. Scope of the Study Area (Source: Drawn by the author)

* Corresponding author: yangdongfeng@dlut.edu.cn

2.1 Formation and Development of the Arc de Triomphe and Surrounding Area

The study area includes 3 km² around the Arc de Triomphe in Paris. The radial road system surrounding the Arc de Triomphe was designed by André Le Nôtre in 1664, with the aim of extending the symbolic power of Louis XIV's palace. At this stage, the radial road network was still in its early stages, with the cross-shaped road layout being the dominant feature (Figure 2). Before the Haussmann renovation, the area around the Arc de Triomphe was on the urban fringe, with only a high-density medieval residential area to the northeast. Southeast of the Arc, the northern land of Passy village, which included farmland, vineyards, and quarries, was divided into residential plots by the Passy Plain Land Company. This area was planned similarly to the European district, with a central circular square located at the intersection of diagonal roads [9].

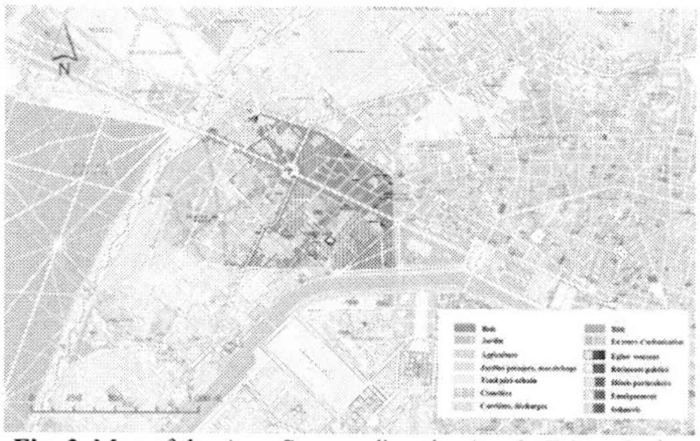

Fig. 2. Map of the Area Surrounding the Arc de Triomphe in Paris, 1850 (Source: [9])

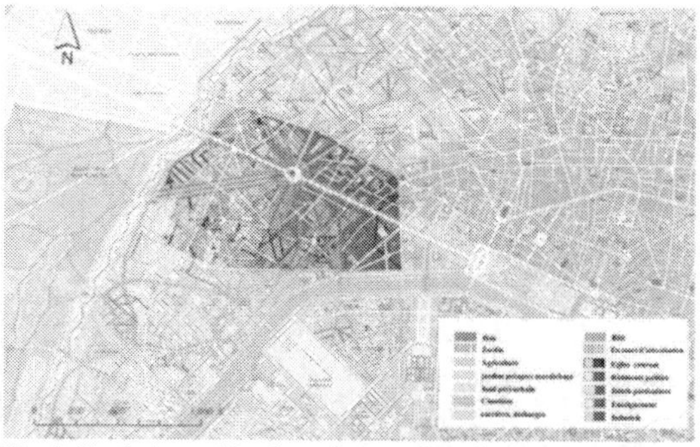

Fig. 3. Evolution of the Area Surrounding the Arc de Triomphe in Paris from 1790 to 1850 (Source: [9])

During the Haussmann renovation of 1860, Paris's urban structure underwent a comprehensive transformation. To address the public health issues at the time, Haussmann initiated a large-scale urban renewal project, including the widening of roads and the development of public facilities [8]. At this time, some of the areas around the Arc de Triomphe had already been developed into independent residences and gardens. Haussmann added new boulevards and public amenities to these areas, which accelerated the urbanization process [9]. Over the period between 1840 and 1910, as the area surrounding the Arc de Triomphe was gradually developed, building density increased, resulting in blocks of varied shapes and sizes. These blocks, which came to be known as "Haussmann blocks," consisted of both new structures and remnants of old ones (Figure 3). These historical and urban contexts make this area an ideal subject for studying the application and impact of the ring-radial layout in a modern city.

2.2 Formation and Development of Zhongshan Square and Surrounding Area in Dalian

In contrast to Paris, the study area in Dalian encompasses 3 km² surrounding Zhongshan Square. Zhongshan Square was first established in 1898, under the direction of Tsar Nicholas II. Engineer Vladimir Sakharov, who had experience in port construction, was responsible for the planning [10]. Dalian's urban layout adopted a Baroque-style structure, utilizing the topography to establish squares at higher elevations, which were connected by a radial street network. The 1900 plan shares many similarities with Paris's Haussmann renovation, such as a main road linking various squares and forming an east-west axis. From the central square, radial roads extended outward in all directions, while similar tree-lined boulevards connected the central square to other areas. However, unlike Paris, the area around Zhongshan Square was zoned primarily for functional buildings, such as offices and banks, leading to lower building density in the surrounding blocks [10].

Due to the outbreak of the Russo-Japanese War, Japan took control of Dalian and inherited the Russian urban plan. By 1940, Zhongshan Square had developed into a typical example of a ring-radial urban layout. The area around the square exhibits a diverse range of architectural styles, including a mix of historic Western-style buildings and modern high-rise structures, which reflect the characteristics of urban development during different historical periods.

3. Morphological Types of Blocks Surrounding Radial Squares

3.1 Methods and Data

Captions should be typed in 9-point Times. They should be centred above the tables and flush left beneath the figures. This study employs factor analysis and principal component analysis (PCA) to study the morphology of the blocks surrounding radial squares, using urban blocks as the smallest unit of analysis. Blocks, as mid-level factors within the hierarchy of urban morphology, effectively reflect the basic layout patterns of urban spaces. However, due to the large number of elements involved in the measurement of urban morphology, it is difficult to ensure accuracy and comprehensiveness. In recent years, some scholars have attempted to quantify the relationships between these elements. For example, Berghauser Pont and Haupt used the Spacematrix method to quantify the relationship between building density and urban texture [11]. Therefore, this study applies factor analysis in combination with principal component analysis to fully account for the various elements of block morphology, clustering the blocks to enable more accurate assessments of their efficiency. Factor analysis is an effective multivariate statistical method that consolidates various elements and extracts key factors, revealing the main characteristics of different blocks, which are then classified based on these characteristics [12].

Table 1. Factor Score Table for Paris

Indicator		Block Level					Building Level								
		Capacity Indicator		Morphology Indicator			Capacity Indicator			Morphology Indicator					
		Block Area	Block Perimeter	Shape Index	Fractal Dimension	Compactness	Floor Area Ratio	Building Density	Number of Buildings	Maximum Ground Coverage	Maximum Floors	Ground Coverage Standard Deviation	Average Ground Coverage	Average Number of Floors	Enclosure Degree
Factor Score	1	-0.43	-0.38	0.41	0.37	-0.26	-0.88	-0.88	-0.50	-0.62	-0.92	-0.33	-0.31	-0.94	-0.85
	2	0.83	0.85	0.58	0.24	-0.61	0.02	0.07	0.70	0.22	0.19	0.07	-0.34	-0.17	0.06
	3	0.27	0.17	-0.14	-0.12	0.17	-0.31	-0.27	0.05	0.31	-0.14	-0.02	0.81	-0.11	-0.29
	4	-0.22	0.03	0.69	0.48	-0.59	0.07	0.02	-0.21	-0.05	-0.03	0.02	0.32	0.21	0.04
	5	0.05	0.00	-0.02	-0.02	0.03	-0.27	-0.38	-0.09	-0.11	-0.08	-0.14	-0.13	0.20	-0.38
	6	0.00	0.00	0.00	0.01	0.03	0.06	-0.01	-0.26	0.30	-0.03	0.17	0.06	0.00	-0.02

Table 2. Factor Score Table for Dalian

Indicator		Block Level					Building Level								
		Capacity Indicator		Morphology Indicator			Capacity Indicator			Morphology Indicator					
		Block Area	Block Perimeter	Shape Index	Fractal Dimension	Compactness	Floor Area Ratio	Building Density	Number of Buildings	Maximum Ground Coverage	Maximum Floors	Ground Coverage Standard Deviation	Average Ground Coverage	Average Number of Floors	Enclosure Degree
Factor Score	1	0.94	0.14	0.03	0.05	0.06	0.20	0.39	-0.16	0.62	0.09	0.33	0.32	0.18	0.19
	2	0.074	0.22	-0.24	-0.14	0.22	0.52	0.85	0.35	0.42	0.44	0.45	0.05	0.32	0.76
	3	0.27	0.17	0.15	0.07	-0.19	-0.17	-0.30	-0.23	0.18	-0.10	0.24	0.27	-0.12	-0.44
	4	0.053	0.05	0.01	-0.01	-0.01	0.69	-0.06	-0.09	-0.02	0.79	0.64	-0.02	0.72	-0.16
	5	0.76	0.76	0.18	-0.01	-0.19	-0.27	-0.02	0.66	0.27	0.01	0.10	-0.00	-0.27	-0.17

3.2 Classification Results

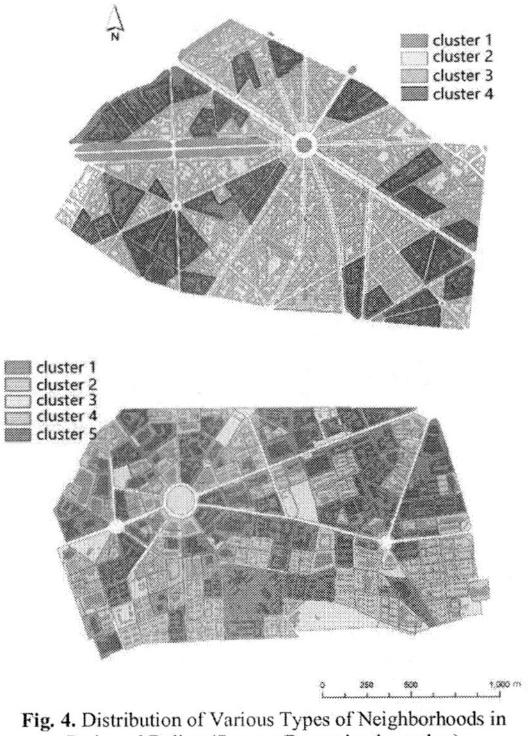

Fig. 4. Distribution of Various Types of Neighborhoods in Paris and Dalian (Source: Drawn by the author)

Factor analysis was applied to the morphological indicators of the blocks, followed by clustering to identify the dominant elements in each dimension for each cluster (Tables 1 and 2). KMO and Bartlett's tests were performed, yielding KMO values of 0.65 for Paris and 0.74 for Dalian, both exceeding the threshold of 0.5, and significant results (sig < 0.05), confirming the suitability of the data for factor analysis. Using the Kaiser criterion and scree plots, six principal components for Paris (explaining 83.5% variance) and five for Dalian (explaining 84.64% variance) were selected. After constructing the factor matrix, Varimax orthogonal rotation clarified the contribution of each variable (Tables 1 and 2). For instance, in Paris, Factor 1 mainly reflects building capacity with FAR (-0.878) and building density (-0.881) as key contributors, while Factor 2 represents block morphology with block area (0.829) and perimeter (0.848) as dominant indicators.

Based on clustering results, Paris and Dalian were classified into four and five categories, respectively, each representing a different type of block (Figure 4). The clusters were visualized in GIS (Figure 7). The classification differences in block types between Paris and Dalian are evident. In Paris, blocks serve as tools of urban construction, with unique block types contributing to the city's high-density characteristics. Furthermore, Paris has a strict planning and control system that includes the "Paris Local Urban Planning" for regulatory urban planning (equivalent to China's control plan for block depth), the "Historic Preservation and Utilization Plan" for historic preservation, and the "Agreement Development Zone Plan" for construction planning.

These three plans complement each other, achieving comprehensive spatial coverage. While each plan varies in target and scope, they all emphasize preserving the cityscape, which is reflected in their content. This leads to a stable structure of buildings and open spaces within blocks, with a unified architectural style. The blocks are often surrounded by continuous facades that are independent of the internal parcel structure, yet vary in geometry—from triangles and squares to more complex polygons. Multiple blocks can even combine into grand parcels. In Paris, block types are mainly determined by area and shape.

Type 1 features low floor area ratio and building density, characterizing "park-type" blocks, with the Foch Avenue as a representative. Type 2, with higher density, compactness, and shape index, is the "compact-type" Haussmann block, which is also the most common and is typical of Haussmann-style areas. Type 3 has lower floor area ratio but a higher average building footprint, representing "landmark-type" blocks. Type 4, with larger block area and perimeter, and lower density than Type 2, is the "loose-type" Haussmann block, mostly found in the southwestern Passy area, a district developed as a unified zone. In contrast, the "compact-type" blocks were primarily formed through redevelopment between 1790 and 1850. Compared to Paris, Dalian's block types are more varied, with relatively lower building density, reflecting the needs of different historical periods and urban functions. Block types in Dalian are determined primarily by building morphology, followed by block shape. Type 1 has fewer buildings but higher structures, characterizing "stand-alone high-rise" blocks, mainly used for small-scale commercial and office purposes. Type 2 has high floor area ratio and building density with few buildings, representing "complex-type" blocks, mainly commercial complexes. Type 3 has the lowest density and floor area ratio, classified as "park-type" blocks. Type 4 has high building density and enclosure with lower buildings, making it the most common type, classified as "residential-type" blocks. Type 5 is large in area, with density similar to Type 4 but with taller buildings, known as "mixed-type" blocks. "Residential-type" blocks are the most widely distributed, characterized by row-style building layouts, mainly constructed in the 1970s and 1980s, with features typical of that era. "Mixed-type" blocks retain early planning features, such as enclosed buildings and low-rise European-style architecture. The "complex-type" and "stand-alone high-rise" blocks are the most recent, located mainly along transportation lines and major squares. Overall, Paris has maintained a strong historical continuity in its planning structure, with minimal structural and architectural changes following intensive construction and renovation. In contrast, Dalian has developed a mix of diverse styles over nearly a century, resulting in a lack of unified cityscape across block types.

4. Efficiency Comparison of Block Types

4.1 Selection of Efficiency Indicators

Efficiency in urban planning and architectural design refers to maximizing functional use, comfort, and sustainability under limited resource and spatial conditions through effective planning, design, and management. Existing studies address various aspects of urban spatial efficiency, such as land use intensity, spatial density in urban design, and building energy efficiency. These studies cover not only physical space utilization, like building density and floor area ratio, but also non-physical space usage, including public space accessibility, service facility distribution, and building sustainability analysis (Table 3).

Table 3. Calculation Methods for Each Indicator

Element Level	Indicator Type	Calculation Method
Spatial Utilization Efficiency	Floor Area Ratio	$FAR = \frac{A_{total}}{A_{land}}$
	Building Density	$D_{build} = \frac{A_{footprint}}{A_{land}} \times 100\%$
	Functional Composite Degree	$H' = -\sum_{i=1}^{n} p_i \ln(p_i)$
Traffic Efficiency	Walkability within a 400-meter Radius	$Accessibility = \frac{1}{m}\sum_{j=1}^{m}\max(\sum_{i=1}^{n}\frac{1}{Depth_{ij}})$
Energy Efficiency	Average Sky Exposure Ratio	$SVF = \frac{1}{n}\sum_{i=1}^{n}\sin(\gamma_i)$
	Spatial Volume Compactness	$C_{volume} = \frac{V_{building}}{V_{bounding}}$

This study focuses on the central radial plaza blocks in Paris and Dalian, where indicator selection must align with the radial layout form. Compared to other layouts, the radial city center requires a higher level of efficiency to optimize resource use and alleviate the central area's heavy load. Starting with block morphology, we analyze three levels—macro, meso, and micro: Spatial Utilization Efficiency: Indicators such as floor area ratio, density, and functional diversity reflect the spatial resource use across blocks, which helps assess the spatial carrying capacity of different blocks within the radial layout. Transit Efficiency: The 400-meter radius pedestrian accessibility serves as an indicator to evaluate whether the block layout facilitates pedestrian activity. The radial layout typically tightly connects the city center with surrounding areas, making this indicator crucial for assessing convenience in movement. Environmental and Energy Efficiency: Indicators like average sky exposure ratio and spatial compactness reveal environmental quality and sustainable energy use. The compactness of buildings in a radial layout can impact sunlight, ventilation, and energy demands, making these indicators essential for a comprehensive evaluation of block morphology's environmental benefits.

By examining these three types of indicators, we assess the correlations between blocks, compactness of layouts, and the comfort and efficiency of buildings within blocks to evaluate the overall efficiency of city center layouts influenced by radial morphology.

4.2 Characteristic Evaluation

Blocks, as mid-level units in urban morphology, play a crucial role in shaping spatial quality. By quantifying key efficiency indicators at the block level, this study offers a detailed comparison of the spatial efficiency of blocks surrounding radial squares in Paris and Dalian, evaluating their performance across various metrics.

(1) FAR: FAR (Floor Area Ratio) reflects the intensity of vertical space usage on a plot of land. It is calculated by dividing the total building area by the plot area. As seen in Figure 5, Dalian exhibits generally higher and more variable FAR values, with the highest concentrated around radial squares and major roads. Paris, by contrast, shows a more uniform FAR distribution in the city center, indicating consistency in planning and development standards (Figure 5).

(2) Building Density: Building density measures the ratio of the building footprint area to the total land area, reflecting horizontal space utilization and the distribution of open spaces. It is calculated by dividing the total building footprint by the plot area. Figure 6 shows that Dalian has lower building density overall compared to Paris, with greater fluctuations. Higher density is found around the central square and certain peripheral zones. Paris shows minimal variation in density, with higher values concentrated in its eastern region (Figure 6).

(3) Spatial Volume Compactness: Spatial volume compactness is the ratio of a building's actual volume to the volume of its bounding box, illustrating howefficiently space is used. It is calculated by dividing the building volume by the minimum enclosing bounding box volume. Figure 7 indicates that while some historical areas near Dalian's radial square and certain mixed-use blocks demonstrate high compactness, overall compactness is low due to dispersed layouts. Paris, with its cohesive urban design, maintains higher compactness, efficiently arranging buildings within available space (Figure 7).

(4) Functional Diversity: Functional diversity is assessed using the Shannon Diversity Index, which measures the variety of different functions within a given system. A higher index suggests greater functional complexity. The data shows that Dalian's radial square area has a higher Shannon Diversity Index compared to Paris, indicating greater functional diversity in its urban layout (Figure 8).

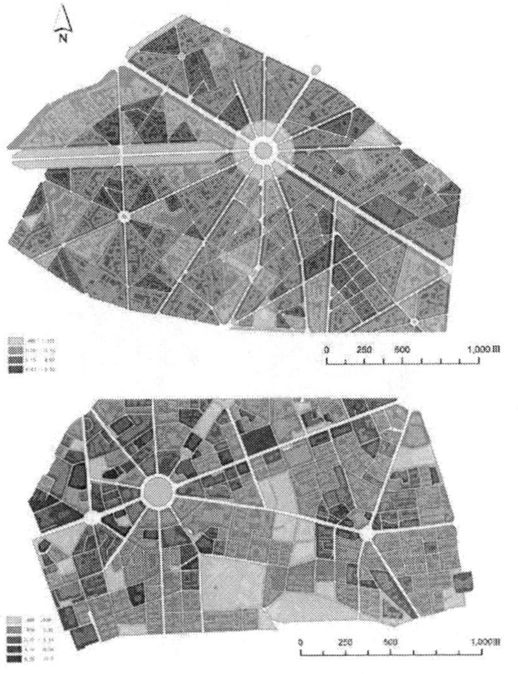

Fig. 5. Paris-Dalian FAR comparison

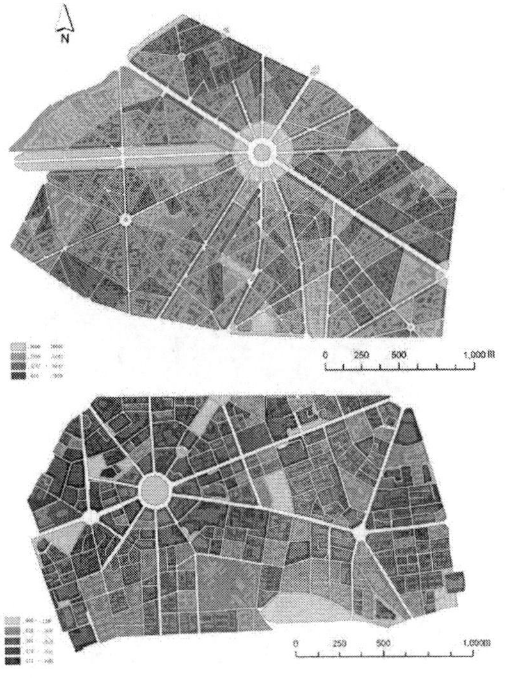

Fig. 6. Paris-Dalian Building Density comparison

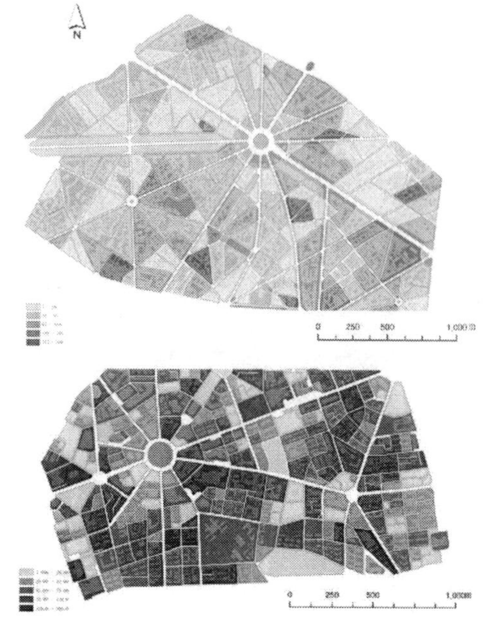

Fig. 7. Paris-Dalian Spatial Volume Compactness comparison

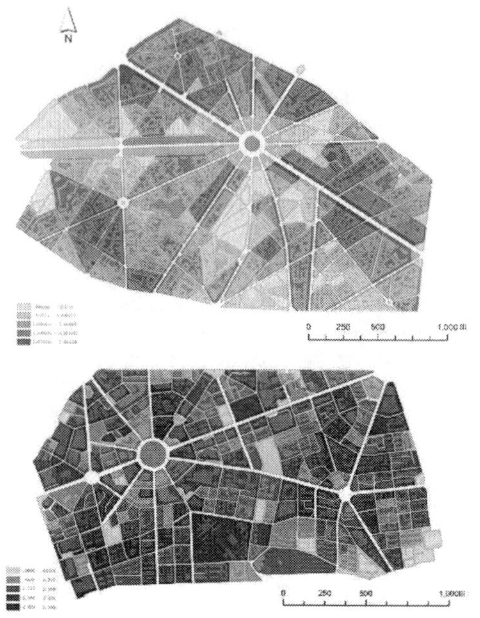

Fig.8. Paris-Dalian Functional Diversity comparison

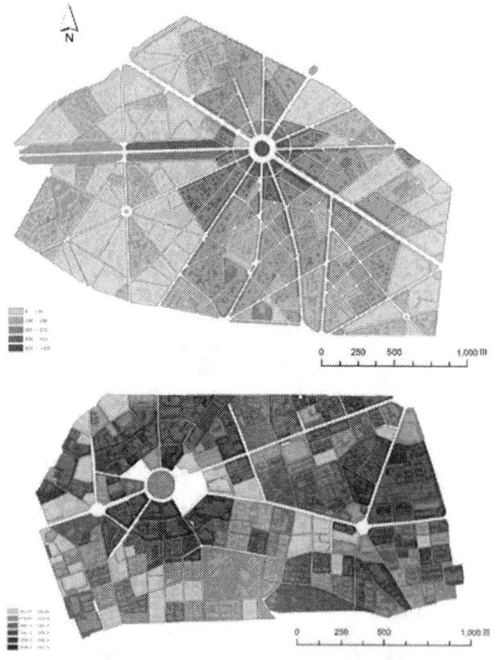

Fig. 9. Paris-Dalian Pedestrian Accessibility within a 400-Meter Radius comparison

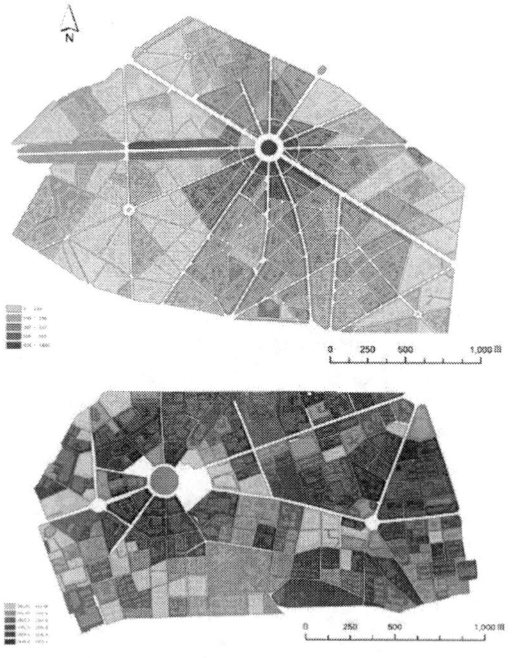

Fig. 10. Paris-Dalian SVF comparison

(5) Pedestrian Accessibility within a 400-Meter Radius: This indicator evaluates walkability by analyzing the street network's complexity and connectivity within a 400-meter radius of each block. A 400-meter radius is considered the maximum distance most people are willing to walk. Data shows that pedestrian accessibility is highest near Paris's radial square, where a dense street network offers convenience for pedestrians. In Dalian, accessibility is also strong near the radial square, but some peripheral areas perform well, suggesting a more complex road network design (Figure 9).

(6) Sky View Factor (SVF): SVF measures the openness of an urban area, reflecting the extent to which buildings obstruct views of the sky. A higher SVF indicates more open environments with better natural light and ventilation. The data indicates that older areas of

Paris have lower SVF values, suggesting that buildings block the sky more, reducing ventilation and lighting. Conversely, some lower-density areas in Dalian show higher SVF values, offering a more open environment with improved ventilation and lighting (Figure 10).

In conclusion, while Paris excels in spatial volume compactness and FAR consistency, Dalian stands out for its functional diversity and pedestrian accessibility in peripheral areas, showcasing distinct approaches to urban block development.

Table 4. Evaluation for Different Types of Neighborhoods

Spatial Utilization Efficiency		
Floor Area Ratio	Building Density	Functional Composite Degree
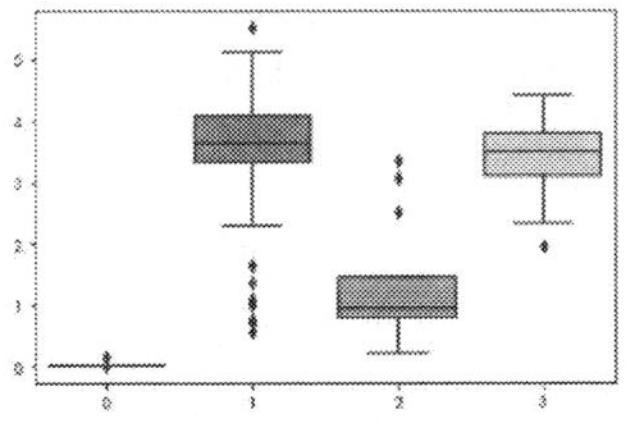	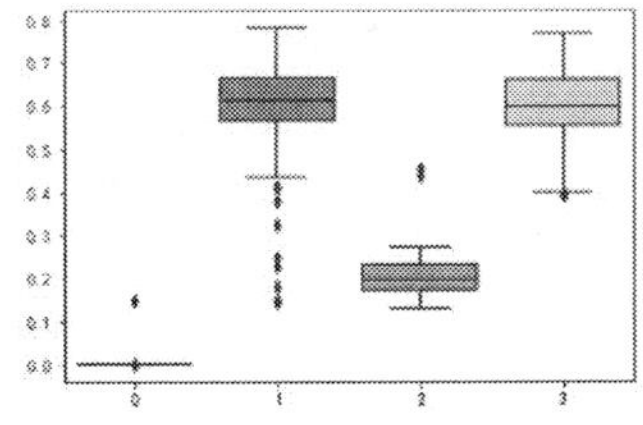	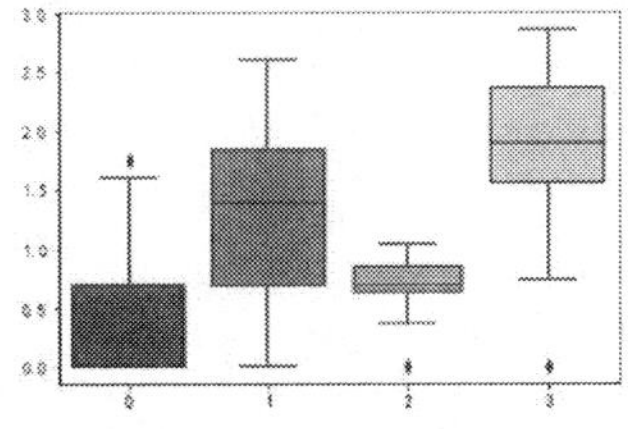
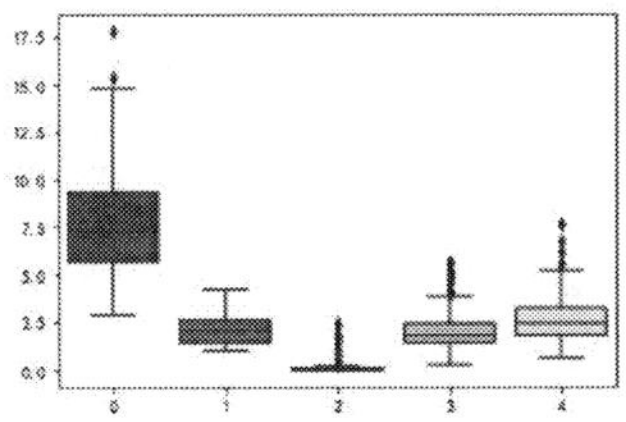	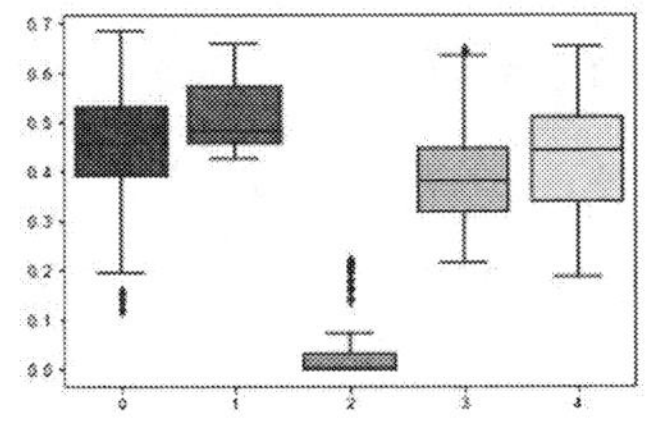	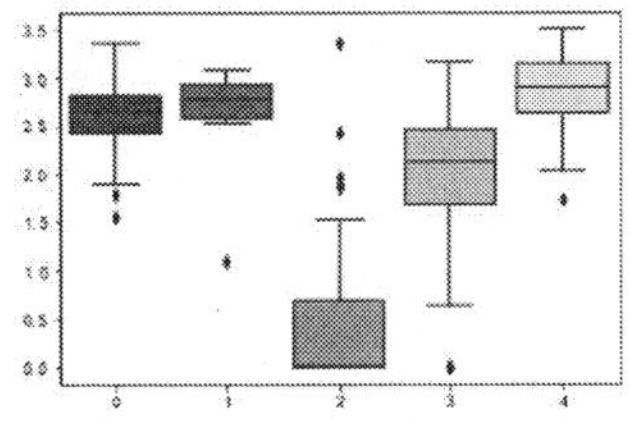
Traffic Efficiency	Energy Efficiency	
Walkability within a 400-meter Radius	Average Sky Exposure Ratio	Spatial Volume Compactness
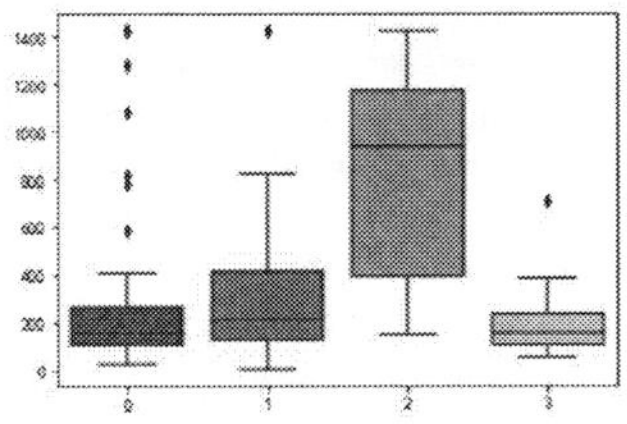	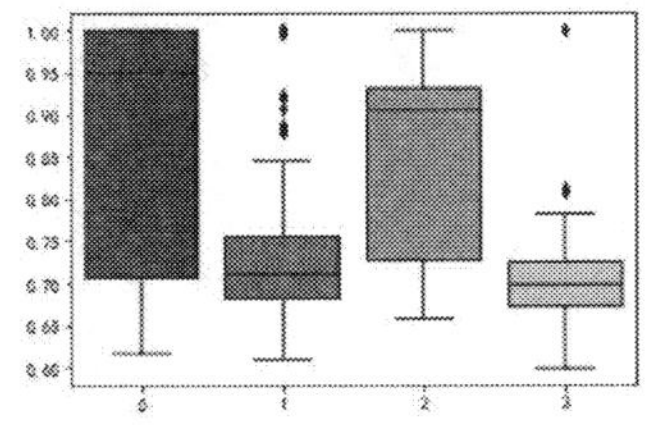	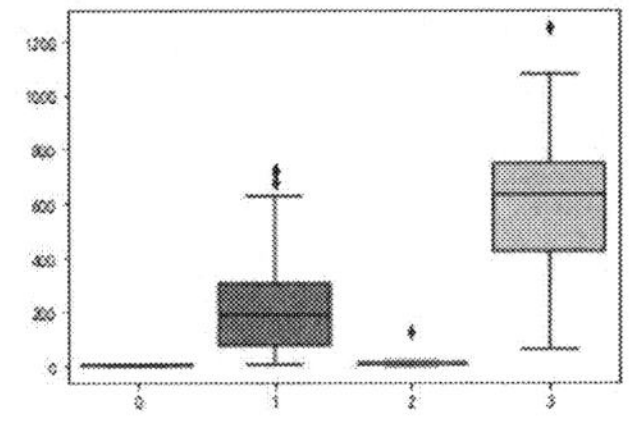
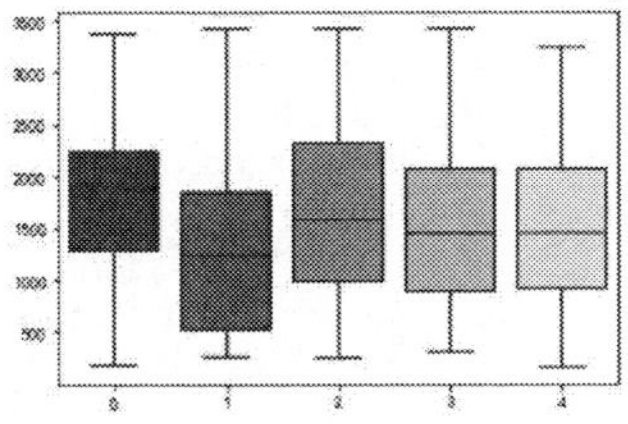	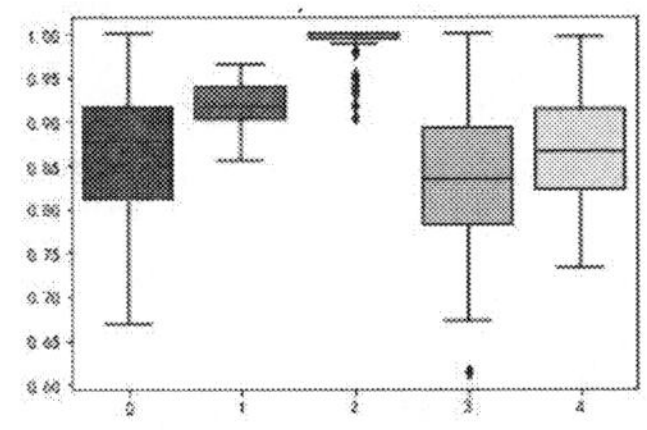	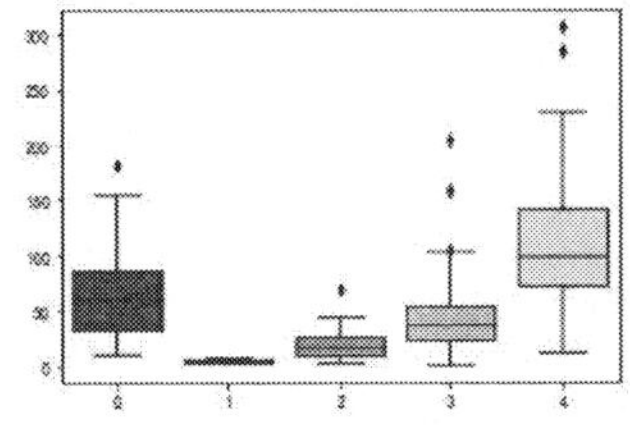

4.3 Type-Characteristic Analysis

After examining the block types in Paris and Dalian, this study further investigated the relationships between these types and key efficiency indicators. Through an analysis of FAR, building density, spatial compactness, functional diversity, pedestrian accessibility within a 400-meter radius, and SVF, the study identified notable differences in the performance of various block types, emphasizing the impact of block morphology on urban spatial efficiency (Table 4).

In terms of spatial utilization, significant differences emerge between Paris and Dalian. In Paris, "compact Haussmann blocks" (Type 2) demonstrate the highest FAR and building density, maximizing space efficiency. In Dalian, the differences in FAR and density across block types are less pronounced. For example, "independent high-rise" blocks (Type 1) feature fewer buildings but achieve relatively high FAR due to building height, compensating for lower density. Regarding functional diversity, Parisian blocks exhibit clear distinctions, showing a strong link between block morphology and functionality. "Loose Haussmann blocks" (Type 4) perform well in supporting multiple functions, enhancing space utilization. Similarly, Dalian's "mixed-use" blocks (Type 5) also exhibit high functional diversity, reflecting their adaptability within the city's urban layout. Pedestrian accessibility also varies significantly between Paris and Dalian. In Paris, "landmark-type" Haussmann blocks (Type 3) excel in accessibility due to their central location within the radial road network. This highlights the strong relationship between block layout and road network design. In contrast, Dalian's block types show a weaker correlation with accessibility, influenced by historical development patterns and functional zoning, which have reduced the relationship between block types and road network accessibility. When it comes to energy efficiency, Paris blocks, with higher spatial compactness, show superior thermal performance. "Compact Haussmann blocks" (Type 2) and "loose Haussmann blocks" (Type 4), despite lower SVF, benefit from compact building structures that reduce heat loss and improve thermal insulation. Conversely, Dalian's blocks, characterized by more dispersed layouts, generally have lower spatial compactness. While Dalian's blocks perform better in terms of SVF, particularly the "independent high-rise" and "park-type" blocks with excellent natural lighting, their looser configurations lead to higher heat loss, reducing overall energy efficiency.

In summary, Parisian blocks outperform in terms of FAR, building density, and spatial compactness, reflecting their highly efficient urban design. Dalian's blocks, however, excel in functional diversity, pedestrian accessibility, and SVF, supporting a "functional efficiency" model. These differences highlight the distinct priorities of the two cities in their urban development strategies.

5. Conclusion and Discussion

5.1 Conclusion

This study focuses on Paris and Dalian, two urban areas with classic ring-radial square layouts, examining the relationship between block morphology and urban spatial efficiency. Using factor analysis and clustering methods, block types in both cities are categorized. The results show that Paris blocks, particularly Haussmann blocks, demonstrate higher FAR, building density, and spatial compactness, indicating efficient land use. In contrast, Dalian's blocks are more diverse in layout and provide better pedestrian accessibility, reflecting different historical influences and urban development needs. Regarding block morphology and efficiency, Paris blocks excel in spatial utilization, highlighting the lasting impact of Haussmann's planning. Meanwhile, Dalian's blocks outperform Paris in functional diversity and SVF, offering advantages in multifunctional use and better living conditions, such as improved natural light and ventilation.

5.2 Extended Discussion

(1) Application of Multi-Source Data and Construction of a Morphology-Efficiency Evaluation System: This study classifies blocks based on morphological types and establishes efficiency indicators to analyze and categorize the relationships between urban block types and morphology. With the advancement of big data and machine learning, the application of multi-source data in the future could provide more accurate, comprehensive, and detailed information on urban morphology. By integrating data from ecological, economic, and social dimensions, such as socioeconomic data, traffic flow data, environmental quality data, and resident behavior data, a multidimensional urban morphology efficiency evaluation model could be constructed. This model could comprehensively assess urban morphology across various dimensions, including spatial utilization efficiency, transit efficiency, energy efficiency, social function diversity, and environmental sustainability. Through such a systematic evaluation, it would be possible to identify high-efficiency blocks within cities and detect primary issues in low-efficiency blocks, offering specific directions for improvement in urban planning and renewal.

(2) Urban Planning Strategies for Morphology Control and Efficiency Enhancement: Efficiency-Based Morphological Design and Control: In urban planning, how to enhance urban efficiency through morphological control is a crucial consideration in current planning practices. Efficiency-based morphological design and control strategies could be developed from several perspectives: Compact Development and Mixed Use—studies show that compact development and mixed functions are key to improving urban spatial efficiency. By increasing building density and floor area ratio, spatial utilization efficiency can be effectively improved. Introducing multiple functions, such as commercial, residential, and office spaces within blocks, can enhance functional diversity, reducing urban functional segregation. This morphological control strategy not only improves land use efficiency but also promotes social interaction, increasing urban vibrancy and sustainability. Accessibility Optimization and Public Space Configuration enhancing block accessibility is essential to improving urban transit efficiency and residents' quality of life. Optimizing street networks and public transportation systems can improve block accessibility, shorten travel times, and increase residents' convenience. Additionally, well-planned public spaces, such as parks and squares, can boost environmental efficiency and promote social interaction, strengthening community cohesion. By introducing an efficiency evaluation system, planners could implement targeted design controls for different block types. For instance, in high-density blocks, compactness and energy-efficient design should be emphasized, while in low-density blocks, more attention could be given to public space configuration and functional diversity. Moreover, in urban renewal and expansion processes, planners could base refined morphological design strategies on evaluation results, ensuring that the efficiency of new areas meets expected targets.

References

1. F. De Lillo, F. Cecconi, G. Lacorata, A. Vulpiani, EPL, **84** (2008)1.Jiao, Y., Lu, H., Sun, Y., et al.

(2011). The exploration and enlightenment of improving the living conditions of migrant workers in typical cities[C]//Transformation and Reconstruction: The Proceedings of Annual National Planning Conference 2011. Nanjing: Southeast University Press.

2. Liu, J. & Deguchi, A. (2022). The Localization of "Imported" Urban Form in Northeast China: The Case Study on Circular-Radial Space in Dalian City. Landscape Architecture Frontiers, 10(6), 32-47.
3. Li, L. & Tang, P. (2015). Preliminary Thoughts on the "Ring + Radial" Road Network. Sichuan Architecture, 35(06), 4-8.
4. Alexander, C. (2017). A City is Not a Tree[M]. Sustasis Press/Off The Common Books.
5. Yang, J., Shi, B., & Yang, Y. (2013). Evaluation Model of Intensive Land Use in Urban Central Areas: A Quantitative Analysis Based on 50 Samples. Journal of Southeast University (Natural Science Edition), 43(04), 877-884.
6. Song, Y., Zhang, Y., & Han, D. (2024). A Method for Measuring Urban Block Morphology Based on "Ingress Structure." Urban Planning, 48(03), 98-106.
7. Han, D., Song, Y., & Ge, X. (2020). Cognition and Design of Morphological Structure in Compact Urban Blocks. Architectural Journal, (11), 79-85.
8. Wang, X., Ye, Y., Yang, L., et al. (2023). A Study on the Measurement of Sustainable Spatial Morphological Elements in New Urban Areas Based on Grid Scale: A Case Study of Yinzhou New City in Ningbo. Tropical Geography, 43(10), 1929-1939.
9. Moulinié, C. & Naudin-Adam, M. (2005). Appréhender la densité. Formes urbaines et densité, Institut d'aménagement et d'urbanisme Île-de-France Note Rapide, 384, 4.
10. Delage, I. & Prévot, C. (2014). Atlas de Paris au temps de Napoléon[M]. Parigramme.
11. Zhang, X. (2016). Study on the Evolution of Urban Morphology in Modern Dalian (1898-1945) [D]. Beijing Jiaotong University.
12. Berghauser Pont, M. Y., & Haupt, P. (2023). Spacematrix: Space, Density and Urban Form - revised edition. TU Delft OPEN.

Jiading New City Green Ecological Urban Planning and Practice — A Case Study of Yuanxiang Lake Central Activity Zone

Guoshu Yang*, Hongyan Pan and Hua Sun

Shanghai Institute of Building Science Co., Ltd, Shanghai, China

Abstract. The "14th Five-Year Plan" represents a critical window for the green and low-carbon transition. Jiading New City, characterized by its pioneering role as a "Low-Carbon Urban Area Leader," focuses on advancing the development of a green ecological urban district and explores the construction of a livable, green, and low-carbon new city. This paper takes the Yuanxiang Lake Central Activity Zone as the subject, elaborating on the green ecological positioning, green ecological indicator system, planning strategies, and the implementation safeguard system. It innovatively proposes the planning and design scheme of a "Green Gene Chain" to activate the Yuanxiang Lake Central Activity Zone, establishing it as a model for the green ecological urban district of Jiading New City and providing a reference for the green ecological planning and construction of other new citys.

1 Introduction

In February 2021, Shanghai issued the "Implementation Opinions on Accelerating the Promotion of New City Planning and Construction during the 14th Five-Year Plan," proposing that new cities are a key strategic space for promoting Shanghai's cluster-based urban development. At the same time, six key sector-specific documents were released, including the "14th Five-Year Plan Special Plan for Environmental Quality and New Infrastructure in New Cities," which mandates that 100% of newly developed areas in new cities adhere to green ecological city standards. In September 2021, the "Shanghai Green Building Management Measures" were announced, specifying that "the construction of Jiading New City, Songjiang New City, Qingpu New City, Fengxian New City, and Nanhui New City should be incorporated into the creation of green ecological cities." In 2022, both the "Shanghai Carbon Peak Implementation Plan" and the "Carbon Peak Action Plan for Urban and Rural Construction in Shanghai" included the development of green ecological cities as a key initiative for transitioning urban and rural construction towards a green, low-carbon model.

Since 2018, Shanghai has continuously advanced the development of green ecological cities. By the end of 2023, a total of 26 green ecological cities had been successfully established across the city, covering approximately 73 square kilometers of land [1]. To further advance the green ecological initiatives within the scope of the new cities, the "Green Ecological Planning and Construction Guidelines" for Jiading, Qingpu, Songjiang, Fengxian, and Nanhui New Cities were issued in 2024. These guidelines, following the principle of "differentiation by new city, by region, and by priority," identified 28 reserve projects for green ecological city development [2], thereby promoting the scaled development of green ecological cities within the new cities. The "14th Five-Year Plan" serves as a critical window for the high-level planning and construction of new cities, emphasizing the importance of maintaining a low-carbon focus, deepening green development, and comprehensively promoting the construction of green ecological cities to contribute to achieving carbon peak and carbon neutrality pathways for the new cities.

2 Achievements in Promoting Green and Low-Carbon Development in Jiading New City

To advance the high-quality development of Jiading New City, the "Jiading New City 14th Five-Year Plan Action Plan" proposed the creation of three exemplary demonstration areas. The Ximen Historical and Cultural District represents Jiading's past, focusing on the protective development of historical and cultural streets and the adaptive reuse of historic buildings [3]. The Yuanxiang Lake Central Activity Zone represents Jiading's present, emphasizing the concept of "integrating the city with nature and bringing nature into the city," to create a city living room [4] that is recognizable, livable, and highly regarded. Jiabao Smart Bay represents Jiading's future, leading with "the integration of production, life, and ecology" and digital transformation, shaping a future city experimental zone [5] characterized by eco-friendly scientific and technological innovation industries.

In 2021, Jiading District released the "Work Plan for Promoting the Creation of a Green Ecological City in

* Corresponding author: yangguoshu@sribs.com

Jiading New City (2021-2023)," which proposed a focused and planned approach to advancing the creation of green ecological city pilot zones in the Yuanxiang Lake Central Activity Zone and Jiabao Smart Bay. The 2022 "Jiading District Carbon Peak Implementation Plan" proposed promoting green ecological city pilots and carrying out pilot projects for low-carbon development zones and low-carbon communities. As of October 2024, two green ecological cities, one low-carbon development practice zone, and two low-carbon communities have been established within the new city's boundaries. In 2023, the Yuanxiang Lake Central Activity Zone was awarded the status of a three-star green ecological city pilot project by Shanghai [6], while Jiabao Smart Bay passed the green ecological city pilot evaluation in June 2024. Additionally, in 2023, the Yuanxiang Lake area was included in Shanghai's Low-Carbon Development Practice Zone [7], and Jiayue Community in Juyuan New District and Zhongfang Community's Defu Second Village in Malu Town were selected as low-carbon community pilots [7-8] in 2022 and 2023, respectively. Through the pioneering creation of Jiading New City as a low-carbon demonstration area, these initiatives have promoted the green ecological development of urban construction and further enhanced the urban ecological capacity of Jiading District.

3 Problems of Ecological Construction in Jiading New City

Building Jiading New City into a high-quality green and low-carbon demonstration area with good ecological foundation and first-mover advantages, but the following shortcomings still exist: First, the land use is not intensive enough, the occupational and residential space is relatively separate, and the occupational/residential ratio is only 0.46[9]. Second, longer commuting distances and a commuting ratio of more than 20%[9], private cars are more than transit trips, and the slow-moving system needs to be upgraded. Third, there is a lack of diversity in public service support and a relative lack of high-quality public space. Fourth, an efficient and low-carbon energy system has yet to be formed. Fifth, the smart city construction model is just getting started.

4 Planning and Practice of Yuanxiang Lake Central Activity Zone

The Yuanxiang Lake Central Activity Zone is positioned as the comprehensive functional center of Jiading New City and a focal area for displaying the city's image. The project focuses on the planning and construction of a green eco-city, with five major aspects of low-carbon, health, resilience, vitality and Intelligence for early and pilot implementation, and explores the construction of a livable, green, and low-carbon new city.

4.1 Green Ecological Positioning and Objectives

The project is located in the core area of Jiading New City, with a total land area of approximately 4.56 square kilometers. As one of the key model demonstration areas of Jiading New City, the overall planning concept for the Yuanxiang Lake Central Activity Zone is "By the Yuanxiang Lake, a New City is Settled." The area is designed with green "DNA" integrated into its development, embedding low-carbon, resilience, spatial, economic, and Intelligence "DNA" chain factors (**Fig. 1**). Through the activation of these "green gene chains"[a] the Yuanxiang Lake Central Activity Zone aims to become a central activity area driven by green genes, and an innovative eco-city that offers a future living experience.

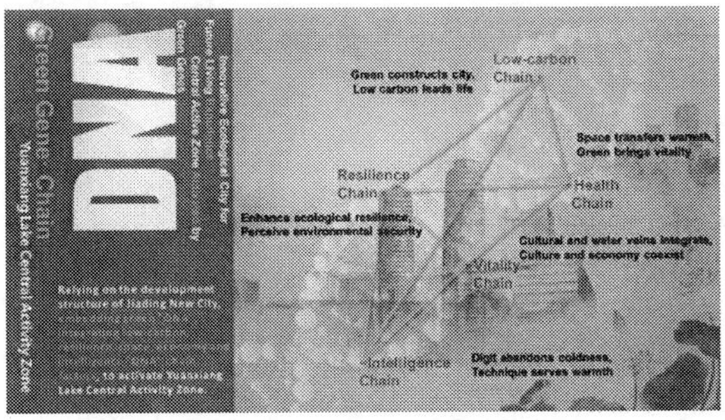

Fig. 1. Project Green Ecological Planning Positioning Diagram.

4.2 Green Ecological Indicator System

The project aims to establish itself as a three-star green ecological city pilot in Shanghai. Based on the "Green DNA" concept, a green ecological indicator system has been constructed, consisting of 5 primary goals, 15 secondary indicators, and 30 indicator items. Among these, 17 are control indicators with rigid constraints, and 13 are guiding indicators that highlight distinctive features (**Table 1**). This balanced approach of combining rigid and flexible green indicators will lead the development of a central activity zone activated by green genes.

Table 1. Green Ecological Indicators of the Project (Partial).

Primary Indicator	Secondary Indicator	Indicator Item	Unit	Indicator Value	Control/ Guiding
"Low-Carbon Chain"	Low-Carbon Emissions	Regional carbon emission intensity reduction rate	%	20	Guiding
	Low-Carbon Buildings	Demonstration zone for ultra-low energy buildings	Unit	1	Control
	Low-Carbon Energy	New buildings adopting at least one renewable energy source	%	100	Control

---
[a] "Green gene chain": interpreted as a key linking factor in green eco-building.

	Low-Carbon Transportation	Proportion of new energy buses	%	100	Control
"Health Chain"	Spatial Layout	15-minute walking coverage rate of community-level public service facilities	%	100	Control
	Environmental Ecology	Length of waterfront greenways	km	18	Guiding
	All-Age-Friendly Design	Child-friendly blocks	Unit	≥2	Guiding
"Resilience Chain"	Environmental Protection	Proportion of water surface area (rivers and lakes)	%	≥12	Control
	Sponge City	Annual runoff control rate	%	≥73	Control
	Solid Waste Resources	Household waste resource utilization rate	%	70	Control
"Vitality Chain"	Low-Carbon Industry	Number of green finance projects	Unit	≥2	Guiding
	Cultural Vitality	Yuanxiang Axis "Heart-Stirring Experience" route	Route	1	Guiding
"Intelligence Chain"	Smart Facilities	Comprehensive pole application coverage rate	%	100	Control
	Smart Scenarios	Number of smart transportation scenarios	Route	≥3	Guiding
	Refined Management	BIM coverage rate for newly constructed buildings above a certain size	%	100	Control

Note: Black-colored indicators represent distinctive features.

4.3 Green Ecological Planning Strategies

4.3.1 "Low-Carbon Chain"—— Green Conservation

Focusing on the goal of a low-carbon city, it aims to create high-quality building spaces, provide sustainable energy support, and develop comfortable, pedestrian-friendly green neighborhoods. The planning fully incorporates high-star-rated green buildings, with 100% of new buildings meeting at least two-star green building standards, and 100% of large public buildings meeting three-star green building standards. A concentrated demonstration zone for ultra-low energy buildings will be built, along with two near-zero energy building demonstration projects, forming a cluster of high-quality green buildings. The plan also promotes the large-scale adoption of renewable energy. At least 30% of the roof space of new buildings will be equipped with photovoltaic installations, with the proportion for school buildings being no less than 50%. The plan envisions a network of comfortable, pedestrian-friendly paths, including a lake-ringed greenway, a Z-axis-themed pathway, and shared bicycle-pedestrian lanes. A total of 18 kilometers of waterfront greenways will be constructed, creating a pleasant lakefront slow-living district.

4.3.2 "Health Chain "——Human-Centered and Livable

Focusing on the concept of a healthy city, the plan emphasizes creating a pleasant neighborhood scale, diverse green ecological spaces, and building an all-age-friendly, healthy district. Centered around Yuanxiang Lake, the plan will establish the "Yuanxiang Cultural Ring" as an urban landmark, ensuring 100% coverage of community-level public service facilities within a 15-minute walking radius. The introduction of "guidance-oriented mixed-use land" will systematically optimize land use layout and structure, and achieving a mixed-use neighborhood ratio of 75%. Focusing on the unique urban spaces of "One Lake and Two Axes" the plan will create a biodiversity-friendly lakeside ecological space. Prioritizing human-centered needs, the project will develop three lakeside healing gardens and two diverse, child-friendly blocks (**Fig. 2**), fostering the development of high-quality, all-age-friendly residential communities.

Fig. 2. Real-World Image of Child-Friendly Area in Yuanxiang Lake Park.

4.3.3 "Resilience Chain"——Safe and Ecological

Focusing on resilient city construction, the plan emphasizes strengthening the governance of ecological systems, enhancing water environment safety, and building sponge city demonstrations to comprehensively improve urban safety capabilities. The plan aims to create multifunctional composite waterfront spaces, with a water surface area ratio of ≥12% and 90% of the waterfront designated for ecological and residential use. Combining the "One Lake and Two Axes" landscape space (**Fig. 3**), it will also establish citizen-friendly walking corridors and linear pathway parks, creating a naturally wild "Green Forest Trail" experience. Adapted to local conditions, the project will establish four types of sponge city demonstration zones: a sponge retrofit demonstration zone, the Yuanxiang Lake Sponge Core Area, a high-quality sponge space demonstration zone, and a rainwater block demonstration zone. The goal is to achieve a regional annual runoff control rate of ≥73%.

Fig. 3. Real-World Image of the Ziqi Donglai Landscape Axis.

4.3.4 "Vitality Chain"——Cultural Sharing

The plan focuses on embedding the essence of a people-oriented city, it aims to invigorate the core area of the new city, promote the aggregation of low-carbon and innovative industries, and preserve the cultural memory of the old city while upgrading the public cultural service experience. The plan revolves around "Three New" key engines: the new empowerment center for the regional economy, the new living base for the scientific and technological community, and the new core for urban service experiences. It seeks to build an integrated structure of industrial and urban functions, transforming the Yuanxiang Lake Central Activity Zone into a future hub of 24/7 vitality. Centered around Yuanxiang Lake, the plan will create the Yuanxiang Axis "Heart-Stirring Experience" route, with 10 experiential nodes along the axis, forming a well-rounded green, ecological, low-carbon experiential space.

4.3.5 "Intelligence Chain"——Efficient and Convenient

Fig. 4. Real-Yumin South Road Smart Road.

The plan is anchored in the direction of smart city development, comprehensively deploy 5G smart facilities and build diverse smart scenarios and a refined smart management platform. It integrates the new city's 5G network and new infrastructure construction, advancing communication network facilities and carrying out smart intersection construction in the Yuanxiang Lake area, thereby enhancing the intelligence of urban facilities. The plan includes the construction of the Yuanxiang Lake "One Ring and Three Roads" (one autonomous driving demonstration ring and three smart roads) as part of the smart transportation application. In 2021 and 2022, upgrades to Baiyin Road and Yumin South Road were completed (**Fig. 4**), transforming them into smart roads, contributing to the creation of a national smart transportation pilot zone.

4.4 Planning Implementation Safeguard System

The project relies on Shanghai's first district-level green ecological city management regulation—the "Jiading District Green Ecological City Construction Management Measures (Trial)"—to explore the establishment of a comprehensive planning implementation safeguard system, covering the entire process from planning to construction and operation management. The system advances the step-by-step implementation of green ecological indicators and low-carbon technologies, ensuring a closed-loop management system for the full life cycle of the project. During the planning phase, indicators will be broken down for both the regional and neighborhood levels, incorporated into the comprehensive planning control platform, and integrated into land transfer agreements. During the construction phase, the development guidelines for primary development and the detailed construction regulations for secondary development will guide low-carbon construction. In the operation management phase, annual and third-party assessments will be conducted regularly to establish a dynamic evaluation and feedback mechanism, thereby achieving the goal of promoting planning and construction through effective management.

5 Conclusion

During the "14th Five-Year Plan" period, Jiading New City has comprehensively enhanced its urban capacity and implemented the national "dual carbon" strategy, focusing on building a green, humanistic, and ecological city through smart transportation and low-carbon building construction. From the initial stages of planning, the project innovatively proposed the concept of a "Green Gene Chain" to activate the design of the Yuanxiang Lake Central Activity Zone. The plan clearly defines the goal of transforming the area into a model for a three-star green ecological city in Shanghai. It is expected that the urban green space rate will reach over 35%, and the per capita park green area will be 31.5 square meters per person. By implementing measures such as low-carbon buildings, low-carbon transportation, and low-carbon energy, it is expected that the regional carbon emission intensity will decrease by more than 20%. It establishing a new low-carbon green landmark and a new model for industrial-urban integration within Jiading New City. At the same time, the project established a rigid and flexible indicator system, a distinctive planning strategy, and a closed-loop management planning and implementation system. It also provides valuable insights for the development of green ecological cities in Shanghai's five new cities and across the nation.

Acknowledgments

1 Shanghai Municipal State-owned Assets Supervision and Administration Commission Technological Innovation and Capacity Enhancement Program “Key Technological Innovation and Industrial Service for Low Carbon Transformation in Construction Field” (2022023).

2 Study on Upgrading the Evaluation Criteria of Green Ecological Urban Areas Based on the New Round of Green and Low-Carbon Construction of High-Quality Cities (KY10000038.20220006).

References

1. Shanghai Green Building Development Report (2023) [R]. Shanghai: Shanghai Municipal Commission of Housing and Urban-Rural Development, 2023.
2. The Five New Citys "Green Ecological Planning and Construction Guidelines" Officially Released (2024) [E]. Shanghai Municipal People’s Government Website: https://www.shanghai.gov.cn/nw31406/20240205/b25ff4a16f804ee2afc6a58f7522b339.html
3. Sun Yiyuan. Jiading Ximen Old Street: Preserving the Roots of Jiading and Revitalizing the Old City’s "Past" [J]. Shanghai State-Owned Assets, 2022(2): 38-40.
4. Sun Yiyuan. The Present: The Reshaping and Enhancement of Jiading Yuanxiang Lake Central Activity Zone [J]. Shanghai State-Owned Assets, 2022(2):41-44.
5. Sun Yiyuan. Jiabao Smart Bay: Sketching the "Future of Jiading" [J]. Shanghai State-Owned Assets, 2022(2):45-48.
6. Approval for Yuanxiang Lake Central Activity Zone as Shanghai Green Ecological City Pilot (Shanghai Construction and Building Materials [2023] No. 308)
7. Notice on Carrying Out 2023 Low-Carbon Demonstration Creation Work by Shanghai Municipal Bureau of Ecology and Environment (Shanghai Environmental Climate [2022] No. 243)
8. Notice on Carrying Out Low-Carbon Demonstration Creation Work by Shanghai Municipal Bureau of Ecology and Environment (Shanghai Environmental Climate [2022] No. 27)
9. Zhang Tianran, Wang Bo, Zi Haibo, et al. Comparative Study on the Spatial Characteristics of Occupational and Residential Spaces in Five New Cities in Shanghai[J]. Shanghai Urban Planning, 2021.

Urban Waterfront Space in Macao: Distribution Characteristics, Existing Problems and Design Strategies

Xuefeng Li, Qi Yue*, Xinchun Sang and Xiangguo Zhang

National Ocean Technology Center, Tianjin, China

Abstract. As a new urban culture emerged, waterfront spaces underwent a historic transformation in terms of humanistic aspects, functionality, and spatiality by shifting their primary functions to include leisure tourism, landscape effects, ecological services among others. By combining Macao's unique developmental history with its environmental topography, a distinctive coastal cityscape belonging to Macao has been created. Through extensive field investigations, the fundamental distribution characteristics of waterfront spaces in Macao have been comprehensively summarized. Furthermore, this study examines the prevailing issues and challenges associated with Macao's urban waterfront space resulting from extensive land reclamation, such as the repercussions on the natural ecological environment, the reinforced susceptibility to ecological catastrophes and the reconfiguration of urban waterfront spatial structure. At last, based on the case studies and public participation, the present study proposes countermeasures and strategies for the design of waterfront spaces in Macao, with the objective of creating urban waterfront areas that are environmentally safe, resilient to climate change, ecologically friendly, convenient, and possess a sense of historical significance and composite attributes.

1 Introduction

Urban waterfront spaces hold significant importance for every city in terms of both time and space. The development, construction, and functional evolution of these spaces have undergone three crucial stages[1]. Initially, during the pre-industrial era, the focus was on drainage, irrigation, and transportation functions. This was followed by the industrial era where industrial water supply and transportation took precedence. Finally, we entered the post-industrial era where leisure tourism became the primary function. The influence of the ocean on urban aesthetics is primarily reflected through its impact on coastal architecture and spatial forms while simultaneously being influenced by the urban texture itself. Macao is a renowned waterfront city that benefits from its coastal location advantage and has effectively utilized the ocean as an important aspect of its development approach. In 2015, the State Council of the People's Republic of China published the new Map of the Administrative Division of Macao, which adjusted and clarified 85 km^2 administrative scope of the sea area managed by Macao. In light of the urban development trend towards the fourth dimension and Macao's strategic positioning as a global hub for tourism and leisure, the waterfront space in Macao has encountered significant opportunities for growth. Thus, it can be asserted that the waterfront space not only serves as the foundation for urban development but also acts as a catalyst for future urban expansion in Macao.

As a coastal city, Macao possesses a substantial amount of waterfront space and shoreline. However, due to its limited land area and high population density, the conflict between people and land is prominent, resulting in insufficient space for urban development. Consequently, land reclamation has become the primary approach to expand land space and promote urban growth. Over nearly a century of such activities, Macao's total land area has reached 33.3 km^2. Nevertheless, these reclamation efforts have altered the natural characteristics of the sea area and brought about various adverse impacts on the urban waterfront spaces. Through extensive field investigations and meticulous data collection, this study aims to comprehensively summarize the fundamental distribution characteristics of waterfront spaces while meticulously analyzing prevailing challenges encountered in these areas. Furthermore, innovative ideas will be proposed for effective planning and design of urban waterfront spaces in Macao with an aim to provide crucial information for managing water and land areas as well as assisting the Macao government in scientifically designing these spaces.

* Corresponding author: 375705336@qq.com

2 The distribution characteristics of Macao urban waterfront space

2.1 The coastline distribution of Macao

To facilitate the effective administration of the sea areas of Macao, the Macao Special Administrative Region government demarcated and revises the coastline. The total length of Macao's coastline is 79.5 km, including 18.4 km of Macao Peninsula, 49.9 km of outlying islands (including Taipa, Cotai Reclamation Area and Coloane Island), 5.7 km of New Urban reclamation and 2.7 km of Hong Kong-Zhuhai-Macao Bridge Macao Port Management Area (Table 1).

Table 1. The coastline length by different regions in Macao[2].

Region	Coastline length (km)	Percentage (%)
Macao Peninsula	18.5	23.3
Outlying Islands	49.5	62.2
New Urban Zone A	5.7	7.2
Macao Port Administration Area	2.7	3.4
New Urban Zone C	3.1	3.9
Total	79.5	100

2.2 The coastline types of Macao

According to the latest statistics in 2023, the coastline of Macao is dominated by artificial coastline, accounting for about 84.3% of the total length of the coastline, which is mainly distributed on the Macao Peninsula and the north and east sides of Taipa Island, with a total length of 67 km (Table 2). The typical artificial slope embankment coastline is the east side of Macao Science Center and artificial vertical embankment coastline is the north bay of Chopsticks for instance. In contrast, the coastline of Macao is supplemented by natural coastline, accounting for about 15.7% of the total length of the coastline (Table 2). Among them, the sandy coastline is mainly distributed in the Black Sand Beach and Cheoc Van Beach of Coloane Island, with a total length of 2.2 km; the rocky coastline is mainly distributed in the south and east parts of Coloane Island, with a total length of 5.2 km; there is no spontaneously recuperated coastline through the latest statistics data; the renovated coastline is distributed in the forest planting area and the ecological protection zone 2 in the west of Taipa Island along the coast, with a total length of 5.1 km.

Table 2. The coastline types by different regions in Macao[3].

Types	Total (km)	Pct. (%)	Macao Peninsula (km)	Outlying Islands (km)	Other areas
Artificial	67.0	84.3	18.4	37.1	11.5
Sandy	2.2	2.8	0.1	2.1	0.0
Rocky	5.2	6.5	0.0	5.2	0.0
Spontaneously recuperated	0.0	0.0	0.0	0.0	0.0
Renovated	5.1	6.4	0.0	5.1	0.0

2.3 The coastline utilization status of Macao

The development and utilization status of Macao's coastline can be divided into industrial coastline, transportation coastline, tourism coastline, urban and rural construction coastline and other coastlines, accounting for 92.6% of the total coastline length (Table 3). Among them, the most important development and utilization types are urban and rural construction shoreline (41.8% of total length), transportation shoreline (26.9% of total length) and tourism shoreline (18% of total length). In addition, the unutilized coastline is mainly rocky coastline and sandy coastline without tourism function, which are mainly distributed in the south of Coloane Island and the north of Taipa Island, accounting for 7.4% of the total coastline length.

Table 3. The Coastline utilization status by different regions in Macao[3].

Status	Total (km)	Pct. (%)	Macao Peninsula (km)	Outlying Islands (km)	Other areas
Industrial	2.7	3.4	0.0	2.7	0.0
Transportation	20.3	26.9	3.6	16.7	0.0
Tourism	14.3	18	4.6	9.7	0.0
Urban and rural construction	33.2	41.8	9.6	12.1	11.5
Other	3.1	3.9	0.0	3.1	0.0
Unutilized	5.9	7.4	0.7	5.2	0.0

3 The existing problems of Macao urban waterfront space

3.1 Repercussions on the natural ecological environment

Due to large-scale land reclamation activities, the natural coastline has been gradually replaced by artificial shoreline. As depicted in Table 1, the proportion of artificial coastline has risen to 84.3%, while the sandy coastline with favorable hydrophilic spatial properties merely accounts for 2.8% and spans a total length of 2.2 km. The shape of the coastline also tends to become regular and flat due to cutting and straightening, altering the marine hydrodynamic environment in nearby waters, diminishing seawater's self-purification capacity, impeding pollutant migration and diffusion, and eventually leading to the deterioration of seawater quality and marine environment pollution[4]. Similarly, land reclamation exerts a profoundly significant impact on marine organisms inhabiting adjacent waters. It encroaches upon vital ecological spaces such as diverse coastal natural wetlands and results in habitat loss or degradation for species survival including fish eggs, larvae, benthic organisms, rare birds and wetland vegetation[4]. For instance, vast areas of mangrove wetlands, possessing crucial ecological value and serving as environmental buffers, are rapidly vanishing. This series of issues imposes immense pressure on the coastal region's ecosystem; henceforth augmenting the proportion of natural restoration alongside shoreline rehabilitation stands as pivotal directions for future urban ecological construction in Macao.

3.2 Reinforced susceptibility to ecological catastrophes

The land reclamation projects have extended the coastline of Macao into the sea by tens of kilometers, resulting in significant alterations to the topography and geomorphology of the original coastal zones. This has led to a substantial reduction in biodiversity and natural landscape resources within inter-tidal areas, while also eradicating the natural barrier between the city and water. Instead, non-ecological hard revetments have replaced the coastline along with diminishing bay-body river channels. Consequently, this has compromised the disaster prevention and mitigation capabilities of Macao's coastal zone, thereby adversely affecting flood control and drainage initiatives. As a result, there has been a notable increase in ecological disasters such as storm surges, water-logging events, and seawater inundation leading to intensified damages. Various regions across Macao experience varying degrees of water-logging disasters throughout different times of the year due to inefficient discharge mechanisms for surface water during heavy rainfall or astronomical spring tides[5]. Additionally, certain natural flood discharge outlets are obstructed which further exacerbates infiltration issues where surface water permeates into underlying ground layers at an increased rate. This subsequently elevates groundwater levels in specific areas causing building foundations to become saturated by groundwater infiltration. Consequently, these circumstances impose additional strain on river channels and drainage networks while significantly impacting urban streets with flooding problems that severely disrupt local residents' livelihoods.

3.3 Reconfiguration of urban waterfront spatial structure

In the initial phase of urban development, the construction of waterfront spaces in Macao was characterized by disorder and excessive development. Numerous reclamation projects resulted in the construction of tall and robust sea walls, creating vast artificial landscapes. While this rapid urbanization met short-term construction needs, it failed to address long-term developmental requirements. The unbalanced emphasis on functional aspects in urban infrastructure disregarded comprehensive considerations, leading to detrimental impacts on natural environments, public spaces, and the overall urban context within the densely populated areas of Macao. Consequently, structural alterations occurred in waterfront spaces with a disconnection from ecosystems, resulting in a loss of their original functions such as self-purification and energy flow transformation between ecosystems. Overall, no stable spatial structure has been established at an urban scale for waterfront areas: wetland patches have suffered severe fragmentation while mangrove wetlands have experienced significant reduction without establishing stable connections with urban ecosystems; green space resources are unevenly distributed throughout the city with scattered locations lacking organic connections; green corridors are interrupted by high-density buildings resulting in fragmented conditions; leisure spaces like riverside parks and green areas exhibit low utilization rates while connectivity among pedestrian systems, landscape interfaces, and ecological bases is inadequate.

4 The design strategies for urban waterfront space in Macao

4.1 Case studies

Barcelona initiated the implementation of the "Open the City to the Sea" development concept in the 1980s. Through significant events such as the 1992 Olympics and the 2004 World Cultural Forum, it has effectively facilitated coastal utilization optimization and urban spatial structure enhancement, thereby enhancing public life in coastal areas. Barcelona has successfully evolved into an enchanting tourist destination and a showcase for its city image. The progressive transformation of coastline utilization, integration of coastal and hinterland functions, demand-driven humanized spaces and facilities, as well as landscape control and distinctive shaping of both overall cityscape and coastal regions represent valuable experiences that can inform future endeavors in transforming coastal utilization methods and spatial reconstruction.

The waterfront space in Copenhagen is divided into four zones from land to sea: the activity zone, the trail zone, the waterfront zone, and the water zone. The "activity zone" serves as a crucial link between land and water, functioning as a pivotal area for public gatherings and social activities. It encompasses diverse functional areas such as waterfront parks and recreational green spaces. Among these four zones, it holds paramount importance. The "trail zone" is dedicated to fitness enthusiasts who enjoy walking and cycling, combining lush landscaping with water features to create a natural fitness area that promotes stress relief, mental relaxation, and connection with nature. The "waterfront zone" functions as the central hub for wetland activities and primarily facilitates water-related endeavors. Alongside constructing relevant water infrastructure platforms, it also encompasses the cultivation and planning of green vegetation on both sides to establish a robust ecological environment. The "water zone" denotes the central covered area of the body of water that currently retains its original natural state. It will serve as a pivotal focal point for future development of multi-dimensional waterfront spaces, encompassing an organic integration of aquatic elements, aquatic communities, wetland ecology, and human habitation environments.

4.2 Public participation

The involvement of the public in urban waterfront spaces is a crucial matter encompassing urban planning, design, management, and public interests. It also plays a pivotal role in shaping the city's waterfront areas. Only by granting Macao citizens a voice in determining the city's development can designs better cater to the aspirations and desires of actual end users. Macao citizens possess a unique and profound understanding of local culture, making it essential to heed their opinions as an integral part of establishing a new identity for waterfront parks. The active participation and cooperation of Macao community residents not only ensure that the design of public spaces aligns with the needs and memories of community members but also foster a sense of engagement and unity within the community. Through collaborative construction projects, Macao community residents gain deeper insight into and involvement in the development of public spaces, transforming them into truly shared and supportive places belonging to the community. Subsequently, through tripartite collaboration among the Macao government, social organizations, and residents, effective governance and management strategies for waterfront spaces are achieved. This participatory governance model employed in designing Macao's urban waterfront areas maximizes each party's roles and advantages while efficiently integrating and utilizing waterfront resources to promote sustainable social development.

4.3 Design strategies

4.3.1 Urban waterfront space with resilience and security

Under the pressure of climate change and flood prevention, enhancing the resilience of urban waterfront spaces has emerged as a prominent research topic[6]. Given Macao's frequent exposure to ocean disasters, constructing resilient and secure urban waterfront spaces has become an imperative practical necessity. Firstly, safeguarding the original topography takes precedence while expanding the shoreline to create a reclamation-based buffer zone. The inter-tidal zones, sandbars, low-lying areas, and wetlands within this region can serve as vital components for flood control and retention purposes while providing habitats and improving water quality. By employing diverse spatial configurations such as shallow hills, depressions, and marshes with meticulous micro-topographical shaping at the water-land interface to establish high-porosity shore habitat space featuring varying flow speeds that adhere to wetland vegetation succession rules gradually forming a more robust waterfront ecosystem with enhanced anti-interference capacity for self-organization and self-recovery. Simultaneously, attention should be directed towards establishing organic connectivity between waterfront buffer zones and urban ecological patches in order to construct a collective flood barrier for cities. Secondly, emphasis should be placed on eco-friendly seawall construction. By implementing a multi-level ecological embankment construction model that integrates beach-fronts, marshlands, coastal protection forests, embankments, elevated roads, ground roads, and architectural overhead designs, the resistance and load-bearing capacity of seawalls against marine disasters can be significantly enhanced. At last, urban waterfront spaces exhibit a linearly connected spatial configuration with relatively weak hierarchy and centrality. Based on factors such as walking space radius, proximity to neighboring communities, and availability of landscape resources, multiple landscape nodes are formed within waterfront areas. These dispersed nodes establish their own distinct spatial domains while maintaining relative independence. As a result, the spatial organization structure adopts a modular form with multiple centers to enhance resilience in the face of disturbances.

4.3.2 Urban waterfront space with resilience and security

Ecological restoration refers to the rehabilitation of urban areas' damaged natural environment and ecosystems, encompassing forests, green spaces, wetlands, and other ecological systems that foster a conducive urban environment. Urban renovation focuses on rectifying functionality and quality issues in public service facilities to enhance urban functions and cultural vitality. Given the substantial amount of hardened urban land in Macao, it is imperative to maximize the utilization of transformable soft land or underutilized hard land for facilitating ecological restoration endeavors. While safeguarding the preservation of the old city texture and World Heritage sites, an optimal strategy should be adopted that minimizes repairs while simultaneously rehabilitating wetlands and green spaces. The restoration process should be conducted incrementally based on landscape ecology's "Patch-Corridor-Matrix" theory to achieve optimization in ecological structure[7]. By reinstating soft shoreline matrices along both sides of corridors, Macao's rigid shoreline can be restored while establishing mangrove wetland patches. Through riverfront ecological corridors connecting dispersed green patches within urban parks, residential areas, and surrounding streets, an all-encompassing urban green network can materialize. Simultaneously integrating this network with water systems and surrounding mountains will establish a comprehensive ecological network system aimed at preserving natural landscapes while perpetuating the existing urban landscape pattern. Based on the restoration of natural habitats, the transformation and development of waterfront spaces are guided by functional integration. For instance, expanding waterfront trail areas towards the water's edge, constructing boardwalks around mangroves, and implementing temporary or permanent floating facilities to cater to people's activity needs. The enhancement of spatial functions essentially perpetuates a mindset of expansion towards the sea and utilization of water space, employing strategies such as bringing activity areas closer to the water zone and integrating trail areas with waterfront zones. Furthermore, both materials and construction methods strictly adhere to ecological standards in order to minimize adverse impacts on aquatic ecosystems while achieving the objective of restoring urban functionality for better fulfilling people's desire for proximity to water.

4.3.3 Urban waterfront space with a sense of history and complexity

The public spaces in waterfront areas of cities should not only serve as recreational parks, but also function as an extension of urban work and residential activities. In terms of coastal utilization, it is essential to consider the potential for coastal development based on historical culture, tangible economy, and industrial considerations. By leveraging the abundant marine location factors, the development of coastal culture and maritime industries can support urban growth by fostering deeper connections between city-ocean-humanity and diversifying interactive modes[8]. Integrating Macao's rich historical and cultural heritage with waterfront spaces involves preserving traditional and modern elements as well as Portuguese and Chinese cultures while showcasing its World Heritage core area that attracts tourists from around the world. To enhance pedestrian systems' readability and cultural connotations, galleries exhibiting Macao's history and Sino-Portuguese culture can be arranged along waterfront promenades. Upgrading maritime industries while adjusting their format will create a sustainable urban ecological leisure service shoreline centered around third sector industries such as tourism industry or service industry. Planning facilities like libraries, community activity centers for residents' needs can be done alongside providing supporting commercial facilities like shopping centers and dining establishments. Sports fields, fitness facilities, children's activities, elderly wellness facilities can also be included to construct a multifaceted leisure walking system that creates a historically significant yet diverse waterfront space for Macao residents.

5 Conclusions

Macao possesses a rich tapestry of cultures, distinctive industries, and abundant waterfront resources. However, the ongoing land reclamation in Macao has led to an absence of comprehensive urban waterfront design, presenting numerous challenges to the current state and future development of the city. The allocation of jurisdiction over 85 km^2 of sea area to Macao in 2015 and the issuance of the Development Plan Outline for Guangdong-Hong Kong-Macao Greater Bay Area in 2019 have brought significant opportunities for enhancing Macao's urban waterfront spaces. Effective planning and design can seamlessly connect the city, its inhabitants, and the ocean through multiple guiding principles that truly orient people towards embracing the ocean as an integral part of their surroundings. Based on the case studies and public participation, this article proposes stringent protection and orderly utilization strategies for Macao's managed sea area by reshaping its urban spatial structure, optimizing designs for urban waterfront spaces, and reinforcing ecological restoration projects along seawalls. These measures aim to ensure well-designed waterfront spaces with livable residential environments while harnessing the ocean's potential to provide residents with high-quality ecological products, picturesque natural settings, and exceptional ecological services. By achieving safety, resilience and friendliness while preserving historical charm combined with modernity within its urban waterfront spaces, Macao will establish a solid foundation towards becoming a world-class tourism leisure center.

Acknowledgments

This research was funded by the National Natural Science Foundation of China International Cooperation Project, grant number 52061160366; High Level Science and Technology Innovation Talents Program of Ministry of Natural Resources, grant number Y2190NF02; and Tianjin "131" High-level Innovative Talents Funding Project, grant number Y2210TF01.

References

1. Y. Tang, L. Li, L. Han, S. Xie. (2022) A review of international research on the transformation and development of urban waterfront space. Prog. Geogr. **41**, 6: 1123-1135.
2. The Macao SAR government Cartography and Cadastre Bureau. (2024) Macao SAR Coastline. https://www.dscc.gov.mo/en/geo_coastline_intro_details/article/geo_coastline_intro.html#scroll.
3. The Macao SAR government Cartography and Cadastre Bureau. (2024) Types and utilization status of Macao SAR coastlines. https://www.dscc.gov.mo/en/geo_coastline_intro_details/article/geo_coastline3.html#scroll.
4. B. Cui, T. Xie, Q. Wang, S. Li, J. Yan, S. Yu, K. Liu, J. Zheng, Z. Liu. (2017) Impact of large-scale reclamation on coastal wetlands and implications for ecological restoration, compensation, and sustainable exploitation framework. Bull. Chin. Acad. Sci. **32**: 418-425.
5. J. Zheng, J. Li, G. Liu, Y. Fei. (2018) Macao's Inner Harbor Waterfront Revitalization Based on the Model of Functional Wetland Reclamation. Chin. Landsc. Archit **34**: 91-97.
6. J. Zou. (2023) Resilience Mechanism and Design Response of Urban Waterfront. Shanghai Urban Plan. Rev. **1**: 40-46.
7. A.A. Ragheb, R.A. EL-Ashmawy. (2020) Urban Waterfront Development for Designing Space in Coastal Cities. Int. J. Sustain. Dev. Plan. **15**: 345-352.
8. D. Üzümcüoğlu, M. Polay. (2023) Enhancing urban waterfront development: a groundbreaking framework for fostering creativity. Geo J. **88**, 6091-6104.

Environmental, Social, and Governance (ESG) Committees and Corporate ESG Information Disclosure: From the Perspective of Organizational Information Processing

Hang Yang*

School of management, Shanghai University, Shanghai, China

Abstract. Based on the organizational information pro- cessing theory, this paper examines the impact and mech- anisms of ESG committees on the level of corporate ESG information disclosure using a sample of Chinese A+H share listed companies from 2016 to 2022. The study finds that establishing an ESG committee significantly improves the level of corporate ESG information disclosure. ESG commit- tees enhance corporate information processing capabilities through three aspects: improving ESG governance structure, increasing board and management attention to ESG informa- tion, and attracting external stakeholders' attention. These improvements contribute to higher levels of ESG information disclosure. Moderation tests show that in companies with higher internationalization, intense industry competition, and high economic policy uncertainty, the increased information processing task encourages more proactive ESG committee activities, thereby significantly improving ESG information disclosure levels.

1. Introduction

The Environmental, Social, and Governance (ESG) con- cept has gained global attention, particularly in China, where rapid economic growth and environmental concerns have heightened demands for corporate ESG performance. This concept supports sustainable development and is essential for achieving the "dual carbon" strategy. The Chinese government has implemented policies to enhance ESG information disclosure, such as the 2016 "Guide- lines for Establishing a Green Financial System" and the 2018 "Corporate Governance Code," which mandates listed companies to report ESG data. In June 2024, new guidelines urged central enterprises to strengthen ESG governance and practices, reflecting the rising importance of sustainable development.

As Chinese companies increasingly focus on ESG infor- mation disclosure, they encounter significant challenges, including information asymmetry [1], incomplete report- ing, and a lack of uniform standards [2]. Although relevant policy frameworks are being established, practical im- plementation remains difficult. The effectiveness of ESG disclosure largely depends on a company's processing capabilities, which involve data collection, application, and sharing. Companies often struggle to collect comprehen- sive ESG data due to inadequate organizational structures and recognition of its importance. Furthermore, the appli- cation of ESG data is limited, as management tends to treat it merely as a compliance tool. In sharing information, some firms may falsify ESG disclosures to protect their interests, leading to misleading data for investors.

To address these issues, many companies are forming ESG committees, dedicated to overseeing ESG matters with clear roles and responsibilities. This governance structure has been shown to positively impact areas such as human rights [3], financial performance [4], and sus- tainable development [5]. However, research on ESG committees' role in enhancing corporate ESG information disclosure, especially in the Chinese context, remains lim- ited and warrants further investigation. In addition, existing literature on corporate ESG information disclosure primar- ily focuses on the impact of factors such as government influence [6], economic development [7], CEO tenure [8], corporate size [9], and ownership nature [10] on ESG disclosure. However, there is limited research exploring the intrinsic mechanisms of corporate ESG information disclosure, particularly regarding the collection, applica- tion, and sharing of ESG data. This paper offers new perspectives and insights through the lens of organizational information processing, thereby enhancing the theoretical research in this field.

This study aims to uncover the mechanisms behind corporate ESG information disclosure by examining three areas: collection, analysis, and sharing capabilities. It also explores how ESG committees affect disclosure levels. Using quantitative methods, the research analyzes A+H share listed companies from 2016 to 2022, focusing on the establishment of ESG

* Corresponding author: *Anna14982@163.com

committees among Chinese firms. The study investigates how these committees influence ESG disclosure and considers factors like corporate internationalization, industry competition, and economic policy uncertainty as moderating effects.

2. Hypothesis

Organizational Information Processing Theory, developed by scholars like Simon [11] and Thompson [12], highlights the relationship between information processing tasks and an organization's capacity to manage informa- tion. "Uncertainty" prompts organizations to improve their information processing capabilities, which include col- lecting, analyzing, and applying information for decision- making [13]. This study identifies three levels of corporate information processing: collection, application, and shar- ing of ESG data.

The ESG committee serves as a governance structure that coordinates communication and oversees sustainable development strategies, enhancing social and environmen- tal responsibility. Its key roles include supervising ESG management and advising on strategy, risk management, and stakeholder communication.

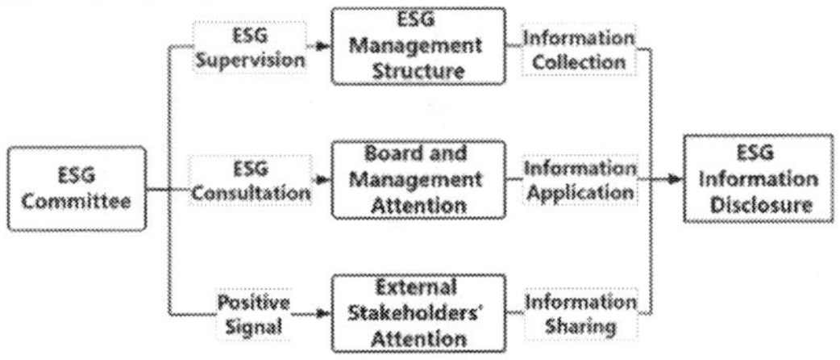

Fig 1. Research hypotheses.

Firstly, the ESG committee enhances the company's governance structure, improving information collection and disclosure. It supervises the integration of ESG prin- ciples into management, ensuring effective data consol- idation and increasing reporting efficiency from various departments.

Secondly, the ESG committee advises the board and management on applying ESG information, improving disclosure. It recommends strategies and stakeholder com- munication, enabling the board to analyze ESG data ef- fectively, which enhances external information disclosure and daily ESG practices.

Finally, the ESG committee attracts external stakehold- ers' attention, improving information sharing and dis- closure. It signals the company's commitment to ESG issues, increasing monitoring and pressure to disclose information, thus enhancing the authenticity and reliability of ESG reporting.

Based on the above analysis, as shown in Figure 1 the following hypothesis is proposed.

H1. Establishing an ESG committee will improve the level of a company's ESG information disclosure.

International enterprises face complex regulatory en- vironments and diverse stakeholder demands [14], increasing uncertainty in ESG information disclosure [15]. ESG committees are essential for unifying policies and balancing stakeholder needs across regions, enhancing information processing capabilities and improving overall ESG disclosure.

H2. ESG committees enhance disclosure more for highly internationalized companies.

In competitive industries, companies face higher exit risks and uncertainty. To establish a competitive advantage, they disclose more positive ESG information, which helps maintain reputation [15], gain consumer recognition [16], and attract investors. ESG committees enhance informa- tion processing and improve disclosure levels, leading to stronger market competitiveness.

H3. ESG committees boost disclosure more in com- petitive industries.

Economic policy uncertainty complicates ESG informa- tion disclosure for companies. Frequent policy changes challenge long-term planning and ESG goals [17]. In response, companies often enhance disclosure to boost credibility and transparency. ESG committees help by de- veloping strategies and improving information processing capabilities.

H4. ESG committees enhance disclosure more under high economic policy uncertainty.

3. Data and methodology

3.1 Sample

This study analyzes A+H-listed companies in China from 2016 to 2022, excluding financial industry data, abnormal samples, and those with missing data. It includes 148 companies with 455 observations. Data on financial and governance come from the CSMAR database, while ESG reports are sourced from Wind. Continuous variables are trimmed at the 1% and 99% percentiles to mitigate ex- treme values. The data encompasses basic characteristics, financial and macroeconomic data, and ESG committee information, with ESG report data compiled manually.

3.2 Variable Definitions

This study measures ESG information disclosure level (ESG_DE) using the ESG disclosure score from Bloomberg, based on corporate ESG and social responsi- bility reports. This score combines environmental, social responsibility, and corporate governance indicators. The explanatory variable is the ESG committee, coded as 1 if established and 0 otherwise. Control variables include company characteristics (size, age, ownership), financial characteristics (growth, leverage, total asset turnover, re- turn on assets), and governance structure (board size, in- dependent director ratio, CEO duality, management share- holding). Time and industry dummy variables are also included for rigor. Detailed definitions and measurement methods are in Table I.

3.3 Model construction

In this study, the following model 1 is proposed for hypothesis testing:

$$ESG\_DE_{it} = \alpha_i + \beta ESGcommittee_{it} + \gamma Controls_{it} + \Sigma\, Year + \Sigma\, Industry + \varepsilon_{it} \quad (1)$$

4. Results

Table II presents the descriptive statistics for the vari- ables in the regression model, showing an average variance inflation factor (VIF) of 1.71, indicating no multicollinear- ity.

To test the effect of ESG committees on corporate ESG information disclosure, Model 1 is tested, with results in Table III. The coefficient for the ESG committee is significantly positive at the 1% level, confirming that establishing ESG committees enhances corporate ESG information disclosure, thereby supporting hypothesis H1.

4.1 Robustness

This study uses a two-stage instrumental variable method to address potential endogeneity issues, selecting the average number of female executives in each province as the instrumental variable. Firms with a focus on gender balance are more likely to establish ESG committees, thereby enhancing ESG management (Column 1 of Ta- ble IV).

Additionally, the propensity score matching method (PSM) is employed to control for sample selection bias, resulting in a significant positive coefficient for the ESG committee (Column 2 of Table IV).

Lagged variable testing shows consistent positive re- sults, confirming robustness (Column 3 of Table IV).

Finally, substituting the dependent variable with data from the Bloomberg database also yields significant posi- tive results, supporting the study's conclusions (Column 4 of Table IV).

4.2 Moderating effect testing

This study categorizes firms by internationalization, defining the top 50% as highly internationalized. Regres- sion results are presented in Table V. In the highly inter- nationalized group (Column 1), the regression coefficient for the ESG committee is significant at the 1% level;

TABLE I. DEFINITION OF VARIABLES

Variable		Measurement
Dependent Variable	ESG_DE	Bloomberg's ESG Disclosure Score
Explanatory Variable	ESG committee	Dummy variable (yes = 1/no = 0) responding the question "Does the company have an ESG committee?"
Control Variables	Size	The total number of employees
	Lev	Total liabilities at the end of the year/total assets at the end of the year.
	Age	ln (current year year - listed year + 1).
	Growth	(Current year operating income/previous year operating income)-1.
	ATO	Operating Income/Average Total Assets.
	ROA	Net profit/Average Balance of Total Assets.
	Board	The number of board of directors takes the natural logarithm.
	Indep	Independent directors/number of directors.
	Dual	Chairperson and general manager is the same person for 1, otherwise 0.
	Mshare	Number of shares held by management/total share capital.
	SOE	The state-owned enterprise holds a controlling stake for 1, otherwise 0.

TABLE II. DESCRIPTIVE STATISTICS

Variable	Observation	Mean	Std.	Min.	Max.	VIF
ESG DE	455	48.59	8.13	27.97	67.64	—
ESG committee	455	0.17	0.38	0.00	1.00	1.44
Size	455	24.99	1.43	21.30	28.50	1.87
Lev	455	0.56	0.17	0.14	0.89	1.67
Age	455	2.86	0.47	0.69	3.47	1.60
Board	455	2.23	0.22	1.79	2.71	1.62
Indep	455	0.40	0.07	0.31	0.67	1.61
Dual	455	0.19	0.40	0.00	1.00	1.42
Growth	455	1.29	9.54	-31.10	75.61	1.66
Mshare	455	0.02	0.08	0.00	0.59	1.72
ATO	455	0.64	0.44	0.07	2.93	1.51
SOE	455	0.69	0.46	0.00	1.00	1.56
ROA	455	0.30	1.98	-10.57	14.86	1.80

TABLE III. ESG COMMITTEES ' BENCHMARK REGRESSION RESULTS ON CORPORATE ESG DISCLOSURE

	(1) ESG DE	(2) ESG DE
ESG	4.4150***	3.5519***
committee		
		(4.19)
Size	(5.13)	1.8134***
		(6.78)
Lev		-5.1390**
		(-2.12)
Age		0.2064
		(0.23)
Board		6.1991***
		(3.66)
Indep		10.4196*
		(1.88)
Dual		-0.6109
		(-0.55)
Growth		0.0134

		(0.40)
Mshare		6.4420
		(1.44)
ATO		-1.2126
		(-1.47)
SOE		0.4942
		(0.58)
ROA		0.3067*
		(1.67)
cons	41.9273***	-20.0703**
	(31.49)	(-2.43)
Year FE	Yes	Yes
Industry FE	Yes	Yes
N	455	455
R2	0.34	0.43
Adj. R2	0.32	0.40

TABLE IV. ROBUSTNESS TEST RESULTS

VARIABLES	(1) IV ESG_DE	(3) PSM ESG_DE	(5) Lagged ESG_DE	(7) Substitution ESG_DE
ESG committee	14.3247**	2.8227**		0.2093***
	(2.28)	(1.83)		(3.83)
L.ESG committee			4.1376***	
			(3.82)	
Constant	-7.5033	-34.6081	-3.1132	-0.4037
	(-0.60)	(-1.25)	(-0.33)	(-1.21)
N	417	118	348	455
Controls	YES	YES	YES	YES
Year FE	YES	YES	YES	YES
Industry FE	YES	YES	YES	YES
Adj. R2	0.23	0.41	0.26	0.15
F-statistic	13.17	—	—	—

whereas, in the less internationalized group (Column 2), the coefficient is not significant. This indicates that highly internationalized firms have a significant ESG committee impact,thereby supporting hypothesis H2.

Using the Herfindahl-Hirschman Index (HHID) to as- sess industry competition, defining the top 50% as high competition groups. As shown in Table V, in the high competition group (Column 3), the regression coefficient for the ESG committee is significant at the 1% level; whereas, in the low competition group (Column 4), the coefficient is not significant.Results show that in high competition, the ESG committee significantly boosts ESG information disclosure, helping firms attract investors and enhance competitive advantages, thereby supporting hy- pothesis H3.

To assess the impact of economic policy uncertainty, the study uses the Baker index to measure China's pol- icy uncertainty, defining the top 50% as high uncertainty groups. The regression results in Table V reveal that in the high uncertainty group (Column 5), ESG committee's regression coefficient is significant at the 1% level; while in the low uncertainty group (Column 6), the coefficient is not significant and is lower than in Column (5). This show that in high uncertainty, ESG committees significantly im- prove information disclosure, enhancing their supervisory and advisory roles, thereby supporting hypothesis H4.

TABLE V. MODERATING EFFECT TESTING

VARIABLES	ESG DE					
	(1) High International	(2) Low International	(3) High Competitive	(4) Low Competitive	(5) High Uncertainty	(6) Low Uncertainty
ESG committee	4.2899***	0.4460	6.4481***	0.3009	5.1171***	3.0933
	(4.55)	(0.27)	(5.07)	(0.25)	(3.72)	(1.20)
Constant	-7.6689	-26.6635***	-15.0188	-23.6159*	-11.0480	-47.8200**
	(-0.51)	(-2.76)	(-1.09)	(-1.89)	(-0.79)	(-2.52)
N	225	227	228	226	180	127
Adj R-squared	0.425	0.465	0.424	0.406	0.242	0.231
Controls	YES	YES	YES	YES	YES	YES
Year FE	YES	YES	YES	YES	YES	YES
Industry FE	YES	YES	YES	YES	YES	YES
P-Value		0.025	0.000			0.033

5. Conclusion

ESG committees are becoming important in corporate governance, especially in emerging markets. However, research on their roles and the quality of ESG disclosure is still limited, particularly in local contexts. This study focuses on Hong Kong and Mainland China to explore how ESG committees impact corporate ESG information disclosure. The findings show that these committees sig- nificantly improve ESG disclosure by enhancing infor- mation collection through better governance frameworks, increasing awareness among boards and management, and boosting external media engagement. These improvements ensure the integrity of ESG disclosures.

Additionally, factors like corporate internationalization, industry competition, and economic uncertainty raise in- formation processing demands. In response, ESG com- mittees help manage these challenges, leading to better disclosures. The analysis also highlights that ESG commit- tees positively influence market expectations and corporate value, signaling confidence to investors and ultimately enhancing market value.

Based on these conclusions, the study offers recom- mendations from corporate, industry, and governmental perspectives:

(1) Corporate level. Companies should establish ESG committees, clarifying their oversight and advisory roles to centralize their function in ESG governance.

They should also refer to Hong Kong's ESG practices to refine ESG governance frameworks, institute effective feedback mech- anisms, and regularly publish high-quality ESG reports detailing targets, measures, and outcomes.

(2) Industry level. Industry organizations should formu- late detailed ESG disclosure guidelines tailored to industry characteristics, establish independent supervisory bodies to regularly review and report on ESG disclosures, and facilitate intra-industry exchange through forums or work- shops to share experiences and best practices. Addition- ally, industry bodies should maintain open communication with government and investors to align industry standards and oversight mechanisms with policy trends and market needs.

(3) Government level. The government should draw from Hong Kong's policy experiences, implementing "comply or explain" principles. Strengthening oversight on ESG disclosures by instituting rigorous review and penalty systems ensures authenticity and completeness of disclosures. Moreover, the government should enhance market incentives for ESG disclosures, encourage financial institutions to develop ESG-related financial products, and stimulate demand and application for ESG information.

References

1. Yuan, X., Li, Z., Xu, J., & Shang, L. (2022). ESG disclosure and corporate financial irregularities: Evidence from Chinese listed firms. Journal of Cleaner Production, 332, 129-992.
2. Sun, Y., Zhao, D., & Cao, Y. (2024). The impact of ESG perfor- mance, reporting framework, and reporting assurance on the tone of ESG disclosures: Evidence from Chinese listed firms. Journal of Cleaner Production, 142-698.
3. Mallin, C., Michelon, G., & Raggi, D. (2013). Monitoring intensity and stakeholders' orientation: How does governance affect social and environmental disclosure? Journal of Business Ethics, 114, 29- 43.
4. Panwar, R., Nawani, S., & Pandey, V. (2018). Legislated CSR: A Brief Introduction. In: Corporate Social Responsibility; Business and Society 360. Emerald Publishing Limited: Bingley, UK, 133- 146.
5. Bifulco, G. M., Savio, R., Paolone, F., & Tiscini, R. (2023). The CSR committee as moderator for the ESG score and market value. Corporate Social Responsibility and Environmental Management, 30(6), 3231-3241.
6. Cordazzo, M., Bini, L., & Marzo, G. (2020). Does the EU directive on non-financial information influence the value relevance of ESG disclosure? Italian evidence. Business Strategy and the Environ- ment, 29(8), 3470-3483.
7. Ali, W., Frynas, J. G., & Mahmood, Z. (2017). Determinants of Corporate Social Responsibility (CSR) disclosure in developed and developing countries: A literature review. Corporate Social Responsibility and Environmental Management, 24(4), 273-294.
8. Bravo, F., & Reguera-Alvarado, N. (2018). Sustainable develop- ment disclosure: Environmental, social, and governance reporting and gender diversity in the audit committee. Business Strategy and the Environment, 28(2), 418-429.
9. Yu, E. P.-Y., Guo, C. Q., & Luu, B. V. (2018). Environmental, social and governance transparency and firm value. Business Strategy and The Environment, 27, 987-1004.
10. Gjergji, R., Vena, L., Sciascia, S., & Cortesi, A. (2020). The effects of environmental, social and governance disclosure on the cost of capital in small and medium enterprises: The role of family business status. Business Strategy and the Environment, 30(1), 683-693.
11. Simon, H. A. (1957). Administrative behavior. New York: The Free Press.
12. Thompson, J. D. (1967). Organizations in action. New York: McGraw-Hill Book Company.
13. Tushman, M. L., & Nadler, D. A. (1978). Information Processing as an Integrating Concept in Organizational Design. Academy of Management Review, 3(3), 613-624.
14. Kolk, A. (2005). Corporate social responsibility in the coffee sector: The dynamics of MNC responses and code development. European Management Journal.
15. Zheng, Q., & Lu, Y. (2018). " Following the Trend" or " Strategic Choice": Response Mechanisms of Corporate Social Responsibility. Nankai Business Review, 21(4), 169-181. (in China)
16. Bardos, K. S., Ertugrul, M., & Gao, L. S. (2020). Corporate social responsibility, product market perception, and firm value. Journal of Corporate Finance, 14(46), 124-135.
17. Zhao, T., Xiao, X., & Zhang, B. (2021). Economic policy uncer- tainty and corporate social responsibility performance: Evidence from China. Sustainability Accounting, Management and Policy Journal, 12(5), 1003-1026.

Spatial Planning and Sustainable Development of Qiangang Village Based on Catalyst Theory

Xiuhong Lin[1,2,*], Yilin Wu[3]

[1]School of Urban Architecture, Guangzhou Huali College, Guangzhou 511325, China
[2]Faculty of Innovation and Design, City University of Macau, Macau 999078, China
[3]School of Architecture, Huaqiao University, Xiamen 361021, China

Abstract. In previous years, construction has proven that the development model of urbanization, large-scale demolition and construction, and pursuit of efficiency are not suitable for rural areas. We need to explore new ways that are more in line with rural development. Catalyst Theory is the process of generating catalytic reactions through small-scale element modification or introduction, driving the activation rate of surrounding large-scale spaces, and striving to create a spatial development model of "small leads to large" with minimal cost investment and minimal spatial intervention. Based on our team's field research data on Qiangang Village over the past six years, we have proposed the spatial planning concept of "Old Village - Catalyst" in Qiangang Village and constructed a planning framework for spatial catalyst, functional catalyst, and cultural catalyst. This article helps to fill the gap in the application of Catalyst Theory in rural planning practice and provides a guidance manual for the restoration and sustainable development of villages in Guangdong Province.

1 Introduction

At present, Chinese society is in a period of rapid development and transformation, and the renewal and development of rural space are facing deeper dilemmas. In the past few years, putting the urbanization development model into the countryside has proved to be too hasty, bringing not only constructive damage to the countryside with commercial demolition and real estate development but also causing cultural intrusion, one-sided villages, ecological damage, and other problems [1]. Unlike high-cost investment, fast-paced construction, and commercial development in urban development, rural areas lack labor and financial support. Therefore, the development model of large-scale demolition and construction in urbanization and the pursuit of efficiency is not suitable for rural protection and renewal.

Considering the slow development and current situation of rural areas, we need to explore new ways that are more suitable for rural spatial development - following the model of "small leading to large, point leading to area" in terms of space; Following the pace of "gradual and progressive development" in terms of time, we strive to drive rural development with minimal cost investment and minimal spatial intervention.

"Catalyst" is a substance that can change the chemical reaction rate of other substances in a chemical reaction, and its quality and chemical properties have not changed before and after the reaction. Its characteristic is that it is small in quantity and efficient. Nowadays, "catalyst" is also used to metaphorically refer to a medium that promotes changes in things. Urban Catalyst Theory was proposed by Wayne Attoe and Donn Logan, who believed that the various components of a city are interrelated. If one element produces a positive effect, it will act as a catalyst, similar to a catalyst in a chemical reaction, and thus affect the changes of other elements [2]. Unlike the traditional top-down spatial design system from macro to micro, Urban Catalyst Theory aims to drive the development of surrounding areas on a large scale through the intervention of micro factors, from bottom to top [3].

The spatial practice of catalyst theory mostly focuses on urban public spaces, and there is relatively little exploration of its application in rural space design. This paper takes the spatial planning of Qiangang Village in Conghua City, Guangdong Province as an example. Based on the field research data of Qiangang Ancient Village in the past six years, we put forward the concept of "Old Village - Catalyst", which means activating the internal and connecting with the external, realizing the effect of catalyst linkage; we also constructed a framework of spatial catalyst, functional catalyst, and cultural vein catalyst; we focus on the feasibility of combining Catalyst Theory with village construction.

2 Literature Review

In the 1980s and 1990s, the old urban areas of European and American cities began to expose numerous problems that urgently needed to be solved. In 1989, Wayne Attoe and Donn Logan proposed that the renewal of buildings or other components in a city could trigger a chain reaction on a larger scale, thus starting to construct the theory system of touch media [4]. In 2006, Jin explored

* Corresponding author: U22092120351@cityu.edu.mo

the "chain reaction" of Catalyst Theory from the perspective of urban design, promoting the continuous and gradual development of urban space according to people's will, and opening up China's exploration of catalyst theory in the field of space [2].

In recent years, the exploration of Catalyst Theory is gradually moving towards the refinement of methods and diversity of objects. Qin et al. explored the modernization and renewal of traditional commercial streets by using micro-renewal as a means to explore the modernization and protection of traditional commercial streets in a mild and low-impact way with public policy orientation from the surface to the inside [5]. Cui et al. take pocket park as a typical representative of micro-space and discuss the selection of catalyst elements, activation and improvement of catalyst, and guidance of catalyst effect in three steps of Catalyst Theory [6]. Chang et al. address the community's economic, social, and humanistic ecological problems, and build a revitalization strategy for the community business field, living field, and ecological field [4]. Duan explores new ideas for urban stock renewal in the post-industrial era through a theoretical study of the catalytic design of abandoned railroad landscapes and the design practice of the 139 Railway [3].

However, although the application of Catalyst Theory in spatial fields has already included landscape renewal, historical districts, residential communities, micro-updates, commercial districts, and other aspects, there is still relatively little exploration in the academic community of using Catalyst Theory in rural spatial design. Therefore, we strive to propose rural spatial design strategies at the micro level, which will trigger a macro-catalytic chain reaction. This article helps to fill the gap in the application of Catalyst Theory in rural planning practice and provides a guide manual for the restoration and sustainable development of ancient villages in Guangdong Province.

3 Overview and Overall Design Concept of Qiangang Village

3.1 Overview of Qiangang Village

Known as the "Pearl on the Tropic of Cancer" and the "Back Garden of Guangzhou," Qiangang Village is situated in Conghua District, Guangzhou, Guangdong Province. It is said that "without Conghua, there was Qiangang first." Qiangang Village was constructed during the Song Dynasty. Qiangang Village is a typical representative of well-preserved Guangfu residential dwellings, and the Guangyu Temple in the village is referred to by experts as a "chronological history" and "living specimen" of architecture. Qiangang Village has rich historical and cultural resources, such as the Negative Emperor's Sea Crossing, Guangyu Temple, Jiangcheng Map, and the Southern Guangdong Ancient Road. The village layout is a barracks chessboard style, with watchhouses in all four directions and well-preserved.

3.2 Overall Design concept of Qiangang Village: Old Village - Catalyst

Based on Catalyst Theory and our field research data, we put forward the concept of "Old Village - Catalyst" for the planning of Qiangang Village, which means to activate the internal and connect the external and drive the development of the surface by the point to realize the effect of the catalyst theory. The plan selects Guangyu Temple, the most famous temple in Qiangang Village, as the center of the plan, and builds a spatial planning framework of spatial catalyst, functional catalyst, and cultural catalyst.

The planning and design of rural areas have regionalism, locality, uniqueness, and ecological sustainability. The concept of "Old Village - Catalyst" follows the key spaces of the village, such as ancestral halls, water bodies, fields, etc. The design concept starts from rural planning, architectural renewal, and folk culture, aiming to maintain the old building community to the greatest extent possible and to maintain the ancient village with a relatively humble approach. From the perspective of architectural space and settlement culture, the humble concession is a respect for ancient villages and history.

In addition, we are also considering how to reconnect the past, present, and future through the latest virtual reality technology. Therefore, the focus of this planning concept is to establish connections between new and old places and spaces. The connection between old and new places is achieved through "spatial catalysts" and "functional catalysts". The old and new spaces refer to the old villages of today and a hundred years ago, achieved through real spaces and virtual old scenes. There are two methods: one is to display the scenery of the old village during the day, and at night, to use digital technology to project light curtains to form a virtual scene; The second is to enter the virtual scene of building the old village through human-computer interaction and setting up VR glasses.

4 Spatial Planning of Qiangang Village Based on Catalyst Theory

4.1 Spatial Catalyst

Spatial catalyst is not only the renovation planning and reconstruction of the internal space, landscape, ecology, and business development of old villages but also the method of closely connecting villages with cities; Through the improvement of business formats and industries, old villages have become important nodes and characteristic areas in urban development. Therefore, the spatial planning of Qiangang Village serves as a spatial catalyst through the integration of spatial structure, traffic flow, landscape axis, and water systems (Fig. 1).

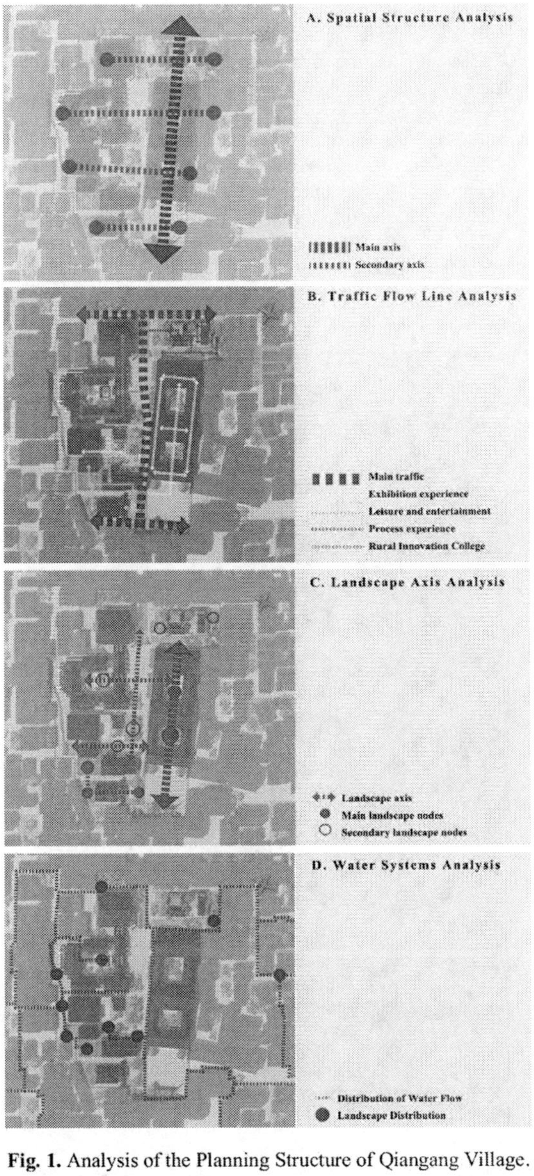

Fig. 1. Analysis of the Planning Structure of Qiangang Village.

4.1.1 Spatial Structure

The main spatial axis of the old village is roughly north-south, facing north and south, with a good landscape orientation. The main axis extends along the axis of Guangyu Temple, and the east-west direction is the secondary axis of the spatial structure. The main transportation axis is the north-south village road, with craft workshops and rural innovation colleges distributed on both sides of the east and west, and transitional stitching between landscape space and outdoor leisure space (Fig. 1A).

4.1.2 Traffic Flow

Retain the main spatial axis of the original old village, expand spatial distribution along the main axis, and plan various types of activity flow lines. According to the spatial function layout, set up tourist flow lines, leisure and entertainment flow lines, craft experience flow lines, and rural innovation college flow lines, and stitch together the missing traffic flow lines (Fig. 1B).

4.1.3 Landscape Axis

The landscape axis unfolds along the spatial main axis, and the main landscape setting area is arranged along the left and right sides of the main passageway, and the landscape space is set up with the leisure area, and the main spaces are the patio of Guangyu Temple, the square in front of the theater, the children's play area behind the theater, the tea room and water view area, and the open-air cafe, etc. (Fig. 1C).

4.1.4 Water Systems

The water channel in Qiangang Village winds and turns, passing through the halls and houses, passing through the Moon Marsh, and finally flowing into the village pond. The water channel passing through the front and back of the house is a major landscape resource in the village. Considering resource utilization, we chose to enlarge the nodes of the local water channel and build a small water pool as a display area for lighting and water features, which is also an important part of improving the village's climate (Fig. 1D).

4.2 Functional Catalyst

Qiangang Village has abundant space and a strong culture of traditional houses, private schools, and academies. Guangyu Temple, the most famous ancient building in Qiangang Village, has attracted scholars to come here for investigation. After conducting field research, we found that only the Guangyu Temple itself has been well repaired near Guangyu Temple. The ancient buildings, landscapes, and spaces with research value nearby have not been restored, and most of the surrounding buildings have collapsed, resulting in poor business conditions.

- As a result, the critical repair space for functional catalysts revolves around Guangyu Temple, and we suggest the following specific directions:
- With virtual reality as the theme and Guangyu Temple as the center, the space is transformed and updated with diffusion to the surrounding areas.
- The renovation and update content includes building posture, spatial form, and landscape form.
- Retain the existing appearance of ancient buildings, utilize existing old sites, and integrate with the development of higher education to implant new functions into the space, such as rural tea rooms, rural living rooms, historical and cultural research institutes, VR equipment experience halls, rural innovation colleges, scholar research institutes, etc.

We strive to use leisure services and cultural spaces, combined with the establishment of rural innovation colleges in universities, as a functional catalyst to drive the development of some spaces through cultural functions (Fig. 2).

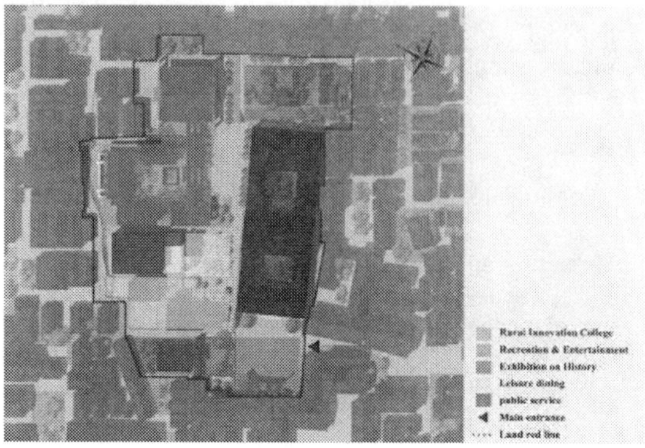

Fig. 2. Aerial View and Functional Zoning Analysis of Qiangang Village.

4.3 Cultural Catalyst

The core cultural area of Qiangang Village is Guangyu Temple and the surrounding areas. The "cultural catalysts" are tested in the abandoned buildings surrounding Guangyu Temple. It is possible to create a space where the virtual and the real, the past and the present, converge by digitizing the creation of virtual scenes and weaving them together with real scenes. The majority of the residents of the old village have relocated to the new village, and those who have remained there are primarily the elderly, according to our research. There are very few leisure activities available to the elderly, who spend most of their days farming. The majority of visitors come to this place to explore the old village's historic construction, and the range of people visiting is very narrow (Fig. 3).

After comparing multiple solutions and ensuring that the original spatial environment is not damaged, simple repairs will be made to the surrounding ancient architectural environment, and new functions will be implanted, as follows:

• The square in front of Guangyu Temple is designed with a stage based on the surrounding architectural style, facing the elderly from the old village.

• Transform the ancient building next to Guangyu Temple into a herbal tea shop, tea room, and light dining bar, catering to tourists and Indigenous people

• Transformation of the surrounding courtyard as a village creation institute, providing workplaces for scholars on study tours and also attracting young people to return to the old village, triggering the young generation's attention to the old village and their care for the old village elders.

• Reinforcement and maintenance of the walls based on the original ruins, not to build new buildings on the basis of the original, but to use holographic digital technology to reproduce the old scenes and use transparent colored organic materials to build virtual reality equipment pavilions.

• For the renovation of the field behind Guangyu Ancestral Hall, the location of the field is re-planned, and the buildings in the yard are remodeled to serve as a base for parent-child agro-entertainment business, which not only broadens the village's audience, but also re-develops the abandoned field as part of the landscape and planting experience, and at the same time drives the development of the village's commercial chain.

In the plan, these types of businesses around Guangyu Temple can not only be developed around Guangyu Temple, but also apply to other areas of the village, such as private schools, academies, and old buildings.

Fig. 3. Partial rendering of Qiangang Village spatial planning.

5 Discussion: Old Village - Catalyst and Rural Sustainable Development

The catalyst theory can not only be applied to rural planning, but also promote the common development of cities and rural areas. The city and the countryside work well together, and the growth of one area affects the other's progress in all directions [7]. In addition to internal catalysts in rural areas, rural catalysts are more important in building connections between rural areas and cities, as well as between rural areas and society. Due to backward development, rural areas often become disconnected from the mainstream society of cities, and both economic and industrial development lag behind urban space to a certain extent [8]. But compared to the city, the countryside has more ecological resources. Therefore, the first step in rural construction is to improve the management of rural ecology, and secondly, to align the basic supporting facilities of rural areas with the level of cities. Only on this basis can we better achieve the subsequent rural catalyst and sustainable development.

However, rural catalysts and the sustainable development of rural and urban areas also face multiple challenges: first, the transportation industry's slow growth in the rural's natural geographic location; second, the majority of industries suitable for the countryside's development are in agriculture and light industry, while the digital industry cannot be developed well in the countryside [9].

Previous experiences and resident feedback are also influencing the creation of functional and cultural catalysts in rural planning. For example, Shanghai's historic districts are being renovated by implementing protection measures, introducing new roles, and leveraging industries to drive the growth of historic areas. Lishui welcomes tourists by developing teaching and cultural businesses in rural areas and universities with excellent natural resources, as well as collaborating between universities and rural colleges. These models have received widespread recognition from both locals and tourists.

6 Conclusion

We have constructed a rural symbiosis catalyst strategy of "taking the small with the big, step by step", and through the spatial planning framework of spatial, functional and cultural catalysts, on the one hand, it has an important practical role in revitalizing the remaining rural space, developing rural resources in an orderly manner, and it has an important academic reference value in the exploration of the rural revitalization and the village spatial development mode; On the other hand, it has also become an effective way to trace the origin of rural culture and protect and develop rural culture as a non-renewable resource.

The new concept of "Old Village - Catalyst", combined with virtual reality to transform the countryside, is undoubtedly a bold way to co-develop the old and the new. Through the catalyst of space, the functionality of the internal space of the countryside will be reproduced, which will lead to a larger effect, thus linking the countryside and the city, and truly realizing minimal intervention and maximizing the benefits.

Acknowledgments

This research was funded by the Guangdong Province Philosophy and Social Sciences 2023 Annual Discipline Co-construction Project (Grant No. GD23XSH15), the Guangzhou Phi-losophy and Social Science Planning 2022 Project (Grant No. 2022GZYB51), and 2021 Guangzhou Huali University Research Project (Grant No. HLKY-2021-SK-1).

References

1. Wei, Y.; Zhang, A.; Ma, Y. A Bibliometric Review of Rural Living Environment Improvement Research in China Based on CNKI Database: 1992–2022. Sustainability 2023, 15 (8), 6561. https://doi.org/10.3390/su15086561.
2. Jin G.; Chen Y. On the Impact of UrbanDesign Projects under "Catalyst Effects" on the Surrounding Environment. Planners 2006, No. 11, 8–12.
3. Duan W. Space Renewal of Abandoned Railway Landscape under Urban Catalyst Effect: Research and Practice on the Renewal Design of Jinzhong 139 Railway Park. Art Research 2022, No. 6, 105–110. https://doi.org/10.13318/j.cnki.msyj.2022.06.007.
4. Chang W.; Zheng K.; YiLizire J.; Xue Y. Live· Network·Field: Research on Construction of Community-Life Circle of Yingping Community in Xiamen Based on Urban Catalyst Theory. Urban Development Studies 2020, 27 (8), 117–124.
5. Qin H.; Hu L. Micro renewal strategy for traditional commercial districts based on urban catalytic effect. Planners 2019, 35 (S1), 81–86. https://doi.org/1006-0022(2019)S0-0081-06.
6. Cui L.; Zhao Y.; Li D. Design Experiment Study on the Correlation between Urban Micro-Space Renewal and Urban Catalyst Effect: Record of the International Competition for Renewal Design of Pocket Park in Nanqiao, Fengxian. Chinese Landscape Architecture 2020, 36 (8), 47–51. https://doi.org/10.19775/j.cla.2020.08.0047.
7. Yao, Q.; An, N.; Gu, G.; Yang, E.; Yang, H.; Li, C.; Yan, K. Research Progress on Features and Characteristics of Rural Settlements: Literature Distribution, Key Issues, and Development Trends. Buildings 2023, 13 (10), 2457. https://doi.org/10.3390/buildings13102457.
8. Zhou, Z.; Zheng, X. A Cultural Route Perspective on Rural Revitalization of Traditional Villages: A Case Study from Chishui, China. Sustainability 2022, 14 (4), 2468. https://doi.org/10.3390/su14042468.
9. Lin, X.; Zhang, Y.; Wu, Y.; Yang, Y.; Yang, W. A Bibliometric Review of Chinese Traditional Defensive Settlement Heritage. Builings 2024, 14 (10). https://doi.org/10.3390/buildings14103107.

A Preliminary Investigation on the Relationship Between the Sustainable Construction Field and the Sustainable Development Goals: A Proposed Research Framework

Zhenquan Zhou[1,2], Deprizon Syamsunur[1,3,*], Xiang Fan[2] and Jiawen Li[4]

[1]Department of Civil Engineering, Faculty of Engineering, Technology and Built Environment, UCSI University, Kuala Lumpur, 56000, Malaysia
[2]School of Engineering and Materials, Tianfu College of SWUFE, Mianyang, 621050, China
[3]Postgraduate Studies, Department of Civil Engineering, Universitas Bina Darma Palembang, Kota Pa-lembang 30111, South Sumatera, Indonesia
[4]School of Business and Law, Taylor's University, Selangor, 47500, Malaysia

Abstract. In recent years, there has been a wide concern regarding climate change and the future development of human beings. Some scholars support the idea that construction activity has been exacerbating the global warming issue, which requires a sustainable transformation of the building industry. This article examines the correlation between the construction sector and the Sustainable Development Goals (SDGs) published by the United Nations. Within the framework of global advocacy for sustainable development, the construction sector is progressively advancing the creation of green buildings to facilitate the achievement of the SDGs by minimising resource use and pollution. Researching convincing academic materials and completing the relevant analysis elucidates the low-degree relations between the construction industry and the relevant goals (1, 2, 4, 10, 14, 16 and 17) and the medium-level correlation with objectives 5 and 8. More importantly, the building sector, especially considering the use of modular constructions, significantly contributes to SDGs 3, 6, 7, 9, 11, 12, 13 and 15 considering its influence on energy efficiency, climate change mitigation, urban infrastructure, clean water preservation and many other aspects. A comprehensive research framework is proposed to explore the challenges of using modular constrictions and improve the building industry's sustainability, facilitating SDG achievement. More first-hand data can be collected in the subsequent investigations, and the multi-criterion analysis method can be employed in further research, improving the accuracy and objectiveness of current research findings.

1 Introduction to the SDGs

The quality of life may be markedly enhanced as society advances by adopting creative solutions across several sectors [1]. The global landscape has experienced a steady enhancement in numerous tangible and intangible factors, including economic growth, resource availability, communication infrastructure, transportation efficiency, educational attainment, level of industrialisation, food supply and diversity, life expectancy, and happiness indices. The likelihood of future global conflicts is intensified by the considerable developmental difference between wealthy nations and emerging sectors [2]. The United Nations, an international institution founded in the last century, is crucial for sustaining global peace, resolving wars, and alleviating regional crises. The principal aims of the United Nations, when considered over a prolonged duration, are to advance regional development and foster international peace. In recent decades, the United Nations has committed to promoting sustainable development to ensure humanity's existence on Earth. In 2015, this esteemed worldwide organisation initiated the 2030 Agenda for SDGs, emphasising sustainable development across many states, regions, and sectors. The effective implementation of the SDGs includes poverty eradication, environmental conservation, peace maintenance, and wealth enhancement. The successful implementation of the SDGs can offer a strategic framework for tackling global difficulties among various human groups and yield significant insights for sustainable development on the planet [3].

2 Definition of Constructions

In light of the UN's SDGs, the construction industry has progressively advocated for sustainable buildings in recent years. Traditionally, the execution of a construction project necessitates the fabrication of building components for onsite assembly, covering the full process from raw material extraction to structural completion. The initial phase of typical cast-in-situ construction involves collecting and acquiring necessary raw materials, including stone, metal, wood, water, cement, concrete,

* Corresponding author: deprizon@ucsiuniversity.edu.my

masonry, and other components [4]. The gathered materials need to be processed and conveyed to the building site. Executing the specified onsite constructions guarantees the effective delivery of the required structures. Moreover, the off-site construction method exhibits minor discrepancies, including prefabricated or modular structures. The subsequent phase involves the production of crucial construction components at designated manufacturing plants and the onsite assembly of the precast modules upon their arrival at the construction site [5]. The process considers material extraction, transportation, manufacture, assembly, and final delivery. This domain permits many interpretations of architectural forms based on investor bids and pertinent plans, culminating in outputs that encompass, but are not confined to, residential dwellings, commercial structures, bridges, tunnels, and railroads. Construction is essential for national progress and urban development, impacting several industries and aspects of life. Buildings are often classified into low-rise, high-rise, and supertall categories, corresponding to the distinct construction codes of various countries and regions. Construction items serve many purposes, including residential, industrial, educational, medical, commercial, governmental, transportation, hydraulic, and national defence responsibilities. Moreover, development can be categorised into public and private structures based on specific attributes of immovable property. Consequently, construction may be categorised into several project categories based on distinct evaluation criteria [3, 5].

3 Relation between SDGs and Buildings

Table 1 presents the correlation between the construction sector and the UN's 2030 agenda concerning the specific SDGs analysed with pertinent perspectives substantiated by published and peer-reviewed academic literature in the subsequent paragraphs [3]. This table offers a succinct overview of particular SDGs and assesses the relationship between the stated objectives and the building sector. Concerning the 17 objectives established by the UN, it is observed that seven targets (namely SDGs 1, 2, 4, 10, 14, 16, and 17) are deemed to possess low relevance for the construction sector [6, 7]. The building sector can fulfil fundamental living requirements, contributing to the realisation of SDGs 1, 2, and 4 by 2030. Specifically, the development of residential structures provides basic housing for inhabitants, hence advancing the attainment of SDG 1. The creation of educational institutions, granaries, and diverse industrial edifices can markedly enhance the conditions related to SDGs 2 and 4. The construction industry is vital for employment generation, economic growth, and the establishment of necessary infrastructure, which may indirectly influence the achievement of low-degree connection SDGs 10, 14, 16, and 17. For instance, essential housing stability and continuous economic growth can enhance the quality of life and mental well-being, diminishing the likelihood of adverse emotional states within communities [8]. Meanwhile, stakeholders can obtain more equitable and adequate work prospects amidst diminishing competitiveness and external discriminatory practices, thereby advancing the achievement of SDGs 10 and 16. Consequently, it can be deduced that the construction sector can generate a ripple effect, wherein active building projects enhance community welfare and bolster the local economy, fostering harmony and a comparatively more equitable society for both essential living and professional advancement [6]. Despite the lower degree of development in island regions relative to inland locations, the building sector significantly influences the environmental sustainability of the ocean and its ecosystem. Insufficient and uncontrolled moisture from polluted building water and atmospheric pollution throughout climatic cycles in island locations may adversely affect marine ecology and jeopardise its biodiversity. The building industry is responsible for influencing the operational efficacy of SDG 14 in many instances. Furthermore, transnational building projects are increasingly prevalent and transcend national and cultural boundaries. The active involvement and efficient implementation of international construction projects can enhance stakeholders' comprehension of varied laws and regulations, economic circumstances, and social viewpoints across different nations and areas. This scenario might greatly enhance international cooperation among states and regions, advancing the outcomes of SDG 17 [3, 6]. Thus, prior research demonstrates that the construction industry pertains to several SDGs, including SDGs 1, 2, 4, 10, 14, 16, and 17. Nonetheless, as shown in Table 1 [6, 7, 9, 10], their interconnectedness degrees are relatively low owing to indirect impacts.

Table 1. The relations between the building industry and SDGs.

Specific SDGs	Relations and Degree	Supportive literatures
1 NO POVERTY	Relevant (Low)	[7]
2 ZERO HUNGER	Relevant (Low)	[7]
3 GOOD HEALTH AND WELL-BEING	Relevant (High)	[6, 9, 10]
4 QUALITY EDUCATION	Relevant (Low)	[7]
5 GENDER EQUALITY	Relevant (Medium)	[7]
6 CLEAN WATER AND SANITATION	Relevant (High)	[6, 7]
7 AFFORDABLE AND CLEAN ENERGY	Relevant (High)	[6, 7, 9, 10]

8 DECENT WORK AND ECONOMIC GROWTH	Relevant (Medium)	[6]
9 INDUSTRY, INNOVATION AND INFRASTRUCTURE	Relevant (High)	[6, 10]
10 REDUCED INEQUALITIES	Relevant (Low)	[7]
11 SUSTAINABLE CITIES AND COMMUNITIES	Relevant (High)	[6, 7, 9, 10]
12 RESPONSIBLE CONSUMPTION AND PRODUCTION	Relevant (High)	[6, 9, 10]
13 CLIMATE ACTION	Relevant (High)	[6, 10]
14 LIFE BELOW WATER	Relevant (Low)	[7]
15 LIFE ON LAND	Relevant (High)	[6, 10]
16 PEACE, JUSTICE AND STRONG INSTITUTIONS	Relevant (Low)	[7]
17 PARTNERSHIPS FOR THE GOALS	Relevant (Low)	[7]

Conversely, several targets are seen as having a moderate correlation with the building industry, including SDGs 5 and 8. The building industry is intricately associated with the aims above, yet the degree of their correlation might be enhanced. The construction sector is considered a vital business in many developing nations, significantly impacting economic growth and national development in alignment with the objectives of SDG 8. The construction sector accounts for almost 50% of DFC in Nigeria [11], while the building industry has been seen as a significant economic catalyst in China in recent decades. The regulation of the construction sector influences its alignment with SDG 8 and may also aid in eradicating gender discrimination in the workforce, which is pertinent to the UN's SDG 5. The construction sector constitutes a crucial business that employs a significant workforce and generates matching earnings, particularly in several developed areas. Hence, dealing with the inequalities in employment opportunities and career development in this sector might significantly impact the promotion of gender equality, as specified by SDG 5. In contrast to the previous SDGs (which exhibit less relevance to the construction sector), the proposed group in this paragraph is more pertinent to advancing the building industry. In other words, the relevant criteria indicate that the construction industry does not substantially impact these objectives (SDG 5 and 8) but is more closely involved. The relationship between the chosen SDGs and the construction industry is assessed to possess a modest degree of correlation.

The building business is intricately linked to SDGs 3, 6, 7, 9, 11, 12, 13, and 15. Assistance from the construction industry is crucial for SDG 3 since developing medical facilities and hospitals is fundamental to supplying important resources for healthcare and wellness systems in communities. The construction industry should efficiently control water pollution and maintain the availability of essential water infrastructure to enhance the sustainability of clean water supply globally, directly linked to SDG 6 [12]. Conversely, the successful implementation of sustainable construction projects offers a viable option for energy consumption patterns and building technologies, fostering progress on SDG 7 in developing and developed countries. Several nations, including China and Japan, have enacted pertinent legislation to promote the energy-efficient transformation and enhancement of the building sector, thereby fostering the utilisation of sustainable raw materials and diminishing the consumption of permissible non-renewable energy. The construction sector is integral to SDG 9 since it develops necessary facilities and durable buildings [3]. In the Industrial 4.0 era, incorporating advanced building technologies, such as modular construction, GIS, and VR techniques, and innovative construction management processes (e.g. smart factories and BIM plus models) can substantially enhance innovation in the building sector and its associated engineering industries. This dynamic promotes industry and innovation in urbanised areas [13].

Following the UN's SDG 11, the construction sector may cultivate inclusive communities and create resilient infrastructures that enhance sustainable living for inhabitants. The advancement of social inclusion and cohesion may be realised through improving public infrastructure, effective transportation connections, and the safe operation of functional systems [3]. Accessible housing, secure and eco-friendly living circumstances, and strong infrastructure may greatly enhance human well-being, promoting the long-term sustainability and inclusiveness of urban regions and communities [3]. The building sector is intrinsically connected to the SDG 12. There will probably be improvements in the energy efficiency of natural and non-renewable resources by 2030, underscoring the crucial need for sustainable transformation and advancement in the construction sector to attain this objective (facilitating the achievement of SDG 12). Contemporary construction methodologies, such as modular construction techniques, can enhance energy efficiency in building development in conjunction with the common smart factory-driven model in the sector. The 4.0 era of industry can enhance the sustainability of manufacturing and production [13, 14]. The construction sector is the predominant energy user and generates significant carbon emissions relative to other industries, particularly concerning SDG 13 and SDG 15 [6]. The increasing prevalence of construction projects associated with rapid urbanisation is anticipated to exacerbate global warming and the associated threats of climate change. Consequently, advancements in energy conservation and emission reduction could substantially enhance climate initiatives focused on global environmental sustainability (SDG 13) and safeguard terrestrial ecosystems (SDG 15) from a long-term developmental perspective [3]. The preceding sections demonstrate that the construction sector profoundly impacts the attainment of SDGs, particularly on SDGs 3, 6, 7, 9, 11, 12, 13, and 15 [6, 7, 9, 10].

4 The Case Study of Modular Buildings

Table 1 of the literature evaluation demonstrates the construction sector's beneficial influence on attaining SDGs. Conversely, the SDGs established by the UN delineate essential criteria for the advancement and execution of the building sector [3,6]. A secondary analysis indicates that the worldwide building industry constitutes around one-third of terminal energy consumption, while construction-related activities represent 40% and 33% of overall energy consumption in the United States and Canada, respectively [15]. The SDGs seek to advance a comprehensive strategy for sustainability across multiple sectors, including the environment, economy, culture, society, and technology. The construction sector should undergo substantial change and enhancement in several areas, particularly in advancing and promoting innovative building technologies. The construction industry has considerable potential to serve as a pivotal catalyst for the effective achievement of the SDGs. An increasing cohort of construction professionals and academic researchers advocates for the extensive adoption of modular construction technology in urbanisation and building development, owing to its considerable benefits in enhancing project sustainability [16]. This sustainable building process aligns with SDGs 3, 6, 7, 9, 11, 12, 13, and 15, fostering environmental sustainability and guaranteeing long-term developmental efficacy [3, 5]. Prior building projects foresee the application of cast-in-place technology for significant structures, including load-bearing components and their interconnections. Alternatively, the secondary structural element may be fabricated within the specified facility, thereby optimising the assembly stage of a building project [17]. During the project lifecycle, it is crucial to examine the elements that promote and obstruct the adoption of prefabrication in construction projects and how its effective implementation can improve environmental sustainability and other aspects [5, 16].

The worldwide construction sector is witnessing a notable increase in applying prefabricated technology in construction projects [5]. This sustainable methodology, originating in London, was implemented in residential design throughout the 19th century [18]. Over the last decade, prefabrication has progressively surpassed cast-in-place construction as the predominant building technology, especially in industrialised countries (SDG 9). The prefabricated building approach heavily relies on the promotion of industrialisation. Using this pattern can also facilitate the industrialised transformation of the building sector and assist in completing the infrastructure establishments. Certain construction projects utilise an offsite methodology; however, most integrate modular building technology with conventional cast-in-situ techniques [5]. A rising number of construction projects are employing this sustainable building technology in the context of regional urbanisation (SDG 11). For example, green building technologies can be effectively used in modular constructions, facilitating sustainable building operations from a long-term perspective. The scientific paper states that prefabrication often involves the production of modules and panels in industrial settings, which are then delivered to the building site [17]. The completed building components are systematically incorporated into the next stage of construction to form the overall structure [16]. Prefabrication is an offsite building method that can diminish the construction timeline and on-site operations, reducing work stress and risks for labourers (SDG 3).In other words, this offsite approach lowers the on-site risks and reduces the health issues on building stakeholders, largely emphasising the safety of operations [19]. Meanwhile, the significant decrease in construction wastes, energy use and polluted water can assist in achieving SDGs 6 and 7 as the green building technologies used in modular constructions can protect drinkable water and promote the use of clean energy to some extent. Based on this condition, adopting modular constructions can benefit the local environment under climate change issues and systematically decrease the negative impacts on the on-land life, which also makes efforts on the SDGs 13 and 15. Although reliance remains on the traditional cast-in-situ method for producing certain building goods [20], many industry professionals have recognised the benefits and significance of modular construction. Considering the significant relationship between the UN's SDGs and the practical application of prefabricated buildings, as discussed before, they have begun promoting its broader implementation, particularly in areas necessitating significant construction and infrastructure improvements [3, 5, 6].

5 Discussions

The analysis indicates that the effective promotion of modular construction can enhance the environmental performance of building projects, aligning with the SDGs. However, certain challenges may impede the successful execution of this advanced construction method [5, 17]. Based on this condition, it is essential to introduce a research approach designed to comprehensively examine the potential opposition to the progress of modular building. Figure 1 delineates the intricate structure of the research process and gives the foundation for addressing the principal investigation inquiry. The initial phase involves a comprehensive assessment of credible literature by establishing clear inclusion and exclusion criteria. A further case study utilising current news articles and interviews may validate or augment the aggregate findings derived from Step 1. The questionnaire survey seeks to assess and validate the state of each influencing factor, encompassing the incentives and obstacles to adopting modular structures in the targeted regions. The integrated results from the systematic review and case study, considered in conjunction with implementing the Unified Theory of Acceptance and Use of Technology (UTAUT) model, can facilitate the development of electronic questionnaires disseminated to various stakeholder groups according to specified inclusion criteria. The Statistical Package for Social Sciences (SPSS) can proficiently analyse the primary data from the questionnaire survey employing the UTAUT model. The third phase involves employing multi-criteria decision analysis in conjunction with the Delphi technique to evaluate the dependability of the results obtained from Phase 2. The findings can be corroborated through in-person discussions with esteemed professionals to assess and improve the contributing elements. Considering this well-rounded and good-quality research framework, all three steps can explore the influencing factors of using modular constructions. In other words, this situation can address the relevant obstacles and effectively promote modular constructions in construction-needed countries and regions.

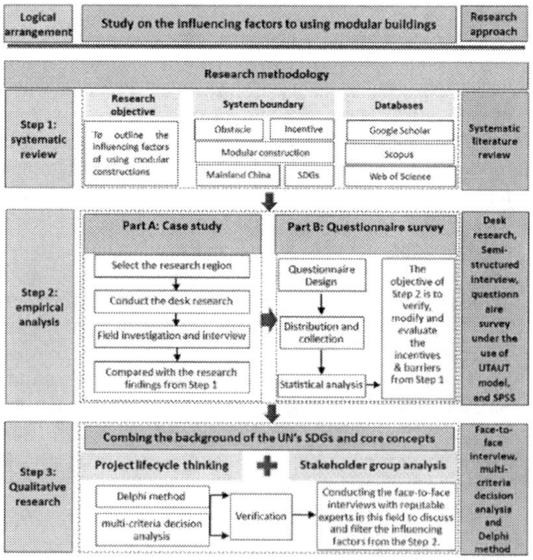

Figure 1. The study flow of further investigations.

6 Conclusions

The UN's SDGs seek to enhance human welfare, foster economic advancement, and safeguard the global environment. As a fundamental component of national growth and urban advancement, the building sector directly or indirectly impacts several SDGs. To some extent, the building sector fulfils basic requirements, generates employment, and stimulates economic growth, contributing to achieving SDGs 1, 2, 4, 10, 14, 16 and 17. The construction sector notably contributes to the objectives of SDG 3, 6, 7, 9, 11, 12, 13 and 15 through the development of clean water and energy resources, as well as the construction of sustainable buildings and infrastructure. It is clear that the building sector greatly contributes to reducing carbon emissions and mitigating climate change. Besides, the construction industry is partly pertinent to attaining SDGs 5 and 8, progress in social inclusion and workplace equality would foster more equitable social development. In other words, the sustainable transformation of the construction industry is crucial for attaining multiple SDGs, requiring coordination across all stakeholders at the policy, technical, and practical levels to promote sustainable growth within the sector. The SDGs established by the UN delineate critical standards for the advancement and execution of the building sector. Consequently, the construction industry should progress and enhance several aspects, particularly in advancing and promoting modular building technology. From this perspective, the wide adoption of modular construction possesses considerable potential as a pivotal catalyst for effectively achieving SDGs, especially for its targets 3, 6, 7, 9, 11, 12, 13 and 15. Thus, many building professionals advocate for the extensive implementation of modular construction technology in urban development and building initiatives, owing to its considerable benefits in enhancing project sustainability and aligning with the UN's SDGs based on long-term development insight. Importantly, exploring the potential hindrances to using modular constructions is essential to ensuring the effective use and wide promotion of this sustainable building approach. If modular constructions can be more widely adopted by building sectors, this situation can positively contribute to achieving the UN's SDGs.

Acknowledgements

It is necessary to express appreciation for the support from the Faculty of Engineering, Technology and Built Environment, UCSI University, Malaysia, as well as many thanks for the assistance from Tianfu College of SWUFE, China.

References

1. Bogner, A. and Mennell, S., 2022. Civilisations, Civilising Processes and Modernity-A Debate. Springer International Publishing.
2. Che, C., Li, S., Yin, Q., Li, Q., Geng, X. and Zheng, H., 2023. Does income inequality have a heterogeneous effect on carbon emissions between developed and developing countries? Evidence from simultaneous quantile regression. Frontiers in Environmental Science, 11, p.1271457.
3. The United Nations 2016, THE 17 GOALS, The Department of Economic and Social Affairs, viewed 16 July 2024, <https://sdgs.un.org/goals>.
4. Domone, P. and Illston, J., 2018. Construction materials: their nature and behaviour. CRC press.
5. Zhou, Z., Syamsunur, D., Wang, L. and Nugraheni, F., 2024. Identification of Impeding Factors in Utilising Prefabrication during Lifecycle of Construction Projects: An Extensive Literature Review. Buildings, 14(6), p.1764.

6. Fei, W., Opoku, A., Agyekum, K., Oppon, J.A., Ahmed, V., Chen, C. and Lok, K.L., 2021. The critical role of the construction industry in achieving the sustainable development goals (SDGs): Delivering projects for the common good. Sustainability, 13(16), p.9112.
7. Goubran, S., 2019. On the role of construction in achieving the SDGs. Journal of sustainability research, 1(2).
8. Barbosa, F., Woetzel, J. and Mischke, J., 2017. Reinventing construction: A route of higher productivity. McKinsey Global Institute.
9. Wen, B., Musa, S.N., Onn, C.C., Ramesh, S., Liang, L., Wang, W. and Ma, K., 2020. The role and contribution of green buildings on sustainable development goals. Building and Environment, 185, p.107091.
10. Omer, M.A. and Noguchi, T., 2020. A conceptual framework for understanding the contribution of building materials in the achievement of Sustainable Development Goals (SDGs). Sustainable Cities and Society, 52, p.101869.
11. Saka, N. and Olanipekun, A.O., 2023. Relationship between the economy, construction sector and imports in Nigeria. International Journal of Construction Management, 23(2), pp.297-306.
12. Belayutham, S., Gonzalez, V.A. and Yiu, T.W., 2016. The dynamics of proximal and distal factors in construction site water pollution. Journal of Cleaner Production, 113, pp.54-65.
13. Maskuriy, R., Selamat, A., Maresova, P., Krejcar, O. and David, O.O., 2019. Industry 4.0 for the construction industry: Review of management perspective. Economies, 7(3), p.68.
14. Zhou, Z., Wang, L., Tian, C. and Syamsunur, D., 2023, December. The development of the smart factory during the Industry 4.0 period under the study of system thinking. In Proceedings of the 2023 International Conference on Artificial Intelligence, Systems and Network Security (pp. 211-215).
15. Kamali, M. and Hewage, K., 2017. Development of performance criteria for sustainability evaluation of modular versus conventional construction methods. Journal of cleaner production, 142, pp.3592-3606.
16. Cao, X., Li, X., Zhu, Y. and Zhang, Z., 2015. A comparative study of environmental performance between prefabricated and traditional residential buildings in China. Journal of cleaner production, 109, pp.131-143.
17. Zhou, Z., Syamsunur, D., Wang, L. and Kit, A.C., 2024. Exploring the feasibility of using modular technology for construction projects in island areas. Journal of Infrastructure, Policy and Development, 8(4), p.3424.
18. Agha, A., Shibani, A., Hassan, D. and Zalans, B., 2021. Modular construction in the United Kingdom housing sector: barriers and implications. Journal of Architectural Engineering Technology, 10(2), p.236.
19. Fard, M.M., Terouhid, S.A., Kibert, C.J. and Hakim, H., 2017. Safety concerns related to modular/prefabricated building construction. International journal of injury control and safety promotion, 24(1), pp.10-23.
20. Philip, A.M. and Kannan, M.R., 2021. Constructability assessment of cast in-situ, precast and modular reinforced concrete structures. Materials Today: Proceedings, 45, pp.6011-6015.

E3S Web of Conferences **617**, 03015 (2025) https://doi.org/10.1051/e3sconf/202561703015
EEUPD 2024

Hefei City Carding in the 21st Century

Menghan Zhu[1], Baoqiang Wang[2,*]

[1] Department of Urban Planning and Design, Faculty of Innovation and Design, City University of Macau, Macau, China
[2] School of Architecture and Urban Planning, Suzhou University of Science and Technology, Suzhou, Jiangsu, China

Abstract. Hefei, the capital of Anhui Province, is situated in central China. This city, with a history spanning over 2,200 years, remained a medium-sized urban area until the late 1970s. However, in the past decade, Hefei has undergone significant spatial expansion and economic growth. This paper examines the city's development trajectory, achievements in socio-economic and spatial growth, and the challenges it faces in the future. Urban researchers have largely overlooked cities in central China; thus, this study aims to enrich the literature in this field, providing readers with a deeper understanding of the urbanization and industrialization efforts in a typical emerging inland city in China.

1. Introduction

1.1 City and Timeframe

The city selected in this article is Hefei City, Anhui Province, and the time selected is the 21st century (Figure 1) (Source: Hefei Municipality, 2015)

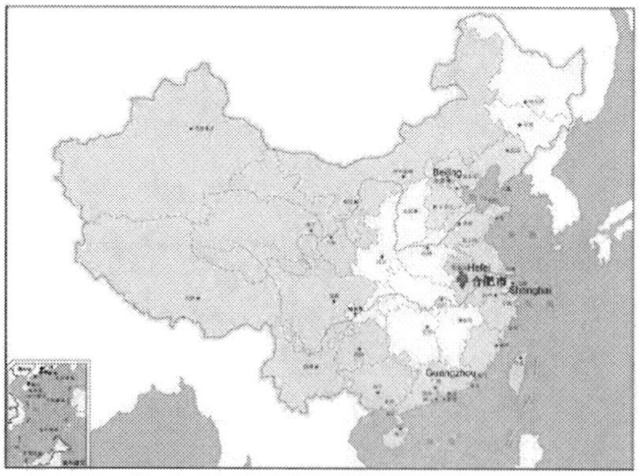

Figure 1. Location of Hefei in China.
(Source: Hefei Municipality, 2015)

1.2 Rationale for Selection

Hefei, the capital of Anhui Province located in central China, is a city with over 2,200 years of history. Despite this long history, it remained a medium-sized city until the late 1970s. However, in the past decade, Hefei has experienced significant spatial expansion and economic growth. Since the beginning of the 21st century, Hefei has leveraged a series of initiatives—such as combining demolition with new construction, fostering an industrially driven urban economy, and prioritizing technological innovation—to overcome inherent disadvantages and achieve rapid progress, ultimately emerging as one of the fastest-developing cities in the 21st century.

Furthermore, urban researchers have largely overlooked cities in central China. Thus, this paper contributes to the literature in this field by offering insights that help readers better understand the urbanization and industrialization efforts of a typical emerging inland city in China.

2. Historical Background Characteristics

2.1 Historical Background

2.1.1 Socioeconomic Background

From 2007 to 2019, Hefei was one of China's fastest-growing economies. The city's total economic output rose from 14.04 billion RMB in 2007 to 62.74 billion RMB in 2016, with an average annual growth rate of 16.5% (Figure 2) (Hefei Bureau of Statistics, 2017). This period is thus referred to as Hefei's "Golden Decade." Industrial and producer services accounted for 85% of the city's total economy, acting as the primary drivers behind its robust economic growth. [1]

As an inland city, Hefei had a weak industrial base, with few large-scale industrial projects allocated to it during the planned economy era. For a long time, Hefei was positioned as a city focused on science and education. However, due to its limited industrial foundation, the city's scientific and technological advances struggled to find local applications. Confronted with the challenge of an imbalance between strong research capacity and weak industry, Hefei launched an industry-oriented development strategy in 2005. Under this strategy, Hefei aimed to bridge research resources with industrial development to foster innovation. As the first step, Hefei applied to become one of China's pilot cities for

* Corresponding author: wbg318@163.com

technological innovation. Simultaneously, the city established a series of institutional reforms and enterprises designed to promote knowledge flow and technology transfer among research institutes. Hefei's goal was to cultivate emerging industrial clusters in fields such as information and communication technology, artificial intelligence equipment, new energy, AI-driven voice technology, and integrated circuits. The city also prioritized modernizing its traditional manufacturing sector. For example, in 2016, Hefei produced 65 million units of the "four major appliances" (air conditioners, refrigerators, televisions, and washing machines), becoming the largest home appliance manufacturing base in the country. Over the past decade, Hefei has developed six major leading industries: automotive, equipment manufacturing, home appliances, LCD technology, new energy, food processing, and agricultural by-products.

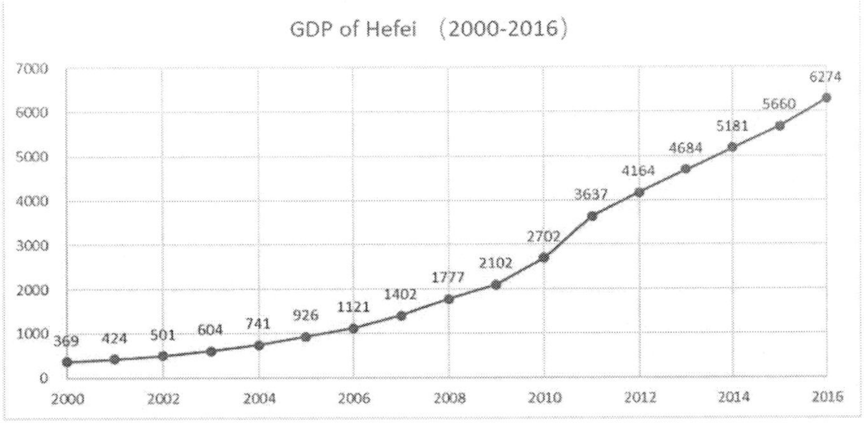

Figure 2. The GDP growth of Hefei from 2000 to 2016 (unit: billion yuan). (Source: Hefei Statistical Bureau, 2017).

2.1.2 Socio-Cultural Background

Since the 21st century, Hefei has developed a distinctive cultural background driven by technological innovation. As one of China's science and technology centers, Hefei has not only thrived in the high-tech industry but has also undergone significant transformations in the cultural sector. Technological innovation has become a leader in the development of urban culture, with the rise of high-tech industries injecting modern technological elements into the city. The construction of representative projects, such as large scientific installations, has closely integrated technology and culture, forming an interactive pattern between technological innovation and cultural development.

With the rise of the cultural and creative industries, Hefei has actively promoted the development of cultural creation, enriching the city's cultural life through various artistic activities, design exhibitions, and creative markets. The emergence of the cultural and creative industries has not only provided local artists and creative talents with extensive development opportunities but has also added unique cultural charm to the city. The construction of urban cultural facilities has become a key element in enhancing cultural soft power. The establishment of museums, art galleries, libraries, and other cultural venues has provided citizens with more opportunities for learning and artistic exchange, further enriching Hefei's cultural heritage.

In the process of modernization, Hefei emphasizes the inheritance and development of local culture. By hosting traditional cultural activities, preserving historical relics, and exploring folk culture, the city maintains a respect for traditional culture, forming a diverse and inclusive cultural landscape. Modern technology and traditional culture coexist and prosper in this city, laying a solid foundation for Hefei's civilizational progress.

2.1.3 Socio-Scientific and Educational Background

Hefei is a city rich in scientific and educational resources. In 1970, one of China's most prestigious universities, the University of Science and Technology of China (USTC), relocated from Beijing to Hefei. With the move of USTC, many national research institutions, including several branches of the Chinese Academy of Sciences (CAS), also settled in Hefei. USTC and CAS represent the highest levels of research in the country, establishing Hefei as a renowned city for science and education. As of 2017, Hefei had more than 60 higher education institutions, with a total enrollment of over 410,000 university students (Hefei Statistical Bureau, 2017).

Additionally, Hefei is home to 564 research institutions and 71 academicians from the Chinese Academy of Sciences and the Chinese Academy of Engineering. As a center for basic scientific research, Hefei has demonstrated strong capabilities in innovation and technology transfer. As early as the 1980s, Hefei developed the country's first microcomputer and the world's first Video CD (VCD) player. Although these technologies had significant market potential, they unfortunately could not be applied locally due to the weak industrial foundation at that time. After the local government implemented numerous measures to enhance industrial support, Hefei's innovation capabilities began to play an increasingly important role in enhancing the competitiveness of the city's industries.

In 2017, the central government approved the construction plan for the Hefei National Comprehensive Science Center, making Hefei the second national-level innovation system base platform after Shanghai. According to this plan, Hefei will focus on interdisciplinary research in the fields of information, energy, health, and environment. By the end of 2016, the number of patent inventions in the city ranked seventh nationwide, the number of national-level high-tech enterprises also ranked seventh, and the ratio of R&D to GDP ranked second in the country (China Review, 2017). Leveraging its innovation capabilities, Hefei has developed several influential industry clusters based on advanced technologies, including a globally leading artificial intelligence voice research cluster and a nationally leading cluster for LEDs and integrated circuits.

2.2 Historical Events

1. Establishment of the National Synchrotron Radiation Laboratory at the Hefei Institute of Physical Science, Chinese Academy of Sciences (2009): The construction of this laboratory marked a significant breakthrough for China in the field of synchrotron radiation, providing strong support for technological innovation in Hefei and the surrounding areas, and promoting the development of high-tech industries.

2. Launch and Expansion of Hefei Metro (2016 onward): The operation of the Hefei Metro not only improved the city's transportation efficiency but also facilitated closer connections among various urban areas, promoting the flow of commerce and people, and having a positive impact on urban development.

3. Construction and Development of the Hefei High-Tech Industrial Development Zone: As the core area of China's Science City, the development zone has attracted a large number of high-tech enterprises and research institutions, driving the upgrading of Hefei's industrial structure and rapid economic growth.

4. Completion of the Anhui Grand Theatre (2012): The construction of the Anhui Grand Theatre injected new vitality into Hefei's cultural industry, serving as a cultural center for hosting various artistic activities and performances, thus fostering the prosperity of local cultural and artistic endeavors.

5. Strategic Cooperation between the University of Science and Technology of China and Hefei (2015): This strategic partnership organically combined the strengths of USTC with the needs of Hefei, promoting deep integration between the university and the city, and facilitating the transformation of research achievements and talent cultivation.

6. Hefei's "14th Five-Year Plan" for the Development of the New Energy Vehicle Industry (2021): This plan mentioned the ambition to establish Hefei as the "Capital of New Energy Vehicles."

These events collectively outline the trajectory of Hefei's development in the 21st century, driving the city's comprehensive transformation from traditional to modern.

3. Urban Layout Characteristics

3.1 Urban Layout

Since entering the 21st century, Hefei's urbanization process has accelerated further, with a dramatic expansion of the urban scale. The 2006 version of the urban master plan designated all construction land in the Hefei urban area as part of the central urban district, leading to an expansion of the city along the banks of Chaohu Lake and a rapid increase in the urban built-up area. In 2012, Hefei began preparing a new version of the urban master plan. Following the administrative division adjustment in 2011, the administrative area of Hefei rapidly expanded, and the urban morphology changed along with the changes in administrative boundaries. This ultimately resulted in the current urban form of Hefei, characterized by a "1-3-3-1" clustered layout. As the speed of urban expansion has outpaced the preparation of urban planning, an analysis using Google Earth satellite imagery from 2000 to 2015 was conducted (see Figure 3). From 1995 to 2005 and then to 2015, Hefei's urban built-up area has expanded dramatically. The northern part of the city has grown along the north-south axis due to the presence of the Dongpu Reservoir and the Dafangying Reservoir, extending along both sides of the reservoirs. However, the scale of this expansion has been limited due to reasons such as the protection of water sources. Simultaneously, the city has rapidly developed toward the southwest, incorporating Dashushan Mountain into the urban area, while the southern part has extended to the shores of Chaohu Lake[7].

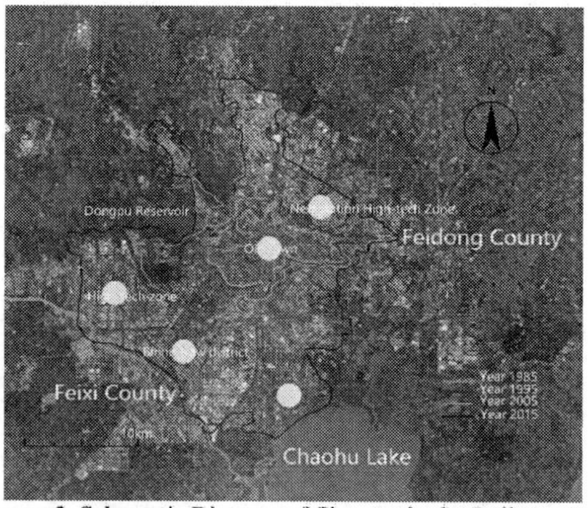

Figure 3. Schematic Diagram of Changes in the Built-up Area of Hefei[2]

3.2 Architectural Styles

3.2.1 Traditional Architectural Style

Hefei is located in the central Anhui region. Based on extensive literature review and practical research, the characteristics of vernacular residential architecture in this area reveal that the convergence of multiple cultures has led to diverse forms of local dwellings. This results in a complex, pluralistic, and inclusive architectural style, which embodies the simplicity and conservatism of

Huizhou architecture while also incorporating the ruggedness of northern architectural styles[6].

(1) **Roof Characteristics**: The architectural forms of residential buildings in the central Anhui region are closely related to the climate. The subtropical, rainy summer climate leads to roofs with a slight slope and no eaves. The predominant roof color is grayish-blue, with some roofs in brick red. Materials commonly used include small blue tiles, laid directly on the rafters in an alternating pattern without mortar in between, creating a double-sloped hard mountain tile roof. Some buildings feature modern flat roofs. The residential architecture in Sanhe Ancient Town has also absorbed elements of Huizhou architecture, such as the horse-head wall, which has characteristics reminiscent of flying eaves and steep cliffs. Unlike the step-like white horse-head walls often seen in southern Anhui, the vernacular houses in central Anhui feature double-sloped roofs covered with small blue tiles.

(2) **Wall Characteristics**: Walls of residential buildings in central Anhui are generally constructed with blue bricks or red bricks, utilizing various masonry techniques, with little plastering to avoid mold and peeling due to the humid climate. Some residences still retain rammed earth walls, which have been reinforced and continue to be used today.

(3) **Decoration Characteristics**: Most residential buildings in central Anhui are brick-structured and range from two to three stories high, displaying a modern style. The residences typically feature double-leaf doors, with door canopies that create a neat and open appearance. The facades of the houses are often adorned with wooden windows. Many buildings include courtyards, exuding rustic tranquility that seamlessly blends with the surrounding natural landscape of mountains and fields.

(4) **Form Characteristics**: The central Anhui region is flat and open, allowing for large building footprints. The blending of various cultures has resulted in a diverse range of architectural forms that integrate characteristics of both southern and northern Anhui residences. By combining courtyards and light wells along a central axis, the buildings are linked in a way that enhances their layered appearance. [3]

3.2.2 Modern Architectural Styles

The period of large-scale construction (2005–2012) marked a significant transformation in the urban image of Hefei. During this time, a multitude of high-quality and aesthetically pleasing modern buildings were completed, enhancing the city's visual appeal and infusing it with new vitality. This era has helped Hefei evolve into a vibrant and culturally rich city with a distinct personality. These outstanding structures exhibit unique characteristics, such as the profound historical sense of the new Anhui Museum (Figure 4), the ecological and energy-efficient design of the Hefei Grand Theatre, the grandeur and high-tech features of the Hefei Olympic Sports Center, the impressive and dynamic presence of the Battle of the Yangtze River Memorial Museum (Figure 5), the unconventional design of the Hefei Art Museum (Figure 6), the Huizhou style of the new Anhui Celebrities Hall, and the dynamic form of the Anhui Broadcasting and Television Center, among others[4].

Figure 4. The new Anhui Museum

Figure 5. Memorial Hall of the Battle of Crossing the River

Figure 6. Hefei Art Museum

3.2.3 The Evolution of Architectural Styles in Hefei: From Tradition to Modernity

The evolution of Hefei's architectural styles, from traditional to modern, profoundly reflects the long-term changes in its socioeconomic landscape. Through the analysis of specific architectural examples, we can better understand the impacts of this transformation.

(1) Traditional Architectural Case: Baogong Temple

Baogong Temple was built during the Qing Dynasty to commemorate the famous Northern Song statesman Bao Zheng. This building features traditional wooden structures and brick carvings, showcasing the architectural style of the Ming and Qing dynasties.Baogong Temple serves not only as a venue for religious and cultural activities but also as a symbol of local politics and social morality. Its existence reflects the dominance of scholar culture and the bureaucratic system in Hefei's societal structure at the time, emphasizing the city's commitment to traditional moral values amid economic development.

(2) Modern Architectural Case: Hefei New Station Complex

The Hefei New Station Complex is a large integrated development that combines commercial, office, and residential functions. Its modern design incorporates glass curtain walls and high-tech materials, exemplifying the functionality and aesthetics of contemporary architecture.

Socioeconomic Background:

With the rapid economic development of Hefei, particularly the rise of high-tech industries, the New Station Complex has emerged as a modern commercial center that meets the growing demands for economic activities. The development of this building has not only boosted local economic growth but also attracted significant investment and talent, marking an acceleration in Hefei's urbanization process.

(3) Comparative Analysis of Traditional and Modern

Architecture

Cultural and Economic Intertwining: Baogong Temple embodies the accumulation of traditional culture and historical legacy, while the Hefei New Station Complex represents the direction of modernization and economic development. The contrast between the two highlights Hefei's varying pursuits of architectural function and form across different historical stages.

Sense of Social Identity and Historical Continuity: As a cultural landmark, Baogong Temple attracts tourists and researchers, enhancing the cultural identity of local residents. Meanwhile, the construction of the Hefei New Station Complex reflects societal aspirations for modernization while promoting economic development. This combination of tradition and modernity showcases Hefei's diversity and inclusivity in urban development.

3.3 Planning Concepts

3.3.1 Recent Land Use Plans

Over the past two decades, Hefei has developed two rounds of land use planning. The first round, approved by the State Council in 1999 (1997-2010), played a crucial role in controlling and guiding land use. However, it failed to anticipate the rapid expansion of the city, leading to a significant imbalance between land supply and demand. The limitations of this land use plan severely constrained urban spatial growth. In response, the government introduced a second round of land use planning (2006-2020) aimed at adjusting and optimizing the land use structure (Figure 7).

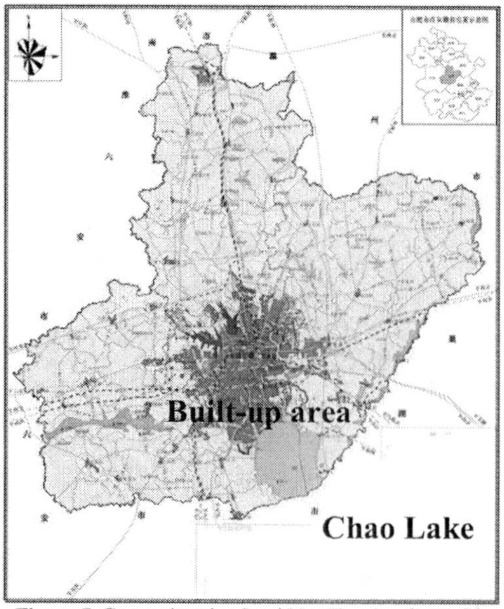

Figure 7. Comprehensive Land Use Planning for Hefei. (Source: Ministry of Land and Resources of China, 2013).

The second round of land use planning adheres to four principles: strict protection of arable land, improving land use efficiency, coordinating different types of land use, and enhancing the ecological environment. By the end of 2005, the total land area of Hefei was 704,701 hectares, of which arable land accounted for 77.98%, construction land for 16.30%, and unused land for 5.72% (Ministry of Land and Resources of China, 2013). Based on the land inventory from 2005, the land use plan set five key objectives to be achieved from 2006 to 2020: (1) the net reduction of arable land and basic farmland should not exceed 3,537 hectares; (2) the total area of land gained through reclamation and redevelopment should reach 22,821 hectares; (3) construction land should not exceed 137,749 hectares, with a net increase of 11,697 hectares in new construction land; (4) per capita construction land should decrease from 226 m^2 to 196 m^2; and (5) priority should be given to the layout of ecological corridors to enhance land use efficiency (Ministry of Land and Resources of China, 2013). By optimizing the land use structure, Hefei's development will become more compact and sustainable.

3.3.2 Development Zone Layout

Since the late 1970s, development zones have played a significant role in promoting industrialization in China. These zones serve as platforms for economic growth, attracting foreign direct investment (FDI), and creating more job opportunities. Hefei has two national-level development zones and approximately ten provincial-level development zones. These areas provide administrative entities with relative autonomy, improving infrastructure, and land reserves, and offering one-stop government services.

The Hefei Economic and Technological Development Zone (ETD) is one of the national-level development zones, established in 1993. Covering an area of 269 km^2, with a built-up area of 83 km^2 (Figure 8), the ETD focuses on the development of manufacturing sectors such as automotive parts, equipment manufacturing, home appliances, light industry, and daily chemicals.

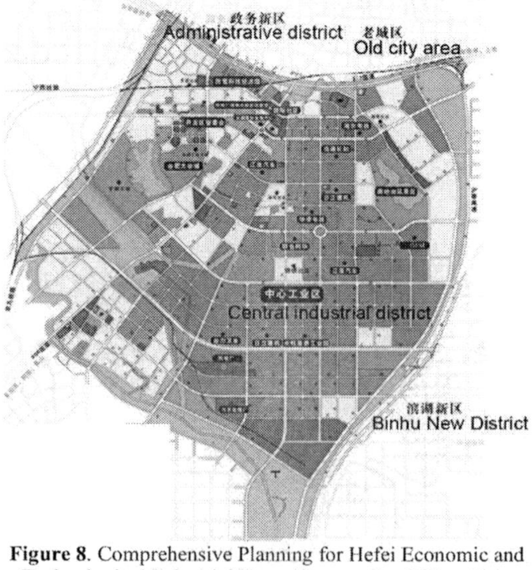

Figure 8. Comprehensive Planning for Hefei Economic and Technological Industrial Zone. (Source: Pearl River Data Industry Alliance, 2010).

https://doi.org/10.1051/e3sconf/202561703015

With strong foreign investment and abundant human resources, the development zone has become the largest manufacturing base in the country for refrigerators, forklifts, excavators, tires, hydraulic presses, and laptops. In 2014, the ETD became the first development zone in Anhui Province to achieve over 10 billion yuan in tax revenue.

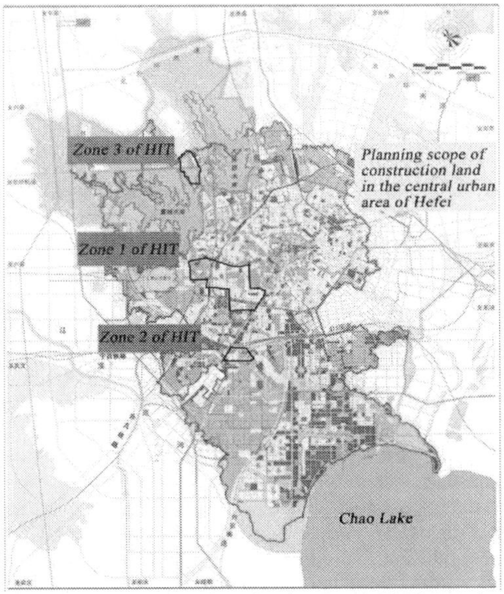

Figure 9. Comprehensive Planning for Hefei High-tech Industrial Zone. (Source: Hefei HTI, 2016).

Hefei's other national-level development zone, Hefei High-tech Industrial Zone (HTI), prioritizes the development of innovation-driven high-tech industries, including electronic information, intelligent voice technology, smart manufacturing, renewable energy, and biomedicine (Figure 9). Established in 1991, HTI was designated as one of China's first high-tech development zones. By 2017, HTI was home to more than 20,000 enterprises and startups (including 603 high-tech companies), ranking sixth in comprehensive competitiveness among 146 national-level high-tech zones in the country. Companies in HTI have developed several groundbreaking technologies, including the world's first optical quantum computer and intelligent voice-interactive banking service robots.

Researchers from the University of Science and Technology of China and other local universities have established many such enterprises. The Hefei Industrial Park, covering an area of 128 square kilometers, has become a sub-center in the western part of Hefei, serving as a platform for the city to engage in global competition through technological innovation.

4. Related Research Introduction

4.1 Scholars and Research

Gu Dazhi, Associate Professor and Master's Supervisor at Hefei University of Technology; Research Focus: Urban Renewal and the Protection of Historical and Cultural Heritage.

Related Research: "A Study on Urban Development and Morphological Changes in Hefei."

5. Status and Influence of the City

5.1 Status

Hefei is a city with a history of over 2,200 years. At the time of the establishment of the People's Republic of China in 1949, Hefei was a small city with a population of around 50,000. By the end of 2016, it had developed into a metropolis with a permanent population of 7.87 million and a land area of 11,408 square kilometers. The city consists of four administrative districts: Yaohai, Luyang, Shushan, and Baohe. The Hefei metropolitan area also includes four additional counties (Feidong, Feixi, Changfeng, and Lujiang) and a county-level city (Chao Lake).

In 2014, the State Council designated Hefei as a sub-center city within the Yangtze River Delta urban region, meaning Hefei would be included in national-level strategies related to regional coordination alongside eastern coastal cities such as Nanjing, Hangzhou, Suzhou, and Ningbo. Hefei serves as the political, economic, cultural, and educational center of Anhui Province, having received honors such as National Garden City, National Sanitary City, and National Excellent Tourist City. It is one of the cities in the Yangtze River Delta National Science Center and a national pilot city for technological innovation and a research and education base, boasting strong research and educational institutions. In 2022, Hefei ranked second globally in the field of quantum information, just behind New York.

5.2 Long-term Influence

As an important city in the Yangtze River Delta, Hefei has a profound and lasting impact on surrounding cities and the entire region.

1. Technology Innovation Leading Regional Development: Hefei's outstanding performance in technological innovation will provide momentum for the development of the Yangtze River Delta. The experience of Hefei's research institutions and high-tech enterprises in promoting technological innovation and industrial upgrading is expected to influence the entire region, driving a transition toward high-end manufacturing and an innovation-driven economy.

2. Talent Attraction and Mobility: The technological development and livable city construction in Hefei will attract more high-level talent, creating a talent agglomeration effect. This is expected to promote talent mobility in surrounding cities, accelerating collaboration and exchange in the Yangtze River Delta. The population flow network within the Yangtze River Delta region exhibits a distinct hierarchical characteristic, structured around a polycentric core-periphery model featuring Shanghai, Nanjing, Hangzhou, and Hefei. The average daily population flows for these cities are as follows:

Shanghai with 674,671, Nanjing with 449,230, Hangzhou with 437,229, and Hefei with 427,525 person-times.

Shanghai serves as the primary center, with dense population flow corridors connecting it to the secondary centers of Nanjing, Hangzhou, and Hefei. Within this framework, Nanjing, Hangzhou, and Hefei experience high-intensity population movements with cities within their respective provinces. Additionally, the Yangtze River Economic Belt between Shanghai and Nanjing facilitates a concentrated corridor of population flow. [8]

3. Industrial Collaborative Development: Hefei's advantageous industries will foster collaborative development with surrounding cities. For example, the construction of high-tech industries and research centers is likely to become a key engine for industrial upgrading in the Yangtze River Delta, facilitating closer ties in the industrial division of labor and cooperation among cities.

4. Transportation Hub Role: Hefei's transportation advantages, particularly the convenient connections provided by high-speed rail and highways, will strengthen its position as a transportation hub in the Yangtze River Delta. This will enhance logistics efficiency across the entire region and promote economic synergy. High-speed rail has effectively improved Hefei's accessibility within the Yangtze River Delta region, reducing intercity rail travel time costs by approximately 50%. Researchers such as Yan Yalei and colleagues analyzed the impact of supplementary travel time using high-speed rail maps via API interfaces. They obtained average travel times between 196 research units based on conventional rail and high-speed rail. Their findings revealed that the average travel time between cities in the Yangtze River Delta decreased from 414 to 1,252 minutes for conventional rail to just 236 to 695 minutes for high-speed rail, indicating that high-speed rail has significantly optimized the accessibility patterns within the region.

5. Cultural and Tourism Exchange: As a city of cultural significance, Hefei's rich historical and cultural heritage will be a highlight of cultural exchange in the Yangtze River Delta. It will promote mutual appreciation of culture among surrounding cities and the broader region, fostering the joint development of the cultural industry.

In summary, Hefei's development in technology, industry, talent, transportation, and culture will bring positive long-term impacts to surrounding cities and the Yangtze River Delta, promoting economic prosperity and social progress across the entire region.

6. Conclusion and Reflections

6.1 Urban Morphology

The importance of urban morphology has often been overlooked in the processes of urban planning and development; however, changes in urban form are closely linked to people's production and daily lives. From Duliang City to the "fan-shaped city" and then to the multi-centered clustered city, Hefei's urban morphological development reflects China's rapid urbanization. As a typical inland city, Hefei's spatial expansion combines intrinsic development with extrinsic growth, facilitating rapid urban morphological expansion. Hefei has consistently pursued its aspiration to connect with rivers and seas. In 2011, through administrative adjustments, Hefei transformed Chao Lake into its "inner lake," utilizing it to connect with the river city belt, achieving a significant breakthrough in accessing the Yangtze River, a "golden waterway." In the future, Hefei's urban form is expected to further expand along the Chao Lake corridor while actively improving the construction of the high-tech zone in the west, linking these two areas to develop a multi-centered clustered city.

6.2 Architectural Style

Hefei's traditional architecture, situated in the central Anhui region, exhibits diverse forms and reflects complex, pluralistic characteristics. It combines the conservative simplicity of Huizhou architecture with the ruggedness typical of northern structures, but the number of existing traditional buildings is limited. Most surviving structures from the Republic of China period are either former residences of celebrities or other historically significant sites. In the context of China's rich traditional architecture, Hefei's local building styles do not stand out; instead, they often utilize the "celebrity story + historical building" approach to repurpose and develop remaining structures.

Hefei's urban image began to transform during the period of large-scale construction, characterized by the completion of numerous high-quality and aesthetically pleasing modern buildings.

6.3 Urban and Environmental Aspects

In recent years, Hefei has explored new pathways for urbanization and industrialization in inland areas, leveraging China's advantages in scientific research, abundant land resources, and low labor costs. However, rapid development has brought several challenges. The first challenge is the increasingly severe environmental pollution, which impacts Hefei's capacity for sustainable development. As a traditional national garden city, Hefei has planned many open spaces and parks within the urban area; however, the city's industrial-oriented policies have contributed to declining environmental quality. Large-scale construction and manufacturing have led to serious air pollution, placing Hefei among the lowest-ranking cities in terms of air quality during the rapid development years of 2015 and 2016 (Tianqi, 2015).

Land space ecological restoration is a major initiative in China's ecological civilization construction, relating to national ecological security and public welfare. In recent years, the provincial government has implemented Xi Jinping's ecological civilization ideas and strengthened ecological construction, increasing efforts in ecological protection and restoration. Hefei has begun to place greater emphasis on ecological conservation and is actively implementing the "Anhui Provincial Land Space Ecological Restoration Plan (2021–2035) [5]."

Sustainable Development of Hefei's Industrial Growth in the Context of Ecological Protection Plans: Recent Data on Hefei's Green and Low-Carbon Transformation, according to the latest data, from January to September 2023, the output value of large-scale enterprises in the new energy and energy-saving environmental protection sectors reached 123.98 billion yuan, reflecting a year-on-year growth of 35.01%. Leveraging its technological innovation advantages, Hefei has carved out a distinctive path for green and low-carbon transformation.

For instance, the China Environmental Valley Exhibition Hall, located in Shushan District, serves as a primary battlefield for Hefei's energy-saving and environmental protection industry, as well as an advanced model for Anhui's environmental industry cluster. It continues to expand its influence in the national environmental sector. Within the valley, Anhui Zhongke Smart Environmental Testing Technology Service Co., Ltd. has independently developed the "Hefei AI Water Quality Automatic Testing Center." This system, centered on automatic detection instruments, can analyze water samples from rivers such as Yangcheng Lake and Baiyangdian, guiding government environmental protection departments in targeted remediation strategies. Currently, it has the capacity to test up to 500 samples per day.

Today, the China Environmental Valley has attracted 30 high-end research platforms in the environmental field at the national and provincial levels. The favorable environment for scientific and technological innovation has further drawn numerous environmental enterprises. Statistics indicate that the China Environmental Valley has gathered over 380 key enterprises in the environmental sector, represented by Sunshine Power (Zero Carbon Base), Hunan Lihe, Landun Optoelectronics, and Sinopec Environmental. Revenue in this sector increased from 5 billion yuan in 2019 to 24.3 billion yuan in 2021, surpassing 35 billion yuan in 2022.

In recent years, Hefei has focused on five key sectors: new energy, efficient energy conservation, advanced environmental protection, resource recycling, and carbon neutrality, carving out a "new runway" for the integration of economic development and environmental protection. The scale of Hefei's new energy and energy-saving environmental protection industries is continuously expanding. In 2022 and the first half of 2023, Hefei ranked first in the province for the comprehensive performance of its dual recruitment and dual attraction initiatives in these sectors. Currently, there are 257 large-scale enterprises and 15 listed companies in the new energy and energy-saving environmental protection sectors, forming a preliminary complete industrial chain. A vertically integrated industrial system has been established, encompassing silicon wafers, battery cells, battery modules, photovoltaic adhesive films, photovoltaic glass, inverters, and photovoltaic power generation applications.

Hefei is now working to create a "fund jungle," promoting the establishment of a 12 billion yuan maternal fund for new energy and energy-saving environmental protection in Anhui Province to be based in Hefei. Additionally, three sub-funds, each exceeding 2 billion yuan in scale, are being prepared for Shushan District, Hefei High-tech Zone, and Feixi County. In the future, Hefei will implement seven major initiatives aimed at cultivating economic entities, developing industrial clusters, and promoting technological products, accelerating toward the goal of a 200 billion yuan industrial cluster.

6.4 Industrial Aspects

Hefei faces challenges in its industrial structure as well. Currently, sectors such as finance, insurance, commerce, logistics, tourism, and e-commerce are relatively weak. Compared to other data sub-center cities in the Yangtze River Delta, such as Hangzhou, Nanjing, and Suzhou, Hefei lacks competitiveness in innovation and needs to catch up with the pace of development in the region. Hefei will require more time to nurture flagship enterprises that can drive economic structural adjustments and industrial upgrades.

6.5 Heritage Conservation Aspects

Balancing urban expansion with cultural heritage preservation in Hefei requires a comprehensive consideration of economic, social, and environmental factors. By formulating integrated planning, encouraging multi-stakeholder participation, strengthening policy support, and learning from successful case studies, Hefei can effectively protect and transmit its rich cultural heritage while pursuing rapid development. This balanced strategy will lay a solid foundation for the sustainable development of the city.

In the process of urban expansion, it is crucial to integrate ecological restoration with cultural heritage conservation. Creating green public spaces can enhance the ecological functions of cultural landscapes; for example, introducing green areas within historical districts can improve environmental quality while preserving historical culture.

Additionally, selecting representative cultural heritage sites for protection and adaptive reuse can serve as demonstrative projects. These initiatives can showcase effective methods for preserving cultural heritage amidst urban growth and provide valuable references for other regions.

1. Baogong Park Conservation and Development:

Baogong Park is an important cultural landmark in Hefei that retains historical relics while attracting visitors through cultural activities and commercial facilities within the park. This project has successfully balanced the needs of cultural heritage preservation with economic development during the city's expansion.

2. Revitalization of Hefei's Old Streets:

The protective renovation of Hefei's old street areas has maintained the character of traditional architecture while incorporating modern commercial elements. This approach not only safeguards cultural heritage but also enhances regional economic vitality and improves residents' quality of life.

Through these strategies, Hefei can achieve a sustainable balance between growth and heritage conservation, ensuring that its cultural legacy is preserved for future generations.

References

1. Tianqi. (2015). Analysis of Air Quality Trends in Major Chinese Cities.
2. Hefei-HTI (2016). The comprehensive plan for the Hefei High-Tech Industrial Zone. Available at http://www.hefei-stip.com.cn/gg/73499.shtml, Accessed date: October 2017.
3. Pearl River Data Industry Alliance. (2010). Comprehensive Planning for Hefei Economic and Technological Industrial Zone.
4. Gu Dazhi. Research on Urban Development and Morphological Changes in Hefei.
5. Anhui Provincial Government. (2021). Anhui Provincial Land Space Ecological Restoration Plan (2021–2035).
6. Zuo Guangzhi & Zhao Wenjun. (2022). Conservation and renewal strategies of vernacular houses in Central Anhui from the perspective of heritage inheritance: A case study of rural living room in Baodian Village, Lujiang County.
7. Gu Dazhi. (2018). Research on Hefei Urban Development and its Morphological Changes (Ph. D. Dissertation, Southeast University).
8. Yan, Y., Yu, T., & Lai, J. (2022). The Impact of High-Speed Rail on Regional Accessibility and Spatial Connectivity Patterns: A Case Study of the Yangtze River Delta Urban Agglomeration. *Shanghai Urban Planning*, (06), 153-158.

Evaluation on the Network Structure of the Park Square System Based on the Social Network Analysis Method —— An Case of Qindu District, Xianyang City, China

Yuyi Chen[a], Tianpeng Wang[*]

School of Architecture and Urban Planning, Lanzhou Jiaotong University, Lanzhou, China

Abstract. In this study, the social network analysis method was used to evaluate the network structure system of Park Square in Qindu District, Xianyang City. The analysis is based on the overall parameters, centrality and cohesive subgroups of the network. The results show that (1) there are obvious differences in connectivity, centrality and aggregation coefficient among the main park squares, and some nodes play a key role in the network. (2) There is a significant cohesive subgroup structure in the park square network, and each node forms its own small group, and the connectivity between the small groups is average. At the same time, by identifying the key nodes and bridging nodes, specific strategies to optimize the network structure are proposed, such as increasing the connectivity between key nodes and improving the accessibility of peripheral nodes.

1 Introduction

Urban parks and plazas are indispensable components in urban construction, serving as important places to regulate the urban environment and improve the living conditions of city residents. Their networked layout is of great significance for achieving sustainable urban development [1-2]. The urban park and plaza system, composed of various levels and types of parks and plazas that are rationally configured, serving as a crucial material carrier for park city construction [3-5].

This paper takes Qindu District, Xianyang City as the research object, constructs a network structure model for the park and plaza system in Qindu District, and leverages social network analysis methods to evaluate the structural characteristics and service effectiveness of the park and plaza system network in Qindu District from multiple dimensions, including network density, clustering coefficient, and centrality.

2 Study methods and data collection

2.1 Social Network Analysis Method and Social Network Model Construction

Social network analysis is a set of theories and methods used to analyze social structures [6]. Based on the perspective of network theory, it views social relations as networks consisting of nodes (individuals, organizations, or units) and edges (relationships, interactions, or connections between nodes). In recent years, social network analysis has gradually demonstrated its unique value and application potential in the research of urban public spaces, particularly in studies of urban park systems, green infrastructure, urban pedestrian networks, and public transportation networks[7-8].

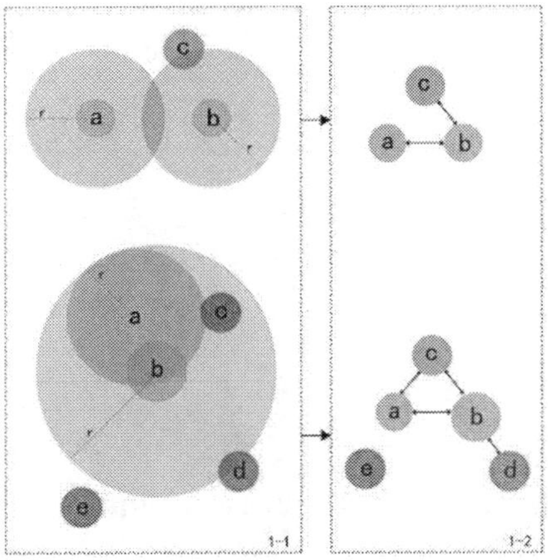

Fig. 1. Expression of Park and Plaza Network Relationships.
Fig. 1-1. The relationships between parks and plazas;
Fig. 1-2. Collaborative Service Network.

In this social network, the "nodes" of the urban park plaza social network and the relationships between these nodes are the park plazas and their service radii, respectively. Firstly, if the service area of urban park plaza A covers park plaza B, then there is a collaborative service relationship between A and B. When multiple parks are in close proximity or have large service areas, park plazas with overlapping service radii will form a collaborative relationship. This type of relationship can enhance the

[*] Corresponding author: [*]wangtp@mail.lzjtu.cn, [a]18691820292@163.com

breadth of park plaza services and the stability of the park plaza network. Individual urban parks or plazas that are isolated from the network have weaker service capabilities to their surroundings. The specific relationship is shown in Figure 1.

2.2 Study objects and data sources

This study takes the parks and plazas (a total of 58) in Qindu District, Xianyang City, as the research objects, including 6 comprehensive parks, 10 specialized parks, 14 community parks, and 28 plazas. The specific locations of the parks and their service radii are shown in Figure 2.

The park and plaza data in this study are sourced from data scraped from Gaode Maps and Baidu Maps; the administrative divisions and boundary data of Qindu District, Xianyang City are sourced from the National Geographic Information Public Service Platform.

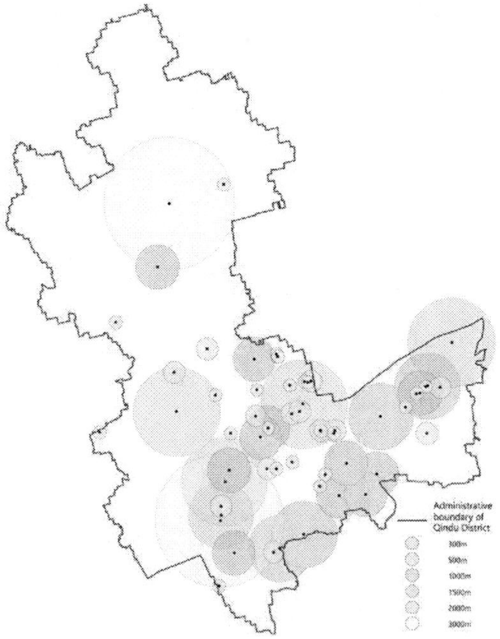

Fig. 2. Distribution and service scope of Qindu District Park Square, Xianyang City.

3 Network structure evaluation of the park square

3.1 Evaluation index of the social network model of the park and square system

The social network of the park and plaza system is structured by the synergistic service relationships among individual nodes and their surrounding nodes. Based on this underlying logic, the evaluation and analysis of the park and plaza system's social network must commence with an examination of the overall structure of the parks and plazas, focusing on assessing the network completeness, which includes network scale and network connectivity. The network scale is analyzed through network density and the overall centralization of the network. The network's connectivity is closely associated with its small-world properties, which are typically evaluated by examining the network's clustering coefficient and the value of its average path length, the analysis of nodes delves into node degree and centrality, with centrality further broken down into degree centrality, betweenness centrality, and closeness centrality [9].

1) Network completeness index

Network scale: The meaning of network scale refers to the number of all nodes in the social network, representing the size of the social network model. The more the number of nodes, the larger the network scale is:

$$p = m/[n(n-1)] \quad (1)$$

where p is the network density, m is the number of actual relationships in the network, and n is the number of nodes in the network.

2) Network connectivity index

Clustering coefficient: The clustering coefficient reflects the coverage breadth and aggregation degree of the network. It is the result of summing the mean values of all nodes after calculating the clustering coefficient of individual nodes. The value range is between 0-1. The calculation formula is:

$$C_i = 1/n \sum_{i=1}^{n} 2E_i/k_i(k_i - 1) \quad (2)$$

where C_i is the overall clustering coefficient of the network, n is the number of nodes in the network, and E_i is the actual number of connections between the k_i adjacency nodes of node i. When C_i =0, there is no connection between two nodes in the network; when C_i =1, any two nodes in the network can be connected. The clustering coefficient in the park square social network represents the connectivity and accessibility between public spaces.

3) Node measurement indicators

Node degree: The node degree represents the number of other nodes in the network that are actually connected to the node. The calculation formula is: The calculation formula is as follows:

$$C_D = d(n_i)/(n-1) \quad (3)$$

where C_D is the point degree, $d(n_i)$ is the number of nodes connected with the node, and n is the total number of nodes in the network. The larger the C_D value is, the more important the position of the node in the network is. In the park Square social network, the node degree represents the spatial node level.

3.2 Analysis of the overall network parameters

3.2.1 Network completeness

Based on the overall network analysis results shown in Figure 3, the social network of parks and plazas in Qindu District, Xianyang City has a network size of 58 with a total of 187 edges, and a network density of 0.0566, which is relatively low. According to the analysis of the social network model of parks and plazas in Qindu District, Xianyang City, the main reasons are that there are too many independent subgroups and isolated nodes in the entire network. Subgroups such as the small group centered around Xixian Central Park and the Wenxian Park group lack connections with the overall network, resulting in poor completeness of the entire network.

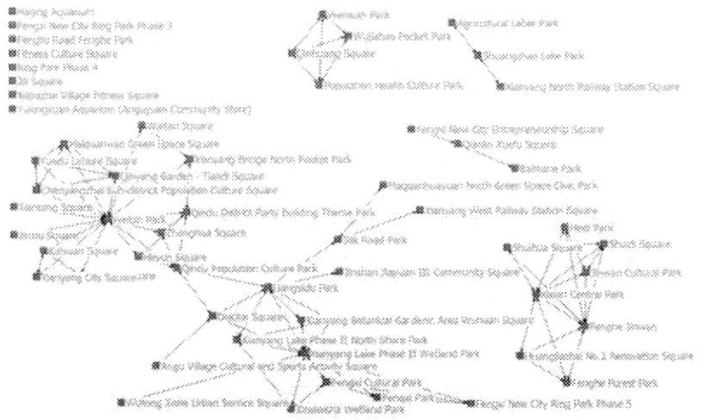

Fig. 3. Social network model of park square system in Qindu District, Xianyang City.

3.2.2 Network connectivity

The calculated clustering coefficient of the social network of parks and plazas in Qindu District, Xianyang City is 0.752, and the weighted overall clustering coefficient is 0.454, indicating a high degree of network coverage and aggregation. A higher clustering coefficient for a single node suggests a higher degree of interconnection among adjacent nodes of that node. The calculated average distance of the entire network is 3.455, indicating that one needs to pass through approximately 3.4555 park and plaza nodes to reach the next desired park or plaza from one. Based on the distance, the cohesion of the network is 0.179, suggesting poor cohesion in the social network of parks and plazas in Qindu District, Xianyang City. In the software UCINET, a random social network with the same number of nodes and network density as the social network of parks and plazas in Qindu District, Xianyang City is generated based on these parameters. Analysis of the clustering coefficient and average path length of the random network reveals a clustering coefficient of 0.084 and an average path length of 3.213. Based on these calculations, the small-world value of the social network of parks and plazas is 8.026. This indicates that the social network of parks and plazas in Qindu District, Xianyang City exhibits significant small-world properties, with the presence of locally clustered groups within the network.

3.3 Centrality analysis

3.3.1 Degree centrality

The degree centrality of the network shows a trend centered around the Weibin Park node and the Xianyang Lake Phase II Wetland Park node, with a ratio of 0.2137, indicating that certain parks in the network are more connected than others and may play important roles in residents' activities. According to Figure 4, it can be seen that the Weibin Park node has the highest degree, with a degree of 15, while the Xianyang Lake Phase II Wetland Park has a degree of 10, suggesting that Weibin Park occupies a core and significant position in the entire network, and its collaborative service capabilities with surrounding parks and plazas hold a notable advantage within the network. Weibin Park is a general comprehensive park with a service radius of 2000m, and it is geographically located in a relatively central position.

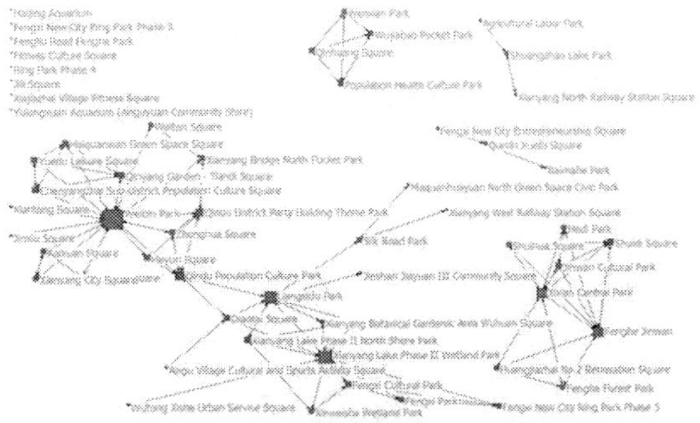

Fig. 4. Central network model of nodes of Park Square system in Qindu District, Xianyang City.

3.3.2 Closeness centrality

Due to the presence of unconnected subgroups within the network, the overall closeness centrality of the network cannot be calculated in the UCINET software, and consequently, a network graph depicting closeness centrality cannot be generated in the NetDraw software either. The results indicate that the node representing Liangsidu Park has the highest closeness centrality in the network. This node is essentially located at the geometric center of the network, minimizing the shortest distance to other nodes within the network.

3.3.3 Betweenness centrality

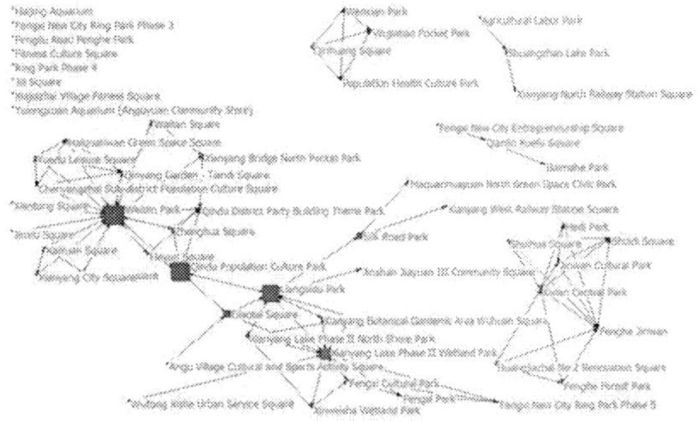

Fig. 5. Intermediate central network model of park square system in Qindu District, Xianyang City.

The betweenness centrality of the network is 0.1906, reflecting that certain parks and plazas play a bridging role in connecting different functional modules within the network, facilitating the flow of information and resources. According to Figure 5, the betweenness centrality of the Qindu Population Cultural Park node is 660.667, which is the highest in the entire network, indicating that multiple small groups within the network are overly reliant on the Qindu Population Cultural Park node for relational transmission, and this node occupies an extremely important position in the overall network.

3.4 Condensate subgroup analysis

Based on the research object of this study, the social network of parks and plazas in Qindu District, Xianyang City, we applied the n-clique analysis from social network analysis' cohesive subgroup analysis. The minimum size of the subgroup was set to 3, meaning that the maximum distance between each park and plaza node within the subgroup was 2. A total of 11 2-clique subgroups were identified under these conditions. These 11 subgroups represent highly tightly connected groups within the entire network, where strong direct or indirect connections exist among the various nodes within these groups. The results indicate that the social network of parks and plazas in Qindu District, Xianyang City, is somewhat hierarchical and comprises multiple relatively independent or semi-independent groups. These groups form a larger-scale community structure. The research findings are presented in Figure 6 and Table 1.

A network analysis of the largest subgroup reveals that Subgroup 1 has a network density of 0.2583 and a network size of 62. Compared to the entire network, most nodes within Subgroup 1 have more direct connections, demonstrating strong internal cohesion. This suggests that Subgroup 1 serves as a core layer within the network, undertaking crucial functions of communication, coordination, and information transmission.

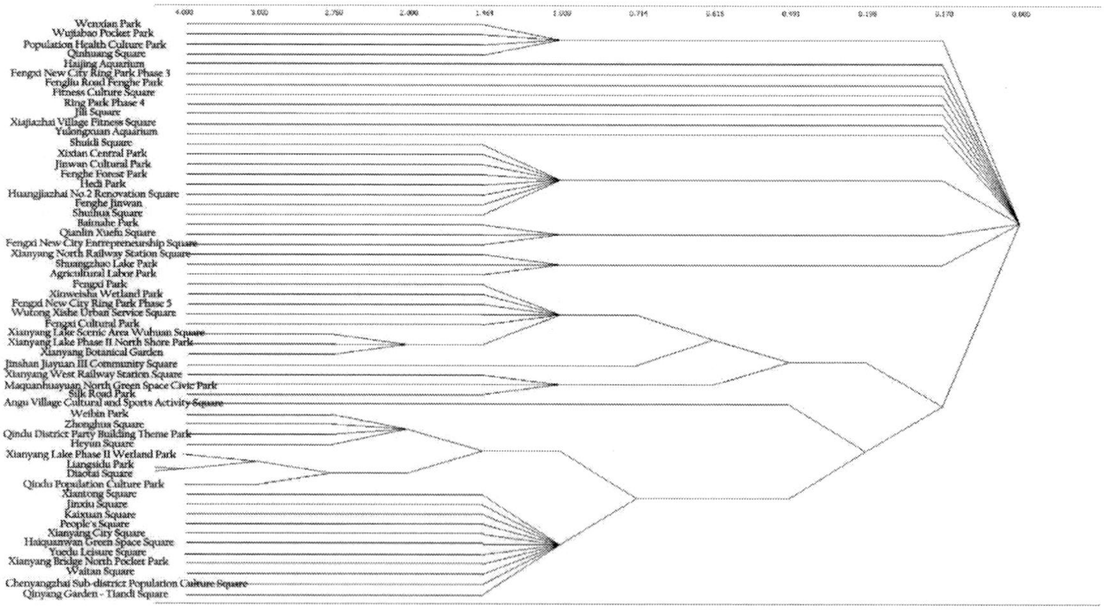

Fig. 6. n-clique analysis results of Park square system in Qindu District, Xianyang City.

Table 1. Results of the condensate subgroup analysis.

subgroup	Park square	quantity
1	Xiantong Square, Jinxiu Square, Kaixuan Square, People's Square, Xianyang City Square, Weibin Park, Zhonghua Square, Haiquanwan Greenland Square, Yuedu Leisure Square, Pocket Park North of Xianyang Bridge, Waitan Square, Qindu District Party Building Theme Park, Heyun Square, Qindu Population Culture Park, Chenyangzhai Subdistrict Population Culture Square, Qinyang Garden - Tiandi Square	16
2	Weibin Park, Zhonghua Square, Diaotai Square, Liangsidu Park, Qindu District Party Building Theme Park, Heyun Square, Qindu Population Culture Park	7
3	Wuhuan Square in Xianyang Lake Scenic Area, North Shore Park of Xianyang Lake Phase II, Wetland Park of Xianyang Lake Phase II, Diaotai Square, Xianyang Botanical Garden, Liangsidu Park, Qindu Population Culture Park, Square of Jinshan Jiayuan District 3, Silk Road Park	9
4	Wetland Park of Xianyang Lake Phase II, Diaotai Square, Cultural and Sports Activity Square of Angu Village, Liangsidu Park, Qindu Population Culture Park	5
5	Wenxian Park, Wujiabao Pocket Park, Population Health Culture Park, Qinhuang Square	4

subgroup	Park square	quantity
6	Water Drop Square, Xi'an-Xixian Central Park, Jinwan Cultural Park, Fenghe Forest Park, Riverbank Park, Renovated Huangjiazhai No.2 Square, Fenghe Jinwan, Water Splash Square	8
7	Fengxi New City Circular Park Phase 3, Fengliu Road Fenghe Park, Circular Park Phase 4, Fengxi Park, Baima River Park, Qianlin Xuefu Square, Fengxi New City Entrepreneurship Square	7
8	Square in Front of Xianyang West Railway Station, Liangsidu Park, Green Space Civic Park on the North Side of Maquan Huayuan, Silk Road Park	4
9	Fengxi New City Circular Park Phase 3, Fengxi Park, Fengxi New City Circular Park Phase 5, Wetland Park of Xianyang Lake Phase II, Fengxi Cultural Park	5
10	Wuhuan Square in Xianyang Lake Scenic Area, North Shore Park of Xianyang Lake Phase II, Fengxi Park, Xinweisha Wetland Park, Fengxi New City Circular Park Phase 5, Wutong Xishe Urban Service Square, Wetland Park of Xianyang Lake Phase II, Fengxi Cultural Park, Diaotai Square, Xianyang Botanical Garden, Liangsidu Park	11
11	Square in Front of Xianyang North Railway Station, Shuangzhao Lake Park, Agricultural Labor Park	3

4 Conclusions

This study systematically evaluates the network structure of parks and squares in Qindu District, Xianyang City through social network analysis. The findings reveal several key points: Firstly, there are evident central nodes in the network of parks and squares in Qindu District, which play crucial hub roles in urban public spaces. Secondly, some peripheral nodes exhibit low connectivity and clustering coefficients, indicating inefficient utilization of public spaces in these areas. Lastly, while the overall network density is moderate, the average path length between some nodes is relatively long, suggesting that the connection between parks and squares in certain areas is not sufficiently tight.

The following optimization strategy is proposed:

1) For peripheral nodes with lower connectivity, measures such as adding footpaths, greenways, or expanding their scale can be implemented to improve the connection between these areas and other nodes, thereby enhancing the overall network density and connectivity.

2) Increasing the density of the park and square network not only helps improve residents' travel and leisure experiences, increasing the utilization rate of public spaces, but also enhances the urban ecological environment and residents' health levels. The increase in density also fosters diversification and refinement in urban infrastructure investments.

3) Through community activities and educational projects, public awareness and utilization of public spaces can be raised. Enhance the utilization rate of public spaces and community satisfaction through public participation, thereby strengthening community cohesion.

References

1. Lin Yuan. (2018) City Park Systems and Human Wellbeing: On Park City for People. J. Chinese Landscape Architecture, 34 (10): 39-44.
2. Wujuan WEN, Yuan LIN. (2023) Urban Green Space Creation in 1920-30s China: through the Induction and Development of Urban Park Systems. J. Chinese Landscape Architecture, 39 (08): 133-138.
3. Yu Hao, Xusheng He, Ningjing Liu, Xiao Han. (2021) International Experience in the Construction and Practice of Urban Park SystemÐA Case Study of London, Tokyo and Toronto. J. Chinese Landscape Architecture, 37 (S1): 34-39.
4. Dennis Hardy. (2001) Regaining Paradise: Englishness and Early Garden City Movement. J. Journal of Historical Geographt, 27 (4): 605-606.
5. Zhu, G, Huang, L, & Zhang, Z. (2020). Optimization strategy of landscape ecological planning in urban green space system. J. IOP Conference Series Earth and Environmental Science, 474, 072005.
6. Márcia Oliveira, Gama J. (2012) An overview of social network analysis. J. Wiley Interdisciplinary Reviews Data Mining & Knowledge Discovery, 2 (2):99-115.
7. Ying Yang, Song Liu. (2023) Evaluation and optimization path of urban park system network structure in Shanghai based on social network analysis. J. Landscape Architecture, 30 (11):51-58.
8. Rong Du, Jieping Chen. (2022) Evaluation of the Green Space Landscape System in the Fringe Area of the Main City of Nanjing Based on Social Network Analysis. J. Chinese Landscape Architecture, 38 (4):68-73.
9. Jun Liu. (2019) Whole Network Approach: a practical guide to UCINET. Gezhi Publishing House, Shanghai.

Research on the Planning Strategy of Child Friendly Commuting Pathways Based on Society Ecosystems Theory: A Case Study of Xicheng Community in Zhuhai

Kang'en Chen [1,*], Helan Zhang [2], Yujuan Pan [3], Tao Xia [4]

[1] Faculty of Innovation and Design, City University of Macau, Macau, China
[2] Zhuhai Institute of Urban Planning and Design, Zhuhai, China
[3] Guangdong Urban-Rural Planning and Design Research Institute Technology Group Co. Ltd., Guangzhou, China
[4] Zhuhai Dechuang Construction Engineering Consulting Limited Company, Zhuhai, China

Abstract. The Society Ecosystems Theory reveals the importance of the influence of government departments, communities, schools, families, peers and other subjects on children, and can provide theoretical guidance and analytical frameworks for the explanation of children's learning behaviors and the construction of child friendly commuting pathways. Based on the review of the Society Ecosystems Theory, this paper optimizes the theoretical framework for the construction of child friendly commuting pathways and uses it to guide the practice of optimizing child friendly commuting pathways in the Xicheng community of Zhuhai. Four main strategies are adopted: to carry out the whole process of public participation with children as the core and parents as the linkage, so as to improve children's subjective initiative and parental participation; to carry out a series of traffic safety education activities with the community and school as the main body, so as to cultivate children's safety awareness; to take government departments as the leading role, to improve the physical environment of the pathway and to create an atmosphere of active access to the pathway; and to take the planner as the linkage to set up a platform for the communication of various system elements and to build consensus.

1. Introduction

In the late 1970s, the Danish municipality of Odense initiated what is considered to be the world's first Safe Passage Path program. The program aimed to reduce the high incidence of child pedestrian accidents by identifying and addressing road hazards through interventions such as new sidewalks, bike lanes, traffic islands, and narrowing roads [1]. With the concern of developed countries about child travel safety, obesity, urban traffic congestion, air pollution and other issues brought about by motorization of commuting, the Safe Commuting Paths program has gradually expanded from Denmark to European regions such as the United Kingdom and to countries such as New Zealand, Australia, Canada and the United States. When the program is implemented, the government formulates policies, communities and schools organize and implement them, and neighborhoods, families and children participate in them; through the adoption of engineering or non-engineering measures, in order to achieve the purpose of providing a relatively independent transportation environment for children to walk or bike to school, changing the attitudes and behaviors of children and their parents, and regulating the traffic behaviors of people who drive, bike, and walk, etc. [2]. Safe routes to school programs are multi-participant, multi-pronged, and ever-expanding, expanding from an initial focus on safety features to later efforts to make walking or biking to school more convenient and fun for children, and thus are collectively referred to as comprehensive programs that create safe, convenient, and fun opportunities for children to walk and bike to school [3]. For the purposes of this paper, these are referred to as Child Friendly Commuting Pathways programs, and the term "Child Walking and Biking Pathways and Safe and Fun Pathways" is collectively referred to as Child Positive Pathways. Overall, the implementation of the Child Friendly Commuting Pathways program has achieved good results, including improving children's physical and mental health, increasing the number of children walking or biking to school, making children's pathways safer, and lowering health care and transportation costs for school districts and families [4].

In China, in recent years, with the in-depth promotion of child-friendly city planning and construction, cities such as Shenzhen and Changsha have explored the construction of child-friendly paths to school and made them the main spatial carriers of child-friendly city construction. At the same time, a series of research results on child friendly commuting pathways have been formed in academia, which mainly cover the technical methods of pathway assessment, behavioral analysis of pathways, pathway planning and design, etc. Articles focusing on the

* Corresponding author: 412975199@qq.com

theoretical analysis framework are not common [5-8], which is difficult to guide the specific practice. In fact, as mentioned above, the construction of child friendly commuting pathways is a comprehensive and systematic project, and the physical and mental characteristics of children determine that to realize positive child learning, it is necessary for government departments, schools, communities, families and children to participate in a coordinated manner and for a number of initiatives to be coordinated and linked. The Society Ecosystems Theory, which is widely used in the fields of developmental psychology, sociology and public health, focuses on describing the relationship between human behavior and the social environment, especially revealing the importance of the influence of all parties on children's behavior, which can provide theoretical guidance and an analytical framework for the explanation of children's learning behaviors and the construction of a child-friendly pathway to schooling, and at the same time facilitate the government's formulation of more precise policies and measures.

2. Literature Review

2.1 The developmental lineage of Society Ecosystems Theory

Society Ecosystems Theory is often referred to as ecological system theory within the field of sociology and social work, and it is a theory used to examine the interactive relationship between human behavior and the social environment [9]. Before the emergence of the Society Ecosystems Theory, people's research on children's psychological development process mainly focuses on a specific environment, for a particular child's behavior to conduct experiments or observations; sometimes in order to make the experiment accurate often on the children's behavior constraints, or to guide them to produce a certain kind of behavior [10], ignoring the children's social attributes. Since the 1970s, the ecological orientation has continued to influence research in the field of developmental psychology, which argues that developmental psychology research should explore the psychological development of individuals in their natural environments and in specific sociocultural contexts [11]. The American psychologist Bronfenbrenner is an important representative of the ecological orientation and was the first to propose the ecological systems theory. He also criticized the traditional laboratory research methods of children's behavior mentioned above, as well as the descriptive methods used by anthropologists, which he considered subjective and inferential in their explanations of cause and effect; and regarded developmental psychology at that time as "the science of studying the strange behaviors of children that occur in unfamiliar situations with unfamiliar adults in the shortest possible period of time (of time spent together) ". To this end, he proposed a new research perspective, the ecology of human development, which he saw as the study of the gradual adaptation between a motivated and growing human being and the changing nature of his or her immediate environment, an adaptation process that is influenced by the relationships between different environments as well as by the larger context in which the environment is situated [12-13]. Bronfenbrenner divided ecosystems into microsystems, mesosystems, exosystems, and macrosystems, with the four systems resembling a set of Russian nesting dolls, focusing on the (child) individual to form a set of concentric circles. Subsequently, almost all graphical depictions of ecosystem theory have relied on variations of this concentric circle model [14]. Since the 1980s, ecosystem theory has been gradually applied to research and practice in the field of social work, and its model has been continuously improved. In particular, at the beginning of this century, American scholar Charles Zastrow elaborated on the multiple systems formed between the individual and the social environment and the interaction between the two, and they regarded the individual's survival environment as a complete ecosystem, and the individual is also a part of the ecosystem [15]; it is believed that the social system (survival environment) is a series of interconnected factors constituting a functional whole, which can be classified into a micro system (individual), mesosystem (family, group, etc.) and macrosystem (community, public institutions, culture, etc.). Individual microsystems are influenced by meso- and macro- systems such as families, groups, organizations, institutions, and communities, and also interact with these systems in a continuous and dynamic manner [16-17].

2.2 Weaknesses of Society Ecosystems Theory

Although the Society Ecosystems Theory can be used as an effective framework for analyzing children's behavior, it also has shortcomings. For one thing, the theory is too general to explain well what kind of influences are produced through what mechanisms among systems and between systems and individuals; for another, it does not take into account enough the subjective initiative of individual children, who are often in a passive position. Therefore, it is necessary to further deepen the theoretical cognition and optimize the model to make it more instructive, taking into account the specific objects of research or practice.

3. Perceptions of Society Ecosystems Theory in the Construction of a Child Friendly Pathways to General Studies

3.1 Key variables influencing child friendly commuting pathways

The research and practice of the path of child friendly commuting in foreign countries show that the influencing factors of children's positive commuting can be summarized in three aspects. First, at the micro level, whether children's own physical and mental development is suitable for active learning. Second, at the meso level, family factors also influence children's active travel,

https://doi.org/10.1051/e3sconf/202561703017

especially parents' attitudes and perceptions of children's commuting. Parents' attitudes towards children's ability to travel actively and independently can be categorized into three "profiles" (Promoters, Pragmatists and Protectors), and different attitudes have different impacts on children's commuting [18]. Interdependence of parents and children's traveling (e.g., parents driving to and from work on the same route as children's passages) [19], among others, also affects children's passages. Therefore, parental attitudes are a factor that cannot be ignored when conducting research on child commuting. Next, at the macro level, external environmental factors (e.g., community, school, government initiatives, etc.) can also have an impact on children's positive generic learning. For example, in neighborhoods, the social environment (community safety, cohesion, and connectedness) and attributes of the child-specific built environment influence children's travel, with children being more motivated to travel independently if there is strong community cohesion and proximity to schools [20].

The author surveyed 745 parents of elementary school children in the community before the construction of child-friendly through-school paths in the Xicheng community, and found that 42% of the children were transported by their parents in their own cars, and 38% of the children walked or biked to school. The top two reasons for transporting children by car were distance from school (some schools outside the community were under construction, resulting in some children having to travel across districts to attend school) and safety hazards of traveling, accounting for 55% and 49%, respectively; followed by commuting to and from work and poor quality of the travel environment, which were generally in line with the findings of the existing studies (see Figure 1 for details).

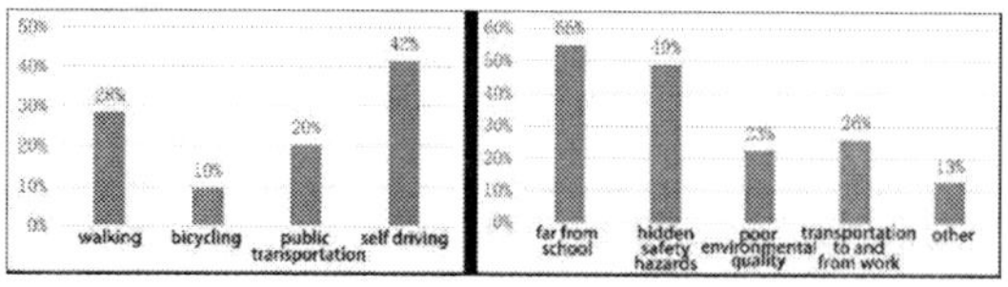

Fig. 1. The Proportion of Children Who Commute to School and the Reasons Why Parents Drive Their Children to School

3.2 Theoretical framework of child friendly commuting

Overseas scholars have constructed an analytical framework for child friendly commuting pathways by applying the Society Ecosystems Theory around the key variables mentioned above. Martin et al. [21] believe that the social-ecological system model includes five levels of influence from individuals to public policies, which can provide a framework reference for the comprehensive strategies adopted by child-friendly pathway programs. Each of these dimensions is divided into infrastructural and non-infrastructural projects, with the former focusing on "hardware" such as the construction of facilities and the creation of a spatial and physical environment, and the latter involving "software" such as coercive, educational, and incentive measures. Mitra [22] summarized a four-factor model of interlocking influences: the external environment, such as the natural environment, policy and social context, influences the urban environment, including the spatial structure of the city, the built environment and the social environment, while the urban environment influences children's travel behavior by influencing the family's economic conditions, attitudes and beliefs that lead to travel decisions and choices. The model also notes the importance of children's physical and mental growth on travel behavior and travel mode choice, and therefore considers children's attitudes, beliefs, and abilities as important factors. Panter et al [23], on the other hand, put more emphasis on the multiple influences of the physical environment, external factors, and the family on children's decision-making process of travel mode choice, especially highlighting the perceptions of the parents and the main variables such as the child's age, gender and the influence of key variables such as distance to destination.

Generally speaking, Society Ecosystems Theory has been better applied in the study of child-friendly pathway to schooling, and a series of analytical frameworks have been formed, but they still need to be further optimized: firstly, these models basically place children at the ultimate level, that is, they believe that the macro-system influences the meso-system, which then influences the micro-system - children's pathway to schooling behaviors. In fact, children's daily behavior is not only directly influenced by parents and school education, but also by the whole community culture, social atmosphere, public opinion and other macro-systems. Secondly, even though the influence of multiple factors on children is taken into account, children are in a passive position in the whole Friendly Pathway program, and whether they are active in learning or not is regarded as the influence of the macro system and the meso system, as well as the result of the corresponding cognition and ability of children in their growth stage, ignoring their subjective initiative, and children are not really involved in the construction of the Friendly Pathway. This also reflects that in the past, in the process of urban planning and governmental decision-making, children's demands were seldom listened to, and children's right to participation was neglected. Third, these models lack the consideration of "communication platform". When the Society Ecosystems Theory is used to guide the construction of child friendly commuting pathways, in addition to emphasizing the mutual influence of various systems, it should also pay attention to the role of communication among various systems, so as to reach a consensus on the outstanding problems and corresponding strategies to be solved in the construction of child friendly commuting pathways, and then provide the government with a more accurate and practical construction plan. Based on this, the author explores the optimal use of Society Ecosystems Theory to construct an analytical framework for the construction of a child-friendly pathway based on existing research results and combining the views of Charles Zastrow and others, with a view to guiding the subsequent practice (see Figure 2 for details).

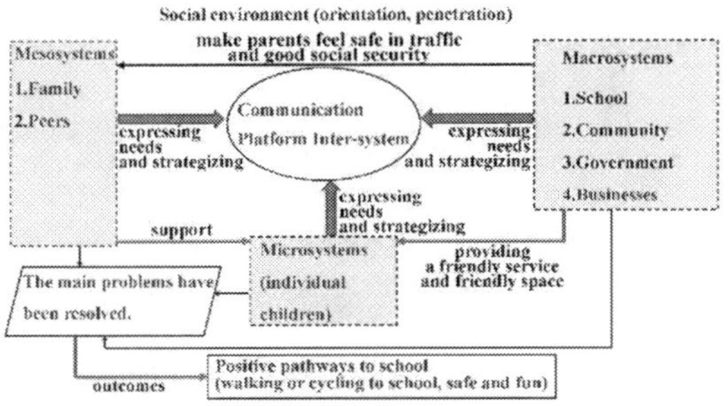

Fig. 2. Modeling the Influence of Society Ecosystems on Child Friendly Commuting

The promotion of children's active learning depends on the social environment, the macro-, meso- and micro-systems, and the role of "intermediaries". The social environment, which includes the formation of a child-friendly atmosphere, good social ethics and a social consensus in support of child-friendly community building, guides and permeates all systems. The macrosystem, including schools, communities, government departments and street-level businesses, provides child-friendly spatial environments (or facilities) and services for children's active learning through engineering and non-engineering measures, which also positively affects the mesosystem, including families and peers, for example, making parents feel safe in transportation and good social security, and thus supporting children's active learning. In addition, the involvement of parents in the construction of access routes and the provision of safety education to children will also have a positive effect. As a micro-system, in addition to good physical and mental development and the basic conditions for positive access, individual children need to exert their subjective initiative and actively participate in the construction of children's access paths in order to realize autonomous positive access. At the same time, there are many elements in each system, which need to be linked up by "intermediaries" to "dialog" on the communication platform, fully express the demands of all parties, and finally reach a consensus. The "intermediary" can be a planner who facilitates the systems to reach consensus and act together.

The public participation approach in this study, with children as central stakeholders, has a different focus when compared to the child-centered urban planning models (see Figure 3 for details). The child-centered public participation approach emphasizes the direct involvement of children in urban planning. Specifically, it includes children's participation in social activities, evaluation of public services, and planning and agenda-setting for public spaces. The child-centered urban planning model, on the other hand, focuses more on the design and planning of urban spaces from the perspective of children's needs, such as safety, accessibility, and fun. The difference between the two is that the former emphasizes more on children's direct participation and expression of opinions, while the latter considers children's needs and interests more from an adult perspective.

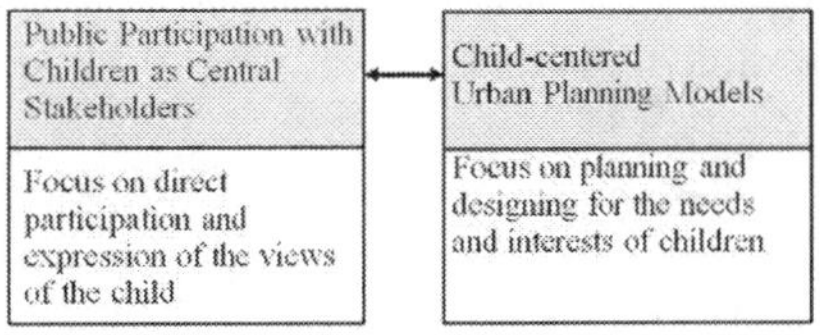

Fig. 3. Comparison: Public Participation with Children as Central Stakeholders & Child-centered Urban Planning Models

4. Practice of Society Ecosystems Theory in Xicheng Community

4.1 Background related to the planning and construction of child friendly commuting pathways in Xicheng Community

Xicheng Community is located in Sanzao Town, Jinwan District, with a core area of about 7 square kilometers, which is the location of Jinwan Center, the western central city of Zhuhai. According to the seventh census data, Xicheng Community has a total population of 8,506 children (0-18 years old), which is the third largest among 332 communities in the city. In 2019, Jinwan District selected Xicheng Community as a pilot site to take the lead in the city in the planning and construction of a child-friendly community (see Figure 4 for details). At the beginning of the work, children reflected problems related to schools, streets, parks, play facilities, traffic safety, reading rooms, mother and baby rooms, etc., which can be categorized into three categories: public service facility space, outdoor open space and street space. In order to facilitate the efficient promotion of the pilot work on child-friendly communities and to form replicable and scalable experiences as soon as possible, after many communications between the government, communities, schools and parents, it was finally decided to select one space in each of the three categories to carry out child-friendly renovation, resulting in the first three key projects: construction of a child-friendly through-school pathway, renovation of Jinshan Park, and construction of a reading space for children in the community.

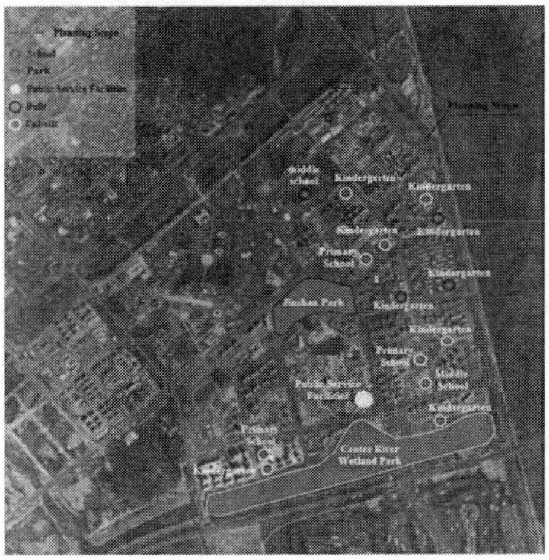

Fig. 4. Core Scope of Xicheng Community

4.2 Top-level design of the construction of child friendly commuting pathways in Xicheng community

Based on the theoretical framework of society ecosystems, the construction of child friendly pathways brings together the macrosystems, mesosystems, micro-systems, and "links". First, in the macrosystems, the government, schools and communities are responsible for enhancing the physical environment of child-friendly pathways through a combination of engineering and non-engineering measures, as well as supporting child-friendly services, systems and cultural environments to promote children's active learning. Secondly, parents in the mesosystems and children in the microsystems, through public participation, put forward their reasonable needs and get them met, so that children will feel proud of the pathway and have a sense of intimacy with it, thus promoting children's independent and active learning. Finally, the planner serves as a "link" to build a communication platform, organically linking the elements of each system, forging a consensus, and contributing to the efficient construction of a child-friendly pathway.

4.3 Strategies for planning child-friendly paths through learning

4.3.1 Conducting public participation activities centered on children in the microsystems and linking parents in the mesosystems to increase children's initiative and family participation

The whole process of participation, linking parents in the mesosystem and children in the microsystem, is divided into four stages. The first stage is the problem identification stage, in which the problems and causes of children's learning pathways were identified through children's free speech, parents' interviews and questionnaires. The second stage is the design stage, in which children's and parents' suggestions for optimizing the pathway through the school are collected through site visits and workshops, and the design of the renovation plan is formed. The third stage was the construction stage, in which children and parents were organized to participate in activities such as painting the paths and learning about the progress of the construction works, taking into account the characteristics of Jinwan Aerotropolis. Finally, in the evaluation stage, children and parents were organized to participate in the satisfaction survey after the path was put into use, and optimization suggestions were put forward, which were used as a reference for the subsequent improvement direction and the work of the child-friendly community in the next year. The whole process (see Figure 5 for details) effectively stimulated children's subjective initiative and improved parents' understanding of and participation in the construction of the Pathway, which played a supportive and encouraging role for children's active access to the Pathway.

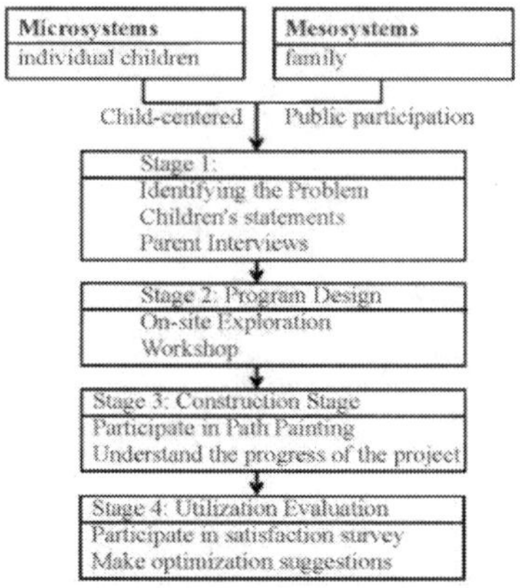

Fig. 5. Public Participation Centered on Children (Microsystems) and Involving Family (Mesosystems)

4.3.2 At the macro system level, the governmental sector upgrades the physical environment of the pathway, while communities and schools conduct a series of traffic safety education activities for children

At the macro system level, the main roles are played by government departments, communities and schools (see Figure 6 for details). At the governmental level, the finance department, the transportation and construction department, and the publicity department all play a role in the construction of friendly through-school paths. First, the finance department is responsible for securing the funds needed for the construction of the pathway. Secondly, the transportation and construction departments are responsible for optimizing and upgrading the environment along the path and its routes (including in front of the school) in cooperation with the school. The scope of optimization and upgrading focuses on the 800-meter service radius of the elementary school in Aviation New Town, and the northern section of Jin Er Road, which is relatively densely populated by residential areas, has been chosen as the first place to carry out the work in the near future. Then, the publicity department is responsible for filming videos, pushing news, and creating a series of child-friendly brand columns to disseminate information to the community about child-friendly paths to school and child-friendly community building, and to create a child-friendly atmosphere.

At the community level, community staff mobilize children in their districts to join the children's councils and, relying on the councils, organize children's participation in the construction of the pathways. At the same time, through organizing a series of activities such as "Discovering the Community", "Learning the Way" painting, green families, "Red Scarf Patrol" and formulating a resident's convention, we have strengthened traffic safety education for children, Green travel education for children.

In schools, in addition to regular traffic safety education for students, school teachers and staff also actively organized students to participate in the District Traffic Safety Knowledge Contest for Primary and Secondary School Students; organized students to take the lead in "Learning the Way" walking experience and mobilized parents to volunteer to maintain traffic order on Tung Choi Road and in front of the school during peak hours, so as to create a safe environment for children to travel to and from school. This creates a safe environment for children to go to school.

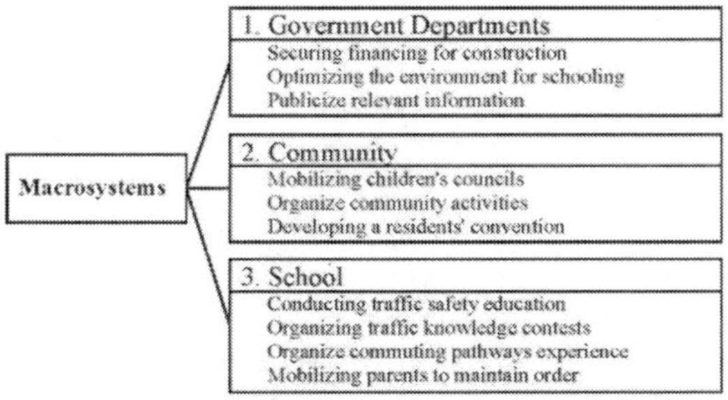

Fig. 6. Macrosystems in Their Respective Roles

4.3.3 Using the planner as a link to build a communication platform for each system element to build consensus

After listening to the needs of children and parents and their opinions on the renovation, the planner carried out a preliminary design, from which a number of renovation nodes were extracted and classified into four major aspects: sidewalk renovation, greening enhancement along the route, optimization of traffic management, and improvement of facilities in front of the school. The planner assisted the lead department to convene all the relevant parties for consultation in the form of coordination meeting and on-site meeting, and formed the final program for the construction of through-school path after repeated communication (see Figure 7 for details).

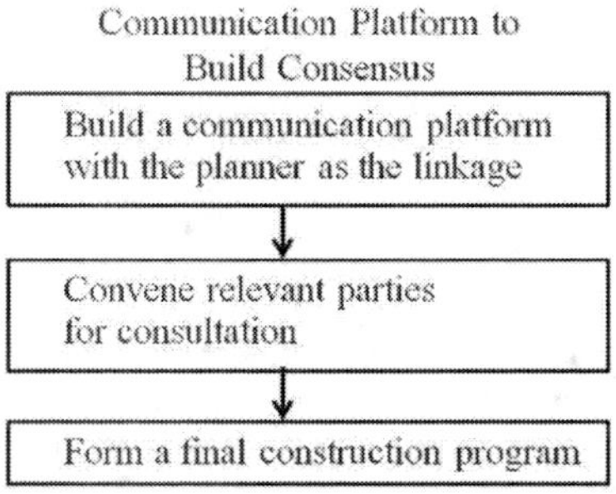

Fig. 7. Planners Build Communication Platform

For example, for the aviation new town elementary school in front of open space but no waiting area, the lack of motor vehicle pick-up and drop-off buffer zone and other issues, the original program proposed the use of open space in front of the school to increase the parking area and waiting area, after the communication of the program changed to increase the waiting area, no parking lot, but the school in front of the green belt to further open up the pick-up and drop-off vehicles to take the "stop and go" way of organizing traffic. "Stop and go" approach to organize traffic (see Table 1 for details).

Table 1. the Construction Program of the Child Friendly Pathways to School

	Problem	Optimization plan
1	Many vehicles illegally parked near the school and made U-turns by pressing double solid lines.	Improve the monitoring system;. Use railings for physical isolation.
2	There are many construction vehicles on Jin'er Road, and the speed is fast.	Add speed cameras. Restricted driving during school hours.
3	Yiyun intersection lacks traffic signal lights.	Add traffic signals and traffic guidance.
4	The zebra crossing on the road is not prominent enough.	Add a love zebra crossing.
5	Pedestrians crossing the street without following traffic rules.	Strengthen traffic rule training and extend the duration of street signal lights.
6	Electric bicycles and motorcycles occasionally enter the sidewalk at high speeds.	Set up a 1.5m wide "learning path" on one side of the sidewalk by paving or drawing lines. Add traffic guidance.
7	Poor physical experience when walking or using non motorized vehicles for travel (some sections of the sidewalk are damaged, there are few sun and rain protection facilities and recreational facilities along the way, the quality of green spaces along the way is poor, and the road is not interesting enough)	Repair the road surface. Add facilities such as seats, sun and rain protection. Timely construction of pocket parks. Aviation themed painting is used on the general education path.
8	The duration of the street signal near the school is too short.	Extend the duration of street signal lights.
9	There is a lack of buffer space for motor vehicle transportation in front of the school.	Further open the green belt in front of the school gate, parents can pick up and drop off immediately.
10	There is an open space in front of the school but no waiting area.	Add a waiting area for parents.

5. Establishment of Child Friendly Commuting Pathways to School and Post-construction Evaluation

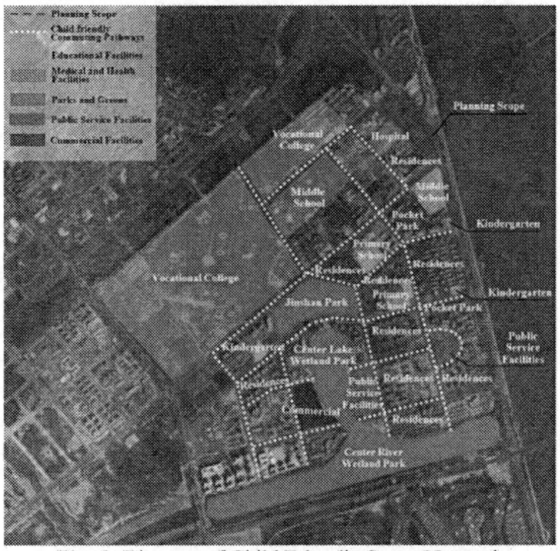

Fig. 8. Diagram of Child Friendly Space Network

With the active participation of children and parents, and the concerted efforts of the government, community, schools and planners, the above mentioned improvement strategies were basically implemented, and the first child-friendly pathway in Zhuhai was built to give children the exclusive right of way to school and to ensure a safe and interesting school journey (see Figure 8 for details). In the post-construction evaluation of the child-friendly pathway, the project team conducted a questionnaire survey and random on-site visits to collect opinions.

According to the results of the survey, users recognized the overall work of enhancing the child-friendly pathways and the surrounding environment very highly. According to the 248 valid questionnaires collected, 186 respondents, or 75% of the respondents, were satisfied with the work, and 56.8% of the respondents believed that the work had facilitated more students to walk to and from school. In terms of the main effects of the construction of child-friendly paths to and from school, about 70% of the respondents recognized that "the safety of the roads around the school has been improved", more than half of the respondents recognized that "the walking environment for going to and from school has become more interesting", and 40% of the respondents recognized that "Parents have a resting place under the sun and out of the rain while waiting".

Of the 248 valid questionnaires, a total of 10 people were dissatisfied with the construction of the pathway to and from school. The main feedback problems are concentrated in the transportation part, mainly "the traffic environment around the school has not been improved", "the lack of pavilions, tables and chairs and other rest facilities in the road". In addition, nearly 30% of the respondents believe that there is a lack of cultural atmosphere, and there is the phenomenon of non-motorized vehicles occupying the school road.

In terms of how to further improve the design of the school paths, 72% of the respondents (178 people) thought that there was a need to increase the number of easy-to-understand wayfinding signs for children and teenagers, and 64% of the respondents (159 people) suggested that traffic control around the schools should be strengthened. In addition, nearly half of the respondents (120 people) proposed to strengthen the cultural atmosphere of the school road and the surrounding environment, as well as 46% of the respondents hope that motor vehicle traffic control by time (to avoid, let the peak period of school). They also cited the need for more pocket parks for short stops for play, rich planting configurations, and refined designs for child-friendly bus stops.

Based on the feedback and suggestions from the respondents, combined with the information obtained from the site survey, the observation of pedestrian flow during school hours and the interviews, the optimization and upgrading of the school route will focus on four aspects: building a child- and youth-friendly signage system, strengthening the construction of a cultural atmosphere, providing a strict means of traffic control, and installing interesting pocket parks along the school route. In terms of longitudinal research, further studies will be conducted on the influence of different types of spaces commonly used by children on the pathway on children's schooling behavior.

6. Discussion: Long-term Mechanisms for the Construction of Child Friendly Commuting Pathways

6.1 Children's Participation Mechanism

Relying on the first batch of key construction projects such as the Child-Friendly Pathway, Xicheng Community has set up the city's first community children's council, and standardized the operation of the children's council in terms of concepts, rules, composition, responsibilities, etc., so as to guide the children's participation in the public affairs of the community in accordance with the rules and regulations. With the Children's Council as a platform, the community has carried out hundreds of activities for children and their families since its establishment, serving thousands of people, and children's participation in community affairs has been greatly enhanced.

6.2 Multi-party communication mechanism

In order to avoid problems of non-cooperation, inefficient communication or unclear authority and responsibility caused by the lack of knowledge and understanding of the relevant parties in the process of project construction, the planner assisted the government-led department in formulating a work program, refining the construction objectives, clarifying specific requirements, responsible departments and timeframes for completion, and setting up a platform for communication between the responsible departments and the community, schools, parents, and children, forming a multi-party communication platform for the community, schools, parents, and children. and children, forming a good mechanism for multi-party consultation.

6.3 Continuous promotion mechanism

First, the financial guarantee mechanism, the pilot phase is guaranteed by special funds from the district finance; in the follow-up maintenance phase, the relevant departments will apply for funds or the community will apply for funds for people's livelihood micro-activities. The second is the personnel guarantee mechanism, which gradually brings into play the role of the community, with the community committee taking the lead, linking volunteers, community residents and other people to participate in the construction of child-friendly communities, and introducing social institutions and organizations to make the relevant services more professional. Finally, the normative level, on the basis of fully studying the experiences and standards of construction of child-friendly passages and child-friendly communities at home and abroad, and combining the practice of Xicheng community, the Guidelines for Construction of Child-Friendly Communities in Jinwan District of Zhuhai City, including the guidelines for the construction of child-friendly passages, have been prepared and issued to further standardize the subsequent construction.

7. Conclusion

The purpose of the construction of child friendly commuting pathways is to promote children's active through schooling, which is a systematic project. As scholars in China have pointed out, the construction of child friendly commuting pathways has not only been limited to the spatial level, but has also been extended to include more flexible services such as "walking buses" by linking up with schools and parent volunteers [24]. Society Ecosystems Theory is undoubtedly one of the most powerful theories to explain the logic behind this phenomenon, which in turn can effectively guide the construction of child friendly commuting pathways, and even act as a "toolkit" to make the guidance more operational after enriching the strategies adopted by each system.

References

1. Appleyard B S. (2003) Planning Safe Routes to School. J. Planning, 69(5):34-37.
2. Jiao J. (2019) Promoting Children's Walking and Biking to School:Successful Experiences and Implications of Safe Routes to School Programs in Europe and the United States. J. Shanghai Urban Planning, 3:6.
3. Foundation R. (2007) Safe Routes to School: 2007 State of the States Report. https://www.policyarchive.org/handle/10207/21559.
4. Mcdonald N C, Yang Y, Abbott S M, et al. (2013) Impact of the Safe Routes to School program on walking and biking: Eugene, Oregon study. J. Transport Policy, 29(SEP.):243-248.
5. WU FW, CHEN MY. (2020) Evaluation and optimization strategy of pedestrian through-school paths from the perspective of children's health and safety. J. Urban Development Research,27(7):20-27.
6. XU Z, LIU AQ, ZHOU ZQ, et al. (2021) Baidu map-based walking paths and scenario analysis of secondary school students in Jianye District, Nanjing. J. Modern Urban Research,36(2):33-40,91.
7. WANG M, CHEN XJ. (2018) Analysis of spatial and temporal characteristics and constraints of elementary school through-school travel in Xi'an city. J. Urban Planning,42(11):142-150.
8. WU ZF, LEI HX. (2022) Research on the connotation analysis and strategy framework of child-friendly streets--Based on the concept of child-friendly cities in China. J. Urban Planning,46(11):32-41,51.
9. SHI HL, FAN YN. (2005) Human Behavior and Social Environment under the Interpretation of Social-Ecological Systems Theory - A New Exploration on Human Behavior and Social Environment by Charles Zastrow in 2004. J. Journal of Capital Normal University (Social Science Edition), (4):94-97.
10. Liu J, Meng HM. (2009) On Bronfenbrenner's ecological systems theory of developmental psychology. J. Chinese Journal of Health Psychology, 17(2):250-252.
11. Zhu, Temperature. (2005) Review and prospect of ecological research in developmental psychology in the past 50 years. J. Psychological Science, 28(4): 922-925.
12. Hicks G L. (2010) Applied Anthropology: The Ecology of Human Development. by Urie Bronfenbrenner. J. American Anthropologist, 83(3): 643-643.
13. Bronfenbrenner, U. (1979) The ecology of human development: Experiments by nature and design. Harvard University Press, Cambridge.
14. Neal J W , Neal Z P . (2013) Nested or Networked? Future Directions for Ecological Systems Theory. J. Social Development, 22(4):722-737.
15. Zhuo CQ. (2013) The Development of Ecological Systems Theory in Social Work. J. Journal of Jianghai,(3):113-119.
16. LIU L, SHI N, HE YL, et al. (2022) Child-friendly city construction practice. J. Urban Planning, 46(1): 44-52.
17. Charles H. Zastrow, Karen K Koester A. (2006) Human Behavior and the Social Environment (Sixth Edition). Renmin University of China Press, Beijing.
18. Juliane S, Julia F, Florian A. (2018) Exploring independent and active mobility in primary school children in Vienna. J. Journal of Transport Geography, 68:31-41 .
19. Weigand L. (2008) A Review of Literature:The Effectiveness of Safe Routes to School and Other Programs to Promote Active Transportation to School. J. Physical fitness.
20. En-Yi Lin, Karen W, Melody O, et al. (2017). Social and Built-environment Factors Related to Children's Independent Mobility: the Importance of Neighbourhood Cohesion and Connectedness. J. Health & Place, 46: 107-113.
21. Moeti R, Pullen-Seufert N . (2009) Implementing Safe Routes to School: Application for the Socioecological Model and Issues to Consider. J. Health Promot Pract, 10(4):606-614.
22. Mitra, Raktim. (2013) Independent Mobility and Mode Choice for School Transportation: a Review and Framework for Future Research. J. Transport Reviews, 33(1):21-43.
23. Panter J R , Jones A P , Sluijs E .(2008) Environmental determinants of active travel in youth: a review and framework for future research. J. International Journal of Behavioral Nutrition and Physical Activity, 5(1):1-14.
24. Liu L, Wu XL, Lei YC, et al. (2022) Child Friendly Cities in China. China Construction Industry Press, Beijing.

A Review of the Applications of CFRP Reinforcements in Civil Engineering

Jue Wang*, Xincheng Su and Quanqing Gao

CCCC Highway Bridges National Engineering Research Centre Co., Ltd., Beijing 100088, China

Abstract. Carbon fiber reinforced polymer (CFRP), as an advanced composite material, has been attracting a great deal of attention from researchers because of its excellent mechanical properties and durability. CFRP reinforcements are increasingly used in the construction and strengthening engineering. In recent years, considering the application of CFRP reinforcements in civil engineering has entered a stage of rapid development, it is necessary to review the development and recent advances in civil engineering. The present paper reviews the applications of CFRP reinforcements in civil engineering, the characteristics and types of CFRP reinforcements and cables, as well as the anchoring system of CFRP cables, are summarized. Furthermore, the applications of CFRP reinforcements and cables in the construction and strengthening engineering are reviewed, especially the developments and applications in bridge engineering. The main goal of this research is to provide a comprehensive review of the application development of CFRP reinforcements in engineering structures. Moreover, the critical issues should be solved for expanding the application scope of them are discussed.

1 Introduction

CFRP is composed of carbon fiber and polymer matrix materials (such as epoxy resin, polyester, nylon resin, etc.), where carbon fiber provides ultra-high strength and high elastic modulus, and the polymer matrix plays a role in bonding, stress transmission, and protecting carbon fiber. CFRP has showed enormous potential in a variety of engineering fields [1].

CFRP was first used in the aerospace field in the 1960s [2]. For the past few years, manufacturing technology has developed rapidly, material technology is constantly advancing, and material costs have also been continuously optimized. These factors have contributed to the gradual expansion of the engineering application scope of CFRP. At present, CFRP has been studied and applied in the fields of automobile manufacturing, ships and submarines, equipment, and construction engineering.

In recent years, with the deepening of research on CFRP and the improvement of production processes, the applications of CFRP cables have gradually increased. CFRP cables can obviously improve the carrying capacity, enhance the durability and seismic performance of structures. Furthermore, compared with steel reinforcements, there are relatively lower carbon emissions of CFRP reinforcements for bridges [3], and also better recycling performance, for instance, the recycled CFRP can be used in the automotive industry [4]. Therefore, CFRP is more environmentally friendly and sustainable, has been extensively applied in engineering.

2 Material properties of CFRP

2.1 Material properties of FRP reinforcements

The commonly used FRP reinforcements include AFRP, GFRP, and CFRP reinforcement. CFRP has the characteristics of ultra-high strength, excellent creep performance, and low relaxation, but the cost is relatively high. The comparison of mechanical properties among these FRP reinforcements and steel reinforcements is shown in Figure 1 [5].

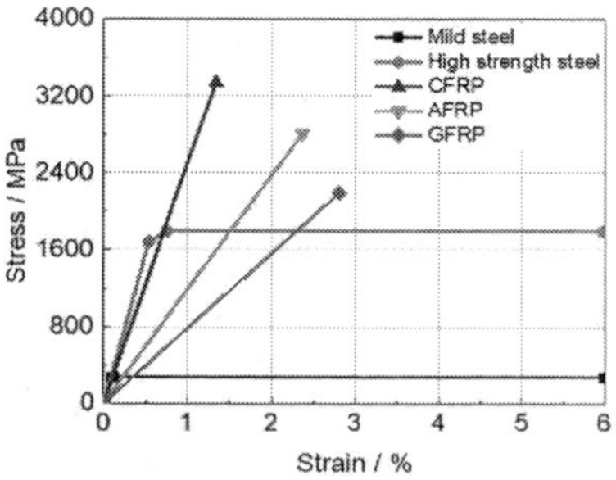

Fig. 1. Strength comparison among FRP reinforcements and steel reinforcements [5].

2.2 Characteristics of CFRP cable

In comparison to traditional cables, CFRP cables have broader application prospects with the characteristics of ultrahigh strength under tensile load, and low weight to

* Corresponding author: wangjue20221126@163.com

volume ratio. Table 1 compares several main characteristics of the two kinds of cables [6, 7].

Table 1. Comparison of characteristics between different cables.

Types of cable	Steel cable	CFRP cable
Tensile strength (MPa)	1860	2500
Elastic modulus (GPa)	195	150
Density (g/cm^3)	7.85	1.5~1.8
Maximum longitudinal strain (%)	4.0	1.3~1.6
Expansion coefficient (10^{-6}/°C)	13.2	-0.9~0.5

The high cost is one of the reasons limiting the widespread application of CFRP reinforcements in engineering structures. However, CFRP reinforcements have superior corrosion resistance, are suitable for complex climatic conditions, and there are few maintenance requirements during the service period of structures [8]. The cost of CFRP reinforcements in the lifecycle and the initial cost of construction can be considered equivalent. However, the cost of traditional materials like steel reinforcements in lifecycle is the sum of the initial cost of construction, the cost of maintenance and repairing of structures in lifecycle, and the indirect cost that caused by the temporary closure when the structures need maintaining, which is always expensive, especially for large-scale projects with the design service life of no less than one hundred years. This makes the application of CFRP reinforcements more advantageous from the perspective of full lifecycle economic performance.

3 The summarize of CFRP reinforcements and cables

3.1 Types of CFRP reinforcements

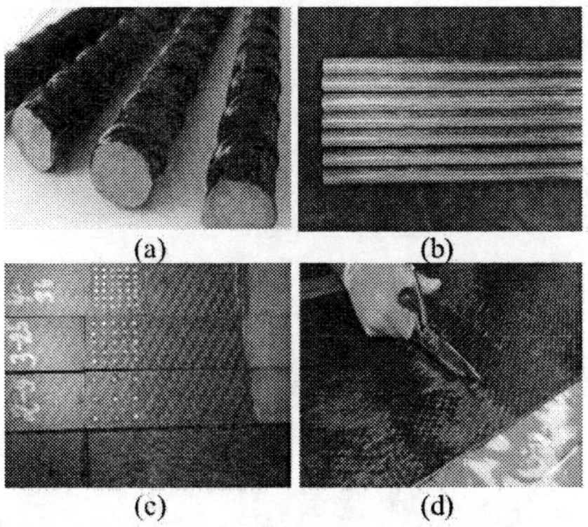

Fig. 2. Different types of CFRP reinforcements: (a) CFRP bars [5]; (b) CFRP rods [9]; (c)CFRP plates [10]; (d) CFRP sheets [11].

See Figure 2, the main types of CFRP reinforcements include CFRP bars [5], CFRP rods [9], CFRP plates [10], and CFRP sheets [11].

3.2 Types of CFRP cables

CFRP cables types mainly include cable of parallel wires [12], cable of twisted wires [13], cable of plates, and cable of nonlaminated straps [14], which are shown in Figure 3. The applications of the above types of CFRP cables in practical engineering will be introduced in the following paragraphs.

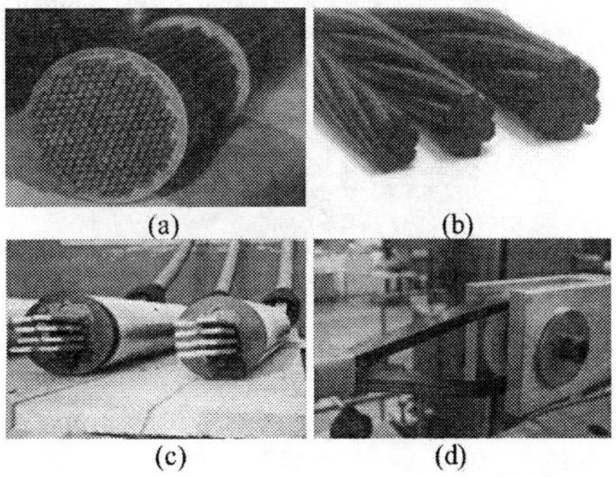

Fig. 3. Types of CFRP Cables: (a) cable of parallel wires [12]; (b) cable of twisted wires [13]; (c) cable of plates [14]; (d) cable of nonlaminated straps [14].

CFRP cables have high axial tensile strength but poor transverse shear resistance due to the aeolotropism of the material, the anchoring methods of steel cables are no longer suitable to CFRP cables. Consequently, one of the most important technical challenges of large-scale applications of CFRP cable is that the anchoring system is still very unripe presently, and the anchoring methods will directly affect the utilization effect of CFRP cables, especially for large-span spatial structures. To avoid the damage in the anchoring area and fully utilize the tensile strength of CFRP cable, various anchoring methods have been designed for CFRP cables, which can be generally divided into mechanical anchoring, adhesive anchoring, wedge anchoring, and composite anchoring methods, according to the anchoring mechanism and anchoring performances, see Figure 4 [15, 16].

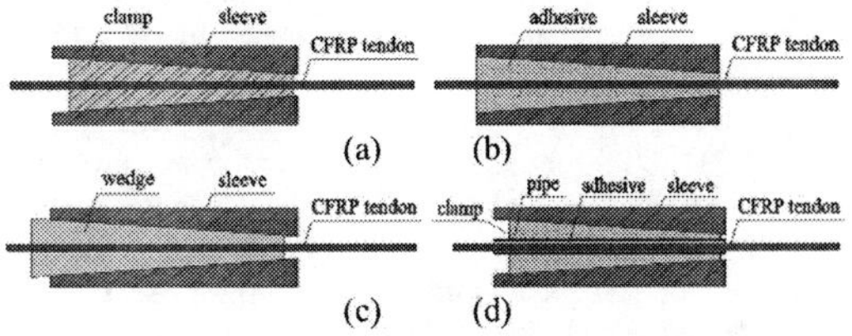

Fig. 4. Types of anchoring methods for CFRP cables: (a)mechanical anchoring; (b)adhesive anchoring; (c)wedge anchoring; (d)composite anchoring method.

4 Research and application of CFRP reinforcement in new construction and strengthening

Because of superior characteristics, CFRP was used as reinforcement materials in CFRP prestressed high-strength concrete components manufactured by prefabricated manufacturer SCACA, including CFRP prestressed electricity pylons, CFRP prestressed HPSCC (high-performance self-compacting concrete) L-shaped beams, and CFRP prestressed HPSCC columns, respectively [17,18], see Figure 5.

(a) (b) (c)

Fig. 5. CFRP reinforcements are applied in construction [25]: (a) CFRP prestressed electricity pylons; (b) CFRP prestressed HPSCC L-shaped beams; (c) CFRP prestressed HPSCC columns.

CFRP reinforcements were also used in strengthening projects. As for the strengthening of buildings, CFRP strips and plates were used to repair the ancient palace which built in 1600 [19], as shown in Figure 6(a), and CFRP rods were used to enhance the aseismic performance of the temple [20], as shown in Figure 6(b).

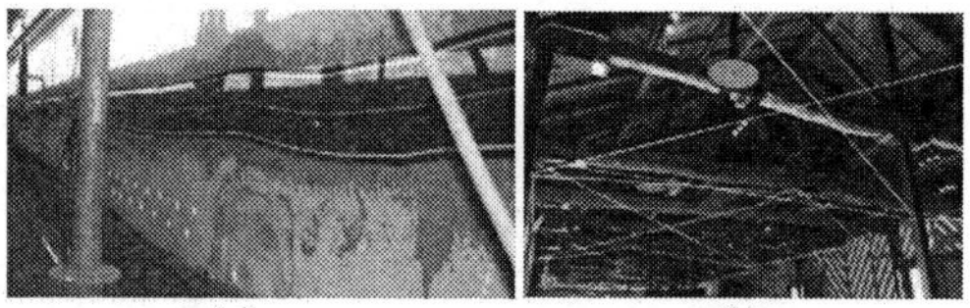
(a) (b)

Fig. 6. CFRP reinforcements are applied in the strengthening of buildings [19,20].

Furthermore, CFRP reinforcements were also applied in strengthening of bridges. A steel truss railway bridge built in Switzerland in 1892, was strengthened using CFRP plates in 2013 [21]. The bridge has a span of 45.2m, a width of 5.0m, and a truss beam height of 6.15m, as shown in Figure 7 (a). There is no need to weld or drill on the existing bridge when using the strengthening system of CFRP reinforcements. Moreover, the system can be disassembled if necessary in future, see Figure 7 (b). The CFRP plate adopted for strengthening work has the elastic modulus of 167.2 GPa and tensile strength of 2710 MPa.

(a) (b)

Fig. 7. The steel truss railway bridge strengthened by CFRP plates [21].

The Aabach Bridge was built in 1928 in Lahern, Switzerland, which is a riveted railway steel bridge [22], as shown in Figure 8 (a). In 2020, a new retrofit system using prestressed CFRP rods was used to strengthen the bridge, see Figure 8 (b). The diameter of the CFRP rod is 8mm, the elastic modulus and tensile strength of CFRP plate are 160 GPa and 2047 MPa, respectively. The maximum prestress load applied to the rod is 50kN.

(a) (b)

Fig. 8. The steel railway bridge strengthened by CFRP rods [22]: (a) the Aabach bridge; (b) the retrofit system.

5 Applications of CFRP cables in civil engineering

The Komatsu Saito office building is located in Ishikawa prefecture, Japan [20]. In 2016, the building was strengthened using the CFRP rods called CABKOMA, by connecting the roof of the building to the external foundation, which a 160m long CFRP rod weighed just 12kg, as shown in Figure 9. A mesh structure made of CFRP rods was also installed inside for strengthening the building.

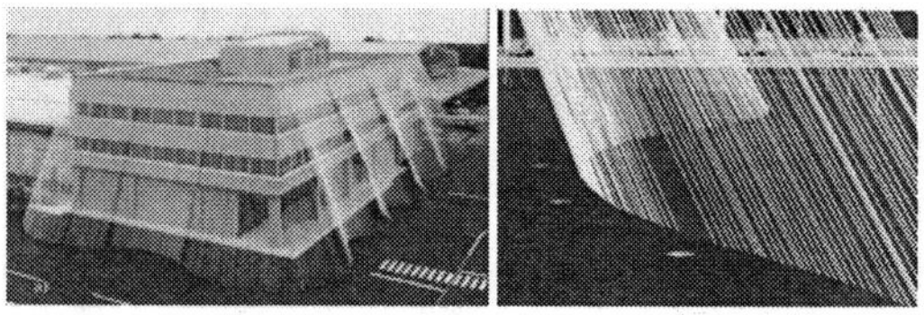
(a) (b)

Fig. 9. Office building strengthened with CFRP rods [20].

The Sanya sports stadium in Hainan province was completed in 2021 (see Figure 10) [23]. The stadium adopted a chamfered pentagonal awning of steel structure, and the roof structure consists of a spokewise cable-truss structure and a membrane structure roof, the cantilever length of the roof is 45m. The spokewise cable-truss structure is composed of fully locked ring cables and CFRP plate cables. The sports stadium has become the first large-span spatial structure which applying CFRP cables in the world.

(a) (b)

Fig. 10. Sanya sports stadium [23]: (a) the exterior of the stadium; (b) the interior of the stadium.

Up to now, although CFRP cables have been applied to some building structures, they are mainly applied in bridge structures. The following examples can be used to show how CFRP cable applications are developed in bridges.

As the first highway cable-stayed bridge using CFRP cables in the word, Stork bridge in Switzerland was opened to traffic in 1996[12], as shown in Figure 11. The bridge has a A-shaped tower with a height of 38 m, and with span arrangement being (63+61) m. Each CFRP cable consists of 241 5mm diameter wires, and had been cyclically loaded over 10 million cycles in the laboratory, with the load reached three times the allowable load for the bridge.

Fig. 11. Stork bridge using CFRP cables [12].

In 1999, a cable-stayed footbridge was available to the public in the vicinity of the railway station in Herning, Denmark (see Figure 12) [24]. The footbridge was constructed across a railway switchyard, and the spans on either side of the pylon are both 40m. The footbridge was constructed for the purpose of researching and developing the new noncorrodible materials that are suitable for bridge. 16 stays were all made of CFRP cables with the polyethylene coats, protective stainless steel sleeves up to 3.5m above deck were installed to prevent vandalism.

Fig. 12. The cable-stayed footbridge using CFRP cables in Herning [24].

In 2002, The Laroin cable-stayed bridge was constructed in France, as shown in Figure 13 [25], which is the first cable-stayed bridge using CFRP cables in France. It is a cable-stayed pedestrian overpass, with steel main beams and inverted V-shaped steel tube bridge towers, and has a span of 110 m. There are 8 CFRP cables distributed at both sides of the bridge, each cable consists of 7 parallel 6mm diameter CFRP rods.

Fig. 13. Laroin cable-stayed pedestrian bridge using CFRP cables [25].

The cable-stayed bridge with single tower and double cable planes located in Jiangsu University was built in May 2005, which is a footbridge with 16 CFRP stays in total, as shown in Figure 14 [26]. There are three types of stays consisting of six, eleven, and sixteen parallel 8mm diameter CFRP rods, respectively.

Fig. 14. The cable-stayed bridge using CFRP stays in Jiangsu University [26].

The first cable-stayed bridge in Maine, Penobscot Narrows bridge (see Figure 15) was opened to traffic on December 2006 [27]. This bridge was constructed as the emergency replacement for the Waldo–Hancock bridge, which suffered corrosion and premature damage due to the geographical location near the coast. For purpose of assessing and evaluating the application effect of CFRP cables in marine environments, six epoxy coated steel strands were replaced with CFRP cables by the Maine Department of Transportation in June 2007, in addition, these steel cables and CFRP cables were monitored.

(a) (b)

Fig. 15. Penobscot Narrows bridge [27].

The three span (72+72+72m) stressed ribbon pedestrian bridge crossing over Júcar river in Cuenca, Spain, was constructed in 2011 to improve the transportation condition of the campus, as shown in Figure 16 (a) [28]. Considering the superior mechanical and durability properties, CFRP cables, see Figure 16 (b), were used in place of conventional cables. The superstructure consists of CFRP cables and precast concrete beam segments, the thickness of bridge deck is 0.34m, and the slenderness ratio of the deck is 211. The bridge is the first application of CFRP cables in stressed ribbon pedestrian bridge.

(a) (b)

Fig. 16. The stressed ribbon pedestrian bridge using CFRP cables [28]: (a) the stressed ribbon pedestrian bridge; (b) the CFRP cables.

The public urban transport operator (SSB) in Stuttgart, Germany, constructed a network arch bridge of 124 m span with CFRP hangers in 2020 (see Figure 17) [29]. It is the first large bridge fully depending on CFRP hangers in the world. The cross-sectional area of CFRP hanger was reduced by 75% compared to what would be required in steel. Moreover, compared with the design with steel hangers, the excellent fatigue resistance makes CFRP hangers a more economical solution.

(a) (b)

Fig. 17. The network arch bridge using CFRP hangers in Stuttgart [29].

In 2023, the through concrete-filled steel tube tied-arch bridge (see Figure 18) named Sanduoxi bridge was constructed in Gaoyou, Jiangsu [30, 31]. The bridge is the first tied-arch highway bridge in China using CFRP hangers, with a span of 92.64m, and the deck width of 12.8m. There are a total of 32 hangers adopted in the bridge, four symmetrically positioned hangers of which are CFRP rods, and the rest of them are steel strand rods.

(a) (b)

Fig. 18. The tied-arch highway bridge using CFRP hanger rods [30, 31].

In 2023, the cable-stayed bridge with single tower and double cable plane on Huangshan road in Jiangyin, Jiangsu (see Figure 19) using the longest CFRP cable in China [32], was completed with the cable-stayed lifting. The bridge has a span arrangement of (30+100+88+30) m, which is composed of separated steel box girders, herringbone bridge towers, and double saddle shaped cable plane. There are a total of 48 cables on the bridge, forty-six of which are 1770 MPa grade high-strength and zinc-aluminum alloy coated steel cables, the other two are CFRP cables with a length of around 68m and an elastic modulus exceeding 170 GPa.

(a) (b)

Fig. 19. The cable-stayed bridge with CFRP cables on Huangshan road in Jiangyin, Jiangsu [32].

The Longmen cross-sea bridge is an important component of the mainline of national highway G228, is also an important highway connecting the Guangxi north bay economic zone [33], which bridge closure was completed in January 2024. The suspension bridge has one span steel box girder of fully floating system, with the main bridge length of 1198m, the main span of 1098m, and the tower height of 174m, as shown in Figure 20, which is the first cross-sea suspension bridge using CFRP cables in the world.

Fig. 20. Guangxi Longmen cross-sea bridge using CFRP cables [33].

The cable-stayed bridge with steel arch tower and double cable plane that cross over Tuhai river in Liaocheng, Shandong. It is the first highway cable-stayed bridge using CFRP cables in China [34, 35], and the span arrangement of main bridge is (100+100) m, see Figure 21. There are two main towers with a height of 55.9 m and one auxiliary tower with a height of 51.9 m, the semi-floating system was adopted in the main bridge. There are 36 pairs of CFRP cables (Φ 7-121) in double cable planes, and are arranged as fan-shaped along the longitudinal direction. The maximum length of these CFRP cables is 73.9 m, and the maximum bearing capacity of the cables exceeds 1000 tons.

(a) (b)

Fig. 21. Tuhai River Bridge using CFRP cables [34, 35].

The Changtai Yangtze river bridge (see Figure 22), with a main span of 1208m, is the largest highway-railway cable-stayed bridge in the world [36]. The bridge is still under construction while the main structure of the bridge located on the Changzhou side had been completed in May 2024. There are twenty-eight CFRP cables (Φ7-127) arranged on the upstream and downstream, namely seven CFRP cables on each side of the beam, the corbel on deck and crossbeam corbel on the mid span side of the main tower are connected by these CFRP cables.

Fig. 22. Changtai Yangtze river bridge using CFRP cables[36].

6 Conclusion

This article provides a comprehensive review of the application and development of CFRP reinforcements in engineering structures, and draws the following conclusions:

1. The existing research and engineering applications have proved that CFRP reinforcements can not only replace the traditional steel reinforcements but also show better mechanical properties and durability. With the continuous maturity of the manufacturing process of CFRP reinforcements, the production efficiency is improved and the cost could be reduced, and the application of CFRP reinforcements in engineering structures will have a larger scope of development.

2. The main factor currently limiting the widespread application of CFRP reinforcements in structures is the lack of support from normative documents. The application of CFRP reinforcements will be greatly promoted by improving relevant design specifications and regulations and providing corresponding design references.

3. The combination with high-performance materials, including high-performance concrete and high-strength steel, makes it possible to expand the application field of CFRP reinforcements from traditional structural systems to new structural systems with better structural and economic performance.

Acknowledgements

This research was supported by the National Key Research and Development Program of China (2021YFB3704400).

References

1. Karataş M A, Gökkaya H. A review on machinability of carbon fiber reinforced polymer (CFRP) and glass fiber reinforced polymer (GFRP) composite materials. Defence Technology, 2018, 14(4): 318-326.
2. Perner M, Algermissen S, Keimer R, et al. Avoiding defects in manufacturing processes: A review for automated CFRP production. Robotics and Computer-Integrated Manufacturing, 2016, 38: 82-92.
3. Liu H Y. Assessment of environmental impact for reinforced concrete bridges and CFRP bridges, World Bridges, 2024, 52 (03): 129-130.
4. He D, Kim H C, Sommacal S, et al. Improving mechanical and life cycle environmental performances of recycled CFRP automotive component by fibre architecture preservation, Composites Part A: Applied Science and Manufacturing, 2023, 175: 107749.
5. Frhaan W K M, Abu Bakar B H, Hilal N, et al. CFRP for strengthening and repairing reinforced concrete: A review, Innovative Infrastructure Solutions, 2021, 6: 1-13.
6. Wang L C. Research on mechanical behaviors of CFRP cables and its connected joint for tension cable roof, Tianjin University, 2017.
7. Zhang P. Study on the mechanical performance and design method of CFRP parallel-rod cables and anchorages, Tsinghua University (2014).
8. Yang Y, Fahmy M F M, Guan S, et al. Properties and applications of FRP cable on long-span cable-supported bridges: A review, Composites Part B: Engineering, 2020, 190: 107934.
9. Liu X, Su Q, Zhu J, et al. The aging behavior and life prediction of CFRP rods under a hygrothermal environment, Polymers, 2023, 15(11): 2490.
10. Arnautov A K, Kulakov V, Andersons J, et al. Experimental investigation on stiffness and strength of single-lap z-pinned joints in a laminated CFRP stress-ribbon strip, The Baltic Journal of Road and Bridge Engineering, 2016, 11(2): 120-126.
11. Hemeda S. Carbon fiber reinforced polymers (CFRP) for strengthening and seismic retrofitting of historic circular masonry stone columns, Advances and Challenges in Structural Engineering: Proceedings of the 2nd GeoM East International Congress and Exhibition on Sustainable Civil Infrastructures, Egypt 2018–The Official International Congress of the Soil-Structure Interaction Group in Egypt (SSIGE). Springer International Publishing, 2019, 114-137.
12. Meier U. Carbon fiber reinforced polymer cables: Why? Why not? What if? Arabian Journal for Science and Engineering, 2012, 37: 399-411.
13. Zdanowicz K, Kotynia R, Marx S. Prestressing concrete members with fibre-reinforced polymer reinforcement: State of research, Structural Concrete, 2019, 20(3): 872-885.
14. Motavalli M, Czaderski C, Pfyl-Lang K. Prestressed CFRP for strengthening of reinforced concrete structures: Recent developments at Empa, Switzerland, Journal of Composites for construction, 2011, 15(2): 194-205.
15. Schmidt J W, Bennitz A, Täljsten B, et al. Mechanical anchorage of FRP tendons–a literature review, Construction and Building Materials, 2012, 32: 110-121.
16. Li Y, Shi T, Qiu Y, et al. Anchorage research for CFRP tendons: a review, Materials, 2024, 17(13): 3208.
17. Terrasi G P, Brönnimann R, Bättig G. CFRP tendon prestress monitoring by resistance measurement in a spun concrete powerline pylon, Proceedings of the first middle east conference on smart monitoring, assessment and rehabilitation of civil structures (SMAR 2011), American University, Dubai, UAE, 2011, 810.
18. Terrasi G P. Prefabricated thin-walled structural elements made from high performance concrete prestressed with CFRP wires, Journal of Materials Science Research, 2013, 2(1): 1.

19. Bastianini F, Corradi M, Borri A, et al. Retrofit and monitoring of an historical building using "Smart" CFRP with embedded fibre optic Brillouin sensors, Construction and building materials, 2005, 19(7): 525-535.
20. Ghafoori E, Motavalli M, Nussbaumer A, et al. Design criterion for fatigue strengthening of riveted beams in a 120-year-old railway metallic bridge using pre-stressed CFRP plates, Composites Part B: Engineering, 2015, 68: 1-13.
21. Heydarinouri H, Nussbaumer A, Motavalli M, et al. Strengthening of steel connections in a 92-year-old railway bridge using prestressed CFRP rods: Multiaxial fatigue design criterion, Journal of Bridge Engineering, 2021, 26(6): 04021023.
22. Franta A, Zając D. Carbon fiber reinforced polymer and tensegrity structures in search of model architectural and engineering solutions, Technical Transactions, 2022, 119(1).
23. Liang C, Zhu Z, Bai G, et al. Design and research on roof structure of the stadium of Sanya International Sports Industry Park, Proceedings of IASS Annual Symposia. International Association for Shell and Spatial Structures (IASS), 2023(4): 1-8.
24. Christoffersen J, Hauge L, Bjerrum J. Footbridge with carbon-fibre-reinforced polymers, Denmark, Structural engineering international, 1999, 9(4): 254-256.
25. Dehmous H, Welemane H. Multi-scale reliability analysis of composite structures–Application to the Laroin footbridge, Engineering Failure Analysis, 2011, 18(3): 988-998.
26. Mei K, Li Y, Lu Z. Application study on the first cable-stayed bridge with CFRP cables in China, Journal of Traffic and Transportation Engineering (English Edition), 2015, 2(4): 242-248.
27. Jay Rohleder Jr W, Tang B, Doe T A, et al. Carbon fiber-reinforced polymer strand application on cable-stayed bridge, Penobscot Narrows, Maine, Transportation research record, 2008, 2050(1): 169-176.
28. Romera L, Hernández S, Baldomir A, et al. Study of pedestrian comfort in a three span stress ribbon footbridge with carbon fibre cables, WIT Transactions on The Built Environment, 2020, 196(14): 139-152.
29. Meier U O, Winistörfer A U, Haspel L. World's first large bridge fully relying on carbon fiber reinforced polymer hangers, Proceedings of the SAMPE Europe Conference, Amsterdam, The Netherlands, 2021, 30.
30. Five bridges flying over the river! The waterway regulation project in Gaoyou of Tongyang line has been fully completed by China Railway Tenth Bureau (2023). https://baijiahao.baidu.com/s?id=1773763638682459669.
31. Liu L, Zhu Y, Huang S, et al. Design and application of CFRP suspender cable for tied arch bridge, Modern Transportation and Metallurgical Materials, 2021, 1 (04): 70-74.
32. The longest carbon fiber composite cable in China made its debut in Jiangyin (2023). https://www.ccia.xin/guonazixun/1399.html.
33. The longest cross-sea bridge in Guangxi has fully entered the superstructure construction stage (2023). https://www.gov.cn/yaowen/2023-05/01/content_5753902.htm.
34. Lu C, Wang X, Ning Y, et al. Numerical investigation on the structural behavior of a short-span cable-stayed bridge with steel and CFRP hybrid cables, Materials, 2024, 17(9): 2032.
35. Classic Project 2: The bridge crossing over Tuhai river on Xinghua road in Liaocheng | The largest carbon fiber cable-stayed bridge in the world (2024). https://www.sdcpd.com/news/491.html.
36. Expected to close in June! The Changzhou side of the Chang-Tai Yangtze river bridge was the first to be connected (2024). https://www.thepaper.cn/newsDetail_forward_27289088.

The Evolution of Chinese Architecture in the 21st Century

Tiantian Yao*

Shanghai Zhongqiao Vocational and Technical University, Shanghai, China

Abstract. With the opening of five ports in 1843, there was a collision of Chinese and Western architecture in terms of design ideas, concepts and styles. However, Chinese architecture in the 20th century introduced a large number of Western architectural styles. While nothing stands still, not even architectural design. In the 21st century, Chinese architecture has gradually evolved into a design concept based on Chinese traditional culture, which dominates the evolution of architectural forms. At the same time, with the global sustainable development and green ecological concepts, the combination of traditional culture and architectural design has formed a unique Chinese architectural style, showing a diversified development. This article will mainly explore the expression and integration of traditional culture in early Chinese architecture, as well as the new concepts of contemporary Chinese architecture in the 21st century.

1. Introduction

In the mid-19th century, as China gradually opened five treaty ports, Western building materials and technologies began to flow into the country. Initially, the structural technology brought by the West was mainly concentrated on masonry buildings, such as British classical, Renaissance or column buildings. From the early 20th century to the mid-20th century, domestic architecture showed a trend of diversified development in style and materials. During this period, European neoclassicism coexisted with traditional Chinese architectural styles, and reinforced concrete and building structures of more than five floors appeared. Buildings such as the Oriental Bank of China Building and the Shanghai General Association Building are more modern in appearance, while also retaining the charm of traditional Chinese culture. The architecture of this period witnessed the integration and innovation of Chinese and Western cultures. From 1925 to 1937, the modernist style of European and American "International Style" architecture was introduced into Shanghai, gradually replacing the simplified classical style and the Baroque architectural style that was integrated with commercial society. The buildings of this period were mainly high-rise buildings, such as the Sassoon Building and the Bank of China Building. These high-rise buildings not only meet the needs of business and residence, but also embody the simplicity and practicality of modernism in design and structure. In the 1950s, the combination of localized Soviet design and traditional Chinese style continued until the 1990s. Although Soviet-style classical architecture has gradually been replaced by the functionality required by the current era, it does not prevent the Sino-Soviet Friendship Building from being a witness and symbol of the era of Sino-Soviet friendship[6]. Therefore, the protection and restoration of historical buildings have become a hot topic of the times.

Turning to 21st century, it was an era of exchange and collision between China and the West, and many architects and wonderful architectural works emerged. New concept, such as green buildings, which refer to high-quality buildings that maximize resource conservation and achieve harmonious coexistence between humans and nature during the life cycle of the building. Sustainable concepts drive the development of Chinese architecture. This article mainly follows the chronological order, and analyzes the future development of Chinese architecture under the concept of global sustainable development from architectural cases in the 21st century.

2. Case Studies

Traditional Chinese architectural elements include horse-head walls, enclosed courtyards, quadrangles, vermilion gates, hollow flower windows, wood carvings, and flying eaves. These elements not only reflect the unique style of traditional Chinese architecture, but also contain rich cultural connotations. However, by the end of the 20th century, Chinese architecture was in an exploratory stage. Traditional architecture has evolved into the "big roof" form, and functionalists also have different opinions on architectural design. In the 21st century, Chinese architects' expression of architecture is not only reflected in the traditional style, but gradually they have their own ideas. In 2006, the Suzhou Museum, which was designed by I.M. Pei. It is a classic example of the combination of tradition and modernity. Decorative lines in white and gray are used to reduce the sense of weight of the building. The overall building materials use glass and concrete instead of traditional wood, which is an innovation. The National Stadium (Bird's Nest) makes excellent use of

* Corresponding author: tiantianyao183@gmail.com

new materials and new structures to show complex and exquisite aesthetics and mechanical properties. At the same time, it gives the building a display of traditional Chinese culture and symbolizes China's innovative ability in the field of architecture, becoming a new landmark in Beijing. In 2010, Shanghai, China held the 41st World Expo, which was also a profound reflection on architectural design. Many types of technologies, such as energy-saving and environmental protection technologies, ecologically sustainable technologies, new materials, etc., are integrated into the architectural design of the China Pavilion. This is a challenge for Chinese architectural design. For example, the building combines passive energy-saving measures such as self-shading, natural ventilation, and lighting, and active ecological energy-saving technologies such as solar roofs and ice storage technologies, to comprehensively improve the environmental awareness and technological content of the China Pavilion[7]. Turning to 2012, Zaha Hadid designed Galaxy SOHO based on the deconstructionist design concept[10]. Its design is inspired by the Wind and Water flow in nature. They can maintain a streamlined shape from any angle and avoid large corners. The open design makes the whole space full of agility and vitality. The streamlined terraces and corridors are interspersed with each other, the spatial layout is reasonable, and the transition between various areas is natural and smooth, which not only meets the needs of commercial functions, but also fully considers the comfort experience of people. Sunlight shines into the huge atrium through the circular glass windows on the roof. The full use of natural light also reduces the use of lights in the building and reduces energy consumption (Galaxy Soho, 2012). The building integrates traditional Chinese cultural elements and modern architectural design. For example, drawing on the idea of Chinese courtyards, four elliptical buildings are enclosed to form a sunken square, which is also an important public space inside the building[5]. However, what is most special is that the building has skylights in each unit. These modules provide gentle light to the interiors of the four building units. This modular structure increases the surface area of the building complex. This structure increases the surface area of the building complex and exposes the interior of the building to more sunlight.

The design of Shanghai Tower embodies the perfect fusion of modernization and sustainability, showing the possibility of super high-rise buildings in the future, similar buildings include Canton Tower, Guangzhou. In the late 20th century, China's reform and opening up entered a climax, and economic development was greatly improved. Many areas in Shanghai have transformed from barren farmland to areas with rapid economic and commercial development[2]. On the one hand, the limited land in the city needs to meet people's demand for high-density buildings, which is a conflict between people and nature and the environment. On the other hand, with the fast-paced city life, people also need to find a trace of green close to nature in the cold concrete buildings. Gensler, the designer of the Shanghai Tower, combines traditional Chinese cultural elements while integrating urban development trends and modern design concepts into the building. The designer was inspired and inspired by the traditional element hutongs. Residential buildings are organized around the public open space in the middle. Similarly, in the Shanghai Tower, these communities are vertical, and each community has its own "sky garden" to promote interaction and create a sense of community. From the perspective of policy and urban development, the Shanghai Municipal Government requires that 30% of the site be green. This building not only meets this requirement, but also draws on the temples, pagodas and palaces in Chinese history in the landscape design of the site. This design is not only an intuitive embodiment of green buildings, but also a fusion of humanistic and social ideas. Stefano Boeri first proposed the vertical forest as an innovative urban greening design concept. Its main feature is to plant a large number of trees and plants on the building structure to form a green vegetation wall, thereby solving environmental problems such as air pollution, heat island effect and ecosystem destruction caused by urbanization. In 2022, China's first vertical forest complex "Huanggang Juran Home" was completed in Huanggang City, Hubei Province. Open and closed balconies integrate a large number of green plants into the interior. This design greatly reduces carbon emissions every year.

Looking around the world, Melbourne's Pixel Building is Australia's first carbon-neutral office building. The developer uses a low-carbon concrete called "Pixelcrete", which is mainly composed of industrial waste, recycled or regenerated materials, reducing the traditional cement content, thereby reducing carbon emissions and achieving carbon neutrality. Singapore has a natural advantage in developing vertical forest buildings, and Oasia Hotel Downtown is a representative practice in this field. The building has a mesh shell on the outside, which is covered with 21 different types of climbing plants, providing food and living space for birds and insects, and creating a unique ecological living experience for nearby urban residents. The Bahrain World Trade Center is located in Manama, the capital of the island nation of Bahrain. This building is also one of the first buildings in the world to have wind turbines built directly into the building structure, providing experience for the development of wind energy buildings in global cities. As cities grow and high-rise buildings increase, Asia has become the center of high-rise buildings, while in Europe and the United States, they pay more attention to manufacturability and efficiency than building height[9]. The Al-Bahar towers in Abu Dhabi is an example of a building whose honeycomb-like appearance is computer-determined. The solar roof allows the building to dissipate heat and keep warm very well as the climate changes. The design saves a lot of electricity energy. On the one hand, houses have not undergone fundamental changes, architecture is traditional. On the other hand, architecture is innovative (Yuldasheva M.K, 2023). For example, France's ECO-house uses natural materials and innovative technologies to maximize energy savings and improve living comfort while causing minimal harm to the environment.

In summary, from a green and sustainable global perspective, architects prefer low-carbon buildings and people's experience inside buildings. This means the application of new materials and new structures. Different from the traditional roofs, beams and columns of Chinese architecture in the 20th century, architects in the 21st century are more inclined to integrate deep traditional cultural concepts into architectural design. At the same time, buildings are combined with urban planning to create landmark buildings. More importantly, the integration of Chinese and Western cultures has become more frequent, and Chinese designers are constantly thinking and combining advanced Western concepts to integrate them into architectural design.

3. Discussion

Rather than attracting renowned overseas architects and multinational design firms, the organizers of the China Pavilion design competition called for participation from architects of Chinese descent around the world. This unconventional eligibility criteria stems from an industry paradox. During the modernization boom of the past few decades, domestic architects have not yet successfully translated the country's rich architectural heritage into an easily recognizable modern architectural language. The process of globalization has also caused a collision between Chinese and Western cultures, including architecture. Many architects and critics oppose the excessive Westernization of domestic architectural design, believing that China has actually become a test site for foreign architects to try new design techniques and materials. Of course, due to different cultural backgrounds, the works of these architects cannot always integrate local characteristics, and the designed buildings seem a little mediocre to Chinese people, both from an aesthetic and functional perspective. This has also inspired Chinese architects to think about how to combine regional culture with traditional characteristics in architecture.From a vertical perspective, reviews the changes of the China Pavilion in eight historical events from the Philadelphia World Expo in 1876 to the Shanghai World Expo in 2010, China, over the past 134 years, showing two common features: traditional vocabulary and color selection[8]. The inspiration for these buildings comes from in-depth designs based on traditional Chinese culture. The choice of color reflects the Chinese preference for red. Through structural expression elements and gorgeous Chinese red, the implemented "dougong" scheme perfectly matches these two traditional attributes. As an innovative combination, this case sets a model for integrating traditional vocabulary into modern architecture. However, in terms of the safety of the building, Juhua Yang (2012) pointed out the need to improve the seismic resistance index of the China Pavilion's building structure through earthquake resistance tests, and proposed improvement plans for weak points[1]. Meanwhile, contemporary people's criticism and demand for Chinese architecture are also increasing. The novel appearance of buildings is gradually unable to meet people's needs for buildings. People are more concerned about how the functions of buildings, the design performance of internal spaces, and the creation of light behind the novel appearance of these buildings serve people. In the face of urban environments, how to combine urban renewal with future development. From the 19th century when Chinese architecture was mainly traditional, to the 21st century when architects constantly think about integrating tradition with modernity and creating great architectural works, this not only reflects the upgrade of design innovation, but also reflects the development of Chinese architectural design. Academician He Jingtang, chief designer of the China Pavilion, also constantly thinks that how to integrate traditional Chinese culture, which has a long history, into architecture and express the essence and connotation of culture with specific images and design language is a huge challenge for future Chinese architecture. The design focuses more on symbolism rather than utilitarianism. However, in the reconstruction process after the Expo, the China Pavilion was considered to have a dual mission. Even with an aesthetic shell, single-purpose spaces are prone to underutilization after the Expo and combining dedicated and multifunctional spaces will provide greater flexibility. By supplementing the iconic roof with a fully functional platform, the possibility of the China Pavilion becoming a complete "useless thing" is minimized, as evidenced by its successful transformation after the Expo. This case shows the possibility of a building achieving a good balance between form shaping and functional adaptability. Moreover, Maryam Farajpour (2024) proposed that the spatial atmosphere in the building space is the key to the quality of the building and an important theme for conveying the sense of place experience[4]. However, there are few studies on the atmosphere factor. Therefore, for the 21st century, architectural design needs to be combined with philosophical thinking to explore the quality of the spatial atmosphere inside the building. At the same time, Kabinesh (2024) summarized the evolution of architecture in the 21st century[3]. He divided the 21st century into three stages. The first stage was the early 21st century, emphasizing green buildings and sustainable bio-compatible design. The second stage was the 2010s, when architecture was combined with cities and architecture was integrated into urban planning. The third stage is the present, combining bio-compatible design with human health and environmental sustainability. Bio-compatibility refers to the interaction between building materials and the biological environment and their impact on organisms. Therefore, bio-compatibility is also a necessary consideration for future architectural design. By combining bio-compatible design with architecture, it not only alleviates environmental problems, but also involves social and economic development. In short, bio-compatibility represents a sustainable design concept and method that promotes harmonious coexistence between people, buildings and the environment. Wherefore, in the latest research, we can see that designers pay attention to the creation of spatial atmosphere and people's experience, as well as the continuous exploration of new sustainable materials. Therefore, in addition to cultural transmission, global sustainability is still the focus of people's architectural design in the 21st century.

4. Conclusion

The term sustainable development has become very popular. According to this concept, buildings should cause minimal harm to the environment. Modern China has been constantly exploring architectural design styles based on this concept. In the 21st century, Chinese designers no longer pile up traditional elements in architecture, but integrate Chinese traditional virtues and cultural essence into architecture. The use of new materials breaks through traditional materials, making buildings more low-carbon and environmentally friendly, and combining green sustainable concepts to promote the organic development of green cities. The value of architecture reflects both cultural value and commercial value, and its functions are more diversified. Unlike Western countries, my country's current situation is highly developed urbanization and dense population. Therefore, future buildings need to meet the functional needs of high-density cities while being low-carbon, such as increasing the height of buildings. This article mainly sorts out the morphological evolution of contemporary Chinese architecture, analyzes and discusses 21st century architecture, critically points out the influence and limitations of different buildings, and lays a case reference for future architectural design trends. It provides case support for studying the evolution of Chinese architectural style in the 21st century. However, this study also has certain limitations, such as the short time span of the selected cases and only one third of the 21st century has passed, which will lead to an incomplete analysis of the evolution of Chinese architecture. In terms of regional selection, cities are mainly highly developed cities, mainly Beijing and Shanghai. Although they are representative and economically developed regions in my country, they still have geographical limitations because high-density cities and ordinary cities have different requirements for building development and urban planning. Overall, future research on Chinese architectural style can use this as a basis to analyze the architecture of the 21st century and the future.

References

1. Juhua, Y.(2012). Shaking table tests on China Pavilion for Expo 2010 Shanghai China. *Journal of Tall and Special Buildings.* 21(4), 265-282.
2. Jun X., Dennis P., and Douglas M.(2010). "Case Study: Shanghai Tower" CTBUH Journal,2010 Issue II. Retrieved from: https://global.ctbuh.org/resources/papers/download/12-case-study-shanghai-tower.pdf
3. Kabinesh, V., Vennila, S., Baranidharan, K., Ravi, R., Krishnamoorthi, S. and Thirunavukkarasu, M.(2024). Sustainable Spaces-The Evolution of Biophilic Design in Modern Architecture: A Review. Asian Journal of Environment & Ecology, 23(5), pp.64-77.
4. Maryam F., Mehrdad M., Nayer T. and Zahra K. (2024). An overview of phenomenological theories and its impact on the atmosphere of architecture from the perspective of architectural thinkers and philosophers. Journal of Urban Management and Energy Sustainability.5(3), pp.60-76.
5. Meinhold, B. (2012). "Zaha Hadad's Light-Filled LEED Silver Galaxy SOHO Opens in Beijing" 14 Nov 2012. Retrieved from: http://inhabitat.com/zaha-hadids-light-filled-leed-silver-galaxy-soho-opens-in-beijing/
6. Xuerui,W. & Xiangning, L. (2023). Intersecting narratives of site: Fragrant Hill Hotel as a reconstruction project. Journal of Asian Architecture and Building Engineering. 22(5), 2506-2522.
7. Xu M.X., Chen F.X., Xie S., and Zhou H.X.(2011). Structural system of China Pavilion at 2010 Shanghai World Expo. Journal of South China University of Technology (Natural Science Edition), 2011(039)009, P82-87.
8. Ying,D.(2013).Conceptualizing mega-event flagships —A case study of China Pavilion of Expo 2010 Shanghai China. Frontiers of Architectural Research, 2(1), 107-115.
9. Yuldasheva M.K and Raxmanova M.B (2023). Architecture Of The Future – Modern Trends And Innovation. Journal Of Arts And Design. pp.314-319.
10. Zaha Hadid Architects (2012). DeZeen Magazine. Retrieved from: http://www.dezeen.com/ 2012/10/29/galaxy-soho-by-zaha-hadid-architects/

A Study on the Age-Friendly Landscape Evaluation System for Rural Communities in Western Zhejiang: A Case Study of Quzhou

Jiahe Xu[a*], Mimi Tian[b]

School of Design and Architecture, Zhejiang University of Technology, Hangzhou, Zhejiang, China

Abstract. With the intensification of population aging, the issue of age-friendly development in rural communities has become increasingly prominent. This study focuses on the age-friendly landscape in rural communities in western Zhejiang, and develops an evaluation system to provide both theoretical support and practical guidance for age-friendly construction. The study identifies and selects objective indicators such as air quality and green coverage, as well as subjective indicators like folk culture and historical relics. Using the Analytic Hierarchy Process (AHP) and entropy method for integrated weighting, the study minimizes the subjective biases and information loss inherent in traditional evaluation methods. Furthermore, by analyzing the weight distribution of various landscape elements, the study reveals the priority for improving natural environments, built infrastructures, and social landscapes, and provides improvement recommendations. The research findings supplement existing evaluation standards for age-friendly communities, specifically in the context of landscape accessibility for the elderly, and offer a scientific basis for the formulation and implementation of future policies.

1. Introduction

1.1 Research Background

As global aging intensifies, creating suitable living environments for older adults has become an urgent issue for countries worldwide. In response, the World Health Organization (WHO) launched the Age-Friendly Cities and Communities (AFCC) initiative in 2007, providing an important framework for the development of age-friendly communities globally [1]. By the end of 2023, the elderly population aged 60 and above in China had reached 297 million, accounting for 21.1% of the total population. In December 2020, China began its selection of exemplary national age-friendly communities[a]. However, the results show that rural aging infrastructure lags significantly behind urban areas, with the situation in Quzhou particularly representative. In rural communities, where the aging population ratio is higher, infrastructure is underdeveloped, and resources are limited, there is an urgent need for a scientific and rational evaluation system to guide improvements and enhance the quality of life for older adults.

The western Zhejiang region is predominantly hilly and basin terrain, with dense villages and a relatively high proportion of elderly residents, creating an urgent demand for age-friendly measures. As a pioneer in rural revitalization, Zhejiang has made significant strides in resource allocation, infrastructure improvement, and the advancement of social services. However, the level of aging-in-place development remains uneven. Focusing on the aging-in-place landscape of rural communities in western Zhejiang not only offers targeted recommendations for the region but also provides valuable reference points for rural areas nationwide, laying a scientific foundation for the planning and development of age-friendly rural communities.

1.2 Literature Review

As global aging intensifies, the construction of age-friendly cities has become a key topic in improving the quality of life for older adults. Research abroad has primarily focused on aging-in-place design in urban environments, encompassing aspects such as transportation, architecture, and housing. In particular, the optimization of transportation and housing design is seen as crucial for promoting older adults' mobility and independent living[2]. Furthermore, high-quality age-friendly landscapes, such as green spaces and public areas, have been shown to improve elderly individuals' mental health and social interactions[3–5].

However, driven by technological advancements, the World Health Organization's (WHO) framework has not fully integrated technological factors and requires expansion to better meet the needs of modern cities [6].

[a] 2020 National Health and Health Commission, National Office on Aging *Notice on Carrying Out the Creation of Model National Age-Friendly Communities, National Standards for Model Urban and Rural Age-Friendly Communities (for Trial Implementation)*

* Corresponding author: [a*] 838080133@qq.com, [b] tmm518@zjut.edu.cn

Meanwhile, research on aging in rural communities has lagged behind. Studies indicate that rural areas face significant deficits in health services and social participation [7,8], particularly in the development of social landscapes, where older adults are confronted with greater challenges due to social isolation [9,10]. Quality aging-in-place landscapes have been demonstrated to reduce symptoms of depression in elderly populations in rural areas[11]. However, there remains a lack of sufficient research on landscape design for mental health in rural settings, and the rural-urban disparity in aging infrastructure remains prevalent.

The development of aging-in-place evaluation systems has evolved based on urban research, primarily focusing on the improvement of infrastructure and service facilities. For example, the Age-Friendly City Criteria (AFCC) theory offers several evaluation guidelines and technical indicators, covering neighborhood environments, housing, and service facilities [12-15]. However, existing evaluation systems have not adequately considered the unique needs of rural communities, making it difficult to directly apply urban aging-in-place results to rural environments.

To address this, scholars both domestically and internationally have proposed methods such as the Fuzzy Analytic Hierarchy Process (FAHP) and hybrid approaches aimed at optimizing the scientific accuracy of evaluation systems[16]. Some studies have constructed multi-level evaluation frameworks using literature analysis and expert consultations, but these methods are mostly focused on urban aging-in-place, lacking applicability to rural settings. Torku et al. analyzed common methods used in AFCC evaluations and found that subjective approaches (e.g., questionnaires, interviews, and focus group discussions) continue to dominate. They suggested the integration of more objective and hybrid methods, emphasizing the inclusion of quantitative indicators based on subjective feedback [17]. Despite the higher proportion of elderly populations in rural areas, research in this domain remains scarce, and comprehensive consideration of multi-dimensional factors such as rural aging-in-place landscapes, artificial landscapes, and social landscapes is still lacking.

In conclusion, current research on aging-in-place landscapes in rural communities is still relatively limited, primarily manifested in insufficient landscape focus, imbalances between urban and rural studies, and the difficulty of directly applying existing evaluation systems to rural environments. Some studies have explored measures to support rural elderly populations by increasing medical resources, but there is inadequate attention to other aspects of artificial and social landscapes, resulting in a lack of comprehensiveness in improving the quality of life for the elderly in rural areas. The existing aging-in-place evaluation systems are largely based on urban experiences and lack applicability to rural settings, while the influence of technological factors and dynamic environmental changes has not been sufficiently incorporated. Therefore, future research should focus more on the unique landscape design needs of rural communities and, by incorporating objective quantitative indicators, develop a scientifically sound and contextually appropriate evaluation framework to comprehensively enhance the quality of life for elderly populations in rural areas.

1.3 Research Subjects

Western Zhejiang primarily refers to the Jinqu Basin and its surrounding areas in the western part of Zhejiang Province, including parts of Tonglu, Chun'an, and Jiande. The region is characterized by hilly and basin terrain, presenting a distinct geographical contrast to other parts of Zhejiang, such as the coastal plains. The undulating topography of western Zhejiang has resulted in a dense layout of villages, most of which are built along the mountains. These villages are home to a large population, with a relatively high proportion of elderly residents. Rural communities in this region are relatively large in scale, with agriculture and traditional handicrafts serving as the primary economic pillars. However, the overall level of economic development remains relatively low.

Despite the dense population and numerous villages, age-friendly facilities in rural western Zhejiang are still underdeveloped. Medical services and accessibility infrastructure fail to meet the needs of the elderly population, and basic infrastructure such as roads, transportation, and age-friendly amenities require urgent improvement. The low quality of life among the elderly in this region is directly impacted by the lack of sufficient age-friendly facilities. Therefore, enhancing the infrastructure of rural communities, particularly the living environments and social support systems for elderly residents, has become an urgent issue that needs to be addressed.

2. Construction of the Age-Friendly Landscape Evaluation System for Rural Communities in Western Zhejiang

2.1 Identification and Selection of Indicators

This section aims to develop an indicator system for age-friendly landscape evaluation in rural communities in Western Zhejiang. Through in-depth analysis of both domestic and international age-friendly evaluation standards and the characteristics of rural communities, the study identifies core indicators that are scientifically sound and applicable. Domestic samples include Tianxianyucun in Beijing, Dayi Xinfuxie Community in Chengdu, Liangzhu Cultural Village in Zhejiang, Japan's "Nonbiri Village" and Germany's Welfare Village. The study identifies five key age-friendly resource elements: 1) Location and transportation conditions, 2) Natural environment, 3) Cultural environment, 4) Facility conditions, and 5) Socioeconomic environment.

The study proceeds by reviewing technical guidelines and literature related to age-friendly construction [b], performing meta-analysis of studies on age-friendly evaluation, rural aging suitability evaluation, AFCC evaluation research, and the factors affecting rural elderly care. Through frequency statistics, the main factors were identified. The five elements were categorized into three types: natural landscapes, artificial landscapes, and social landscapes. Combining the national rating criteria for model age-friendly communities and age-friendly factors from the literature[c], the study uses correlation analysis to select 24 evaluation indicators for rural community landscape aging, adjusting some indicators' focus areas, as indicated in Table 1.

Table 1 Indicators and Evaluation Types

Constituent Elements	Evaluation Indicators	Evaluation Types
Age-Friendly Natural Landscape	Green Coverage Rate	Objective
	Climate Attributes	Objective
	Air Quality	Objective
	Water Resource Conditions	Integrated Evaluation
	Topographical Attributes	Objective
	Noise Situation	Objective
Age-Friendly Artificial Landscape	Traffic Volume	Objective
	Elderly-Friendly Housing Conditions	Integrated Evaluation
	Environmental Sanitation Conditions	Objective
	Road and Lighting Conditions	Objective
	Public Transportation Facilities	Objective
	Medical Support Facilities	Integrated Evaluation
	Educational Support Facilities	Integrated Evaluation
	Leisure and Recreation Facilities	Integrated Evaluation
	Historical Sites and Building Preservation	Integrated Evaluation
Age-Friendly Social Landscape	Folk Culture Preservation	Integrated Evaluation
	Village Cultural Attributes	Objective
	Historical Figures and Events	Objective
	Public Security Conditions	Subjective
	Social Elderly Care Services	Subjective
	Elderly Social Participation Level	Subjective
	Pension Insurance Coverage	Objective
	Medical Insurance Coverage	Objective
	Per Capita Disposable Monthly Income	Objective

2.2 Hierarchical Model and Weight Distribution

This study adopts a combined multi-evaluation method of Analytic Hierarchy Process (AHP) and Entropy Weight Method to assess the age-friendliness of rural community landscapes. By integrating these two methods, the study introduces both expert subjective judgment and objective data, ensuring the comprehensiveness and scientific validity of the evaluation results.

First, a hierarchical structure for evaluation indicators is established based on the three major components identified in the previous section and the twenty-four key evaluation indicators. The relative importance between these categories and specific indicators is then determined. Given that landscape age-friendliness in rural communities involves multiple domains, to ensure the scientific accuracy and feasibility of the results, the study invited 31 experts and scholars from universities, government, and industry to participate in weight judgment. Experts performed pairwise comparisons of each indicator's impact on the quality of life for the elderly, using Saaty's "1-9 scale of comparison" for value assignment[18], as indicated in Table 2. The comparison results formed a judgment matrix, reflecting the relative importance of each indicator.

Table 2 Weight Judgment Matrix

Am	B1	B2	...	Bn
B1	b11	b12	...	b1n
B2	b21	b22	...	b2n
...	...	...	...	...
Bn	bn1	bn2	...	bnn
Scale	Meaning			
1	i and j are equally important			
3	i is slightly more important than j			
5	i is obviously more important than j			
7	i is strongly more important than j			
9	i is extremely more important than j			
2,4,6,8	Compromise is used when necessary			

After repeated comparisons and adjustments, the study achieved a consistency check result of less than 0.1, which ensured the reliability of the pairwise comparisons. This process yielded the first-round subjective weights for all indicators.

The Entropy Weight Method was used to assign weights to objective indicators. The core principle of this method is to determine the weight of each indicator based on the degree of data dispersion; the greater the variation in the data for a specific indicator, the more information it contains, and thus the higher its weight.

For the objective weighting process, 14 rural communities in the Quzhou region were selected as samples, including natural villages, traditional villages, historical and cultural villages, and national model age-friendly communities. During data processing, the first-round subjective weights were denoted as:

[b] 2007 World Health Organization *Lobal Guidelines for the Construction of Age-Friendly Cities*
2009 Office of the Chinese Committee on Aging *Guidelines for the Construction of Livable Communities for the Elderly*
2013 Shanghai Municipal Committee on Aging *Guidelines for the Construction of Age-Friendly Cities in Shanghai (for Trial Implementation)*
2016 Office of the Chinese Committee on Aging *Guiding Opinions on Promoting the Construction of Livable Environment for the Elderly.*

[c] 2020 National Health and Health Commission, National Office on Aging *Notice on Carrying Out the Creation of Model National Age-Friendly Communities*

$$W_{initial} = (w_1, w_2 \ldots, w_n) \quad (1)$$

For the specific groups of objective indicators, secondary weighting was applied to obtain new weights for these indicators, which are recorded as:

$$W_{objective} = \left(w_i', w_j', \ldots, w_k'\right) \quad (2)$$

The secondary weights of the objective indicators, denoted as $W_{objective}$ were normalized to ensure that the total sum of the weights equals 1. Here, mmm represents the number of objective indicators that underwent secondary weighting:

$$w_i'' = \frac{w_i'}{\sum_{i=1}^{m} w_i'} \quad (3)$$

A coefficient β=0.5\beta = 0.5β=0.5 was defined, indicating that the weights obtained through the AHP method and the Entropy Method each contribute equally (with a proportion of 0.5) to the final combined weight. The final weight for the secondary weighted objective indicators is calculated as follows:

$$w_i^{final} = \beta \times w_i'' + (1 - \beta) \times w_i \quad (4)$$

For the indicators that did not undergo secondary weighting, the initial weights obtained from the subjective weighting process are directly used:

$$w_j^{final} = w_j \quad (5)$$

All the final weights were combined and normalized to ensure that the sum of the final weights equals 1. After adjustment, the comprehensive weight values were calculated and are presented in the following Table 3:

Table 3 Indicator Weights

Goal Layer	Criterion Layer	Criterion layer weights	Indicator Layer	First-round subjective weights	Second-round integrated weights	Composite Weights
A Age-Friendly Landscape in Western Zhejiang Rural Communities	**B1** Age-Friendly Natural Landscape	0.305	**C1** Green Coverage Rate	0.183	0.181	0.055205
			C2 Climate Attributes	0.114	0.111	0.033855
			C3 Air Quality	0.191	0.189	0.057645
			C4 Water Resource Conditions	0.179	0.176	0.05368
			C5 Topographical Attributes	0.154	0.157	0.047885
			C6 Noise Situation	0.188	0.186	0.05673
	B2 Age-Friendly Artificial Landscape	0.424	**C7** Traffic Volume	0.102	0.094	0.039856
			C8 Elderly-Friendly Housing Conditions	0.119	0.134	0.056816
			C9 Environmental Sanitation Conditions	0.113	0.114	0.048336
			C10 Road and Lighting Conditions	0.117	0.123	0.052152
			C11 Public Transportation Facilities	0.115	0.112	0.047488
			C12 Medical Support Facilities	0.124	0.136	0.057664
			C13 Educational Support Facilities	0.091	0.089	0.037736
			C14 Leisure and Recreation Facilities	0.112	0.113	0.047912
			C15 Historical Sites and Building Preservation	0.107	0.085	0.03604
	B3 Age-Friendly Social Landscape	0.271	**C16** Folk Culture Preservation	0.112	0.098	0.026558
			C17 Village Cultural Attributes	0.106	0.094	0.025474
			C18 Historical Figures and Events	0.102	0.099	0.026829
			C19 Public Security Conditions	0.107	0.122	0.033062
			C20 Social Elderly Care Services	0.121	0.138	0.037398
			C21 Elderly Social Participation Level	0.113	0.109	0.029539
			C22 Pension Insurance Coverage	0.118	0.119	0.032249
			C23 Medical Insurance Coverage	0.116	0.117	0.031707
			C24 Per Capita Disposable Monthly Income	0.105	0.104	0.028184

The weighting results demonstrate that the relative importance of different indicators in the age-friendly evaluation was reasonably differentiated. The comprehensive weighting method effectively balanced subjective opinions and objective data, ensuring a fair and scientifically robust evaluation process.

2.3 Coring Criteria and Basis

Using the 24 age-friendly landscape evaluation indicators constructed in the previous section, each evaluation target was scored individually. A four-point scale was adopted: "Excellent" received 4 points, "Good" 3 points, "Average" 2 points, and "Poor" 1 point. Weighted scores for each indicator and criterion layer were then calculated.

To ensure practicality, the score thresholds for the 24 indicators were set based on relevant standards and literature. Additionally, dimensionality reduction was applied to some high-dimensional indicators. A score of 60% (i.e., 2.4 points) was established as the benchmark for assessment. Communities with a total score of 2.4 or higher were considered to have the potential for developing an age-friendly landscape, while those scoring below this threshold were deemed in need of improvement in key elements.

 https://doi.org/10.1051/e3sconf/202561703020

3. Research Application and Conclusion

3.1 Data Sources

This section validates the scientific rigor and practical applicability of the constructed evaluation system through the application and analysis of actual data from rural communities in western Zhejiang. The research data were primarily obtained from public data provided by relevant government departments, field visits, and on-site surveys.

Data for the natural environment, such as green coverage rates and climate attributes, were sourced from the Forestry Bureau and Meteorological Bureau, respectively. Environmental factors such as air quality were based on meteorological data and verified through field inspections.

For the artificial landscape and social landscape, data related to indicators such as the adaptability of housing for the elderly, medical facilities, educational resources, and recreational facilities were collected through on-site surveys to ensure data authenticity and relevance.

Additionally, data for the social landscape, including village cultural attributes and public security conditions, were provided by village committees and supplemented with feedback from residents, further enriching the data foundation of the evaluation system.

3.2 Application and Discussion

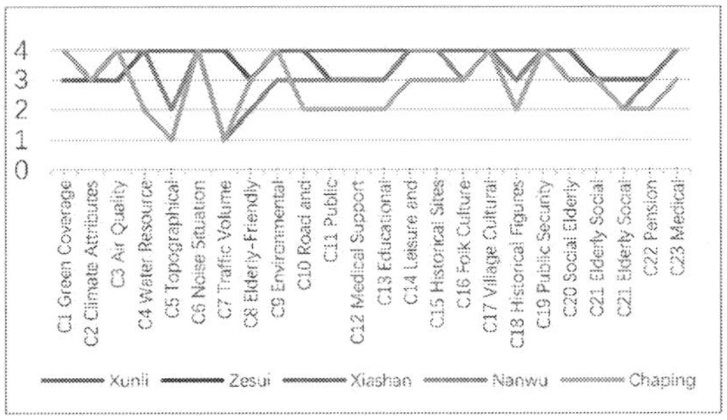

Fig. 1. Score chart

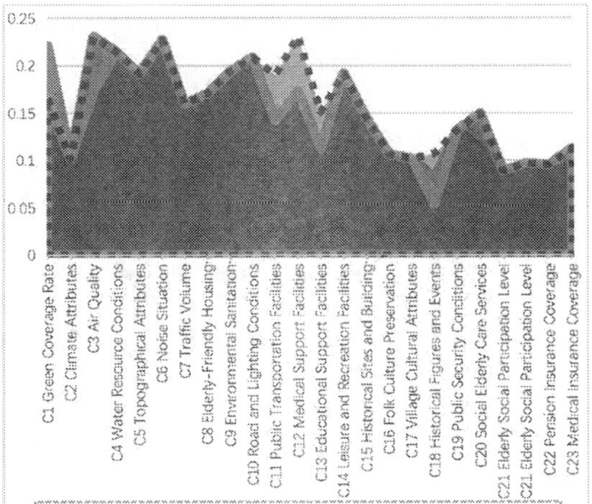

Fig. 2. Weight comparison chart

1) Analysis of Rural Age-Friendliness Differences: The Imbalance of Regional Development

As shown in Figure 1,2, this study found significant diversity and imbalance in the age-friendly level of rural areas. The differences between villages reflect disparities in regional development and resource allocation, with scores ranging from 3.76 to 2.76, indicating a clear gap in natural conditions, resource investment, and infrastructure improvement.

Specifically, Xunli Village, as part of the 28 Dufu Ancient Town, benefits from convenient transportation, good cultural preservation, and well-developed infrastructure. After tourism development, it has attracted a large number of visitors, indirectly driving the development of age-friendly facilities in the village. Zesui Village and Xiashan Village have a high proportion of original residents and elderly people, with relatively well-developed transportation and medical facilities, making them overall suitable for elderly living, though there is room for improvement in community activities. Nanwu Village retains a good appearance and is a quiet, pleasant place to live, but lacks sufficient transportation and eldercare facilities. Chaoping Village scored the lowest due to its complex terrain, poor transportation, and lack of age-friendly facilities, highlighting the major shortcomings in rural age-friendly construction.

The evaluation system in this study is based on the natural environment and climate conditions of the Western Zhejiang region, so the current weights are applicable to similar regions. For regions with significant differences in climate, terrain, and economic conditions, the existing weights may not be fully applicable. To enhance the adaptability of the evaluation system, future research should readjust the weights based on the natural landscape elements of different regions.

2) Comparison with Existing Evaluation Systems: The study shows that there are significant differences in the goals and focuses of the national elderly-friendly community evaluation standards and the evaluation system in this research. The national standards primarily focus on the functionality and convenience of infrastructure, while this study expands on this by emphasizing landscape design and overall livability, stressing the long-term positive impact of the environment on elderly physical and mental health.

Specifically, the core elements of the national standards, such as household water access and road paving, focus on improving basic conditions. For example, Chaoping Village was designated as a national exemplary elderly-friendly community, yet its age-friendly landscape score was low, revealing the insufficient attention to overall livability in the national standards.

In terms of policy implementation, environmental factors such as air quality, green coverage, and cultural preservation should be included in the evaluation system, emphasizing the role of comprehensive livability in promoting the mental health of the elderly. By integrating multidimensional indicators of natural, artificial, and social landscapes, the comprehensive construction of age-friendly rural communities can be promoted, reducing elderly social isolation, enhancing their mental health, and increasing social participation.

3) Resource and Time Constraints: Core Challenges in Rural Age-Friendly Development

Rural age-friendly construction faces dual constraints of resources and time, particularly in the transformation of natural, artificial, and social landscapes, where high economic costs and long construction cycles are prominent issues that affect the comprehensive advancement of age-friendly facilities.

Regarding natural landscapes, respondents generally agree that although natural landscapes play an important role in age-friendly evaluation, their transformation (such as terrain optimization) comes with significantly higher economic and time costs compared to other elements, especially in mountainous villages. For artificial landscapes, respondents emphasized the importance of transportation systems, with the main improvements being road widening and surface repairs aimed at reducing travel time between rural areas and surrounding towns. Regarding social landscapes, economic conditions were identified as a key constraint in rural age-friendly construction. Respondents pointed out that insufficient economic resources in rural communities slow down infrastructure construction and maintenance, directly hindering the achievement of age-friendly goals.

To address these issues, a phased implementation strategy could be adopted. Prioritizing the transformation of transportation and infrastructure, especially improving the connectivity between rural areas and cities, is crucial. More detailed and specific standards for leisure and entertainment facilities should be developed to ensure clear implementation directions and standards for community construction. For natural landscapes, gradual terrain optimization measures could be employed to reduce the difficulty of transformation in high-cost areas, especially in mountainous and resource-scarce regions. With limited resources, rural age-friendly levels can be gradually improved, enhancing elderly living environments and promoting better integration of the elderly into social activities, thus improving their physical and mental health.

3.3 Conclusions

This study, through an investigation and evaluation of the current status of age-friendly landscapes in rural communities of western Zhejiang, highlights deficiencies in the existing landscape development, particularly in its support and enhancement of elderly residents' physical and mental health. The research indicates that while there are national standards for age-friendly communities, their applicability to rural environments remains limited. By constructing an evaluation system for age-friendly landscapes in rural communities, the study demonstrates that combining the Analytic Hierarchy Process (AHP) and Entropy Method effectively balances subjective and objective indicators, overcoming the limitations of traditional evaluations that overly rely on subjective weighting. This approach enhances the scientific rigor and reliability of the evaluation. The results not only accurately reflect the overall performance of communities in terms of age-friendliness but also validate the effectiveness of the constructed evaluation system.

Additionally, this research provides a framework for evaluating age-friendly landscapes based on the rural communities of western Zhejiang, addressing the insufficient attention to landscape factors in the existing evaluation standards for age-friendly communities. By integrating natural landscapes, artificial infrastructure, and social landscapes into a comprehensive set of indicators, this framework offers theoretical support and empirical evidence for policymakers seeking to improve the living environments of elderly residents.

Despite its contributions, this study has limitations. First, it focuses solely on the rural areas of western Zhejiang, and the generalizability of the evaluation system requires further validation in a broader geographic context. Second, the study does not delve deeply into the impact of economic conditions and policy factors on rural age-friendly development, which are critical aspects in rural settings. Future research should expand the sample scope and incorporate economic factors, dynamic assessment mechanisms, and real-time monitoring technologies. Integrating digital tools such as GIS and IoT will ensure that the evaluation system can adapt to the dynamic changes in rural environments while maintaining broad applicability and flexibility. This will provide more scientific and effective support for rural age-friendly development.

Acknowledgments

This paper is one of the phased achievements of the General Project of the National Social Science Fund of China: "Research on Models and Approaches for Establishing a Protection and Inheritance System for Historical and Cultural Towns and Villages in the Yangtze River Delta in the New Era (22BSH043)." Supported by the Fundamental Research Funds for the Provincial Universities of Zhejiang (GB202301005).

References

1. Dikken, J., van den Hoven, R. F. M., van Staalduinen, W. H., Hulsebosch-Janssen, L. M. T. & van Hoof, J. (2020). How Older People Experience the Age-Friendliness of Their City: Development of the Age-Friendly Cities and Communities Questionnaire. *Int. J. Environ. Res. Public. Health* 17, 6867.
2. Ann, B., Denise. (2001). Elder-friendly plans and planners' effort to involve older citizens in the plan-making process. (University of New Orleans.).
3. Chen Chongxian, Luo Weijing, Li Haiwei, Xia Yu, & Xie Chao. (2021). Research Progress on the Relationship Between Residential Area Landscape Environment and Elderly Health. Southern Architecture, 22–28.
4. Zhuang Yini & Chen Dan. (2019). The Impact of Residential Area Plant Landscapes on Elderly Neighbors' Social Interaction—A Case Study of Hongkou District, Shanghai. Journal of Shanghai

Jiao Tong University (Agricultural Science Edition), 37, 70–75.

5. Cheng Xuefei, Wang Shufen, & Wu Qingru. (2017).Research and Application on the Correlation Between Community Landscape Environment and Outdoor Activities of Elderly People. In Sustainable Development and Rational Planning—Proceedings of the 2017 China Urban Planning Annual Conference (Housing Construction Planning) (ed. Chinese Society of Urban Planning), 22 (China Architecture & Building Press, Dongguan, Guangdong, China).
6. Marston, H. R. & van Hoof, J. (2019). "Who Doesn't Think about Technology When Designing Urban Environments for Older People?" A Case Study Approach to a Proposed Extension of the WHO's Age-Friendly Cities Model. *Int. J. Environ. Res. Public. Health* **16**, 3525.
7. Wang Yiming, Li Xue, & Li Bin. (2021). Research on the Functional Integration Strategies of Community Day Care Facilities Based on the Needs of the Elderly. Modern Urban Research, 65–71, 105.
8. Hao Liang. (2016). Building a Community-Based Home Care Service Supply System—A Demand-Oriented Approach for the Elderly. Modern Business, 190–191.
9. Conway, S. F., Farrell, M., McDonagh, J. & Nolan, N. (2024). Creating an Age-Friendly Environment in Farming – The Farmer's Yards Approach. *J. Agromedicine*.
10. Gallagher, E., Menec, D. V. & Keefe, D. J. (2006). Age-friendly rural and remote communities: A guide. *Public Health Agency Can.*
11. Liu, Y., Pan, Z., Liu, Y. & Li, Z. (2023). Can living in an age-friendly neighbourhood protect older adults' mental health against functional decline in China? *Landsc. Urban Plan.* **240**, 104897.
12. Yu Yifan, Zhu Feiyang, Jia Shuying, Guo Yuting, & Hu Yuting. (2020). Research on the Evaluation System of Elderly-Friendly Communities. Shanghai Urban Planning, 1–6.
13. Guo Yawen, Fu Hua, Tang Yuyang, Lü Zhe, & Ding Xiaocang. (2016). Survey Results Analysis of Core Indicators for Elderly-Friendly Cities. Shanghai Journal of Preventive Medicine, 28, 717–723, 751.
14. Li Shan & Yang Zhongzhen. (2012). Research on the Connotation and Evaluation System of Urban Elderly-Friendly Communities. Northwest Population, 33, 17–21, 26.
15. An Haoyuan. (2014). Research on the Evaluation System of Elderly-Friendliness in Urban Community Housing. (Tianjin University).
16. Mei, W.-B., Hsu, C.-Y. & Ou, S.-J. (2020). Research on Evaluation Indexes and Weights of the Aging-Friendly Community Public Environment under the Community Home-based Pension Model. *Int. J. Environ. Res. Public. Health* **17**, 2863.
17. Torku, A., Chan, A. P. C. & Yung, E. H. K. (2021). Age-friendly cities and communities: a review and future directions. *Ageing Soc.* **41**, 2242–2279.
18. Saaty, T. L. (2005). Analytic Hierarchy Process. in *Encyclopedia of Biostatistics* (John Wiley & Sons, Ltd).

Comprehensive Performance Evaluation for Sustainable Development of Small Towns: A Case Study of Tianhuangping Town, Anji County

Chao Li, Feng Yang*, Maomao Yan and Ying Qu

China National Institute of Standardization, 100191 Beijing, China

Abstract. With the rapid development of urbanization in China, a large number of small towns have emerged. These small towns face problems such as limited resources, backward industries, and aging infrastructure. How to achieve the sustainable development of small towns is crucial for China to implement its sustainable development goals. Based on the status and existing research of small towns, this paper develops a performance evaluation index system encompassing five key dimensions, and employs the evaluation index system, along with the maturity assessment, to appraise the sustainable development of small towns. Applying this process to Tianhuangping Town, the results highlight key areas for improvement and provide targeted suggestions for sustainable growth. Our findings contribute to the understanding of sustainability metrics in small towns and offer practical recommendations for policymakers.

1 Introduction

Small towns are an important part of World's urbanization and a crucial support for its economic and social development [1,2]. Despite their significance, there is no universally accepted definition of small towns [3,4]. The size and definition of small towns varies according to the criteria of different countries or regions. In China, small towns were definited as those with a permanent resident population of less than 500,000 and legally established administrative townships [5]. In recent years, the scale of small towns has been steadily increasing [2].

Small towns play an irreplaceable role in the process of achieving sustainable development [6]. Small towns have certain advantages in sustainable development due to their large development potential, diverse economic structure, prominent spatial characteristics, flexible resource utilization, and relatively small population size [7]. However, small towns face significant challenges in achieving sustainable development. These challenges include limited resource and environmental carrying capacity, low levels of economic development, inadequate infrastructure and public services, and weak social governance and cultural construction [8].

To promote the sustainable development of small towns, it is essential to conduct quantitative performance evaluations of their sustainable development and promptly identify any issues. Although research on urban sustainable development evaluation models has made some progress and has proposed multiple indicator evaluation systems around the economic, social, and environmental dimensions of cities [9,10], the continuously enriching connotations of sustainable development in recent years [4,11] have shown that these three dimensions cannot fully cover the evaluation of sustainable development, adversely affecting its evaluation effectiveness. Furthermore, most sustainable development research mainly focuses on medium and large cities, with relatively few studies on sustainable development evaluation models for small towns [9,12]. The unique characteristics, development orientation, and current development stage of small towns give new meanings and content to their sustainable development. Therefore, when analyzing and evaluating the sustainable development of small towns, it is necessary to fully consider the unique factors of small towns.

To address this gap, this study introduces a systematic approach to evaluating sustainable development performance. It incorporates a multidimensional index system and complete evaluation process. By applying this framework to Tianhuangping Town in Anji County, Zhejiang Province, we can assess the town's sustainable development performance, identify key areas for improvement, and offer practical recommendations. This study aims to enhance sustainability in small towns, promoting continuous improvement and effective management.

2 Study Area and Methodology

2.1 Study Area

This paper takes a small town, Tianhuangping Town under the jurisdiction of Anji County, as an example. Tianhuangping Town has a population of 22,000, which conforms to the scope of small towns [5]. At the same time, as the birthplace of ecological civilization, it has

* Corresponding author: yangfeng@cnis.ac.cn

strong representativeness and comparability, making it suitable as the research object of this paper.

2.2 Indicator Setting

With the continuous enrichment of the connotations of sustainable development in recent years [4,13], economic, social, environment, culture and governance have become increasingly important, making them indispensable parts of the evaluation of sustainable development [5]. Furthermore, the selection process should be determined based on objectives and the characteristics of the indicators. Characteristics of Key Performance Indicators should be representativeness, measurability, comparability, manageability, independence [14,15].

2.3 Indicator Confirmation and Screening

In this study, a comprehensive approach that integrates literature analysis, expert consultation and field research was employed. First, the study determines the preliminary evaluation indicators based on the five dimensions of sustainable development, referencing relevant research and practices, especially ISO 37111 and GB/T 43155, as well as the connotations of sustainable development in small towns. Second, the Delphi method is used to further screen and optimize the indicator system. Ten experts from different fields, including urban and rural planning, economic management, sociology, environmental science, and public policy, were invited to rate the importance, feasibility, and dynamism of each indicator. After receiving the reply from the experts, this study screened the indicators, and adjusted and optimized the indicator system. On this basis, we identified several small towns and, through communication with local government officials and experts, as well as trial completion, reviewed and confirmed the indicators. The final indicator system is presented in Table 1.

2.4 Setting Weights and Determining Benchmark Values

Considering that data related to small towns is relatively scarce, this paper recommends employing the Analytic Hierarchy Process (AHP) method to determine the weights of the indicators. We have considered that sustainable development requires a balanced development of all five aspects, hence the weights of the indicators should not differ too much.

We employ the historical optimum value method for indicators with relatively continuous data, and complement our benchmark values with local government planning and expert consultation methods. This study selected the period from 2017 to 2021 for Tianhuangping Town, and used data from statistical yearbooks and other sources to determine the optimum value of each indicator as the benchmark value, to calculate the sustainable development level of Tianhuangping town in 2023, as shown in Table 1.

2.5 Data Dimensionless Normalization Processing

Considering the incomplete data in small towns and the inability to ensure a normal distribution, while also aiming to simplify the workflow and make the calculation process concise and understandable, this study employs the range method for dimensionless normalization processing.

Table 1. Indicators and related values for sustainable development of Tianhuangping town.

Primary index	Secondary index	Three-level index	Weight	Reference value	Data	Normalized
Economy	Economic growth	Per capita GDP	0.03	50,000	52,000	0.6
		Growth rate	0.03	8%	8.5%	0.425
	Economic structure	Primary industry	0.02	10%	9%	0.1
		Secondary industry	0.02	40%	42%	0.8
		Service sector	0.02	50%	49%	0.1
	Economic efficiency	Energy consumption intensity	0.02	0.5	0.4	0.4
		Labor productivity	0.02	100 thousand	110 thousand	0.4
	Economic welfare	Resident income	0.03	40000	42,000	0.5
		Poverty rate	0.02	5%	4%	0.4
Society	Social stability	Population density	0.02	1000	1300	0.3
		Population aging	0.02	15%	16%	0.2
		Criminal rate	0.02	1%	1%	0
	Social equity	Educational level	0.03	90%	88%	0.6
		Medical level	0.03	80%	78%	0.3
		Social security	0.03	70%	68%	0.1
	Social participation	A sense of citizenship	0.02	60%	55%	0.25
		Social organizations	0.02	50%	45%	0
		Informatization level	0.02	40%	35%	0.25
Environment	Environment quality	Air quality	0.03	80	75	0.25
		Water quality	0.03	90	85	0.5
		Sound pollution	0.02	50	55	0.5
	Environmental resources	Land use	0.02	60%	55%	0.25
		Water resource	0.02	70%	65%	0.5
		Biodiversity	0.02	80%	75%	0.75

	Environmental service	Ecosystem services	0.02	90%	85%	1
		Environmental capacity	0.02	100%	95%	1
		Environmental risk	0.02	10%	15%	0.5
Culture	Cultural heritage	Historical and cultural heritage	0.02	10	8	0.8
		Intangible cultural heritage	0.02	20	18	0.9
		Local culture	0.02	30	28	1
	Cultural development	Cultural sector	0.03	40%	35%	0.25
		Cultural innovation	0.03	50%	45%	0.5
		Cultural exchange	0.02	60%	55%	0.75
	Cultural vitality	Cultural activities	0.02	70%	65%	1
		Cultural facilities	0.02	80%	75%	1
		Cultural satisfaction	0.02	90%	85%	1
Govern	Governance ability	Administrative efficiency	0.02	80%	75%	0.25
		Public service level	0.02	80%	75%	0.25
		Emergency management ability	0.02	80%	75%	0.25
	Governance level	Rule of law construction	0.02	80%	75%	0.25
		Government trust	0.02	80%	75%	0.25
		Social supervision	0.02	80%	75%	0.25
	Governance mechanism	Government-led	0.02	50%	50%	0
		Market regulation	0.02	50%	50%	0
		Social participation	0.02	50%	50%	0

2.6 The comprehensive index of sustainable development

The indicators for sustainable development in small towns exhibit relative independence, with no significant interactions between them; hence, this study uses the weighted average method to calculate the comprehensive index of sustainable development. Based on the weight of each indicator, this paper calculates the comprehensive index of the five primary indicators of economy, society, environment, culture, and governance of small towns, and then calculates the comprehensive index of sustainable development of small towns based on the weight of the primary indicators.

2.7 Maturity Assessment

After calculating the comprehensive index of sustainable development for the town, it is necessary to assess the maturity of sustainable development in the town to reflect its level and issues [16]. Given that this study has obtained benchmark values and considering the principles of intuitiveness and ease of use, this study employs the percentage deviation method to construct a maturity model for maturity assessment.

By using the benchmark values, the normalized values of each indicator are compared with the benchmark values to obtain the deviation for each primary indicator and the overall system. The formula is as follows:

$$D = \frac{S-B}{B}$$

where D is the deviation of sustainable development performance for the overall system, S is the comprehensive index of sustainable development for the overall system, and B is the benchmark value for the overall system.

Subsequently, a maturity classification needs to be conducted. This study refers to previous research [17] and, considering the current situation of small towns, adopts an equal division method to provide reference thresholds. The formula is as follows:

$$M = \begin{cases} \text{Primary stage,} & \text{If } D < -0.2 \\ \text{Development stage,} & \text{If } -0.2 \le D < 0 \\ \text{Mature stage,} & \text{If } 0 \le D < 0.2 \\ \text{Optimization stage,} & \text{If } 0.2 \le D < 0.4 \\ \text{Leading stage,} & \text{If } D \ge 0.4 \end{cases}$$

where M is the maturity level of overall sustainable development performance, and D is the deviation of overall sustainable development performance.

3 Results

Table 2. Comprehensive index and maturity grade of sustainable development of small towns.

Town	Primary index	Weight	Comprehensive index	Deviation	Maturity grade
Tianhuanping town	Economy	0.32	0.58	0.16	Mature stage
	Society	0.28	0.54	0.08	Mature stage
	Environment	0.24	0.68	0.33	Optimization stage
	Culture	0.08	0.84	0.63	Leading stage
	Govern	0.08	0.38	-0.25	primary stage
	Overall	1	0.58	0.10	Mature stage

From the maturity levels of each primary indicator (Table 2), the town has reached the optimization or leading stage in terms of environment and culture, and have reached mature stage in economy and society. This indicates that the town has a high level of sustainable development performance in these areas, surpassing the baseline values, and has certain advantages and exemplary effects. However, the governance aspect of the town is still in the primary stage, indicating a relatively low level of sustainable development performance, below the baseline values.

4 Discussion

4.1 Results Analysis

Our findings are consistent with several studies that highlight the environmental and economic strengths of small towns while identifying social and govern sustainability as a common challenge [8,18].

The high scores in environmental sustainability reflect effective waste management practices and the availability of green spaces in Tianhuangping Town. These findings suggest that the town has successfully implemented policies aimed at environmental protection, which aligns with SDGs. Economic sustainability, driven by robust local businesses and employment rates, further supports the town's overall sustainability performance. This indicates a healthy economic base that can support further development initiatives. The lower scores in govern sustainability highlight areas needing attention. Community participation and social equity, in particular, were identified as weak points. These findings resonate with studies by [19], which noted similar challenges in social integration and community engagement in rural China. Addressing these social sustainability issues is crucial for achieving balanced development.

Therefore, the town should strengthen governance capacity and level. It should improve administrative efficiency, the level of public services, emergency management capabilities, rule of law, government trust, and social supervision. It should optimize governance mechanisms, achieve an organic combination of government leadership, market regulation, and social participation, enhance the transparency, fairness, and effectiveness of governance, and thus improve the overall sustainable development performance of the town.

4.2 Methodological Analysis

This study delves into the detailed assessment pathway for the sustainable development level of small towns, examining each step in the construction of indicators, calculation of weights, benchmark values, and maturity evaluation. We have selected suitable approaches in line with the characteristics of small towns at each stage.

In the process of constructing the indicator system, although the comprehensive index system is complex, it is sufficiently comprehensive to better encompass the connotations of sustainable development. The selection of indicators should not only consider as many indicators as possible but also integrate field research and the actual conditions of small towns to select relevant indicators that reflect their situation. The setting of weights and benchmark values should be more aligned with the availability of data in small towns. Regarding the Data Dimensionless Normalization process and the establishment of comprehensive indicators, we have chosen methods based on maturity, reliability, and the ability to compare. In terms of maturity assessment, this paper adopts the fixed-level maturity model, which is popular and easy to implement [20].

While this study provides a robust framework for evaluating the sustainability of small towns, there are several limitations to consider:

Subjective Expert Opinions: The reliance on the Delphi method for indicator selection introduces subjectivity. Future research could incorporate a broader range of stakeholders, including local residents, to balance these perspectives.

Static Analysis: The evaluation provides a snapshot of the town's sustainability at a single point in time. Longitudinal studies are needed to understand the dynamics of sustainability and the impact of implemented policies over time.

Generalizability: The case study approach, while detailed, limits the generalizability of the findings. Applying the evaluation framework to other small towns in different regions can validate its applicability and identify regional variations.

5 Conclusion

This paper proposes a systematic comprehensive evaluation index system and maturity assessment model to evaluate the sustainable development level of small towns in terms of economy, society, environment, culture, and governance. This study analyzes the evaluation pathway of the sustainable development of small towns, and selects the appropriate method according to the characteristics of small towns at each stage. The application of this evaluation process to Tianhuangping Town, demonstrates its practical utility, revealing strengths in environmental and cultural dimensions and areas for improvement in governance. Despite certain limitations, including the subjectivity inherent in the Delphi method and the static nature of our analysis, our approach provides valuable insights and practical guidance for policymakers and stakeholders. Future research should aim to extend this framework to small towns with diverse characteristics and regional backgrounds, enhancing its applicability and uncovering regional disparities in sustainable development performance. By doing so, we can further promote the sustainable development of small towns across various dimensions and regions.

Acknowledgments

This research was funded by the Central Public-Interest Scientific Institution Basal Research Fund, CNIS [Grant No.512024Y-11450], National Key Research and Development Program [Grant No.2022YFC3802901].

References

1. Chen, W.H., Chen, X.L. (2010) Study on Indicators of Evaluating Sustainable Development in Small - Sized Towns. Ecological Economy: 114-116.
2. Gu, C.,Li, Y., Han, S.S. (2015) Development and transition of small towns in rural China. Habitat International, 50: 110-119. doi:10.1016/j.habitatint. 2015.08.017.
3. Zhai, Y. (2008) Several thoughts on the construction of small towns in China. City: 99-103.
4. Guin, D. (2019) Contemporary perspectives of small towns in India: A review. Habitat International, 86: 19-27. doi:10.1016/j.habitatint.2019.02.003.
5. SAC. (2023) GB/T 43155-2023 Sustainable cities and communities – Management and implementation guidance for sustainable development of small towns.
6. Crowhurst Lennard, S.H. (2010) Small town sustainability: Economic, social, and environmental innovation. URBAN DESIGN International, 15: 189-190. doi:10.1057/udi.2010.11.
7. Bell, D., Jayne, M. (2009) Small Cities? Towards a Research Agenda. International Journal of Urban and Regional Research, 33: 683-699. doi:10.1111/j.1468-2427.2009.00886.x.
8. Shen, L.,Ren, Y.,Xiong, N.,Li, H., Chen, Y. (2018) Why small towns can not share the benefits of urbanization in China? Journal of Cleaner Production, 174: 728-738. doi:10.1016/j.jclepro.2017.10.150.
9. Pearce, D. Blueprint 3: Measuring sustainable development, Routledge: 2014.
10. Li, C., Tang, X. Research on Connotation and Assessment of Small Towns' Sustainable Development: A Case Study on Hengdian Town. In Proceedings of the 2008 International Conference on Intelligent Computation Technology and Automation (ICICTA), 20-22 Oct. 2008, 2008; pp. 753-757.
11. ISO. (2016) ISO 37101:2016 Sustainable development in communities – Management system for sustainable development – Requirements with guidance for use.
12. Visvaldis, V.,Ainhoa, G., Ralfs, P. (2013) Selecting Indicators for Sustainable Development of Small Towns: The Case of Valmiera Municipality. Procedia Computer Science, 26: 21-32. doi:10.1016/j.procs. 2013.12.004.
13. ISO. (2024) ISO 37111:2024 Sustainable cities and communities – Urban settlements– Guidance for a flexible approach to phased implementation of ISO 37101.
14. Sun, X.,Liu, X.,Li, F.,Tao, Y., Song, Y. (2017) Comprehensive evaluation of different scale cities' sustainable development for economy, society, and ecological infrastructure in China. Journal of Cleaner Production, 163: S329-S337. doi:10.1016/j.jclepro. 2015.09.002.
15. Haiyirete, X.,Xu, Q.,Wang, J.,Liu, X., Zeng, K. (2024) Comprehensive Evaluation of the Development Level of China's Characteristic Towns under the Perspective of an Urban–Rural Integration Development Strategy. Land, 13. doi:10.3390/ land13071069.
16. Aragão, F.V.,Chiroli, D.M.,Zola, F.C.,Aragão, E.V.,Marinho, L.H.,Correa, A.L., Colmenero, J.C. (2023) Smart Cities Maturity Model—A Multicriteria Approach. Sustainability, 15. doi:10.3390/su15086695.
17. Chen, S.,Shao, C.,Gao, J.,Zhao, R., Yang, L. (2022) Evaluation technology and application of sustainable development in resource-dependent cities based on SDGs: A case study of Chenzhou. Acta Ecologica Sinica, 42: 4807-4822.
18. Liu, L.,Wu, T.,Li, S.,de Jong, M., Sun, Y. (2017) The drivers of local environmental policy in China: An analysis of Shenzhen's environmental performance management system, 2007–2015. Journal of Cleaner Production, 165: 656-666. doi:10.1016/j.jclepro. 2017.07.165.
19. Zhang, M.,Wu, W.,Zhong, W.,Zeng, G., Wang, S. (2017) The reshaping of social relations: Resettled rural residents in Zhenjiang, China. Cities, 60: 495-503. doi:10.1016/j.cities.2016.06.007.
20. Santos-Neto, J.B.S.d., Costa, A.P.C.S. (2019) Enterprise maturity models: a systematic literature review. Enterprise Information Systems, 13: 719-769. doi:10.1080/17517575.2019.1575986.

Experimental Study on the Strength Characteristics of Loess Reinforced with Calcium-based Mixture, Chitosan and Sisal Fiber

Lei Lei[1,2], Hao Wan [1,2], Xiaolong Wei [1,2], Jun Liang [3], Han Zhang [3], Xiaoli Liu [4], Zhuo Chen [5,*], and Kanliang Tian [5]

[1] Electric Power Research Institute, State Grid Shaanxi Electric Power Company, Xi'an, China
[2] State Grid (Xi'an) Environmental Protection Technology Center Co., Xi'an, China
[3] State Grid Shaanxi Electric Power Company, Xi'an, China
[4] Electric Power Research Institute of State Grid Shaanxi Electric Power Company, Xi'an, China
[5] Institute of Soil and Water Conservation, Northwest A& F University, Yangling, China

Abstract. Loess has engineering characteristics such as loose structure, strong collapsibility, and rapid decrease in strength when exposed to water, which can easily cause environmental and engineering problems. Calcium-based mixture, chitosan and sisal fiber were introduced to improve soil. Loess samples with different additive contents and sisal fiber contents were prepared, and the liquid limit, plastic limit and unconfined compressive strength test were carried out. The effect and mechanism of each material on improving the liquid limit, plastic limit and unconfined compressive strength of loess were analyzed. The test results showed that: calcium-based mixture chould significantly improve the liquid limit, plastic limit and strength of fiber reinforced loess. The optimum content of mixture wass 9 %, which increased the liquid limit, plastic limit and unconfined compressive strength to 1.35,1.35 and 3.81 times of untreated soil. Chitosan chould significantly change the liquid limit and strength of fiber reinforced loess, but had no significant effect on the plastic limit. The optimum dosage of chitosan was 0.48 %, which increased the liquid limit and unconfined compressive strength to 1.30 and 5.50 times that of untreated soil. The sisal fiber chould significantly improve the unconfined compressive strength of the treated soil, but had no significant effect on the liquid plastic limit. The optimum content of sisal fiber was 0.75 %. The combined use of biological, natural materials and short fibers chould enhance the interaction between soil particles and formed an overall three-dimensional interlocking grid structure, which significantly improved the engineering mechanical properties of loess.

1. Introduction

Loess is a kind of soil with vertical joints, loose structure, strong collapsibility and rapid decrease in strength when exposed to water. In engineering practice, the mineral components in loess are easily dissolved under high water content, which leads to the decrease of soil strength. [1]. Under the action of external force, it is easy to cause soil fracturing failure, shear failure, overall settlement and slip, which have adverse effects on the environment and engineering safety [2]. Therefore, engineers and scholars often use natural materials, chemical synthetic materials, biological materials to improve soil strength characteristics and stability [3-6]. Studies have shown that the addition of a variety of natural materials and chemical synthetic materials can change the soil microstructure and interface characteristics, improve soil physical properties and mechanical properties, and affect the activity of soil microorganisms and enzymes [4]. The improvement of chemical synthetic materials is a kind of soil enhancement method which is widely used at present, and usually has good effect. However, there are also some shortcomings, such as high improvement cost, insufficient research on material degradation process and biological toxicity, and unclear impact on soil microorganisms and ecosystems [5, 6]. At present, more attention has been paid to biological materials such as biological mycorrhiza, microbial broth and biological reaction products with better ecological and environmental compatibility, which are considered to be environmentally friendly and reliable soil reinforcement materials [5-10]. A number of studies have shown that carrageenan, agar gel, chitosan, carrageenan and other molecular biopolymers have a good effect on improving the compressive strength, cohesion, water resistance and stability of soil [7-10]. After the application of these materials, the performance of soil specimens has been effectively improved under adverse conditions. Biomaterials have been gradually applied in environmental remediation, soil solidification, sand enhancement, soil erosion control and dust suppression. From an environmental point of view, natural materials and biomaterials have better ecological and environmental effects than traditional reinforcing materials [7-11].

In recent years, the simultaneous use of short fibers and natural and biological materials to enhance soil has become an important research direction. The physical enhancement effect of short fibers and the improvement

* Corresponding author: 450832762@qq.com

effect of soil microstructure of natural and biological materials make them achieve better results when used in combination [12]. People used glass fiber, rubber fiber, flax fiber, organic fiber and other short fibers combined with brick kiln dust, xanthan gum and other materials to enhance the soil, which improves the shear strength, cohesion, load deformation resistance and California bearing ratio of different soils [9-15]. The above research proves the potential application value of using composite materials to improve soil mechanical properties and engineering properties. However, there were few studies on the use of biopolymers combined with natural fibers to enhance the properties of loess. In this study, a series of liquid limit, plastic limit and unconfined compressive strength (UCS) tests were carried out by using chitosan (CB), calcium-based mixture (CM) and sisal fiber (SF) as the reinforcement materials of loess. The effects of different materials on the characteristics of liquid limit, plastic limit and unconfined compressive strength of loess were studied, which provided new ideas for the application of soil reinforcement in practical engineering.

2. Materials and methods

2.1 Materials

(1) Loess: The loess was taken from Jiaxian County, Yulin City, Shaanxi Province. It is yellow-brown and contains a small amount of fine sand. After sampling, the impurities were removed, dried, crushed, and passed through a 2mm sieve to obtain the loess. The basic parameters of soil were determined according to "Standard for geotechnical testing method" [16], and the soil particle size was measured by SYNC laser particle size analyzer. The test results are shown in Table 1 and Figure 1.

Table 1. Basic soil parameters.

Proportion	Plastic limit	Liquid limit	Maximum dry density
2.65	18.36%	29.05%	1.74g/cm^3

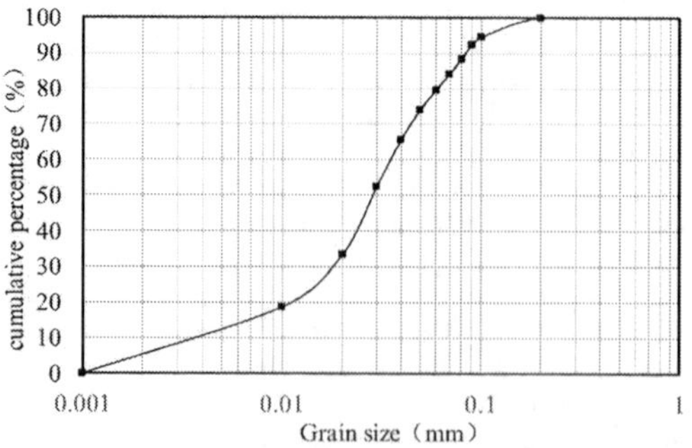

Fig. 1. Soil particle composition.

(2) Chitosan Biopolymer (CB): The CB was extracted and processed from crab shells and shrimp shells. There are five main processing steps: grinding, decalcification, deproteinization, dehydration, and deacetylation [17]. It is a macromolecular polysaccharide biopolymer with excellent toughness, excellent tensile strength, and good hydrophilicity. The molecular formula is $(C_6H_{11}NO_4)_n$. The characterization of chitosan is shown in table 2.

Table 2. Chitosan characterization.

Colour	Water-solubility	Viscosity	Density
Pale yellow	Insoluble	150MPa·s	0.25g/cm^3

(3) Calcium-based mixture (CM): The CM was made by mixing 50% by mass of calcium based bentonite and 50% by mass of hydrated lime. The calcium based bentonite has significant water absorption and expansion properties. The main production steps are: crushing of bentonite ore, addition of sodium carbonate, mixing and extrusion, drying and grinding, air classification, and material collection. Calcium based bentonite and hydrated lime were purchased from a Mineral products processing plant in Lingshou County, China. The characterization of materials is shown in table 3.

Table 3. Materials characterization.

Calcium based bentonite	Hydrated lime
Fraction of montmorillonite: 70%	Major constituent: $Ca(OH)_2$
water absorption rate: 180%	Mass fraction of calcium: 95%
wet compressive strength: 20KPa	Density: 2.24g/ml
expansion ratio: 6ml/g~12ml/g	Water-solubility: 1.65g/L
the cement: ≥350ml/15g	Alkaline and dry

(4) Sisal fiber (SF): SF is a kind of natural fiber extracted from sisal. It is mainly composed of cellulose, lignin, pectin and so on. It is generally light white, with high elastic modulus and specific surface area, strong toughness, wear resistance and water resistance. The main production steps are : sisal leaf cleaning, crude fiber stripping, boiling treatment, fiber rinsing, drying and combing. The basic sisal fiber parameters are shown in table 4. In this study, the fibers were cut into short fiber strips of 1.5~2 cm.

Table 4. Basic sisal fiber parameters.

Fiber bundles width	tensile strength	Density
9μm~423μm	450MPa	1.45g/cm^3

2.2 Test methods

All tests were carried out according to the "standard of soil test method". The mass ratio of the treated soil is shown in Table 5.

Table 5. Proportioning scheme.

Material type and dosage (%)	Sisal fiber content (%)
CM:3、6、9、12	0.25、0.5、0.75、1
CB:0.12、0.24、0.35、0.48	

The main process of the experiment is as follows:

(1) Material treatment: group and uniformly mix the materials (Fig.2.A) according to the scheme.

(2) Liquid limit (LL) and plastic limit (PL) test: take 200g of mixed material with the same ratio and divide it into three parts. Add pure water to adjust the moisture content, so that they reach the moisture content close to the liquid limit, intermediate state, and plastic limit in sequence. Mix the mixed material thoroughly and place it in a test dish, then scrape off any excess soil. Operate the liquid plastic limit combined determination device for testing. Read and calculate the results based on the data.

The device is the liquid plastic limit combined determination device (Fig. 2.B).

(3) Samples preparation and curing: Make different proportions of samples (diameter 6.18cm, height 8cm, moisture content 17, dry density 1.5g/cm^3) with three-lobe lightweight compaction device (Fig. 2.C). Curing the samples in a curing box for 7 days, with controlled conditions of 20 degrees Celsius and humidity greater than 90% inside the curing box (Fig. 2.D).

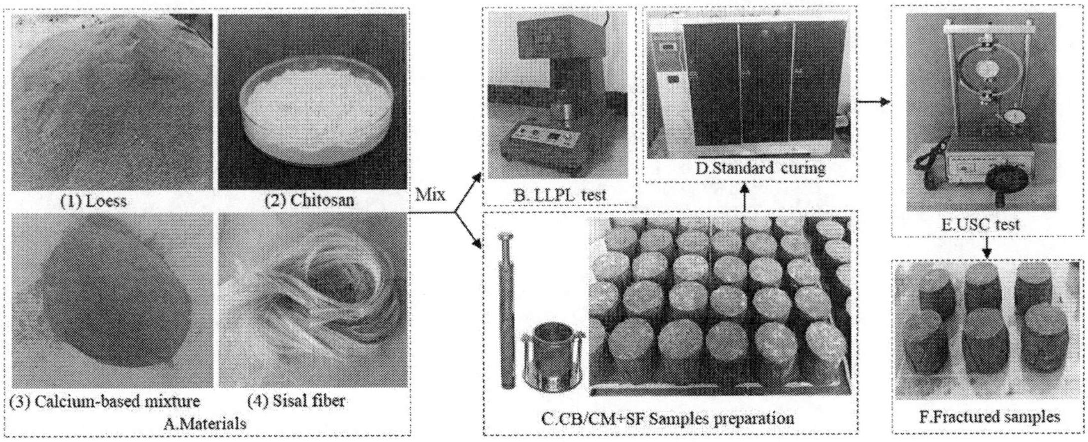

Fig. 2. Test process.

(4) Unconfined compressive strength (UCS) test: check and zeroing the strain controlled unconfined compression instrument equipment (Fig.2.E). Take out the test sample from the curing box and put it into the test board. Start the device and adjust the rising speed of the test plate to 1%~3%. Record the test data until the strain reaches 20%. Calculate the compressive strength according to the following formula.

$$\sigma = \frac{CR}{A} \times 10 \qquad (1)$$

σ is the axial stress(KPa);

C is the dynamometer coefficient (N/0.01mm);

R is the dynamometer reading (0.01mm);

A is the bearing area of the specimen (cm^2).

The main process is shown in the following figure:

3. Result and analysis

3.1 The influence of different materials on liquid limit and plastic limit

Liquid and plastic limits are important parameters to determine soil state, classify soil types and evaluate soil engineering properties. Fig.3 and Fig.4 show the changes of liquid and plastic limits of soil treated with different ratios.

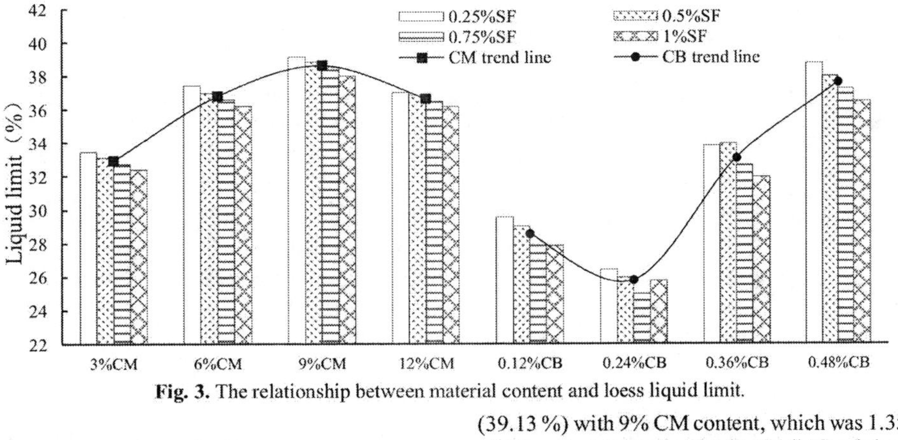

Fig. 3. The relationship between material content and loess liquid limit.

3.1.1 The influence on liquid limit

The liquid limit of untreated loess was 29.05 %. In the calcium-based mixture (CM) and sisal fiber (SF) mixed group, when the CM content changed from 3 % to 12 %, the liquid limit of the treated soil increased firstly and then decreased. The average liquid limit reached a peak (39.13 %) with 9% CM content, which was 1.35 times that of the untreated soil. The liquid limit of the 12 % CM content group was less than that of the 6 % CM content group, which indicated that the CM could effectively improve the loess liquid limit when the SF content is constant. However, when the CM content exceeded a certain value, the liquid limit decreased instead. 9 % CM content was the optimal addition ratio in the ratio group.

In the chitosan (CB) and sisal fiber (SF) mixed group, the liquid limit of the treated soil decreased firstly and then increased with the CB content changed from 0.12 % to 0.48 %. When the CB content was 0.24 %, the average liquid limit was as low as 25.78 %, which was 0.89 times that of the untreated soil. The average liquid limit increased to 37.63 % with 0.48 % CB content, which was 1.30 times that of the untreated soil.

When the CM or CB content of the treated soil was constant and the SF content changed from 0.25 % to 1 %, the plastic limit of the treated soil in different groups increased firstly and then decreased. The reduction of the CB group was higher than that of the CM group. The liquid limit of the 0.48 % CB group decreased the most. The liquid limit decreased from 38.76 % to 36.51 % with the SF content changed from 0.25 % to 1 %.

3.1.2 The influence on plastic limit

The plastic limit of untreated loess was 18.36 %. In the group of CM and SF, when the content of CM in the treated soil changed from 3 % to 12 %, the plastic limit of the treated soil increased firstly and then decreased. The average plastic limit of the treated soil reached the peak (24.94 %) with 9 % CM content, which was 1.36 times that of the untreated soil. The plastic limit of the 0.5 % fiber content 0.5 % group was the highest, which could reach 25.02 %. This showed that when the content of SF was constant, the CM could effectively improve the plastic limit of loess. However, when the CM content exceeded a certain value, the plastic limit decreased instead. 9 % CM content was the optimal addition ratio in the ratio group.

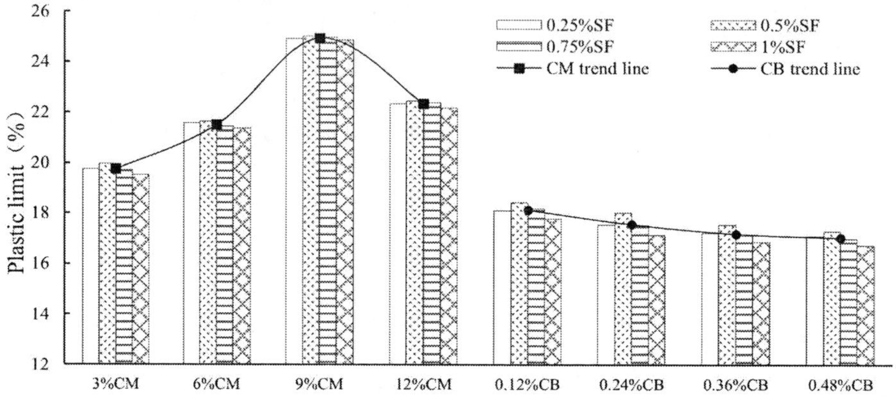

Fig. 4. The relationship between material content and loess Plastic limit.

In the mixed group of CB and SF, when the content of CB changed from 0.12 % to 0.48 %, the liquid limit of the treated soil decreased slowly, reaches the lowest value with the 0.48 % CB content, and the average liquid limit was as low as 17.03 %, which was only 1.33 % lower than that of the untreated soil. This showed that the content of CB had little effect on the plastic limit.

When the CM or CB content of the treated soil was constant and the SF content changed from 0.25 % to 1 %, the plastic limit of the treated soil in different groups increased firstly and then decreased. The reduction in the CB group was higher than that in the CM group. The plastic limit in the 0.24 % CB content group changed the most. The liquid limit decreased from 18.03 % to 17.15 % with the SF content changed from 0.5 % to 1 %.

3.1.3 Mechanism analysis

In general, the change law of the plastic limit of the treated soil with the CM content was basically consistent with the change law of the liquid limit. 9 % CM content is the optimal addition ratio in the ratio group, which could effectively improve the liquid plastic limit of the loess. There was no significant change in the plastic limit with CB content increased. 0.48 % CB content was the optimal addition ratio in the ratio group. When the content of CM and CB was constant and the content of SF increased, there was no significant change in the liquid and plastic limits of the treated soil.

In engineering practice, the lower the moisture content of loess, the higher its mechanical strength. The increase of liquid limit and plastic limit indicates the improvement of engineering characteristics of loess. There are two main reasons why CM increase the plastic limit of processed soil liquid. The first reason is that CM have good water absorption and expansion properties. When it comes into contact with water, montmorillonite particles absorb a large amount of water between the loess skeleton and expand several times or even more than ten times. The particles formed by expansion reduce the free water between soil particles and lower soil pores, increasing the friction and cohesion between soil particles, making the soil more difficult to liquefy and displace. The second reason is that hydrated lime absorb carbon dioxide and produces calcium carbonate crystals in the soil, reducing soil pores and cementing soil particles while increasing particle friction.

However, montmorillonite and hydrated lime consumed a large amount of water and carbon dioxide between soil particles. When the material in the pores was insufficient and the mixing amount continues to increase (CM content≥9%), part of the CM existed in an uneven powder state between the soil particles, which actually reduced the liquid plastic limit of the soil.

The reason why CB have a certain impact on liquid limit and plastic limit is that: there are a large number of hydrophilic groups on the surface of CB molecules. When the CB content is low (0.12 %, 0.24 %), CB absorbed some free water in the soil to form a smaller biopolymer aggregate in the soil particles [18]. This makes it easier for displacement to occur between soil particles, which led to a decrease in liquid limit. When the content continued to increase (0.36 %, 0.48 %), the biopolymer aggregates would cement each other to form structures with a certain volume and integrity (Fig. 5.). The structures reduced soil pores and displacement, causing an increase in liquid limit. CB had little effect on plastic limit because it did not significantly affect the main mineral components of soil. Fig. 5 shows that chitosan reduced soil pores.

SF had little effect on soil liquid plastic limit, because SF mainly improved the overall strength of soil by fiber soil particle friction interface. However, it does not significantly alter the properties of soil particles and the soil itself [14].

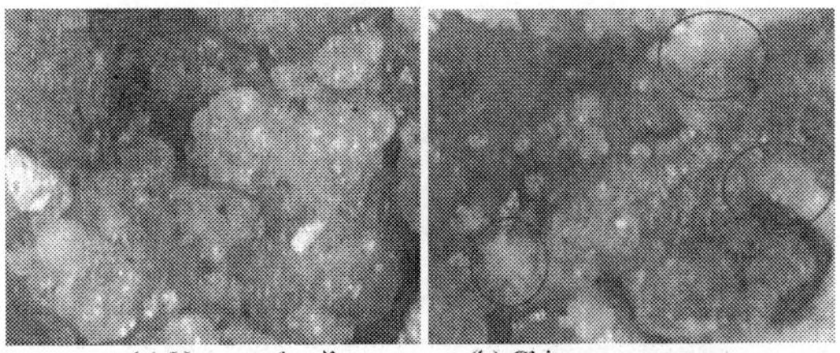

Fig. 5. 300x optical microscopy images Chitosan

3.2 The influence of different materials on unconfined compressive strength

Unconfined compressive strength (UCS) is an important strength index to describe the stability and bearing capacity of soil. Fig. 6 shows the change of UCS of soil treated with different ratios.

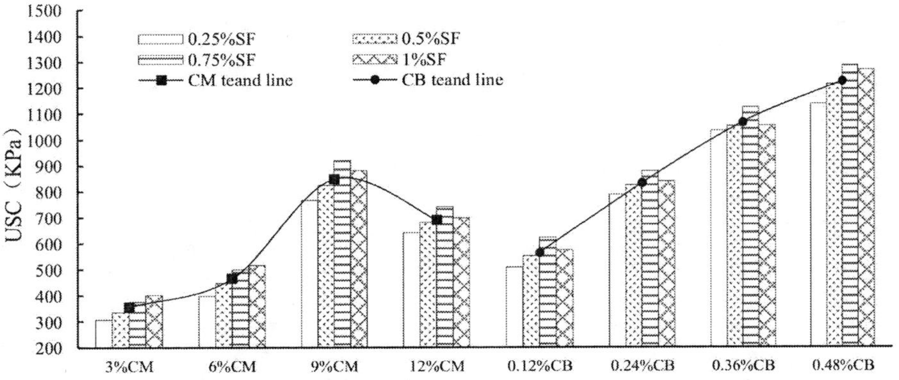

Fig. 6. The relationship between material content and unconfined compressive strength.

3.2.1 The influence of CM and CB on USC

It can be seen from Fig.6 that the incorporation of CM, CB and SF significantly improved the UCS of loess samples. The UCS of untreated soil was 223 KPa. In the group of CM and SF, when the CM content changes from 3 % to 12 %, the UCS of the treated soil increased firstly and then decreased. The average UCS reached the peak (849 KPa) with 9% CM content, which was 3.81 times that of the untreated sample. When CM content was 9 % and SF content was 0.75 %, the UCS reached 921 KPa, which is 4.13 times that of the untreated sample. This showed that CM effectively improved the UCS of loess with a constant SF content. But USC decreased after the content exceeded a certain value. 9 % CM content is the optimal addition ratio in the ratio group.

The mechanism of CM enhancing USC is the same as that of enhancing liquid limit and plastic limit. Montmorillonite and hydrated lime absorb water and carbon dioxide from soil particles. The generated calcium carbonate and expanding particles reduce soil pores, free water, and increase soil cohesion, friction, and cohesion.

In the group of CB and SF, when the content of CB changed from 0.12 % to 0.48 %, the UCS of the treated soil increased steadily. The average UCS can reach 1226KPa with 0.48 % CB content, which was 5.50 times that of the untreated sample. When the content of CB was 0.48 % and the content of SF was 0.75 %, the UCS can reach 1285KPa, which is 5.76 times that of the untreated sample.

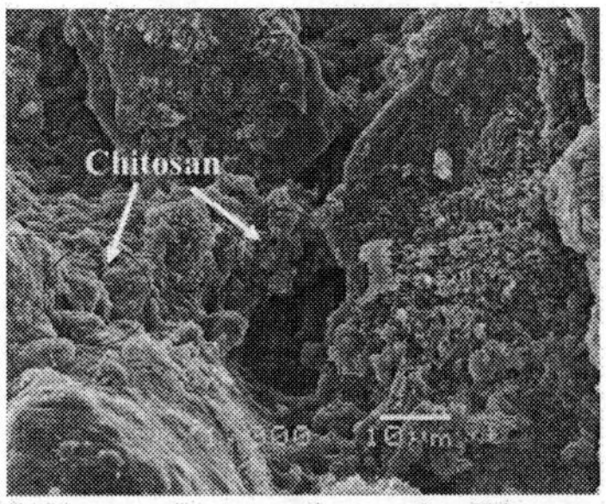

Fig. 7. SEM image of loess surface under Chitosan treatment.

0.48 % CB content was the optimal addition ratio in the ratio group. The mechanism of CB enhancing USC is that CB molecules usually have positive charges and a large number of hydrophilic groups on the surface, which will attract free water and negatively charged ion groups in the soil to form biopolymer aggregates with a certain volume and strength. They reduce soil pores and connect soil skeleton structure with soil particle (Fig. 7). Thus, the cohesion, internal friction, stability and overall strength of the sample are improved [19].

3.2.2 The influence of SF on USC

When the SF content changed from 0.25 % to 1 %, the UCS of groups with 3 % and 6 % CM content increased steadily, but the UCS of groups with CM 9 % and 12 % increased firstly and then decreased. The reason may be that the effect of fibers changed under higher test block strength. The effect changed from filling soil pores and preventing soil particle movement to separating soil particles, reducing the contact area and friction between soil particles.

When the admixture of the treated sample was CB and SF content changed from 0.25 % to 1 %, the UCS of the treated sample increased firstly and then decreased. The UCS of the treated soil reached a peak with 0.75 % SF content. In addition, in different treatment groups, the effect of SF on increasing UCS increased with the improve of the sample strength. 0.75 % SF content was the optimal addition ratio in the ratio group.

The mechanism of SF increasing UCS is that the rough surface texture of the fiber can prevent the sliding displacement and rearrangement of the sand particles by providing interface friction and interlocking. At the same time, the friction interface of the fiber and the soil particles can form a three-dimensional interlocking structure (Fig.8.), thereby improving the overall toughness and strength of the specimen. When SF and CM or CB were mixed into reinforced soil, the agglomeration formed by CM or CB was further in full contact with the fiber surface, rubbing, cementing and locking each other to improve the mechanical enhancement effect [18-20].

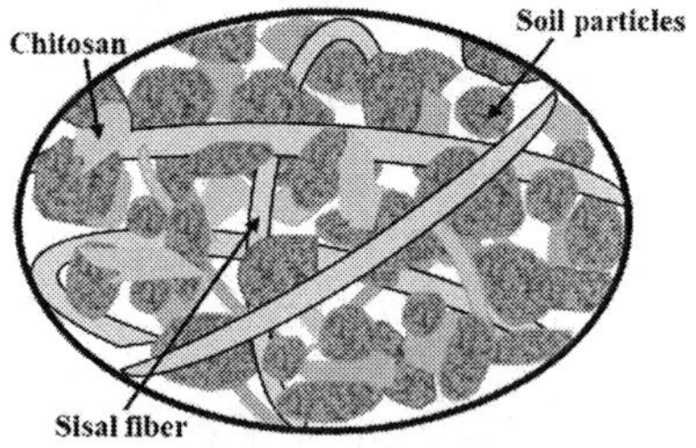

Fig. 8. Schematic diagram of 3d interlocking structure.

3.3 The relationship between liquid and plastic limits and UCS of treated soil

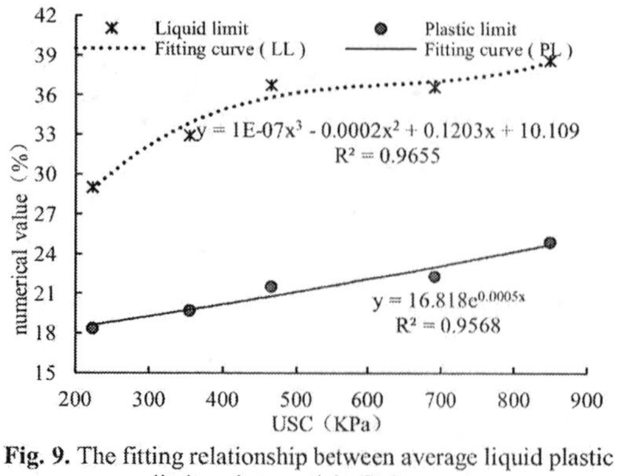

Fig. 9. The fitting relationship between average liquid plastic limit and strength in CM group.

Fig. 9 is plotted based on the average liquid and plastic limits and UCS data of each group of materials, showing the relationship and fitting line between the liquid and plastic limits and UCS of different groups.

It can be seen from Fig. 9 and Fig. 10 that the UCS of the CM-SF group is positively correlated with the liquid limit and the plastic limit. The liquid-plastic limits of the CM-SF treated soil was improved, and the UCS was also increased. The plastic limit of CB-SF group is negatively correlated with the UCS. The UCS increased when the plastic limit of CB-SF group decreased.

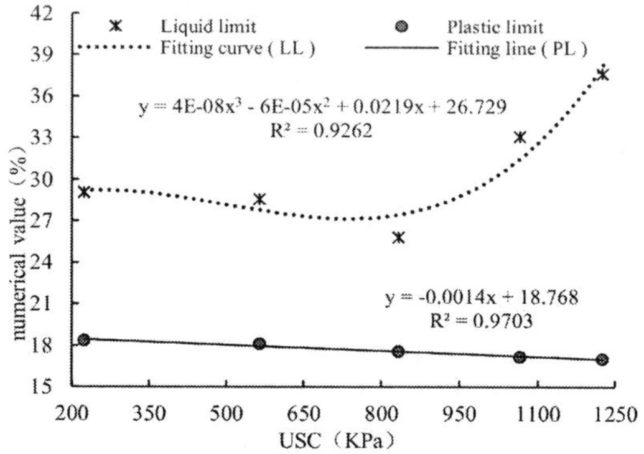

Fig. 10. The fitting relationship between average liquid plastic limit and strength in CB group.

The liquid limit of CB-SF is nonlinearly related to the UCS. The UCS might increase or decrease when the liquid limit was increased within a certain range. The reason for the significant difference in correlation between the two composite materials may be due to the different mechanisms of action of CM and CB in soil.

The enhancement mechanism of CM is to increase soil cohesion and soil particle friction by absorbing water, expanding, generating calcium carbonate and reducing pores. Therefore, the correlation among liquid limit, plastic limit and unconfined strength is very simple. In addition to the above functions, CB also forms "line" and "film" connecting soil particles with certain mechanical strength between particles, forming a finishing polymer bonding structure (Fig.7. and Fig.8). This leads to the fact that the reinforcement effect of low content CB on reinforced soil is better than that of CM, and the effect of different content of CB on liquid limit is different.

4. Discussion

Our results confirm the enhancing effect of three materials on soil engineering properties. Generally speaking, the strength of soil depends on the effects of friction and cohesion, and the main mechanism of reinforcement is based on the friction between particles. They have different characteristics.

The mechanism of BM reinforcement is to absorb water, expand, reduce free water and pores, and increase cohesion. Its mechanism is simple, but its effect is reliable. The materials can be purchased at a low cost, with good durability and high production capacity, making it suitable for large-scale soil treatment applications. But it also has significant limitations. Firstly, the enhancement efficiency of this material is relatively low and requires a large dosage to achieve good results. Secondly, this material has weak alkalinity and is not suitable for most areas with high salinity, which may affect the growth of alkali tolerant vegetation.

Chitosan is a typical biopolymer that can significantly improve strength with low dosage and high cost-effectiveness. It is a very promising modifier. Studies have shown that chitosan has the ability to promote plant growth and inhibit bacterial damage. Similar polymers, including guar gum, xanthan gum, etc., can enhance soil erosion resistance while significantly reducing hydraulic conductivity [15, 21]. Scholars used guar gum in sandy soil with a dosage of 4% to increase USC to 5800KPa, and also used 1.5% xanthan gum to increase USC to 1500KPa [22]. However, these two types of adhesives are produced by microorganisms through biochemical reactions, which limits their production efficiency and application. The degradable properties of chitosan make it have good environmental protection effects, but the impact of degradation time and process on engineering soil properties has not been fully studied at present [9].

Strengthening soil in SF is a mechanical and physical reinforcement method. SF has the characteristics of a wide range of production materials, mechanical strength, and strong corrosion resistance [9]. Scholars have used SF for soil reinforcement, expansion inhibition, and improvement of frost resistance, resulting in good ecological effects. There are also studies indicating that the mycelium formed by kenaf microorganisms and the microbial fragments present in the soil have an inherent ability to consolidate soil particles. This enhances the soil's resistance to disintegration and minimizes the possibility of pore air and water pressure development. Other commonly used fibers include coconut shell fiber, jute fiber, glass fiber, polypropylene fiber, etc. They can mainly be divided into two categories: natural and synthetic, but both have limitations. For example, polypropylene fibers have good mechanical properties and are reasonably priced, but chemical synthesis production consumes a large amount of energy and generates carbon dioxide. Huangma fiber has a high yield and excellent properties, but its soil reinforcement effect is weaker than that of SF.

Coconut shell fiber has strong strength and stress resistance, but the fiber width is relatively thick, which has various limitations. The degradation products of sisal fiber also have certain bactericidal effects. SF can be used as fertilizer, which not only provides the nutrients needed by plants, but also improves soil structure. Therefore, it has excellent ecological and environmental protection. It has good effects on supplementing soil nutrients, reducing runoff erosion, improving strength and ecological effects.

This article believes that the combination of natural fibers and natural biopolymers has great potential for application. The basic principle is shown in Figure 8, which forms a good three-dimensional locking structure, strengthens the soil and has a good ecological effect, improving both the internal properties of the soil and the external friction conditions.

5. Conclusions

(1) Calcium-based mixture can enhance the liquid limit, plastic limit and UCS of loess. With the increase of CM content, the liquid limit, plastic limit and UCS of the treated soil increase firstly and then decrease. The optimum dosage of CM is 9 %. Under this dosage, the average liquid limit, plastic limit and UCS of the CM-SF group are 39.13 %, 24.94 % and 849KPa, which are 1.35,1.36 and 3.81 times that of the untreated soil, respectively.

(2) Chitosan can affect the liquid limit of loess and enhance the UCS of loess. With the increase of CB content, the liquid limit of the treated soil decreases firstly and then increases, the plastic limit does not change significantly, and the UCS increases rapidly. The optimum dosage of CB mixture is 0.48 %. Under this dosage, the average liquid limit and UCS of CB-SF treated soil are 37.63 and 1226 KPa, which are 1.30 and 5.50 times that of untreated soil, respectively.

(3) Sisal fiber can enhance the UCS of loess but has little effect on the liquid plastic limit. With the increase of SF content, the UCS of the treated soil will increase firstly and then decrease. The best SF content is 0.75 %. Excessive incorporation will reduce the UCS of the treated soil.

(4) In the case of low CB dosage (0.12 %~0.48 %), the soil strength can still be greatly improved, and the soil enhancement effect is much stronger than that of CM. The use of biomaterials such as CB combined with short fibers to enhance soil has a good application prospect. When combined, it can simultaneously enhance the interaction between soil particles and form an overall three-dimensional interlocking grid structure. It has certain application potential in foundation reinforcement, slope protection and soil and water conservation.

Acknowledgments

This paper is one of the initial achievements of the scientific and technological project of State Grid Shaanxi Electric Power Company ' Research and Application of Key Technologies for Risk Identification and Control of Soil Erosion and Secondary Disasters in Northern Shaanxi Power Grid Project '(5226KY22000K).

References

1. CHENG L H, NIE R S, LIU F. (2010) An experimental study of the uniaxial compressive strength of fractured loess. HYDROGEOLOG & ENGINEE RING GEOLOGY, 44(5): 7. DOI: 10.16030/j.cnki.issn.1000-3665.2017.05.13.
2. CHENG L Q, CHEN G H, LIU Y Y, et al. (2022) Formation mechanism and disaster-generating model of Loess landslide and collapse in Northern Shaanxi Province. GEOLOGY OF SHAANXI, 2022, 40(02): 65-69.
3. LIAO W Y. (2022) Experimental Study on Soil Structure Improved by Biomaterial. NANCHANG

UNIVERSITY, DOI: 10.27232/d.cnki.gnchu.2022.003292.

4. LI F Z. (2021) Research and application of improved materials for raw soil rapid maturation. China University of Mining & Technology, Beijing, DOI: 10.27624/d.cnki.gzkbu.2021.000054.
5. ZHANG R Q. (2019) Advances in Research on Effects of Improvers on Soil Properties of Reclaimed Cultivated Land in Hollow Village. Journal of Agricultural Sciences, 9(6): 7. DOI: 10.12677/HJAS.2019.96072.
6. ZHANG C B, WANG Q J, QIN R J, et al. (2023) Natural hematite as low-cost auxiliary material for improving soil remediation by in-situ microbial community.Environmental Science and Pollution Research, 2023(35): 30.
7. Khatami H R, O'Kelly B C. (2013) Improving mechanical properties of sand using biopolymers. Journal of Geotechnical & Geoenvironmental Engineering, 139(8): 1402-1406. DOI: 10.1061/(ASCE)GT.1943-5606.0000861.
8. Nakamatsu J, Kim S, Ayarza J, et al. (2017) Eco-friendly modification of earthen construction with carrageenan: Water durability and mechanical assessment. Construction and Building Materials, 139 (Complete): 193-202. DOI: 10.1016/j.conbuildmat. 2017.02.062.
9. Montoya B M, Dejong J T, Boulanger R W. (2013) Dynamic response of liquefiable sand improved by microbial-induced calcite precipitation. Geotechnique, 63(4): 302-312. DOI: 10.1680/geot. SIP13.P.019.
10. Wade E, Zowada R, Foudazi R. (2021) Alginate and guar gum spray application for improving soil aggregation and soil crust integrity. 2021. DOI: 10.1016/J.CARPTA.2021.100114.
11. Young, I, M, Crawford J.W. (2004)Interactions and Self-Organization in the Soil-Microbe Complex. Science, 304(5677): 1634-1637. DOI: 10.1126/science.1097394.
12. ZHU X F, ZHENG J Q, WEI J H, et al. (2023) Experimental study on shear strength of fiber-water soluble polymer compo. Advances in Science and Technology of Water Resources, 43 (06): 50-59.
13. Yohanna P, Etim R K, Ekene N I, et al. (2024) Case evaluation of structural strength improvement of cement stabilized lateritic soil reinforced with sisal fibers and plastic waste strips. Discover Civil Engineering, 1(1): 1-18.
14. HAO S F, YU Y X, SONG J L, et al. (2023) Study on direct shear strength properties of sand mixed with polyurethane prepolymer and sisal fiber[J]. Environmental earth sciences, 82(19): 436.1-436.14.
15. Ma Q, Wu J, Xiao B H. (2024) Effect of Xanthan Gum on the mechanical properties of fiber reinforced sandy soil. Bulletin of engineering geology and the environment. 83(5): 177.1-177.14. DOI: 10.1007/s10064-024-03673-2.
16. Ministry of Housing and Urban-Rural Development. (2019) GB/T50123-2019 Standard for geotechnical testing method. China plans publishing house.
17. Hataf N, Ghadir P, Ranjbar N. (2017) Investigation of soil stabilization using chitosan biopolymer. Journal of Cleaner Production, 170(jan.1): 1493-1500. DOI: 10.1016/j.jclepro.2017.09.256.
18. Sujatha E R, O'Kelly B C. (2023) Biopolymer Based Soil Treatment for Geotechnical Engineering Applications. 2023. DOI: 10.1007/978-981-16-6603-2_22-1.
19. Ilman B, Balkis A. (2023) Sustainable biopolymer stabilized earthen: Utilization of chitosan biopolymer on mechanical, durability, and microstructural properties. Journal of Building Engineering, 2023. DOI: 10.1016/j.jobe. 2023. 107220.
20. Ren G, Gao X, Wen X, et al. (2024) Experimental study on the long-term performance of sisal fiber reinforced high-performance concrete subjected to drying-wetting cycles[J]. Case Studies in Construction Materials, 21. DOI: 10.1016/j.cscm.2024.e03911.
21. Yohanna P, Etim R K, Ekene N I, et al. (2024) Case evaluation of structural strength improvement of cement stabilized lateritic soil reinforced with sisal fibers and plastic waste strips. Discover Civil Engineering, 1(1): 1-18. DOI: 10.1007/s44290-024-00069-9.
22. Pydi R, Yadu L, Chouksey S K. (2024) A Novel Prediction of Strength Properties of Biopolymer-Treated Soil: A Response Surface Approach. Transportation infrastructure geotechnology, 2024(4): 11. DOI: 10.1007/s40515-024-00384-9.

Study on Mechanical Properties of Recycled Aggregate Concrete with Cement-Based Crystalline Materials

Jinhao Ou[a], Jiang Feng[b], Jiaqi Liu[*]

Rocket Force University of Engineering, Xi'an 710025, China

Abstract. The acceleration of urbanization has led to the increasing scale of construction waste, which has caused a lot of waste of resources, and recycled aggregate concrete technology can effectively utilize these wastes. Due to the existence of complex systems such as mortar and microcracks at the interface of recycled aggregate, the mechanical properties of recycled aggregate concrete are insufficient. Based on this, this paper proposes a permeable crystalline material XYPEX containing a large number of active components, and adds different amounts of XYPEX. The working performance and mechanical properties of recycled aggregate concrete were measured. The conclusions are as follows: the incorporation of XYPEX can improve the slump expansion of recycled aggregate concrete mixture and enhance the fluidity of concrete. With the increase of XYPEX content, the mechanical strength of recycled aggregate concrete also increases, and the best performance is achieved when the content is 2.0 %. XYPEX can make recycled aggregate concrete generate more crystals insoluble in water, fill in its pores, enhance its mechanical properties, and meet the application requirements of recycled aggregate concrete.

1 Introduction

In recent years, with the acceleration of urbanization, the amount of solid waste generated by construction and demolition operations has also continued to grow on a large scale. According to the growth trend, the scale of these solid wastes will be close to 3 billion tons in 2025[1-3]. These wastes not only occupy a large area of valuable land resources, but also easily lead to a series of serious pollution problems if not handled properly, which is a great hidden danger to China 's economy, environment and society.

Recycled aggregate concrete technology is an innovative method. It uses concrete, bricks and stones in abandoned buildings as raw materials. It is processed into recycled aggregates, and then these aggregates are mixed with cementitious materials to re-produce concrete. The application of this technology can effectively reduce the demand for non-renewable natural resources and double carbon emissions, and reduce energy consumption. Therefore, since the advent of recycled aggregate concrete technology, it has been widely praised and recognized by all walks of life[4-7] .Compared with ordinary concrete, recycled aggregate concrete still has many shortcomings in terms of workability and mechanical properties. C.S.Poon et al. have shown that the compressive strength of recycled aggregate concrete prepared by replacing natural coarse aggregate with recycled coarse aggregate will be significantly reduced [8]; xiao et al.[9] found that with the increase of recycled aggregate replacement rate, the uniaxial tensile strength of concrete will gradually decrease, and Talamona et al. ' s research has also obtained similar results [10]. Sun Yue-dong et al. have shown that the old cement mortar contained in recycled aggregate will make the mechanical properties and durability of recycled concrete insufficient [11].

In order to optimize the mechanical properties of recycled aggregate concrete, this paper considers the use of XYPEX permeable crystalline material as a concrete admixture. The permeable crystalline material contains a large number of active components. After these components are dissolved in water, they can react with calcium ions (Ca2 +) in aggregate cement slurry, and then form complexes insoluble in water. These generated complexes will migrate with water as a medium within the consolidated body. When they encounter unhydrated cement with high activity, the active chemical substances in the complex will be replaced by more stable aluminate ions, silicate ions, etc. Subsequently, calcium ions react with these ions again to form crystals that are insoluble in water. These crystals will effectively fill the cracks and pores of concrete, thereby improving its overall performance[12-14].

In summary, in this paper, recycled aggregate is used to replace natural aggregate, and different amounts of XYPEX (0 %, 0.5 %, 1.0 %, 1.5 %, 2.0 %, 2.5 %, 3.0 %) are added to prepare modified recycled aggregate concrete. The macroscopic properties of recycled aggregate concrete were tested, and the slump expansion, T500 time, J-ring expansion, compressive strength and splitting tensile strength of concrete with different curing time and different XYPEX content were measured. The effect of different XYPEX content on the mechanical strength of recycled aggregate concrete was studied.

[*] Corresponding author: [*]15129569792@163.com; [a]670385212@qq.com; [b]watermelon0724@163.com

Table 1. Performance indicators of cement.

fineness /% (80μm)	standard consistency water consumption (%)	final setting time (h:min)	28 day flexure strength (MPa)	28 day compressive Strength (MPa)
3.7	28.2	3：12	7.6	46.7

Table 2. Performance index of coarse aggregate.

diameter (mm)	apparent density (kg/m3)	soil content (%)	mud content (%)	flat elongated particles content (%)
5～20	2713	0.51	0.4	5.2

Table 3. Performance index of fine aggregate.

apparent density (kg/m3)	bulk density (kg/m3)	soil content (%)	porosity (%)
2688	1533	2.5	39.5

Table 4. Performance index of water reducing agent.

water reduction (%)	bleeding rate (%)	gas content (%)	28d compressive strength ratio (%)
35	30	4.0	165

Table 5. Grade I fly ash performance index.

bulk density (g/cm)	density (g/cm)	moisture content (%)	chloridion (%)	calcium oxide (%)	sulfur trioxide (%)
1.23	2.64	0.88	0.014	5.2	2.6

Table 6. Performance index of XYPEX permeable crystalline material.

appearance	water content (%)	chloride ion content (%)	total alkali (%)	gas content (%)	28d shrinkage ratio (%)
evenness	0.6	0.017	12.56	2.5	105

2 Test scheme and raw materials

2.1 Test raw materials

The recycled aggregate concrete grade is C40, and the test uses PO42.5 ordinary Portland cement. The relevant performance indicators are shown in Table 1. Coarse and fine aggregates are processed from a construction waste in Baqiao District, Xi 'an City, with reference to the national standard JGJ52-2006 ' Standard for quality and inspection methods of sand and stone for ordinary concrete ' [15]. The specific indicators are shown in Table 2 and Table 3;

The polycarboxylate superplasticizer was used in the test, and the performance index should meet the requirements of GB50119-2013 ' Technical Specification for Application of Concrete Admixtures '[16]. The performance index is shown in Table 4. Refer to the national standard GB / T1596-2017 ' Fly ash used in cement and concrete ' [17], the relevant performance indicators of fly ash used in this test are shown in Table 5. The XYPEX permeable crystalline material is used, and its specific performance indicators are shown in Table 6.

2.2 Test scheme

2.2.1 Working performance test

Refer to the standard ' Ordinary concrete mixture performance test method standard ' (GB / T 50080-2016)[18], the expansion time T500, slump expansion test and gap passing test were carried out on the recycled aggregate concrete after mixing, so as to evaluate the workability and fluidity of recycled aggregate concrete.

2.2.2 Strength performance test

The compressive strength and splitting strength of recycled aggregate concrete after 3 days, 7 days and 28 days of curing in constant temperature and humidity curing room with different XYPEX content were tested by rock triaxial shear rheological testing machine. The size of the specimen is 150mm x 150mm x 150mm.

3 Analysis of test results

3.1 Analysis of filling results

The relationship between the slump expansion of recycled aggregate concrete with XYPEX material, T500 and XYPEX content is shown in Figure 1.

According to Figure 1, when XYPEX material is added to cement, the slump expansion and T500 value of concrete mixture show significant changes with the increase of dosage. When the content of XYPEX is 1.0 %, the slump expansion of concrete mixture reaches 591 mm, which is 1.2 % higher than that of the reference group without XYPEX. When the content of XYPEX increases to 2.5 %, the slump expansion reaches the maximum value of 648 mm, which is 1.4 % higher than that of the reference group. However, if the content of XYPEX continues to increase on this basis, the slump expansion of concrete mixture shows a trend of decreasing first and then stabilizing.

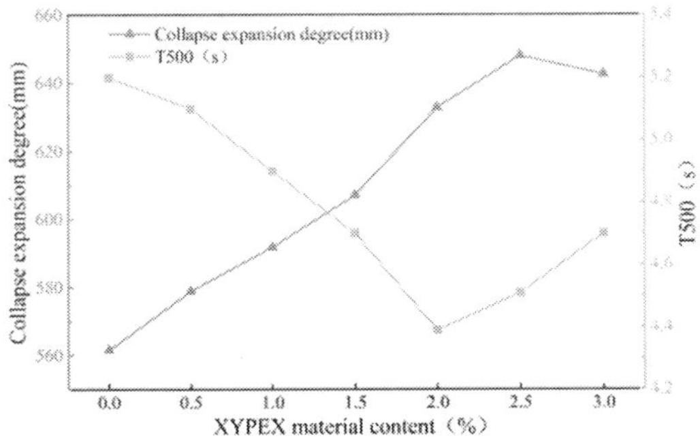

Figure 1. The relationship curves of slump expansion, T500 and XYPEX content of recycled aggregate concrete.

In the absence of XYPEX, the time required for the concrete mixture to flow to a distance of 500 mm is 5.2 seconds. As the XYPEX content increases to 2.0 %, the T500 time is gradually shortened to 4.3 seconds, and the flow rate is significantly accelerated, which is 20 % shorter than the reference group. When the content of XYPEX continues to increase to 3.0 %, the T500 time increases from 4.3 seconds to 4.7 seconds, and the time increases by 9 % compared with the content of 2.0 %. At this time, the flow rate of concrete mixture begins to slow down, the cohesiveness is enhanced, and the filling property is reduced. Despite this, the T500 time was reduced by 9 % compared to the baseline group. This shows that the addition of XYPEX has a significant effect on the fluidity of concrete mixture. Therefore, with the addition of XYPEX material, the filling performance of concrete mixture is improved. Especially when the content of XYPEX is 2.0 % to 3.0 %, the filling performance of concrete is better.

3.2 Analysis of clearance passing results

By calculating the difference between the slump spread and the J-ring spread, the gap passing ability of concrete is characterized. The test results are shown in Figure 2.

According to the curve shown in Figure 2, it can be observed that the effect of XYPEX material content on the J-ring expansion of concrete mixture shows a trend of increasing first and then decreasing. When the content of XYPEX material increases from 0 % to 2.5 %, the expansion degree of J ring increases gradually, and reaches the peak value of 648 mm when the content is 2.5 %, which is 15.5 % higher than that of the reference group without XYPEX material.

In order to further study the performance of the concrete mixture, the difference between the slump expansion and the J-ring expansion was calculated. As shown in Figure 2, the difference decreases first and then increases with the increase of XYPEX material content. When the content of XYPEX material is 0.5 %, the difference is almost unchanged compared with the reference group; however, as the content further increased to 2.0 %, the difference gradually decreased, and reached the minimum value of 34.5 mm when the content was 2.0 %, which was 15.8 % lower than that of the reference group. However, when the content exceeds 2.0 %, the difference begins to increase gradually. The above results show that the addition of XYPEX material has a significant improvement effect on the gap passing performance of concrete. Especially when the content is 2.0 %, the gap performance of concrete reaches the optimal level.

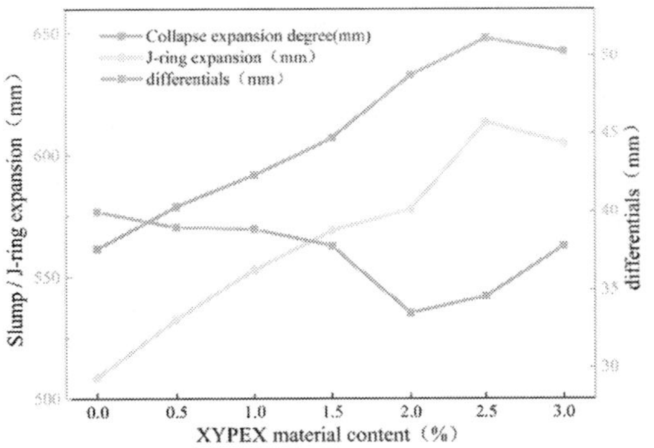

Figure 2. The relationship curves of slump expansion, J-ring expansion difference and XYPEX material content.

In summary, the incorporation of XYPEX material can effectively improve the filling performance and gap passing performance of concrete mixture. Combined with the two test results of J ring expansion and the difference between slump expansion and J ring expansion, it can be concluded that when the content of XYPEX material is 2.0 %, the improvement effect on the working performance of concrete mixture is the best. However, it is worth noting that excessive incorporation of XYPEX materials will lead to a decrease in the filling performance and clearance performance of concrete mixtures. This may be due to the fact that the total amount of water in the slurry remains unchanged, and the excessive XYPEX material requires more adsorbed water, resulting in a decrease in free water content, which in turn affects the fluidity and working performance of the concrete.

3.3 The influence of the content of permeable crystalline material on the compressive strength of recycled aggregate concrete

The strength of concrete reflects the ability of concrete to resist external loads and is one of the key indicators for evaluating the performance of concrete. The compressive strength of recycled aggregate concrete with different XYPEX content was tested, and the test results are shown in Figure 3.

From the change trend of Figure 3 curve, it can be seen that under the same curing age conditions, with the continuous increase of XYPEX content, the compressive strength of recycled aggregate concrete specimens shows a change rule of increasing first and then stabilizing. When the curing age is 3 days, the compressive strength of concrete specimens increases gradually in the range of 0 % ~ 2.5 % XYPEX content, and reaches the maximum value of 31 MPa when the content is 2.5 %. Compared with the reference group without XYPEX, its strength is increased by 7.67 %. When the content is further increased to 3.0 %, the compressive strength of the specimen decreases slightly, but it is still 6.2 % higher than that of the reference group. For the recycled aggregate concrete specimens with a curing age of 7 days, the change trend of compressive strength is similar to that of the specimens with a curing age of 3 days. In the range of XYPEX content from 0 % to 2.0 %, the compressive strength gradually increased, and reached a peak of 34 MPa at a content of 2.0 %, which was 7.9 % higher than that of the reference group. When the curing age is extended to 28 days, the compressive strength of concrete specimens shows a more significant growth trend with the increase of XYPEX content. Compared with the specimens of 3 days and 7 days curing age, the growth rate of compressive strength of the specimens of 28 days curing age is obviously accelerated when the content of XYPEX increases. When the content of XYPEX reached 2.0 %, the growth of compressive strength gradually stabilized, and reached the maximum value of 42.96 MPa when the content was 2.5 %, which was 13.89 % higher than that of the reference group.

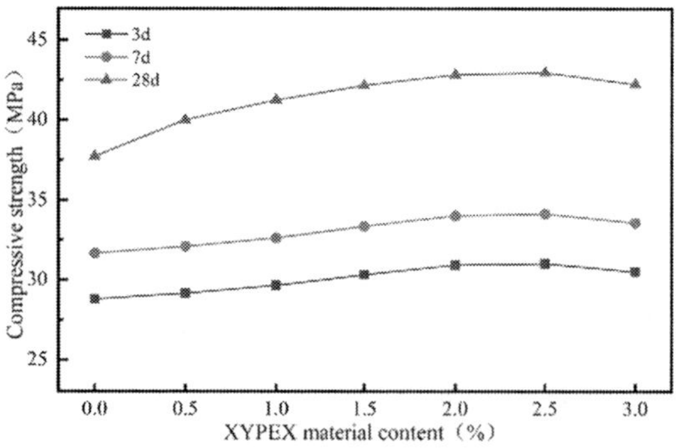

Figure 3. The compressive strength curve of recycled aggregate concrete with different XYPEX content and different ages.

In summary, the incorporation of XYPEX material has a certain improvement effect on the compressive strength of concrete at each curing age. However, for concrete specimens with curing ages of 3 days and 7 days, the incorporation of XYPEX has a relatively small increase in concrete strength. When the curing age reaches 28 days, with the increase of XYPEX content, the increase of concrete compressive strength increases compared with the early stage, and gradually tends to a stable value. These results provide a valuable reference for us to further study and optimize the application of XYPEX in concrete.

3.4 The effect of the content of permeable crystalline material on the splitting tensile strength of recycled aggregate concrete

The splitting tensile strength of concrete is also an important index of its mechanical properties. The splitting tensile strength test was carried out on concrete specimens with different curing time and different content of XYPEX permeable crystalline material. The test results are shown in Figure 4.

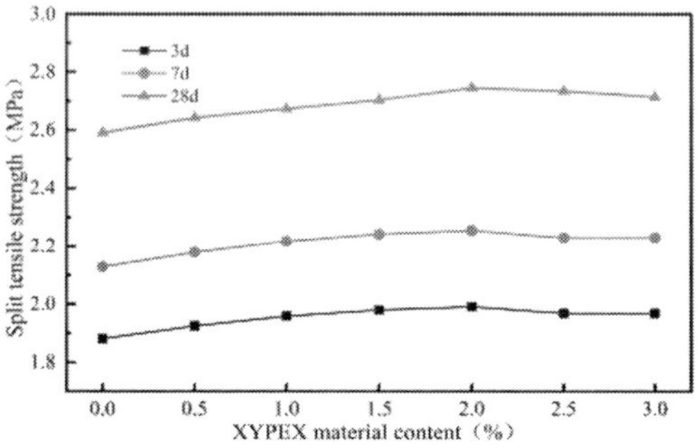

Figure 4. The variation curve of splitting tensile strength of recycled aggregate concrete with different XYPEX content and different ages.

It can be seen from Figure 4 that under the same curing age, with the gradual increase of XYPEX content, the splitting tensile strength of concrete specimens shows a slow but continuous growth trend. When the content of XYPEX is in the range of 0 % ~ 2.0 %, the splitting tensile strength of concrete specimens with curing age of 3 days, 7 days or 28 days gradually increases with the increase of XYPEX content. When the content of XYPEX reaches 2.0 %, the splitting tensile strength of concrete specimens at each curing age reaches the maximum value, which is 1.99 MPa (3d), 2.53 MPa (7d) and 2.74 MPa (28d), respectively. Compared with the reference group, it increased by 5.85 %, 6.13 % and 6.29 %, respectively ; when the content of XYPEX exceeds 2.0 %, the splitting tensile strength of concrete at each curing age tends to be stable, and no longer changes significantly with the increase of the content.

Based on the results of the above compressive test and splitting tensile test of recycled aggregate concrete, the following conclusions can be drawn: Under standard curing conditions, the incorporation of XYPEX material has a certain improvement effect on the mechanical properties of concrete. Especially when the content of XYPEX is 2.0 %, the mechanical properties of concrete in each curing age are excellent, and both compressive strength and splitting tensile strength reach a high level.

The reason why the incorporation of XYPEX can improve the mechanical properties of concrete is that XYPEX material has excellent self-healing ability, which is mainly due to its active material catalyst. These catalysts can produce water-insoluble dendritic crystals in concrete, which can be combined with concrete to form a whole, thereby blocking water and other liquids from all directions. When micro-cracks appear in concrete, the active ingredients in the XYPEX material can be reactivated to produce new crystals that automatically repair cracks. Especially for micro-cracks below 0.4 mm, XYPEX material can self-heal without additional treatment. These crystals are filled into the cracks and tiny pores of the consolidation body, and some spaces inside the consolidation body are blocked, so that the consolidation body itself is more compact. With the increase of compactness, the mechanical properties of the concrete itself will also be improved accordingly.

4 Conclusion

In this paper, the influence of XYPEX on the working performance and mechanical strength of recycled aggregate concrete is studied by changing the content of XYPEX, and the reasons for XYPEX to improve the performance of recycled aggregate concrete are analyzed from the perspective of the action mechanism of permeable crystalline materials. The main conclusions are as follows:

(1) The incorporation of XYPEX can improve the slump expansion of concrete, reduce the T500 time and the difference between the slump expansion and the J ring expansion, thereby enhancing the working performance of concrete; when the content of XYPEX is 2.0 % ~ 2.5 %, the above indexes of recycled aggregate concrete reach the best value, and its working performance reaches the best state.

(2) Compared with the mechanical properties of ordinary recycled aggregate concrete, the compressive strength and splitting tensile strength of recycled aggregate concrete increase with the increase of XYPEX content, and when the content is 2.0 %, the mechanical properties of recycled aggregate concrete are improved best and meet the requirements of the specification.

(3) The incorporation of XYPEX accelerates the hydration rate of cement in concrete, and some water-insoluble crystals are formed inside the consolidation body. These crystals can be combined with concrete to form a whole, thus blocking water and other liquids from all directions, making the structure dense and improving its mechanical properties.

References

1. Wang B, Yan L, Fu Q, et al. Resources conservation and recycling, 2021, 171.
2. Ha J, Shin J, Chung L, et al. Performance Evaluation of Recycled Aggregate Concrete Made of Recycled Aggregate Modified by Carbonation[J]. Journal of the Korea Concrete Institute, 2016, 28(4):445-454.
3. Xiao Jian-zhuang, Hao Lu-cen, Cao Wan-zhi, et al., Journal of Tongji University (Natural Science Edition), 2022, 50 (3), 378.
4. von Greve-Dierfeld S, Lothenbach B, Vollpracht A, et al. Materials and structures, 2020, 53(6).
5. Maduabuchukwu Nwakaire C, Poh Yap S, Chuen Onn C, et al. Construction & building materials, 2020, 235:117444.
6. Sousa V, Bogas J A. Journal of cleaner production, 2021, 306:127277.
7. Xiao J, Xiao Y, Liu Y, et al. Structural concrete, 2020.
8. Poon, C.S., Kou, S.C., Lam, L. Use of recycled aggregates in molded concrete bricks and blocks [J]. Construction and Building Materials. 2002, 16 (5): 281-289.
9. Xiao Jian-zhuang, Li Jia-bin, Sun Zhen-ping, et al. Study on compressive strength of recycled concrete [J]. Journal of Tongji University: Natural Science Edition, 2004, 32 (12): 4.
10. Talamona, D., Tan, M.H. Properties of recycled aggregate concrete for sustainable urban built environment [J]. Journal of Sustainable Cement-based Materials. 2012, 1 (4): 202-210.
11. Sun Yue-dong, Xiao Jian-zhuang. Recycled concrete aggregate [J]. Concrete, 2004.
12. ZHANG C C, GUAN X C, LI J L, et al. Coupling effect of cementitious capillary crystalline waterproof material and exposure environments on self-healing properties of engineered cementitious composites (ECC). Journal of Building Engineering, 2023, 63, 105471.
13. Zhang Li-qing, Yu Jia-le, Wang Yun-yang, et al. [J / OL]. Material introduction, 1-30 [2024-07-18].
14. Yang Xiao-hua, Zheng Kun-long and Xu Li-xiao.Experiment on the influence of permeable crystalline material additives on the performance of cement slurry [J]. China Journal of Highway, 2019, 32 (07): 129-135 + 157.
15. China Academy of Building Research. Standard for quality and test methods of sand and gravel for ordinary concrete [M]. Industry standard-Building industry. 2006: 118P.; A4.
16. China Academy of Building Research, China General Research Institute of Building Materials Science, Ministry of Railways Product Quality Supervision and Inspection Center, etc. Technical Specifications for Application of Concrete Admixtures [M]. National Standard of the People 's Republic of China. 2013: 114p: A4.
17. China General Research Institute of Building Materials Science, Changjiang Water Resources Commission, Changjiang Academy of Sciences, Shanghai Academy of Building Research, etc. Fly ash used in cement and concrete [M]. General Administration of Quality Supervision, Inspection and Quarantine of the People 's Republic of China; China National Standardization Administration. 2017: 16.
18. Standard for performance test method of ordinary concrete mixture [M]. National standard of the People's Republic of China. 2003: 66p: A4.

E3S Web of Conferences 617, 03024 (2025) https://doi.org/10.1051/e3sconf/202561703024
EEUPD 2024

How Digital Empowerment Promotes Rural Autonomous Governance? — A Case Study of Houwei Village, Xingdian Street, Pukou District, Nanjing

Xiu Li[1,2,*], Zhixuan Yang[3] and Yinglu Liu[4]

[1] Shinawatra University, Faculty of Psychology, 12160 Pathunm Thanni, Thailand
[2] Jinken College of Technology, Faculty of Architecture, 210000 Nanjing, China
[3] INTI International University, Faculty of Engineering and Quantity Surveying, 71800 Nilai Negeri Sembilan, Malaysia
[4] University of New South Wales, Faculty of Science, 2052 Sydney, Australia

Abstract. This paper uses Houwei Village, Xingdian Street, Pukou District, Nanjing, as a case study to explore how agricultural IoT, smart community services, and e-commerce platforms empower rural governance and enhance villagers' subjectivity and governance effectiveness. By leveraging these digital tools, the paper also examines their contribution to building sustainable rural communities, in line with the global goal of promoting sustainable cities and communities.

1 Introduction

In recent years, the implementation of China's rural revitalization strategy has introduced new requirements for traditional rural governance models, pushing governance systems from being human-driven to more intelligent and scientific directions [1]. In this context, digital technology has become an important tool to solve rural governance problems due to its efficiency, transparency, and inclusivity [2]. The use of digital technology to empower grassroots governance in the era of big data has become a new trend. Current academic research on rural digital governance has made some progress but generally faces two limitations: first, an over-focus on the technical functions themselves, while neglecting the complex interactions between technology and rural society, culture, and ecosystems; second, the research often focuses on policy interpretation or case descriptions, lacking theoretical analysis of the deep governance mechanisms enabled by digital technology.

This paper, using Houwei Village in Pukou District, Nanjing, as a case study, combines the theories of autonomous governance and technology-society systems to explore how digital technology can be integrated into rural governance systems, empowering autonomous governance and enhancing governance effectiveness. The paper also identifies prominent issues in the current practice of rural digital governance and proposes targeted improvement suggestions, providing academic references and practical insights for the modernization of governance under the rural revitalization strategy.

2 Theoretical Framework

2.1 Autonomous Governance Theory

Autonomous governance theory asserts that the foundation of effective rural governance lies in fully unleashing the initiative of villagers as active governance agents. It advocates for the achievement of co-governance goals through negotiation, cooperation, and active participation [3]. By decentralizing authority and fostering a more inclusive governance structure, this theory emphasizes the importance of empowering local communities to engage directly in decision-making processes.

The introduction of digital technology provides new and diverse avenues for villagers to participate in governance, offering tools that facilitate enhanced information transparency, resource sharing, and collaborative decision-making. These technological innovations create opportunities for deeper, more meaningful engagement in public affairs, which in turn strengthens the synergy of grassroots governance. Digital platforms enable villagers to engage more directly with governance processes, reducing barriers to participation and ensuring that all voices are heard. This broadens the scope of collective cooperation and mutual support, facilitating more inclusive and responsive governance.

Research has shown that digital empowerment can overcome the limitations of traditional governance models by enhancing communication, increasing transparency, and promoting greater collaboration among villagers. Furthermore, it stimulates the agency of governance subjects, encouraging individuals to take an active role in shaping their community's future [4]. In this context, digital tools serve as powerful enablers that

* Corresponding author: eli851929@gmail.com

enhance both the participation and empowerment of villagers, allowing them to become more engaged and invested in the governance process.

2.2 Technology-Society Systems Theory

The Technology-Society Systems Theory posits that the introduction of technology does not merely alter governance tools but fundamentally reshapes the underlying power structures, resource distribution, and social interactions within a given system. In rural contexts, the success and effectiveness of digital technology are heavily influenced by local social, cultural, and ecological factors. For technology to be successfully integrated into rural governance systems, it is essential that technological innovations align closely with the unique needs and dynamics of the rural environment. Thus, the success of technology adoption is determined not only by the functionality of the digital tools themselves but also by their ability to harmonize with the socio-ecological fabric of rural governance [5].

"Reducing the perceived risk of digital technologies: Management personnel must enhance risk awareness. Over-trusting and over-developing emerging technologies without adequate risk assessment can lead to insufficient early warning systems, potentially resulting in serious consequences. For example, algorithmic errors or malicious network attacks could cause business systems to generate erroneous feedback, leading to uncontrollable outcomes" [6]. This perspective on managing perceived risk is especially crucial in rural communities adopting digital tools, as heightened perceptions of risk can severely hinder the acceptance and effective utilization of these technologies. Addressing these concerns and building trust in digital systems is essential to ensure their widespread adoption and maximize their governance potential [7].

3 Case Study of Houwei Village

3.1 Village Overview and Governance Background

Houwei Village is located in Xingdian Street, Pukou District, Nanjing, covering an area of 295.69 hectares, with a permanent population of 777 people. The village has gradually developed a predominantly aquatic farming and rural tourism economy, taking advantage of its favorable natural resources, such as abundant water bodies suitable for aquaculture and scenic areas conducive to rural tourism. However, despite these advantages, Houwei Village faces several governance challenges. The aquatic farming industry is hindered by a lack of branding, leading to limited market competitiveness and low product recognition. Additionally, agricultural non-point source pollution has exacerbated water quality issues, threatening the sustainable development of the aquaculture industry. Moreover, the public service infrastructure at the village level remains insufficient, and there is a low level of villager participation in public affairs and decision-making processes. These challenges have created a fertile ground for the integration of digital technologies, which can help address the pressing issues and improve overall governance efficiency.

3.2 Introduction and Practice of Digital Technology

Houwei Village has embraced digital innovation by integrating agricultural IoT (Internet of Things), smart community services, and e-commerce platforms to enhance governance effectiveness across various dimensions, as shown in Fig. 1. These technologies have enabled the village to tackle its challenges while creating new economic opportunities for its residents.

Fig. 1. Agricultural IoT Platform.

3.2.1 Deployment of Agricultural IoT Technology

To improve aquaculture management, Houwei Village deployed a water quality monitoring and dynamic control system covering over 10,000 acres of aquatic farming areas. The system employs a network of sensors that monitor parameters such as water temperature, dissolved oxygen levels, and PH levels in real-time (As shown in Fig. 2). Using remote control technology, the system optimizes water circulation and adjusts environmental conditions to improve aquaculture productivity. The deployment of this technology has led to a 15% increase in the success rate of aquaculture, and the per-acre economic benefit has risen from 4,000 RMB to over 5,000 RMB. This integration of IoT technology has injected substantial value into traditional aquaculture practices, making them more sustainable and efficient.

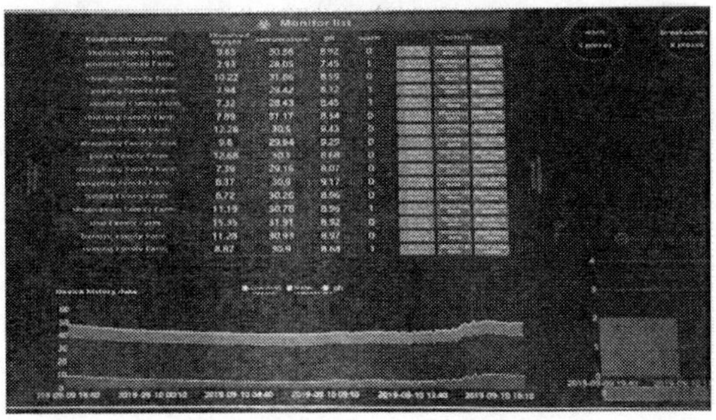

Fig. 2. Intelligent monitoring system.

 https://doi.org/10.1051/e3sconf/202561703024

3.2.2 Promotion of Smart Community Services

To enhance the living environment and provide better services to its residents, Houwei Village installed smart streetlights that integrate monitoring, Wi-Fi, and environmental protection functions. This initiative has significantly improved public safety and convenience in the village. Furthermore, the establishment of an online public service platform has streamlined government services, allowing villagers to process administrative tasks more efficiently. According to a satisfaction survey, more than 90% of the villagers expressed approval of the smart community services, highlighting the positive impact of these digital innovations on daily life.

3.2.3 Application of E-commerce Platforms

Houwei Village has also capitalized on digital tools to enhance its agricultural economy by promoting local products through e-commerce platforms. By branding its signature product, "Hong Crab," the village has been able to reach a broader consumer base. In 2019, e-commerce sales of "Hong Crab" exceeded 5 million RMB, with over 70% of the sales coming from online orders. This shift to online sales has not only expanded the market reach but also helped strengthen the regional agricultural brand and improve the income of local farmers.

3.3 Innovation in Digital Governance Mechanisms

The digitalization of Houwei Village has been accompanied by several innovations in governance mechanisms, which have improved the overall effectiveness of local administration. On the governance front, the village has implemented a grid-based party building management system, enabling more efficient guidance and supervision of grassroots affairs by the party organization. This system has streamlined decision-making processes and improved transparency in governance. It has enhanced the villagers' sense of well-being and acquisition, raised their participation and concern in rural public affairs, and in turn strengthened their right to make major village decisions and participate in public affairs.

Additionally, the village has adopted a collective economic shareholding model, where villagers can invest their land and assets in the collective economy and receive dividends based on their contributions. This system has resulted in an average annual dividend growth rate of 5.8%, significantly enhancing villagers' economic well-being.

Moreover, Houwei Village has integrated its natural and cultural resources to further enhance rural tourism development. By protecting the ChuHe Wetland and promoting the village's unique "Hong Culture," the village has successfully combined ecological conservation with cultural tourism. This integration has opened up new avenues for sustainable rural tourism development, benefiting both the local economy and the environment. As shown in Fig. 3, the Golden Autumn aquatic Harvest Festival was held in Houwei Village, Combine ecological protection and culture.

Fig. 3. Combined ecological conservation with cultural tourism.

4 Outcomes of Digital Empowerment in Rural Autonomous Governance

Houwei Village's digital governance practice has yielded significant results in economic, social, and ecological areas. Economically, the synergetic development of aquaculture and rural tourism has greatly enhanced the added value of the village's industries. Socially, the application of digital tools has strengthened villagers' sense of participation and satisfaction. Ecologically, the application of agricultural IoT technology has reduced non-point source pollution (As shown in Fig. 4, the beautiful ecological environment after treatment), while the development of "Hong Culture" has enhanced villagers' cultural identity and the village's external appeal [8].

Fig. 4. Ecological effectiveness of smart governance.

5 Conclusions

While significant strides have been made in the digital governance practices of Houwei Village, several challenges persist that could hinder the full realization of its digital transformation goals. These challenges include the insufficient participation of social capital, a shortage of highly skilled technical personnel, and the difficulties associated with maintaining and upgrading digital infrastructure over the long term. Addressing these challenges is crucial to ensuring the continued success and scalability of digital governance in the village.

To mitigate these issues, this paper proposes the following key recommendations:

Building a Diversified Investment Mechanism: To enhance the financial sustainability of digital governance initiatives, it is critical to attract social capital through Public-Private Partnership (PPP) models and cooperative funding mechanisms. This approach would not only broaden the sources of governance funding but also ensure a more diversified and resilient financial base for digital projects, facilitating long-term investment in rural development [9].

Strengthening Digital Talent Training: A strategic collaboration between universities, technology companies, and local government bodies to provide targeted digital skills training is essential. This collaboration would empower both villagers and local management staff with the necessary competencies to effectively leverage digital tools in governance [10]. By building local technical capacity, Houwei Village can ensure that digital solutions are not only adopted but also maintained and expanded with local expertise.

Improving Long-term Facility Management Mechanisms: The long-term sustainability of digital infrastructure requires dedicated funding plans for ongoing maintenance and periodic upgrades. A well-established maintenance framework is essential to ensuring that digital systems continue to function optimally and provide lasting benefits to the village. Additionally, it is vital to implement strategic planning for infrastructure renewal to prevent technological obsolescence and to adapt to evolving governance needs [11].

As the case of Houwei Village demonstrates, digital technology holds immense potential for advancing the efficiency and sustainability of rural governance. By optimizing resource allocation, minimizing waste, and improving overall operational efficiency, digital technologies can significantly enhance governance capacity while delivering tangible economic and ecological benefits to rural communities. These advancements are not merely technical; they lay the groundwork for a more sustainable, resilient, and equitable rural future. With the continued integration of digital tools, rural areas like Houwei Village can foster environments that are both adaptive and forward-thinking, contributing to the broader goals of rural revitalization and sustainable development. However, the realization of digital empowerment of rural governance model promotion and construction is not a one-day thing, in the context of the comprehensive promotion and implementation of the rural revitalization strategy, the use of modern digital technology to empower rural governance has been a general trend, can be gradually optimized and seek new breakthroughs in order to achieve the modernization of rural governance.

References

1. Chen, X. (2018). The Rural Revitalization Strategy in China. China Rural Economy, 34(5), 3-12. https://doi.org/10.3969/j.issn.1006-420X.2018.05.001 (in Chinese)
2. Yang, J., & Li, X. (2023). Empowering Rural Communities through Digital Technologies: A Case Study of Rural Revitalization in Jiangsu Province. Technology in Society, 78, 101-112. https://doi.org/10.1016/j.techsoc.2023.101205
3. Guo, Z., & Wang, L. (2023). Digital Transformation in Rural Governance: A Comparative Study of Rural Development Projects in China and Europe. International Journal of Rural Planning, 37(2), 88-104. https://doi.org/10.1016/j.ruralplan.2023.01.014
4. MR Uddin, AKM Hye, and MM Habib, "A Study on Digital Transformation of Hangzhou Garden Company", International Supply Chain Technology Journal, vol. 9, no. 12, 2023, doi: https://doi.org/10.20545/isctj.v9i12.323
5. Wong, F., & Lee, S. (2020). Harnessing digital tools for effective governance: A case study of rural community development. Journal of International Public Administration, 42 (3), 234 - 245. https://doi.org/10.1177/0095399720901839
6. Xu Zheng and Tippawan Lertatthakornkit, "Willingness of Digital Transformation of Chinese (Xiamen) Enterprises", Asia Pacific Journal of Religions and Cultures, Vol. 8, No. 1 (2024), 107-122.https://so06.tcithaijo.org/index.php/ajrc/article/view/265585
7. Xu, Y., & Zhang, H. (2022). The Role of Big Data in Enhancing Rural Governance: Lessons from Rural China. Journal of Rural Development and Planning, 50(1), 101-118. https://doi.org/10.1108/JRDP-12-2021-0115
8. Mergel, I., et al. (2018). Digital governance and its impact on rural society: Lessons from rural development projects. Journal of Public Administration Research and Theory, 28(4), 500-512. https://doi.org/10.1093/jopart/muy046
9. Li, P., & Zhang, M. (2019). Digital Technology and the Capacity Building of Rural Governance. Urban Planning Journal, 34(7), 44-52. https://doi.org/10.3969/j.issn.1006-3831.2019.07.044 (in Chinese)
10. Li, S., & Zhao, W. (2021). Exploring the Impact of Smart Cities on Rural Governance: Evidence from China. Sustainable Cities and Society, 70, 102-112. https://doi.org/10.1016/j.scs.2021.102412
11. Wang, D. (2020). The Role of Agricultural Internet of Things in Promoting Rural Revitalization. Agricultural Science and Technology Communications, 33(4), 56-62. https://doi.org/10.3969/j.issn.1000-3617.2020.04.056 (in Chinese)

AUTHOR INDEX

Bai, Shujuan........250
Budthimedhee, Kanjanee........157
Cai, Jinxing........173
Cao, Qi-Guang........62
Chen, Guanghua........235
Chen, Huafeng........250
Chen, Kang'en........352
Chen, Liang........62
Chen, Shuaiming........235
Chen, Tiantian........228
Chen, Yuyi........347
Chen, Zhuo........383
Chen,........235
Cheng, Wei........292
Cheng, Yufei........55
Chu, Xiaoming........298
Ding, Yunxia........173
Dong, Weiwei........107
Fan, Sijun........204
Fan, Xiang........332
Feng, Jiang........391
Feng, Yang........145
Fu, Guomeng........119
Fu, Haiyan........50
Fu, Hao........145
Fu, Wantong........242
Gao, Quanqing........360
Gao, Ziqi........145
Gong, Fangbing........115
Gu, Xueji........76
Guo, Anni........216
Guo, Fa........141
Guo, KaiZhao........26
Guo, Liangjie........87
Guo, Xinning........303
He, Li........33
Hu, Yuanyuan........235
Huang, Juan........76
Huang, Shan........274
Huang, Zheng........180
Kang, Xinwei........162
Lai, Xinghua........288
Lei,........383
Li, Chao........378
Li, Haiyang........288
Li, Jiawen........332
Li, Jiaxing........250
Li, Jun........62
Li, Linlin........250
Li, Song........62
Li, Xiu........254, 396
Li, Xuefeng........123, 317
Li, Yuanzhi........1
Li, Zhaojie........204
Liang, Jun........383
Liang, Xiaohong........137
Lin, Guilan........107
Lin, Hongru........152
Lin, Xiuhong........327
Liu, Hanxin........267
Liu, Hao........180
Liu, Jiaqi........391
Liu, Jing........228
Liu, Pingping........11
Liu, Tuo........141
Liu, Wei........87
Liu, Xiaoli........383
Liu, Yanhong........259
Liu, Yanlin........254
Liu, Yaolai........69
Liu, Yinglu........396
Liu, Yong........167
Liu, Yumeng........76
Liu, Zhifan........50
Li-Xing, Nan........62
Long, Song-Hua........40
Long, Yetian........235
Lu, Meng........191
Lu, Peng........62
Lu, Shuming........235
Lu, Yilei........191
Luo, Jiajun........50
Ma, Yingying........15
Ma, Zongkai........33
Mao, Hui........186
Mao, Weiyun........26
Meng, Yayun........69
Nan, Yanyun........82
Ou, Jinhao........391
Ou, Menglin........50
Pan, Hongyan........284, 312
Pan, Yujuan........352
Peng, Shuairan........69
Qiu, Binbin........33
Qiu, Cai-Sheng........40
Qiu, Hua-Jiao........40

Qiu, Xilan ... 173, 284
Qu, Ying ... 378
Sang, Xinchun ... 123, 317
Shan, Xianjin ... 235
Shang, Anfan ... 145
Shangguan, Yufei ... 33
Shao, Ze ... 119
Shen, Chen ... 26
Shi, Jiayi ... 292
Shi, Mingqi ... 292
Song, Zhangpan ... 288
Su, Xincheng ... 360
Sun, Hua ... 312
Sun, Kexin ... 224
Sun, Xiang-Ping ... 40, 45
Sun, Yinrui ... 21
Sung, Wook Won ... 40
Syamsunur, Deprizon ... 332
Tan, Junjie ... 267
Tang, Chunling ... 115
Teng, Xin ... 123
Tian, Kanliang ... 383
Tian, Mimi ... 371
Tian, Yao ... 204
Wan, Hao ... 383
Wang, Baoqiang ... 274, 338
Wang, Chao ... 21
Wang, Chen ... 204
Wang, Hongfeng ... 119
Wang, Hongqijie ... 11
Wang, Jing ... 228
Wang, Jiuyi ... 119
Wang, Jue ... 360
Wang, Junde ... 55
Wang, Junyue ... 87
Wang, Liang ... 145
Wang, Purong ... 93, 100
Wang, Tianpeng ... 347
Wang, Xianfei ... 69
Wang, Ya ... 26
Wang, Zhuo ... 45
Wei, Xiaolong ... 383
Wei, Yanhua ... 191
Wu, Gang ... 267
Wu, Jingyun ... 180
Wu, Yilin ... 327
Xia, Tao ... 352
Xia, Zhouzhenyang ... 50
Xiao, Rong ... 26
Xiao, Wenjun ... 137
Xiao, Yiqing ... 228
Xiao-Zhang, Hui ... 62
Xie, Guo-Li ... 62
Xie, Shizhe ... 82
Xu, Duoxun ... 141
Xu, Guoyin ... 93, 100
Xu, Jiahe ... 371
Xu, Jing ... 167
Xu, Xuan ... 115
Xuan, Zhaojie ... 224
Yan, Jin ... 82
Yan, Maomao ... 378
Yan, Yuchen ... 141
Yang, Dongfengt ... 303
Yang, Dong-Qing ... 62
Yang, Feng ... 378
Yang, Guoshu ... 216, 284, 312
Yang, Hang ... 322
Yang, Shengfei ... 141
Yang, Shuai ... 204
Yang, Ting ... 55
Yang, Xin ... 267
Yang, Xingmei ... 259
Yang, Yi ... 224
Yang, Yu ... 298
Yang, Yunsong ... 129
Yang, Zhixuan ... 396
Yao, Sucheng ... 157
Yao, Tiantian ... 367
Ye, Xiangfei ... 69
Yin, Qifeng ... 259
Yin, Weiye ... 204
Yu, Lan ... 254
Yu, Xingchen ... 119
Yu, Yunxiao ... 298
Yuan, Li ... 69
Yue, Qi ... 317
Zeng, Jia ... 115
Zeng, Yuchen ... 7
Zhai, Yongbo ... 87
Zhang, Dongsheng ... 26
Zhang, Gaijing ... 186, 216
Zhang, Han ... 383
Zhang, Helan ... 352
Zhang, Kefu ... 26
Zhang, Liang ... 76
Zhang, Lina ... 186
Zhang, Pei ... 141
Zhang, Xiangguo ... 123, 317
Zhang, Xiaoyan ... 267
Zhang, Yong ... 167
Zhao, Chuanting ... 76
Zhao, Jing ... 119
Zhao, Ping ... 167

Zhao, Wenbin 26
Zhao, Xin-Ling 40
Zhao, Zhi-Min 40
Zheng, Ming-Yue 62
Zheng, Yuchen 21
Zhong, Guo-Qian 40, 45
Zhong, Wook Won 45
Zhou, Linkang 211
Zhou, Shuai 204
Zhou, Yiying 191
Zhou, Zhenhai 211
Zhou, Zhenquan 332
Zhu, Hua 288
Zhu, Menghan 338
Zhuang, Yuan 198
Zhuo, Wang 40
Zuo, Yanlei 87